WELCOME TO CALCULUS

Calculus is the mathematics of CHANGE and almost everything in our world is changing.

Calculus is among the most important and useful developments of human thought, and even though it is over 300 years old, it is still considered the beginning and cornerstone of modern mathematics. It is a wonderful and beautiful and useful set of ideas and techniques.

You will see the fundamental ideas of this course over and over again in future courses in mathematics, the sciences (physical, biological and social) as well as in economics, engineering and others.

But calculus is an intellectual step up from your previous mathematics courses. Many of the ideas are more carefully defined, and they have both a functional and a graphical meaning.. Some of the algorithms are more complicated, and in many cases you will need to decide on the appropriate algorithm to use. And there is a huge variety of applications, too many to be able to discuss each of them in detail.

What this means for you, the student

Probably more than in your previous mathematics classes you need

* to think about the concepts as well as the techniques
* to think about the patterns as well as the individual steps
* to think about the meaning of the concepts and the techniques in the context of particular applications
* to think about how the ideas and techniques apply to functions given by graphs and tables and words as well as by formulas, and
* to spend enough time (1 to 2 hours a day) doing problems to sort out the concepts and to master the techniques and to get better and more efficient with the algebra skills that are vital to you success.

Sometime all of this mental stretching can seem overwhelming, but stick with it (and do a lot of problems). It an even become fun.

So welcome to calculus.

CONTEMPORARY CALCULUS: **Contents**

Each section contains Practice Problems throughout the section. The solutions to these Practice Problems are at the end of that section, after the Problem Set for the section.

Each section also contains a Problem Set. The solutions to the odd problems of each Problem Set are at the end of each chapter.

How to Succeed in Beginning Calculus

Calculus takes time. Do not get behind. Use the textbook intelligently. Working the problems. Work together. Study with a friend. Work in small groups.

Chapter 4: The Integral

How to Succeed in Beginning Calculus

The following comments are based on over thirty years of watching students succeed and fail in calculus courses at colleges and universities and of listening to their comments as they went through their study of calculus. This is the best advice we can give to help you succeed.

Calculus takes time. Almost no one fails calculus because they lack sufficient "mental horsepower". Most people who do not succeed are unwilling (or unable) to devote the necessary time to the course. The "necessary time" depends on how smart you are, what grade you want to earn and on how competitive the calculus course is. Most calculus teachers and successful calculus students agree that 2 (or 3) hours every weeknight and 5 or 6 hours each weekend is a good way to begin if you seriously expect to earn an A or B grade. If you are only willing to devote 5 or 10 hours a week to calculus outside of class, you should consider postponing your study of calculus.

Do NOT get behind. The brisk pace of the calculus course is based on the idea that "if you are in calculus, then you are relatively smart, you have succeeded in previous mathematics courses, and you are willing to work hard to do well." It is terribly hard to **catch up** and **keep up** at the same time. A much safer approach is to work very hard for the first month and then evaluate your situation. If you do get behind, spend a part of your study time catching up, but spend most of it trying to follow and understand what is going on in class.

Go to class, every single class. We hope your calculus teacher makes every idea crystal clear, makes every technique obvious and easy, is enthusiastic about calculus, cares about you as a person, and even makes you laugh sometimes. If not, you still need to attend class. You need to hear the vocabulary of calculus spoken and to see how mathematical ideas are strung together to reach conclusions. You need to see how an expert problem solver approaches problems. You need to hear the announcements about homework and tests. And you need to get to know some of the other students in the class. Unfortunately, when students get a bit behind or confused, they are most likely to miss a class or two (or five). That is absolutely the worst time to miss classes. Come to class anyway. Ask where you can get some outside tutoring or counseling. Ask a classmate to help you for an hour after class. If you must miss a class, ask a classmate what material was covered and skim those sections before the next class. Even if you did not read the material, come back to class as soon as possible.

Work together. Study with a friend. Work in small groups. It is much more fun and is very effective for doing well in calculus. Recent studies, and our personal observations, show that students who **regularly** work together in small groups are less likely to drop the course and are more likely to get A's or B's. You need lots of time to work on the material alone, but study groups of 3–5 students, working together 2 or 3 times a week for a couple hours, seem to help everyone in the group. Study groups offer you a way to get and give help on the material and they can provide an occasional psychological boost ("misery loves company?"). They are a place to use the mathematical language of the course, to trade mathematical tips, and to "cram" for the next day's test. Students in study groups are less likely to miss important points in the course, and they get to know some very nice people, their classmates.

Use the textbook effectively. There are a number of ways of using a mathematics textbook:

i. to gain an overview of the concepts and techniques,
ii. to gain an understanding of the material,
iii. to master the techniques, and
iv. to review the material and see how it connects with the rest of the course.

The first time you read a section, just try to see what problems are being discussed. Skip around, look at the pictures, and read some of the problems and the definitions. If something looks complicated, skip it. If an example looks interesting, read it and try to follow the explanation. This is an exploratory phase. Don't highlight or underline at this stage — you don't know what is important yet and what is just a minor detail.

The next time through the section, proceed in a more organized fashion, reading each introduction, example, explanation, theorem and proof. This is the beginning of the "mastery" stage. If you don't understand the explanation of an example, put a question mark (in pencil) in the margin and go on. Read and try to understand each step in the proofs and ask yourself why that step is valid. If you don't see what justified moving from one step to another in the proof, pencil in question marks in the margin. This second phase will go more slowly than the first, but if you don't understand some details just keep going. Don't get bogged down yet.

Finally, worry about the details. Go quickly over the parts you already understand, but slow down and try to figure out the parts marked with question marks. Try to solve the example problems before you refer to the explanations. If you now understand parts that were giving you trouble, cross out the question marks. If you still don't understand something, put in another question mark and **write down** your question to ask your teacher, tutor, or classmate.

Finally it is time to try the problems at the end of the section. Many of them are similar to examples in the section, but now you need to solve them. Some of the problems are more complicated than the examples, but they still require the same basic techniques. Some of the problems will require that you use concepts and facts from earlier in the course, a combination of old and new concepts and techniques. Working lots of problems is the "secret" of success in calculus.

Working the Problems: Many students read a problem, work it out and check the answer in the back of the book. If their answer is correct, they go on to the next problem. If their answer is wrong, they manipulate (finagle, fudge, massage) their work until their new answer is correct, and then they go on to the next problem. **Do not try the next problem yet!** Before going on, spend a short time, just half a minute, thinking about what you have just done in solving the problem. Ask yourself, "What was the point of this problem?" , "What big steps did I have to take to solve this problem?" , "What was the **process**?" Do not simply review every single step of the solution process. Instead, look at the outline of the solution, the **process**. If your first answer was wrong, ask yourself, "What about this problem should have suggested the right process the first time?" As much learning and retention can take place in the 30 seconds you spend reviewing the **process** as took place in the 10 minutes you took to solve the problem. A correct answer is important, but a **correct process, carefully used, will get you many correct answers**.

There is one more step which too many students omit. **Go back and quickly look over the section one more time.** Don't worry about the details, just try to understand the overall logic and layout of the section. Ask yourself, "What was I expected to learn in this section?" Typically this last step, a review and overview, goes quickly, but it is very valuable. It can help you see and retain the important ideas and connections.

Good luck. Work hard. Have Fun.

Calculus is beautiful and powerful.

A recent interesting article in the New York Times makes some additional suggestions:

* Instead of always studying in the same location, simply alternating the room where a person studies improves retention.
* Studying distinct but related skills or concepts in one sitting, rather then focusing intensely on a single thing, also improves retention.
* Cognitive scientists see testing itself – or practice tests and quizzes – as a powerful tool for learning, rather than merely assessment. The process of retrieving an idea seems to fundamentally alter the way the information is subsequently stored, making it far more accessible in the future.

(* "Learning styles" & "left brain – right brain": In a recent review of relevant research a team of psychologists found almost zero support for such ideas. Hoffman: Use all of your senses and all of your brain!)

(On-line at: http://www.nytimes.com/2010/09/07/health/views/07mind.html)

CHAPTER 0

WELCOME TO CALCULUS

Calculus was first developed more than three hundred years ago by Sir Isaac Newton and Gottfried Leibniz to help them describe and understand the rules governing the motion of planets and moons. Since then, thousands of other men and women have refined the basic ideas of calculus, developed new techniques to make the calculations easier, and found ways to apply calculus to problems besides planetary motion. Perhaps most importantly, they have used calculus to help understand a wide variety of physical, biological, economic and social phenomena and to describe and solve problems in those areas.

The discovery, development, and application of calculus is a great intellectual achievement, and now you have the opportunity to share in that achievement. You should feel exhilarated. You may also be somewhat concerned, a common reaction of students just beginning to study calculus. You need to be concerned enough to work to master calculus and confident enough to keep going when you don't understand something at first.

Part of the beauty of calculus is that it is based on a few very simple ideas. Part of the power of calculus is that these simple ideas can help us understand, describe, and solve problems in a variety of fields. This book tries to emphasize both the beauty and the power.

In Section 0.1 (Preview) we will look at the main ideas which will continue throughout the book: the problems of tangent lines and areas. We will also consider a process that underlies both of those problems, the limiting process of approximating a solution and then getting better and better approximations until we finally get an exact solution.

Sections 0.2 (Lines), 0.3 (Functions), and 0.4 (Combinations of Functions) contain review material which you need to recall before we begin calculus. The emphasis in these sections is on material and skills you will need to succeed in calculus. You should have worked with most of this material in previous courses, but the emphasis and use of the material in these sections may be different than in those courses.

Section 0.5 (Mathematical Language) discusses a few key mathematical phrases. It considers their use and meaning and some of their equivalent forms. It will be difficult to understand the meaning and subtleties of calculus if you don't understand how these phrases are used and what they mean.

0.1 PREVIEW OF CALCULUS

Two Basic Problems

Beginning calculus can be thought of as an attempt, a historically successful attempt, to solve two fundamental problems. In this section we will start to examine geometric forms of those two problems and some fairly simple ways to attempt to solve them. At first, the problems themselves may not appear very interesting or useful, and the methods for solving them may seem crude, but these simple problems and methods have led to one of the most beautiful, powerful, and useful creations in mathematics: Calculus.

First Problem: Finding the Slope of a Tangent Line

Suppose we have the graph of a function $y = f(x)$, and we want to find the equation of the line which is **tangent** to the graph at a particular point P on the graph (Fig. 1). (We will give a precise definition of **tangent** in Section 1.0. For now, think of the tangent line as the line which touches the curve at the point P and stays close to the graph of $y = f(x)$ near P.) We know that the point P is on the tangent line, so if the x–coordinate of P is $x = a$, then the y–coordinate of P must be $y = f(a)$ and $P = (a, f(a))$. The only other information we need to find the equation of the tangent line is its slope, m_{tan}, and that is where the difficulty arises. In algebra, we needed two points in order to determine a slope, and so far we only have the point P. Lets simply pick a second point, say Q, on the graph of $y = f(x)$. If the x–coordinate of Q is b (Fig. 2), then the y–coordinate is f(b), so $Q = (b, f(b))$. The slope of the line through P and Q is

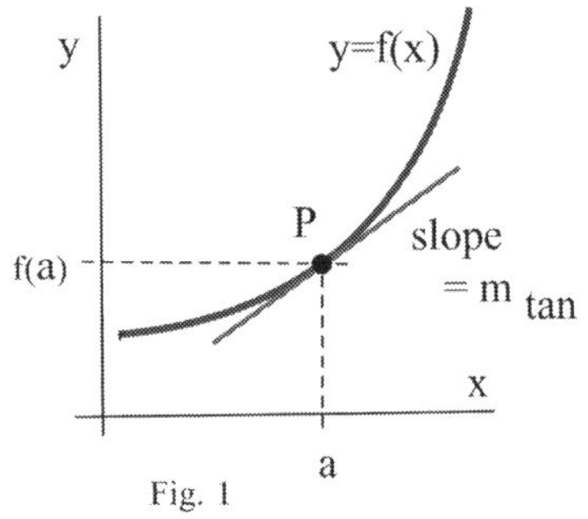

Fig. 1

$$m_{PQ} = \frac{\text{rise}}{\text{run}} = \frac{f(b) - f(a)}{b - a} .$$

If we drew the graph of $y = f(x)$ on a wall, put nails at the points P and Q on the graph, and laid a straightedge on the nails, then the straightedge would have slope m_{PQ} (Fig. 2). However, the slope m_{PQ} can be very different from the value we want, the slope m_{tan} of the tangent line. The key idea is that if the point Q is **close to** the point P, then the slope m_{PQ} is **close to** the slope we want, m_{tan}. Physically, if we slide the nail at Q along the graph towards the fixed point P, then the slope, $m_{PQ} = \frac{f(b) - f(a)}{b - a}$, of the straightedge gets closer and closer to the slope, m_{tan}, of the tangent line. If the value of b is very close to a, then the point Q is very close to P, and the value of m_{PQ} is very close to the value of m_{tan}. Rather than defacing walls with graphs and nails, we can calculate

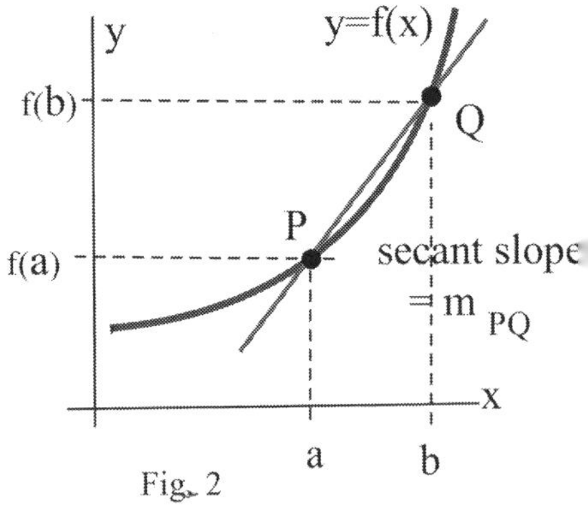

Fig. 2

$$m_{PQ} = \frac{f(b) - f(a)}{b - a}$$

and examine the values of m_{PQ} as b gets closer and closer to a. We say that m_{tan} is the limiting value of m_{PQ} as b gets very close to a, and we write

$$m_{tan} = \lim_{b \to a} \frac{f(b) - f(a)}{b - a} .$$

The slope m_{tan} of the tangent line is called the **derivative** of the function f(x) at the point P, and this part of calculus is called **differential calculus**. Chapters 2 and 3 begin differential calculus.

The slope of the tangent line to the graph of a function will tell us important information about the function and will allow us to solve problems such as:

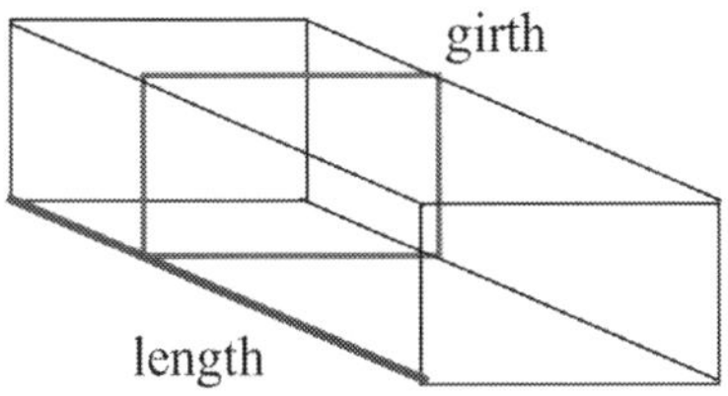

Fig. 3

"The US Post Office requires that the length plus the girth (Fig. 3) of a package not exceed 84 inches. What is the largest volume which can be mailed in a rectangular box?"

An oil tanker was leaking oil, and a 4 inch thick oil slick had formed. When first measured, the slick had a radius 200 feet and the radius was increasing at a rate of 3 feet per hour. At that time, how fast was the oil leaking from the tanker?

Derivatives will even help us solve such "traditional" mathematical problems as finding solutions of equations like $x^2 = 2 + \sin(x)$ and $x^9 + 5x^5 + x^3 + 3 = 0$.

Second Problem: Finding the Area of a Shape

Fig. 4

Suppose we need to find the area of a leaf (Fig. 4) as part of a study of how much energy a plant gets from sunlight. One method for finding the area would be to trace the shape of the leaf onto a piece of paper and then divide the region into "easy" shapes such as rectangles and triangles whose areas we could calculate. We could add all of the "easy" areas together to get the area of the leaf. A modification of this method would be to trace the shape onto a piece of graph paper and then count the number of squares completely inside the edge of the leaf to get a lower estimate of the area and count the number of squares that touch the leaf to get an upper estimate of the area. If we repeat this process with smaller squares, we have to do more counting and adding, but our estimates are closer together and closer to the actual area of the leaf. (This area can also be approximated using a sheet of paper, scissors and an accurate scale. How?)

Each square is 1 sq. cm
totally inside = 1
partially inside = 18
$1 \le$ number ≤ 19
1 sq.cm $\le$ area $\le$ 19 sq.cm

Each square is 1/4 sq/ cm
totally inside = 16
partially inside = 34
$16 \le$ number ≤ 50
4 sq.cm $\le$ area $\le$ 12.5 sq.cm

We can calculate the area A between the graph of a function y = f(x) and the x–axis (Fig. 5) by using similar methods. We can divide the area into strips of width w and determine the lower and upper values of y = f(x) on each strip. Then we can approximate the area of each rectangle and add all of the little areas together to get A_w, an approximation of the exact area. The key idea is that if w is small, then the rectangles are narrow, and the approximate area A_w is very

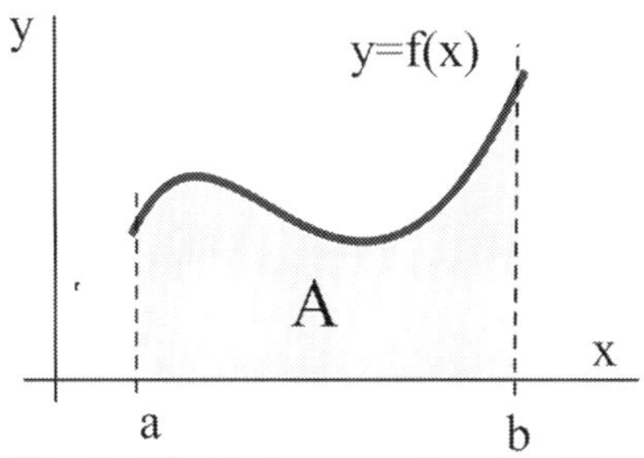

Fig. 5: What is the area of region A?

close to the actual area A. If we take narrower and narrower rectangles, the approximate areas get closer and closer to the actual area: $A = \lim_{w \to 0} A_w$.

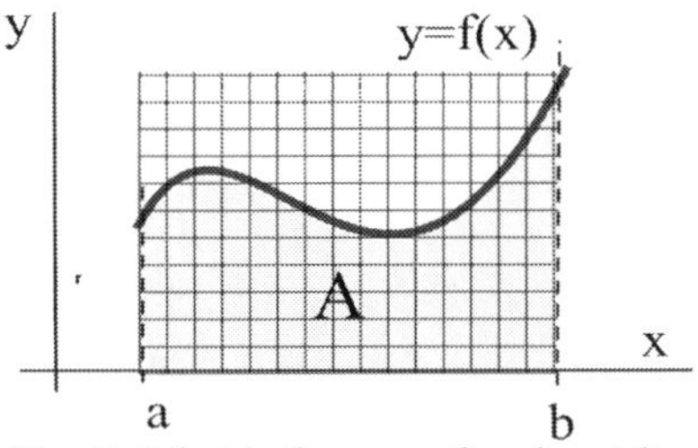

Fig. 5: What is the area of region A?

The process we used is the basis for a technique called **integration**, and this part of calculus is called **integral calculus**. Integral calculus and integration will begin in Chapter 4.

The process of taking the limit of a sum of "little" quantities will give us important information about the function and will also allow us to solve problems such as:

"Find the length of the graph of $y = \sin(x)$ over one period (from $x = 0$ to $x = 2\pi$)."

"Find the volume of a torus ("doughnut") of radius 1 inch which has a hole of radius 2 inches. (Fig. 6)"

"A car starts at rest and has an acceleration of $5 + 3\sin(t)$ feet per second per second in the northerly direction at time t seconds. Where will the car be, relative to its starting position, after 100 seconds?"

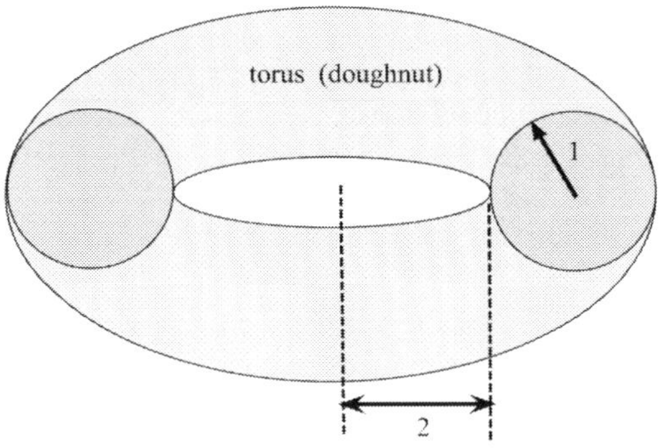

Fig.6: What is the volume of the torus?

A Unifying Process: Limits

We used a similar processes to "solve" both the tangent line problem and the area problem. First, we found a way to get an approximate solution, and then we found a way to improve our approximation. Finally, we asked what would happen if we continued improving our approximations "forever", that is, we "took a limit." For the tangent line problem, we let the point Q get closer and closer and closer to P, the limit as b approached a. In the area problem, we let the widths of the rectangles get smaller and smaller, the limit as w approached 0. Limiting processes underlie derivatives, integrals, and several other fundamental topics in calculus, and we will examine limits and their properties in Chapter 1.

Two Sides Of The Same Coin: Differentiation and Integration

Just as the set–up of each of the two basic problems involved a limiting process, the solutions to the two problems are also related. The process of differentiation for solving the tangent line problem and the process of integration for solving the area problem turn out to be "opposites" of each other: each process undoes the effect of the other process. The Fundamental Theorem of Calculus in Chapter 4 will show how this "opposite" effect works.

Extensions of the Main Problems

The first 5 chapters present the two key ideas of calculus, show "easy" ways to calculate derivatives and integrals, and examine some of their applications. And there is more. In later chapters, new functions will be examined and ways to calculate their derivatives and integrals will be found. The approximation ideas will be extended to use "easy" functions, such as polynomials, to approximate the values of "hard" functions such as $\sin(x)$ and e^x. And the notions of "tangent lines" and "areas" will be extended to 3–dimensional space as "tangent planes" and "volumes".

Success in calculus will require time and effort on your part, but such a beautiful and powerful field is worth that time and effort.

PROBLEMS

(Solutions to odd numbered problems are given at the back of the book.)

Problems 1 – 4 involve estimating slopes of tangent lines.

1) Sketch the lines tangent to the curve shown in Fig. 7 at $x = 1, 2$ and 3. Estimate the slope of each of the tangent lines you drew.

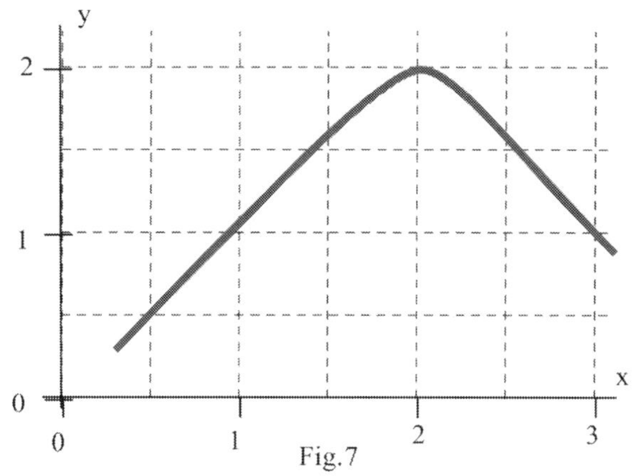

Fig.7

2) Fig. 8 shows the weight of a "typical" child from age 0 to age 24 months. (Each of your answers should have the units "kilograms per month.")

(a) What was the average rate of weight gain from month 0 to month 24?

(b) What was the average weight gain from month 9 to month 12? from month 12 to month 15?

(c) Approximately how fast was the child gaining weight at age 12 months? at age 3 months?

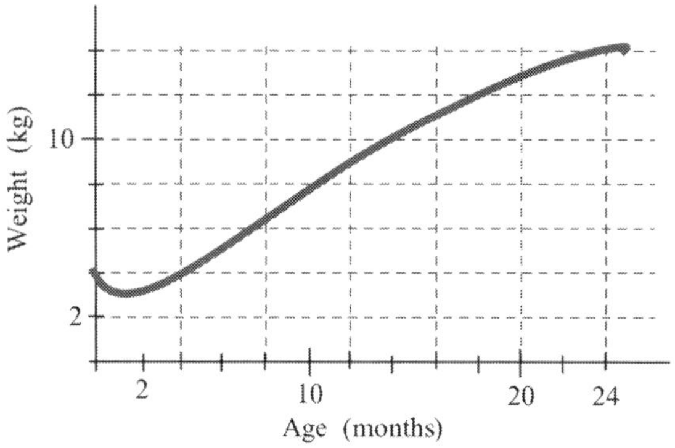

Fig.8

3) Fig. 9 shows the temperature of a cup of coffee during a ten minute period. (Each of your answers in (a) – (c) should have the units "degrees per minute.")

(a) What was the average rate of cooling from minute 0 to minute 10?

(b) What was the average rate of cooling from minute 7 to minute 8? from minute 8 to minute 9?

(c) What was the rate of cooling at minute 8? at minute 2?

(d) When was the cold milk added to the coffee?

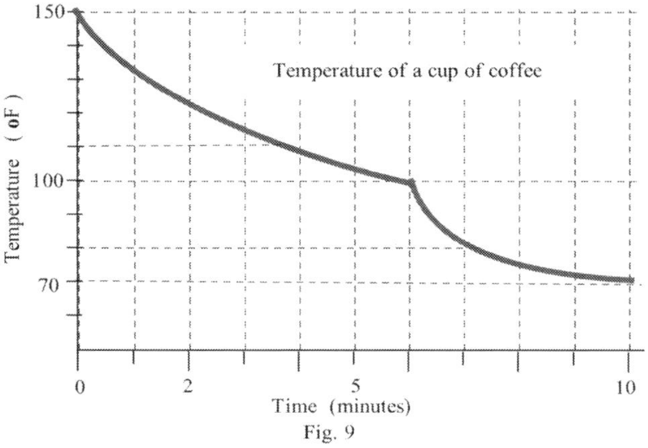

Fig. 9

4) Describe a method for determining the slope at the middle of a steep hill on campus

(a) using a ruler, a long piece of string, a glass of water and a loaf of bread.

(b) using a protractor, a piece of string and a helium–filled balloon.

Problems 5 and 6 involve approximating areas.

5) Approximate the area of the leaf in Fig. 4 .

6) Fig. 10 shows temperatures during a month.

(a) Approximate the shaded area between the temperature curve and the 65° line from Nov. 15 to Nov. 25.

(b) The area of the "rectangle" is (base)(height) so what are the units of your answer in part (a)?

(c) Approximate the shaded area between the temperature curve and the 65° line from Nov. 5 to Nov. 30.

(d) Who might use or care about these results?

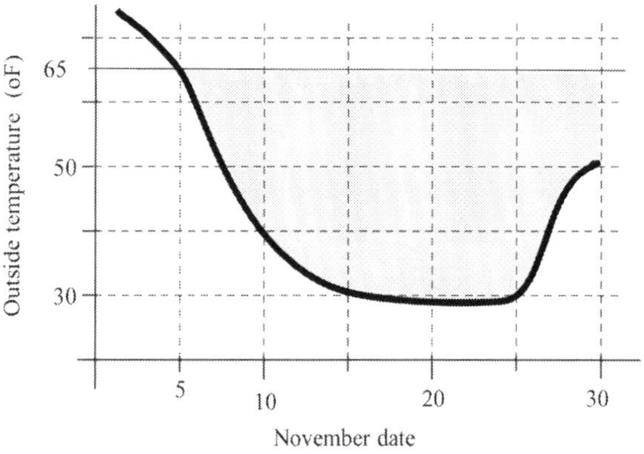

Fig. 10

7) Describe a method for determining the volume of a standard incandescent light bulb using a ruler, a tin coffee can, a scale, and a jug of wine.

0.2 LINES IN THE PLANE

The first graphs and functions you encountered in algebra were straight lines and their equations. These lines were easy to graph, and the equations were easy to evaluate and to solve. They described a variety of physical, biological and business phenomena such as **d = rt** relating the distance d traveled to the rate r and time t of travel, and $\mathbf{C = \frac{5}{9}(F - 32)}$ for converting the temperature in Fahrenheit degrees (F) to Celsius (C).

The first part of calculus, differential calculus, will deal with the ideas and techniques and applications of tangent **lines** to the graphs of functions, so it is important that you understand the graphs and properties and equations of straight lines.

The Real Number Line

The real numbers (consisting of all integers, fractions, rational and irrational numbers) can be represented as a line, called the **real number line** (Fig. 1). Once we have selected a starting location, called the **origin**, a positive direction (usually up or to the right), and unit of length, then every number can be located as a point on the number line. If we move from a point $x = a$ to point $x = b$ on the line (Fig. 2), then we will have moved an **increment** of $b - a$. This increment is denoted by the symbol $\mathbf{\Delta x}$ (read "delta x").

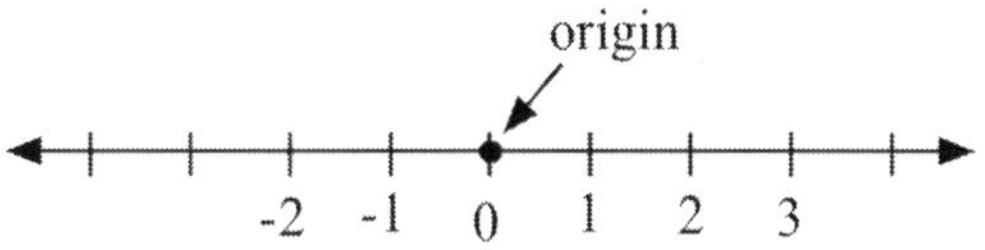

Fig. 1

The Greek capital letter delta, Δ, will appear often in the future and will represent the "change" in something. If b is larger than a, then we will have moved in the positive direction, and $\Delta x = b - a$ will be positive. If b is smaller than a, then $\Delta x = b - a$ will be negative and we will have moved in the negative direction. Finally, if $\Delta x = b - a$ is zero, then a=b and we did not move at all.

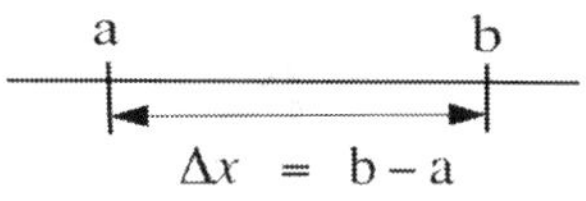

Fig. 2

We can also use the Δ notation and absolute values to write the **distance** that we have moved. On the number line, the distance from $x = a$ to $x = b$ is

$$\mathbf{dist(a,b)} = \begin{cases} b - a & \text{if } b \geq a \\ a - b & \text{if } b < a \end{cases} \quad \text{or simply,} \quad \text{dist(a,b)} = |b - a| = |\Delta x| = \sqrt{(\Delta x)^2} .$$

The **midpoint** of the segment from $x = a$ to $x = b$ is the point $M = \frac{a+b}{2}$ on the number line.

Example 1: Find the length and midpoint of the interval from $x = -3$ to $x = 6$.

Solution: $\text{Dist}(-3,6) = |6 - (-3)| = |9| = 9$. The midpoint is at $\frac{(-3) + (6)}{2} = 3/2$.

Practice 1: Find the length and midpoint of the interval from $x = -7$ to $x = -2$.

(Note: Solutions to Practice Problems are given at the end of each section, after the Problems.)

The Cartesian Plane

A real number **plane** (Fig. 3) is determined by two perpendicular number lines, called the **coordinate axes**, which intersect at a point, called the **origin of the plane** or simply the origin. Each point P in the plane can be described by an **ordered pair** (x,y) of numbers which specify how far, and in which directions, we must move from the origin to reach the point P. The point P = (x,y) can then be located in the plane by starting at the origin and moving x units horizontally and then y units vertically. Similarly, each point in the plane can be labeled with the ordered pair (x,y) which directs us how to reach that point from the origin. In this book, a point in the plane will be labeled either with a name, say P, or with an ordered pair (x,y), or with both P = (x,y). This coordinate system is called the **rectangular coordinate system** or the **Cartesian coordinate system** after Rene Descartes, and the resulting plane is called the Cartesian Plane. The coordinate axes divide the plane into four **quadrants** which are labeled quadrants I, II, III and IV as in Fig. 4 We will often call the horizontal axes the **x-axis** and the vertical axis the **y-axis** and then refer to the plane as the **xy-plane**. This choice of x and y as labels for the axes is simply a common choice, and we sometimes prefer to use different labels and even different units of measure on the two axes.

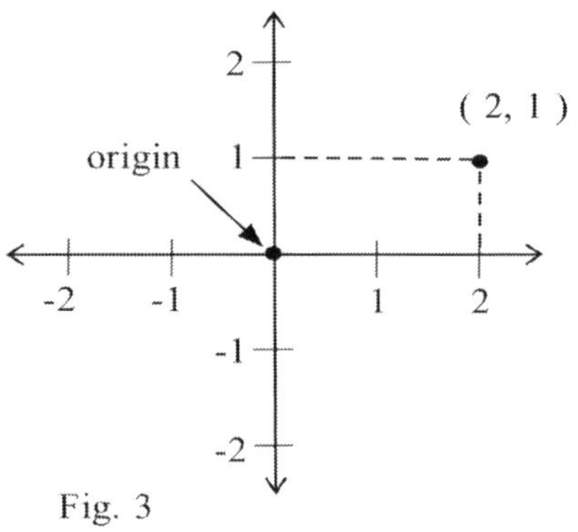

Fig. 3

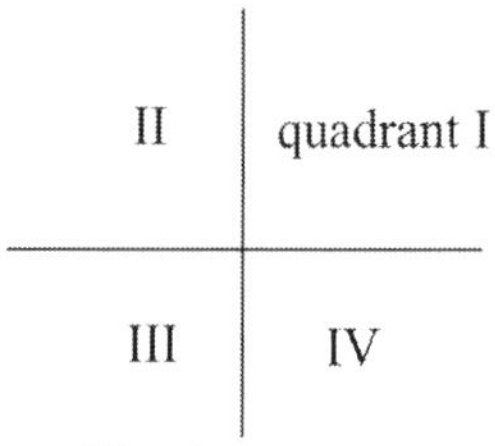

Fig. 4

Increments and Distance Between Points In The Plane

If we move from a point $P = (x_1,y_1)$ to a point $Q = (x_2,y_2)$ in the plane, then we will have two **increments** or changes to consider. The increment in the x or horizontal direction is $x_2 - x_1$ which is denoted by $\mathbf{\Delta x = x_2 - x_1}$. The increment in the y or vertical direction is $\Delta y = y_2 - y_1$. These increments are shown in Fig. 5 . Δx does not represent Δ times x, it represents the difference in the x coordinates: $\Delta x = x_2 - x_1$.

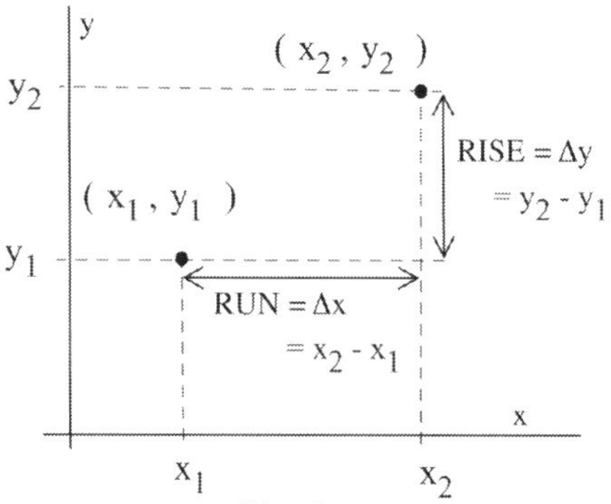

Fig. 5

The distance between the points $P = (x_1,y_1)$ and $Q = (x_2,y_2)$ is simply an application of the Pythagorean formula for right triangles, and

$$\textbf{dist(P,Q)} = \sqrt{(\Delta x)^2 + (\Delta y)^2} = \sqrt{(x_2-x_1)^2 + (y_2-y_1)^2} \;.$$

The **midpoint** M of the line segment joining P and Q is $M = \left(\frac{x_1 + x_2}{2} , \frac{y_1 + y_2}{2} \right)$.

Example 2: Find an equation describing the points P = (x,y) which are equidistant from Q = (2,3) and R = (5,–1). (Fig. 6)

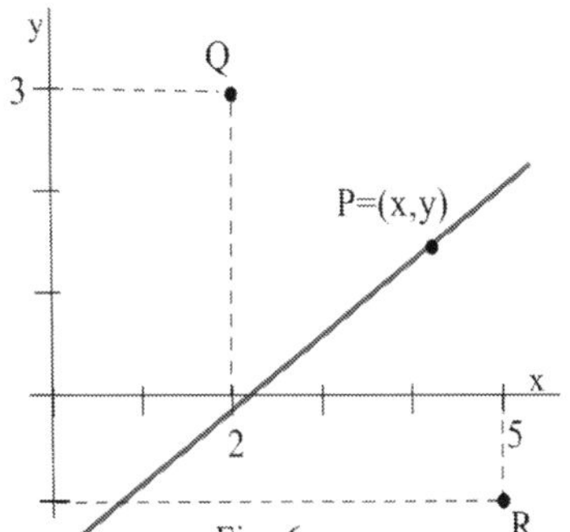

Fig. 6

Solution: The points P=(x,y) must satisfy dist(P,Q) = dist(P,R) so

$$\sqrt{(x-2)^2+(y-3)^2} = \sqrt{(x-5)^2+(y-(-1))^2} \ .$$

By squaring each side we get $(x-2)^2+(y-3)^2 = (x-5)^2+(y+1)^2$.

Then $x^2 - 4x + 4 + y^2 - 6y + 9 = x^2 - 10x + 25 + y^2 + 2y + 1$

so $-4x - 6y + 13 = -10x + 2y + 26$ and $y = .75x - 1.625$, a straight line. Every point on the line $y = .75x - 1.625$ is equally distant from Q and R.

Practice 2: Find an equation describing all points P = (x,y) equidistant from Q = (1,–4) and R = (0,–3).

A **circle** with radius r and center at the point C = (a,b) consists of all points P = (x,y) which are at a distance of r from the center C: the points P which satisfy dist(P,C) = r .

Example 3: Find the equation of a circle with radius r = 4 and center C = (5,–3). (Fig. 7)

Solution: A circle is the set of points P=(x,y) which are at a fixed distance r from the center point C, so this circle will be the set of points P=(x,y) which are at a distance of 4 units from the point C = (5,–3). P will be on this circle if dist(P,C) = 4. Using the distance formula and simplifying,

$\sqrt{(x-5)^2 + (y+3)^2} = 4$ so $(x-5)^2 + (y+3)^2 = 16$ or

$x^2 - 10x + 25 + y^2 + 6y + 9 = 16$.

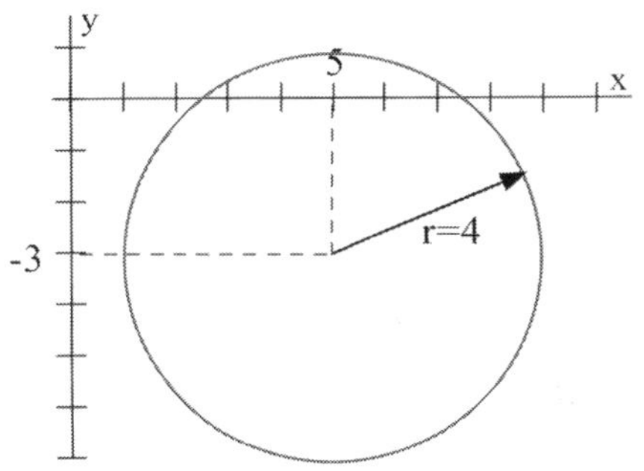

Fig. 7

Practice 3: Find the equation of a circle with radius r = 5 and center C = (–2,6).

The Slope Between Points In The Plane

In one dimension on the number line, our only choice was to move in the positive direction (so the x–values were increasing) or in the negative direction. In two dimensions in the plane, we can move in infinitely many directions and a precise means of describing direction is needed. The **slope** of the line segment joining $P = (x_1, y_1)$ to $Q = (x_2, y_2)$, is

$$m = \{ \text{ slope from P to Q } \} = \frac{\text{rise}}{\text{run}} = \frac{y_2 - y_1}{x_2 - x_1} = \frac{\Delta y}{\Delta x} \ .$$

In Fig. 8, the slope of a line measures how fast we rise or fall as we move from left to right along the line. It measures the rate of change of the y-coordinate with respect to changes in the x-coordinate. Most of our work will occur in 2 dimensions, and **slope** will be a very useful concept which will appear often.

If P and Q have the same x coordinate, then $x_1 = x_2$ and $\Delta x = 0$. The line from P to Q is **vertical** and the slope $m = \Delta y/\Delta x$ is **undefined** because $\Delta x = 0$. If P and Q have the same y coordinate, then $y_1 = y_2$ and $\Delta y = 0$, so the line is **horizontal** and the slope is $m = \Delta y/\Delta x = 0/\Delta x = 0$ (assuming $\Delta x \neq 0$).

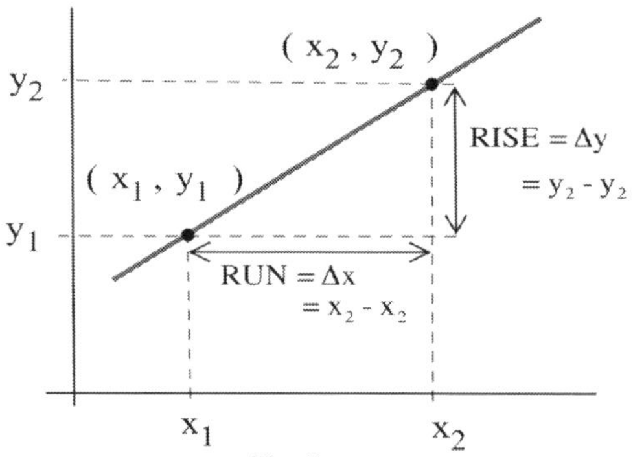

Fig. 8

Practice 4: For $P = (-3,2)$ and $Q = (5,-14)$, find Δx, Δy, and the slope of the line segment from P to Q.

If the coordinates of P or Q contain variables, then the slope m is still given by $\Delta y/\Delta x$, but we will need to use algebra to evaluate and simplify m.

Example 4: Find the slope of the line segment from $P = (1,3)$ to $Q = (1+h, 3 + 2h)$. (Fig. 9)

Solution: $y_1 = 3$ and $y_2 = 3 + 2h$ so $\Delta y = (3 + 2h) - (3) = 2h$. $x_1 = 1$ and $x_2 = 1 + h$ so $\Delta x = (1 + h) - (1) = h$, and the slope is $m = \frac{\Delta y}{\Delta x} = \frac{2h}{h} = 2$. In this example, the value of m is the constant 2 and does not depend on the value of h.

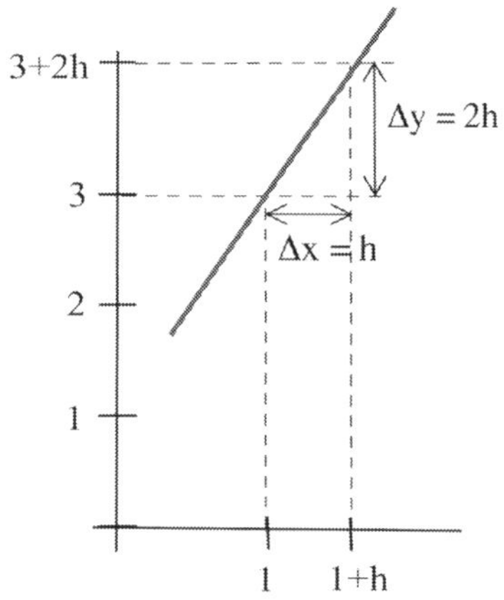

Fig. 9

Practice 5: Find the slope and midpoint of the line segment from $P = (2,-3)$ to $Q = (2 + h, -3 + 5h)$.

Example 5: Find the slope between the points $P = (x, x^2 + x)$ and $Q = (a, a^2 + a)$ for $a \neq x$.

Solution: $y_1 = x^2+x$ and $y_2 = a^2+a$ so $\Delta y = (a^2 + a) - (x^2 + x)$. $x_1 = x$ and $x_2 = a$ so $\Delta x = a-x$ and

the slope is $$m = \frac{\Delta y}{\Delta x} = \frac{(a^2+a) - (x^2+x)}{a-x} = \frac{a^2 - x^2 + a - x}{a - x}$$

$$= \frac{(a-x)(a+x) + (a-x)}{a - x} = \frac{(a-x) \cdot \{(a+x) + 1\}}{a - x} = (a + x) + 1.$$

In this example, the value of m depends on the values of both a and x.

Practice 6: Find the slope between $P = (x, 3x^2 + 5x)$ $Q = (a, 3a^2 + 5a)$ for $a \neq x$.

In application problems it is important to read the information and the questions very carefully. Including the units of measurement of the variables can help you avoid "silly" answers.

Example 6: In 1970 the population of Houston was 1,233,535 and in 1980 it was 1,595,138. Find the slope of the line through the points (1970, 1233535) and (1980, 1595138).

Solution: $m = \dfrac{\Delta y}{\Delta x} = \dfrac{1595138 - 1233535}{1980 - 1970} = \dfrac{361603}{10} = 36{,}160.3$

But 36,160.3 is just a number which may or may not have any meaning to you. If we include the units of measurement along with the numbers we will get a more meaningful result:

$$m = \frac{\Delta y}{\Delta x} = \frac{1595138 \text{ people} - 1233535 \text{ people}}{\text{year } 1980 - \text{year } 1970}$$

$$= \frac{361603 \text{ people}}{10 \text{ years}} = 36{,}160.3 \text{ people/year}$$

which says that during the decade from 1970 to 1980 the population of Houston grew at an average rate of 36,160 people per year.

If the x–unit is time in hours and the y-unit is distance in kilometers, then m is $\dfrac{\Delta y \text{ kilometers}}{\Delta x \text{ hours}}$, so the units for m are kilometers/hour ("kilometers per hour"), a measure of velocity, the rate of change of distance with respect to time. If the x-unit is the number of employees at a bicycle factory and the y-unit is the number of bicycles manufactured, then m is $\dfrac{\Delta y \text{ bicycles}}{\Delta x \text{ employees}}$, and the units for a are bicycles/employee ("bicycles per employee"), a measure of the rate of production per employee.

EQUATIONS OF LINES

Every line has the property that the slope of the segment between any two points on the line is the same, and this constant slope property of straight lines leads to ways of finding equations to represent nonvertical lines.

Point–Slope Equation

In calculus, we will usually know a point on the line and the slope of the line so the point–slope form will be the easiest to apply, and the other forms of equations for straight lines can be derived from the point–slope form.

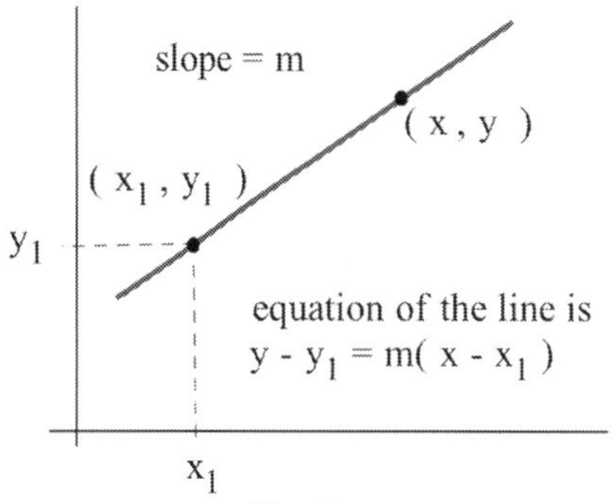

Fig. 10

If L is a nonvertical line through a known point $P = (x_1, y_1)$ with a known slope m (Fig. 10), then the equation of the line L is

Point-Slope: $y - y_1 = m(x - x_1)$.

Example 7: Find the equation of the line through (2,–3) with slope 5.

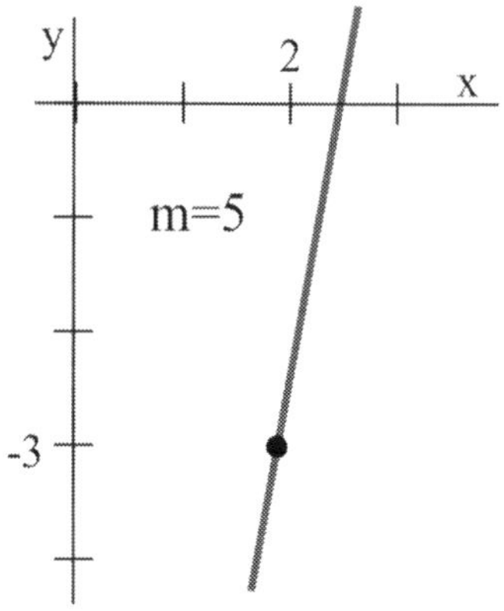

Fig. 11

Solution: The solution is simply a matter of knowing and using the point–slope formula. $m = 5$, $y_1 = -3$ and $x_1 = 2$ so

$y - (-3) = 5(x - 2)$. This equation simplifies to $y = 5x - 13$ (Fig. 11).

The equation of a **vertical line** through a point $P = (a,b)$ is $x = a$. The only points $Q = (x,y)$ on the vertical line through the point P have the same x–coordinate as P.

Two–Point and Slope–Intercept Equations

If two points $P = (x_1,y_1)$ and $Q = (x_2,y_2)$ are on the line L, then we can calculate the slope between them and use the first point and the point–slope equation to get the equation of L:

Two Points: $y - y_1 = m(x - x_1)$ **where** $m = \dfrac{y_2 - y_1}{x_2 - x_1}$.

Once we have the slope m, it does not matter whether we use P or Q as the point. Either choice will give the same simplified equation for the line.

It is common practice to rewrite the equation of the line in the form $y = mx + b$, the **slope-intercept** form of the line. The line $y = \mathbf{m}x + \mathbf{b}$ has slope **m** and crosses the y-axis at the point (0, **b**).

Practice 7: Use the $\Delta y/\Delta x$ definition of slope to calculate the slope of the line $y = mx + b$.

The point-slope and the two-point formulas are usually more useful for finding the equation of a line, but the slope-intercept form is usually the most useful form for an answer because it allows us to easily picture the graph of the line and to quickly calculate y-values.

Angles Between Lines

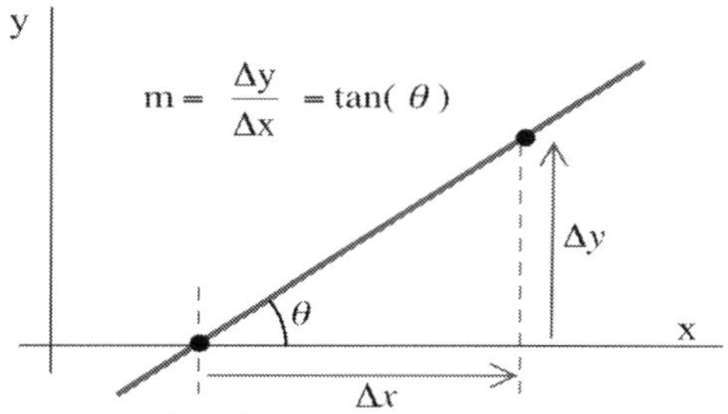

Fig. 12

The **angle of inclination** of a line with the x-axis is the smallest angle θ which the line makes with the positive x-axis as measured from the x-axis counterclockwise to the line (Fig. 12). Since the slope $m = \Delta y/\Delta x$ and since $\tan(\theta)$ = opposite/adjacent, we have that

$$\mathbf{m = \tan(\theta)}\ .$$

The slope of the line is the tangent of the angle of inclination of the line.

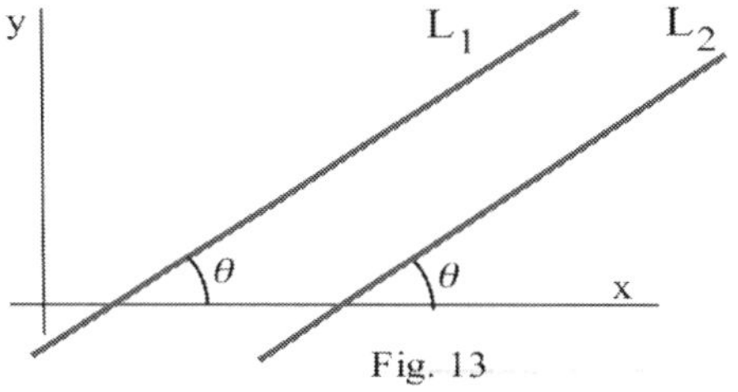

Fig. 13

Parallel and Perpendicular Lines

Two parallel lines L_1 and L_2 make equal angles with the x-axis so their angles of inclination will be equal (Fig. 13) and so will their slopes. Similarly, if their slopes m_1 and m_2 are equal, then the equations of the lines will always differ by a constant:

$$\begin{aligned} y_1 - y_2 &= \{m_1x+b_1\} - \{m_2x+b_2\} \\ &= (m_1-m_2)x + (b_1-b_2) \\ &= b_1 - b_2 \end{aligned}$$

which is a constant so the lines will be parallel. These two ideas can be combined into a single statement:

> **Two nonvertical lines L_1 and L_2 with slopes m_1 and m_2 are parallel if and only if $m_1 = m_2$.**

Practice 8: Find the equation of the line in Fig. 14 which contains the point (–2,3) and is parallel to the line 3x + 5y = 17.

If two lines are perpendicular and neither line is vertical, the situation is a bit more complicated (Fig. 15).

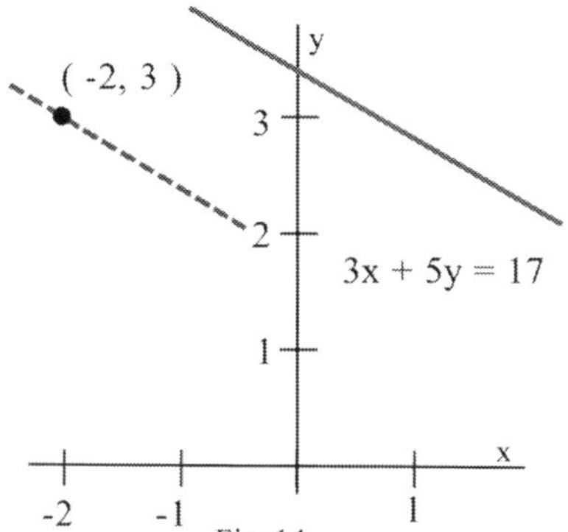

Fig. 14

Assume L_1 and L_2 are two nonvertical lines that intersect at the origin (for simplicity) and that $P = (x_1,y_1)$ and $Q = (x_2,y_2)$ are points away from the origin on L_1 and L_2 , respectively. Then the slopes of L_1 and L_2 will be $m_1 = y_1/x_1$ and $m_2 = y_2/x_2$. The line connecting P and Q forms the third side of the triangle OPQ , and this will be a right triangle if and only if L_1 and L_2 are perpendicular. In particular, L_1 and L_2 are perpendicular if and only if the triangle OPQ satisfies the Pythagorean theorem:

$\{\text{dist}(O,P)\}^2 + \{\text{dist}(O,Q)\}^2 = \{\text{dist}(P,Q)\}^2$ or

$$(x_1-0)^2 + (y_1-0)^2 + (x_2-0)^2 + (y_2-0)^2$$

$$= (x_1 - x_2)^2 + (y_1 - y_2)^2 .$$

By squaring and simplifying, this last equation reduces to

$0 = -2x_1x_2 - 2y_1y_2$ so $y_2/x_2 = -x_1/y_1$ and

$$m_2 = y_2/x_2 = -x_1/y_1 = -\frac{1}{(y_1/x_1)} = -\frac{1}{m_1} .$$

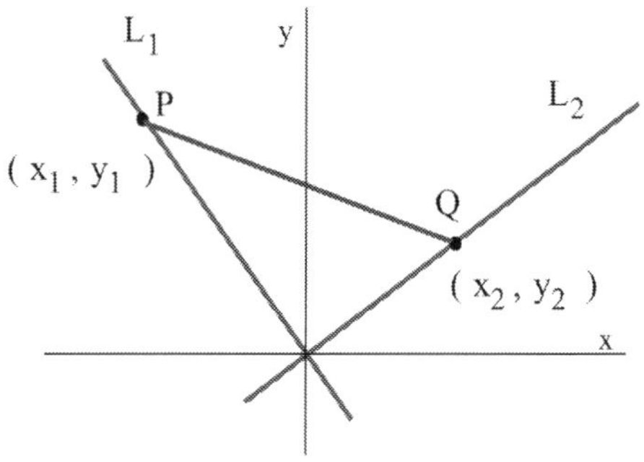

Fig. 15

We have just proved the following result:

Two nonvertical lines L_1 and L_2 with slopes m_1 and m_2 are perpendicular if and only if their slopes are negative reciprocals of each other: $m_2 = -\frac{1}{m_1}$.

Practice 9: Find the line which goes through the point (2,–5) and is perpendicular to the line $3y - 7x = 2$.

Example 8: Find the distance (the shortest distance) from the point (1,8) to the line L: $3y - x = 3$.

Solution: This is a sophisticated problem which requires several steps to solve. First we need a picture of the problem (Fig. 16). We will find the line L* through the point (1,8) and perpendicular to L. Then we will find the point P where L and L* intersect, and, finally, we will find the distance from P to (1,8).

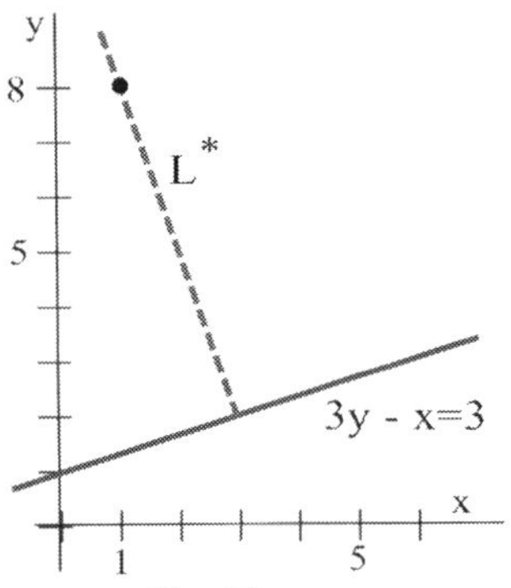

Fig. 16

(i) L has slope 1/3 so L* has slope $m = -\frac{1}{1/3} = -3$, and

L* has the equation $y - 8 = -3(x - 1)$ which simplifies to $y = -3x + 11$.

(ii) We can find the point of intersection of L and L* by replacing the y in the equation for L with the y from L* so

$3(\mathbf{-3x + 11}) - x = 3$. Then $x = 3$ so $y = -3x + 11 = -3(3) + 11 = 2$,

so L and L* intersect at $P = (3,2)$.

(iii) Finally, the distance from L to (1,8) is just the distance from the point (1,8) to the point $P = (3,2)$ which is $\sqrt{(1-3)^2 + (8-2)^2} = \sqrt{40} \approx 6.325$.

Angle Formed by Intersecting Lines

If two lines which are not perpendicular intersect at a point and neither line is vertical, then we can use some geometry and trigonometry to determine the angles formed by the intersection of the lines (Fig. 17). Since θ_2 is an exterior angle of the triangle ABC, θ_2 is equal to the sum of the two opposite interior angles so $\theta_2 = \theta_1 + \theta$ and $\theta = \theta_2 - \theta_1$. Then, from trigonometry,

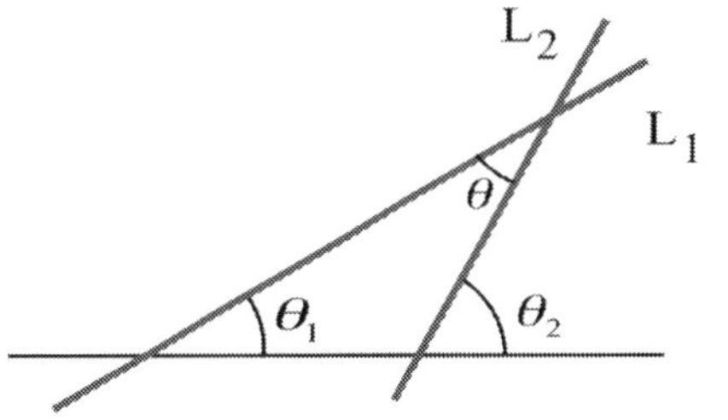

Fig. 17

$$\tan(\theta) = \tan(\theta_2 - \theta_1) = \frac{\tan(\theta_2) - \tan(\theta_1)}{1 + \tan(\theta_2)\tan(\theta_1)} = \frac{m_2 - m_1}{1 + m_2 m_1} .$$

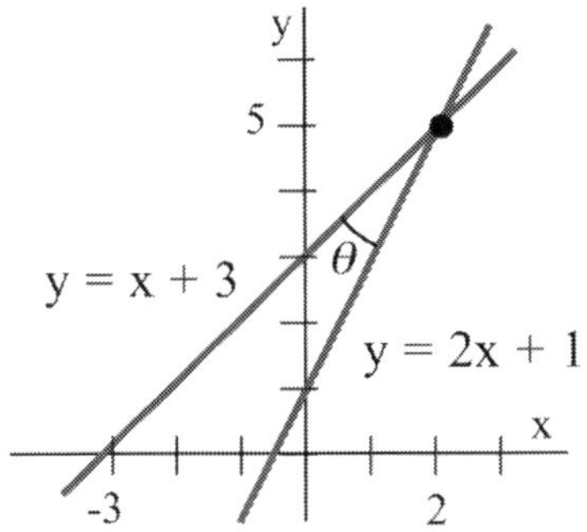

Fig. 18

The inverse tangent of an angle is between $-\pi/2$ and $\pi/2$ (-90° and 90°)

so $\theta = \arctan\left(\frac{m_2 - m_1}{1 + m_2 m_1} \right)$ always gives the smaller of the angles.

The larger angle is $\pi - \theta$ or $180^{\circ} - \theta^{\circ}$.

The smaller angle θ formed by two nonperpendicular lines with slopes m_1 and m_2 is

$$\theta = \arctan\left(\frac{m_2 - m_1}{1 + m_2 m_1} \right).$$

Example 9: Find the point of intersection and the angle between $y = x + 3$ and $y = 2x + 1$. (Fig. 18)

Solution: Solving the first equation for y and then substituting into the second equation, $(x + 3) = 2x + 1$ so $x = 2$. Putting this back into either equation, we get $y = 5$. Each of the lines is in the slope–intercept form so it is easy to see that $m_1 = 1$ and $m_2 = 2$. Then

$$\tan(\theta) = \frac{m_2 - m_1}{1 + m_2 m_1} = \frac{2 - 1}{1 + (2)(1)} = 1/3 \text{ and}$$

$$\theta = \arctan(1/3) = .322 \text{ radians} \approx 18.435^{\circ} .$$

PROBLEMS

1. Estimate the slope of each line in Figure 19.

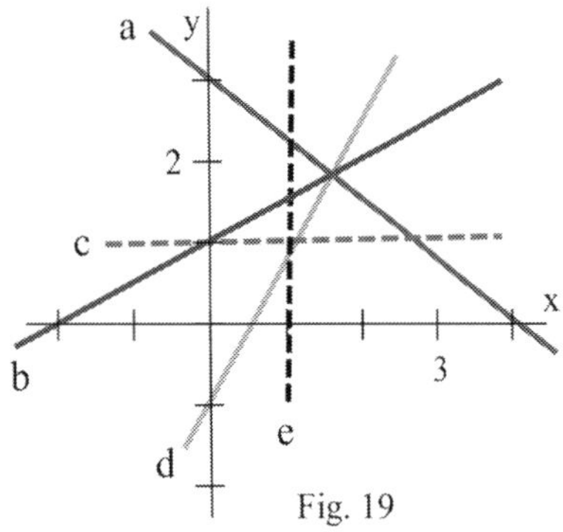

Fig. 19

2. Estimate the slope of each line in Figure 20.

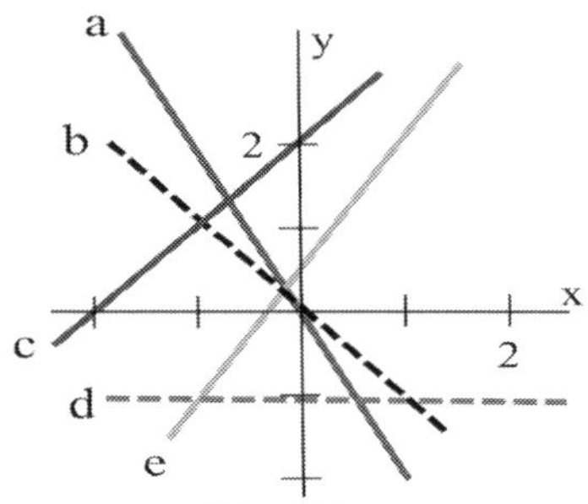

Fig. 20

3. Calculate the slope of the line through each pair of points:
   a) (2,4) , (5,8) b) (–2,4) , (3,–5) c) (2,4) , (x, x^2)
   d) (2,5) , (2+h, $1+(2+h)^2$) e) (x, x^2+3) , (a, a^2+3)

4. Calculate the slope of the line through each pair of points:
   a) (5,–2) , (3,8) b) (–2,–4) , (5,–3) c) (x, 3x+5), (a, 3a+5)
   d) (4,5), (4+h, 5–3h) e) (1,2), (x, $1+x^2$) f) (2,–3), (2+h, $1-(1+h)^2$)
   g) (x, x^2), (x+h, $x^2+2xh+h^2$) h) (x, x^2), (x–h, $x^2-2xh+h^2$)

5. A small airplane at an altitude of 5000 feet is flying East at 300 feet per second (a bit over 200 miles per hour), and you are watching it with a small telescope as it passes directly overhead. (Fig. 21)
   a) What is the **slope** of the telescope 5, 10 and 20 seconds after the plane passes overhead?
   b) What is the **slope** of the telescope t seconds after the plane passes overhead?
   c) After it passes overhead, is the slope of the telescope increasing, decreasing, or staying the same?

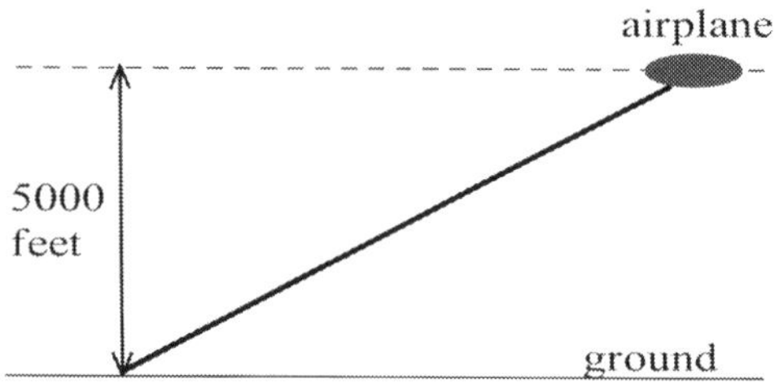

Fig. 21

6. You are at the origin (0,0) and are watching a small bug at the point (t, $1+t^2$) at time t seconds.
   a) What is the **slope** of your line of vision when t = 5, 10 and 15 seconds?
   b) What is the **slope** of your line of vision at any time t?

7. The blocks in a city are all perfect squares. A friend gives you the following directions to a good restaurant; "go north 3 blocks, turn east and go 5 blocks, turn south and go 7 blocks, turn west and go 3 blocks." How far away (straight line distance) is the restaurant?

8. Suppose the directions in problem 7 had been "go north 5 blocks, turn right and go 6 blocks, turn right and go 3 blocks, turn left and go 2 blocks." How far away is the restaurant?

9. How far up a wall will a 20 foot long ladder reach if the bottom must be at least 4 feet from the bottom of the wall? What will be the slope of the ladder if the bottom is 4 feet from the wall? What angle will the ladder make with the ground?

10. Let $P = (1, -2)$ and $Q = (5, 4)$
   (a) Find the point midpoint R on the line segment from P to Q.
   (b) Find the point T which is 1/3 of the way from P to Q: $\text{Dist}(P,T) = (1/3)\cdot\text{Dist}(P,Q)$.
   (c) Find the point S which is 2/5 of the way from P to Q: $\text{Dist}(P,S) = (2/5)\cdot\text{Dist}(P,Q)$.

11. Let $P = (2, 3)$ and $Q = (8, 11)$. Verify that if $0 \le a \le 1$, then the point $R = (x,y)$ with $x = 2a + 8(1-a)$ and $y = 3a + 11(1-a)$ is on the line from P to Q and $\text{Dist}(P,R) = (1-a)\cdot\text{Dist}(P,Q)$.

12. What is the longest straight stick which fits into a rectangular box which is 24 inches long, 18 inches wide and 12 inches high? What angle, in degrees, does the stick make with the base of the box?

C13. The lines $y = x$ and $y = 4 - x$ intersect at the point $(2, 2)$.
   a) Use slopes to show that the lines are perpendicular.
   b) Graph them together on your calculator using the "window" $-10 \le x \le 10,\ -10 \le y \le 10.$
   Why do the lines not appear to be perpendicular on the calculator display?
   c) Find a suitable window for the graphs so the lines so that they do appear perpendicular.

C14. a) Find equations for two lines that both go through the point $(1, 2)$, one with slope 3 and one with slope -1/3.
   b) Choose a suitable window so the lines will appear perpendicular, and graph them together on your calculator.

15. Sketch each line which has slope=m and which goes through the point P. Find the equation of each line.
   a) $m = 3$, $P = (2,5)$ b) $m = -2$, $P = (3,2)$ c) $m = -1/2$, $P=(1,4)$

16. Sketch each line which has slope=m and which goes through the point P. Find the equation of each line.
   a) $m = 5$, $P = (2,1)$ b) $m = -2/3$, $P = (1,3)$ c) $m = \pi$, $P = (1,-3)$

17. Find the equation of each of the following lines.
   a) L_1 goes through the point (2, 5) and is parallel to $3x - 2y = 9$.
   b) L_2 goes through the point (–1,2) and is perpendicular to $2x = 7-3y$.
   c) L_3 goes through the point (3, –2) and is perpendicular to $y = 1$.

18. a) Find a value for A so that the line $y = 2x + A$ goes through the point (3,10).
   b) Find a value for B so that the line $y = Bx + 2$ goes through the point (3,10).
   c) Find a value for D so that the line $y = Dx + 7$ crosses the y–axis at $y = 4$.
   d) Find values for A and B so that the line $Ay = Bx + 1$ goes through the points (1,3) and (5,13).

19. Find the shortest distance between the circles with centers $C_1 = (1, 2)$ and $C_2 = (7, 10)$ and radii
   a) $r_1 = 2$ and $r_2 = 4$ b) $r_1 = 2$ and $r_2 = 7$ c) $r_1 = 5$ and $r_2 = 8$
   d) $r_1 = 3$ and $r_2 = 15$ e) $r_1 = 12$ and $r_2 = 1$

20. Find the equation of the circle with center C and radius r when
a) $C = (2,7)$ $r = 4$ b) $C = (3,-2)$ $r = 1$ c) $C = (-5,1)$ $r = 7$ d) $C = (-3,-1)$ $r = 4$

21. Explain how you can determine, without graphing, whether a given point $P = (x,y)$ is inside, on, or outside the circle with center $C = (h,k)$ and radius r.

22. A box with a height of 2 cm and a width of 8 cm is definitely big enough to hold two semicircular rods with radii of 2 cm (Fig. 22). Will these same two rods fit into a box 2 cm high and 7.6 cm wide? Will they fit in a box 2 cm high and 7.2 cm wide? (Suggestion: turn one of the rods over.)

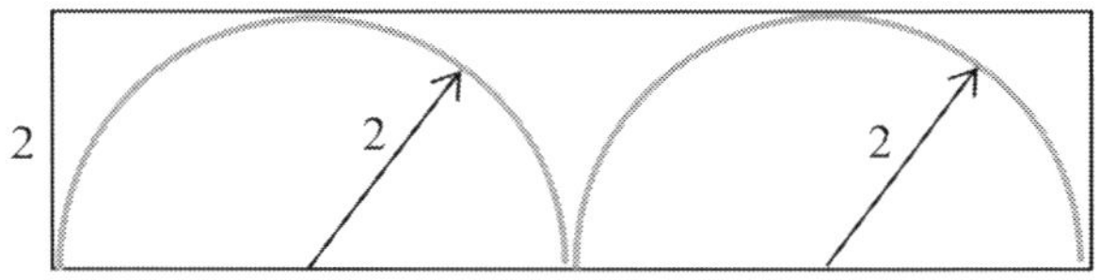

Fig. 22

23. Show that the equation of the circle with center $C = (h,k)$ and radius r is $(x - h)^2 + (y - k)^2 = r^2$.

24. Find the equation of the line which is tangent to the circle $x^2 + y^2 = 25$ at the point P when
a) $P = (3,4)$ b) $P = (-4,3)$ c) $P = (0,5)$ d) $P = (-5,0)$

25. Find the slope of the line which is tangent to the circle with center $C = (3,1)$ at the point P when
a) $P = (8,13)$ b) $P = (-10,1)$ c) $P = (-9,6)$ d) $P = (3,14)$

26. Find the center $C = (h,k)$ and the radius r of the circle which goes through the three points
a) (0,1) , (1,0) , and (0,5) b) (1,4) , (2,2) , and (8,2) c) (1,3) , (4,12) , and (8,4)

27. a) How close does the line $3x - 2y = 4$ come to the point (2,5)?
b) How close does the line $y = 5 - 2x$ come to the point (1,–2)?
c) How close does the circle with radius 3 and center at (2,3) come to the point (8,3)?

28. a) How close does the line $2x - 5y = 4$ come to the point (1,5)?
b) How close does the line $y = 3 - 2x$ come to the point (5,–2)?
c) How close does the circle with radius 4 and center at (4,3) come to the point (10,3)?

29. a) Show that the line L given by $Ax + By = C$ has slope $m = -A/B$. (Fig. 23)

b) Find the equation of the line L* through (0,0) which is perpendicular to line L in part (a).

c) Show that the lines L and L* intersect at the point

$$(x, y) = \left(\frac{AC}{A^2 + B^2} , \frac{BC}{A^2 + B^2} \right).$$

d) Show that the distance from the origin to the point (x,y) in part (c) is

$$\frac{|C|}{\sqrt{A^2 + B^2}} .$$

Steps (a) – (d) show that the distance from the origin to the line $Ax + By = C$ is $\dfrac{|C|}{\sqrt{A^2 + B^2}}$.

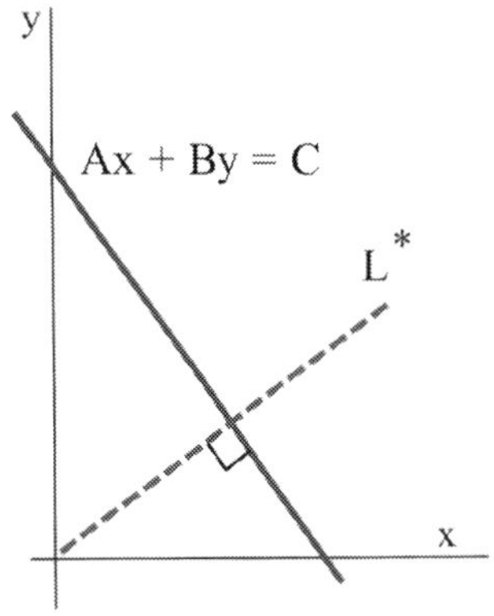

Fig. 23

30. Show that the distance from the point (p,q) to the line $Ax + By = C$ is $\dfrac{|Ap + Bq - C|}{\sqrt{A^2 + B^2}}$.

(The steps will be similar to those in problem 29, but the algebra will be more complicated.)

Section 0.2 **PRACTICE Answers**

Practice 1: Length = Dist(–7, –2) = | (–7) – (–2) | = | –5 | = **5**.

The midpoint is at $\frac{(-7)+(-2)}{2} = \frac{-9}{2} = \mathbf{-4.5}$.

Practice 2: Dist(P,Q) = Dist(P,r) so $\sqrt{(x-1)^2 + (y+4)^2} = \sqrt{(x-0)^2 + (y+3)^2}$.

Squaring each side and simplifying, we eventually have $\mathbf{y = x - 4}$.

Practice 3: The point P = (x , y) is on the circle when it is 5 units from the center C = (–2, 6) so Dist(P,C) = 5. Then Dist((x,y) , (–2,6)) = 5 so

$\sqrt{(x+2)^2 + (y-6)^2} = 5$ or $\mathbf{(x+2)^2 + (y-6)^2 = 25}$.

Practice 4: $\Delta x = 5 - (-3) = \mathbf{8}$, $\Delta y = -14 - 2 = \mathbf{-16}$, and slope = $\frac{\Delta y}{\Delta x} = \frac{-16}{8} = \mathbf{-2}$.

Practice 5: slope = $\frac{\Delta y}{\Delta x} = \frac{(-3+5h)-(-3)}{(2+h)-2} = \frac{5h}{h} = \mathbf{5}$.

The midpoint is at $\left(\frac{(2)+(2+h)}{2} , \frac{(-3+5h)+(-3)}{2} \right) = \left(\mathbf{2 + \frac{h}{2}} , \mathbf{-3 + \frac{5h}{2}} \right)$.

Practice 6: slope = $\frac{\Delta y}{\Delta x} = \frac{(3a^2 + 5a) - (3x^2 + 5x)}{a - x}$

$= \frac{3(a^2 - x^2) + 5(a-x)}{a-x} = \frac{3(a+x)(a-x) + 5(a-x)}{a-x} = \mathbf{3(a+x) + 5}$.

Practice 7: Let $y_1 = mx_1 + b$ and $y_2 = mx_2 + b$. Then

slope = $\frac{\Delta y}{\Delta x} = \frac{(mx_2 + b) - (mx_1 + b)}{x_2 - x_1} = \frac{m(x_2 - x_1)}{x_2 - x_1} = \mathbf{m}$.

Practice 8: The line 3x + 5y = 17 has slope $\frac{-3}{5}$ so the slope of the parallel line is $m = \frac{-3}{5}$.

Using the form $y = \frac{-3}{5}x + b$ and the point (–2, 3) on the line, we have

$3 = \frac{-3}{5}(-2) + b$ so $b = \frac{9}{5}$ and

$\mathbf{y = \frac{-3}{5}x + \frac{9}{5}}$ or $\mathbf{5y + 3x = 9}$ ..

Practice 9: The line 3y – 7x = 2 has slope $\frac{7}{3}$ so the slope of the perpendicular line is $m = \frac{-3}{7}$.

Using the form $y = \frac{-3}{7}x + b$ and the point (2, –5) on the line, we have

$-5 = \frac{-3}{7}(2) + b$ so $b = \frac{-29}{7}$ and

$\mathbf{y = \frac{-3}{7}x + \frac{-29}{7}}$ or $\mathbf{7y + 3x = -29}$.

0.3 FUNCTIONS AND THEIR GRAPHS

When you prepared for calculus, you learned to manipulate functions by adding, subtracting, multiplying and dividing them, as well as calculating functions of functions (composition). In calculus, we will still be dealing with functions and their applications. We will create new functions by operating on old ones. We will derive information from the graphs of the functions and from the derived functions. We will find ways to describe the point–by–point behavior of functions as well as their behavior "close to" some points and also over entire intervals. We will find tangent lines to graphs of functions and areas between graphs of functions. And, of course, we will see how these ideas can be used in a variety of fields.

This section and the next one are a review of information and procedures you should already know about functions before we begin calculus.

What is a function?

Definition of Function:

A **function** from a set X to a set Y is a **rule** for assigning to each element of the set X a single element of the set Y. A function assigns a unique (exactly one) output element in the set Y to each input element from the set X.

The **rule** which defines a function is often given by an equation, but it could also be given in words or graphically or by a table of values. In practice, functions are given in all of these ways, and we will use all of them in this book.

In the definition of a function, the set X of all inputs is called the **domain** of the function. The set Y of all outputs produced from these inputs is called the **range** of the function. Two different inputs, elements in the domain, can be assigned to the same output, an element in the range, but one input cannot lead to 2 different outputs.

Most of the time we will work with functions whose domains and ranges are real numbers, but there are other types of functions all around us. Final grades for this course is an example of a function. For each student, the instructor will assign a final grade based on some rule for evaluating that student's performance. The **domain** of this function consists of all students registered for the course, and the **range** consists of the letters A, B, C, D, F, and perhaps W (withdrawn). Two students can receive the same final grade, but only one grade will be assigned to each student.

Function Machines

Functions are abstract structures, but sometimes it is easier to think of them in a more concrete way. One way is to imagine that a function is a special purpose computer, a machine which accepts inputs, does something to those inputs according to the defining rule, and produces an output. The output is the value of the function for the given input value. If the defining rule for a function f is "multiply the input by itself", f(input) = (input)(input), then Fig. 1 shows the results of putting the inputs x, 5, a, c + 3 and x + h into the machine f.

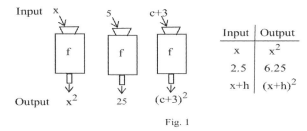

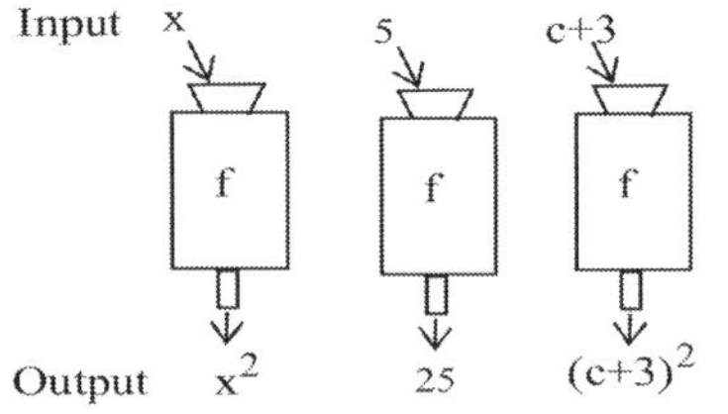

Input	Output
x	x^2
2.5	6.25
x+h	$(x+h)^2$

Fig. 1

Practice 1: If we have a function machine g whose rule is "divide 3 by the input and add 1", $g(x) = 3/x + 1$, what outputs do we get from the inputs x, 5, a, c + 3 and x + h ? What happens if we put 0 into the machine g?

You expect your calculator to behave as a function: each time you press the same input sequence of keys you expect to see the same output display. In fact, if your calculator did not produce the same output each time you would need a new calculator. (On many calculators there is a key which does not produce the same output each time you press it. Which key is that?)

Functions Defined by Equations

If the domain consists of a collection of real numbers (perhaps all real numbers) and the range is a collection of real numbers, then the function is called a **numerical function**. The rule for a numerical function can be given in several ways, but it is usually written as a formula. If the rule for a numerical function, f, is "the output is the input number multiplied by itself", then we could write the rule as $f(x) = x \cdot x = x^2$. The use of an "x" to represent the input is simply a matter of convenience and custom. We could also represent the same function by $f(a) = a^2$, $f(€\#) = \#^2$ or $f(\text{ input }) = (\text{ input })^2$.

For the function f defined by $f(x) = x^2 - x$, we have that $f(3) = 3^2 - 3 = 6$, $f(.5) = (.5)^2 - (.5) = -.25$, and $f(-2) = (-2)^2 - (-2) = 6$. Notice that the two different inputs, 3 and –2, both lead to the output of 6. That is allowable for a function. We can also evaluate f if the input contains variables. If we replace the "x" with something else in the notation "f(x)", then we must replace the "x" with the same thing everywhere in the equation:

$f(\mathbf{c}) = \mathbf{c}^2 - \mathbf{c}$, $f(\mathbf{a+1}) = (\mathbf{a+1})^2 - (\mathbf{a+1}) = (a^2 + 2a + 1) - (a + 1) = a^2 + a$,

$f(\mathbf{x+h}) = (\mathbf{x+h})^2 - (\mathbf{x+h}) = (x^2+2xh+h^2) - (x+h)$, and, in general, $f(\textbf{input}) = (\textbf{input})^2 - (\textbf{input})$.

For more complicated expressions, we can just proceed step–by–step:

$$\frac{f(\mathbf{x+h}) - f(x)}{h} = \frac{\{(\mathbf{x+h})^2 - (\mathbf{x+h})\} - \{x^2 - x\}}{h} = \frac{\{(x^2+2xh+h^2) - (x+h)\} - \{x^2 - x\}}{h}$$

$$= \frac{2xh + h^2 - h}{h} = \frac{h(2x + h - 1)}{h} = 2x + h - 1 \ .$$

Practice 2: For the function g defined by $g(t) = t^2 - 5t$, evaluate $g(1)$, $g(-2)$, $g(w+3)$, $g(x+h)$, $g(x+h) - g(x)$, and $\dfrac{g(x+h) - g(x)}{h}$.

Functions Defined by Graphs and Tables of Values

The **graph** of a numerical function f consists of a plot of ordered pairs (x,y) where x is in the domain of f and $y = f(x)$. A **table of values** of a numerical function consists of a list of some of the ordered pairs (x,y) where $y = f(x)$. Fig. shows a graph of $f(x) = \sin(x)$ for $-4 \le x \le 9$.

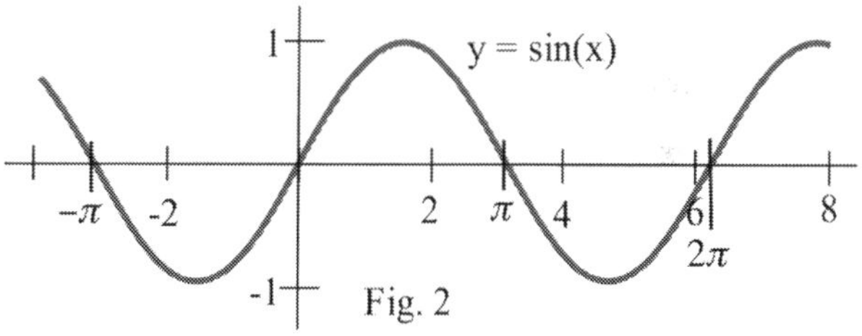

Fig. 2

A function can be defined by a graph or by a table of values, and these types of definitions are common in applied fields. The outcome of an experiment will depend on the input, but the experimenter may not know the "rule" for predicting the outcome. In that case, the experimenter usually represents the experiment function as a table of measured outcome values verses input values or as a graph of the outcomes verses the inputs. The table and graph in Fig. 3 show the deflections obtained when weights are loaded at the end of a wooden stick. The graph in Fig. 4 shows the temperature of a hot cup of tea as a function of the time as it sits in a 68° F room. In these experiments, the "rule" for the function is that f(input) = actual outcome of the experiment.

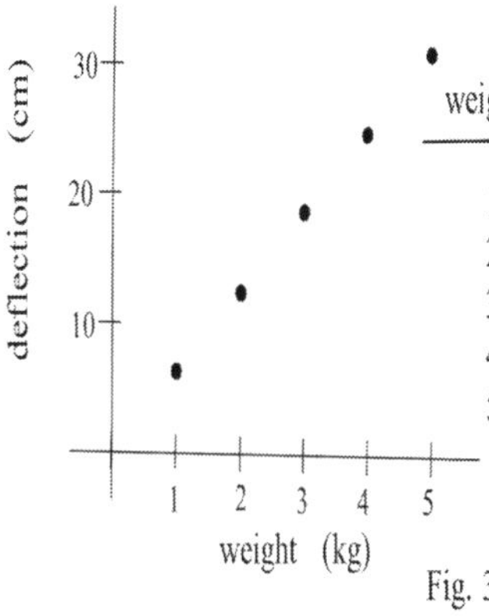

weight	deflection
1	6.2
2	12.4
3	18.6
4	24.8
5	31.0

Fig. 3

Tables have the advantage of presenting the data explicitly, but it is often difficult to detect patterns simply from lists of numbers. Graphs tend to obscure some of the precision of the data, but patterns are much easier to detect visually — we can actually see what is happening with the numbers.

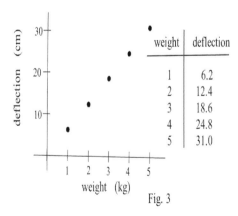

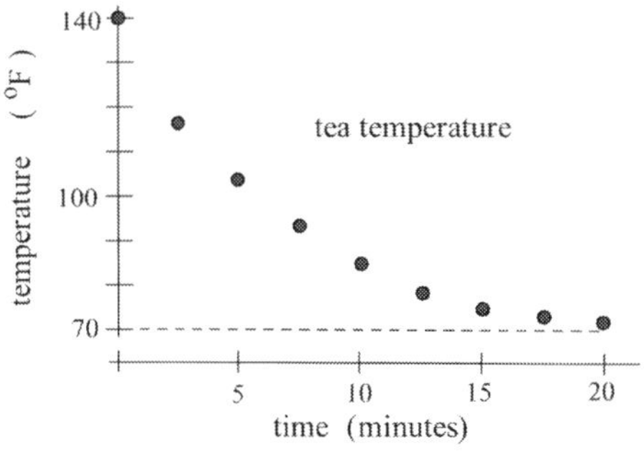

Fig. 4

Creating Graphs of Functions

Most people understand and can interpret pictures more quickly than tables of data or equations, so if we have a function defined by a table of values or by an equation, it is often useful and necessary to create a picture of the function, a graph.

A Graph from a Table of Values

If we have a table of values of the function, perhaps consisting of measurements obtained from an experiment, then we can simply plot the ordered pairs in the xy–plane to get a graph which consists of a collection of points.

Fig. 5 shows the lengths and weights of trout caught (and released) during several days of fishing. It also shows a line which comes "close" to the plotted points. From the graph, you could estimate that a 17 inch trout would weigh slightly more than one pound.

length (in)	weight (pds)
13.5	0.4
14.5	0.9
15.0	0.7
16.0	0.9
18.0	1.2
18.5	1.6
19.5	1.5
20.5	1.7
20.5	2.1

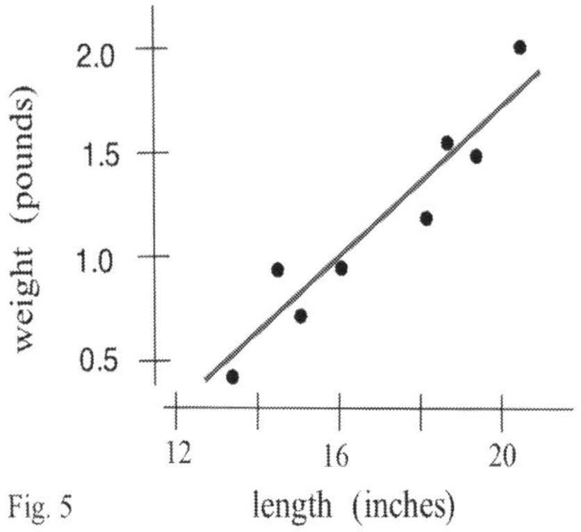

Fig. 5

A Graph from an Equation

Creating the graph of a function given by an equation is similar to creating one from a table of values — we need to plot enough points (x,y) where y = f(x) so we can be confident of the shape and location of the graph of the entire function. We can find a point (x,y) which satisfies y = f(x) by picking a value for x and then calculating the value for y by evaluating f(x). Then we can enter the (x,y) value in a table or simply plot the point (x,y) .

If you recognize the form of the equation and know something about the shape of graphs of that form, you may not have to plot many points. If you do not recognize the form of the equation then you will have to plot more points, maybe 10 or 20 or 234: it depends on how complicated the graph appears and on how important it is to you (or your boss) to have an accurate graph. Evaluating y = f(x) at a lot of different values for x and then plotting the points (x,y) is usually not very difficult, but it can be very time–consuming. Fortunately, there are now calculators and personal computers which will do the evaluations and plotting for you.

Is every graph the graph of a function?

The definition of function requires that each element of the domain, each input value, be sent by the function to exactly one element of the range, to exactly one output value, so for each input x-value there will be exactly one output y–value, y = f(x). The points (x, y_1) and (x,y_2) cannot both be on the graph of f unless $y_1 = y_2$. The graphic interpretation of this result is called the Vertical Line Test.

> **Vertical Line Test for a Function:**
> A graph is the graph of a function if and only if a vertical line drawn through any point <u>in the domain</u> intersects the graph at <u>exactly one</u> point.

Fig. 6(a) shows the graph of a function. Figs. 6(b) and 6(c) show graphs which are not the graphs of functions, and vertical lines are shown which intersect those graphs at more than one point. Non–functions are not "bad", and sometimes they are necessary to describe complicated phenomena.

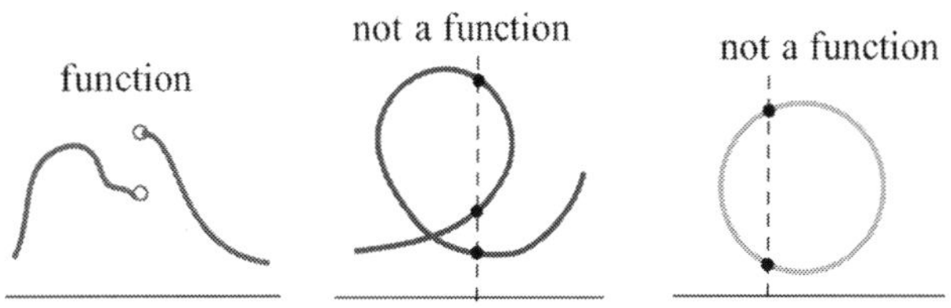

Fig. 6

Reading Graphs Carefully

Calculators and computers can help students, reporters, business people and scientific professionals create graphs quickly and easily, and because of this, graphs are being used more often than ever to present information and justify arguments. And this text takes a distinctly graphical approach to the ideas and meaning of calculus. Calculators and computers can help us create graphs, but we need to be able to read them carefully. The next examples illustrate some types of information which can be obtained by carefully reading and understanding graphs.

Example 1: A boat starts from St. Thomas and sails due west with the velocity shown in Fig. 7

(a) When is the boat traveling the fastest?

(b) What does a negative velocity away from St. Thomas mean?

(c) When is the boat the farthest from St. Thomas?

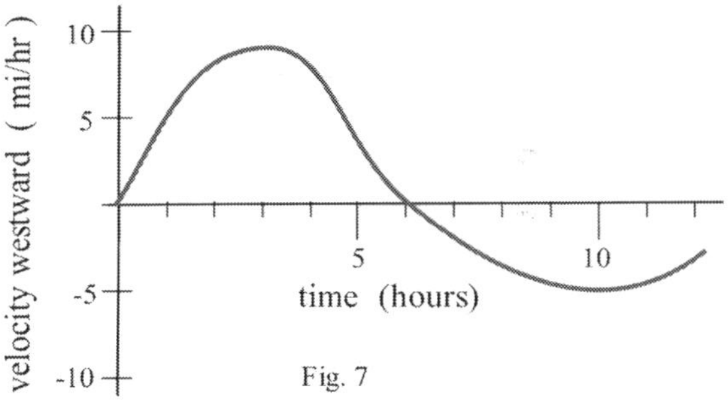

Fig. 7

Solution: (a) The greatest speed is 10 mph at $t = 3$ hours.

(b) It means that the boat is heading back toward St. Thomas.

(c) The boat is farthest from St. Thomas at $t = 6$ hours. For $t < 6$ the boat's velocity is positive, and the distance from the boat to St. Thomas is increasing. For $t > 6$ the boat's velocity is negative, and the distance from the boat to St. Thomas is decreasing.

Practice 3: You and a friend start out together and hike along the same trail but walk at different speeds (Fig. 8).

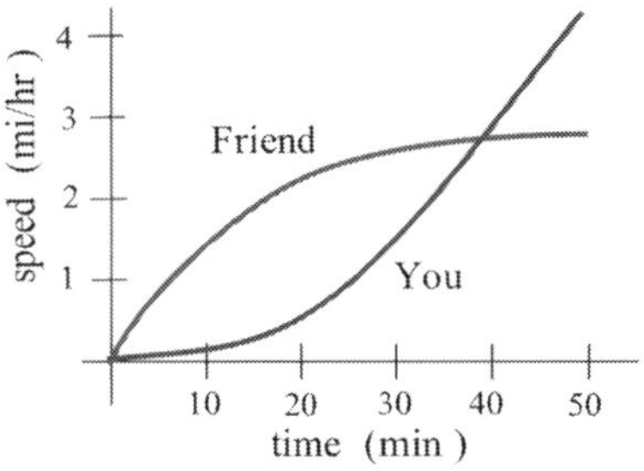

Fig. 8

(a) Who is walking faster at $t = 20$?

(b) Who is ahead at $t = 20$?

(c) When are you and your friend farthest apart?

(d) Who is ahead when $t = 50$?

Example 2: In Fig. 9, which has the largest slope: the line through the points A and P, the line through B and P, or the line through C and P?

Solution: The line through C and P has the largest slope: $m_{PC} > m_{PB} > m_{PA}$.

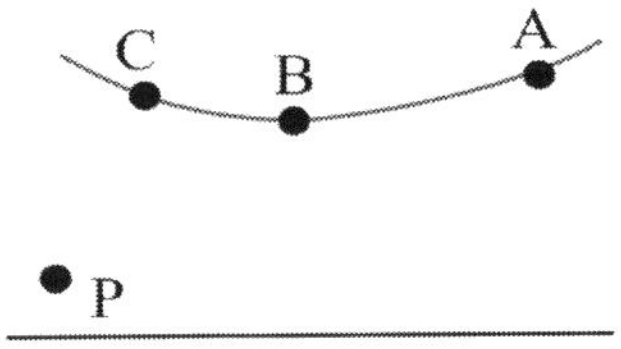

Fig. 9

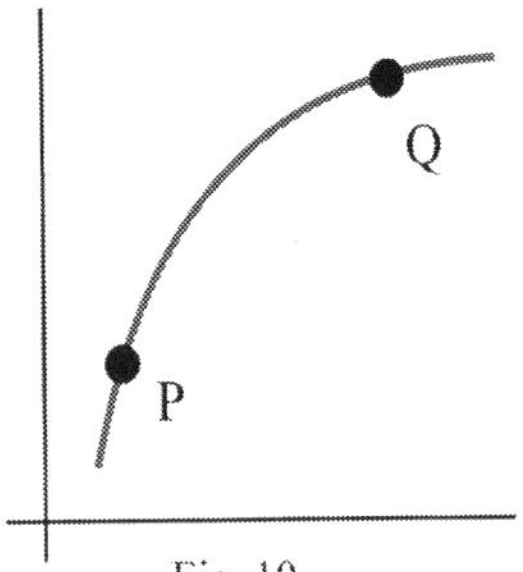

Fig. 10

Practice 4: In Fig. 10, the point Q on the curve is fixed, and the point P is moving to the right along the curve toward the point Q. As P moves toward Q:

(a) the x–coordinate of P is Increasing, Decreasing, Remaining constant, or None of these.

(b) the x–increment from P to Q is Increasing, Decreasing, Remaining constant, or None of these

(c) the slope from P to Q is Increasing, Decreasing, Remaining constant, or None of these.

Example 3: The graph of $y = f(x)$ is shown in Fig. 11. Let $g(x)$ be the **slope** of the line tangent to the graph of $f(x)$ at the point $(x,f(x))$.

(a) Estimate the values $g(1)$, $g(2)$ and $g(3)$.

(b) When does $g(x) = 0$? (c) At what value(s) of x is $g(x)$ largest? (d) Sketch the graph of $y = g(x)$.

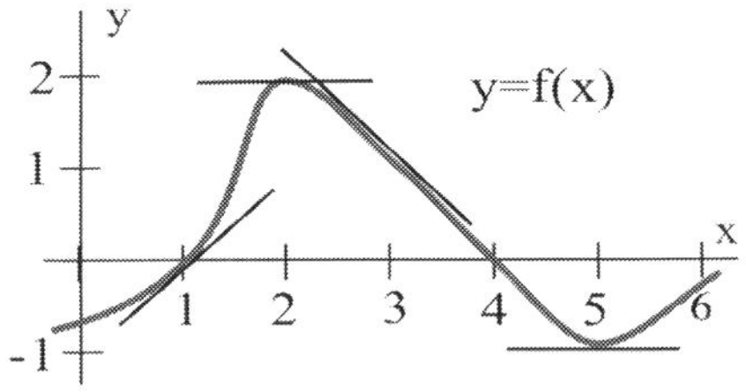

Fig. 11

Solution: (a) Fig. 11 shows the graph $y = f(x)$ with several tangent lines to the graph of f. From Fig. 11 we can estimate that $g(1)$ (the <u>slope</u> of the line tangent to the graph of f at $(1,0)$) is approximately equal to 1. Similarly, $g(2) \approx 0$ and $g(3) \approx -1$.

(b) The slope of the tangent line appears to be horizontal (slope = 0) at $x = 2$ and at $x = 5$.

(c) The tangent line to the graph appears to have greatest slope (be steepest) near $x = 1.5$.

(d) We can build a table of values of $g(x)$ and then sketch the graph of these values.

x	f(x)	g(x) = tangent slope at (x, f(x))
0	–1	.5
1	0	1
2	2	0
3	1	–1
4	0	–1
5	–1	0
6	–.5	.5

The graph y = g(x) is given in Fig. 12.

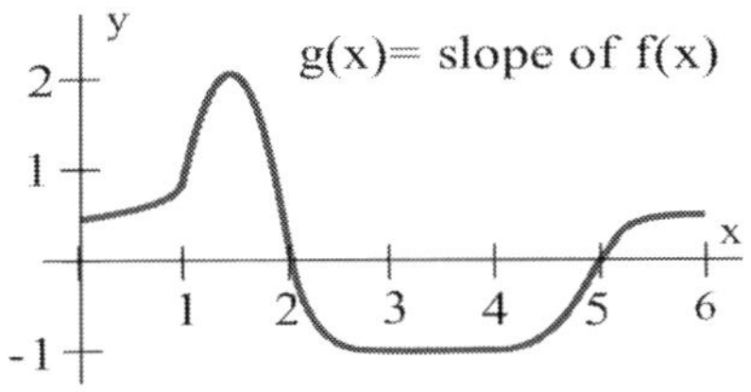

Fig. 12

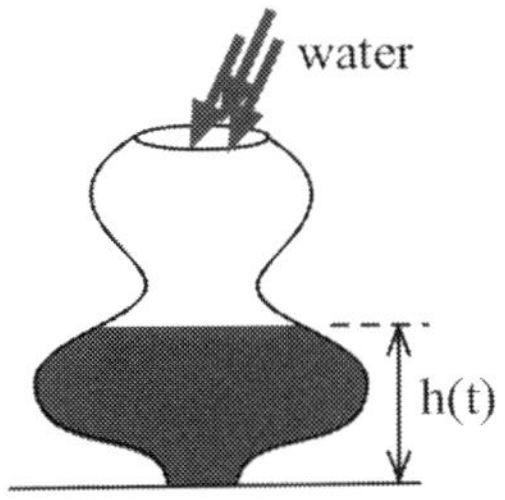

Fig. 13

Practice 5: Water is flowing into a container (Fig. 13) at a constant rate of 3 gallons per minute. Starting with an empty container, sketch the graph of the height of the water in the container as a function of time.

Problems

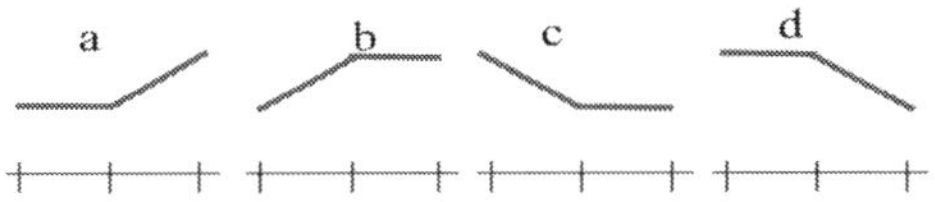

Fig. 14

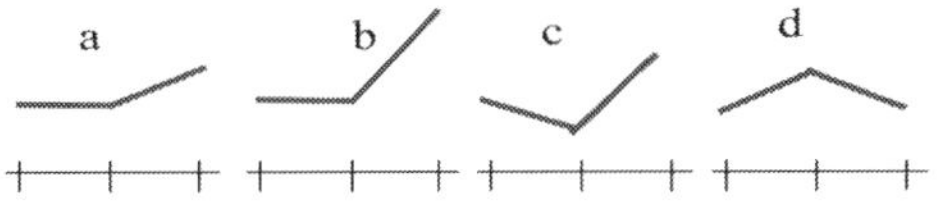

Fig. 15

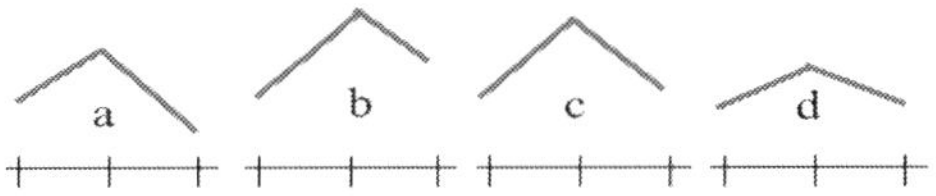

Fig. 16

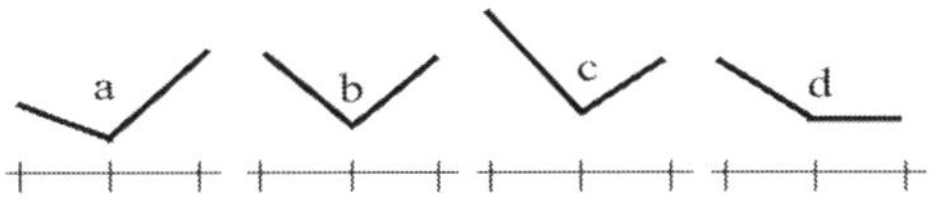

Fig. 17

In problems 1 – 4, use the shapes and slopes of the data to match the given numerical triples to the graphs in the figures. (For example, A: 3, 3, 6 in Problem 1. is "over and up" so it matches graph (a) in Fig. 14. B is "down and over" so it matches graph (c) in Fig. 14.)

1. Fig. 14. Data: A: 3, 3, 6 B: 12, 6, 6 C: 7, 7, 3 D: 2, 4, 4

2. Fig. 15. Data: A: 7, 10, 7 B: 17.3, 17.3, 25 C: 4, 4, 8 D: 12, 8, 16

3. Fig. 16. Data: A: 7, 14, 10 B: 23, 45, 22 C: 0.8, 1.2, 0.8 D: 6, 9, 3

4. Fig. 17. Data: A: 6, 3, 9 B: 18, 10, 10 C: 12, 6, 9 D: 3.7, 1.9, 3.6

5. Water is flowing into each of the bottles in Fig. 18 at a steady rate. Match each bottle shape with the graph of the height of the water as a function of time.

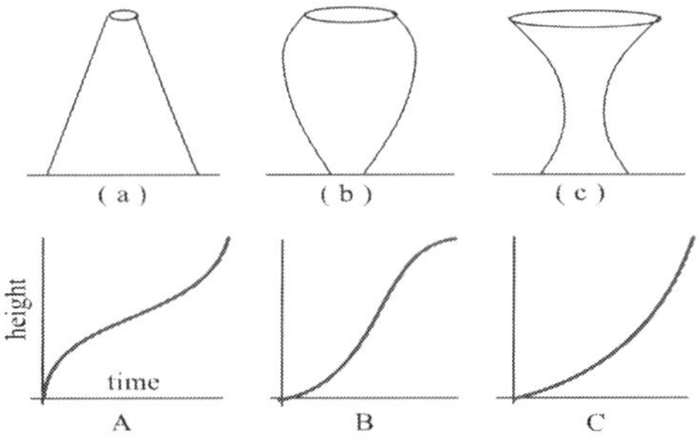

Fig. 18

6. Sketch the shapes of bottles which will have the water height versus time graphs in Fig. 19.

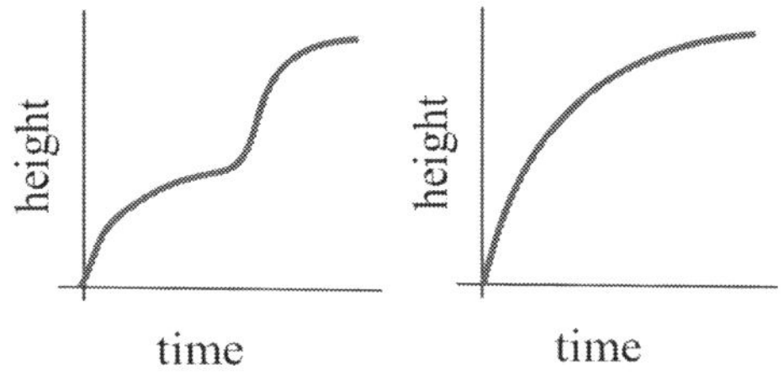

Fig. 19

7. $f(x) = x^2 + 3$, $g(x) = \sqrt{x-5}$, and $h(x) = \frac{x}{x-2}$

(a) evaluate f(1), g(1) and h(1)

(b) graph f(x), g(x) and h(x) for $-5 \le x \le 10$

(c) evaluate f(3x), g(3x) and h(3x)

(d) evaluate f(x+h), g(x+h) and h(x+h)

8. Find the slope of the line through the points P and Q when

(a) P = (1,3), Q = (2,7)

(b) $P = (x, x^2 + 2)$, $Q = (x+h, (x+h)^2 + 2)$

(c) P = (1,3), $Q = (x, x^2 + 2)$ What are the values of these slopes in (c) if x = 2, x = 1.1, x = 1.002?

9. Find the slope of the line through the points P and Q when

(a) P = (1,5), Q = (2,7)

(b) $P = (x, x^2 + 3x - 1)$, $Q = (x+h, (x+h)^2 + 3(x+h) - 1)$

(c) P = (1,3), $Q = (x, x^2 + 3x - 1)$ What are the values of these slopes in (c) if x = 1.3, x = 1.1, x = 1.002?

10. $f(x) = x^2 + x$ and g(x) = 3/x. Evaluate and simplify $\frac{f(a+h) - f(a)}{h}$ and $\frac{g(a+h) - g(a)}{h}$ when a = 1, a = 2, a = –1, a = x.

11. $f(x) = x^2 - 2x$ and $g(x) = \sqrt{x}$. Evaluate and simplify $\frac{f(a+h) - f(a)}{h}$ and $\frac{g(a+h) - g(a)}{h}$ when a = 1, a = 2, a = 3, a = x.

12. The temperatures in Fig. 20 were recorded during a 12 hour period in Chicago.

(a) At what time was the temperature the highest? Lowest?

(b) How fast was the temperature rising at 10 am? At 1 pm?

(c) What could have caused the drop in temperature between 1 pm and 3 pm?

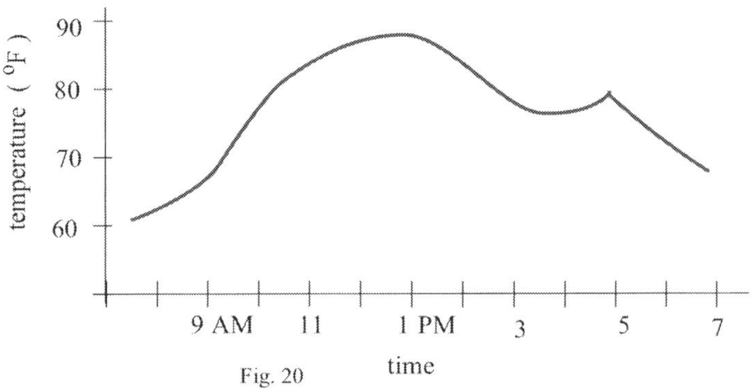

Fig. 20

13. The graph in Fig. 21 shows the distance of an airplane from an airport during a several hour flight.
    (a) How far was the airplane from the airport at 1 pm? At 2 pm?
    (b) How fast was the distance changing at 1 pm?
    (c) How could the distance from the plane to the airport remain unchanged from 1:45 pm until 2:30 pm without the airplane falling?

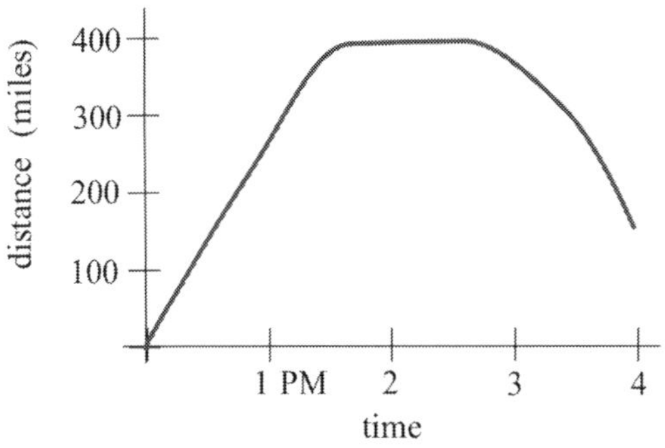

Fig. 21

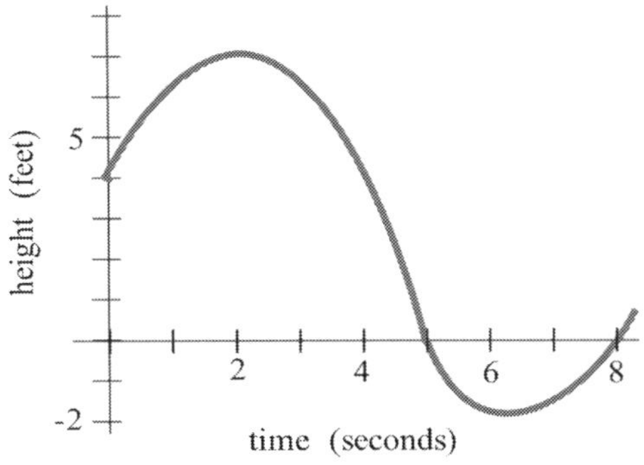

Fig. 22

14. The graph in Fig. 22 shows the height of a diver above the water level at time t seconds.
    (a) What was the height of the diving board?
    (b) When did the diver hit the water?
    (c) How deep did the diver get?
    (d) When did the diver return to the surface?

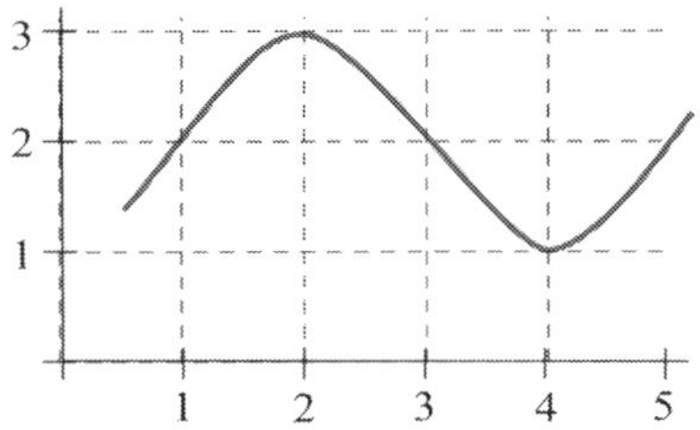

Fig. 23

15. a) Sketch the lines tangent to the curve in Fig. 23 at $x = 1, 2, 3, 4,$ and 5.
    b) For what value(s) of x is the value of the function largest? Smallest?
    c) For what value(s) of x is the slope of the tangent line largest? Smallest?

16. Fig. 24 shows the height of the water (above and below mean sea level) at a Maine beach.
    a) At which time(s) was the most beach exposed? The least?
    b) At which time(s) was the current the strongest?

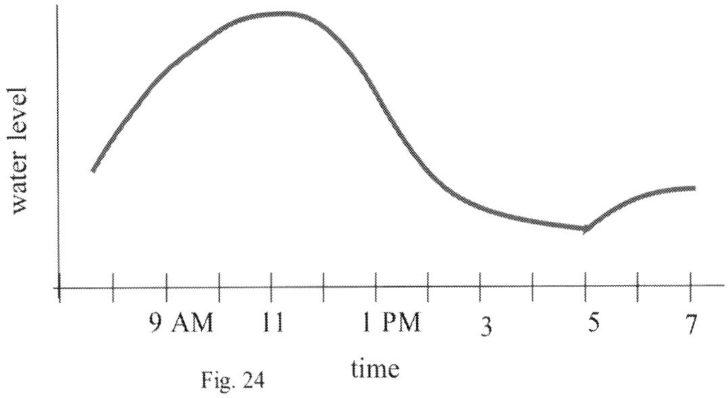

Fig. 24

17. Imagine that you are ice skating, from **left to right**, along the path in Fig. 25. Sketch the path you will follow if you fall at points A, B, and C.

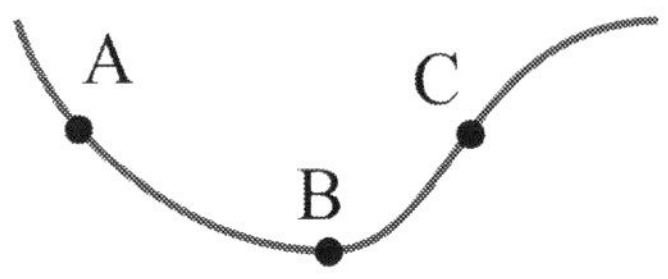

Fig. 25

18. Define s(x) to be the **slope** of the line through the points (0,0) and (x, f(x)) in Fig. 26 . (For example, s(3) = { **slope** of the line through (0,0) and (3, f(3)) } = 4/3.)

a) Evaluate s(1), s(2), and s(4). b) For which integer value of x between 1 and 7 is s(x) smallest?

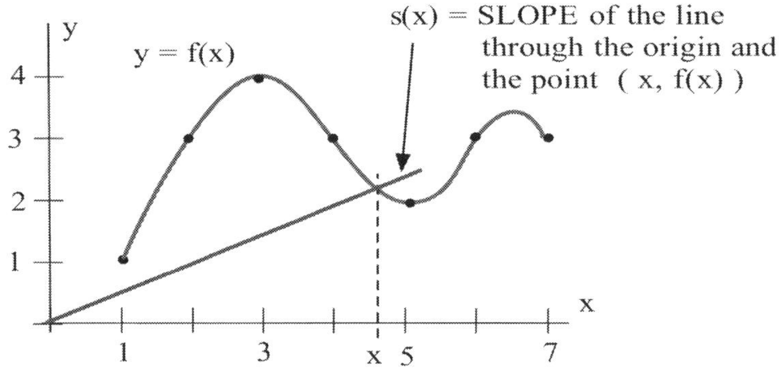

Fig. 26

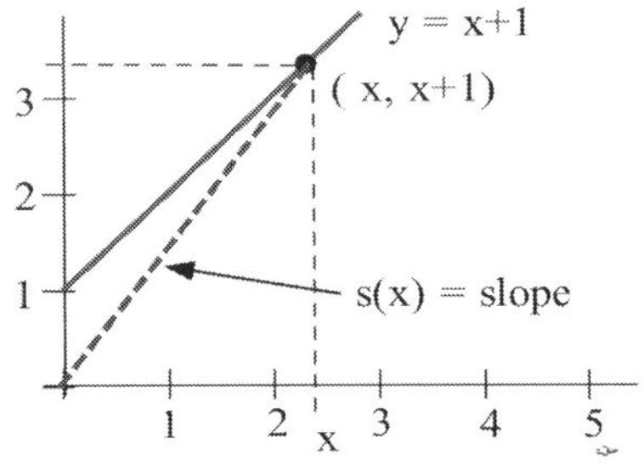

Fig. 27

19. Let f(x) = x + 1 and define s(x) to be the **slope** of the line through the points (0,0) and (x, f(x)) in Fig. 27 . (For example, s(2) = { slope of the line through (0,0) and (2,3) } = 3/2.)

a) Evaluate s(1), s(3) and s(4).

b) Find an equation for s(x) in terms of x.

20. Define A(x) to be the area of the rectangle bounded by the axes, the line y = 2, and a vertical line at x as in Fig. 28. (For example, A(3) = area of a 2 by 3 rectangle = 6.)

a) Evaluate A(1), A(2) and A(5).

b) Find an equation for A(x) in terms of x.

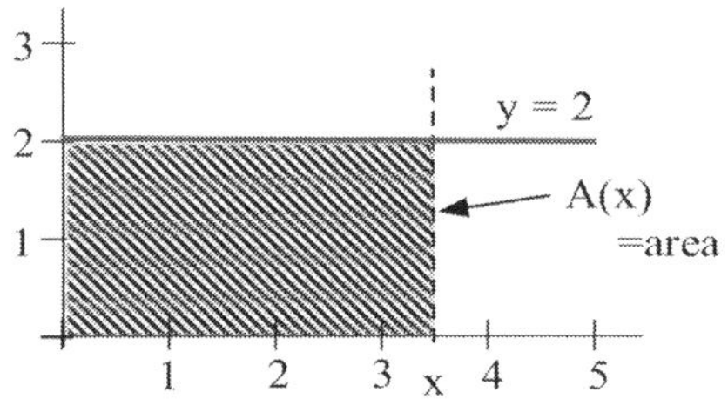

Fig. 28

21. Use the graph of $y = f(x)$ in Fig. 29 to complete the table. (You will have to estimate the values of the slopes.)

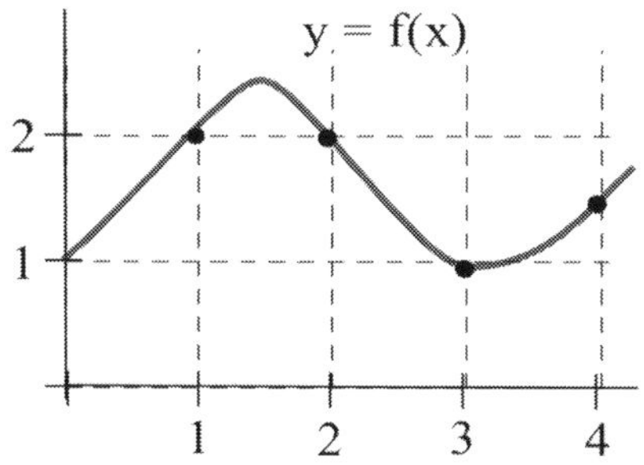

Fig. 29

x	f(x)	**slope** of the line **tangent** to the graph of f at (x, f(x))
0	1	1
1		
2		
3		
4		

22. Sketch the graphs of water height versus time for water pouring into a bottle shaped like:

(a) a milk carton (b) a spherical glass vase
(c) an oil drum (cylinder) lying on its side
(d) a giraffe (e) you.

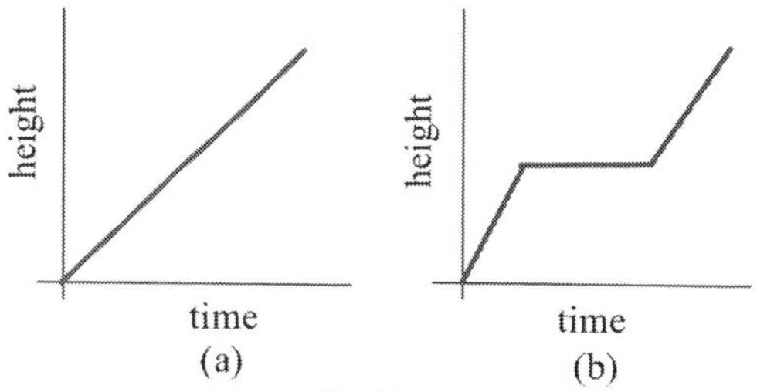

Fig. 30

23. Design bottles whose graphs of (highest) water height versus time will look like those in Fig. 30.

Section 0.3 **PRACTICE Answers**

Practice 1:

Input	Output
x	$\frac{3}{x} + 1$
5	$\frac{3}{5} + 1 = 1.6$
a	$\frac{3}{a} + 1$
0	$g(0) = \frac{3}{0} + 1$ which is not defined because of division by 0.

Input	Output
c + 3	$\frac{3}{c+3} + 1$
x + h	$\frac{3}{x+h} + 1$

Practice 2: $g(t) = t^2 - 5t$.

$g(1) = 1^2 - 5(1) = \mathbf{-4}$. $g(-2) = (-2)^2 - 5(-2) = \mathbf{14}$.

$g(w + 3) = (w + 3)^2 - 5(w + 3) = w^2 + 6w + 9 - 5w - 15 = \mathbf{w^2 + w - 6}$.

$g(x + h) = (x + h) - 5(x + h) = \mathbf{x^2 + 2xh + h^2 - 5x - 5h}$.

$g(x + h) - g(x) = (x^2 + 2xh + h^2 - 5x - 5h) - (x^2 - 5x) = \mathbf{2xh + h^2 - 5h}$.

$$\frac{g(x + h) - g(x)}{h} = \frac{2xh + h^2 - 5h}{h} = \mathbf{2x + h - 5} .$$

Practice 3: (a) **Friend** (b) **Friend**

(c) At $\mathbf{t = 40}$. Before that your friend is walking faster and increasing the distance between you. Then you start to walk faster than your friend and start to catch up.

(d) **Friend**. You are walking faster than your friend at $t = 50$, but you still have not caught up.

Practice 4: (a) The x–coordinate is **increasing**. (b) The x–increment Δx is **decreasing**.

(c) The slope of the line through P and Q is **decreasing**.

Practice 5: See Fig. 31.

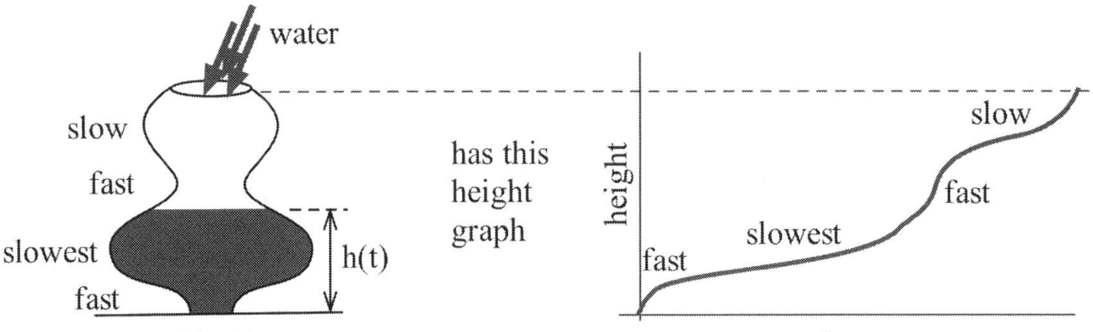

Fig. 13

Fig. 31

0.4 COMBINATIONS OF FUNCTIONS

Multiline Definitions of Functions -- Putting Pieces Together

Sometimes a physical or economic situation behaves differently depending on circumstances, and a more complicated function may be needed to describe the situation.

Sales Tax: Some states have different rates of sale tax depending on the type of item purchased. A "luxury item" may be taxed at 12%, food may have no tax, and all other items may have a 6% tax. We could describe this situation by using a **multiline** function, a function whose defining rule consists of several pieces. Which piece of the rule we need to use will depend on what we buy. In this example we could define the tax T on an item which costs x to be

$$T(x) = \begin{cases} 0 & \text{if x is the cost of a food} \\ 0.12x & \text{if x is the cost of a luxury item} \\ 0.06x & \text{if x is the cost of any other item.} \end{cases}$$

To find the tax on a \$2 can of stew, we would use the first piece of the rule and find that the tax is 0. To find the tax on a \$30 pair of earrings, we would use the second piece of the rule and find that the tax is \$3.60 . The tax on a \$20 book requires using the third rule, and the tax is \$1.20 .

Wind Chill Index: The rate at which a person's body loses heat depends on the temperature of the surrounding air and on the speed of the air. You lose heat more quickly on a windy day than you do on a day with little or no wind. Scientists have experimentally determined this rate of heat loss as a function of temperature and wind speed, and the resulting function is called the Wind Chill Index, WCI . The WCI is the temperature on a still day (no wind) at which your body would lose heat at the same rate as on the windy day. For example, the WCI value for 30° F air moving at 15 miles per hour is 9° F: your body loses heat as quickly on a 30° F day with a 15 mph wind as it does on a 9° F day with no wind.

If T is the Fahrenheit temperature of the air and v is the speed of the wind in miles per hour, then the WCI is a multiline function of the wind speed v (and of the temperature T):

$$\text{WCI} = \begin{cases} T & \text{if } 0 \le v \le 4 \\ 91.4 - \dfrac{10.45 + 6.69\sqrt{v} - 0.447v}{22}(91.5 - T) & \text{if } 4 \le v \le 45 \\ 1.60T - 55 & \text{if } v > 45 \end{cases}$$

The WCI value for a still day ($0 \le v \le 4$ mph) is just the air temperature. The WCI values for wind speeds above 45 mph are the same as the WCI value for a wind speed of 45 mph. The WCI values for wind speeds between 4 mph and 45 mph decrease as the wind speeds increase.

This WCI function depends on two variables, the temperature and the wind speed. However, if the temperature is constant, then the resulting formula for the WCI will only depend on the speed of the wind. If the air temperature is 30° F ($T = 30$), then the formula for the Wind Chill Index is

$$WCI_{30} = \begin{cases} 30^o & \text{if } 0 \le v \le 4 \text{ mph} \\ 62.19 - 18.70\sqrt{v} + 1.25v & \text{if } 4 \le v \le 45 \text{ mph} \\ -7^0 & \text{if } 45 \le v \text{ mph} \end{cases}$$

The graphs of the the Wind Chill Indices are shown on Fig. 1 for temperatures of 40° F, 30° F and 20° F. (From UMAP Module 658, Windchill by William Bosch and L.G. Cobb, 1984.)

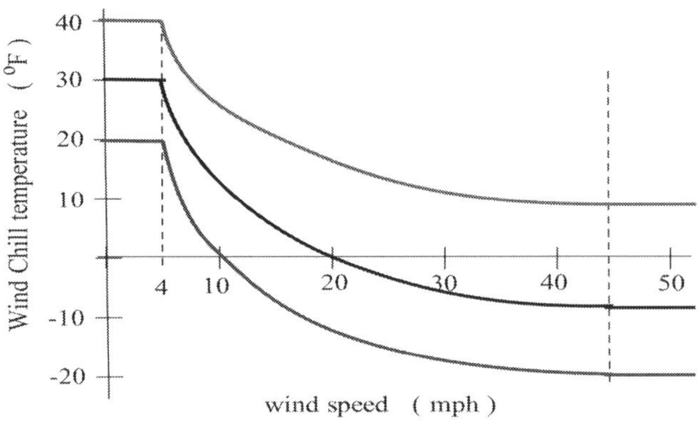

Fig. 1

Practice 1: A motel charges $50 per night for a room during the tourist season from June 1 through September 15, and $40 per night otherwise. Define a multiline function which describes these rates.

Example 1: Define $f(x) = \begin{cases} 2 & \text{if } x < 0 \\ 2x & \text{if } 0 \le x < 2 \\ 1 & \text{if } 2 < x \end{cases}$

Evaluate $f(-3), f(0), f(1), f(4)$ and $f(2)$. Graph $y = f(x)$ for $-1 \le x \le 4$.

Solution: To evaluate the function for different values of x, we must first decide which line of the rule applies. If $x = -3 < 0$, then we need to use the first line of the rule, and $f(-3) = 2$. When $x = 0$ or $x = 1$, we need the second line of the function definition, and then $f(0) = 2(0) = 0$ and $f(1) = 2(1) = 2$. At $x = 4$ the third line is needed, and $f(4) = 1$. Finally, at $x = 2$, none of the lines apply: the second line requires $x < 2$ and the third line requires $2 < x$, so $f(2)$ is undefined. The graph of $f(x)$ is given in Fig. 2. Note the "hole" above $x = 2$ since $f(2)$ is not defined by this rule for f.

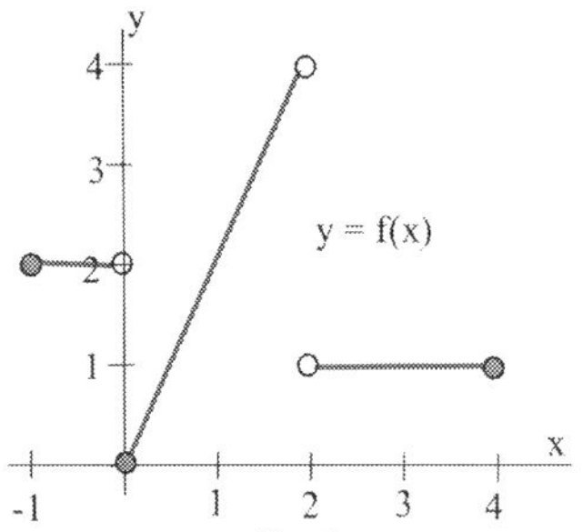

Fig. 2

Practice 2: Define $g(x) = \begin{cases} x & \text{if } x < -1 \\ 2 & \text{if } -1 \le x < 1 \\ -x & \text{if } 1 < x \le 3 \\ 1 & \text{if } 4 < x. \end{cases}$ Graph $y = g(x)$ for $-3 \le x \le 6$ and

evaluate $g(-3), g(-1), g(0), g(1/2), g(1), g(\pi/3), g(2), g(3), g(4)$ and $g(5)$.

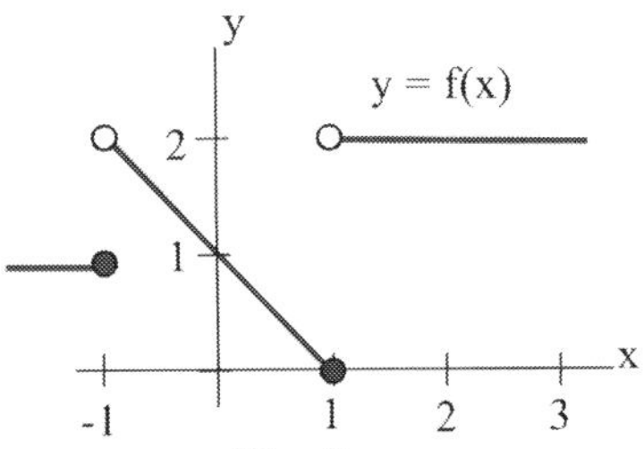

Fig. 3

Practice 3: Write a multiline function definition for the function whose graph is given in Fig. 3.

We can think of a multiline function definition as a machine which first examines the input value to decide which line of the function rule to apply (Fig. 4).

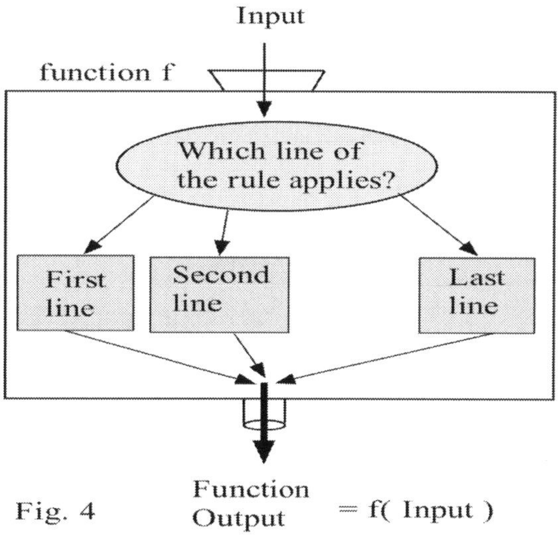

Fig. 4

Composition of Functions — Functions of Functions

Basic functions are often combined with each other to describe more complicated situations. Here we will consider the composition of functions, functions of functions.

> **Definition:** The **composite** of two functions f and g , written f∘g , is **f∘g(x) ≡ f(g(x))**.

The domain of the composite function f∘g(x) = f(g(x)) consists of those x–values for which g(x) and f(g(x)) are both defined — we can evaluate the composition of two functions at a point x only if each step in the composition is defined.

If we think of our functions as machines, then composition is simply a new machine consisting of an arrangement of the original machines. The composition f∘g of the function machines f and g shown in Fig. 5(a) is an arrangement of the machines so that the original input x goes into machine g , the output from machine g becomes the input into machine f , and the output from machine f is our final output. The composition of the function machines f∘g(x) = f(g(x)) is only valid if x is an allowable input into g (x is in the domain of g) and if g(x) is then an allowable input into f . The composition g∘f involves arranging the machines so the original input goes into f , and the output from f then becomes the input for g (Fig. 5(b)).

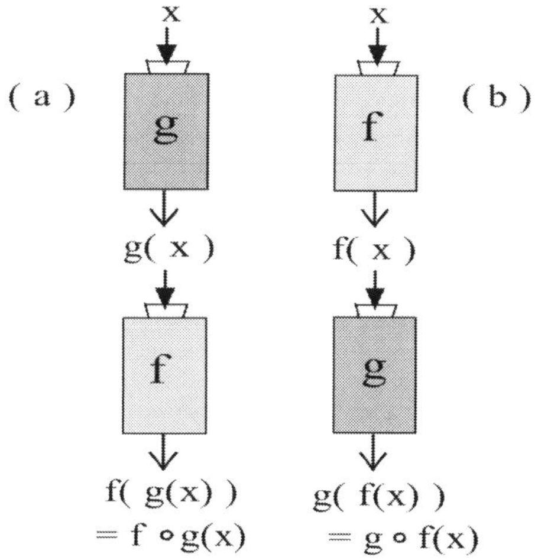

Fig. 5

Example 2: For $f(x) = \sqrt{x-2}$, $g(x) = x^2$, and $h(x) = \begin{cases} 3x & \text{if } x < 2 \\ x-1 & \text{if } 2 \le x \end{cases}$,

evaluate $f\circ g(3)$, $g\circ f(6)$, $f\circ h(2)$ and $h\circ g(-3)$. Find the equations and domains of $f\circ g(x)$ and $g\circ f(x)$.

Solution: $f\circ g(3) = f(\,g(3)\,) = f(\,3^2\,) = f(\,9\,) = \sqrt{9-2} = \sqrt{7} \approx 2.646$

$g\circ f(6) = g(\,f(6)\,) = g(\,\sqrt{6-2}\,) = g(\,\sqrt{4}\,) = g(\,2\,) = 2^2 = 4$

$f\circ h(2) = f(\,h(2)\,) = f(\,2-1\,) = f(\,1\,) = \sqrt{1-2} = \sqrt{-1}$ which is undefined

$h\circ g(-3) = h(\,g(-3)\,) = h(\,9\,) = 9 - 1 = 8$.

$f\circ g(x) = f(\,g(x)\,) = f(\,x^2\,) = \sqrt{x^2-2}$, and the domain of $f\circ g$ is those x–values for which $x^2 - 2 \ge 0$ so the domain of $f\circ g$ is all x such that $x \ge \sqrt{2}$ or $x \le -\sqrt{2}$.

$g\circ f(x) = g(\,f(x)\,) = g\left(\sqrt{x-2}\right) = \{\sqrt{x-2}\}^2 = x - 2$, but we can evaluate the first piece, f, of the composition only if $f(x) = \sqrt{x-2}$ is defined, so the domain of $g\circ f$ is all $x \ge 2$.

Practice 4: For $f(x) = \frac{x}{x-3}$, $g(x) = \sqrt{1+x}$, and $h(x) = \begin{cases} 2x & \text{if } x \le 1 \\ 5-x & \text{if } 1 < x. \end{cases}$

Evaluate $f\circ g(3)$, $f\circ g(8)$, $g\circ f(4)$, $f\circ h(1)$, $f\circ h(3)$, $f\circ h(2)$ and $h\circ g(-1)$. Find the equations for $f\circ g(x)$ and $g\circ f(x)$.

Shifting and Stretching Graphs

Some compositions are relatively common and easy, and you should recognize the effect of the composition on the graphs of the functions.

Example 3: Fig. 6 shows the graph of $y = f(x)$.

Graph (a) $2 + f(x)$, (b) $3\cdot f(x)$, and (c) $f(x-1)$.

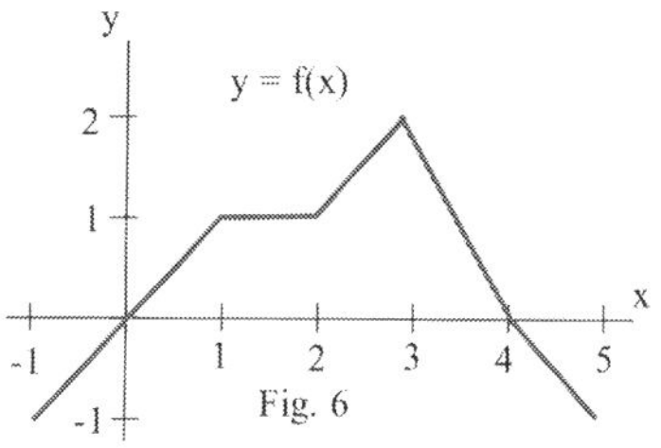

Fig. 6

Solution: All of the new graphs are shown below in Fig. 7 .

(a) Adding 2 to all of the values of f(x) **rigidly shifts** the graph of f(x) **2 units upward**.

(b) Multiplying all of the values of f(x) by 3 leaves all of the roots of f fixed: if x is a root of f then $f(x) = 0$ and $3f(x) = 3(0) = 0$ so x is also a root of $3\cdot f(x)$. If x is not a root of f, then the graph of 3f(x) looks like the graph of f(x) **stretched vertically** by a factor of 3.

(c) The graph of $f(x-1)$ is the graph of f(x) **rigidly shifted 1 units to the right**.

We could also get these results by examining the graph of $y = f(x)$, creating a table of values for f(x)

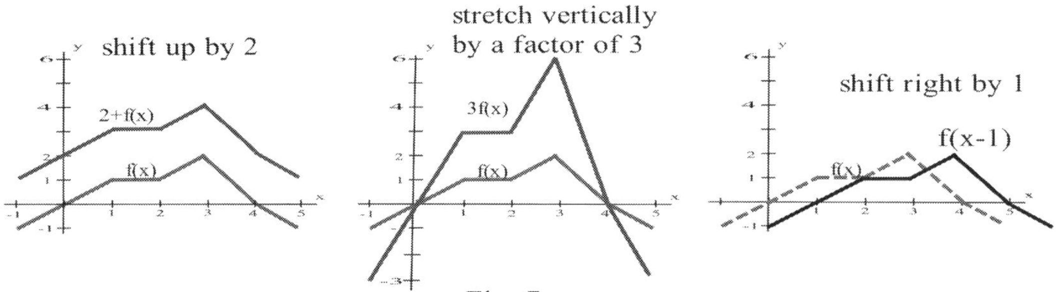

Fig. 7

and the new functions, and then graphing the new functions.

x	f(x)	2 + f(x)	3f(x)	x–1	f(x–1)
–1	–1	1	–3	–2	f(–2) not defined
0	0	2	0	–1	f(0–1) = –1
1	1	3	3	0	f(1–1) = 0
2	1	3	3	1	f(2–1) = 1
3	2	4	6	2	f(3–1) = 1
4	0	2	0	3	f(4–1) = 2
5	–1	1	–3	4	f(5–1) = 0

If k is a positive constant, then

- the graph of $y = \mathbf{k + f(x)}$ will be the graph of $y = f(x)$ rigidly **shifted up by k units**,
- the graph of $y = \mathbf{kf(x)}$ will have the same roots as the graph of f(x) and will be the graph of $y = f(x)$ **vertically stretched by a factor of k**,
- the graph of $y = \mathbf{f(x - k)}$ will be the graph of $y = f(x)$ rigidly **shifted right by k units**,
- the grah of $y = \mathbf{f(x + k)}$ will be the graph of $y = f(x)$ rigidly **shifted left by k units**.

Practice 5: Fig. 8 is the graph of g(x).

Graph (a) 1+g(x), (b) 2g(x), (c) g(x–1) and (d) –3g(x).

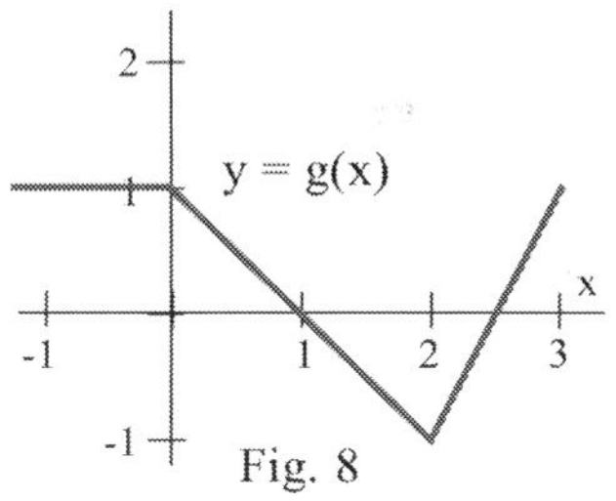

Fig. 8

Iteration of Functions

There are applications which feed the output from a function machine back into the same machine as the new input. Each time through the machine is called an **iteration** of the function.

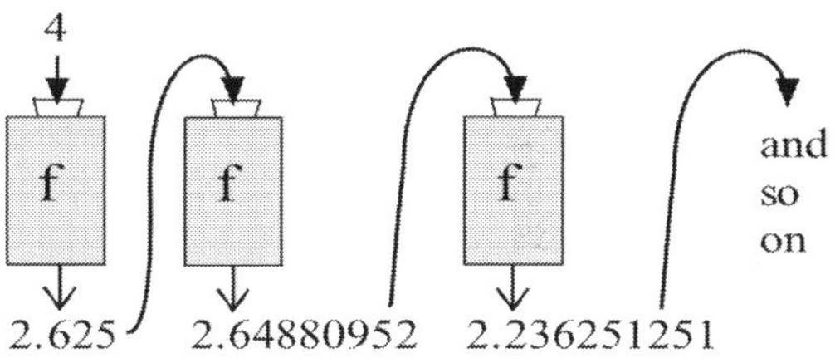

Fig. 9

Example 4: Suppose $f(x) = \dfrac{5/x + x}{2}$, and we start with the input x = 4 and repeatedly feed the output from f back into f (Fig. 9). What happens?

Solution:

Iteration	Input	Output
1	4	$f(4) = \dfrac{5/4 + 4}{2} = 2.625$
2	2.625	$f(\ f(4)\ ) = \dfrac{5/2.625 + 2.625}{2} = 2.264880952$
3	2.264880952	f(f(f(4))) = 2.236251251
4	2.236251251	2.236067985
5	2.236067985	2.236067977
6	2.236067977	2.236067977

Once we have obtained the output 2.236067977, we will just keep getting the same output. You might recognize this output value as $\sqrt{5}$. This algorithm always finds $\pm\sqrt{5}$. If we start with any positive input, the values will eventually get as close to $\sqrt{5}$ as we want. Starting with any negative value for the input will eventually get us to $-\sqrt{5}$. We cannot start with $x = 0$, since 5/0 is undefined.

Practice 6: What happens if we start with the input value $x = 1$ and iterate the function $f(x) = \frac{9/x + x}{2}$ several times? Do you recognize the resulting number? What do you think will happen to the iterates of $g(x) = \frac{A/x + x}{2}$? (Try several positive values of A.)

Two Useful Functions: Absolute Value and Greatest Integer

These two functions have useful properties which let us describe situations in which an object abruptly changes direction or jumps from one value to another value. Their graphs will have corners and breaks.

Absolute Value Function: $|x|$

The **absolute value** function of a number x, $y = f(x) = |x|$, is the distance between the number x and 0. If x is greater than or equal to 0, then $|x|$ is simply $x - 0 = x$. If x is negative, then $|x|$ is $0 - x = -x = -1 \cdot x$ which is positive since $-1 \cdot$(negative number) = a positive number. On some calculators and in some computer programming languages, the absolute value function is represented by **ABS(x)** .

Definition of $|x|$: $|x| = \begin{cases} x & \text{if } x \geq 0 \\ -x & \text{if } x < 0 \end{cases}$ or $|x| = \sqrt{x^2}$.

The domain of $y = f(x) = |x|$ consists of all real numbers. The range of $f(x) = |x|$ consists of all numbers larger than or equal to zero, all non–negative numbers. The graph of $y = f(x) = |x|$ (Fig. 10) has no holes or breaks, but it does have a sharp corner at $x = 0$. The absolute value will be useful later for describing phenomena such as reflected light and bouncing balls which change direction abruptly or whose graphs have corners. The absolute value function has a number of properties which we will use later.

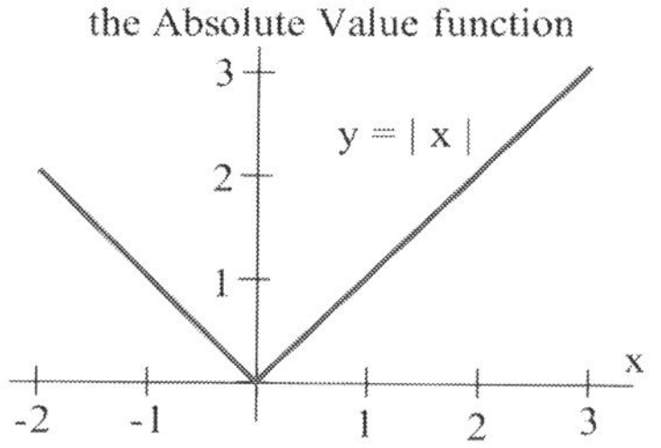

Fig. 10

Properties of $|\ |$: For all real numbers a and b:

(a) $|a| \geq 0$. $|a| = 0$ if and only if $a = 0$.

(b) $|ab| = |a||b|$

(c) $|a + b| \leq |a| + |b|$

Taking the absolute value of a function has an interesting effect on the graph of the function. Since

$|x| = \begin{cases} x & \text{if } x \geq 0 \\ -x & \text{if } x < 0 \end{cases}$, then for any function f(x) we have $|f(x)| = \begin{cases} f(x) & \text{if } f(x) \geq 0 \\ -f(x) & \text{if } f(x) < 0. \end{cases}$

In other words, if $f(x) \geq 0$, then $| f(x) | = f(x)$ so the graph of $| f(x) |$ is the same as the graph of $f(x)$. If $f(x) < 0$, then $| f(x) | = -f(x)$ so the graph of $| f(x) |$ is just the graph of $f(x)$ "flipped" about the x–axis, and it lies above the x–axis. The graph of $| f(x) |$ will always be on or above the x–axis.

Example 5: Fig. 11 shows the graph of $f(x)$. Graph (a) $| f(x) |$, (b) $| 1 + f(x) |$ and (c) $1 + | f(x) |$.

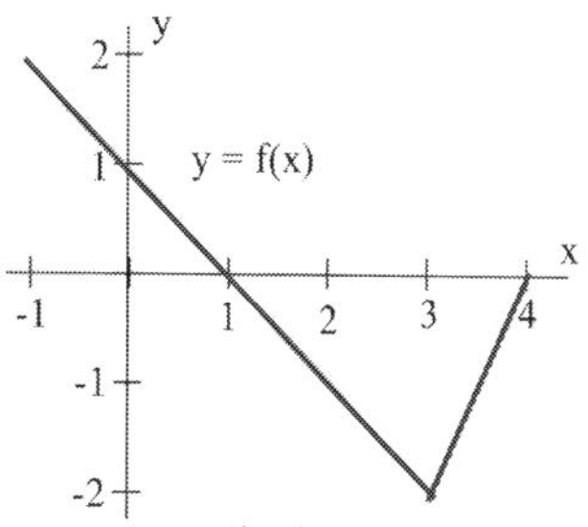

Fig. 11

Solution: The graphs are given in Fig. 12. In (b) we shift the graph of f up 1 unit before taking the absolute value. In (c) we take the absolute value before shifting the graph up 1 unit.

Practice 7: Fig. 13 shows the graph of $g(x)$. Graph (a) $| g(x) |$, (b) $| g(x - 1) |$, and (c) $g(| x |)$.

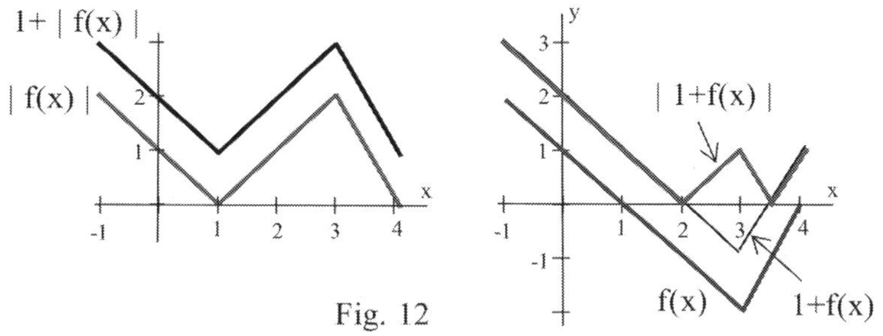

Fig. 12

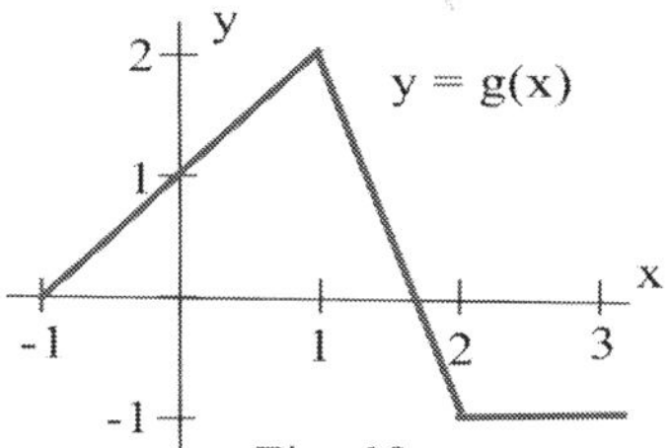

Fig. 13

Greatest Integer Function: [x] or INT(x)

The **greatest integer** function of a number x , $y = f(x) = [x]$, is the largest **integer** which is less than or equal to x . The value of [x] is always an integer and [x] is always less than or equal to x. For example, [3.2] = 3, [3.9] = 3, and [3] = 3. If x is positive, then [x] truncates x (drops the fractional part of x) to get [x]. If x is negative, the situation is different: $[-4.2] \neq -4$ since -4 is **not** less than or equal to -4.2 : $[-4.2] = -5$, $[-4.7] = -5$ and $[-4] = -4$. On some calculators and in many programming languages the square brackets [ ] are used for grouping objects or for lists, and the greatest integer function is represented by **INT(x)** .

Definition of [x]: [x] = **the largest integer which is less than or equal to x**

$$= \begin{cases} \textbf{x} & \textbf{if x is an integer} \\ \textbf{largest integer strictly less than x} & \textbf{if x is NOT an integer.} \end{cases}$$

The domain of The f(x) = [x] is all real numbers. The range of f(x) = [x] is only the integers. The graph of y = f(x) = [x] is shown in Fig. 14. It has a jump break, a step, at each integer value of x, and f(x) = [x] is called a **step function**. Between any two consecutive integers, the graph is horizontal with no breaks or holes. The greatest integer function is useful for describing phenomena which change values abruptly such as postage rates as a function of the weight of the letter ("26¢ for the first ounce and 13¢ additional for each additional half ounce"). It can also be used for functions whose graphs are "square waves" such as the on and off of a flashing light.

the Greatest Integer function

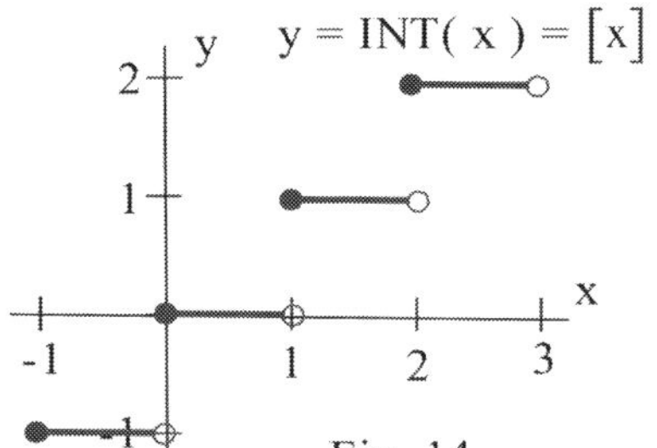

Fig. 14

Example 6: Graph f(x) = INT(1 + .5 sin(x)).

Solution: One way to create this graph is to first graph y = 1 + 0.5sin(x) , the thin curve in Fig. 15, and then apply the greatest integer function to y to get the thicker "square wave" pattern.

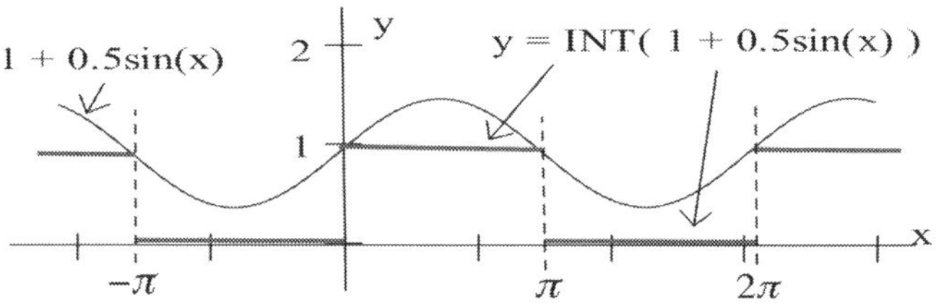

Fig. 15

Practice 8: Sketch the graph of $y = \text{INT}(x^2)$ for $-2 \le x \le 2$.

A Really "Holey" Function

The graph of the greatest integer function has a break or jump at each integer value, but how many breaks can a function have? The next function illustrates just how broken or "holey" the graph of a function can be.

$$\text{Define } h(x) = \begin{cases} 2 & \text{if } x \text{ is a rational number} \\ 1 & \text{if } x \text{ is an irrational number} \end{cases}$$

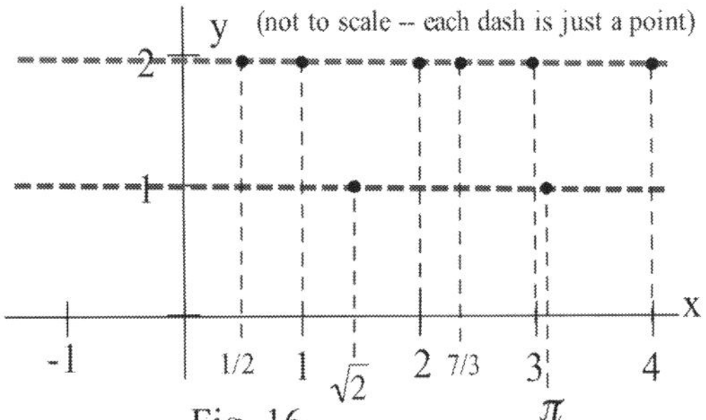

Fig. 16

Then h(3) = 2, h(5/3) = 2 and h(–2/5) = 2 since 3, 5/3 and –2/5 are all rational numbers. h(π) = 1, h($\sqrt{7}$) = 1 , and h($\sqrt{2}$) = 1 since π, $\sqrt{7}$ and $\sqrt{2}$ are all irrational numbers. These and some other points are plotted in Fig. 16 .

In order to analyze the behavior of h(x) the following fact about rational and irrational numbers is useful.

Fact: "Every interval contains both rational and irrational numbers" or, equivalently,

"If a and b are real numbers and a < b, then there is

(i) a rational number R between a and b ($a < R < b$), and

(ii) an irrational number I between a and b ($a < I < b$)."

The Fact tells us that between any two places where the $y = h(x) = 1$ (because x is rational) there is a place where $y = h(x)$ is 2 because there is an irrational number between any two distinct rational numbers. Similarly, between any two places where $y = h(x) = 2$ (because x is irrational) there is a place where $y = h(x) = 1$ because there is a rational number between any two distinct irrational numbers. The graph of $y = h(x)$ is impossible to actually draw since every two points on the graph are separated by a hole. This is also an example of a function which your computer or calculator can not graph because in general it can not determine whether an input value of x is irrational.

Example 7: Sketch the graph of

$$g(x) = \begin{cases} 2 & \text{if } x \text{ is a rational number} \\ x & \text{if } x \text{ is an irrational number} \end{cases}$$

(not to scale -- each dash is just a point)

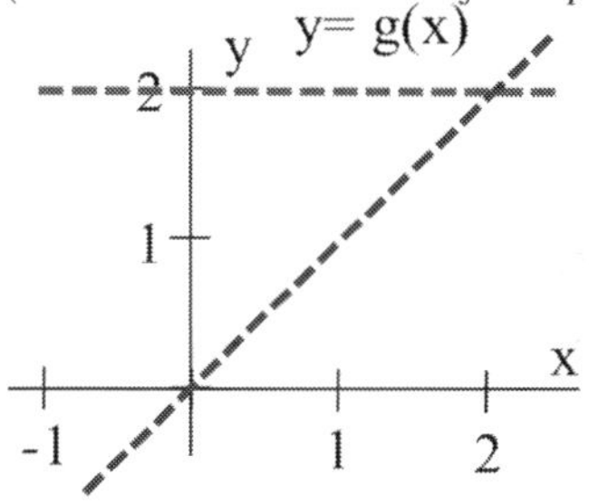

Fig. 17

Solution: A sketch of the graph of $y = g(x)$ is shown in Fig. 17 . When x is rational, the graph of $y = g(x)$ looks like the "holey" horizontal line $y = 2$. When x is irrational, the graph of $y = g(x)$ looks like the "holey" line $y = x$.

Practice 9: Sketch the graph of $r(x) = \begin{cases} 2 & \text{if } x \text{ is a rational number} \\ x & \text{if } x \text{ is an irrational number} \end{cases}$

PROBLEMS

1. If T is the Celsius temperature of the air and v is the speed of the wind in kilometers per hour, then

$$\text{WCI} = \begin{cases} T & \text{if } 0 \le v \le 6.5 \\ 33 - \dfrac{10.45 + 5.29\sqrt{v} - 0.279v}{22}(33 - T) & \text{if } 6.5 \le v \le 72. \\ 1.6T - 19.8 & \text{if } 72 < v \end{cases}$$

Determine the Wind Chill Index (a) for a temperature of 0° C and a wind speed of 49 km/hr and

(b) for a temperature of 11° C and a wind speed of 80 km/hr.

(c) Write a multiline function definition for the WCI if the temperature is 11° C.

2. Use the graph of $y = f(x)$ in Fig. 18 to evaluate $f(0), f(1), f(2), f(3), f(4)$ and $f(5)$. Write a multiline function definition for f.

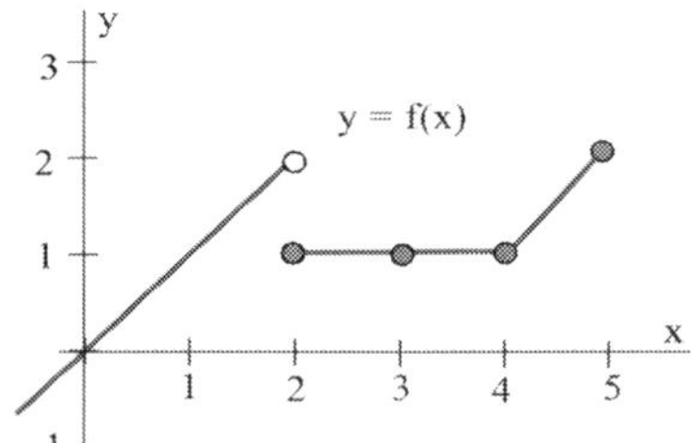

Fig. 18

3. Use the graph of $y = g(x)$ in Fig. 19 to evaluate $g(0), g(1), g(2), g(3), g(4)$ and $g(5)$. Write a multiline function definition for g.

4. Use the values given in the table and $h(x) = 2x + 1$ to determine the values of $f \circ g$, $g \circ f$ and $h \circ g$.

x	f(x)	g(x)	f∘g(x)	g∘f(x)	h∘g(x)
–1	2	0			
0	1	2			
1	–1	1			
2	0	2			

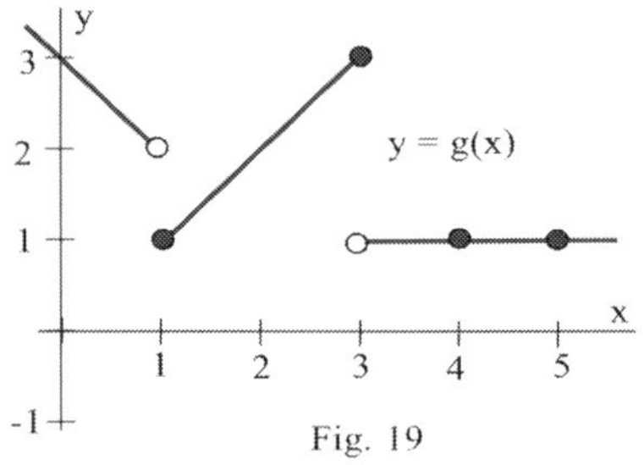

Fig. 19

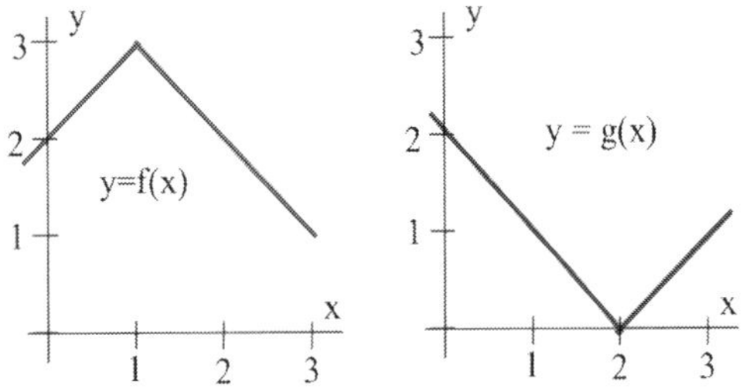

Fig. 20

5. Use the graphs in Fig. 20 and the equation $h(x) = x - 2$ to determine the values of

(a) $f(\,f(1)\,), f(\,g(2)\,), f(\,g(0)\,), f(\,g(1)\,)$

(b) $g(\,f(2)\,), g(\,f(3)\,), g(\,g(0)\,), g(\,f(0)\,)$

(c) $f(\,h(3)\,), f(\,h(4)\,), h(\,g(0)\,), h(\,g(1)\,)$

6. Use the graphs in Fig. 21 and the equation $h(x) = 5 - 2x$ to determine the values of

(a) $h(\,f(0)\,), f(\,h(1)\,), f(\,g(2)\,), f(\,f(3)\,)$

(b) $g(\,f(0)\,), g(\,f(1)\,), g(\,h(2)\,), h(\,f(3)\,)$

(c) $f(\,g(0)\,), f(\,g(1)\,), f(\,h(2)\,), h(\,g(3)\,)$

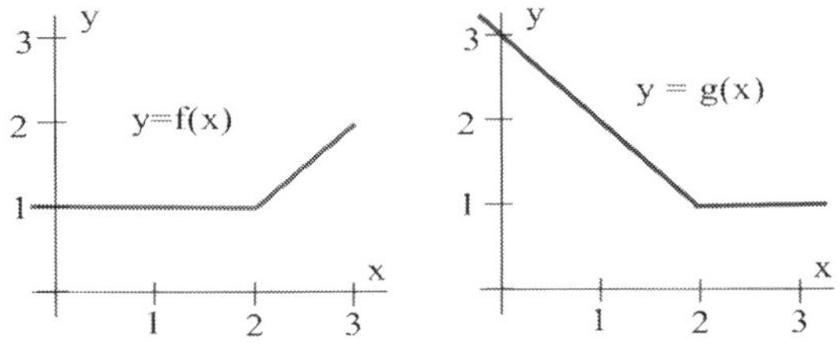

Fig. 21

7. $f(x) = \begin{cases} 3 & \text{if } x < 1 \\ x-2 & \text{if } 1 \le x < 3 \\ 1 & \text{if } 3 \le x \end{cases}$ $\quad g(x) = \begin{cases} x^2 - 3 & \text{if } x < 0 \\ \text{INT}(x) & \text{if } 0 \le x \end{cases}$ $\quad h(x) = x - 2.$

(a) Evaluate $f(x)$, $g(x)$, and $h(x)$ for $x = -1, 0, 1, 2, 3,$ and 4.

(b) Evaluate $f(\,g(1)\,), f(\,h(1)\,), h(\,f(1)\,), f(\,f(2)\,), g(\,g(3.5)\,)$.

(c) Graph $f(x)$, $g(x)$ and $h(x)$ for $-5 \le x \le 5$.

8. $f(x) = \begin{cases} x+1 & \text{if } x<1 \\ 1 & \text{if } 1 \le x < 3 \\ 2-x & \text{if } 3 \le x \end{cases}$ $\quad g(x) = \begin{cases} |x+1| & \text{if } x<0 \\ 2x & \text{if } 0 \le x \end{cases}$ $\quad h(x) = 3.$

   (a) Evaluate f(x), g(x), and h(x) for $x = -1, 0, 1, 2, 3,$ and 4 .

   (b) Evaluate f(g(1)), f(h(1)), h(f(1)), f(f(2)), g(g(3.5)) .

   (c) Graph f(x), g(x) and h(x) for $-5 \le x \le 5$.

9. You are planning to take a one week vacation in Europe, and the tour brochure says that Monday and Tuesday will be spent in England, Wednesday in France, Thursday and Friday in Germany, and Saturday and Sunday in Italy. Let L(d) be the location of the tour group on day d and write a multiline function definition for L(d).

10. A state has just adopted the following state income tax system: no tax on the first \$10,000 earned, 1% of the next \$10,000 earned, 2% of the next \$20,000 earned, and 3% of all additional earnings. Write a multiline function definition for T(x), the state income tax due on earnings of x dollars.

11. Write a multiline function definition for the curve y = f(x) in Fig. 22.

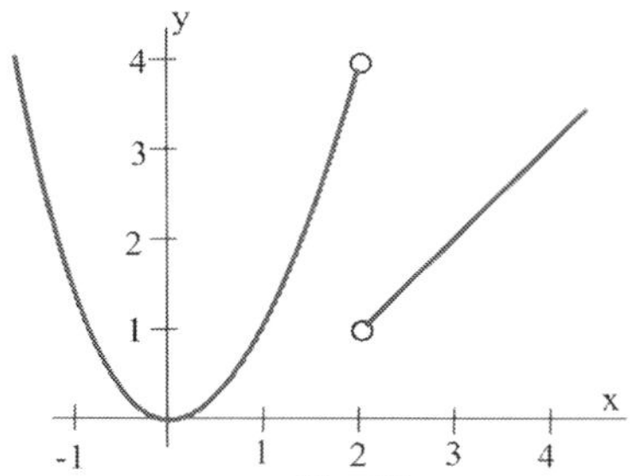

Fig. 22

12. Define B(x) to be the **area** of the rectangle whose lower left corner is at the origin and whose upper right corner is at the point (x, f(x)) for the function f in Fig. 23. Then, for example, B(3)=6. Evaluate B(1), B(2), B(4) and B(5)

13. Define B(x) to be the **area** of the rectangle whose lower left corner is at the origin and whose upper right corner is at the point (x, 1/x).

   a) Evaluate B(1), B(2) and B(3).

   b) Show that B(x) = 1 for all x > 0.

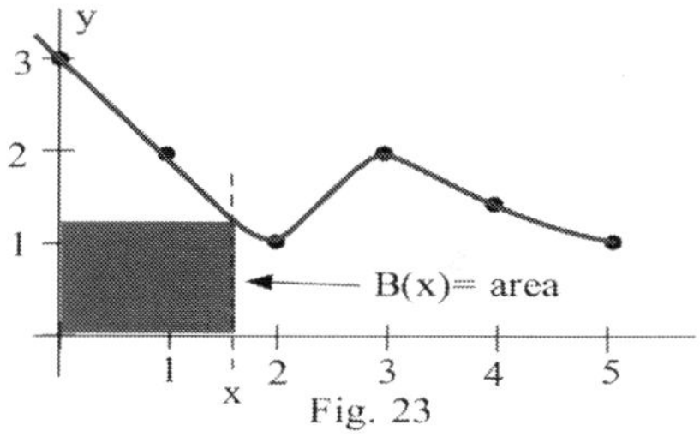

Fig. 23

14. For $f(x) = |9 - x|$ and $g(x) = \sqrt{x-1}$,

   (a) evaluate f∘g(1), f∘g(3), f∘g(5), f∘g(7), f∘g(0)

   (b) evaluate f∘f(2), f∘f(5), f∘f(–2). Does f∘f(x) = |x| for all values of x ?

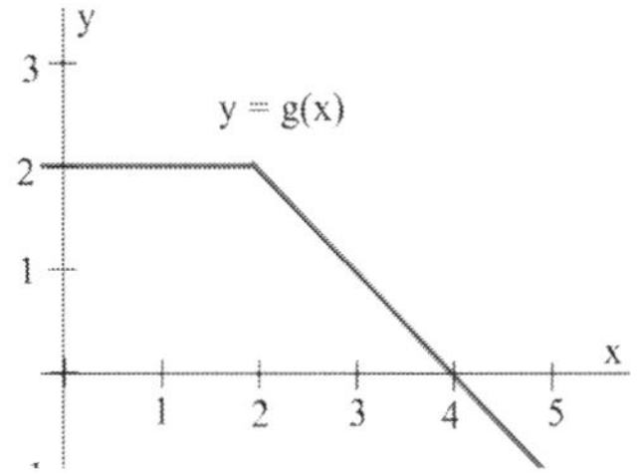

15. Fig. 24 is the graph of g(x). Graph (a) g(x) – 1, (b) g(x–1), (c) | g(x) |, and (d) [g(x)] .

16. Fig. 25 is the graph of f(x). Graph (a) f(x) – 2, (b) f(x – 2), (c) | f(x) |, and (d) [f(x)] .

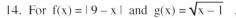

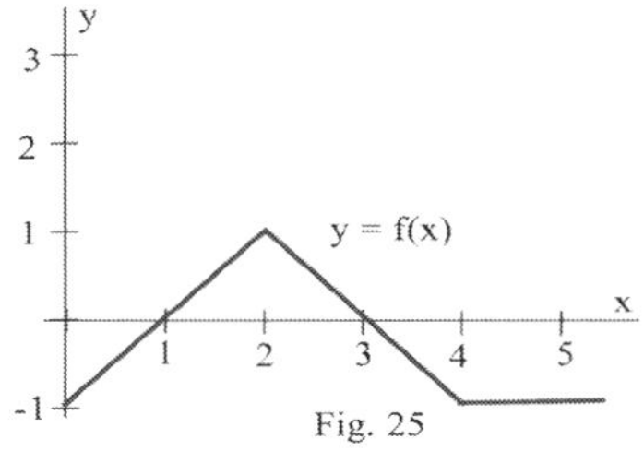

Fig. 25

17. (a) Let f(x) = 3x + 2 and g(x) = 2x + A. Find a value for A so that f(g(x)) = g(f(x)).

   (b) Let f(x) = 3x + 2 and g(x) = Bx – 1. Find a value for B so that f(g(x)) = g(f(x)).

18. (a) Let $f(x) = Cx + 3$ and $g(x) = Cx - 1$. Find a value for C so that $f(g(x)) = g(f(x))$.

(b) Let $f(x) = 2x + D$ and $g(x) = 3x + D$. Find a value for D so that $f(g(x)) = g(f(x))$.

19. Graph $y = f(x) = x - \text{INT}(x)$ for $-1 \le x \le 3$. This function is called the "fractional part of x" and is an example of a "sawtooth" graph.

20. $f(x) = \text{INT}(x + 0.5)$ rounds off x to the NEAREST integer. $g(x) = \dfrac{\text{INT}(10x + 0.5)}{10}$ rounds off x to the nearest tenth, the first decimal place. What function will round off x to (a) the nearest hundredth (2 decimal places)? (b) the nearest thousandth (3 decimal places)?

21. Modify the function in example 6 to produce a "square wave" graph with a "long on, short off, long on, short off" pattern.

22. Some versions of the computer language BASIC contain a "signum" or "sign" function defined by

$$\text{SGN}(x) = \begin{cases} 1 \text{if } x > 0 \\ 0 \text{if } x = 0 \\ -1 \text{if } x < 0 \end{cases}.$$

(a) Graph SGN(x) (b) Graph SGN(x – 2) (c) Graph SGN(x – 4)

(d) Graph SGN(x – 2) SGN(x – 4) (e) Graph 1 – SGN(x – 2) SGN(x – 4)

(f) For a < b, describe the graph of 1 – SGN(x – a) SGN(x – b)

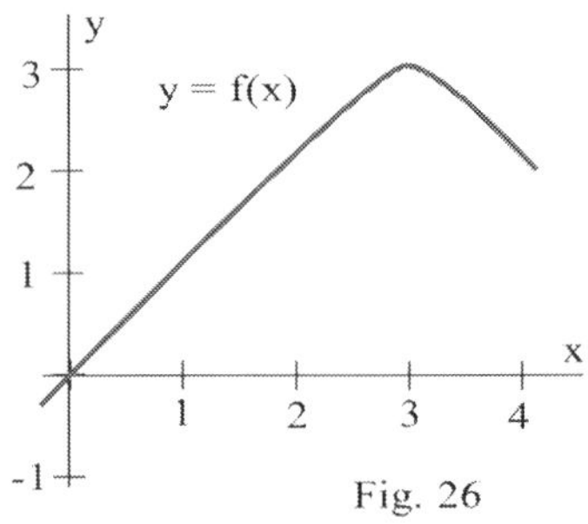

Fig. 26

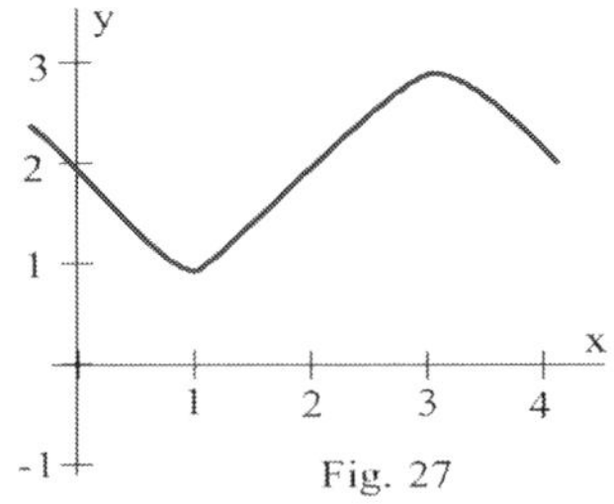

Fig. 27

23. Define g(x) to be the **slope** of the line tangent to the graph of y = f(x) in Fig. 26 at (x,y).

(a) Estimate g(1), g(2), g(3) and g(4).

(b) Graph y = g(x) for $0 \le x \le 4$.

24. Define h(x) to be the **slope** of the line tangent to the graph of y = f(x) in Fig. 27 at (x,y).

(a) Estimate h(1), h(2), h(3) and h(4).

(b) Graph y = h(x) for $0 \le x \le 4$.

C25. Pressing the COS (cosine) button on your calculator several times will produce iterates of f(x) = cos(x). What number will the iterates approach if you start with x = 1 and press the COS button 20 or 30 times? What happens if you start with x = 2 or x = 10? (Be sure your calculator is in radian mode.)

C26. Let f(x) = 1 + sin(x). What happens if you start with x = 1 and repeatedly feed the output from f back into f ? What happens if we start with x = 2 and examine the iterates of f ? (Be sure your calculator is in radian mode.)

C27. Starting with $x = 1$, do the iterates of $f(x) = \frac{x^2+1}{2x}$ approach a number? What happens if you start with $x = .5$ or $x = 4$?

C28. Let $f(x) = \frac{x}{2} + 3$. (a) What are the iterates of f if you start with $x = 2$? 4? 6?

(b) Find a number c so that $f(c) = c$. This value of c is called a **fixed point** of f.

(c) Find a fixed point of $g(x) = \frac{x}{2} + A$.

C29. Let $f(x) = \frac{x}{3} + 4$. (a) What are the iterates of f if you start with $x = 2$? 4? 6?

(b) Find a number c so that $f(c) = c$. (c) Find a fixed point of $g(x) = \frac{x}{3} + A$.

Some iterative procedures are geometric rather than numerical.

30. Start with an equilateral triangle with sides of length 1 (Fig, 28a).

(i) Remove the middle third of each line segment.

(ii) Replace the removed portion with 2 segments with the same length as the removed segment.

The first two iterations of this procedure are shown in Fig. 28b and Fig. 28c. Repeat steps (i) and (ii) several more times, each time removing the middle third of each line segment and replacing it with two new segments. What happens to the length of the shape with each iteration? (The result of iterating over and over with this procedure is called Koch's Snowflake, named for Helga von Koch)

Koch;s Snowflake

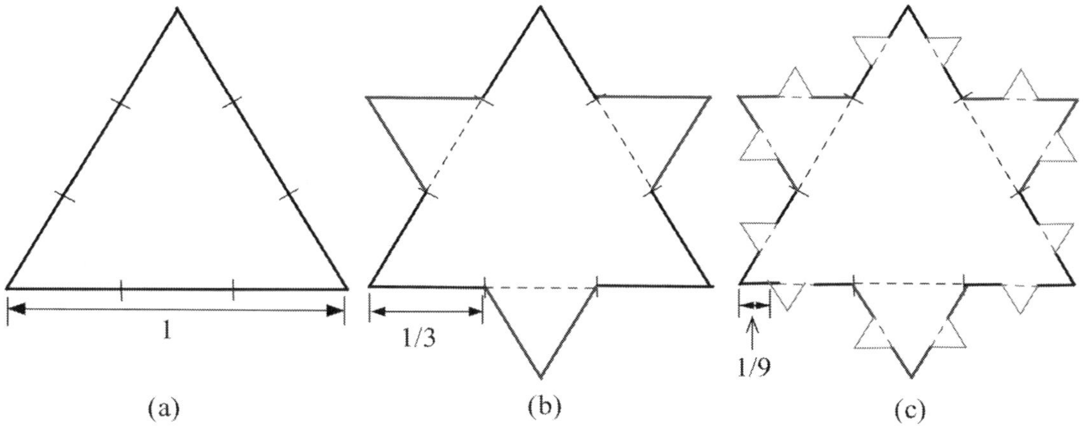

Fig. 28

31. (Optional) Sketch the graph of $p(x) = \begin{cases} 3 - x & \text{if } x \text{ is a rational number} \\ 1 & \text{if } x \text{ is an irrational number} \end{cases}$

32. (Optional) Sketch the graph of $q(x) = \begin{cases} x^2 & \text{if } x \text{ is a rational number} \\ x + 1 & \text{if } x \text{ is an irrational number} \end{cases}$

Section 0.4 **PRACTICE Answers**

Practice 1: C(x) is the cost for one night on date x .

$$C(x) = \begin{cases} \$50 & \text{if } x \text{ is between June 1 and September 15} \\ \$40 & \text{if } x \text{ is any other date} \end{cases}$$

Practice 2:

See Fig. 29

x	g(x)
−3	−3
−1	2
0	2
1/2	2
1	undefined

x	g(x)
π/3	−π/3
2	−2
3	−3
4	undefined
5	1

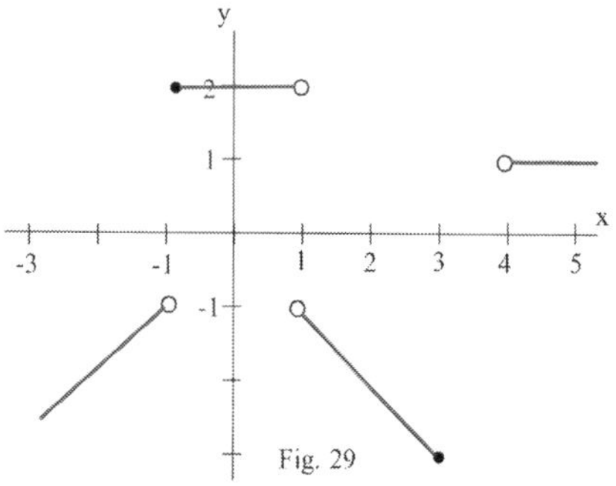

Fig. 29

Practice 3: $f(x) = \begin{cases} 1 & \text{if } x \le -1 \\ 1 - x & \text{if } -1 < x \le 1 \\ 2 & \text{if } 1 < x \end{cases}$

Practice 4: $f \circ g(3) = f(2) = 2/{-1} = \mathbf{-2}$ $f \circ g(8) = f(3)$ is **undefined** $g \circ f(4) = g(4) = \mathbf{\sqrt{5}}$

$f \circ h(1) = f(2) = 2/{-1} = \mathbf{-2}$ $f \circ h(3) = f(2) = \mathbf{-2}$ $f \circ h(2) = f(3)$ is **undefined**

$h \circ g(-1) = h(0) = \mathbf{0}$ $f \circ g(x) = f(\sqrt{1+x}\,) = (\sqrt{1+x}\,)/(\sqrt{1+x} - 3)$, $g \circ f(x) = g(\frac{x}{x-3}) = \sqrt{1 + \frac{x}{x-3}}$

Practice 5: See Fig. 30.

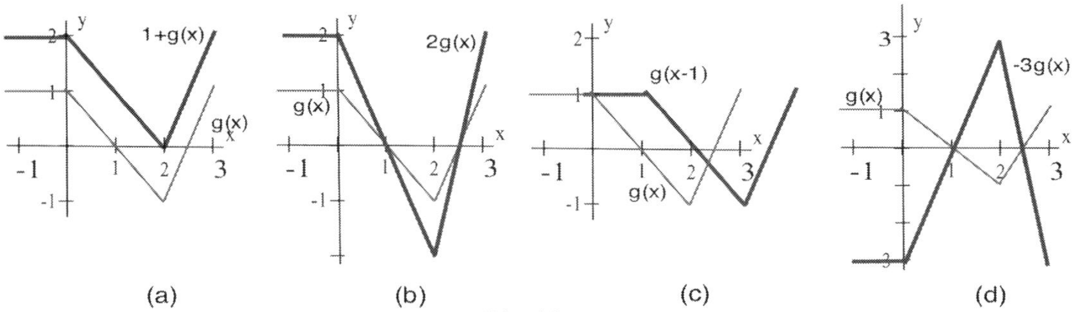

Fig. 30

Practice 6: $f(x) = \dfrac{9/x + x}{2}$. $f(1) = \dfrac{9/1 + 1}{2} = 5$, $f(5) = \dfrac{9/5 + 5}{2} = 3.4$, $f(3.4) \approx 3.023529412$,

$f(3.023529412) \approx 3.000091554$, and $f(3.000091554) \approx 3.000000001$.

These values are approaching 3, the square root of 9.

Putting A = 6, then $f(x) = \dfrac{6/x + x}{2}$.

$f(1) = \dfrac{6/1 + 1}{2} = 3.5$, $f(3.5) = \dfrac{6/3.5 + 3.5}{2} = 2.607142857$,

$f(2.607142857) \approx 2.45425636$, $f(2.45425636) \approx 2.449494372$,

$f(2.449494372) \approx 2.449489743$.

$f(2.449489743) \approx 2.449489743$ (the output is the same as the input for 9 decimal places)

These values are approaching 2.449489743, the square root of 6.

For any positive value A, the iterates of $f(x) = \dfrac{A/x + x}{2}$ (starting with any positive x) will approach $\sqrt{A}$.

Practice 7: Fig. 31 shows some of the intermediate steps and final graphs.

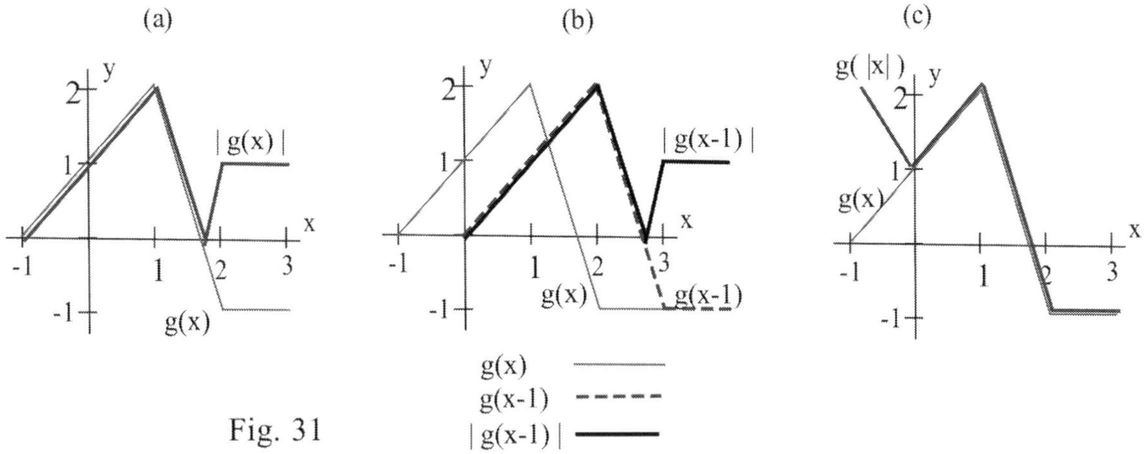

Fig. 31

Practice 8: Fig. 32 shows the graph of $y = x^2$ and the graph (thicker) of $y = INT(x^2)$.

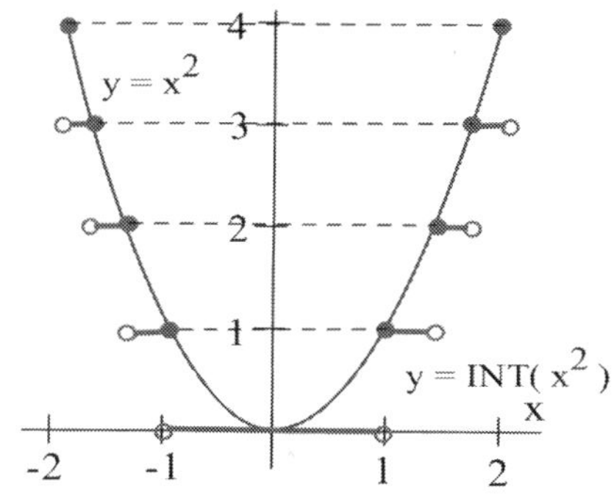

Fig. 32

Practice 9: Fig. 33 shows the "holey" graph of $y = x$ with a hole at each rational value of x and the "holey" graph of $y = \sin(x)$ with a hole at each irrational value of x. Together they form the graph of r(x) .

(This is a very crude image since we can't really see the individual holes which have zero width.)

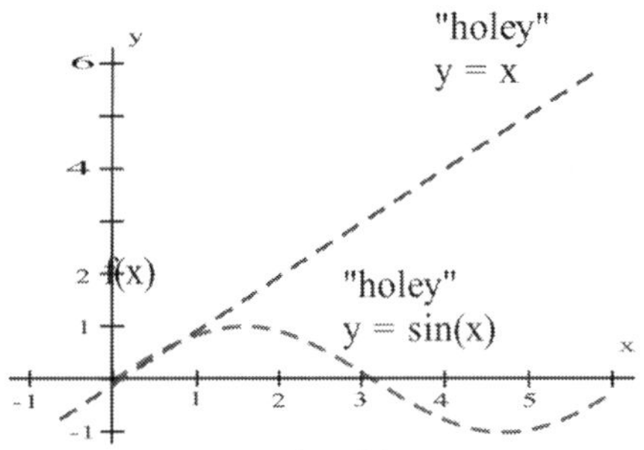

Fig. 33

0.5 MATHEMATICAL LANGUAGE

The calculus concepts we will explore in this book are simple and powerful, but sometimes subtle. To succeed in calculus you will have to master some techniques, but, more importantly, you will have to understand ideas and be able to work with the ideas in words and pictures -- very clear words and pictures. You also need to understand some of the common linguistic constructions used in mathematics. In this section we will discuss a few of the most common mathematical phrases, the meaning of these phrases and some of their equivalent forms.

Your calculus teacher is going to use these types of statements, and it is very important that you understand exactly what the teacher means. You have reached the level in mathematics where the precise use of language is important.

EQUIVALENT STATEMENTS

Two statements are **equivalent** if they always have the same logical value (a logicl value is either "true" or "fale", that is, if they are both true or are both false. The statements "$x = 3$" and "$x + 2 = 5$" are equivalent statements because if one of them is true then so is the other, and if one of them is false then so is the other. The statements "$x = 3$" and "$x^2 - 4x + 3 = 0$" are not equivalent since $x = 1$ makes the second statement true but the first one false.

AND and OR

> The compound statement "A **and** B are true" is equivalent to "**both of** A and B are true."

If A or if B or if both are false, then the statement "A and B are true" is false. The statement "$x^2 = 4$ and $x > 0$" is true when $x = 2$ and is false for every other value of x.

> The compound statement "A **or** B is true" is equivalent to "**at least one** of A or B is true."

If both A and B are false, then the statement "A or B is true" is false. The statement "$x^2 = 4$ or $x > 0$" is true if $x = -2$ or x is any positive number. The statement is false when $x = -3$ and for lots of other values of x.

Practice 1: Which values of x make each statement true?

(a) "$x < 5$" (b) "$x + 2 = 6$" (c) "$x^2 - 10x + 24 = 0$" (d) "(a) and (b)" (e) "(a) or (c)"

NEGATION OF A STATEMENT

For some simple statements we can construct the negation just by adding the word "not."

Statement	Negation of the Statement
x is equal to 3 ($x = 3$)	x is **not** equal to 3 ($x \neq 3$)
x is less than 5 ($x < 5$)	x is **not** less than 5 ($x \not< 5$) x is greater than or equal to 5 ($x \geq 5$)

When the statement contains words such as "all", "no", or "some," then its negation is more complicated.

Statement	Negation of the Statement
All x satisfy A. Every x satisfies A.	At least one x does not satisfy A. There is an x which does not satisfy A. Some x does not satisfy A.
No x satisfies A. Every x does not satisfy A.	At least one x satisfies A. Some x satisfies A.
There is an x which satisfies A. At least one x satisfies A. Some x satisfies A.	No x satisfies A. Every x does not satisfy A.

We can also negate compound statements containing "and" and "or".

Statement	Negation of the Statement
A and B are both true.	At least one of A or B is not true.
A and B and C are all true.	At least one of A or B or C is not true.
A or B is true.	Both A and B are not true.

Practice 2: Write the negation of each of these statements.

(a) $x + 5 \geq 3$ (b) All prime numbers are odd. (c) $x^2 < 4$

(d) x divides 2 and x divides 3. (e) No mathematician can sing well.

IF ... THEN ... : A Very Common Structure in Mathematics

The most common and basic structure used in mathematical language is the

"**If** (some hypothesis) **then** (some conclusion)"

sentence. Almost every result in mathematics can be stated using one or more "If ... then ... " sentences.

> "If A then B" means that when the hypothesis A is true, then the conclusion B must also be true.

If the hypothesis is false, then the "If ... then ... " sentence makes no claim about the truth or falsity of the conclusion — the conclusion may be either true or false.

Even in everyday life you have probably encountered "If ... then ..." statements for a long time. A parent might try to encourage a child with a statement such as " If you clean your room then I will buy you an ice cream cone."

To show that an "If . . . then . . . " statement is not valid (not true), all we need to do is find a **single** example where the hypothesis is true and the conclusion is false. Such an example with a true hypothesis and false conclusion is called a **counterexample** for the "If . . . then . . . " statement. A valid "If . . . then . . . " statement has no counterexample.

> A **counterexample** to the statement "If A then B" is an example in which A is true and B is false.

The only way for the statement " If you clean your room then I will buy you an ice cream cone" to be false is if the child cleaned the room and the parent did not buy the ice cream cone. If the child did not clean the room but the parent still bought the ice cream cone we would say that the statement was true.

The statement "If n is a positive integer, then $n^2 + 5n + 5$ is a prime number" has hypotheses "n is a positive integer" and conclusion "$n^2 + 5n + 5$ is a prime number." This "If ... then" statement is false since replacing n with the number 5 will make the hypothesis true and the conclusion false. The number 5 is a counterexample for the statement. Every invalid "If . . . then . . . " statement has at least one counterexample, and the most convincing way to show that a statement is not valid is to find a counterexample to the statement.

A number of other language structures can be translated into the "If ... then ..." form. The statements below all mean the same as "If (A) then (B)" :

"All (A) are (B)."	"Every (A) is (B)."	"Each (A) is (B)."
"Whenever (A), then (B)."	"(B) whenever (A)."	"(A) only if (B)."
"(A) implies (B)."	"(A) $\Rightarrow$ (B)" (the symbol " $\Rightarrow$ " means "implies")	

Practice 3: Restate "**If** (a shape is a square) **then** (the shape is a rectangle)" as many ways as you can.

"If … then … " statements occur hundreds of times in every mathematics book, including this one. It is important that you are able to recognize the various forms of "If … then … " statements and that you are able to distinguish the hypotheses from the conclusions.

Contrapositive Form of an "If ... then ..." Statement

The statement "If (A) then (B)" means that if the hypothesis A is true, then the conclusion B is guaranteed to be true.

Suppose we know that in a certain town the statement

"If (a building is a church) then (the building is green)"

is a true statement,. What can we validly conclude about a red building? Based on the information we have, we can validly conclude that the red building is "not a church" since every church is green. We can also conclude that a blue building is not a church. In fact, we can conclude that every "not green" building is "not a church." That is, if the conclusion of a valid "If ... then ... " statement is **false**, then the hypothesis must also be **false**.

> The **contrapositive** form of "If (A) then (B)" is
> **"If (negation of B) then (negation of A)"** or **"If (B is false) then (A is false)."**

> The statement "If (A) then (B)" and its contrapositive "If (not B) then (not A)" are **equivalent**.

What about a green building in this town? The green building may or may not be a church – perhaps every post office is also painted green. Or perhaps every building in town is green, in which case the statement "If (a building is a church) then (the building is green)" is certainly true.

Practice 4: Write the contrapositive form of each of the following statements.

(a) If a function is differentiable then it is continuous. (b) All men are mortal.

(c) If (x equals 3) then ($x^2 - 5x + 6$ equals 0) (d) If (2 divides x and 3 divides x) then (6 divides x).

Converse of an "If ... then ..." Statement

If we switch the hypotheses and the conclusion of an "If A then B" statement we get the converse "If B then A."

The converse of an "If ... then ... " statement is a new statement with the hypothesis and conclusion switched: the converse of "If (A) then (B)" is "If (B) then (A)." For example, the converse of "If (a building is a church) then (the building is green)" is "If (a building is green) then (the building is a church)." The converse of an "If ... then ... " statement **is not equivalent** to the original "If ... then ... " statement. The statement "If $x = 2$, then $x^2 = 4$" is true, but the converse statement "If $x^2 = 4$, then $x = 2$" is not true because $x = -2$ makes the hypothesis of the converse true and the conclusion false.

The **converse** of "If (A) then (B)" is "If (B) then (A)."

The statement "If (A) then (B)" and its converse "If (B) then (A)" are **not equivalent**.

Wrap–up

The precise use of language by mathematicians (and mathematics books) is an attempt to clearly communicate ideas from one person to another, but that requires that both people understand the use and rules of the language. If you don't understand this usage, the communication of the ideas will almost certainly fail.

PROBLEMS

In problems 1 and 2, let $A = \{1,2,3,4,5\}$, $B = \{0,2,4,6\}$, and $C = \{-2,-1,0,1,2,3\}$. Which values of x satisfy each statement.

1. a) x is in A **and** x is in B. b) x is in A **or** x is in C. c) x is not in B **and** x is in C.
2. a) x is not in B **or** C. b) x is in B and C but not in A. c) x is not in A but is in B or C.

In problems 3 – 5, list or describe all the values of x which make each statement true.

3. a) $x^2 + 3 > 1$ b) $x^3 + 3 > 1$ c) $[\,x\,] \le |\,x\,|$
4. a) $\dfrac{x^2 + 3x}{x} = x + 3$ b) $x > 4$ and $x < 9$ c) $|\,x\,| = 3$ and $x < 0$
5. a) $x + 5 = 3$ or $x^2 = 9$ b) $x + 5 = 3$ and $x^2 = 9$ c) $|\,x + 3\,| = |\,x\,| + 3$

In problems 6 – 8, write the **contrapositive** of each statement. If the statement is false, give a **counterexample**.

6. a) If $x > 3$ then $x^2 > 9$. b) Every solution of $x^2 - 6x + 8 = 0$ is even.
7. a) If $x^2 + x - 6 = 0$ then $x = 2$ or $x = -3$. b) All triangles have 3 sides.
8. a) Every polynomial has at least one zero. b) If I exercise and eat right then I will be healthy.

In problems 9 – 11, write the **contrapositive** of each statement. If necessary, first write the original statement in the "If . . . then . . ." form.

9. a) If your car is properly tuned, it will get at least 24 miles per gallon.
   b) You can have dessert if you eat your vegetables.
10. a) A well–prepared student will miss less than 15 points.
    b) I feel good when I jog.
11. a) If you love your country, you will vote for me.
    b) If guns are outlawed then only outlaws will have guns.

In problems 12 – 15, write the **negation** of each statement.

12. a) It is raining. b) Some equations have solutions. c) f(x) and g(x) are polynomials.

13. a) f(x) or g(x) is positive. b) x is positive. c) 8 is a prime number.

14. a) Some months have 6 Mondays. b) All quadratic equations have solutions.

 c) The absolute value of a number is positive.

15. a) For all numbers a and b, $|a + b| = |a| + |b|$. b) All snakes are poisonous.

 c) No dog can climb trees.

16. Write an "If . . . then . . . " statement which is true but whose converse is false.

17. Write an "If . . . then . . . " statement which is true and whose converse is true.

18. Write an "If . . . then . . . " statement which is false and whose converse is false.

In problems 19 – 22, state whether each statement is true or false. If the statement is false, give a counterexample.

19. a) If a and b are real numbers then $(a + b)^2 = a^2 + b^2$.

 b) If $a > b$ then $a^2 > b^2$. c) If $a > b$ then $a^3 > b^3$.

20. a) For all real numbers a and b, $|a + b| = |a| + |b|$

 b) For all real numbers a and b, $[a] + [b] \le [a + b]$ ([ ] represents the greatest integer function.)

 c) If f(x) and g(x) are linear functions then f(g(x)) is a linear function.

21. a) If f(x) and g(x) are linear functions then f(x) + g(x) is a linear function.

 b) If f(x) and g(x) are linear functions then f(x)g(x) is a linear function.

 c) If x divides 6 then x divides 30.

22. a) If x divides 50 then x divides 10. b) If x divides yz then x divides y or z.

 c) If x divides a^2 then x divides a.

In problems 23 – 26, rewrite each statement as an "If ... then ... " statement and state whether it is true or false. If the statement is false, give a counterexample.

23. a) The sum of two prime numbers is a prime. b) The sum of two prime numbers is never a prime.

 c) Every prime number is odd. d) Every prime number is even.

24. a) Every square has 4 sides. b) All 4–sided polygons are squares.

 c) Every triangle has 2 equal sides. d) Every 4–sided polygon with equal sides is a square.

25. a) Every solution of x+5=9 is odd. b) Every 3–sided polygon with equal sides is a triangle.

 c) Every calculus student studies hard. d) All (real number) solutions of $x^2 - 5x + 6 = 0$ are even.

26. a) Every straight line intersects the x–axis. b) Every (real number) solution of $x^2 + 3 = 0$ is even.

 c) All birds can fly. d) No mammal can fly.

Section 0.5 **PRACTICE Answers**

Practice 1: (a) All values of x less than 5. (b) $x = 4$

(c) Both $x = 4$ and $x = 6$. (d) $x = 4$

(e) $x = 6$ and all x less than 5.

Practice 2: (a) $x + 5 < 3$.

(b) At least one prime number is even.

There is an even prime number.

(c) $x^2 \ge 4$.

(d) x does not divide 2 or x does not divide 3.

(e) At least one mathematician can sing well.

There is a mathematician who can sing well.

Practice 3: Here are several ways to restate "**If** (a shape is a square) **then** (the shape is a rectangle)."

All squares are rectangles.

Every square is a rectangle.

Each square is a rectangle.

Whenever a shape is a square, then it is a rectangle.

A shape is a rectangle whenever it is a square.

A shape is a square only if it is a rectangle.

A shape is a square implies that it is a rectangle.

Being a square implies being a rectangle.

Practice 4: (a) statement "If a function is differentiable then it is continuous."

contrapositive "If a function is not continuous then it is not differentiable."

(b) statement "All men are mortal."

contrapositves "All immortals are not men."

"If a thing is not mortal then it is not human."

(c) statement "If (x equals 3) then ($x^2 - 5x + 6$ equals 0)."

contrapositive "If ($x^2 - 5x + 6$ does not equal 0) then (x does not equal 3)."

(d) statement "If (2 divides x and 3 divides x) then (6 divides x)."

contrapositive "If (6 does not divide x) then (2 does not divide x **or** 3 does not divide x)."

Chapter Zero Solutions to Odd Numbered Problems

Important Note about Precision of Answers:

In many of the problems in this book you are required to read information from a graph and to calculate with that information. You should take reasonable care to read the graphs as accurately as you can (a small straightedge is helpful), but even skilled and careful people make slightly different readings of the same graph. That is simply one of the drawbacks of graphical information. When answers are given to graphical problems, the answers should be viewed as the best approximations we could make, and they usually include the word "approximately" or the symbol "≈" meaning "approximately equal to." Your answers should be close to the given answers, but you should not be concerned if they differ a little. (Yes those are vague terms, but it is all we can say when dealing with graphical information.)

Section 0.1

1. approx. 1, 0, –1

3. (a) Approx. $\frac{70-150 \text{ deg.}}{10-0 \text{ min}} = -8$ deg/min. Avg. rate of cooling ≈ 8 deg/min. (b) Approx. 6 deg/min cooling, and 5 deg/min cooling. (c) Approx. 5.5 deg/min cooling, and 10 deg/min cooling. (d) When t = 6 min.

5. We estimate that the area is approximately (very approximate) 9 cm^2.

7. Method 1: Measure the diameter of the coffee can, then fill it about half full of wine and measure the height of the wine and calculate the volume. Submerge the bulb, measure the height of the wine again, and calculate the new volume. The volume of the bulb is the difference of the two calculated volumes.
Method 2: Fill the can completely full of wine and weigh the full can. Submerge the bulb (displacing a volume of wine equal to the volume of the bulb), remove the bulb, and weigh the can again. By subtracting, find the weight of the displaced wine and then use the fact that the density of wine is approximately 1 gram per 1 cubic centimeter to determine the volume of the bulb.

Section 0.2

1. (a) –3/4 (b) 1/2 (c) 0 (d) 2 (e) undefined

3. (a) $\frac{4}{3}$ (b) $\frac{-9}{5}$ (c) $x+2$ (if $x \neq 2$) (d) $4+h$ (if $h \neq 0$) (e) $a+x$ (if $a \neq x$)

5. (a) $t=5$: $\frac{5000}{1500}=\frac{10}{3}$, $t=10$: $\frac{5000}{3000}=\frac{5}{3}$, $t=20$: $\frac{5000}{6000}=\frac{5}{6}$ (b) any $t>0$: $\frac{5000}{300t}=\frac{50}{3t}$
(c) decreasing, since the numerator remains constant at 5000 while the denominator increases.

7. The restaurant is 4 blocks south and 2 blocks east. The distance is $\sqrt{4^2+2^2}=\sqrt{20} \approx 4.47$ blocks.

9. $y=\sqrt{20^2-4^2}=\sqrt{384} \approx 19.6$ feet, $m=\frac{\sqrt{384}}{4} \approx 4.9$. $\tan(q)=\frac{\sqrt{384}}{4} \approx 4.9$ so $q \approx 1.37$ ($\approx 78.5^\circ$).

11. The equation of the line through $P=(2,3)$ and $Q=(8,11)$ is $y-3=\frac{8}{6}(x-2)$ or $6y-8x=2$. Substituting $x=2a+8(1-a)=8-6a$ and $y=3a+11(1-a)=11-8a$ into the equation for the line, we get $6(11-8a)-8(8-6a)=66-48a-64+48a$ which equals 2 for every value of a, so the point with $x=2a+8(1-a)$ and $y=3a+11(1-a)$ is on the line through P and Q for every value of a.
The $\text{Dist}(P,Q)=\sqrt{6^2+8^2}=10$. $\text{Dist}(P,R)=\sqrt{(8-6a-2)^2+(11-8a-3)^2}$
$=\sqrt{(6-6a)^2+(8-8a)^2}=\sqrt{6^2(1-a)^2+8^2(1-a)^2}=\sqrt{100(1-a)^2}=10\cdot|1-a|=|1-a|\cdot\text{Dist}(P,Q)$.

13. (a) $m_1 \cdot m_2 = (1)(-1) = -1$ so the lines are perpendicular. (b) Because 20 units of x-values are physically wider on the screen than 20 units of y-values.
(c) Set the window so (xmax – xmin) $\approx$1.7(ymax – ymin).

15. (a) $y - 5 = 3(x - 2)$ or $y = 3x - 1$. (b) $y - 2 = -2(x - 3)$ or $y = 8 - 2x$ (c) $y - 4 = -\frac{1}{2}(x - 1)$ or $y = -\frac{1}{2}x + \frac{9}{2}$

17. (a) $y - 5 = \frac{3}{2}(x-2)$ or $y = \frac{3}{2}x + 2$ (b) $y - 2 = \frac{3}{2}(x+1)$ or $y = \frac{3}{2}x + \frac{7}{2}$ (c) $x = 3$.

19. The distance between the centers is $\sqrt{6^2 + 8^2} = 10$. (a) 10–2–4 = 4 (b) 10–2–7 = 1
(c) 0 (they intersect) (d) 15–10–3 = 2 (e) 12–10–1 = 1.

21. Find $\text{Dist}(P,C) = \sqrt{(x-h)^2 + (y-k)^2}$, and compare the value to r:

$$P \text{ is } \begin{cases} \text{inside the circle} & \text{if Dist}(P,C) < r \\ \text{on the circle} & \text{if Dist}(P,C) = r \\ \text{outside the circle} & \text{if Dist}(P,C) > r \end{cases}$$

23. A point P =(x,y) lies on the circle if and only if its distance from C = (h,k) is r : Dist(P,C) = r. So P is on the circle if and only if $\sqrt{(x-h)^2 + (y-k)^2} = r$ or $(x-h)^2 + (y-k)^2 = r^2$.

25. (a) slope is $-\frac{5}{12}$ (b) undefined (vertical line) (c) $\frac{12}{5}$ (d) 0 (horizontal line)

27. (a) distance ≈ 2.22 . (b) Distance ≈ 2.24 .
(c) (by inspection) 3 units which occurs at the point (5, 3).

29. (a) If $B \neq 0$, we may solve for y: $y = -\frac{A}{B}x + \frac{C}{B}$. The slope is the coefficient of x: $m = -\frac{A}{B}$.
(b) The required slope is B/A (the negative reciprocal of –A/B) so the equation is $y = \frac{B}{A}x$ or $Bx - Ay = 0$.
(c) Solve { Ax + By = C, Bx – Ay = 0 } to get $x = \frac{AC}{A^2 + B^2}$ and $y = \frac{BC}{A^2 + B^2}$.
(d) Distance $= \sqrt{\left(\frac{AC}{(A^2 + B^2)}\right)^2 + \left(\frac{BC}{A^2 + B^2}\right)^2} = \sqrt{\frac{A^2C^2}{(A^2 + B^2)^2} + \frac{B^2C^2}{(A^2 + B^2)^2}}$
$= \sqrt{\frac{(A^2 + B^2)C^2}{(A^2 + B^2)^2}} = \sqrt{\frac{C^2}{A^2 + B^2}} = \frac{|C|}{\sqrt{A^2 + B^2}}$

Section 0.3

1. A–a, B–c, C–d, D–b 3. A–b, B–c, C–d, D–a

5. (a)–C, (b)–A, (c)–B

6. The bottles are sketched in Fig. 0.3P6.

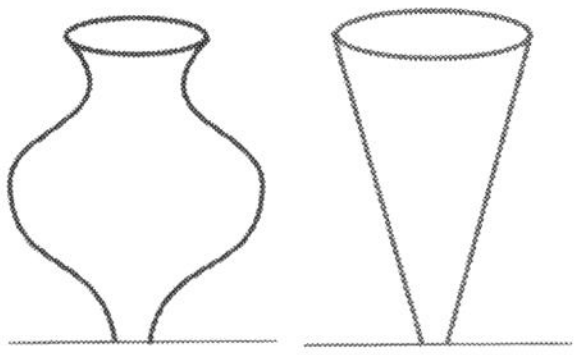

Fig. 0.3P6

7. $f(x) = x^2 + 3$, $g(x) = \sqrt{x - 5}$, $h(x) = \frac{x}{x - 2}$
(a) f(1) = 4, g(1) is undefined , h(1) = –1.
(b) Graphs of f, g and s are shown in Fig. 0.3P7.
(c) $f(3x) = (3x)^2 + 3 = 9x^2 + 3$, $g(3x) = \sqrt{3x - 5}$ (for $x \geq 5/3$) $h(3x) = \frac{3x}{3x - 2}$
(d) $f(x+h) = (x+h)^2 + 3 = x^2 + 2xh + h^2 + 3$, $g(x+h) = \sqrt{x + h - 5}$, $h(x+h) = \frac{x+h}{x+h-2}$

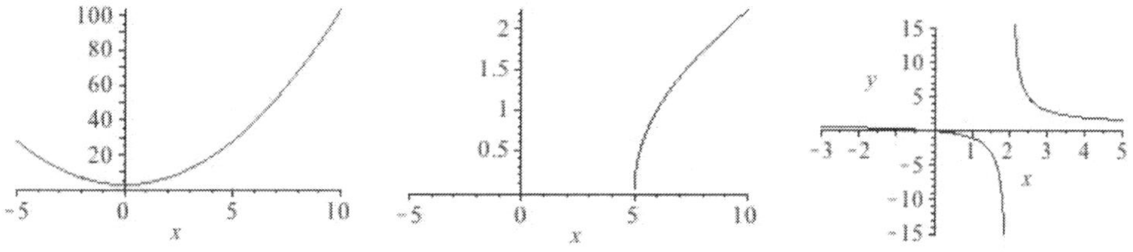

Fig. 0.3P7

9. (a) $m = 2$ (b) $m = 2x + 3 + h$. (c) $m = x + 4$ (if $x \neq 1$)
If $x = 1.3$, then $m = 5.3$. If $x = 1.1$, then $m = 5.1$. If $x = 1.002$, then $m = 5.002$.

11. $f(x) = x^2 - 2x$, $g(x) = \sqrt{x}$.
$m = \dfrac{f(a+h) - f(a)}{h} = 2a + h - 2$ $(h \neq 0)$. If $a = 1$, then $m = h$. If $a = 2$, then $m = 2+h$.
If $a = 3$, then $m = 4 + h$. If $a = x$, then $m = 2x + h - 2$.

$m = \dfrac{g(a+h) - g(a)}{h} = \dfrac{1}{\sqrt{a+h} + \sqrt{a}} = \dfrac{\sqrt{a+h} - \sqrt{a}}{h}$. If $a = 1$, then $m = \dfrac{\sqrt{1+h} - 1}{h}$.
If $a = 2$, then $m = \dfrac{\sqrt{2+h} - \sqrt{2}}{h}$. If $a = 3$, then $m = \dfrac{\sqrt{3+h} - \sqrt{3}}{h}$.
If $a = x$, then $m = \dfrac{\sqrt{x+h} - \sqrt{x}}{h}$.

13. (a) Approx. 250 miles, 375 miles. (b) Approx. 200 miles/hour.
(c) By flying along a circular arc about 375 miles from the airport.

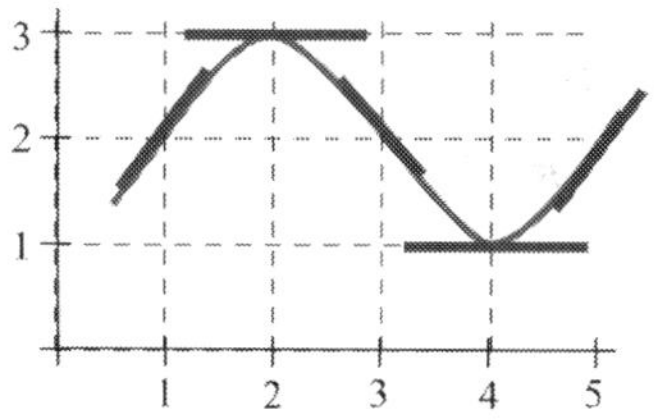

Fig. 0.3P15

15. (a) See Fig. 0.3P15. (b) Max at $x = 2$. Min at $x = 4$.
(c) Largest at $x = 5$. Smallest at $x = 3$.

17. The path of the slide is a straight line tangent to the graph of the path at the point of fall. See Fig. .

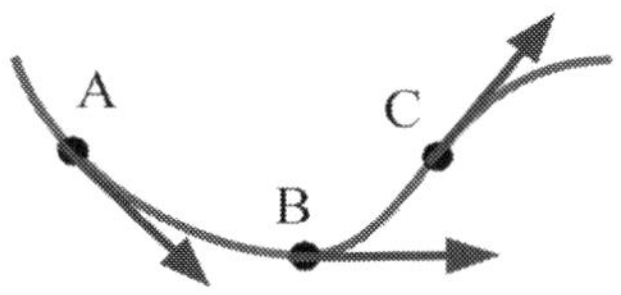

Fig. 0.3P17

19. (a) $s(1) = 2$, $s(3) = 4/3$. $s(4) = 5/4$. (b) $s(x) = \dfrac{x+1}{x}$.

21.

x	f(x)	**slope** of the line **tangent** to the graph of f at (x, f(x))
0	1	1
1	2	1
2	2	–1
3	1	0
4	1.5	0.5

23. On your own.

Section 0.4 Answers

1. (a) ≈ -18 . (b) -2.2

(c) If T = 11°C, WCI = $\begin{cases} 11 & \text{if } 0 \le v \le 6.5 \\ 33 - \dfrac{10.45 + 5.29\sqrt{v} - 0.279\,v}{22}(22) & \text{if } 6.5 < v \le 72 \\ -2.2 & \text{if } 72 < v \end{cases}$.

3. g(0)=3, g(1)=1, g(2)=2, g(3)=3, g(4)=1, g(5)=1. $g(x) = \begin{cases} 3 - x & \text{if } x < 1 \\ x & \text{if } 1 \le x \le 3 \\ 1 & \text{if } x > 3 \end{cases}$

5. (a) f(f(1)) = 1, f(g(2)) = 2, f(g(0)) = 2, f(g(1)) = 3
(b) g(f(2)) = 0, g(f(3)) = 1, g(g(0)) = 0, g(f(0)) = 0
(c) f (h(3)) = 3, f(h(4)) = 2, h(g(0)) = 0, h(g(1)) = –1

7. (a)

x	–1	0	1	2	3	4
f(x)	3	3	–1	0	1	1
g(x)	–2	0	1	2	3	4
h(x)	–3	–2	–1	0	1	2

(b) f(g(1)) = –1, f(h(1)) = 3, h(f(1)) = –3, f(f(2)) = 3, g(g(3.5)) = 3

(c) See Fig. 0.4P7 for the graphs of f, g, and h.

9. L(d) = $\begin{cases} \text{England} & \text{if } d = \text{Mon. or Tue.} \\ \text{France} & \text{if } d = \text{Wed.} \\ \text{Germany} & \text{if } d = \text{Thur. or Fri.} \\ \text{Italy} & \text{if } d = \text{Sat. or Sun.} \end{cases}$

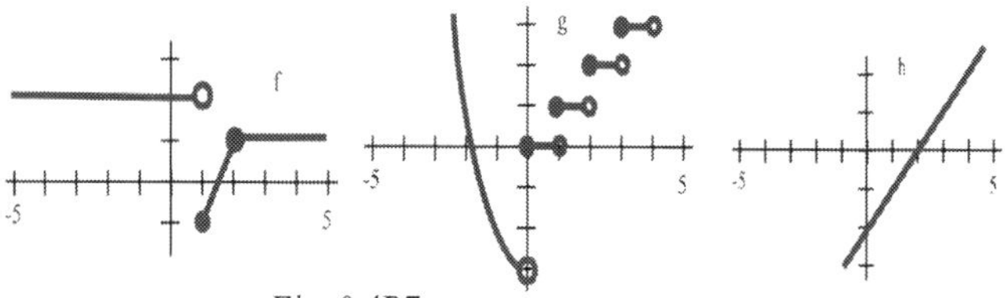

Fig. 0.4P7

11. $f(x) = \begin{cases} x^2 & \text{if } x < 2 \\ x - 1 & \text{if } x > 2 \end{cases}$

13. (a) $B(1) = 1 \cdot f(1) = 1 \cdot \frac{1}{1} = 1$,

$B(2) = 2 \cdot f(2) = 2 \cdot \frac{1}{2} = 1$,

$B(3) = 3 \cdot f(3) = 3 \cdot \frac{1}{3} = 1$.

(b) For x > 0, $B(x) = x \cdot f(x) = x \cdot \frac{1}{x} = 1$.

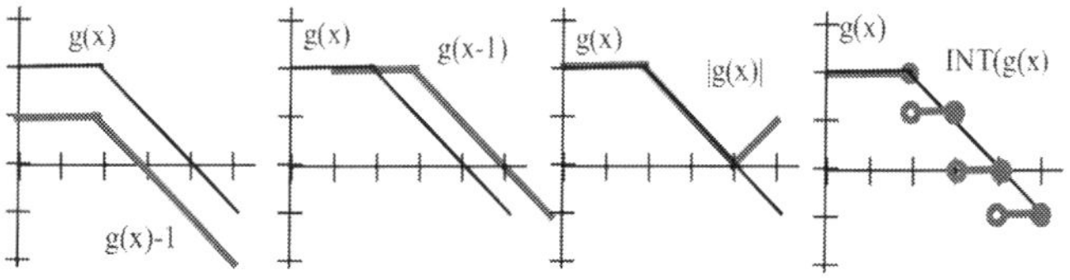

Fig. 0.4P15

15. See Fig. 0.4P15.

17. (a) f(g(x)) = 6x + 2 + 3A, g(f(x)) = g(3x+2) = 6x + 4 + A. If f(g(x)) = g(f(x)), then A = 1.
(b) f(g(x)) = 3Bx – 1, g(f(x)) = 3Bx + 2B – 1. If f(g(x)) = g(f(x)), then B = 0.

19. See Fig. 0.4P19 for the graph of f(x) = x – [x] = x – INT(x).

21. $f(x) = \lfloor 1.3 + 0.5 \cdot \sin(x) \rfloor$ works. The value of $0.5 < A < 1.5$ in $f(x) = \lfloor A + 0.5 \cdot \sin(x) \rfloor$ determines the relative lengths of the long and short parts of the pattern..

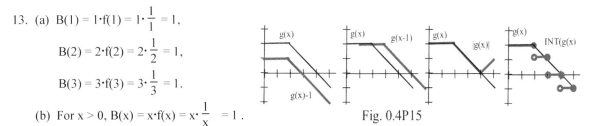

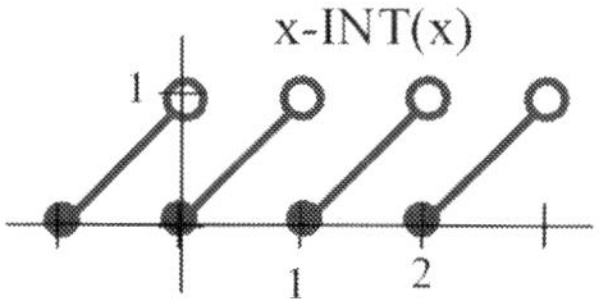

Fig. 0.4P19

23. (a) g(1) = 1, g(2) = 1, g(3) = 0, g(4) = –1. Now graph g

25. ≈ 0.739 starting with x = 1, 2, 10, or any value.

27. $f(x) = (x^2 + 1)/(2x)$. (note that this is the corrected version of the function f)
$f(1) = 2/2 = 1$.
$f(0.5) = 1.25$, $f(1.25) = 1.025$, $f(1.025) \approx 1.0003049$, $f(1.0003049) \approx 1.000000046$, ...
$f(4) = 2.125$, $f(2.125) \approx 1.297794$, $f(1.297794) \approx 1.034166$, $f(1.034166) \approx 1.000564$, ...

29. (a) $f(2) = 14/3 \approx 4.7$, $f(14/3) = 50/9 \approx 5.6$, $f(50/9) = 158/27 \approx 5.85$, $f(158/27) = 482/81 \approx 5.95$
$f(4) = 16/3 \approx 5.3$, $f(16/3) = 52/9 \approx 5.8$, $f(52/9) = 160/27 \approx 5.93$, $f(160/27) = 484/81 \approx 5.975$
$f(6) = 6$.
(b) $c = 6$.
(c) Solve $c = g(c) = c/3 + A$ to get $3c = c + 3A$ and $2c = 3A$ so $c = \frac{3A}{2}$ is a fixed point of g.

31. On your own.

Section 0.5 Answers

1. (a) $x = 2, 4$ (b) $x = -2, -1, 0, 1, 2, 3, 4, 5$ (c) $x = -2, -1, 1, 3$

3. (a) all x (all real numbers) (b) $x > \sqrt[3]{-2}$ (c) all x

5. (a) $x = -2, -3, 3$ (b) no values of x (c) $x \geq 0$

7. (a) If $x \neq 2$ and $x \neq -3$, then $x^2 + x - 6 \neq 0$. True.
(b) If an object does not have 3 sides, then it is not a triangle. True.

9. (a) If your car does not get at least 24 miles per gallon, then it is not tuned properly.
(b) If you can not have dessert, then you did not eat your vegetables.

11. (a) If you will not vote for me, then you do not love your country.
(b) If not only outlaws have guns, then guns are not outlawed. (poor English)
If someone legally has a gun, then guns are not illegal.

13. (a) Both f(x) and g(x) are not positive. (b) x is not positive. ($x \leq 0$)
(c) 8 is not a prime number.

15. (a) For some numbers a and b, $| a+b | \neq | a | + | b |$. (b) Some snake is not poisonous.
(c) Some dog can climb trees.

17. If x is an integer, then 2x is an even integer. True.
Converse: If 2x is an even integer, then x is an integer. True.
(It is not likely that these were the statements you thought of. There are lots of other examples.)

19. (a) False. Put $a = 3$ and $b = 4$. Then $(a + b)^2 = (7)^2 = 49$, but $a^2 + b^2 = 3^2 + 4^2 = 9 + 16 = 25$.
(b) False. Put $a = -2$ and $b = -3$. Then $a > b$, but $a^2 = 4 < 9 = b^2$.
(c) True.

21. (a) True. (b) False. Put $f(x) = x + 1$ and $g(x) = x + 2$. Then $f(x)g(x) = x^2 + 3x + 2$ is not a linear function.
(c) True.

23. (a) If a and b are prime numbers, then a + b is prime. False: take $a = 3$ and $b = 5$.
(b) If a and b are prime numbers, then a + b is not prime. False: take $a = 2$ and $b = 3$.
(c) If x is a prime number, then x is odd. False: take $x = 2$. (this is the only counterexample)
(d) If x is a prime number, then x is even. False: take $x = 3$ (or 5 or 7 or ...)

25. (a) If x is a solution of $x + 5 = 9$, then x is odd. False: take $x = 4$.
(b) If a 3–sided polygon has equal sides, then it is a triangle. True. (We also have nonequilateral triangles .)
(c) If a person is a calculus student, then that person studies hard. False (unfortunately), but we won't mention names.
(d) If x is a (real number) solution of $x^2 - 5x + 6 = 0$, then x is even. False: take $x = 3$.

1.0 TANGENT LINES, VELOCITIES, GROWTH

In section 0.2, we estimated the slope of a line tangent to the graph of a function at a point. At the end of section 0.3, we constructed a new function which was the slope of the line tangent to the graph of a function at each point. In both cases, before we could calculate a slope, we had to **estimate** the tangent line from the graph of the function, a method which required an accurate graph and good estimating. In this section we will start to look at a more precise method of finding the slope of a tangent line which does not require a graph or any estimation by us. We will start with a nonapplied problem and then look at two applications of the same idea.

The Slope of a Line Tangent to a Function at a Point

Our goal is to find a way of exactly determining the slope of the line which is tangent to a function (to the graph of the function) at a point in a way which does not require us to have the graph of the function.

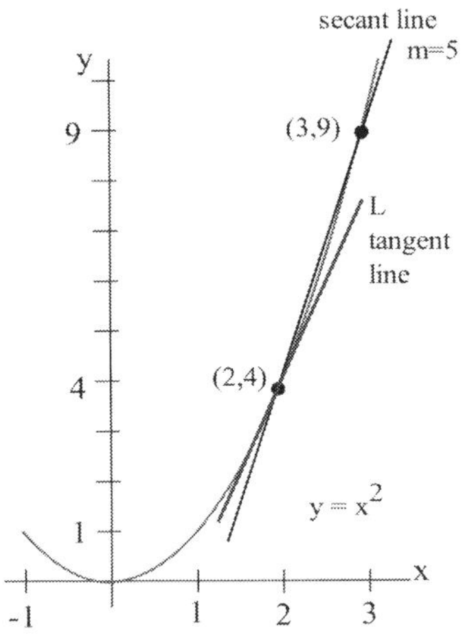

Fig. 1

Let's start with the problem of finding the slope of the line L (Fig. 1) which is tangent to $f(x) = x^2$ at the point (2,4). We could estimate the slope of L from the graph, but we won't. Instead, we can see that the line through (2,4) and (3,9) on the graph of f is an approximation of the slope of the tangent line, and we can calculate that slope exactly: $m = \Delta y/\Delta x = (9-4)/(3-2) = 5$. But $m = 5$ is only an estimate of the slope of the tangent line and not a very good estimate. It's too big. We can get a better estimate by picking a second point on the graph of f which is closer to (2,4) — the point (2,4) is fixed and it must be one of the points we use. From Fig. 2, we can see that the slope of the line through the points (2,4) and (2.5,6.25) is a better approximation of the slope of the tangent line at (2,4): $m = \Delta y/\Delta x = (6.25 - 4)/(2.5 - 2) = 2.25/.5 = 4.5$, a better estimate, but still an approximation. We can continue picking points closer and closer to (2,4) on the graph of f, and then calculating the slopes of the lines through each of these points and the point (2,4):

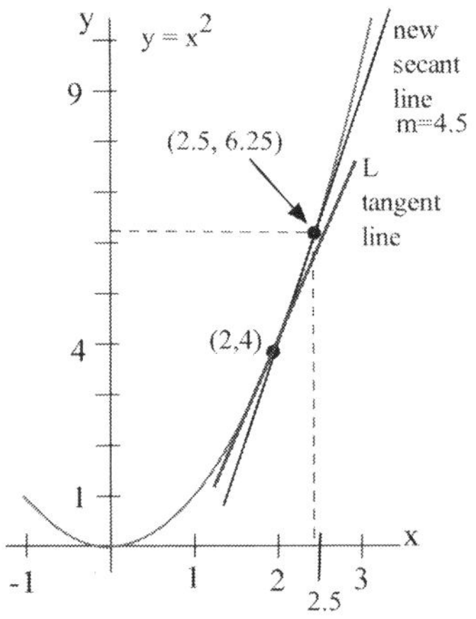

Fig. 2

Points to the left of (2,4)

x	$y = x^2$	slope of line through (x,y) and (2,4)
1.5	2.25	3.5
1.9	3.61	3.9
1.99	3.9601	3.99

Points to the right of (2,4)

x	$y = x^2$	slope of line through (x,y) and (2,4)
3	9	5
2.5	6.25	4.5
2.01	4.0401	4.01

The only thing special about the x–values we picked is that they are numbers which are close, and very close, to x = 2. Someone else might have picked other nearby values for x. As the points we pick get closer and closer to the point (2,4) on the graph of $y = x^2$, the slopes of the lines through the points and (2,4) are better approximations of the slope of the tangent line, and these slopes are getting closer and closer to 4.

Practice 1: What is the slope of the line through (2,4) and (x, y) for $y = x^2$ and x = 1.994? x = 2.0003?

We can bypass much of the calculating by not picking the points one at a time: let's look at a general point near

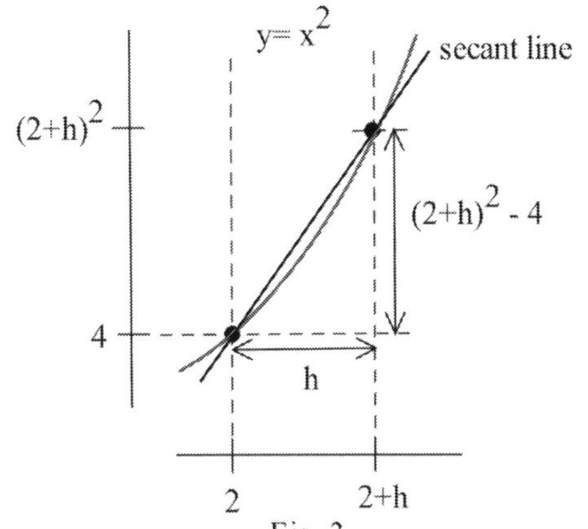

Fig. 3

(2,4). Define x = 2 + h so h is the increment from 2 to x (Fig. 3). If h is small, then x = 2 + h is close to 2 and the point (2+h, f(2+h)) = (2+h, $(2+h)^2$) is close to (2,4). The slope m of the line through the points (2,4) and (2+h, $(2+h)^2$) is a good approximation of the slope of the tangent line at the point (2,4):

$$m = \frac{\Delta y}{\Delta x} = \frac{(2+h)^2 - 4}{(2+h) - 2}$$

$$= \frac{\{4 + 4h + h^2\} - 4}{h} = \frac{4h + h^2}{h} = \frac{h(4+h)}{h} = 4 + h .$$

If h is very small, then m = 4 + h is a very good approximation to the slope of the tangent line, and m = 4 + h is very close to the value 4. The value m = 4 + h is called the slope of the **secant line** through the two points (2,4) and (2+h, $(2+h)^2$). The limiting value 4 of m = 4 + h as h gets smaller and smaller is called the **slope of the tangent line** to the graph of f at (2,4).

Example 1: Find the slope of the line tangent to $f(x) = x^2$ at the point (1,1) by evaluating the slope of the secant line through (1,1) and (1+h, f(1+h)) and then determining what happens as h gets very small (Fig. 4).

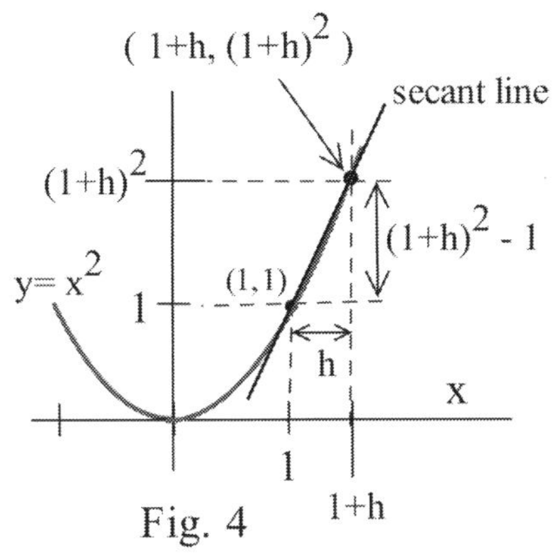

Fig. 4

Solution: The slope of the secant line through the points (1,1) and (1+h, f(1+h)) is

$$m = \frac{f(1+h) - 1}{(1+h) - 1} = \frac{(1+h)^2 - 1}{h} = \frac{\{1 + 2h + h^2\} - 1}{h}$$

$= \frac{2h + h^2}{h} = 2 + h$. As h gets very small, the value of m approaches the value 2, the slope of tangent line at the point (1,1).

Practice 2: Find the slope of the line tangent to the graph of $y = f(x) = x^2$ at the point (–1,1) by finding the slope of the secant line, m_{sec} , through the points (–1,1) and (–1+h, f(–1+h)) and then determining what happens to m_{sec} as h gets very small.

FALLING TOMATO

Suppose we drop a tomato from the top of a 100 foot building (Fig. 5) and time its fall.

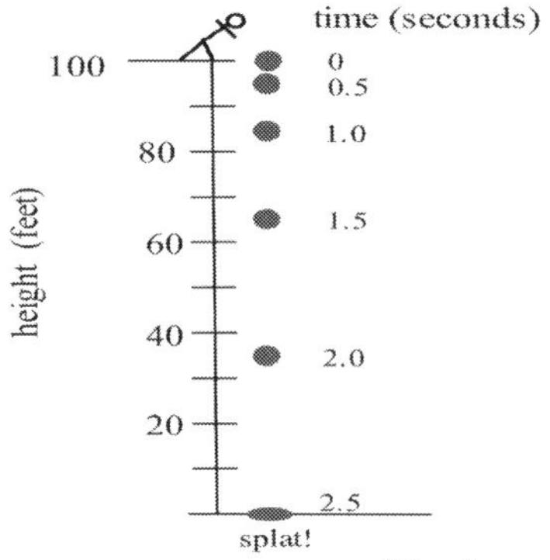

Fig. 5

time (s)	height (ft)
0	100
0.5	96
1.0	84
1.5	64
2.0	36
2.5	0

Some questions are easy to answer directly from the table:

(a) How long did it take for the tomato to drop 100 feet? (2.5 seconds)

(b) How far did the tomato fall during the first second? (100 – 84 = 16 feet)

(c) How far did the tomato fall during the last second? (64 – 0 = 64 feet)

(d) How far did the tomato fall between t =.5 and t = 1? (96 – 84 = 12 feet)

Some other questions require a little calculation:

(e) What was the average velocity of the tomato during its fall?

$$\textbf{Average velocity} = \frac{\textbf{distance fallen}}{\textbf{total time}} = \frac{\boldsymbol{\Delta}\textbf{position}}{\boldsymbol{\Delta}\textbf{time}} = \frac{-100\text{ ft}}{2.5\text{ s}} = -40\text{ ft/s}.$$

(f) What was the average velocity between t=1 and t=2 seconds?

$$\text{Average velocity} = \frac{\Delta\text{position}}{\Delta\text{time}} = \frac{36\text{ ft} - 84\text{ ft}}{2\text{ s} - 1\text{ s}} = \frac{-48\text{ ft}}{1\text{ s}} = -48\text{ ft/s}.$$

Some questions are more difficult.

(g) How fast was the tomato falling 1 second after it was dropped?

This question is significantly different from the previous two questions about average velocity. Here we want the **instantaneous velocity,** the velocity at an instant in time. Unfortunately the tomato is not equipped with a speedometer so we will have to give an approximate answer.

One crude approximation of the instantaneous velocity after 1 second is simply the average velocity during the entire fall, –40 ft/s . But the tomato fell slowly at the beginning and rapidly near the end so the "–40 ft/s" estimate may or may not be a good answer.

We can get a better approximation of the instantaneous velocity at t=1 by calculating the average velocities over a short time interval near t = 1 . The average velocity between t = 0.5 and t = 1 is $\frac{-12 \text{ feet}}{0.5 \text{ s}} = -24$ ft/s, and the average velocity between t = 1 and t = 1.5 is $\frac{-20 \text{ feet}}{.5 \text{ s}} = -40$ ft/s so we can be reasonably sure that the instantaneous velocity is between –24 ft/s and –40 ft/s.

In general, the shorter the time interval over which we calculate the average velocity, the better the average velocity will approximate the instantaneous velocity. The average velocity over a time interval is $\frac{\Delta \text{position}}{\Delta \text{time}}$, which is the slope of the **secant line** through two points on the graph of height versus time (Fig. 6). The instantaneous velocity at a particular time and height is the slope of the **tangent line** to the graph at the point given by that time and height.

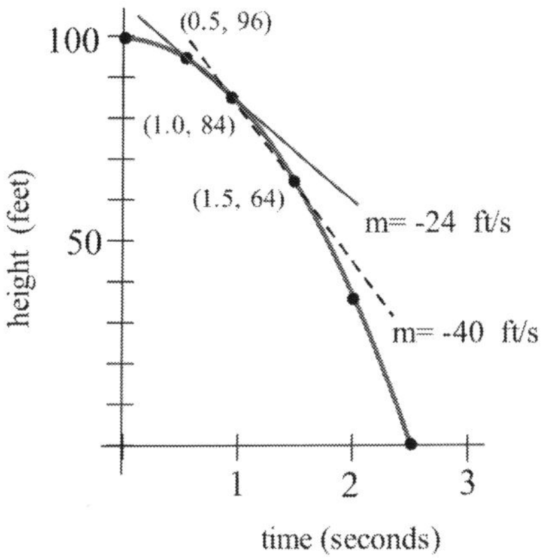

Fig. 6

Average velocity $= \frac{\Delta \textbf{position}}{\Delta \textbf{time}}$

= slope of the secant line through 2 points.

Instantaneous velocity = slope of the line tangent to the graph.

Practice 3: Estimate the velocity of the tomato 2 seconds after it was dropped.

GROWING BACTERIA

Suppose we set up a machine to count the number of bacteria growing on a petri plate (Fig. 7). At first there are few bacteria so the population grows slowly. Then there are more bacteria to divide so the population grows more quickly. Later, there are more bacteria and less room and nutrients available for the expanding population, so the population grows slowly again. Finally, the bacteria have used up most of the nutrients, and the population declines as bacteria die.

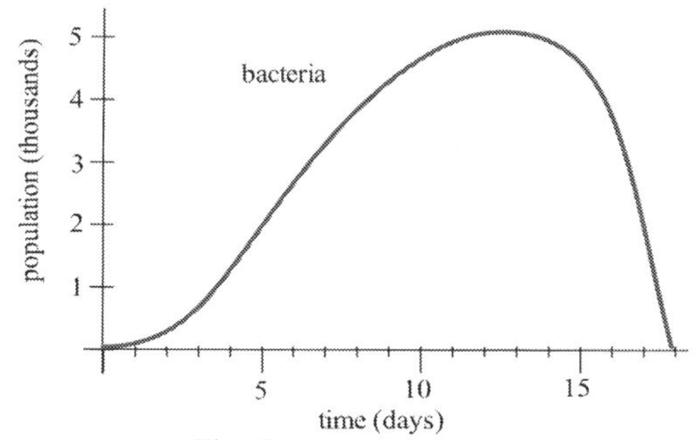

Fig. 7

The population graph can be used to answer a number of questions.

(a) What is the bacteria population at time t = 3 days?
(Answer: about 500 bacteria)

(b) What is the population increment from t = 3 to t = 10 days? (about 4000 bacteria)

(c) What is the **rate** of population growth from $t = 3$ to $t = 10$ days? (Fig. 7)

Solution: The rate of growth from $t = 3$ to $t = 10$ is the average change in population during that time:

$$\text{average change in population} = \frac{\text{change in population}}{\text{change in time}} = \frac{\Delta\text{population}}{\Delta\text{time}}$$

$$= \frac{4000 \text{ bacteria}}{7 \text{ days}} \approx 570 \text{ bacteria/day} .$$

This is the slope of the secant line through the two points (3, 500) and (10, 4500).

(d) What is the **rate** of population growth on the third day, at $t = 3$?

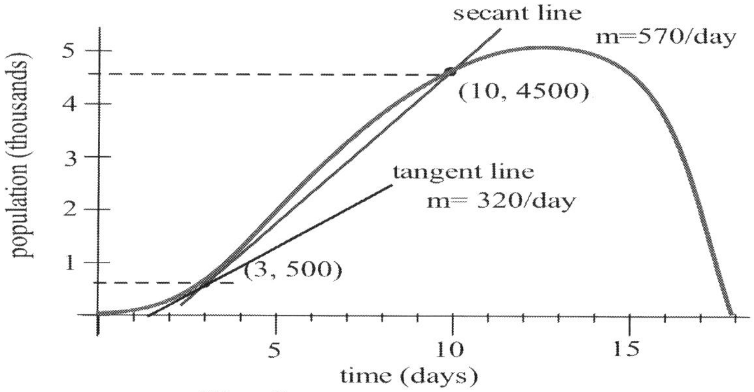

Fig. 8

Solution: This question is asking for the **instantaneous** rate of population change, the slope of the line which is **tangent** to the population curve at (3, 500). If we sketch a line approximately tangent to the curve at (3, 500) and pick two points near the ends of the tangent line segment (Fig. 8), we can estimate that instantaneous rate of population growth is approximately 320 bacteria/day .

> **Average population growth rate** $= \dfrac{\Delta\textbf{population}}{\Delta\textbf{time}}$
>
> **= slope of the secant line through 2 points.**
>
> **Instantaneous population growth rate = slope of the line tangent to the graph.**

Practice 4: Approximately what was the average change in population between $t = 9$ and $t = 13$? Approximately what was rate of population growth at $t = 9$ days?

The tangent line problem, the instantaneous velocity problem and the instantaneous growth rate problem are all similar. In each problem we wanted to know how rapidly something was **changing at an instant in time**, and each problem turned out to be finding the **slope of a tangent line**. The approach in each problem was also the same: find an approximate solution and then examine what happened to the approximate solution over shorter and shorter intervals. We will often use this approach of finding a limiting value, but before we can use it effectively we need to describe the concept of a limit with more precision.

PROBLEMS

1. What is the slope of the line through (3,9) and (x, y) for $y = x^2$ and $x = 2.97$? $x = 3.001$? $x = 3+h$? What happens to this last slope when h is very small (close to 0)? Sketch the graph of $y = x^2$ for x near 3.

2. What is the slope of the line through (–2,4) and (x, y) for $y = x^2$ and $x = -1.98$? $x = -2.03$? $x = -2+h$? What happens to this last slope when h is very small (close to 0)? Sketch the graph of $y = x^2$ for x near –2.

3. What is the slope of the line through (2,4) and (x, y) for $y = x^2 + x - 2$ and $x = 1.99$? $x = 2.004$? $x = 2+h$? What happens to this last slope when h is very small? Sketch the graph of $y = x^2 + x - 2$ for x near 2.

4. What is the slope of the line through (–1,–2) and (x, y) for $y = x^2 + x - 2$ and $x = -.98$? $x = -1.03$? $x = -1+h$? What happens to this last slope when h is very small? Sketch the graph of $y = x^2 + x - 2$ for x near –1.

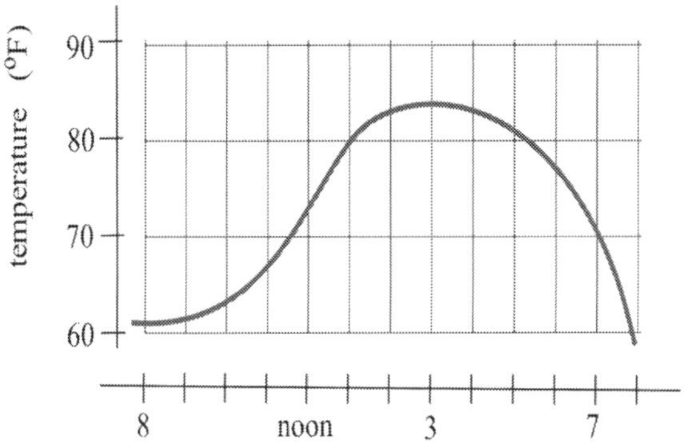

Fig. 9

5. Fig. 9 shows the temperature during a day in Ames.
   (a) What was the average change in temperature from 9 am to 1 pm?
   (b) Estimate how fast the temperature was rising **at** 10 am and **at** 7 pm?

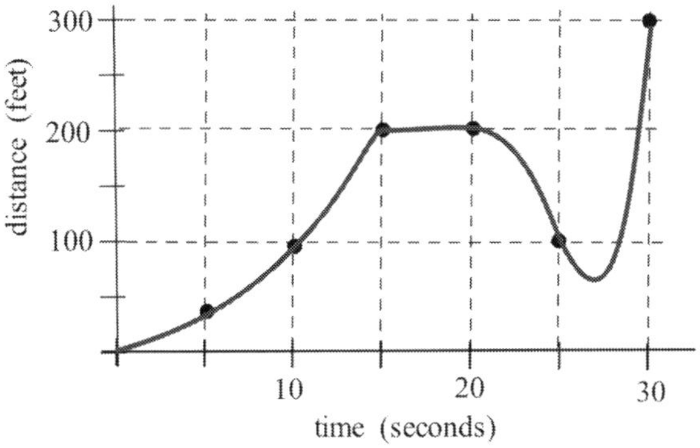

Fig. 10

6. Fig. 10 shows the distance of a car from a measuring position located on the edge of a straight road.
   (a) What was the average velocity of the car from t = 0 to t = 30 seconds?
   (b) What was the average velocity of the car from t = 10 to t = 30 seconds?
   (c) About how fast was the car traveling **at** t = 10 seconds? **at** t = 20 s ? **at** t = 30 s ?
   (d) What does the horizontal part of the graph between t = 15 and t = 20 seconds mean?
   (e) What does the negative velocity at t = 25 represent?

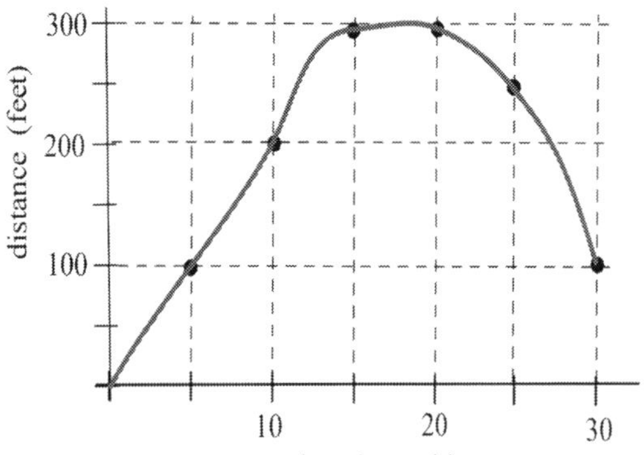

Fig. 11

7. Fig. 11 shows the distance of a car from a measuring position located on the edge of a straight road.

   (a) What was the average velocity of the car from $t = 0$ to $t = 20$ seconds?

   (b) What was the average velocity from $t = 10$ to $t = 30$ seconds?

   (c) About how fast was the car traveling **at** $t = 10$ seconds? **at** $t = 20$ s ? **at** $t = 30$ s ?

8. Fig. 12 shows the composite developmental skill level of chess masters at different ages as determined by their performance against other chess masters. (From "Rating Systems for Human Abilities", by W.H. Batchelder and R.S. Simpson, 1988. UMAP Module 698.)

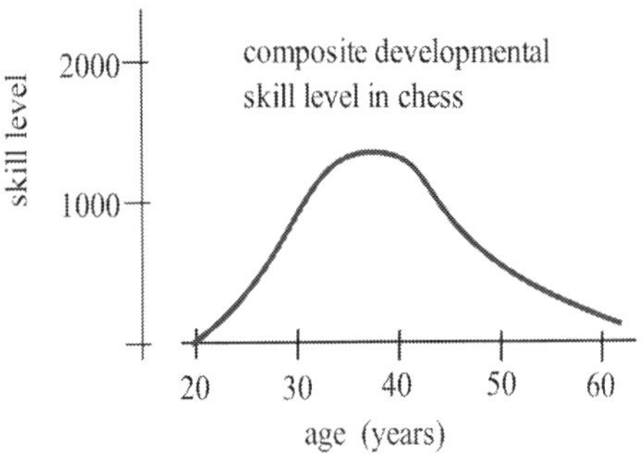

Fig. 12

   (a) At what age is the "typical" chess master playing the best chess?

   (b) At approximately what age is the chess master's skill level increasing most rapidly?

   (c) Describe the development of the "typical" chess master's skill in words.

   (d) Sketch graphs which you **think** would reasonably describe the performance levels versus age for an athlete, a classical pianist, a rock singer, a mathematician, and a professional in your major field.

Problems 9 and 10 define new functions A(x) in terms of AREAS bounded by the functions $y = 3$ and $y = x + 1$. This may seem a strange way to define a functions A(x), but this idea will become important later in calculus. We are just getting an early start here.

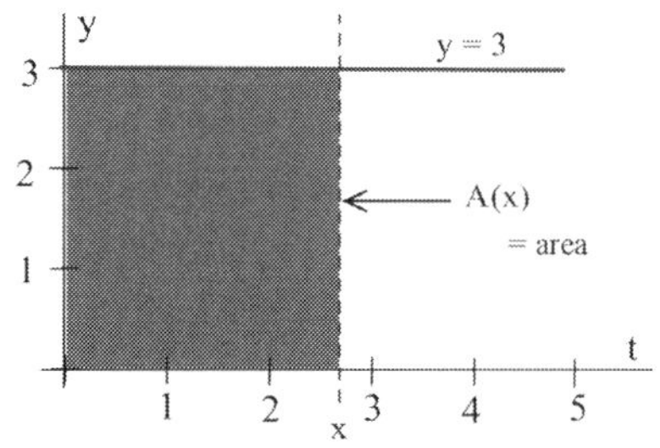

Fig. 13

9. Define A(x) to be the **area** bounded by the t (horizontal) and y axes, the horizontal line $y = 3$, and the vertical line at x (Fig. 13). For example, $A(4) = 12$ is the area of the 4 by 3 rectangle.

   a) Evaluate $A(0), A(1), A(2), A(2.5)$ and $A(3)$.

   b) What area would $A(4) - A(1)$ represent in the figure?

   c) Graph $y = A(x)$ for $0 \le x \le 4$.

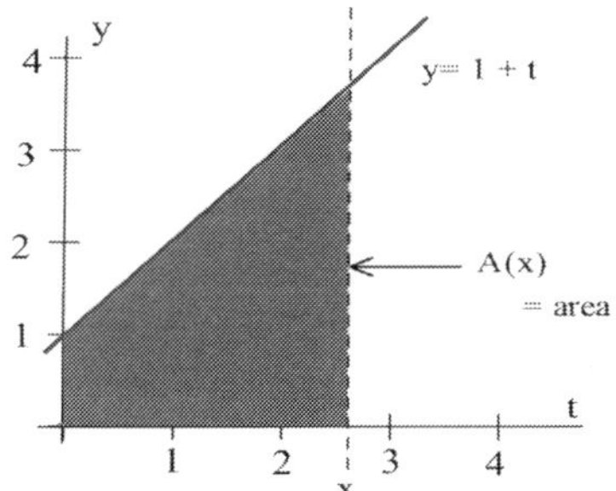

Fig. 14

10. Define A(x) to be the **area** bounded by the t (horizontal) and y axes, the line y = t + 1, and the vertical line at x (Fig. 14). For example, A(4) = 12.

a) Evaluate A(0), A(1), A(2), A(2.5) and A(3).

b) What area would A(3) – A(1) represent in the figure?

c) Graph y = A(x) for $0 \le x \le 4$.

Section 1.0 **PRACTICE Answers**

Practice 1: $y = x^2$

If x = 1.994, then y = 3.976036 so the slope between (2, 4) and (x ,y) is

$$\frac{4-y}{2-x} = \frac{4-3.976036}{2-1.994} = \frac{0.023964}{0.006} \approx 3.994 .$$

If x = 2.0003, then $y \approx 4.0012$ so the slope between (2, 4) and (x ,y) is

$$\frac{4-y}{2-x} = \frac{4-4.0012}{2-2.0003} = \frac{-0.0012}{0.0003} \approx 4.0003 .$$

Practice 2: $m_{sec} = \frac{f(-1+h)-(1)}{(-1+h)-(-1)} = \frac{(-1+h)^2-1}{h} = \frac{1-2h+h^2-1}{h} = \frac{h(-2+h)}{h} = \mathbf{-2+h}$

As $h \to 0$, $m_{sec} = -2 + h \to \mathbf{-2}$.

Practice 3: The average velocity between t = 1.5 and t = 2.0 is $\frac{36-64 \text{ feet}}{2.0-1.5 \text{ sec}} = -56$ feet per second.

The average velocity between t = 2.0 and t = 2.5 is $\frac{0-36 \text{ feet}}{2.5-2.0 \text{ sec}} = -72$ feet per second.

The velocity **at** t = 2.0 is somewhere between –56 ft/sec and –72 ft/sec, probably about the middle of this interval: $\frac{(-56)+(-72)}{2} = \mathbf{-64 \text{ ft/sec}}$.

Practice 4: (a) When t = 9 days, the population is approximately P = 4,200 bacteria. When t = 13, $P \approx 5{,}000$. The average change in population is approximately

$$\frac{5000-4200 \text{ bacteria}}{13-9 \text{ days}} = \frac{800 \text{ bacteria}}{4 \text{ days}} = \mathbf{200 \text{ bacteria per day}}.$$

(b) To find the rate of population growth at t = 9 days, sketch the line tangent to the population curve at the point (9, 4200) and then use (9, 4200) and another point on the tangent line to calculate the slope of the line. Using the approximate values(5, 2800) and (9, 4200), the slope of the tangent line at the point (9, 4200) is approximately

$$\frac{4200 - 2800 \text{ bacteria}}{9 - 5 \text{ days}} = \frac{1400 \text{ bacteria}}{4 \text{ days}} \approx \textbf{350 bacteria/day} .$$

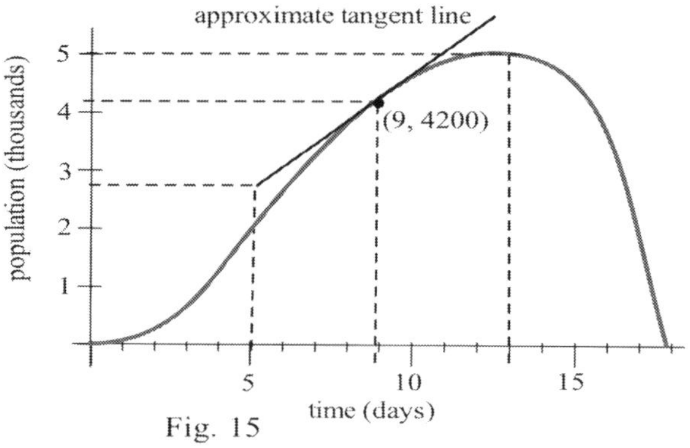

Fig. 15

1.1 THE LIMIT OF A FUNCTION

THE IDEA, Informally

Calculus has been called the study of continuous change, and the **limit** is the basic concept which allows us to describe and analyze such change. An understanding of limits is necessary to understand derivatives, integrals and other fundamental topics of calculus.

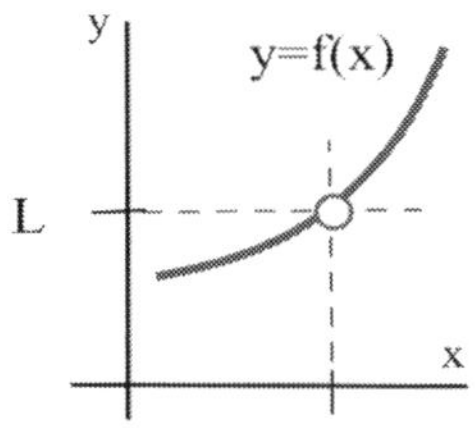

Fig. 1

The limit of a function describes the behavior of the function when the variable is near, **but does not equal** , a specified number (Fig. 1). If the values of f(x) get closer and closer , as close as we want, to one number L as we take values of x very close to (but not equal to) a number c, then we

> say **"the limit of f(x), as x approaches c, is L "** and we
>
> write " $\lim_{x \to c}$ **f(x) = L."** (The symbol " → " means "approaches" or "gets very close to.")

f(c) is a single number that describes the behavior (value) of f **AT** the point x = c.

$\lim_{x \to c}$ **f(x)** is a single number that describes the behavior of f **NEAR, <u>BUT NOT AT</u>,** the point x = c.

If we have a graph of the function near x = c , then it is usually easy to determine $\lim_{x \to c} f(x)$.

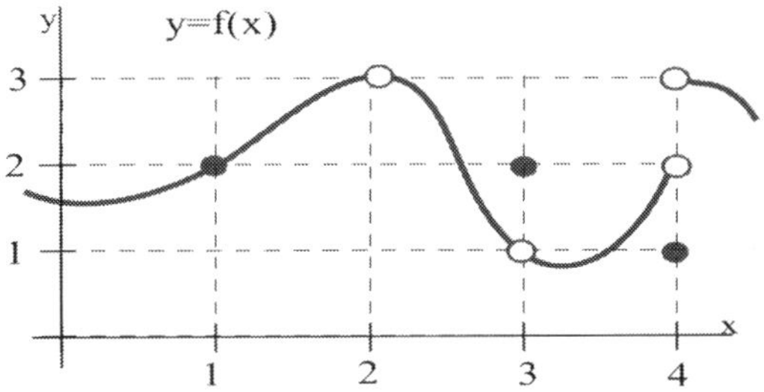

Fig. 2

Example 1: Use the graph of y = f(x) in Fig. 2 to determine the following limits:

(a) $\lim_{x \to 1} f(x)$ (b) $\lim_{x \to 2} f(x)$

(c) $\lim_{x \to 3} f(x)$ (d) $\lim_{x \to 4} f(x)$

Solution: (a) $\lim_{x \to 1} f(x) = 2$. When x is very close to 1, the values of f(x) are very close to y = 2. In this example, it happens that f(1) = 2, but that is <u>irrelevant</u> for the limit. The only thing that matters is what happens for x close to 1 but $x \neq 1$.

(b) f(2) is undefined, but we only care about the behavior of f(x) for x <u>close to</u> 2 and <u>not equal to</u> 2. When x is close to 2, the values of f(x) are close to 3. If we restrict x close enough to 2, the values of y will be as close to 3 as we want, so $\lim_{x \to 2} f(x) = 3$.

(c) When x is close to 3 (or as x approaches the value 3), the values of f(x) are close to 1 (or approach the value 1), so $\lim_{x \to 3} f(x) = 1$. For this limit it is completely irrelevant that f(3) = 2, We only care about what happens to f(x) for x close to and not equal to 3.

(d) This one is harder and we need to be careful. When x is close to 4 and slightly **less than** 4 (x is just to the left of 4 on the x–axis), then the values of f(x) are close to 2. But if x is close to 4 and slightly **larger than** 4 then the values of f(x) are close to 3. If we only know that x is very close to 4, then we cannot say whether y = f(x) will be close to 2 or close to 3 — it depends on whether x is on the right or the left side of 4. In this situation, the f(x) values are not close to a single number so we say $\lim_{x \to 4} f(x)$ **does not exist**. It is irrelevant that f(4) = 1. The limit, as x approaches 4, would still be undefined if f(4) was 3 or 2 or anything else.

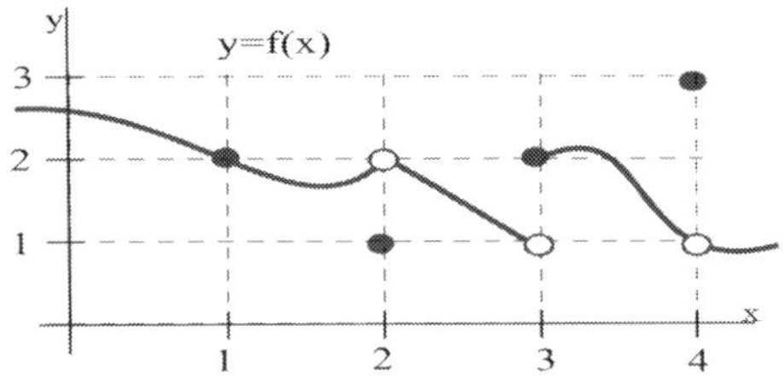

Fig. 3

Practice 1: Use the graph of y = f(x) in Fig. 3 to determine the following limits:

(a) $\lim_{x \to 1} f(x)$ (b) $\lim_{t \to 2} f(t)$

(c) $\lim_{x \to 3} f(x)$ (d) $\lim_{w \to 4} f(w)$

Example 2: Determine the value of $\lim_{x \to 3} \dfrac{2x^2 - x - 1}{x - 1}$.

Solution: We need to investigate the values of f(x) when x is close to 3. If the f(x) values get arbitrarily close to or even equal some number L, then L will be the limit. One way to keep track of both the x and the f(x) values is to set up a table and to pick several x values which are closer and closer (but not equal) to 3. We can pick some values of x which approach 3 from the left, say x = 2.91, 2.9997, 2.999993, and 2.9999999 , and some values of x which approach 3 from the right, say x = 3.1, 3.004, 3.0001, and 3.000002 . The only thing important about these particular values for x is that they get closer and closer to 3 without equaling 3. You should try some other values "close to 3" to see what happens. Our table of values is

x	f(x)
2.9	6.82
2.9997	6.9994
2.999993	6.999986
2.9999999	6.9999998
↓	↓
3	**7**

x	f(x)
3.1	7.2
3.004	7.008
3.0001	7.0002
3.000002	7.000004
↓	↓
3	**7**

As the x values get closer and closer to 3, the f(x) values are getting closer and closer to 7. In fact, we can get f(x) as close to 7 as we want ("arbitrarily close") by taking the values of x very close ("sufficiently close") to 3. $\lim_{x \to 3} \dfrac{2x^2 - x - 1}{x - 1} = 7$.

Instead of using a table of values, we could have graphed y = f(x) for x close to 3, Fig. 4, and used the graph to answer the limit question. This graphic approach is easier, particularly if you have a calculator or computer do the graphing work for you, but it is really very similar to the "table of values" method: in each case you need to evaluate y = f(x) at many values of x near 3.

You might have noticed that if we just evaluate f(3), then we get the correct answer 7. That works for this particular problem, but it often fails. The next example illustrates the difficulty.

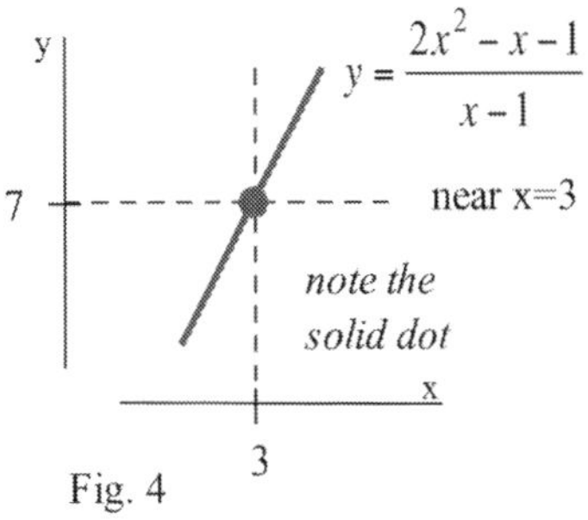

Fig. 4

Example 3: Find $\lim_{x \to 1} \frac{2x^2 - x - 1}{x - 1}$. (Same as Example 2 but with x→1.)

Solution: You might try to evaluate $f(x) = \frac{2x^2 - x - 1}{x - 1}$ at x = 1, but f is not defined at x = 1. It is tempting, **but wrong**, to conclude that this function does not have a limit as x approaches 1.

<u>Table Method</u>: Trying some "test" values for x which get closer and closer to 1 from both the left and the right, we get

x	f(x)	x	f(x)
0.9	2.82	1.1	3.2
0.9998	2.9996	1.003	3.006
0.999994	2.999988	1.0001	3.0002
0.9999999	2.9999998	1.000007	3.000014
↓	↓	↓	↓
1	**3**	1	**3**

The function f is not defined at x = 1, but when x is **close to** 1, the values of f(x) are getting very close to 3. We can get f(x) as close to 3 as we want by taking x very close to 1 so $\lim_{x \to 1} \frac{2x^2 - x - 1}{x - 1} = 3$.

<u>Graph Method</u>: We can graph $y = f(x) = \frac{2x^2 - x - 1}{x - 1}$ for x close to 1, Fig. 5, and notice that whenever x is close to 1, the values of y = f(x) are close to 3. f is not defined at x = 1, so the graph has a hole above x = 1, but we only care about what f(x) is doing for x close to but **not equal to** 1.

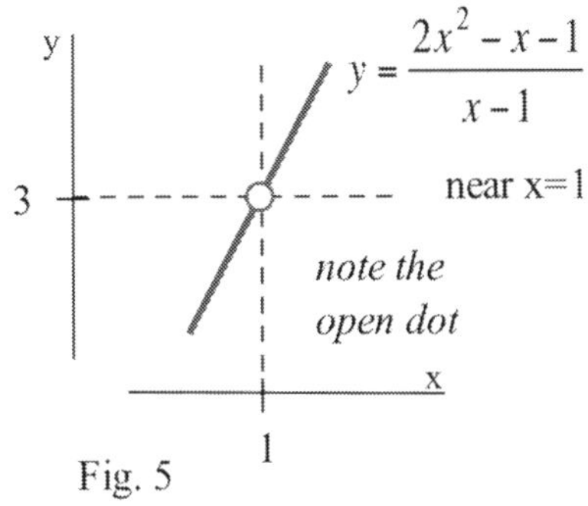

Fig. 5

<u>Algebra Method</u>: We could have found the same result by noting that $f(x) = \frac{2x^2 - x - 1}{x - 1} = \frac{(2x+1)(x-1)}{x-1} = 2x+1$ as long as $x \neq 1$. (If $x \neq 1$, then $x-1 \neq 0$ so it is valid to divide the numerator and denominator by the factor x–1.) The "x→1" part of the limit means that x is close to 1 but **not equal to 1**, so our division step is valid and

$$\lim_{x \to 1} \frac{2x^2 - x - 1}{x - 1} = \lim_{x \to 1} 2x + 1 = 3$$, the correct answer.

THREE METHODS FOR EVALUATING LIMITS

The Algebra Method

The algebra method involves algebraically simplifying the function before trying to evaluate its limit. Often, this simplification just means factoring and dividing, but sometimes more complicated algebraic or even trigonometric steps are needed.

The Table and Graph Methods

To evaluate a limit of a function f(x) as x approaches c, the table method involves calculating the values of f(x) for "enough" values of x very close to c so that we can "confidently" determine which value f(x) is approaching. If f(x) is well–behaved, we may not need to use very many values for x. However, this method is usually used with complicated functions, and then we need to evaluate f(x) for lots of values of x.

> A computer or calculator can often make the function evaluations easier, but their calculations are subject to "round off" errors. The result of any computer calculation which involves both large and small numbers should be viewed with some suspicion. For example, the function
>
> $$f(x) = \frac{\{ (0.1)^x + 1 \} - 1}{(0.1)^x} = \frac{(0.1)^x}{(0.1)^x} = 1$$ for every value of x , and my calculator gives the correct answer for some values of x : $f(3) = \frac{\{ (0.1)^3 + 1 \} - 1}{(0.1)^3} = 1$, and f(8) and f(9) both equal 1.
>
> But my calculator says $\{ (0.1)^{10} + 1 \} - 1 = 0$ so it evaluates f(10) to be 0, definitely an **incorrect** value. Your calculator may evaluate f(10) correctly, but try f(35) or f(107).
>
> **Calculators are too handy to be ignored, but they are too prone to these types of errors to be believed <u>uncritically</u>. Be careful.**

The graph method is closely related to the table method, but we create a graph of the function instead of a table of values, and then we use the graph to determine which value f(x) is approaching.

Which Method Should You Use?

In general, the algebraic method is preferred because it is precise and does not depend on which values of x we chose or the accuracy of our graph or precision of our calculator. **If you can evaluate a limit algebraically, you should do so.** Sometimes, however, it will be very difficult to evaluate a limit algebraically, and the table or graph methods offer worthwhile alternatives. Even when you can algebraically evaluate the limit of a function, it is still a good idea to graph the function or evaluate it at a few points just to verify your algebraic answer.

The table and graph methods have the same advantages and disadvantages. Both can be used on very complicated functions which are difficult to handle algebraically or whose algebraic properties you don't know. Often both methods can be easily programmed on a calculator or computer. However, these two methods are very time–consuming by hand and are prone to round off errors on computers. You need to know how to use these methods when you can't figure out how to use the algebraic method, but you need to use these two methods warily.

Example 4: Evaluate (a) $\lim\limits_{x\to 0} \dfrac{x^2+5x+6}{x^2+3x+2}$ and (b) $\lim\limits_{x\to -2} \dfrac{x^2+5x+6}{x^2+3x+2}$.

Solution: The function in each limit is the same but x is approaching a different number in each of them.

(a) Since $x\to 0$, we know that x is getting closer and closer to 0 so the values of the x^2, 5x and 3x terms get as close to 0 as we want. The numerator approaches 6 and the denominator approaches 2, so the values of the whole function get arbitrarily close to 6/2 = 3, the limit.

(b) As x approaches –2, the numerator and denominator approach 0, and a small number divided by a small number can be almost anything — the ratio depends on the size of the top compared to the bottom. <u>More investigation is needed</u>.

<u>Table Method</u>: If we pick some values of x close to (but not equal to) –2, we get the table

x	x^2+5x+6	x^2+3x+2	$\dfrac{x^2+5x+6}{x^2+3x+2}$
–1.97	0.0309	–0.0291	–1.061856
–2.005	–0.004975	0.005025	–0.990050
–1.9998	0.00020004	–0.00019996	–1.00040008
–2.00003	–0.00002999	0.0000300009	–0.9996666
↓	↓	↓	↓
–2	0	0	**–1**

Even though the numerator and denominator are each getting closer and closer to 0, their ratio is getting arbitrarily close to –1 which is the limit.

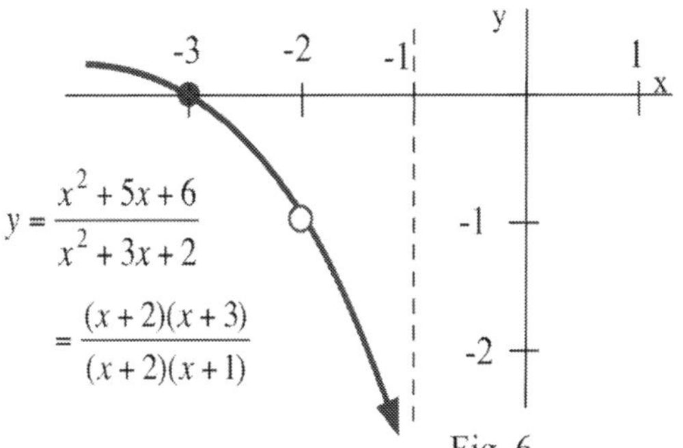

Fig. 6

<u>Graph Method</u>: The graph of $y = f(x) = \dfrac{x^2+5x+6}{x^2+3x+2}$ in ________Fig. 6 shows that the values of f(x) are very close to –1 when the x–values are close to –2.

<u>Algebra Method</u>: $f(x) = \dfrac{x^2+5x+6}{x^2+3x+2} = \dfrac{(x+2)(x+3)}{(x+2)(x+1)}$.

We know $x \to -2$ so $x \neq -2$, and we can divide the top and bottom by (x+2). Then f(x) = (x+3)/(x+1) so $f(x) \to 1/-1 = -1$ as $x \to -2$.

If $\lim_{x \to c} \left\{ \frac{\text{polynomial}}{\text{polynomial}} \right\}$ approaches $\frac{0}{0}$, try dividing the top and bottom by x – c.

Practice 2: Evaluate (a) $\lim_{x \to 2} \frac{x^2 - x - 2}{x - 2}$ (b) $\lim_{t \to 0} \frac{t \cdot \sin(t)}{t^2 + 3t}$ (c) $\lim_{w \to 2} \frac{w - 2}{\ln(w/2)}$.

ONE–SIDED LIMITS

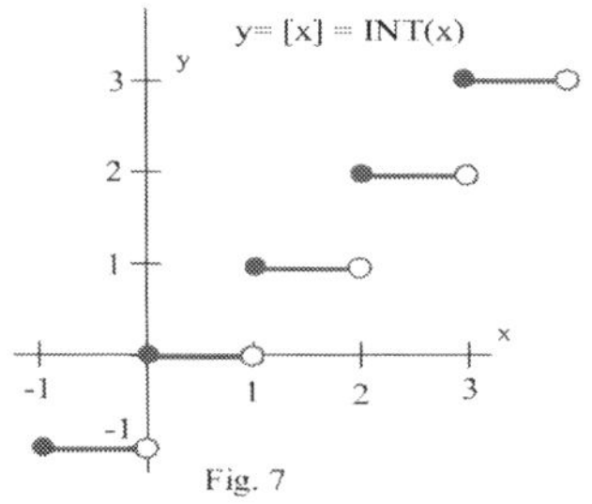

Fig. 7

Sometimes, what happens to us at a place depends on the direction we use to approach that place. If we approach Niagara Falls from the upstream side, then we will be 182 feet higher and have different worries than if we approach from the downstream side. Similarly, the values of a function near a point may depend on the direction we use to approach that point. If we let x approach 3 from the left (x is close to 3 and x < 3), then the values of [x] = INT(x) equal 2 (Fig. 7). If we let x approach 3 from the right (x is close to 3 and x > 3), then the values of [x] = INT(x) equal 3.

On the number line we can approach a point from the left or right, and that leads to **one–sided limits**.

Definition of Left and Right Limits:

The **left limit** as x approaches c of f(x) is L if the values of f(x) get as close to L as we want when x is very close to and **left of c**, x < c: $\lim_{x \to c^-} f(x) = L$.

The **right limit**, written with $\mathbf{x \to c^+}$, requires that x lie to the **right of c**, x > c.

Example 5: Evaluate $\lim_{x \to 2^-} (x - [x])$ and $\lim_{x \to 2^+} (x - [x])$.

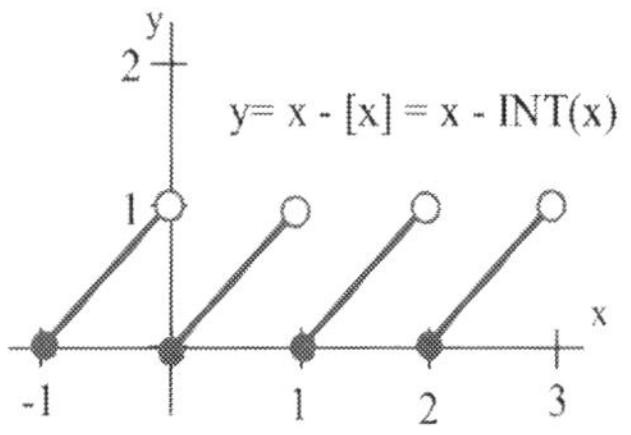

Fig. 8

Solution: The left–limit notation $x \to 2^-$ requires that x be close to 2 and that x be to the left of 2, so x < 2.

If 1 < x < 2, then [x] = 1 so $\lim_{x \to 2^-} (x - [x]) = 2 - 1 = 1$.

If x is close to 2 and is to the right of 2, then 2 < x < 3 so [x] = 2 and $\lim_{x \to 2^+} (x - [x]) = 2 - 2 = 0$.

The graph of f(x) = x – [x] is shown in Fig. 8 .

If the left and right limits have the same value, $\lim_{x \to c^-} f(x) = \lim_{x \to c^+} f(x) = L$, then the value of f(x) is close to L whenever x is close to c, and it does not matter if x is left or right of c so $\lim_{x \to c} f(x) = L$. Similarly, if $\lim_{x \to c} f(x) = L$, then f(x) is close to L whenever x is close to c and less than c and whenever x is close to c and greater than c, so $\lim_{x \to c^-} f(x) = \lim_{x \to c^+} f(x) = L$. We can combine these two statements into a single theorem.

One–Sided Limit Theorem:

$\lim_{x \to c} f(x) = L$ if and only if $\lim_{x \to c^-} f(x) = \lim_{x \to c^+} f(x) = L$.

Corollary: If $\lim_{x \to c^-} f(x) \neq \lim_{x \to c^+} f(x)$, then $\lim_{x \to c} f(x)$ does not exist.

One–sided limits are particularly useful for describing the behavior of functions which have steps or jumps.

To determine the limit of a function involving the greatest integer or absolute value or a multiline definition, definitely consider both the left and right limits.

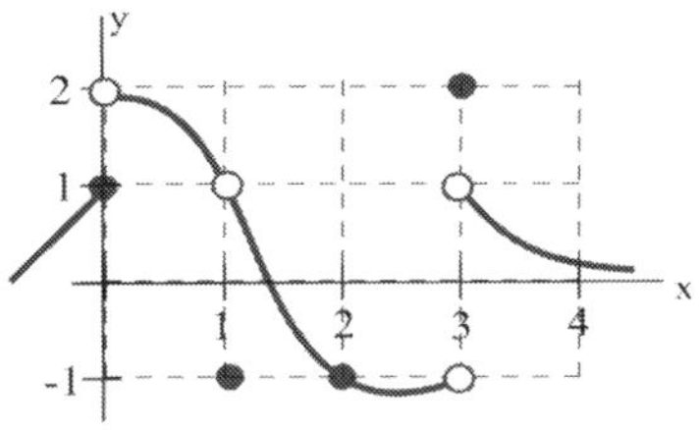

Fig. 9

Practice 3: Use the graph in Fig. 9 to evaluate the one and two–sided limits of f at x = 0, 1, 2, and 3.

Practice 4: Let $f(x) = \begin{cases} 1 & \text{if } x < 1 \\ x & \text{if } 1 < x < 3 \\ 2 & \text{if } 3 < x \end{cases}$.

Find the one and two–sided limits of f at 1 and 3.

PROBLEMS

1. Use the graph in Fig. 10 to determine the following limits.

(a) $\lim_{x \to 1} f(x)$ (b) $\lim_{x \to 2} f(x)$

(c) $\lim_{x \to 3} f(x)$ (d) $\lim_{x \to 4} f(x)$

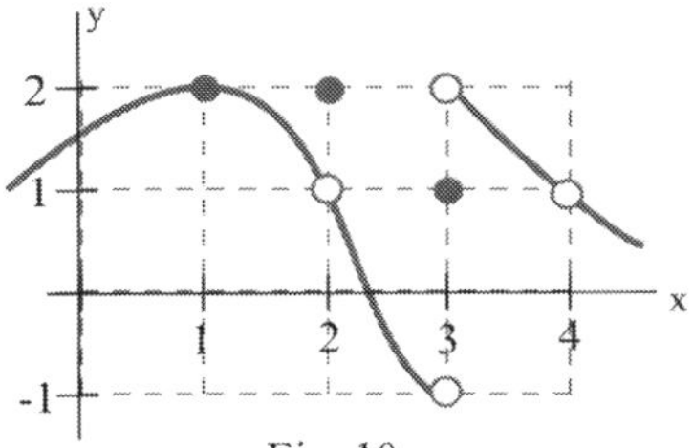

Fig. 10

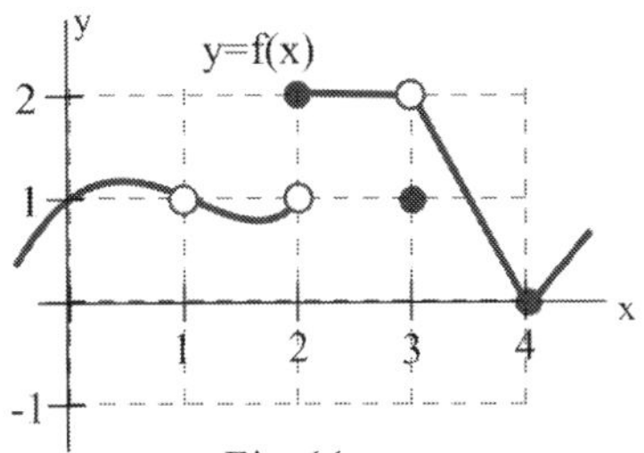

Fig. 11

2. Use the graph in Fig. 11 to determine the following limits.

(a) $\lim_{x \to 1} f(x)$ (b) $\lim_{x \to 2} f(x)$

(c) $\lim_{x \to 3} f(x)$ (d) $\lim_{x \to 4} f(x)$

3. Use the graph in Fig. 12 to determine the following limits.

(a) $\lim_{x \to 1} f(2x)$ (b) $\lim_{x \to 2} f(x-1)$

(c) $\lim_{x \to 3} f(2x-5)$ (d) $\lim_{x \to 0} f(4+x)$

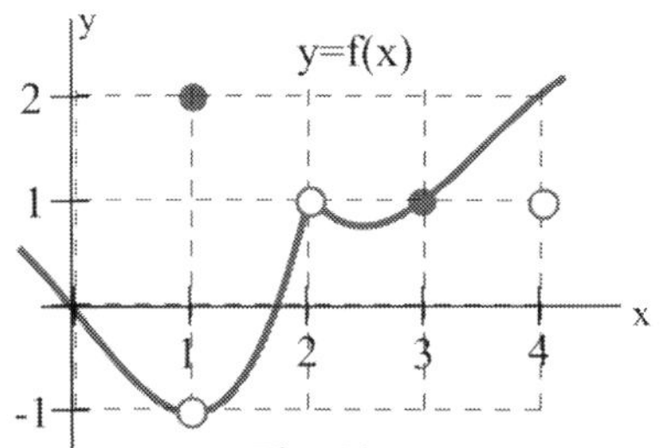

Fig. 12

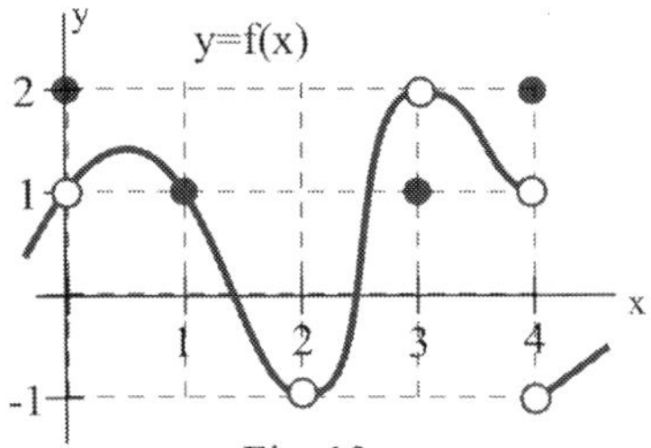

Fig. 13

4. Use the graph in Fig. 13 to determine the following limits.

(a) $\lim_{x \to 1} f(3x)$ (b) $\lim_{x \to 2} f(x+1)$

(c) $\lim_{x \to 3} f(2x-4)$ (d) $\lim_{x \to 0} | f(4+x) |$

5. Evaluate (a) $\lim_{x \to 1} \frac{x^2 + 3x + 3}{x - 2}$ (b) $\lim_{x \to 2} \frac{x^2 + 3x + 3}{x - 2}$

6. Evaluate (a) $\lim_{x \to 0} \frac{x + 7}{x^2 + 9x + 14}$ (b) $\lim_{x \to 3} \frac{x + 7}{x^2 + 9x + 14}$

(c) $\lim_{x \to -4} \frac{x + 7}{x^2 + 9x + 14}$ (d) $\lim_{x \to -7} \frac{x + 7}{x^2 + 9x + 14}$

7. Evaluate (a) $\lim_{x \to 1} \frac{\cos(x)}{x}$ (b) $\lim_{x \to \pi} \frac{\cos(x)}{x}$ (c) $\lim_{x \to -1} \frac{\cos(x)}{x}$

8. Evaluate (a) $\lim_{x \to 7} \sqrt{x - 3}$ (b) $\lim_{x \to 9} \sqrt{x} - 3$ (c) $\lim_{x \to 9} \frac{\sqrt{x} - 3}{x - 9}$

9. Evaluate (a) $\lim_{x \to 0^-} | x |$ (b) $\lim_{x \to 0^+} | x |$ (c) $\lim_{x \to 0} | x |$

10. Evaluate (a) $\lim_{x \to 0^-} \frac{| x |}{x}$ (b) $\lim_{x \to 0^+} \frac{| x |}{x}$ (c) $\lim_{x \to 0} \frac{| x |}{x}$

11. Evaluate (a) $\lim_{x \to 5} | x - 5 |$ (b) $\lim_{x \to 3} \frac{| x - 5 |}{x - 5}$ (c) $\lim_{x \to 5} \frac{| x - 5 |}{x - 5}$

12. $f(x) = \begin{cases} x & \text{if } x < 0 \\ \sin(x) & \text{if } 0 < x \le 2 \\ 1 & \text{if } 2 < x \end{cases}$. Find the one and two–sided limits of f as $x \to 0, 1,$ and 2.

13. $g(x) = \begin{cases} 1 & \text{if } x \le 2 \\ 8/x & \text{if } 2 < x < 4 \\ 6 - x & \text{if } 4 < x \end{cases}$. Find the one and two–sided limits of g as $x \to 1, 2, 4,$ and 5.

In problems 14 – 17 use a calculator or computer to get approximate answers accurate to 2 decimal places.

14. (a) $\lim_{x \to 0} \frac{2^x - 1}{x}$ (b) $\lim_{x \to 1} \frac{\log_{10}(x)}{x - 1}$ 15. (a) $\lim_{x \to 0} \frac{3^x - 1}{x}$ (b) $\lim_{x \to 1} \frac{\ln(x)}{x - 1}$

16. (a) $\lim_{x \to 5} \frac{\sqrt{x-1} - 2}{x - 5}$ (b) $\lim_{x \to 0} \frac{\sin(3x)}{5x}$ 17. (a) $\lim_{x \to 16} \frac{\sqrt{x} - 4}{x - 16}$ (b) $\lim_{x \to 0} \frac{\sin(7x)}{2x}$

18. Define A(x) to be the **area** bounded by the x and y axes, the bent line in Fig. 14, and the vertical line at x. For example, A(4) = 10.

a) Evaluate A(0), A(1), A(2), and A(3).

b) Graph y = A(x) for $0 \le x \le 4$.

c) What area does A(3) – A(1) represent?

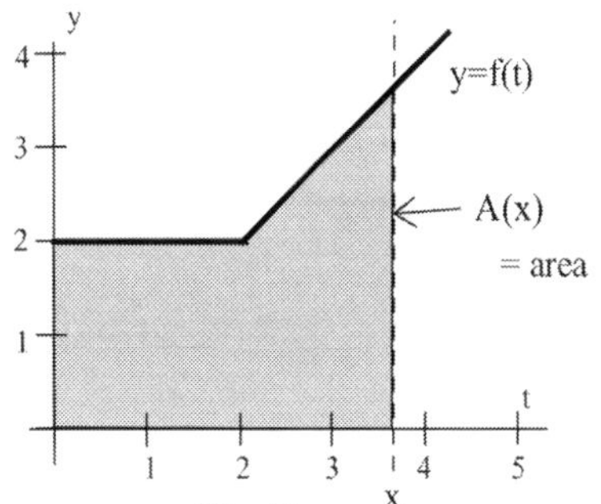

Fig. 14

19. Define A(x) to be the **area** bounded by the x and y axes, the line $y = \frac{1}{2}x + 2$, and the vertical line at x. (Fig. 15).

For example, A(4) = 12.

a) Evaluate A(0), A(1), A(2), and A(3).

b) Graph y = A(x) for $0 \le x \le 4$.

c) What area does A(3) – A(1) represent?

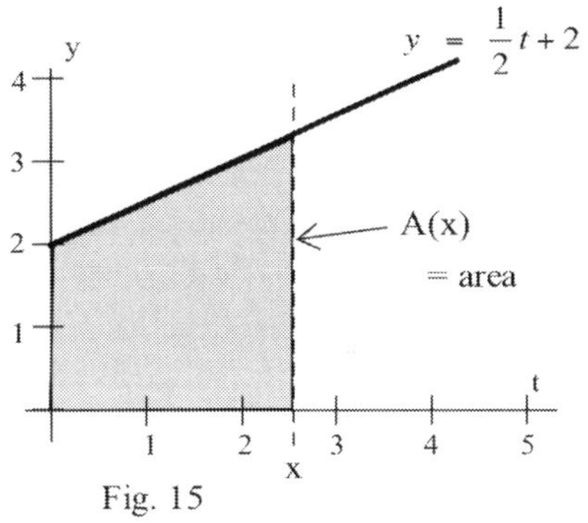

Fig. 15

20. Sketch $f(x) = \sqrt{4x - x^2}$ for $0 \le x \le 4$ (this is a semicircle) .

Define A(x) to be the **area** bounded by the x and y axes, the graph y = f(x) , and the vertical line at x.

a) Evaluate A(0), A(2), and A(4).

b) Graph y = A(x) for $0 \le x \le 4$.

c) What area does A(3) – A(1) represent?

Section 1.1 PRACTICE Answers

Practice 1: (a) **2** (b) **2** (c) **does not exist** (no limit) (d) **1**

Practice 2: (a) $\lim_{x \to 2} \frac{(x+1)(x-2)}{x-2} = \lim_{x \to 2} (x+1) = 3$

(b) $\lim_{t \to 0} \frac{t \cdot \sin(t)}{t(t+3)} = \lim_{t \to 0} \frac{\sin(t)}{t+3} = \frac{0}{3} = 0$

(c) $\lim_{w \to 2} \frac{w-2}{\ln(w/2)} = 2$. Try this one numerically or using a graph.

w	$\frac{w-2}{\ln(w/2)}$
2.2	2.098411737
2.01	2.004995844
2.003	2.001499625
2.0001	2.00005

w	$\frac{w-2}{\ln(w/2)}$
1.9	1.949572575
1.99	1.994995823
1.9992	1.999599973
1.9999	1.99995

Practice 3:

$\lim_{x \to 0^-} f(x) = \mathbf{1}$ $\lim_{x \to 0^+} f(x) = \mathbf{2}$ $\lim_{x \to 0} f(x)$ **does not exist**

$\lim_{x \to 1^-} f(x) = \mathbf{1}$ $\lim_{x \to 1^+} f(x) = \mathbf{1}$ $\lim_{x \to 1} f(x)$ **1**

$\lim_{x \to 2^-} f(x) = \mathbf{-1}$ $\lim_{x \to 2^+} f(x) = \mathbf{-1}$ $\lim_{x \to 2} f(x) = \mathbf{-1}$

$\lim_{x \to 3^-} f(x) = \mathbf{-1}$ $\lim_{x \to 3^+} f(x) = \mathbf{1}$ $\lim_{x \to 3} f(x)$ **does not exist**

Practice 4:

$\lim_{x \to 1^-} f(x) = \mathbf{1}$ $\lim_{x \to 1^+} f(x) = \mathbf{1}$ $\lim_{x \to 1} f(x) = \mathbf{1}$

$\lim_{x \to 3^-} f(x) = \mathbf{3}$ $\lim_{x \to 3^+} f(x) = \mathbf{2}$ $\lim_{x \to 3} f(x)$ **does not exist**

1.2 PROPERTIES OF LIMITS

This section presents results which make it easier to calculate limits of combinations of functions or to show that a limit does not exist. The main result says we can determine the limit of "elementary combinations" of functions by calculating the limit of each function separately and recombining these results for our final answer.

Main Limit Theorem:

If $\lim_{x\to a} f(x) = L$ and $\lim_{x\to a} g(x) = M$,

then

(a) $\lim_{x\to a} \{f(x) + g(x)\} = \lim_{x\to a} f(x) + \lim_{x\to a} g(x) = L + M$

(b) $\lim_{x\to a} \{f(x) - g(x)\} = \lim_{x\to a} f(x) - \lim_{x\to a} g(x) = L - M$

(c) $\lim_{x\to a} k\,f(x) = k \lim_{x\to a} f(x) = kL$

(d) $\lim_{x\to a} f(x)\cdot g(x) = \{\lim_{x\to a} f(x)\}\cdot\{\lim_{x\to a} g(x)\} = L\cdot M$

(e) $\lim_{x\to a} \dfrac{f(x)}{g(x)} = \dfrac{\lim_{x\to a} f(x)}{\lim_{x\to a} g(x)} = \dfrac{L}{M}$ (if $M \neq 0$).

(f) $\lim_{x\to a} \{ f(x) \}^n = \left\{ \lim_{x\to a} f(x) \right\}^n = L^n$

(g) $\lim_{x\to a} \sqrt[n]{f(x)} = \sqrt[n]{\lim_{x\to a} f(x)} = \sqrt[n]{L}$ (if $L > 0$ when n is even)

The Main Limit Theorem says we get the same result if we first perform the algebra and then take the limit or if we take the limits first and then perform the algebra: e.g., (a) the limit of the sum equals the sum of the limits. A proof of the Main Limit Theorem is not inherently difficult, but it requires a more precise definition of the limit concept than we have given, and it then involves a number of technical difficulties.

Practice 1: For $f(x) = x^2 - x - 6$ and $g(x) = x^2 - 2x - 3$, evaluate the following limits:

(a) $\lim_{x\to 1} \{f(x) + g(x)\}$ (b) $\lim_{x\to 1} f(x)g(x)$ (c) $\lim_{x\to 1} f(x)/g(x)$ (d) $\lim_{x\to 3} \{f(x) + g(x)\}$

(e) $\lim_{x\to 3} f(x)g(x)$ (f) $\lim_{x\to 3} f(x)/g(x)$ (g) $\lim_{x\to 2} \{ f(x) \}^3$ (h) $\lim_{x\to 2} \sqrt{1 - g(x)}$

Limits of Some Very Nice Functions: Substitution

As you may have noticed in the previous example, for some functions f(x) it is possible to calculate the limit as x approaches a simply by substituting x = a into the function and then evaluating f(a), but sometimes this method does not work. The Substitution Theorem uses the following Two Easy Limits and the Main Limit Theorem to partially answer when such a substitution is valid.

Two Easy Limits: $\lim_{x \to a} k = k$ and $\lim_{x \to a} x = a$.

Substitution Theorem For Polynomial and Rational Functions:

If P(x) and Q(x) are **polynomials** and a is any number,

then $\lim_{x \to a} P(x) = P(a)$ and $\lim_{x \to a} \frac{P(x)}{Q(x)} = \frac{P(a)}{Q(a)}$ if $Q(a) \neq 0$.

The Substitution Theorem says that we can calculate the limits of polynomials and rational functions by substituting as long as the substitution does not result in a division by zero.

Practice 2: Evaluate (a) $\lim_{x \to 2} 5x^3 - x^2 + 3$ (b) $\lim_{x \to 2} \frac{x^3 - 7x}{x^2 + 3x}$ (c) $\lim_{x \to 2} \frac{x^2 - 2x}{x^2 - x - 2}$

Limits of Other Combinations of Functions

So far we have concentrated on limits of single functions and elementary combinations of functions. If we are working with limits of other combinations or compositions of functions, the situation is slightly more difficult, but sometimes these more complicated limits have useful geometric interpretations.

Example 1: Use the function defined by the graph in Fig. 1 to evaluate

(a) $\lim_{x \to 1} \{ 3 + f(x) \}$ (b) $\lim_{x \to 1} f(2+x)$ (c) $\lim_{x \to 0} f(3-x)$ (d) $\lim_{x \to 2} f(x+1) - f(x)$

Solution: (a) $\lim_{x \to 1} \{ 3 + f(x) \}$ is a straightforward application of part (a) of the Main Limit Theorem:

$$\lim_{x \to 1} \{ 3 + f(x) \} = \lim_{x \to 1} 3 + \lim_{x \to 1} f(x) = 3 + 2 = 5.$$

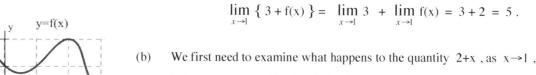

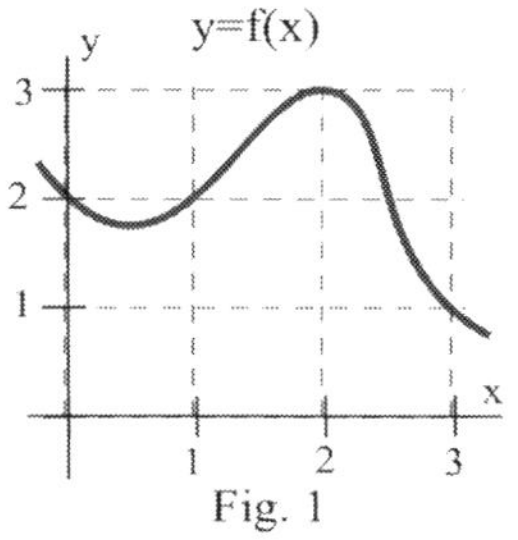

Fig. 1

(b) We first need to examine what happens to the quantity 2+x, as $x \to 1$, before we can consider the limit of f(2+x). When x is very close to 1, the value of 2+x is very close to 3, so the limit of f(2+x) as $x \to 1$ is equivalent to the limit of f(w) as $w \to 3$ (w=2+x), and it is clear from the graph that $\lim_{w \to 3} f(w) = 1$: $\lim_{x \to 1} f(2+x) = \lim_{w \to 3} f(w) = 1$ (w represents 2+x).

In most cases it is not necessary to formally substitute a new variable w for the quantity 2+x, but it is still necessary to think about what happens to the quantity 2+x as $x \to 1$.

(c) As $x \to 0$, the quantity 3–x will approach 3 so we want to know what happens to the values of f when the variable is approaching 3: $\lim_{x \to 0} f(3-x) = 1$.

(d) $\lim_{x \to 2} \{ f(x+1) - f(x) \} = \lim_{x \to 2} f(x+1) - \lim_{x \to 2} f(x)$ replace x+1 with w

$= \lim_{w \to 3} f(w) - \lim_{x \to 2} f(x) = 1 - 3 = -2$.

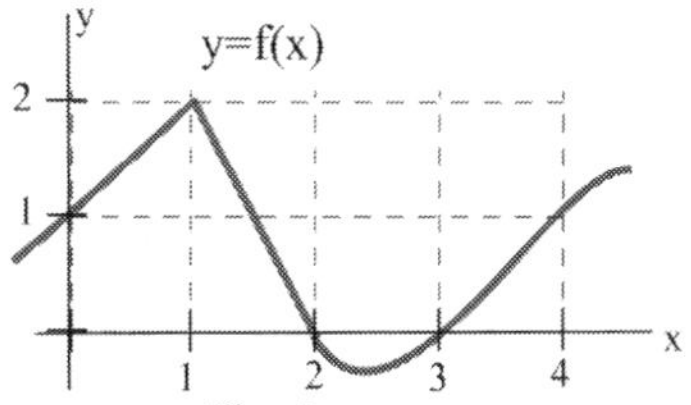

Fig. 2

Practice 3: Use the function defined by the graph in Fig. 2 to evaluate

(a) $\lim_{x \to 1} f(2x)$ (b) $\lim_{x \to 2} f(x-1)$

(c) $\lim_{x \to 0} 3 \cdot f(4+x)$ (d) $\lim_{x \to 2} f(3x-2)$.

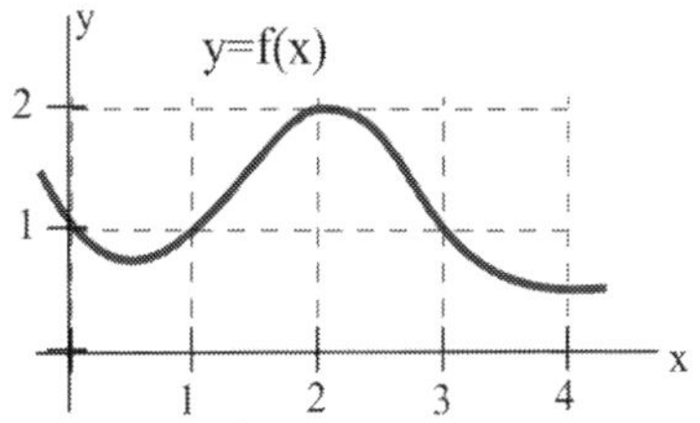

Fig. 3

Example 2: Use the function defined by the graph in Fig. 3 to evaluate

(a) $\lim_{h \to 0} f(3+h)$ (b) $\lim_{h \to 0} f(3)$

(c) $\lim_{h \to 0} \{ f(3+h) - f(3) \}$ (d) $\lim_{h \to 0} \dfrac{f(3+h)-f(3)}{h}$

Solution: Part (d) is a common form of limit, and parts (a) – (c) are the steps we need to evaluate (d).

(a) As $h \to 0$, the quantity w = 3+h will approach 3 so $\lim_{h \to 0} f(3+h) = \lim_{x \to 3} f(w) = 1$.

(b) f(3) is the constant 1 and f(3) does not depend on h in any way so $\lim_{h \to 0} f(3) = 1$.

(c) The limit in part (c) is just an algebraic combination of the limits in (a) and (b):

$$\lim_{h \to 0} \{ f(3+h) - f(3) \} = \lim_{h \to 0} f(3+h) - \lim_{h \to 0} f(3) = 1 - 1 = 0 .$$

The quantity f(3+h) – f(3) also has a geometric interpretation — it is the change in the y–coordinates, the Δy, between the points (3,f(3)) and (3+h,f(3+h)). (Fig. 4)

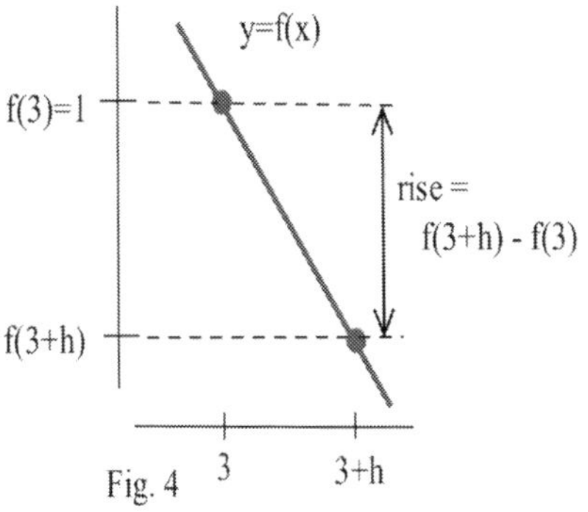

Fig. 4

(d) As $h \to 0$, the numerator and denominator of $\dfrac{f(3+h) - f(3)}{h}$ both approach 0 so we cannot immediately determine the value of the limit. But if we recognize that $f(3+h) - f(3) = \Delta y$ for the two points (3, f(3)) and (3+h, f(3+h)) and that $h = \Delta x$ for the same two points, then we can interpret $\dfrac{f(3+h) - f(3)}{h}$ as $\dfrac{\Delta y}{\Delta x}$ which is the slope of the secant line through the two points. So

$$\lim_{h \to 0} \frac{f(3+h) - f(3)}{h} = \lim_{\Delta x \to 0} \frac{\Delta y}{\Delta x} = \lim_{\Delta x \to 0} \{ \text{ slope of the secant line } \}$$

$$= \text{slope of the tangent line at } (3, f(3)) \approx -2 .$$

This limit, representing the slope of line tangent to the graph of f at the point (3, f(3)) , is a pattern we will see often in the future.

Tangent Lines as Limits

If we have two points on the graph of a function, (x, f(x)) and (x+h, f(x+h)), then $\Delta y = f(x+h) - f(x)$ and $\Delta x = (x+h) - (x) = h$ so the slope of the secant line through those points is $m_{secant} = \frac{\Delta y}{\Delta x}$ and the slope of the line tangent to the graph of f at the point (x, f(x)) is, by definition,

$$m_{tangent} = \lim_{\Delta x \to 0} \{ \text{ slope of the secant line } \} = \lim_{\Delta x \to 0} \frac{\Delta y}{\Delta x} = \lim_{h \to 0} \frac{f(x+h) - f(x)}{h} .$$

Example 3: Give a geometric interpretation for the following limits and **estimate** their values for the function in Fig. 5: (a) $\lim_{h \to 0} \frac{f(1+h) - f(1)}{h}$ (b) $\lim_{h \to 0} \frac{f(2+h) - f(2)}{h}$

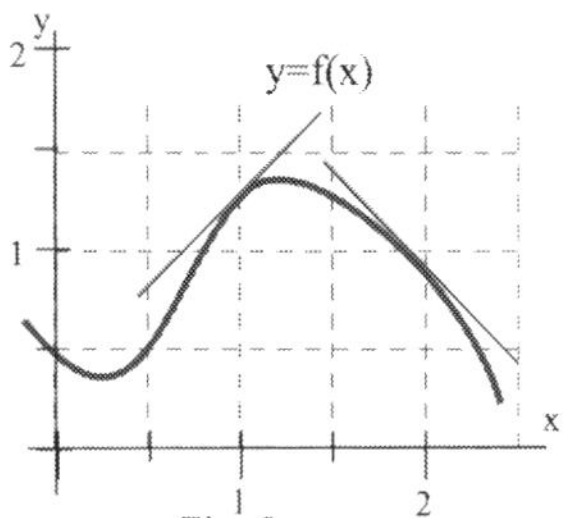

Fig. 5

Solution: Part (a) represents the slope of the line tangent to the graph of f(x) at the point (1, f(1)) so

$\lim_{h \to 0} \frac{f(1+h) - f(1)}{h} \approx 1$. Part (b) represents the slope of the line tangent to the graph of f(x) at the point (2, f(2)) so $\lim_{h \to 0} \frac{f(2+h) - f(2)}{h} \approx -1$.

Practice 4: Give a geometric interpretation for the following limits and estimate their values for the function in Fig. 6:

$$\lim_{h \to 0} \frac{g(1+h) - g(1)}{h} \quad \lim_{h \to 0} \frac{g(3+h) - g(3)}{h} \quad \lim_{h \to 0} \frac{g(h) - g(0)}{h} .$$

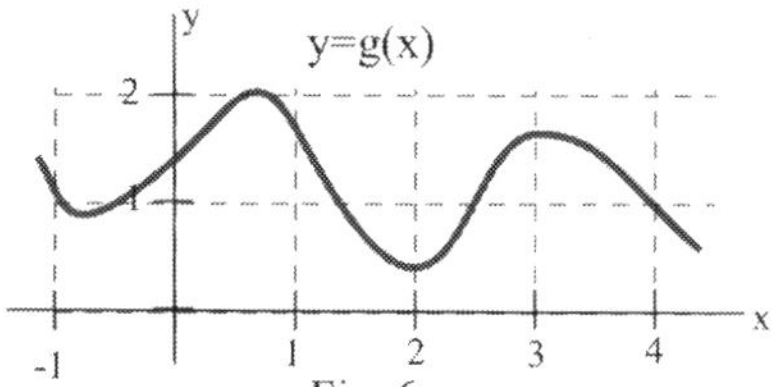

Fig. 6

Comparing the Limits of Functions

Sometimes it is difficult to work directly with a function. However, if we can compare our difficult function with easier ones, then we can use information about the easier functions to draw conclusions about the difficult one. If the complicated function is always between two functions whose limits are equal, then we know the limit of the complicated function.

> **Squeezing Theorem** (Fig. 7):
>
> If $g(x) \le f(x) \le h(x)$ for all x near c (for all x close to but not equal to c)
>
> and $\lim_{x \to c} g(x) = \lim_{x \to c} h(x) = L$
>
> then for x near c, f(x) will be squeezed between g(x) and h(x), and $\lim_{x \to c} f(x) = L$.

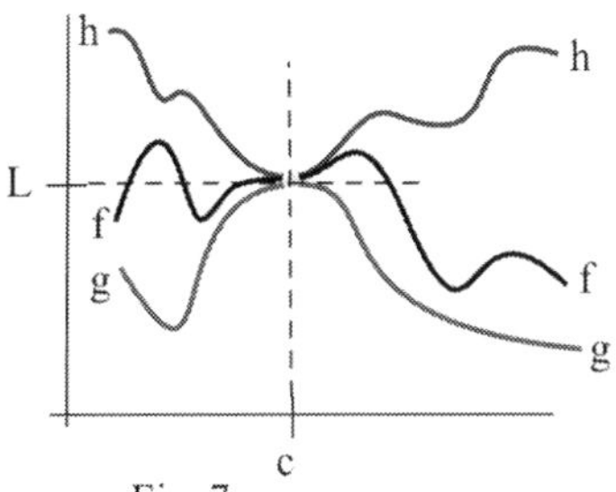

Fig. 7

Example 4: Use the inequality $-|x| \le \sin(x) \le |x|$ to determine $\lim_{x \to 0} \sin(x)$ and $\lim_{x \to 0} \cos(x)$.

Solution: $\lim_{x \to 0} |x| = 0$ and $\lim_{x \to 0} -|x| = 0$ so, by the Squeezing Theorem, $\lim_{x \to 0} \sin(x) = 0$. If $-\pi/2 < x < \pi/2$ then $\cos(x) = +\sqrt{1 - \sin^2(x)}$ so

$$\lim_{x \to 0} \cos(x) = \lim_{x \to 0} +\sqrt{1 - \sin^2(x)} = +\sqrt{1 - 0} = 1 .$$

Example 5: Evaluate $\lim_{x \to 0} x \cdot \sin(\frac{1}{x})$.

Solution: The graph of $y = \sin(\frac{1}{x})$ for values of x near 0 is shown in Fig. 8. The y–values of this graph change very rapidly for values of x near 0, but they all lie between –1 and +1: $-1 \le \sin(\frac{1}{x}) \le +1$. The fact that $\sin(\frac{1}{x})$ is bounded between –1 and +1 implies that $x \sin(\frac{1}{x})$ is stuck between –x and +x , so the function we are interested in , $x \cdot \sin(\frac{1}{x})$, is squeezed between two "easy" functions, –x and x (Fig. 9). Both "easy" functions approach 0 as $x \to 0$, so $x \cdot \sin(\frac{1}{x})$ must also approach 0 as $x \to 0$: $\lim_{x \to 0} x \cdot \sin(\frac{1}{x}) = 0$.

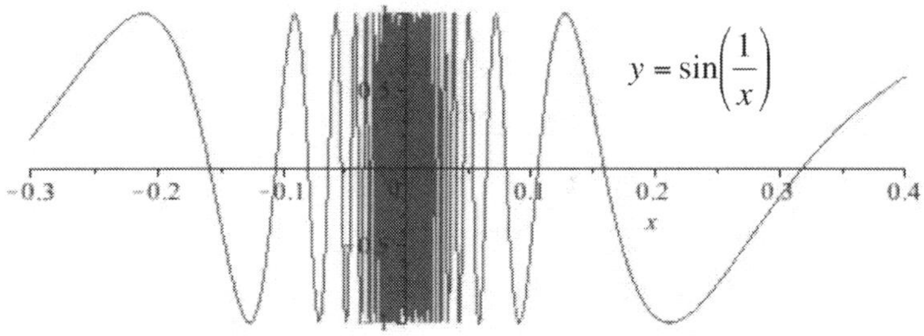

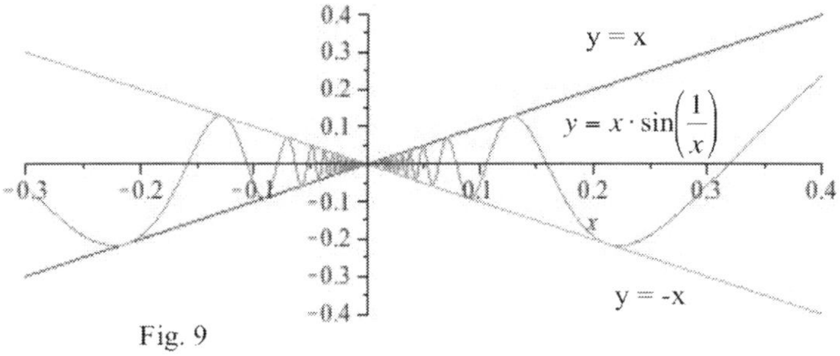

Fig. 9

Practice 5: If f(x) is always between $x^2 + 2$ and $2x + 1$, then $\lim_{x \to 1} f(x) = ?$

Practice 6: Use the relation $\cos(x) \le \frac{\sin(x)}{x} \le 1$ to show that $\lim_{x \to 0} \frac{\sin(x)}{x} = 1$. (The steps for deriving the inequalities are shown in problem 19.)

List Method for Showing that a Limit Does Not Exist

If the limit, as x approaches c, exists and equals L, then we can guarantee that the values of f(x) are as close to L as we want by restricting the values of x to be very, very close to c. To show that a limit, as x approaches c, does not exist, we need to show that no matter how closely we restrict the values of x to c, the values of f(x) are not all close to a single, finite value L. One way to demonstrate that $\lim_{x \to c} f(x)$ does not exist is to show that the left and right limits exist but are not equal.

Another method of showing that $\lim_{x \to c} f(x)$ does not exist is to find two infinite lists of numbers, $\{a_1, a_2, a_3, a_4, \ldots\}$ and $\{b_1, b_2, b_3, b_4, \ldots\}$, which approach arbitrarily close to the value c as the subscripts get larger, but so that the lists of function values, $\{f(a_1), f(a_2), f(a_3), f(a_4), \ldots\}$ and $\{f(b_1), f(b_2), f(b_3), f(b_4), \ldots\}$, approach two different numbers as the subscripts get larger.

Example 6: For $f(x) = \begin{cases} 1 & \text{if } x < 1 \\ x & \text{if } 1 < x < 3 \\ 2 & \text{if } 3 < x \end{cases}$, show that $\lim_{x \to 3} f(x)$ does not exist.

Solution: We can use one–sided limits to show that this limit does not exist, or we can use the list method by selecting values for one list to approach 3 from the right and values for the other list to approach 3 from the left.

One way to define values of $\{a_1, a_2, a_3, a_4, \ldots\}$ which approach 3 from the right is to define $a_1 = 3 + 1$, $a_2 = 3 + \frac{1}{2}$, $a_3 = 3 + \frac{1}{3}$, $a_4 = 3 + \frac{1}{4}$ and, in general, $a_n = 3 + \frac{1}{n}$. Then $a_n > 3$ so $f(a_n) = 2$ for all subscripts n, and the values in the list $\{f(a_1), f(a_2), f(a_3), f(a_4), \ldots\}$ are approaching 2. In fact, all of the $f(a_n) = 2$.

We can define values of $\{b_1, b_2, b_3, b_4, \ldots\}$ which approach 3 from the left by $b_1 = 3 - 1$, $b_2 = 3 - \frac{1}{2}$, $b_3 = 3 - \frac{1}{3}$, $b_4 = 3 - \frac{1}{4}$ and, in general, $b_n = 3 - \frac{1}{n}$. Then $b_n < 3$ so $f(b_n) = b_n = 3 - \frac{1}{n}$ for each subscript n, and the values in the list $\{f(b_1), f(b_2), f(b_3), f(b_4), \ldots\} = \{2, 2.5, 2.67, 2.75, 2.8, \ldots, 3 - \frac{1}{n}, \ldots\}$ approach 3.

Since the values in the lists $\{f(a_1), f(a_2), f(a_3), f(a_4), \ldots\}$ and $\{f(b_1), f(b_2), f(b_3), f(b_4), \ldots\}$ approach two different numbers, we can conclude that $\lim_{x \to 3} f(x)$ does not exist.

Example 7: Let $h(x) = \begin{cases} 2 & \text{if } x \text{ is a **rational** number} \\ 1 & \text{if } x \text{ is an **irrational** number} \end{cases}$ be the "holey" function introduced in Section 0.4 . Use the list method to show that $\lim_{x\to 3} h(x)$ does not exist.

Solution: Let $\{ a_1, a_2, a_3, a_4, \ldots \}$ be a list of rational numbers which approach 3, for example, $a_1 = 3 + 1$, $a_2 = 3 + 1/2, \ldots, a_n = 3 + 1/n$. Then $f(a_n)$ always equals 2 so $\{ f(a_1), f(a_2), f(a_3), f(a_4), \ldots \} = \{ 2, 2, 2, \ldots \}$ and the $f(a_n)$ values "approach" 2. If $\{b_1, b_2, b_3, b_4, \ldots \}$ is a list of irrational numbers which approach 3, for example, $b_1 = 3 + \pi$, $b_2 = 3 + \pi/2, \ldots, b_n = 3 + \pi/n$. then $\{ f(b_1), f(b_2), f(b_3), f(b_4), \ldots \} = \{ 1, 1, 1, \ldots \}$ and the $f(b_n)$ "approach" 1. Since the $f(a_n)$ and $f(b_n)$ values approach different numbers, the limit as $x \to 3$ does not exist.

A similar argument will work as x approaches any number c, so for every c we have that $\lim_{x\to c} h(x)$ does not exist. The "holey" function does not have a limit as x approaches any value c.

PROBLEMS

1. Use the functions f and g defined by the graphs in Fig. 10 to determine the following limits.

   (a) $\lim_{x\to 1} \{ f(x) + g(x) \}$ (b) $\lim_{x\to 1} f(x)\cdot g(x)$

   (c) $\lim_{x\to 1} f(x)/g(x)$ (d) $\lim_{x\to 1} f(g(x))$

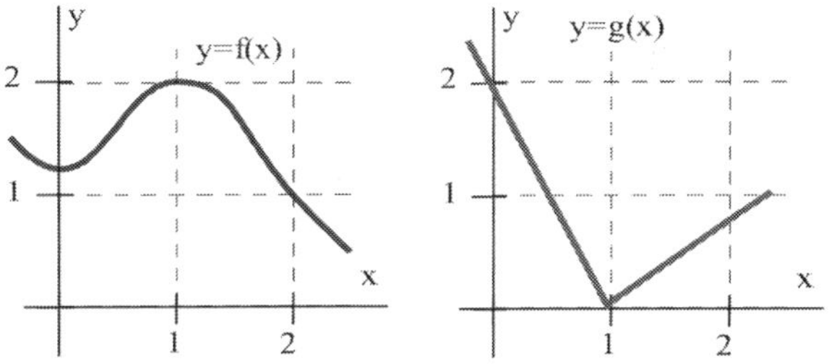

Fig. 10

2. Use the functions f and g defined by the graphs in Fig. 10 to determine the following limits.

   (a) $\lim_{x\to 2} \{ f(x) + g(x) \}$ (b) $\lim_{x\to 2} f(x)\cdot g(x)$

   (c) $\lim_{x\to 2} f(x)/g(x)$ (d) $\lim_{x\to 2} f(g(x))$

3. Use the function h defined by the graph in Fig. 11 to determine the following limits.

   (a) $\lim_{x\to 2} h(2x - 2)$ (b) $\lim_{x\to 2} \{ x + h(x) \}$

   (c) $\lim_{x\to 2} h(1 + x)$ (d) $\lim_{x\to 3} h(x/2)$

4. Use the function h defined by the graph in Fig. 11 to determine the following limits.

   (a) $\lim_{x\to 2} h(5 - x)$ (b) $\lim_{x\to 2} x\cdot h(x - 1)$

   (c) $\lim_{x\to 0} \{ h(3 + x) - h(3) \}$ (d) $\lim_{x\to 0} \dfrac{h(3+x) - h(3)}{x}$

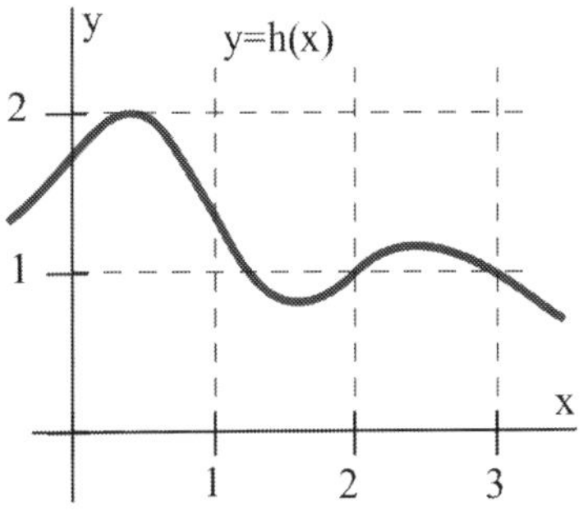

Fig. 11

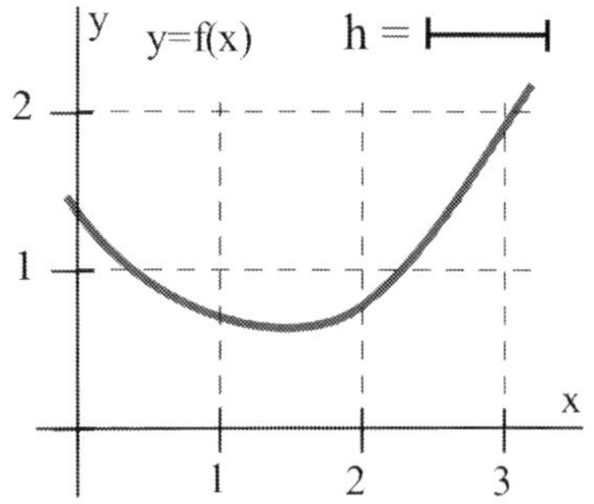

Fig. 12

5. Label the parts of the graph of f (Fig. 12) which are described by

(a) 2 + h (b) f(2) (c) f(2 + h)

(d) f(2 + h) – f(2) (e) $\frac{f(2+h)-f(2)}{(2+h)-(2)}$ (f) $\frac{f(2-h)-f(2)}{(2-h)-(2)}$

6. Label the parts of the graph of f (Fig. 13) which are described by

(a) a + h (b) g(a) (c) g(a + h)

(d) g(a + h) – g(a) (e) $\frac{g(a+h)-g(a)}{(a+h)-(a)}$ (f) $\frac{g(a-h)-g(a)}{(a-h)-(a)}$

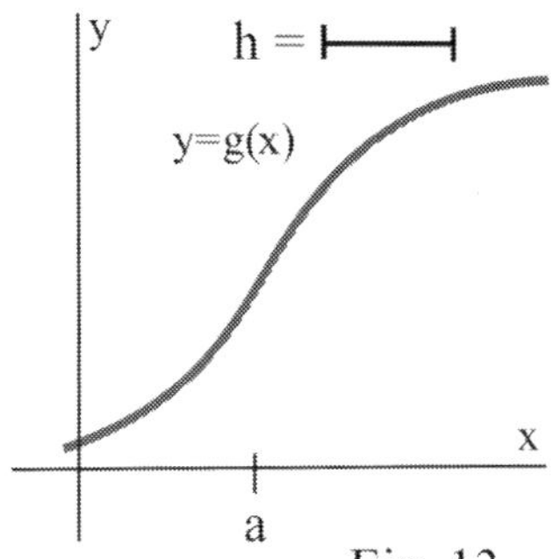

Fig. 13

7. Use the function f defined by the graph in Fig. 14 to determine the following limits.

(a) $\lim_{x \to 1^+} f(x)$ (b) $\lim_{x \to 1^-} f(x)$ (c) $\lim_{x \to 1} f(x)$

(d) $\lim_{x \to 3^+} f(x)$ (e) $\lim_{x \to 3^-} f(x)$ (f) $\lim_{x \to 3} f(x)$

(g) $\lim_{x \to -1^+} f(x)$ (h) $\lim_{x \to -1^-} f(x)$ (i) $\lim_{x \to -1} f(x)$

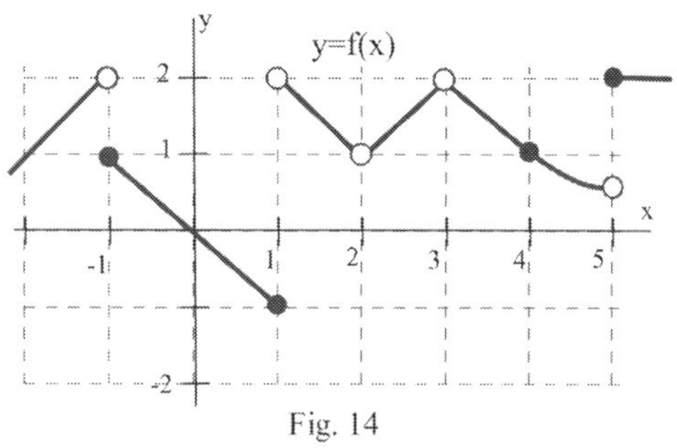

Fig. 14

8. Use the function f defined by the graph in Fig. 14 to determine the following limits.

(a) $\lim_{x \to 2^+} f(x)$ (b) $\lim_{x \to 2^-} f(x)$ (c) $\lim_{x \to 2} f(x)$

(d) $\lim_{x \to 4^+} f(x)$ (e) $\lim_{x \to 4^-} f(x)$ (f) $\lim_{x \to 4} f(x)$

(g) $\lim_{x \to -2^+} f(x)$ (h) $\lim_{x \to -2^-} f(x)$ (i) $\lim_{x \to -2} f(x)$

9. The Lorentz Contraction Formula in relativity theory says the length L of an object moving at v miles per second with respect to an observer is $L = A \cdot \sqrt{1 - \frac{v^2}{c^2}}$ where c is the speed of light (a constant).

a) Determine the "rest length" of the object (v = 0). b) Determine $\lim_{v \to c^-} L$.

10. (a) $\lim_{x \to 2^+} INT(x)$ (b) $\lim_{x \to 2^-} INT(x)$ (c) $\lim_{x \to -2^+} INT(x)$ (d) $\lim_{x \to -2^-} INT(x)$

(e) $\lim_{x \to -2.3} INT(x)$ (f) $\lim_{x \to 3} INT(x/2)$ (g) $\lim_{x \to 3} INT(x)/2$ (h) $\lim_{x \to 0^+} \frac{INT(2+x)-INT(2)}{x}$

11. $f(x) = \begin{cases} 1 & \text{if } x < 1 \\ x & \text{if } 1 < x \end{cases}$ and $g(x) = \begin{cases} x & \text{if } x \neq 2 \\ 3 & \text{if } x = 2 \end{cases}$.

(a) $\lim_{x \to 2} \{ f(x) + g(x) \}$ (b) $\lim_{x \to 2} f(x)/g(x)$ (c) $\lim_{x \to 2} f(\, g(x)\,)$

(d) $\lim_{x \to 0} g(x)/f(x)$ (e) $\lim_{x \to 1} f(x)/g(x)$ (f) $\lim_{x \to 1} g(\, f(x)\,)$

Problems 12 – 15 require a calculator.

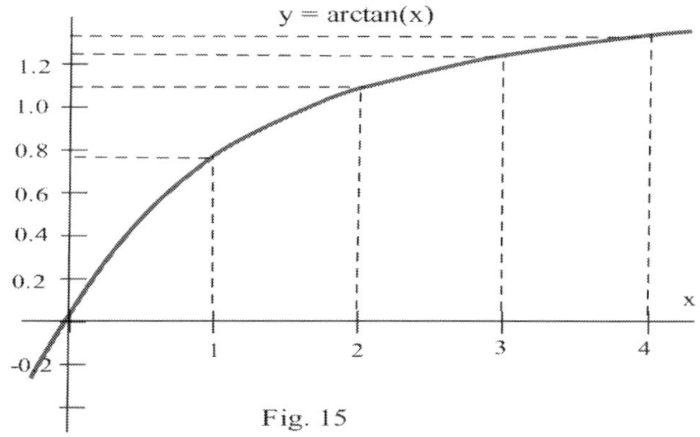

Fig. 15

12. Give geometric interpretations for the following limits and use a calculator to estimate their values.

(a) $\lim_{h \to 0} \dfrac{\arctan(0 + h) - \arctan(0)}{h}$

(b) $\lim_{h \to 0} \dfrac{\arctan(1 + h) - \arctan(1)}{h}$

(c) $\lim_{h \to 0} \dfrac{\arctan(2 + h) - \arctan(2)}{h}$

13. (a) What does $\lim_{h \to 0} \dfrac{\cos(h) - 1}{h}$ represent on the graph of $y = \cos(x)$?

(It may help to recognize that $\dfrac{\cos(h) - 1}{h} = \dfrac{\cos(0 + h) - \cos(0)}{h}$)

(b) Graphically and using your calculator, determine $\lim_{h \to 0} \dfrac{\cos(h) - 1}{h}$.

14. (a) What does the ratio $\dfrac{\ln(1 + h)}{h}$ represent on the graph of $y = \ln(x)$?

(It may help to recognize that $\dfrac{\ln(1 + h)}{h} = \dfrac{\ln(1 + h) - \ln(1)}{h}$.)

(b) Graphically and using your calculator, determine $\lim_{h \to 0} \dfrac{\ln(1 + h)}{h}$.

15. Use your calculator (to generate a table of values) to help you estimate

(a) $\lim_{h \to 0} \dfrac{e^h - 1}{h}$ (b) $\lim_{c \to 0} \dfrac{\tan(1 + c) - \tan(1)}{c}$ (c) $\lim_{t \to 0} \dfrac{g(2 + t) - g(2)}{t}$ when $g(t) = t^2 - 5$.

16. (a) For $h > 0$, find the slope of the line through the points $(h, | h |)$ and $(0,0)$.

(b) For $h < 0$, find the slope of the line through the points $(h, | h |)$ and $(0,0)$.

(c) Evaluate $\lim_{h \to 0^-} \dfrac{|h|}{h}$, $\lim_{h \to 0^+} \dfrac{|h|}{h}$, and $\lim_{h \to 0} \dfrac{|h|}{h}$

17. Describe the behavior of the function y = f(x) in Fig. 16 at each integer using one of the phrases:

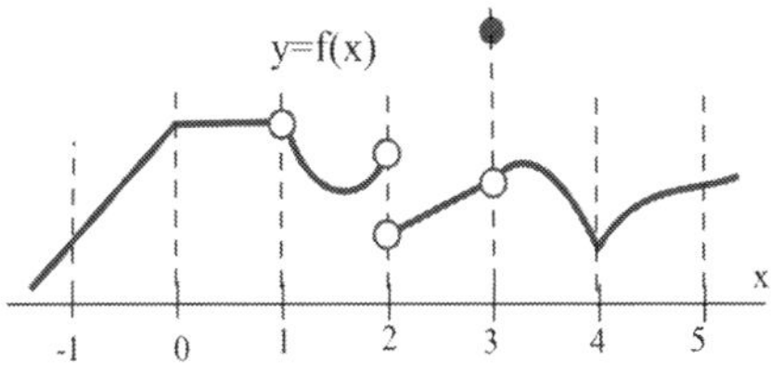

Fig. 16

(a) "connected and smooth", (b) "connected with a corner",

(c) "not connected because of a simple hole which could be plugged by adding or moving one point", or

(d) "not connected because of a vertical jump which could not be plugged by moving one point."

18. Describe the behavior of the function y = f(x) in Fig. 17 at each integer using one of the phrases:
(a) "connected and smooth", (b) "connected with a corner",
(c) "not connected because of a simple hole which could be plugged by adding or moving one point", or
(d) "not connected because of a vertical jump which could not be plugged by moving one point."

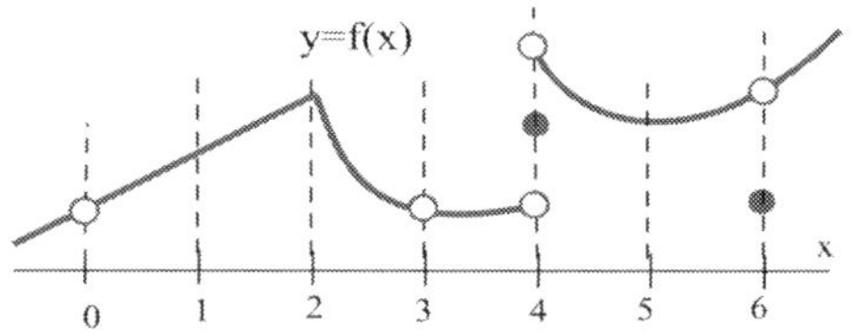

Fig. 17

19. This problem outlines the steps of a proof that $\lim_{\theta \to 0^+} \frac{\sin(\theta)}{\theta} = 1$. Statements (a) – (h) below refer to Fig. 18. Assume that $0 < \theta < \frac{\pi}{2}$ and justify why each statement is true.

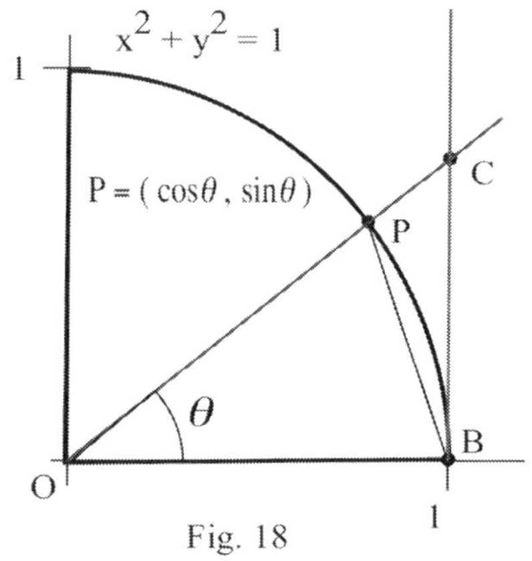

Fig. 18

(a) Area of $\Delta OPB = \frac{1}{2}$ (base)(height) $= \frac{1}{2} \sin(\theta)$.

(b) $\dfrac{\text{area of the sector (the pie shaped region) OPB}}{\text{area of the whole circle}}$

$= \dfrac{\text{angle defining sector OPB}}{\text{angle of the whole circle}} = \dfrac{\theta}{2\pi}$

so (area of the sector OPB) $= \frac{\theta\pi}{2\pi} = \frac{\theta}{2}$.

(c) The line L through the points (0,0) and P = (cos(θ),sin(θ)) has slope $m = \frac{\sin(\theta)}{\cos(\theta)}$, so

$C = (1, \frac{\sin(\theta)}{\cos(\theta)})$ and the area of $\Delta OCB = \frac{1}{2}$ (base)(height) $= \frac{1}{2} (1) \frac{\sin(\theta)}{\cos(\theta)}$.

(d) Area of ΔOPB < area of sector OPB < area of ΔOCB .

(e) $\frac{1}{2} \sin(\theta) < \frac{\theta}{2} < \frac{1}{2} (1) \frac{\sin(\theta)}{\cos(\theta)}$ and $\sin(\theta) < \theta < \frac{\sin(\theta)}{\cos(\theta)}$.

(f) $1 < \frac{\theta}{\sin(\theta)} < \frac{1}{\cos(\theta)}$ and $1 > \frac{\sin(\theta)}{\theta} > \cos(\theta)$.

(g) $\lim_{\theta \to 0^+} 1 = 1$ and $\lim_{\theta \to 0^+} \cos(\theta) = 1$.

(h) $\lim_{\theta \to 0^+} \frac{\sin(\theta)}{\theta} = 1$.

20. Use the list method to show that $\lim_{x \to 2} \frac{|x-2|}{x-2}$ does not exist .

21. Show that $\lim_{x \to 0} \sin(\frac{1}{x})$ does not exist.

(Suggestion: Let $f(x) = \sin(\frac{1}{x})$ and pick $a_n = \frac{1}{n\pi}$ so $f(a_n) = \sin(\frac{1}{a_n}) = \sin(n\pi) = 0$ for every n.

Then pick $b_n = \frac{1}{2n\pi + \pi/2}$ so $f(b_n) = \sin(\frac{1}{b_n}) = \sin(2n\pi + \frac{\pi}{2}) = \sin(\frac{\pi}{2}) = 1$ for every n.)

Section 1.2 **PRACTICE Answers**

Practice 1: (a) **–10** (b) **24** (c) **3/2** (d) **0**
(e) **0** (f) **5/4** (g) **–64** (h) **2**

Practice 2: (a) **39** (b) **–3/5** (c) **2/3**

Practice 3: (a) **0** (b) **2** (c) **3** (d) **1**

Practice 4: (a) slope of the line tangent to the graph of g at the point (1, g(1)): estimated slope ≈ **–2**

(b) slope of the line tangent to the graph of g at the point (3, g(3)): estimated slope ≈ **0**

(c) slope of the line tangent to the graph of g at the point (0, g(0)): estimated slope ≈ **1**

Practice 5: $\lim_{x \to 1} x^2 + 2 = 3$ and $\lim_{x \to 1} 2x + 1 = 3$ so $\lim_{x \to 1} f(x) = \mathbf{3}$.

Practice 6: $\lim_{x \to 0} \cos(x) = 1$ and $\lim_{x \to 0} 1 = 1$ so $\lim_{x \to 0} \frac{\sin(x)}{x} = \mathbf{1}$.

1.3 CONTINUOUS FUNCTIONS

In section 1.2 we saw a few "nice" functions whose limits as $x \to a$ simply involved substituting a into the function: $\lim_{x\to a} f(x) = f(a)$. Functions whose limits have this substitution property are called **continuous functions**, and they have a number of other useful properties and are very common in applications. We will examine what it means graphically for a function to be continuous or not continuous. Some properties of continuous functions will be given, and we will look at a few applications of these properties including a way to solve horrible equations such as $\sin(x) = \frac{2x+1}{x-2}$.

DEFINITION AND MEANING OF CONTINUOUS

Definition of Continuity at a Point

A function **f is continuous at x = a** if and only if $\lim_{x\to a} \mathbf{f(x) = f(a)}$.

The graph in Fig. 1 illustrates some of the different ways a function can behave at and near a point, and Table 1 contains some numerical information about the function and its behavior. Based on the information in the table, we can conclude that f is continuous at 1 since $\lim_{x\to 1} f(x) = 2 = f(1)$.

We can also conclude from the information in the table that f is not continuous at 2 or 3 or 4, because

$\lim_{x\to 2} f(x) \neq f(2)$, $\lim_{x\to 3} f(x) \neq f(3)$, and $\lim_{x\to 4} f(x) \neq f(4)$.

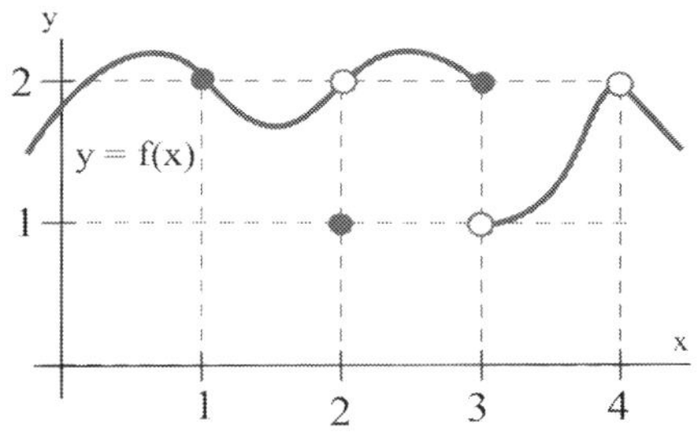

Fig. 1

a	f(a)	$\lim_{x\to a} f(x)$
1	2	2
2	1	2
3	2	does not exist
4	undefined	2

Table 1

Graphic Meaning of Continuity

When x is close to 1, the values of f(x) are close to the value f(1), and the graph of f in Fig. 1 does not have a hole or break at x=1 . The graph of f is connected at x=1 and can be drawn without lifting your pencil. At x=2 and x=4 the graph of f has holes, and at x=3 the graph has a break. The function f is also continuous at 1.7 (why?) , and at every point shown **except** at 2, 3, and 4.

Informally: A function is **continuous** at a point if the graph of the function is **connected** there.

A function is **not continuous** at a point if its graph **has a hole or break** at that point.

Sometimes the definition of continuous (the substitution condition for limits) is easier to use if we break it into several smaller pieces and then check whether or not our function satisfies each piece.

$\{$ f is continuous at a $\}$ if and only if $\{ \lim_{x \to a} f(x) = f(a) \}$ if and only if

(i) f is defined at a,

(ii) the limit of f(x) , as $x \to a$, exists (so the left limit and right limits exist and are equal)

and (iii) the value of f at a equals the value of the limit as $x \to a$: $\lim_{x \to a} f(x) = f(a)$.

If f satisfies conditions (i), (ii) and (iii) , then f is continuous at a. If f does not satisfy one or more of the three conditions at a , then f is not continuous at a.

For the function in Fig. 1, at a = 1, all 3 conditions are satisfied, and f is continuous at 1. At a = 2, conditions (i) and (ii) are satisfied but not (iii), so f is not continuous at 2. At a = 3, condition (i) is satisfied but (ii) is violated, so f is not continuous at 3. At a = 4, condition (i) is violated, so f is not continuous at 4.

A function is **continuous on an interval** if it is continuous at every point in the interval. A function f is **continuous from the left** at a if $\lim_{x \to a^-} f(x) = f(a)$, and is **continuous from the right** if $\lim_{x \to a^+} f(x) = f(a)$.

Example 1: Is $f(x) = \begin{cases} x + 1 & \text{if } x \le 1 \\ 2 & \text{if } 1 < x \le 2 \\ 1/(x-3) & \text{if } x > 2 \end{cases}$ continuous at 1, 2, 3 ?

Solution: We could answer these questions by examining the graph of f(x), but lets try them without the graph. At a = **1**, f(**1**) = 2 and the left and right limits are equal,

$f(x) = \lim_{x \to 1^-} (x+1) = 2$ and $\lim_{x \to 1^+} f(x) = \lim_{x \to 1^+} 2 = 2$, so f is continuous at **1**.

At a = **2**, f(**2**) = 2, but the left and right limits are not equal,

$\lim_{x \to 2^-} f(x) = \lim_{x \to 2^-} 2 = 2$ and $\lim_{x \to 2^+} f(x) = \lim_{x \to 2^+} 1/(x-3) = -1$, so f fails condition (ii) and is not continuous at **2**. f is continuous from the left at **2**, but not from the right.

At a = **3**, f(**3**) = 1/0 which is undefined so f is not continuous at **3** because it fails condition (i).

Example 2: Where is $f(x) = 3x^2 - 2x$ continuous?

Solution: At every point. By the Substitution Theorem for Polynomials, every polynomial is continuous everywhere.

Example 3: Where are $g(x) = \frac{x+5}{x-3}$ and $h(x) = \frac{x^2 + 4x - 5}{x^2 - 4x + 3}$ continuous?

Solution: g is a rational function so by the Substitution Theorem for Polynomials and Rational Functions it is continuous everywhere except where its denominator is 0: g is continuous everywhere except 3. The graph of g (Fig. 2) is connected everywhere except at 3 where it has a vertical asymptote.

$h(x) = \dfrac{(x-1)(x+5)}{(x-1)(x-3)}$ is also continuous everywhere except where its denominator is 0: h is continuous everywhere except 3 and 1 . The graph of h (Fig. 3) is connected everywhere except at 3 where it has a vertical asymptote and at 1 where it has a hole: f(1) = 0/0 is undefined.

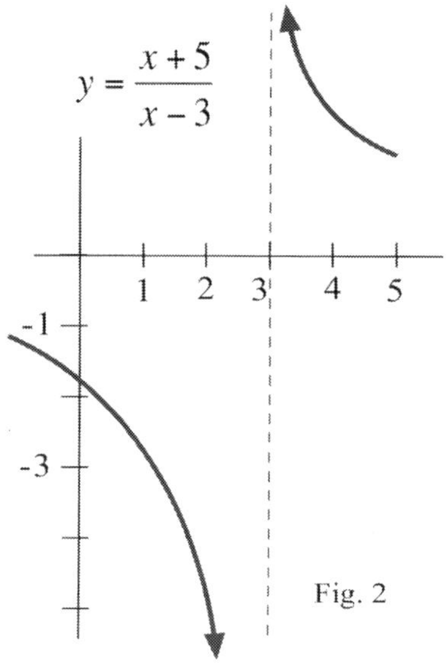

Fig. 2

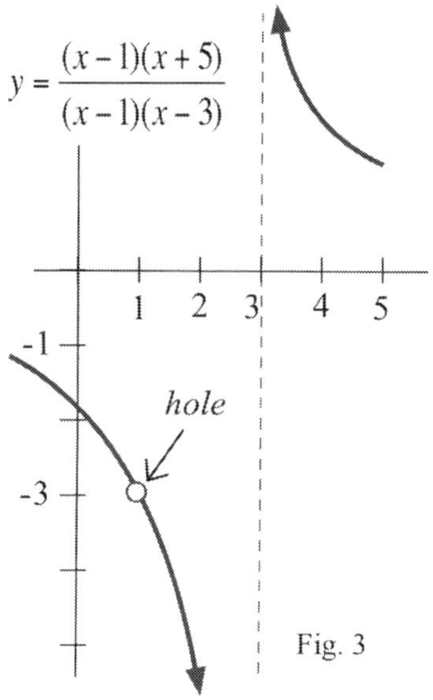

Fig. 3

Example 4: Where is f(x) = INT(x) continuous?

Solution: The graph of y = INT(x) seems to be connected except at each integer, and at each integer there is a "jump" (Fig. 4).

If a is an **integer**, then

$\lim\limits_{x\to a^-} \text{INT}(x) = a-1$, and $\lim\limits_{x\to a^+} \text{INT}(x) = a$, so $\lim\limits_{x\to a} \text{INT}(x)$ is undefined, and INT(x) is not continuous.

If a is **not an integer**, then the left and right limits of INT(x) , as x→ a, both equal INT(a) so

$\lim\limits_{x\to a} \text{INT}(x) = \text{INT}(a) = f(a)$ and INT(x) is continuous.

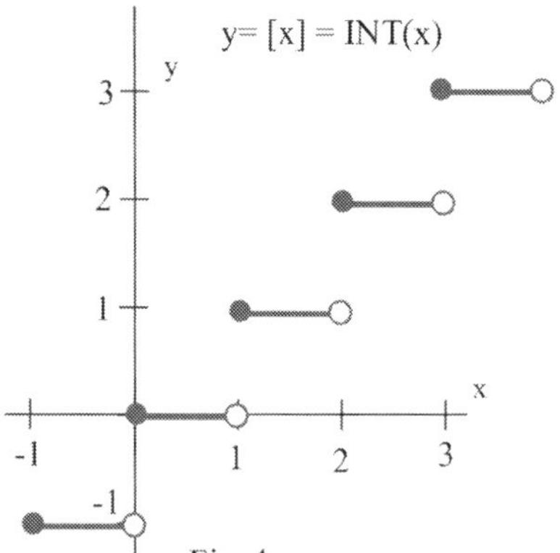

Fig. 4

f(x) = INT(x) is continuous except at the integers.

In fact, f(x) = INT(x) is continuous from the right everywhere and is continuous from the left everywhere except at the integers.

Practice 1: Where is f(x) = |x|/x continuous?

Why do we care whether a function is continuous?

There are several reasons for us to examine continuous functions and their properties:

- Most of the applications in engineering, the sciences and business are continuous and are modeled by continuous functions or by pieces of continuous functions.

- Continuous functions have a number of useful properties which are not necessarily true if the function is not continuous. If a result is true of all continuous functions and we have a continuous function, then the result is true for our function. This can save us from having to show, one by one, that each result is true for each particular function we use. Some of these properties are given in the rest of this section.

- Differential calculus has been called the study of **continuous** change, and many of the results of calculus are guaranteed to be true only for continuous functions. If you look ahead into Chapters 2 and 3, you will see that many of the theorems have the form "If f is **continuous** and (some additional hypothesis), then (some conclusion)".

Combinations of Continuous Functions

Theorem: If f(x) and g(x) are continuous at a, and k is any constant,
then the elementary combinations of f and g
($k \cdot f(x)$, $f(x) + g(x)$, $f(x) - g(x)$, $f(x) \cdot g(x)$, and $f(x)/g(x)$ $(g(a) \neq 0)$)
are continuous at a.

The continuity of a function is defined in terms of limits, and all of these results about simple combinations of continuous functions follow from the results about simple combinations of limits in the Main Limit Theorem. Our hypothesis is that f and g are both continuous at a, so we can assume that

$\lim_{x \to a} f(x) = f(a)$ and $\lim_{x \to a} g(x) = g(a)$ and then use the appropriate part of the Main Limit Theorem.

For example, $\lim_{x \to a} \{ f(x) + g(x) \} = \left\{ \lim_{x \to a} f(x) \right\} + \left\{ \lim_{x \to a} g(x) \right\} = f(a) + g(a)$, so $f + g$ is continuous at a.

Practice 2: Prove: If f and g are continuous at a, then kf and $f - g$ are continuous at a. (k a constant.)

Composition of Continuous Functions

If g is continuous at a and f is continuous at g(a),

then $\lim_{x \to a} \{ f(g(x)) \} = f(\lim_{x \to a} g(x)) = f(g(a))$ so $f \circ g(x) = f(g(x))$ is continuous at a.

This result will not be proved here, but the proof just formalizes the following line of reasoning:

The hypothesis that "g is continuous at a" means that if x is close to a then g(x) will be close to g(a). Similarly, "f is continuous at g(a)" means that if g(x) is close to g(a) then f(g(x)) = f∘g(x) will be close to f(g(a)) =f∘g(a) . Finally, we can conclude that if x is close to a, then g(x) is close to g(a) so f∘g(x) is close to f∘g(a), and therefore f∘g is continuous at x = a.

The next theorem presents an alternate version of the limit condition for continuity, and we will use this alternate version occasionally in the future.

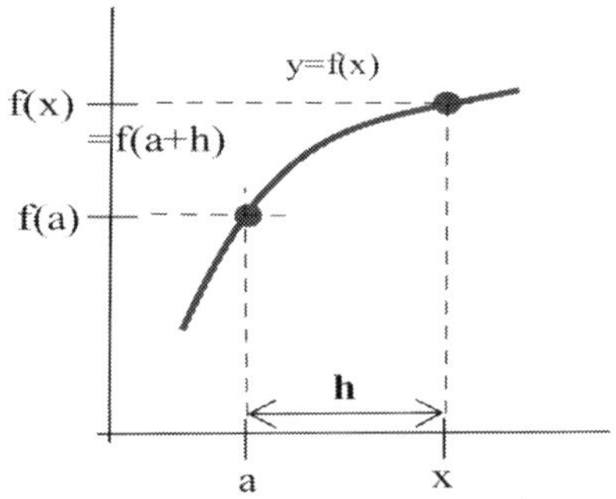

Fig. 5

Theorem: $\lim_{x \to a} f(x) = f(a)$ if and only if $\lim_{h \to 0} f(a+h) = f(a)$.

Proof: Let's define a new variable h by h = x – a so x = a + h (Fig. 5). Then x → a if and only if h = x – a → 0 , so

$\lim_{x \to a} f(x) = \lim_{h \to 0} f(a+h)$, and $\lim_{x \to a} f(x) = f(a)$ if and only if $\lim_{h \to 0} f(a+h) = f(a)$.

A function f is continuous at a if and only if $\lim_{h \to 0} f(a+h) = f(a)$.

Which Functions Are Continuous?

Fortunately, the situations which we encounter most often in applications and the functions which model those situations are either continuous everywhere or continuous everywhere except at a few places, so any result which is true of all continuous functions will be true of most of the functions we commonly use.

Theorem: The following functions are continuous everywhere, at every value of x:
(a) polynomials, (b) sin(x) and cos(x), and (c) | x | .

Proof: (a) This follows from the Substitution Theorem for Polynomials and the definition of continuity.

(b) The graph of y = sin(x) (Fig. 6) clearly indicates that sin(x) does not have any holes or breaks so sin(x) is continuous everywhere. Or we could justify that result analytically:

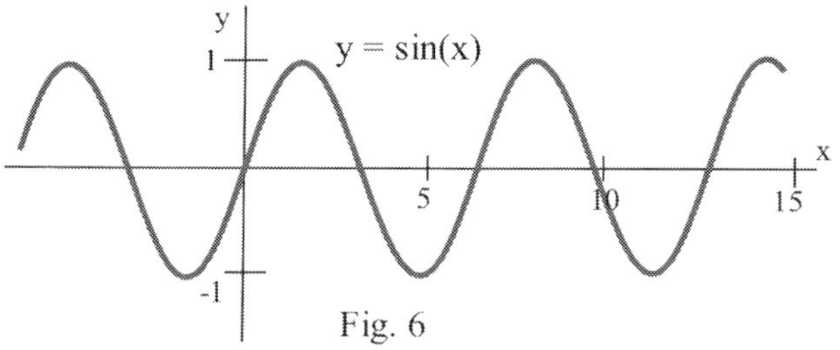

Fig. 6

for every real number a,

$$\lim_{h\to 0} \sin(a+h) = \lim_{h\to 0} \sin(a)\cos(h) + \cos(a)\sin(h)$$

$$= \lim_{h\to 0} \sin(a) \cdot \lim_{h\to 0} \cos(h) + \lim_{h\to 0} \cos(a) \cdot \lim_{h\to 0} \sin(h)$$

(recall from section 1.2 that $\lim_{h\to 0} \cos(h) = 1$ and $\lim_{h\to 0} \sin(h) = 0$)

$$= \lim_{h\to 0} \sin(a) \cdot \mathbf{1} + \lim_{h\to 0} \cos(a) \cdot \mathbf{0} = \sin(a),$$

so $f(x) = \sin(x)$ is continuous at every point. The justification of $f(x) = \cos(x)$ is similar.

(c) $f(x) = |x|$. When $x > 0$, then $|x| = x$ and its graph (Fig. 7) is a straight line and is continuous since x is a polynomial function. When $x < 0$, then $|x| = -x$ and it is also continuous. The only questionable point is the "corner" on the graph when $x = 0$, but the graph there is only bent, not broken:

$$\lim_{h\to 0^+} |x| = \lim_{h\to 0^+} x = 0$$

and $\lim_{h\to 0^-} |x| = \lim_{h\to 0^-} -x = 0$ so $\lim_{h\to 0} |x| = 0 = |0|$,

and $f(x) = |x|$ is also continuous at 0.

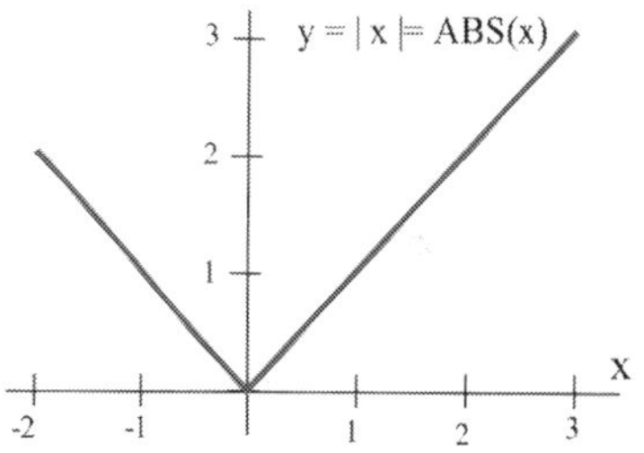

Fig. 7

A continuous function can have corners but not holes or breaks (jumps).

Several results about limits of functions can be written in terms of continuity of those functions. Even functions which fail to be continuous at some points are often continuous most places.

Theorem: (a) A rational function is continuous **except** where the denominator is 0.

(b) Tangent, cotangent, secant and cosecant are continuous **except** where they are undefined.

(c) The greatest integer function $[\ x\ ] = \text{INT}(x)$ is continuous **except** at each integer.

(d) But the "holey" function $h(x) = \begin{cases} 2 & \text{if } x \text{ is a } \textbf{rational} \text{ number} \\ 1 & \text{if } x \text{ is an } \textbf{irrational} \text{ number} \end{cases}$

is discontinuous **everywhere**.

INTERMEDIATE VALUE PROPERTY OF CONTINUOUS FUNCTIONS

Since the graph of a continuous function is connected and does not have any holes or breaks in it, the values of the function can not "skip" or "jump over" a horizontal line (Fig. 8). If one value of the continuous function is below the line and another value of the function is above the line, then **somewhere** the graph will cross the line. The next theorem makes this statement more precise. The result seems obvious, but its proof is technically difficult and is not given here.

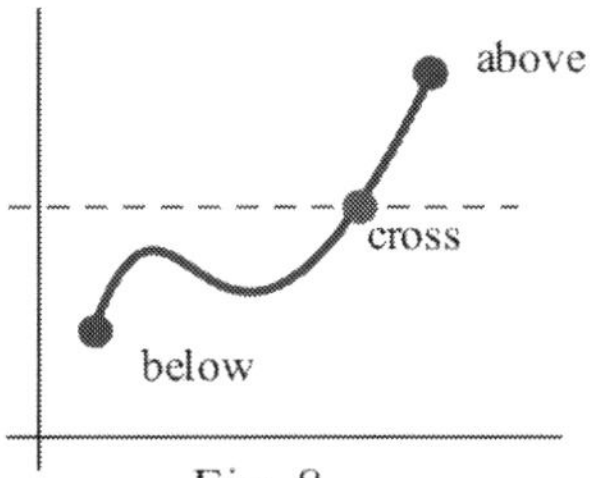

Fig. 8

> **Intermediate Value Theorem for Continuous Functions**
>
> If f is continuous on the interval [a,b] and V is any value between f(a) and f(b),
>
> then there is a number c between a and b so that f(c) = V
>
> (that is, f actually takes each intermediate value between f(a) and f(b).)

If the graph of f connects the points (a, f(a)) and (b, f(b)) and V is any number between f(a) and f(b), then the graph of f must cross the horizontal line y = V **somewhere between x = a and x = b** (Fig. 9). Since f is continuous, its graph cannot "hop" over the line y = V.

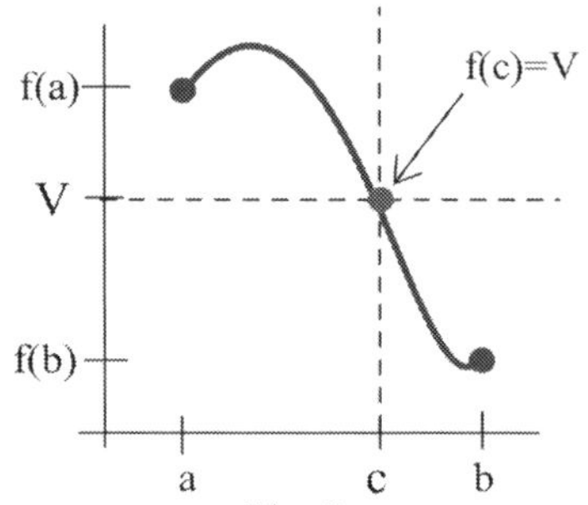

Fig. 9

Most people take this theorem for granted in some common situations:

- If a child's temperature rose from 98.6^{o} to 101.3^{o}, then there was an instant when the child's temperature was exactly 100°. In fact, every temperature between 98.6^{o} and 101.3^{o} occurred at some instant.
- If you dove to pick up a shell 25 feet below the surface of a lagoon, then at some instant in time you were 17 feet below the surface. (Actually, you want to be at 17 feet twice. Why?)
- If you started driving from a stop (velocity = 0) and accelerated to a velocity of 30 kilometers per hour, then there was an instant when your velocity was exactly 10 kilometers per hour.

But we cannot apply the Intermediate Value Theorem if the function is not continuous:

- In 1987 it cost 22¢ to mail a letter first class inside the United States, and in 1990 it cost 25¢ to mail the same letter, but we cannot conclude that there was a time when it cost 23¢ or 24¢ to send the letter. Postal rates did not increase in a continuous fashion. They jumped directly from 22¢ to 25¢ .
- Prices, taxes and rates of pay change in jumps, discrete steps, without taking on the intermediate values.

The Intermediate Value Property can help us finds roots of functions and solve equations. If f is continuous on [a,b] and f(a) and f(b) have opposite signs (one is positive and one is negative), then 0 is an intermediate value between f(a) and f(b) so f will have a root between x = a and x= b.

Bisection Algorithm for Approximating Roots

The Intermediate Value Theorem is an example of an "existence theorem" because it concludes that something exists: a number c so that f(c) = V. Many existence theorems do not tell us how to find the number or object which exists and are of no use in actually finding those numbers or objects. However, the Intermediate Value is the basis for a method commonly used to approximate the roots of continuous functions, the Bisection Algorithm.

Bisection Algorithm for Finding a Root of f(x)

(i) Find two values of x , say a and b, so that f(a) and f(b) have opposite signs

(then f(x) has a root between a and b, a root in the interval [a,b].)

(ii) Calculate the midpoint (bisection point) of the interval [a,b], m = (a+b)/2, and evaluate f(m).

(iii) (a) If f(m) = 0, then m is a root of f , and we are done.

(b) If $f(m) \neq 0$, then f(m) has the sign opposite one of f(a) or f(b):

if f(a) and f(m) have opposite signs, then f has a root in [a,m] so put b = m

if f(b) and f(m) have opposite signs, then f has a root in [m,b] so put a = m

(iv) Repeat steps (ii) and (iii) until a root is found exactly or is approximated closely enough.

The length of the interval known to contain a root is cut in half each time through steps (ii) and (iii) so the Bisection Algorithm quickly "squeezes" in on a root (Fig. 10).

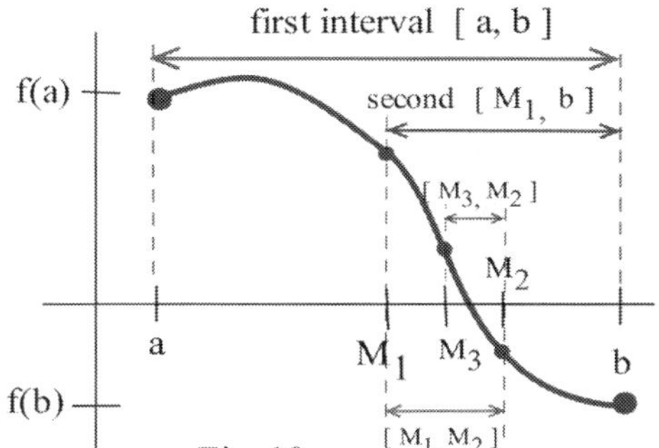

Fig. 10

The steps of the Bisection Algorithm can be done "by hand", but it is tedious to do very many of them that way. Computers are very good with this type of tedious repetition, and the algorithm is simple to program.

Example 7: Find a root of $f(x) = x - x^3 + 1$.

Solution: f(0) = 1 and f(1) = 1 so we cannot conclude that f has a root between 0 and 1. f(1) = 1 and f(2) = –5 have opposite signs, so by the Intermediate Value Property of continuous functions (this function is a polynomial so it is continuous everywhere) we know that there is a number c between 1 and 2 such that f(c) = 0 (Fig. 11). The midpoint of the interval [1,2] is m = (1+2)/2 = 3/2 = 1.5 , and f(3/2) = –7/8 so f changes sign between 1 and 1.5 and we can be sure that there is a root between 1 and 1.5 . If we repeat the operation for the interval [1, 1.5], the midpoint is m = (1+1.5)/2 = 1.25, and f(1.25) = 19/64 > 0 so f changes sign between 1.25 and 1.5 and we know f has a root between 1.25 and 1.5 .

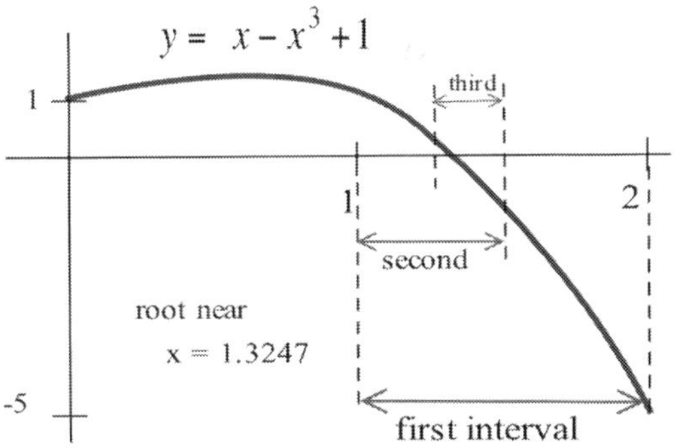

Fig. 11

Repeating this procedure a few more times, we get that

a	b	m = (b+a)/2	f(a)	f(b)	f(m)	root between	
1	2		1	–5		1	2
1	2	1.5	1	–5	–0.875	1	1.5
1	1.5	1.25	1	–0.875	0.2969	1.25	1.5
1.25	1.5	1.375	0.2969	–0.875	–0.2246	1.25	1.375
1.25	1.375	1.3125	0.2969	–0.2246	0.0515	1.3125	1.375
1.3125	1.375	1.34375					

If we continue the table, the interval containing the root will squeeze around the value 1.324718 .

The Bisection Algorithm has one major drawback — there are some roots it does not find. The algorithm requires that the function be both positive and negative near the root so that the graph actually crosses the x–axis. The function $f(x) = x^2 - 6x + 9 = (x-3)^2$ has the root $x = 3$ but is never negative (Fig. 12). We cannot find two starting points a and b so that f(a) and f(b) have opposite signs, and we cannot use the Bisection Algorithm to find the root x = 3. In Chapter 2 we will see another method, Newton's Method, which does find roots of this type.

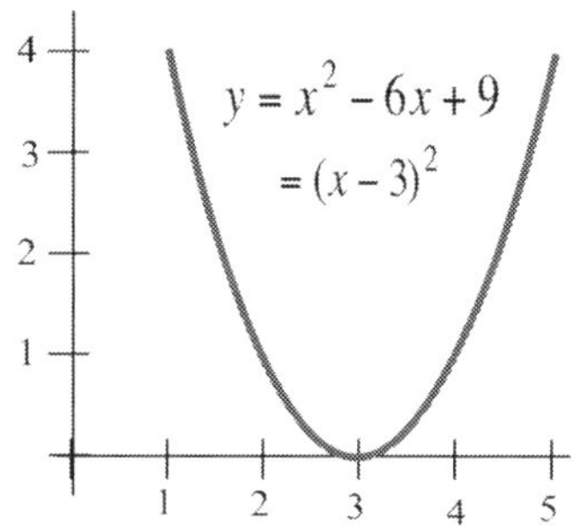

Fig. 12

The Bisection Algorithm requires that we supply two starting points a and b, two x–values at which the function has opposite signs. These values can often be found with a little "trial and error", or we can examine the graph of the function and use it to help pick the two values.

Finally, the Bisection Algorithm can also be used to solve equations because the solution of any equation can always be transformed into an equivalent problem of finding roots by moving everything to one side of the equal sign. For example, the problem of solving the equation $x^3 = x + 1$ can be transformed into the equivalent problem of solving $x + 1 - x^3 = 0$ or of finding the roots of $f(x) = x + 1 - x^3$ which we did in the previous example.

Example 8: Find all of the solutions of $\sin(x) = \dfrac{2x+1}{x-2}$. (x is in radians.)

Solution: We can convert this problem of solving an equation to the problem of finding the roots of $f(x) = \sin(x) - \dfrac{2x+1}{x-2} = 0$. The function f(x) is continuous everywhere except at x = 2, and the graph of f(x) in Fig. 13 can help us find two starting values for the Bisection Algorithm. The graph shows that f(–1) is negative and f(0) is positive, and we know f(x) is continuous on the interval [–1,0]. Using the algorithm with the starting interval [–1,0], we get that the root is contained in the shrinking intervals:

$[-.5,0], [-.25,0], [-.25, -.125], \ldots,$
$[-.238281, -.236328], \ldots, [-.237176, -.237177]$
so the root is approximately $-.237177$.

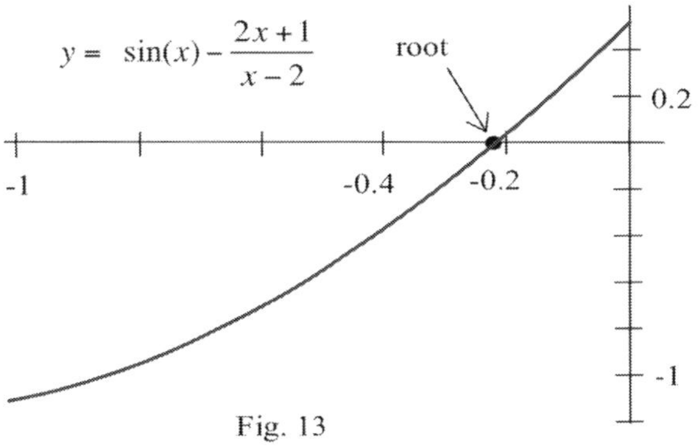

Fig. 13

We might also notice that $f(0) = 0.5$ is positive and $f(\pi) = 0 - \frac{2\pi + 1}{\pi - 2} \approx -6.38$ is negative. Why is it wrong to conclude that $f(x)$ has another root between $x = 0$ and $x = \pi$?

PROBLEMS

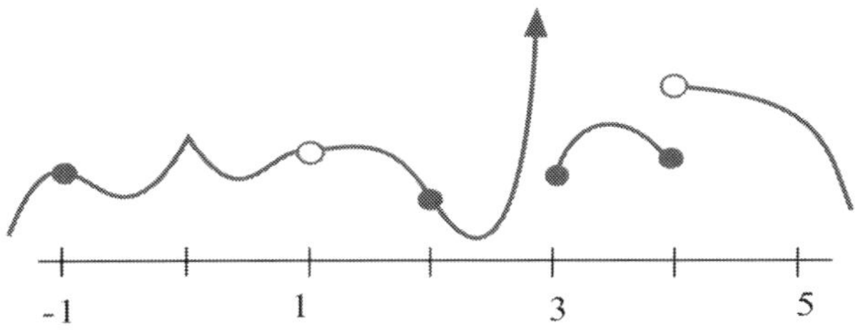

Fig. 14

1. At which points is the function in Fig. 14 discontinuous?

2. At which points is the function in Fig. 15 discontinuous?

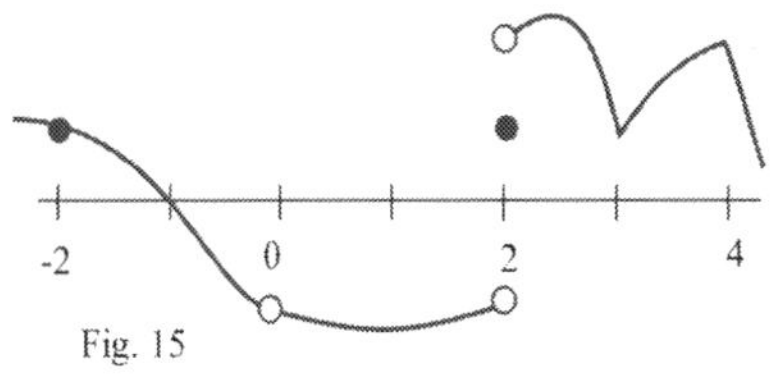

Fig. 15

3. Find at least one point at which each function is not continuous and state which of the 3 conditions in the definition of continuity is violated at that point.

(a) $\frac{x+5}{x-3}$ (b) $\frac{x^2+x-6}{x-2}$ (c) $\sqrt{\cos(x)}$

(d) $\text{INT}(x^2)$ (e) $\frac{x}{\sin(x)}$ (f) $\frac{x}{x}$

(g) $\ln(x^2)$ (h) $\frac{\pi}{x^2 - 6x + 9}$ (i) $\tan(x)$

4. Which <u>three</u> of the following functions are not continuous. Use the appropriate theorems of this section to justify that each of the other functions is continuous.

(a) $\frac{7}{\sqrt{2 + \sin(x)}}$ (b) $\cos(x^5 - 7x + \pi)$ (c) $\frac{x^2 - 5}{1 + \cos^2(x)}$

(d) $\frac{x^2 - 5}{1 + \cos(x)}$ (e) $\text{INT}(3 + 0.5\sin(x))$ (f) $\text{INT}(0.3\sin(x) + 1.5)$

(g) $\sqrt{\cos(\sin(x))}$ (h) $\sqrt{x^2 - 6x + 10}$ (i) $\sqrt[3]{\cos(x)}$

(j) $2^{\sin(x)}$ (k) $\log(|x|)$ (l) $1 - 3^{-x}$

5. A continuous function f has the values given below:

x	0	1	2	3	4	5
f(x)	5	3	–2	–1	3	–2

(a) f has at least ___ roots between 0 and 5. (b) $f(x) = 4$ at least ___ times between 0 and 5.

(c) $f(x) = 2$ at least ___ times between 0 and 5. (d) $f(x) = 3$ at least ___ times between 0 and 5.

(e) Is it possible for f(x) to equal 7 for some x values between 0 and 5?

6. A continuous function g has the values given below:

x	1	2	3	4	5	6	7
g(x)	–3	1	4	–1	3	–2	–1

(a) g has at least ___ roots between 1 and 5. (b) $g(x) = 3.2$ at least ___ times between 1 and 7.

(c) $g(x) = -0.7$ at least ___ times between 3 and 7. (d) $g(x) = 1.3$ at least ___ times between 2 and 6.

(e) Is it possible for g(x) to equal π for some value(s) of x between 5 and 6?

7. This problem asks you to verify that the Intermediate Value Theorem is true for some particular functions, intervals and intermediate values. In each problem you are given a function f, an interval [a,b] and a value V. Verify that V is between f(a) and f(b) and find a value of c in the interval so that $f(c) = V$.

(a) $f(x) = x^2$ on $[0,3]$, $V = 2$. (b) $f(x) = x^2$ on $[-1,2]$, $V = 3$.

(c) $f(x) = \sin(x)$ on $[0,\pi/2]$, $V = 1/2$. (d) $f(x) = x$ on $[0,1]$, $V = 1/3$.

(e) $f(x) = x^2 - x$ on $[2,5]$, $V = 4$. (f) $f(x) = \ln(x)$ on $[1,10]$, $V = 2$.

8. Two students claim that they both started with the points $x = 1$ and $x = 9$ and applied the Bisection Algorithm to the function in Fig. 16. The first student says that the algorithm converged to the root near $x = 8$, but the second claims that the algorithm will converge to the root near $x = 4$. Who is right?

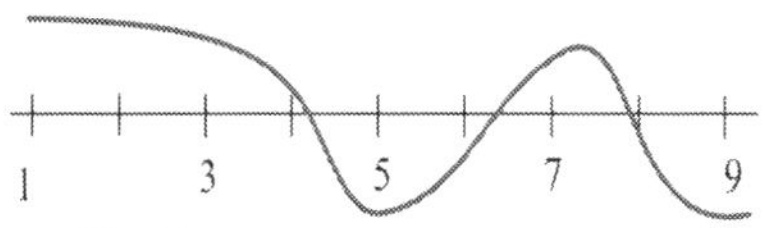

Fig. 16

9. Two students claim that they both started with the points $x = 0$ and $x = 5$ and applied the Bisection Algorithm to the function in Fig. 17. The first student says that the algorithm converged to the root labeled A, but the second claims that the algorithm will converge to the root labeled B. Who is right?

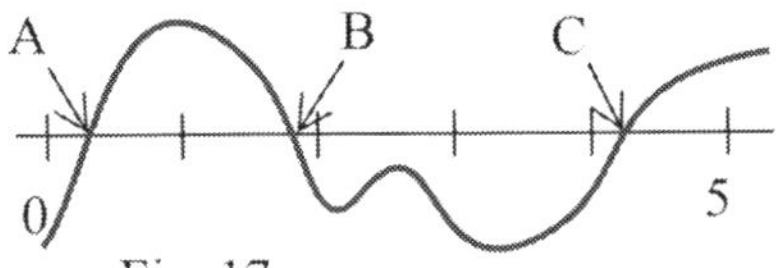

Fig. 17

10. If you apply the Bisection Algorithm to the function in Fig. 18 and use the given starting points, which root does the algorithm find? (a) starting points 0 and 9. (b) starting points 1 and 5. (c) starting points 3 and 5.

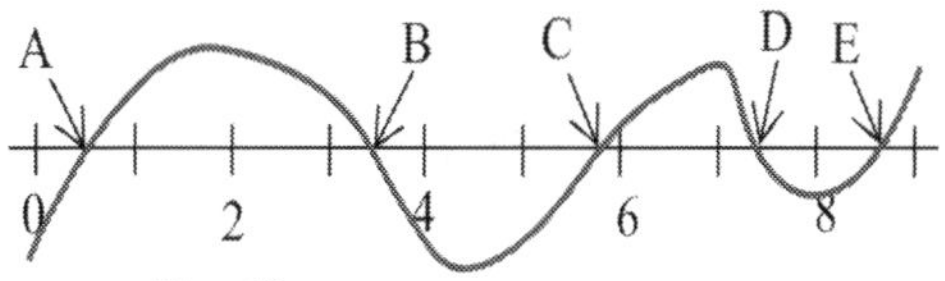

Fig. 18

11. If you apply the Bisection Algorithm to the function in Fig. 19 and use the given starting points, which root does the algorithm find?
(a) starting points 3 and 7. (b) starting points 5 and 6.
(c) starting points 1 and 6.

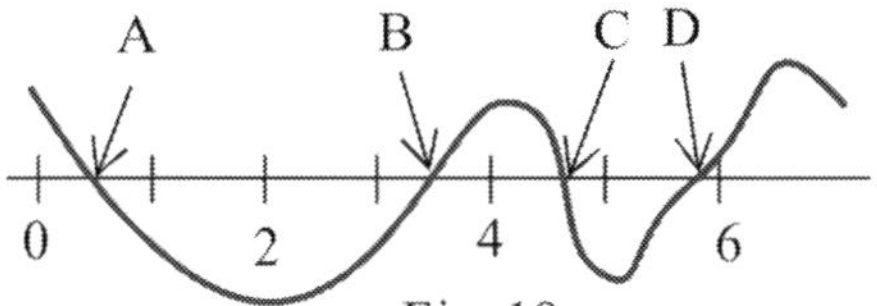

Fig. 19

In problems 12 – 17 , use the Intermediate Value Theorem to verify that each function has a root in the given interval(s). Then use the Bisection Algorithm to narrow the location of that root to an interval of length less than or equal to 0.1 .

12. $f(x) = x^2 - 2$ on $[0, 3]$.

13. $g(x) = x^3 - 3x^2 + 3$ on $[-1, 0]$, $[1, 2]$, $[2, 4]$.

14. $h(t) = t^5 - 3t + 1$ on $[1, 3]$.

15. $r(x) = 5 - 2^x$ on $[1, 3]$.

16. $s(x) = \sin(2x) - \cos(x)$ on $[0, \pi]$

17. $p(t) = t^3 + 3t + 1$ on $[-1, 1]$

18. What is wrong with this reasoning: "If $f(x) = 1/x$ then $f(-1) = -1$ and $f(1) = 1$. Because $f(-1)$ and $f(1)$ have opposite signs, f has a root between $x = -1$ and $x = 1$."

19. Each of the following statements is false for some functions. For each statement, sketch the graph of a counterexample.
   a) If $f(3) = 5$ and $f(7) = -3$, then f has a root between $x = 3$ and $x = 7$.
   b) If f has a root between $x = 2$ and $x = 5$, then $f(2)$ and $f(5)$ have opposite signs.
   c) If the graph of a function has a sharp corner, then the function is not continuous there.

20. Define $A(x)$ to be the **area** bounded by the x and y axes, the curve $y = f(x)$, and the vertical line at x (Fig. 20). From the figure, it is clear that $A(1) < 3$ and $A(3) > 3$.
Do you think there is a value of x, between 1 and 3, so $A(x) = 3$? If so, justify your conclusion and estimate the location of the value of x so $A(x) = 3$.
If not, justify your conclusion.

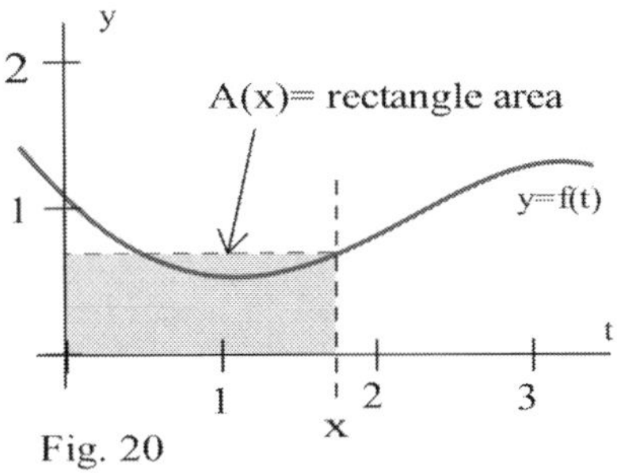

Fig. 20

21. Define A(x) to be the **area** bounded by the x and y axes, the curve y = f(x), and the vertical line at x (Fig. 21).

a) Shade the part of the graph represented by A(2.1) – A(2) and estimate the value of $\frac{A(2.1) - A(2)}{0.1}$.

b) Shade the part of the graph represented by A(4.1) – A(4) and estimate the value of $\frac{A(4.1) - A(4)}{0.1}$.

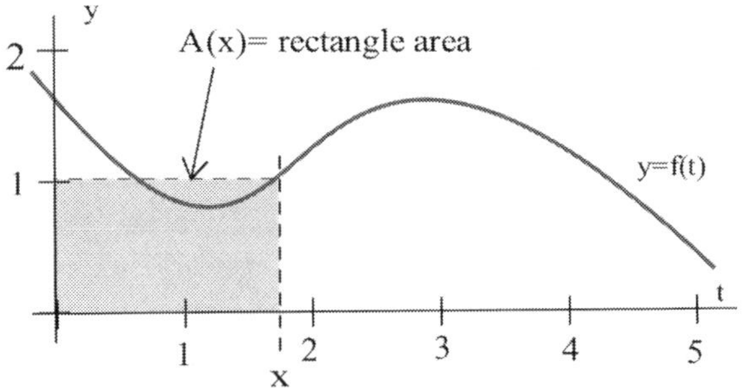

Fig. 21

22. (a) A square sheet of paper has a straight line drawn on it from the lower left corner to the upper right corner. Is it possible for you to start on the left edge of the sheet and draw a connected line to the right edge that does not cross the diagonal line?

(b) Prove: If f is continuous on the interval [0,1] and $0 \le f(x) \le 1$ for all x, then there is a number c, $0 \le c \le 1$, such that f(c) = c. (The number c is called a "fixed point" of f because the image of c is the same as c — f does not move c.)

Hint: Define a new function g(x) = f(x) – x and start by considering the values g(0) and g(1).

(c) What does part (b) have to do with part (a) of this problem?

(d) Is the theorem in part (b) true if we replace the closed interval [0,1] with the open interval (0,1)? True/False: "If f is continuous on the interval (0,1) and $0 < f(x) < 1$ for all x, then there is a number c, $0 < c < 1$, such that f(c) = c."

23. A piece of string is tied in a loop and tossed onto quadrant I enclosing a single region (Fig. 22).

(a) Is it always possible to find a line L which goes through the origin so that L divides the region into two equal areas? (Justify your answer.)

(b) Is it always possible to find a line L which is parallel to the x–axis so that L divides the region into two equal areas? (Justify your answer.)

(c) Is it always possible to find 2 lines, L parallel to the x–axis and M parallel to the y–axis, so L and M divide the region into 4 equal areas? (Justify your answer.)

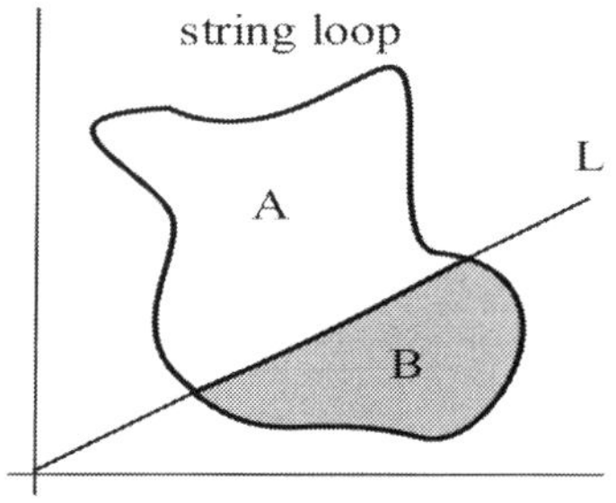

Fig. 22

Section 1.3 **PRACTICE Answers**

Practice 1: $f(x) = \frac{|x|}{x}$ (Fig. 23) is continuous everywhere **except at x = 0** where this function is not defined.

If $a > 0$, then $\lim_{x \to a} \frac{|x|}{x} = 1 = f(a)$ so f is continuous at a.

If $a < 0$, then $\lim_{x \to a} \frac{|x|}{x} = -1 = f(a)$ so f is continuous at a.

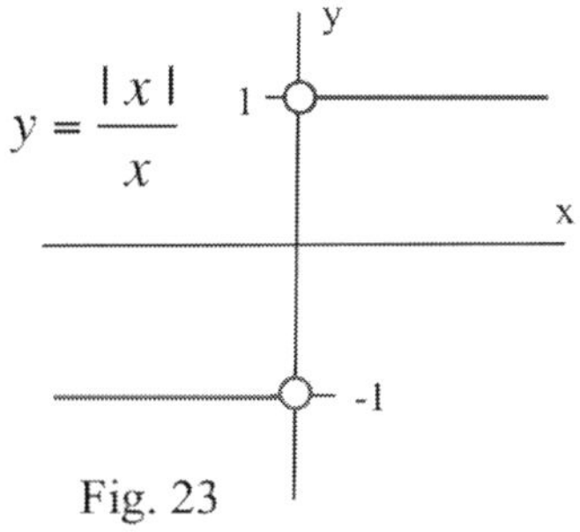

Fig. 23

f(0) is not defined., $\lim_{x \to 0^-} \frac{|x|}{x} = -1$ and $\lim_{x \to 0^+} \frac{|x|}{x} = +1$ so

$\lim_{x \to 0} \frac{|x|}{x}$ does not exist.

Practice 2: (a) To prove that kf is continuous at a, we need to prove that kf satisfies the definition of continuity at a: $\lim_{x \to a} kf(x) = kf(a)$.

Using results about limits, we know

$\lim_{x \to a} kf(x) = k \lim_{x \to a} f(x) = k\,f(a)$ (since f is assumed to be continuous at a) so kf is continuous at a.

(b) To prove that f – g is continuous at a, we need to prove that f – g satisfies the definition of continuity at a: $\lim_{x \to a} (f(x) - g(x)) = f(a) - g(a)$.

Again using information about limits,

$\lim_{x \to a} (f(x) - g(x)) = \lim_{x \to a} f(x) - \lim_{x \to a} g(x) = f(a) - g(a)$ (since f and g are both continuous at a) so f – g is continuous at a.

1.4 DEFINITION OF LIMIT

It may seem strange that we have been using and calculating the values of limits for awhile without having a precise definition of limit, but the history of mathematics shows that many concepts, including limits, were successfully used before they were precisely defined or even fully understood. We have chosen to follow the historical sequence in this chapter and to emphasize the intuitive and graphical meaning of limit because most students find these ideas and calculations easier than the definition. Also, this intuitive and graphical understanding of limit was sufficient for the first hundred years of the development of calculus (from Newton and Leibniz in the late 1600's to Cauchy in the early 1800's), and it is sufficient for using and understanding the results in beginning calculus.

Mathematics, however, is more than a collection of useful tools, and part of its power and beauty comes from the fact that in mathematics terms are precisely defined and results are rigorously proved. Mathematical tastes (what is mathematically beautiful, interesting, useful) change over time, but because of these careful definitions and proofs, the results remain true, everywhere and forever. Textbooks seldom give all of the definitions and proofs, but it is important to mathematics that such definitions and proofs exist.

The goal of this section is to provide a precise definition of the limit of a function. The definition will not help you calculate the values of limits, but it provides a precise statement of what a limit is. The definition of limit is then used to verify the limits of some functions, and some general results are proved.

The Intuitive Approach

The precise ("formal") definition of limit carefully defines the ideas that we have already been using graphically and intuitively. The following side–by–side columns show some of the phrases we have been using to describe limits, and those phrases, particularly the last ones, provide the basis to building the definition of limit.

A Particular Limit	General Limit
$\lim_{x \to 3} 2x - 1 = 5$	$\lim_{x \to a} f(x) = L$
"as the values of x approach 3, the values of 2x–1 approach (are arbitrarily close to) 5"	"as the values of x approach a, the values of f(x) approach (are arbitrarily close to) L"
"when x is close to 3 (but not equal to 3), the value of 2x–1 is close to 5"	"when x is close to a (but not equal to a), the value of f(x) is close to L"
"we can guarantee that the values of f(x) = 2x–1 are as close to 5 as we want by starting with values of x sufficiently close to 3 (but not equal to 3)"	"we can guarantee that the values of f(x) are as close to L as we want by starting with values of x sufficiently close to a (but not equal to a)"

Let's examine what the last phrase ("we can ...") means for the Particular Limit.

Example 1: We know $\lim_{x \to 3} 2x - 1 = 5$. Show that we can guarantee that the values of $f(x) = 2x - 1$ are as close to 5 as we want by starting with values of x sufficiently close to 3.

(a) What values of x guarantee that $f(x) = 2x - 1$ is within 1 unit of 5 ? (Fig. 1a)

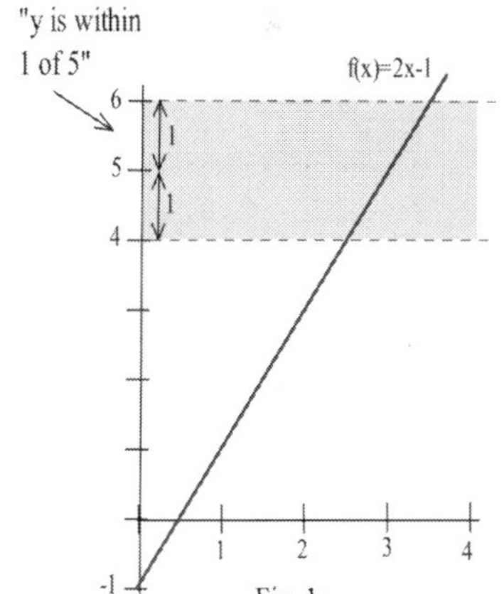

Fig. 1a

Solution: "within 1 unit of 5" means between $5-1 = 4$ and $5+1 = 6$, so the question can be rephrased as "for what values of x is $y = 2x - 1$ between 4 and 6: $4 < 2x - 1 < 6$?" We want to know which values of x put the values of $y = 2x - 1$ into the shaded band in Fig. 1a. The algebraic process is straightforward: solve $4 < 2x - 1 < 6$ for x to get $5 < 2x < 7$ and $2.5 < x < 3.5$. We can restate this result as follows: "If x is within **0.5** units of 3, then $y = 2x-1$ is within 1 unit of 5." (Fig. 1b)

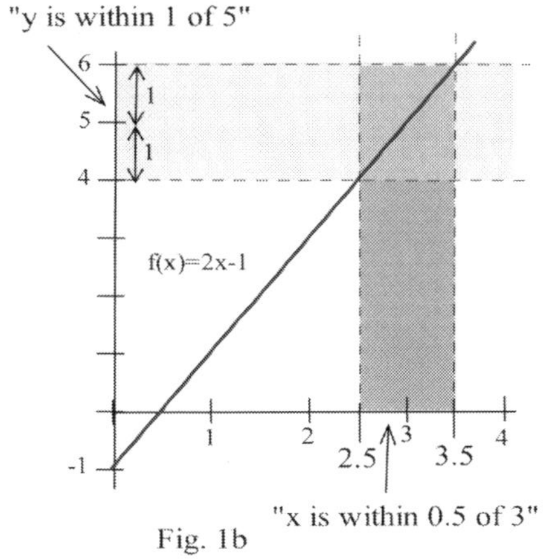

Fig. 1b

Any smaller distance also satisfies the guarantee: e.g., "If x is within **0.4** units of 3, then $y = 2x-1$ is within 1 unit of 5." (Fig. 1c)

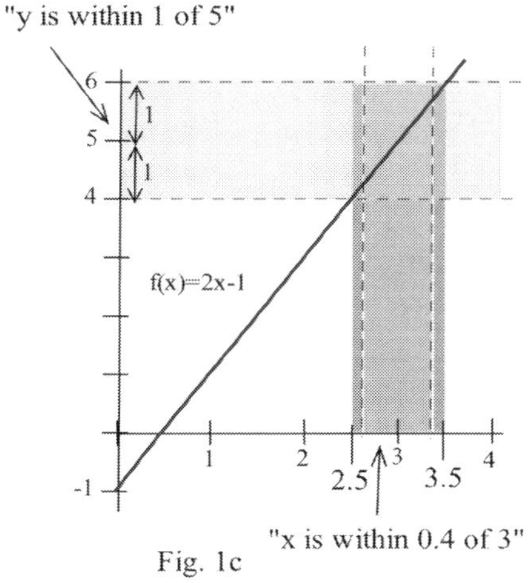

Fig. 1c

(b) What values of x guarantee the $f(x) = 2x - 1$ is within 0.2 units of 5? (Fig. 2a)

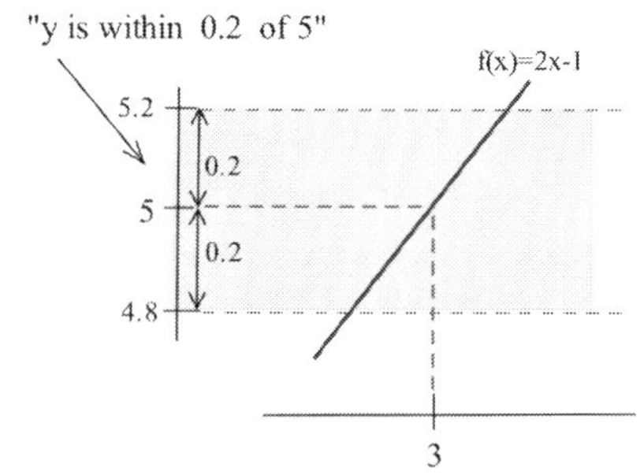

Fig. 2a

Solution: "within 0.2 units of 5" means between $5-0.2 = 4.8$ and $5+0.2 = 5.2$, so the question can be rephrased as "for what values of x is $y = 2x - 1$ between 4.8 and 5.2: $4.8 < 2x - 1 < 5.2$?" Solving for x, we get $5.8 < 2x < 6.2$ and $2.9 < x < 3.1$. "If x is within 0.1 units of 3, then $y = 2x-1$ is within 0.2 units of 5." (Fig. 2b) Any smaller distance also satisfies the guarantee.

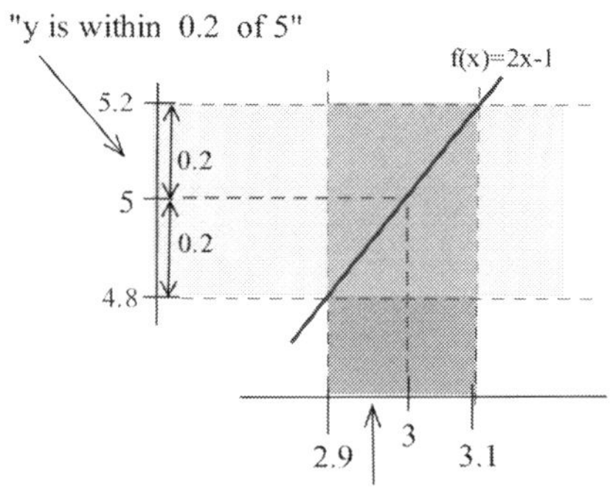

Fig. 2b "x is within 0.1 of 3"

Rather than redoing these calculations for every possible distance from 5, we can do the work once, generally:

(c) What values of x guarantee that $f(x) = 2x - 1$ is within E units of 5? (Fig. 3a)

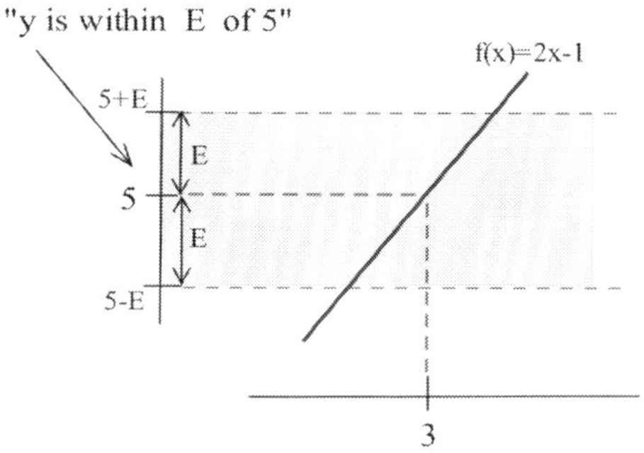

Fig. 3a

Solution: "within E unit of 5" means between 5–E and 5+E , so the question is "for what values of x is $y = 2x - 1$ between 5–E and 5+ε: $5-E < 2x - 1 < 5+E$?" Solving $5-E < 2x - 1 < 5+E$ for x, we get $6-E < 2x < 6+E$ and $3 - E/2 < x < 3 + E/2$. "If x is within E/2 units of 3, then $y = 2x-1$ is within E units of 5." (Fig. 3b) Any smaller distance also satisfies the guarantee.

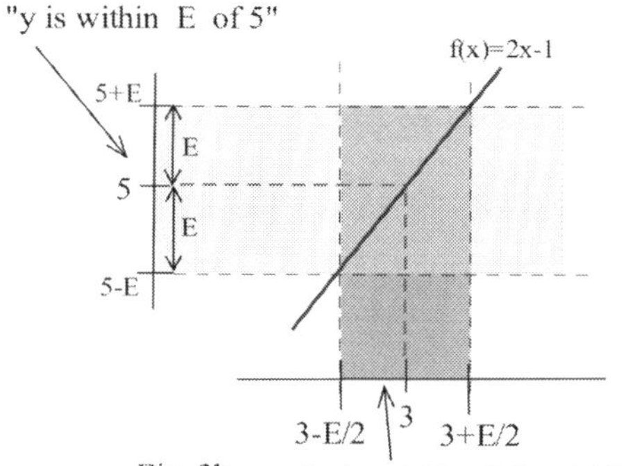

Fig. 3b **"x is within E/2 of 3"**

Part (c) of Example 1 illustrates a little of the power of general solutions in mathematics. Rather than doing a new set of similar calculations every time someone demands that $f(x) = 2x - 1$ be within some given distance of 5, we did the calculations once. And then we can respond for any given distance. For the question "What values of x guarantee that $f(x) = 2x - 1$ is within 0.4, 0.1 and 0.006 units of 5?", we can answer "If x is within 0.2 (= 0.4/2), 0.05 (=0.1/2) and 0.003 (=0.006/2) units of 3."

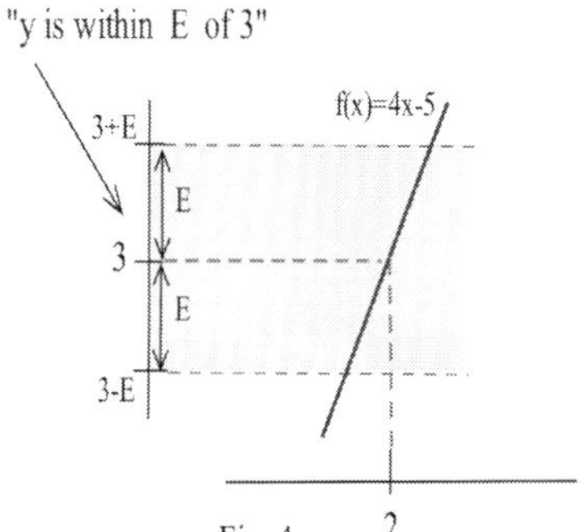

Fig. 4

Practice 1: $\lim_{x \to 2} 4x - 5 = 3$.

What values of x guarantee that $f(x) = 4x - 5$ is within

(a) 1 unit of 3?

(b) 0.08 units of 3? (c) E units of 3? (Fig. 4)

The same ideas work even if the graphs of the functions are not straight lines, but the calculations are more complicated.

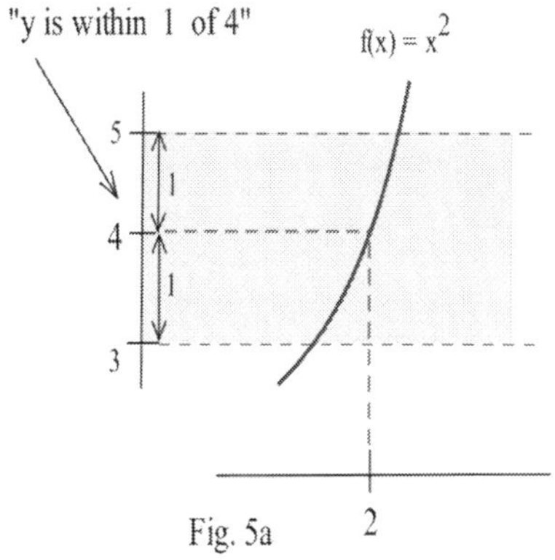

Fig. 5a

Example 2: $\lim_{x \to 2} x^2 = 4$. (a) What values of x guarantee that $f(x) = x^2$ is within 1 unit of 4? (b) Within 0.2 units of 4? (Fig. 5a) State each answer in the form "If x is within ______ units of 2, then f(x) is within 1 (or 0.2) unit of 4."

Solution: (a) If x^2 is within 1 unit of 4, then $3 < x^2 < 5$ so $\sqrt{3} < x < \sqrt{5}$ or $1.732 < x < 2.236$. The interval containing these x values extends from $2 - \sqrt{3} \approx 0.268$ units to the left of 2 to $\sqrt{5} - 2 \approx 0.236$ units to the right of 2. Since we want to specify a single distance on each side of 2, we can pick the **smaller** of the two distances, 0.236 . (Fig. 5b)

"If x is within 0.236 units of 2,

then f(x) is within 1 unit of 4."

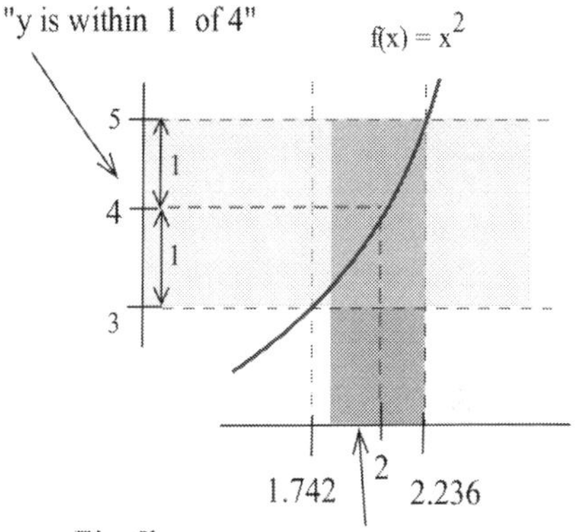

Fig. 5b **"x is within 0.236 of 3"**

(b) Similarly, if x^2 is within 0.2 units of 4, then $3.8 < x^2 < 4.2$ so $\sqrt{3.8} < x < \sqrt{4.2}$ or $1.949 < x < 2.049$. The interval containing these x values extends from $2 - \sqrt{3.8} \approx 0.051$ units to the left of 2 to $\sqrt{4.2} - 2 \approx 0.049$ units to the right of 2. Again picking the smaller of the two distances, "If x is within 0.049 units of 2, then f(x) is within 1 unit of 4."

The situation in Example 2 of different distances on the left and right sides is very common, and we **always** pick our single distance to be the **smaller** of the distances to the left and right. By using the smaller distance, we can be certain that if x is within that smaller distance on either side, then the value of f(x) is within the specified distance of the value of the limit.

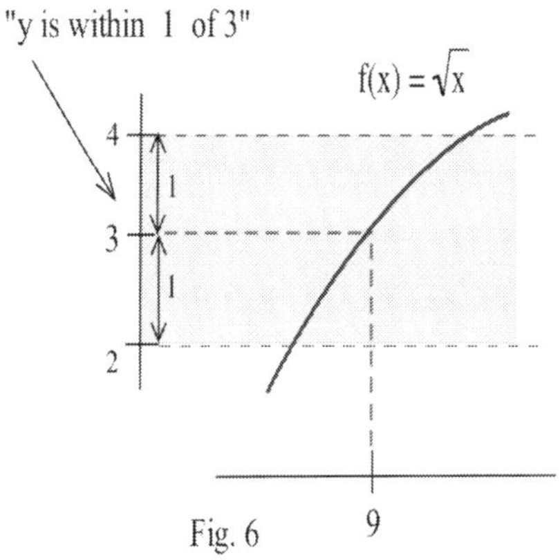

Fig. 6

Practice 2: $\lim_{x \to 9} \sqrt{x} = 3$. What values of x guarantee that $f(x) = \sqrt{x}$ is within 1 unit of 3? Within 0.2 units of 3? (Fig. 6) State each answer in the form "If x is within ______ units of 2, then f(x) is within 1 (or 0.2) unit of 4."

The same ideas can also be used when the function and the specified distance are given graphically, and in that case we can give the answer graphically.

Example 3: In Fig. 7, $\lim_{x \to 2} f(x) = 3$. What values of x guarantee that y = f(x) is within E units (given graphically) of 3? State your answer in the form "If x is within ______ (show a distance D graphically) of 2, then f(x) is within E units of 3."

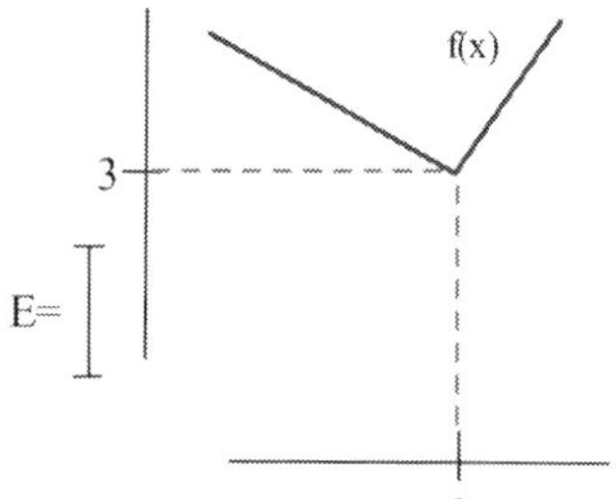

Fig. 7

Solution: The solution process requires several steps as illustrated in Fig. 8:

i. Use the given distance E to find values 3 – E and 3 + E on the y–axis.

ii. Sketch the horizontal band which has its lower edge at y = 3 – E and its upper edge at y = 3 + E.

iii. Find the first locations to the right and left of x = 2 where the graph of y = f(x) crosses the lines y = 3 – E and y = 3 + E, and at these locations draw vertical lines to the x–axis.

iv. On the x–axis, graphically determine the distance from 2 to the vertical line on the left (labeled D_L) and from 2 to the vertical line on the right (labeled D_R).

v. Let the length D be the smaller of the lengths D_L and D_R.

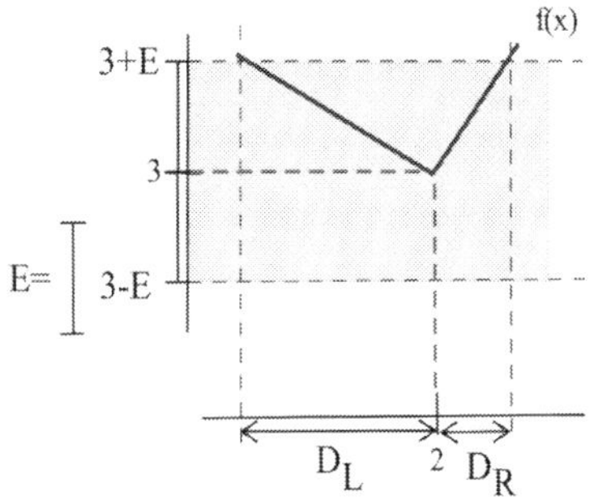

Fig. 8a: steps i and ii

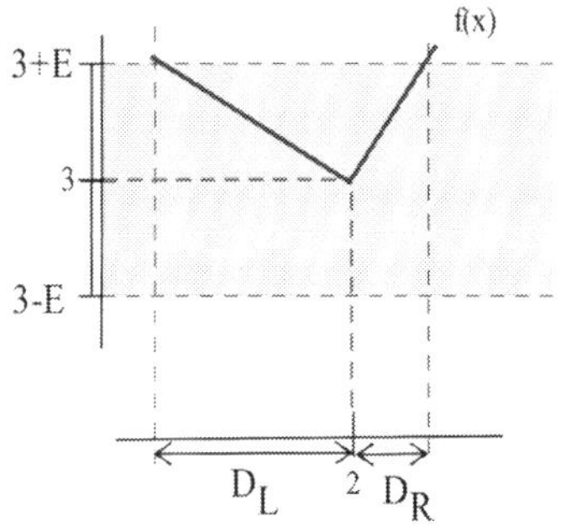

Fig. 8b: steps iii and iv

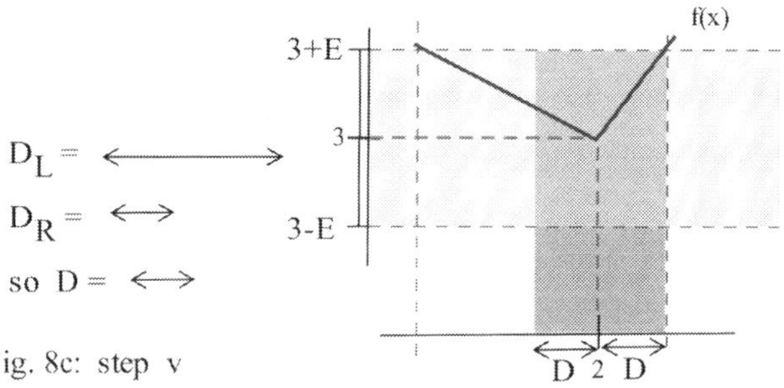

Fig. 8c: step v

Practice 3: In Fig. 9, $\lim_{x \to 3} f(x) = 1.8$. What values of x guarantee that y = f(x) is within E units of 1.8?

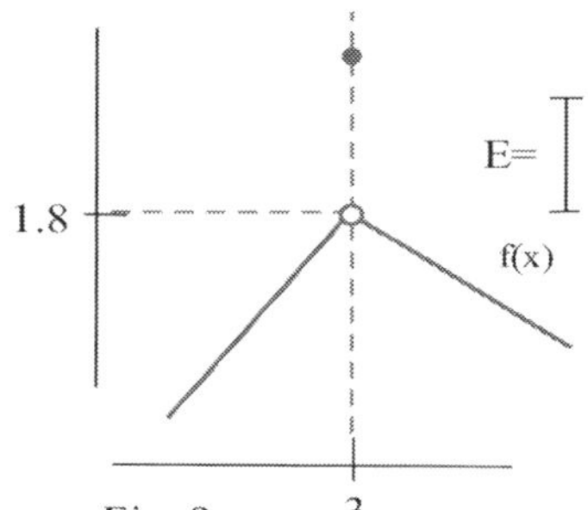

Fig. 9

The Formal Definition of Limit

The ideas of the previous examples and practice problems can be stated for general functions and limits, and they provide the basis for the definition of limit which is given in the box. The use of the lower case Greek letters **ε** (epsilon) and **δ** (delta) in the definition is standard, and this definition is sometimes called the "epsilon–delta" definition of limit.

> **Definition of $\lim_{x \to a} f(x) = L$:**
>
> $\lim_{x \to a} f(x) = L$ means
>
> for every given $\varepsilon > 0$ there is a $\delta > 0$ so that (Fig. 10)
>
> if x is within δ units of a (and $x \neq a$)
>
> then f(x) is within ε units of L.
>
> (Equivalently: $| f(x) - L | < \varepsilon$ whenever $0 < | x - a | < \delta$.)

In this definition, ε represents the given distance on either side of the limiting value y = L, and δ is the distance on each side of the point x = a on the x–axis that we have been finding in the previous examples. This definition has the form of a a "challenge and reponse:" for any positive challenge ε (make f(c) within ε of L), there is a positive response δ (start with x within δ of a and $x \neq a$).

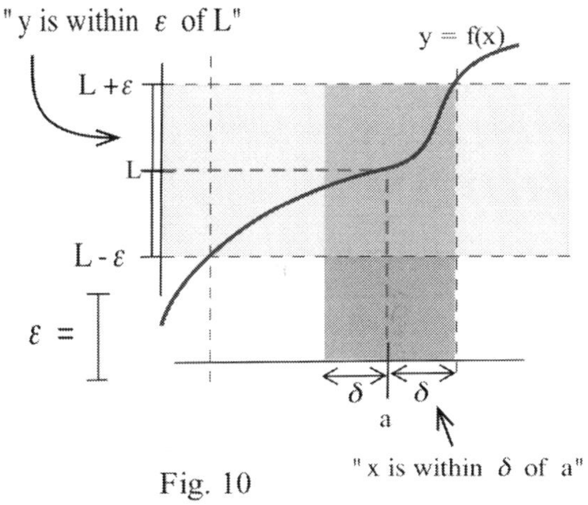

Fig. 10

Example 4: In Fig. 11a, $\lim_{x \to a} f(x) = L$, and a value for ε is given graphically as a length. Find a length for δ that satisfies the definition of limit (so "if x is within δ of a (and $x \neq a$), then f(x) is within ε of L").

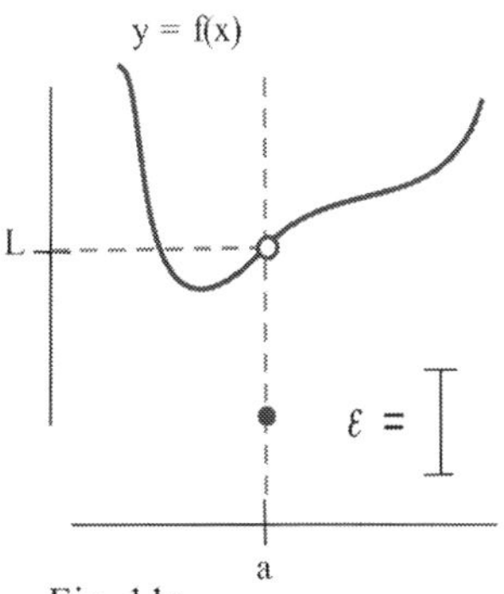

Fig. 11a

Solution: Follow the steps outlined in Example 3. The length for δ is shown in Fig. 11b, and any shorter length for δ also satisfies the definition.

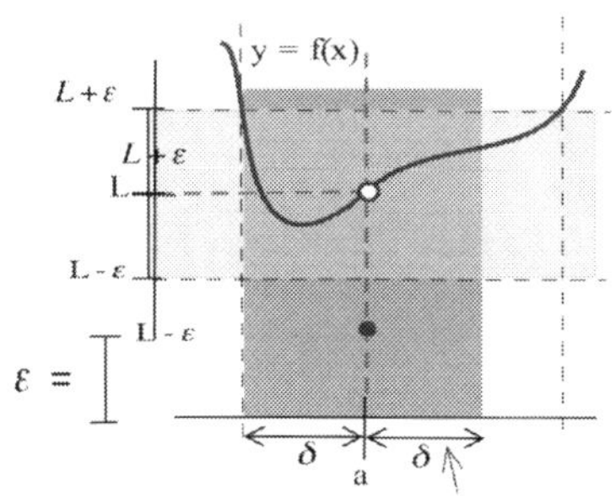

Fig. 11b "x is within δ of a"

Practice 4: In Fig. 12, $\lim_{x \to a} f(x) = L$, and a value for ε is given graphically as a length. Find a length for δ that satisfies the definition of limit

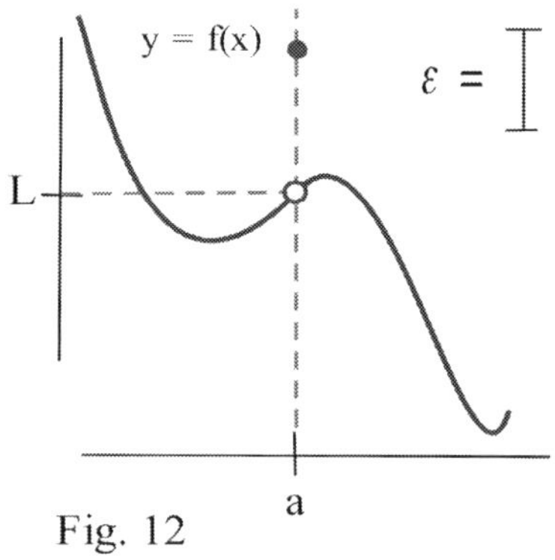

Fig. 12

Example 5: Prove that $\lim_{x \to 3} 4x - 5 = 7$.

Solution: We need to show that

"for every given $\varepsilon > 0$ there is a $\delta > 0$ so that

if x is within δ units of 3 (and $x \neq 3$)

then $4x - 5$ is within ε units of 7."

Actually there are two things we need to do. First, we need to find a value for δ (typically depending on ε), and, second, we need to show that our δ really does satisfy the "if – then" part of the definition.

i. Finding δ is similar to part (c) in Example 1 and Practice 1: assume $4x - 5$ is within ε units of 7 and solve for x. If $7 - \varepsilon < 4x - 5 < 7 + \varepsilon$, then $12 - \varepsilon < 4x < 12 + \varepsilon$ and $3 - \varepsilon/4 < x < 3 + \varepsilon/4$, so x is within $\varepsilon/4$ units of 3. Put $\delta = \varepsilon/4$.

ii. To show that $\delta = \varepsilon/4$ satisfies the definition, we merely reverse the order of the steps in part i.

Assume that x is within δ units of 3. Then $3 - \delta < x < 3 + \delta$ so

$3 - \varepsilon/4 < x < 3 + \varepsilon/4$ (replacing δ with $\varepsilon/4$),

$12 - \varepsilon < 4x < 12 + \varepsilon$ (multiplying by 4), and

$7 - \varepsilon < 4x - 5 < 7 + \varepsilon$ (subtracting 5), so

we can conclude that $f(x) = 4x - 5$ is within ε units of 7. This formally verifies that $\lim_{x \to 3} 4x - 5 = 7$.

Practice 5: Prove that $\lim_{x \to 4} 5x + 3 = 23$.

The method used to prove the values of the limits for these particular linear functions can also be used to prove the following general result about the limits of linear functions.

Theorem: $\lim_{x \to a} mx + b = ma + b$

Proof: Let $f(x) = mx + b$.

Case 1: $m = 0$. Then $f(x) = 0x + b = b$ is simply a constant function, and any value for $\delta > 0$ satisfies the definition. Given any value of $\varepsilon > 0$, let $\delta = 1$ (any positive value for δ works). If x is is within 1 unit of a, then $f(x) - f(a) = b - b = 0 < e$, so we have shown that for any $\varepsilon > 0$, there is a $\delta > 0$ which satisfies the definition.

Case 2: $m \neq 0$. Then $f(x) = mx + b$. For any $\varepsilon > 0$, put $\delta = \frac{\varepsilon}{|m|} > 0$. If x is within $\delta = \frac{\varepsilon}{|m|}$ of a, then

$$a - \frac{\varepsilon}{|m|} < x < a + \frac{\varepsilon}{|m|} \quad \text{so} \quad -\frac{\varepsilon}{|m|} < x - a < \frac{\varepsilon}{|m|} \quad \text{and} \quad |x - a| < \frac{\varepsilon}{|m|} .$$

Then the distance between f(x) and L = ma + b is

$| f(x) - L | = | (mx + b) - (ma + b) | = |m| \cdot | x - a | < |m| \frac{\varepsilon}{|m|} = \varepsilon$

so f(x) is within ε of L = ma + b. (Fig. 13)

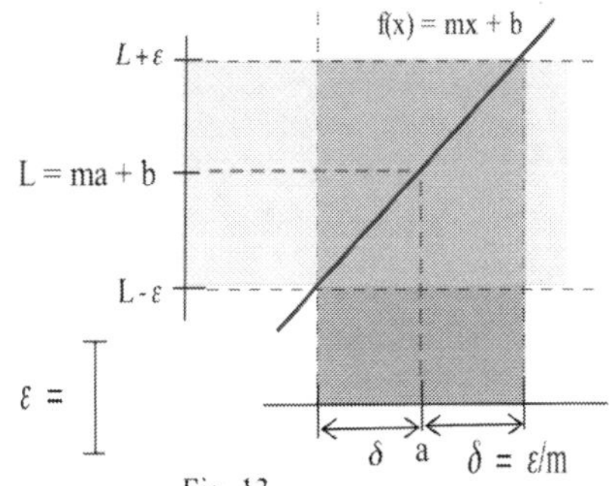

Fig. 13

In each case, we have shown that "given any $\varepsilon > 0$, there is a $\delta > 0$" that satisfies the rest of the definition is satisfied.

If there is even a single value of ε for which there is no δ, then the function does not satisfy the difinition, and we say that the limit "does not exist.:

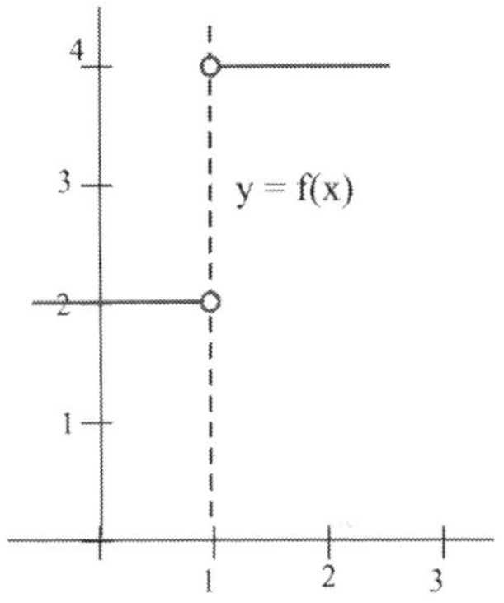

Fig. 14

Example 6: Let $f(x) = \begin{cases} 2 & \text{if } x < 1 \\ 4 & \text{if } x > 1 \end{cases}$ as is shown in Fig. 14 .

Use the definition to prove that $\lim_{x \to 1} f(x)$ does not exist.

Solution: One common proof technique in mathematics is called "proof by contradiction," and that is the method we use here. Using that method in this case, (i) we assume that the limit does exist and equals some number L, (ii) we show that this assumption leads to a contradiction, and (iii) we conclude that the assumption must have been false. Therefore, we conclude that the limit does not exist.

(i) Assume that the limit exists: $\lim_{x \to 1} f(x) = L$ for some value for L. Let $\varepsilon = \frac{1}{2}$. (The definition says "for every ε" so we can pick this value. Why we chose this value for ε shows up later in the proof.) Then, since we are assuming that the limit exists, there is a $\delta > 0$ so that if x is within δ of 1 then f(x) is within ε of L.

(ii) Let x_1 be between 1 and 1 + δ. Then $x_1 > 1$ so $f(x_1) = 4$. Also, x_1 is within δ of 1 so $f(x_1) = 4$ is within $\frac{1}{2}$ of L, and L is between 3.5 and 4.5: $\mathbf{3.5 < L < 4.5}$.

Let x_2 be between 1 and 1 – δ. Then $x_2 < 1$ so $f(x_2) = 2$. Also, x_2 is within δ of 1 so $f(x_2) = 2$ is within $\frac{1}{2}$ of L, and L is between 1.5 and 2.5: $\mathbf{1.5 < L < 2.5}$.

(iii) The two inequalities in bold print provide the contradiction we were hoping to find. There is no value L that simultaneously satisfies $3.5 < L < 4.5$ **and** $1.5 < L < 2.5$, so we can conclude that our assumption was false and that f(x) does not have a limit as $x \to 1$.

Practice 6: Use the definition to prove that $\lim_{x \to 0} \frac{1}{x}$ does not exist (Fig. 15).

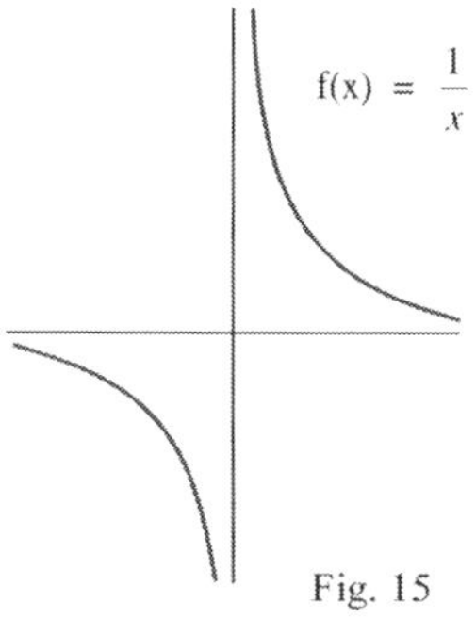

Fig. 15

Proofs of Two Limit Theorems

The theorems and their proofs are included here so you can see how such proofs proceed — you have already used these theorems to evaluate limits of functions.. There are rigorous proofs of all of the other limit properties, but they are somewhat more complicated than the proofs given here.

Theorem: If $\lim_{x \to a} f(x) = L$, then $\lim_{x \to a} k \cdot f(x) = k \cdot L$.

Proof: Case k = 0: The Theorem is true but not very interesting: $\lim_{x \to a} 0 \cdot f(x) = \lim_{x \to a} 0 = 0 \cdot L$.

Case $k \neq 0$: Since $\lim_{x \to a} f(x) = L$, then, by the definition, for every $\varepsilon > 0$ there is a $\delta > 0$ so that $| f(x) - L | < \varepsilon$ whenever $| x - a | < \delta$. For any $\varepsilon > 0$, we know $\frac{\varepsilon}{|k|} > 0$ and pick a value of δ that satisfies $| f(x) - L | < \frac{\varepsilon}{|k|}$ whenever $| x - a | < \delta$. When

$| x - a | < \delta$ ("x is within δ of a") then

$| f(x) - L | < \frac{\varepsilon}{|k|}$ ("f(x) is within $\frac{\varepsilon}{|k|}$ of L") so

$|k| \cdot | f(x) - L | < \varepsilon$ (multiplying each side by $|k| > 0$) and

$| k \cdot f(x) - k \cdot L | < \varepsilon$ ($k \cdot f(x)$ is within ε of $k \cdot L$).

Theorem: If $\lim_{x \to a} f(x) = L$ and $\lim_{x \to a} g(x) = M$, then $\lim_{x \to a} f(x) + g(x) = L + M$.

Proof: Assume that $\lim_{x \to a} f(x) = L$ and $\lim_{x \to a} g(x) = M$. Then, given any $\varepsilon > 0$, we know $\varepsilon/2 > 0$ and that there are deltas for f and g, δ_f and δ_g, so that

if $| x - a | < \delta_f$, then $| f(x) - L | < \varepsilon/2$ ("if x is within δ_f of a, then f(x) is within $\varepsilon/2$ of L", and

if $| x - a | < \delta_g$, then $| g(x) - M | < \varepsilon/2$ ("if x is within δ_g of a, then g(x) is within $\varepsilon/2$ of M").

Let δ be the smaller of δ_f and δ_g. If $| x - a | < \delta$, then $| f(x) - L | < \varepsilon/2$ and $| g(x) - M | < \varepsilon/2$ so

$| (f(x) + g(x)) - (L + M)) | = | (f(x) - L) + (g(x) - M) |$ (rearranging the terms)

$\leq | f(x) - L | + | g(x) - M |$ (by the Triangle Inequality for absolute values)

$< \frac{\varepsilon}{2} + \frac{\varepsilon}{2} = \varepsilon$. (by the definition of the limits for f and g).

Problems

In problems 1–4, state each answer in the form "If x is within _____ units of . . . "

1. $\lim_{x \to 3} 2x + 1 = 7$. What values of x guarantee that $f(x) = 2x + 1$ is (a) within 1 unit of 7?

   (b) within 0.6 units of 7? (c) within 0.04 units of 7? (d) within ε units of 7?

2. $\lim_{x \to 1} 3x + 2 = 5$. What values of x guarantee that $f(x) = 3x + 2$ is within 1 unit of 5 ?

   (b) within 0.6 units of 5? (c) within 0.09 units of 5? (d) within ε units of 5?

3. $\lim_{x \to 2} 4x - 3 = 5$. What values of x guarantee that $f(x) = 4x - 3$ is within 1 unit of 5 ?

   (b) within 0.4 units of 5? (c) within 0.08 units of 5? (d) within ε units of 5?

4. $\lim_{x \to 1} 5x - 3 = 2$. What values of x guarantee that $f(x) = 5x - 3$ is within 1 unit of 2 ?

   (b) within 0.5 units of 5? (c) within 0.01 units of 5? (d) within ε units of 5?

5. For problems 1 – 4, list the slope of each function f and the δ (as a function of ε). For these linear functions f, how is δ related to the slope?

6. You have been asked to cut two boards (exactly the same length after the cut) and place them end to end. If the combined length must be within 0.06 inches of 30 inches, then each board must be within how many inches of 15?

7. You have been asked to cut three boards (exactly the same length after the cut) and place them end to end. If the combined length must be within 0.06 inches of 30 inches, then each board must be within how many inches of 10?

8. $\lim_{x \to 3} x^2 = 9$. What values of x guarantee that $f(x) = x^2$ is within 1 unit of 9? within 0.2 units?

9. $\lim_{x \to 2} x^3 = 8$. What values of x guarantee that $f(x) = x^3$ is within 0.5 unit of 8? within 0.05 units?

10. $\lim_{x \to 16} \sqrt{x} = 4$. What values of x guarantee that $f(x) = \sqrt{x}$ is within 1 unit of 4? Within 0.1 units?

11. $\lim_{x \to 3} \sqrt{1+x} = 2$. What values of x guarantee that $f(x) = \sqrt{1+x}$ is within 1 unit of 2? Within 0.0002 units?

12. You have been asked to cut four pieces of wire (exactly the same length after the cut) and form them into a square. If the area of the square must be within 0.06 inches of 100 inches, then each piece of wire must be within how many inches of 10?

13. You have been asked to cut four pieces of wire (exactly the same length after the cut) and form them into a square. If the area of the square must be within 0.06 inches of 25 inches, then each piece of wire must be within how many inches of 5?

In problems 14 – 17, $\lim_{x \to a} f(x) = L$ and the function f and a value for ε are given graphically. Find a length for δ that satisfies the definition of limit for the given function and value of ε.

14. f and ε as shown in Fig. 16.

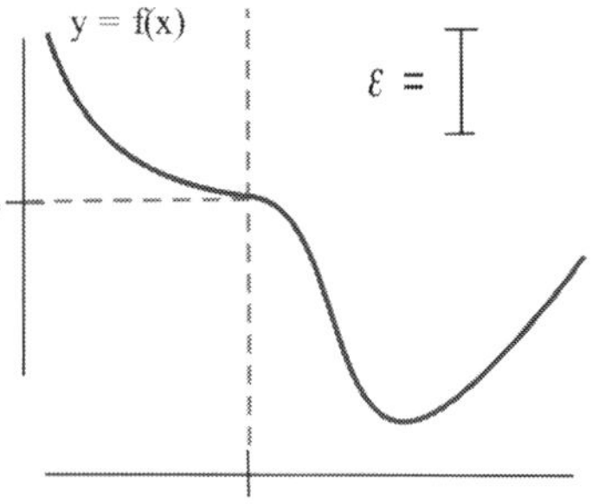

Fig. 16

15. f and ε as shown in Fig. 17.

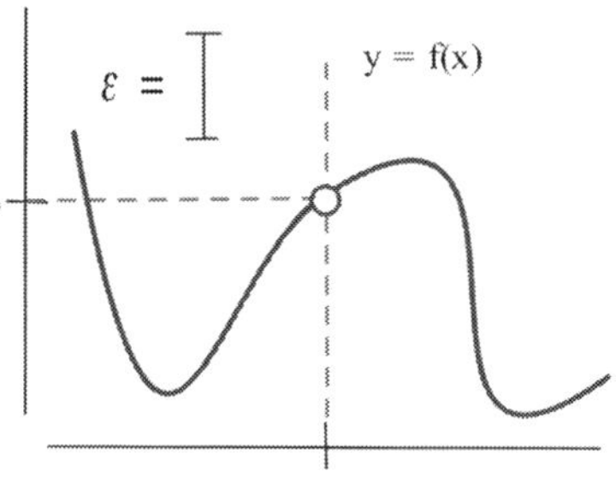

Fig. 17

16. f and ε as shown in Fig. 18.

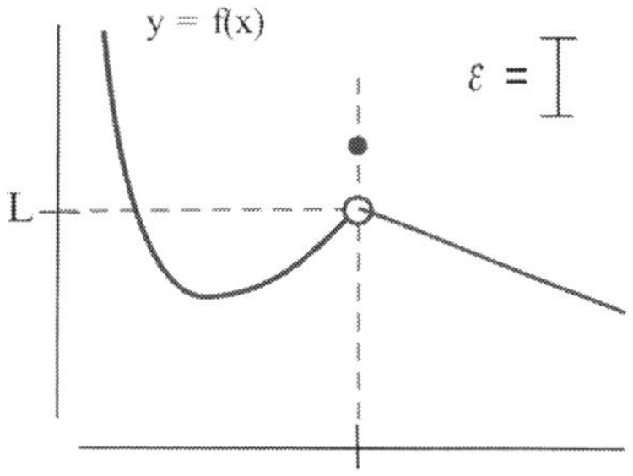

Fig. 18

17. f and ε as shown in Fig. 19.

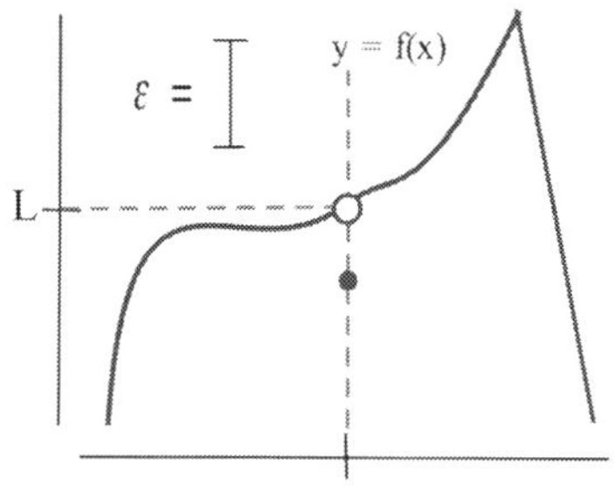

Fig. 19

18. Redo each of problems 14 – 17 taking a new value of ε to be half the value of ε given in the problem.

In problems 19–22, use the definition to prove that the given limit does not exist. (Find a value for ε > 0 for which there is no δ that satisfies the definition.)

19. $f(x) = \begin{cases} 4 & \text{if } x < 2 \\ 3 & \text{if } x > 2 \end{cases}$ as is shown in Fig. 20.

Show $\lim_{x \to 2} f(x)$ does not exist.

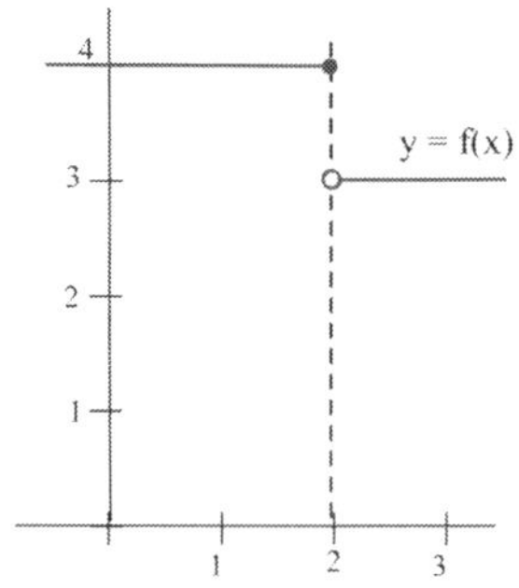

Fig. 20

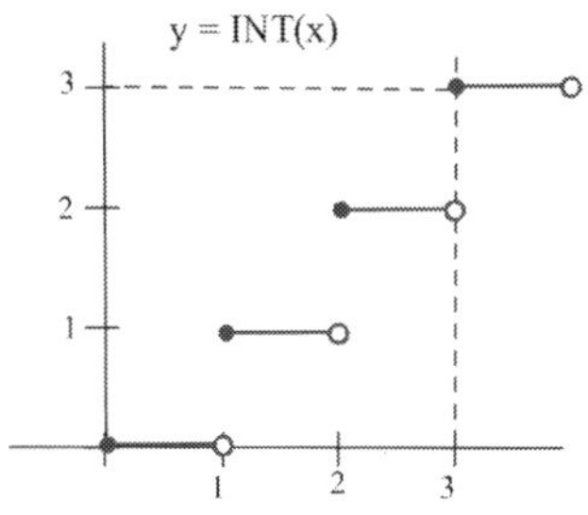

Fig. 21

20. f(x) = INT(x) as is shown in Fig. 21.

Show $\lim_{x \to 3} f(x)$ does not exist.

21. $f(x) = \begin{cases} x & \text{if } x < 2 \\ 6-x & \text{if } x > 2 \end{cases}$. Show $\lim_{x \to 2} f(x)$ does not exist.

22. $f(x) = \begin{cases} x+1 & \text{if } x < 2 \\ x^2 & \text{if } x > 1 \end{cases}$. Show $\lim_{x \to 2} f(x)$ does not exist.

23. Prove: If $\lim_{x \to a} f(x) = L$ and $\lim_{x \to a} g(x) = M$, then $\lim_{x \to a} f(x) - g(x) = L - M$.

Section 1.4 **PRACTICE Answers**

Practice 1: (a) $3 - 1 < 4x - 5 < 3 + 1$ so $7 < 4x < 9$ and $1.75 < x < 2.25$: "x within 1/4 unit of 2."

(b) $3 - 0.08 < 4x - 5 < 3 + 0.08$ so $7.92 < 4x < 8.08$ and $1.98 < x < 2.02$: " x within 0.02 units of 2."

(c) $3 - E < 4x - 5 < 3 + E$ so 8 _ $E < 4x < 8 + E$ and $2 - \frac{E}{4} < x < 2 + \frac{E}{4}$: "x within E/4 units of 2."

Practice 2: "within 1 unit of 3": If $2 < \sqrt{x} < 4$, then $4 < x < 16$ which extends from 5 units to the left of 9 to 7 units to right of 9. Using the **smaller** of these two distances from 9, "If x is within 5 units of 9, then $\sqrt{x}$ is within 1 unit of 3."

"within 0.2 units of 3": If $2.8 < \sqrt{x} < 3.2$, then $7.84 < x < 10.24$ which extends from 1.16 units to the left of 9 to 1.24 units to the right of 9. "If x is within 1.16 units of 9, then $\sqrt{x}$ is wqithin 0.2 units of 3.

Practice 3: See Fig. 22.

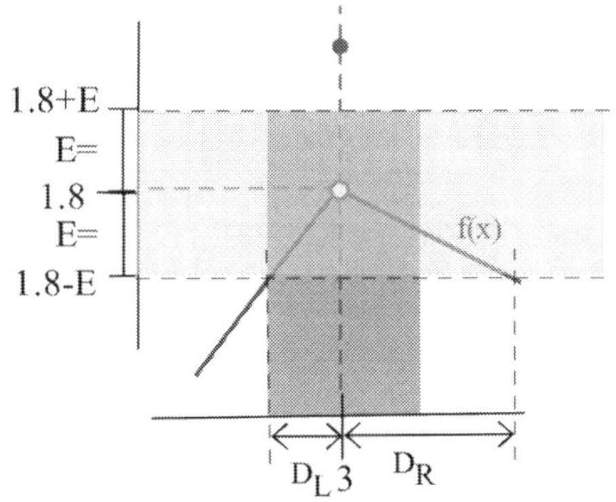

D = smaller of D_L and D_R

D = ⟷

Fig. 22 (based on Fig. 9)

Practice 4: See Fig. 23

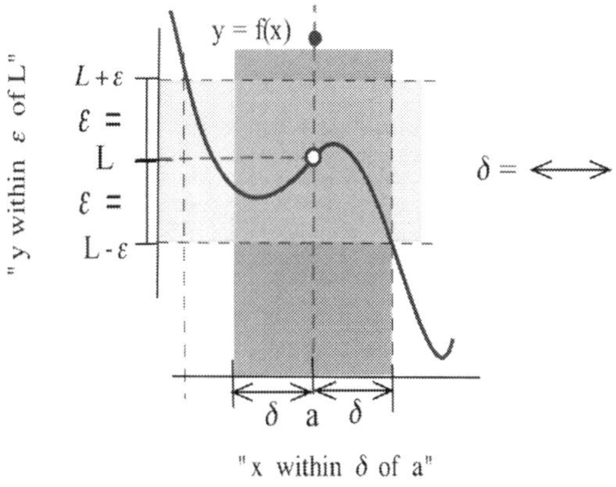

Fig. 23 (based on Fig. 12)

Practice 5: Given any $\varepsilon > 0$, take $\delta = \varepsilon/5$.

If x is within $\delta = \varepsilon/5$ of 4, then

$4 - \varepsilon/5 < x < 4 + \varepsilon/5$ so

$-\varepsilon/5 < x - 4 < \varepsilon/5$ (subtracting 4)

$-\varepsilon < 5x - 20 < \varepsilon$ (multiplying by 5)

$-\varepsilon < (5x + 3) - 23 < \varepsilon$ (rearranging to get the form we want)

so, finally, $f(x) = 5x + 3$ is within ε of $L = 23$.

We have shown that "for any $\varepsilon > 0$, there is a $\delta > 0$ (namely $\delta = \varepsilon/5$)" so that the rest of the definition is satisfied.

Practice 6: This is a much more sophisticated (= harder) problem.

Using "proof by contradiction" as outlined in the solution to Example 6.

(i) Assume that the limit exists: $\lim_{x \to 0} \frac{1}{x} = L$ for some value for L. Let $\varepsilon = 1$. (The definition says "for every ε" so we can pick this value. For this limit, the definition fails for every $\varepsilon > 0$.) Then, since we are assuming that the limit exists, there is a $\delta > 0$ so that if x is within δ of 0 then $f(x) = \frac{1}{x}$ is within $\varepsilon = 1$ of L.

(ii) (See Fig. 24) Let x_1 be between 0 and $0 + \delta$ and also require that $x_1 < \frac{1}{2}$. Then $0 < x_1 < \frac{1}{2}$ so $f(x_1) = \frac{1}{x_1} > 2$. Since x_1 is within δ of 0, $f(x_1) > 2$ is within $\varepsilon = 1$ of L, so L is greater than $2 - \varepsilon = 1$: $\mathbf{1 < L}$.

Let x_2 be between 0 and $0 - \delta$ and also require that $x_2 > -\frac{1}{2}$. Then $0 > x_2 > \frac{1}{2}$ so $f(x_2) = \frac{1}{x_2} < -2$. Since x_2 is within δ of 0, $f(x_2) < -2$ is within $\varepsilon = 1$ of L, so L is less than $-2 + \varepsilon = -1$: $\mathbf{-1 > L}$.

(iii) The two inequalities in bold print provide the contradiction we were hoping to find. There is no value L that satisfies

BOTH **1 < L and L < –1**,

so we can conclude that our assumption was false and that $f(x) = \frac{1}{x}$ does not have a limit as $x \to 0$.

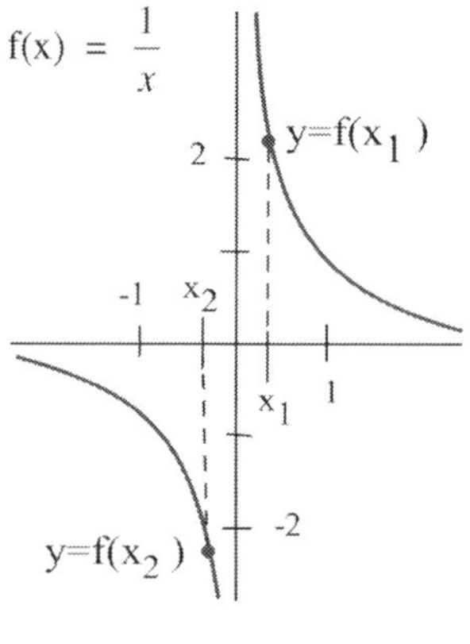

Fig. 23 (based on Fig. 15)

Chapter One

Section 1.0

1. $m = \frac{y-9}{x-3}$. If x = 2.97, then $m = \frac{-0.1791}{-0.03} = 5.97$. If x = 3.001, then $m = \frac{0.006001}{0.001} = 6.001$.

   If x = 3 + h, then $m = \frac{(3+h)^2 - 9}{(3+h) - 3} = \frac{9 + 6h + h^2 - 9}{h} = 6 + h$. When h is very small (close to 0), 6 + h is very close to 6.

3. $m = \frac{y-4}{x-2}$. If x = 1.99, then $m = \frac{-0.0499}{-0.01} = 4.99$. If x = 2.004, then $m = \frac{0.020016}{0.004} = 5.004$.

   If x = 2 + h, then $m = \frac{\{(2+h)^2 + (2+h) - 2\} - 4}{(2+h) - 2} = \frac{4+4h+h^2 + 2+h - 2 - 4}{h} = 5 + h$. When h is very small, 5 + h is very close to 5.

5. All of these answers are **approximate**. Your answers should be close to these numbers.

   (a) average rate of temperature change $\approx \frac{80^o - 64^o}{1\text{ pm} - 9\text{ am}} = \frac{16^o}{4\text{ hours}} = 4^o$ per hour.

   (b) at 10 am, temperature was rising about 5° per hour.

   at 7 pm, temperature was rising about -10° per hour (**falling** about 10° per hour).

7. All of these answers are **approximate**. Your answers should be close to these numbers.

   (a) average velocity $\approx \frac{300\text{ ft} - 0\text{ ft}}{20\text{ sec} - 0\text{ sec}} =$ 15 feet per second.

   (b) average velocity $\approx \frac{100\text{ ft} - 200\text{ ft}}{30\text{ sec} - 10\text{ sec}} =$ –5 feet per second.

   (c) at t = 10 seconds, velocity ≈ 30 feet per second (between 20 and 35 ft/s).

   at t = 20 seconds, velocity ≈ –1 feet per second.

   at t = 30 seconds, velocity ≈ –40 feet per second.

9. (a) A(0) = 0, A(1) = 3, A(2) = 6, A(2.5) = 7.5, A(3) = 9.

   (b) the area of the rectangle bounded below by the x–axis, above by the line y = 3, on the left by the vertical line x = 1, and on the right by the vertical line x = 4.

   (c) Graph of y = A(x) = 3x.

Section 1.1

1. (a) 2 (b) 1 (c) DNE (does not exist) (d) 1

3. (a) 1 (b) –1 (c) –1 (d) 2

5. (a) –7 (b) (13/0) DNE

7. (a) 0.54 (remember, we are using radian mode) (b) –0.318 (c) –0.54

9. (a) 0 (b) 0 (c) 0 10. (a) –1 (b) +1 (c) DNE (does not exist)

11. (a) 0 (b) –1 (c) DNE

13. $\lim_{h\to 0^-} g(x) = 1$ $\lim_{x\to 0^+} g(x) = 1$ $\lim_{h\to 0} g(x) = 1$

    $\lim_{h\to 2^-} g(x) = 1$ $\lim_{x\to 2^+} g(x) = 4$ $\lim_{h\to 2} g(x)$ does not exist

$\lim_{h \to 4^-} g(x) = 2$ $\lim_{x \to 4^+} g(x) = 2$ $\lim_{h \to 4} g(x) = 2$

$\lim_{h \to 5^-} g(x) = 1$ $\lim_{x \to 5^+} g(x) = 1$ $\lim_{h \to 5} g(x) = 1$

15. (a) 1.0986 (b) 1 17. (a) 0.125 (b) 3.5

19. (a) A(0) = 0, A(1) = 2.25, A(2) = 5, A(3) = 8.25
 (b) $A(x) = 2x + x^2/4$
 (c) the area of the region bounded below by the x–axis, above by the line y = x/2 + 2, on the left by the vertical line x = 1, and on the right by the vertical line x = 3.

Section 1.2

1. (a) 2 (b) 0
 (c) DNE (does not exist) (d) 1.5

3. (a) 1 (b) 3 (c) 1 (d) ≈ 0.8

5. See Fig. 1.2P5 .

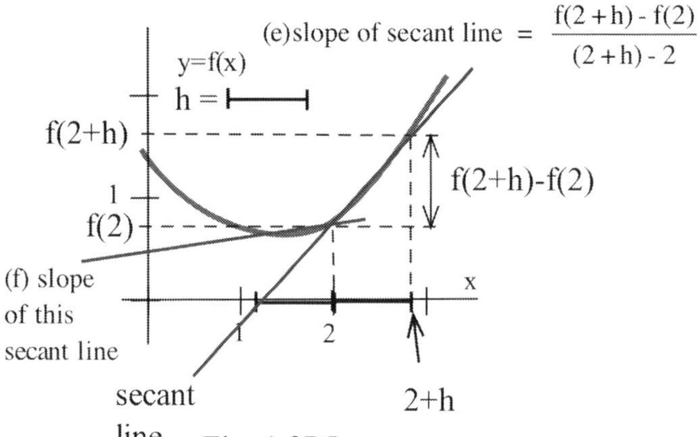

Fig. 1.2P5

7. (a) 2 (b) –1 (c) DNE (d) 2
 (e) 2 (f) 2 (g) 1 (h) 2 (i) DNE

9. (a) When v = 0, L = A .

 (b) $\lim_{v \to c^-} A\sqrt{1 - \frac{v^2}{c^2}} = 0$

11. (a) 4 (b) 1 (c) 2 (d) 0 (e) 1 (f) 1

13. (a) Slope of the line tangent to the graph of y = cos(x) at the point (0,1). (b) Slope = 0.

15. (a) ≈ 1 (b) ≈ 3.43 (c) ≈ 4

17. at x = –1: a at x = 0: b at x = 1: c at x = 2: d
 at x = 3: c at x = 4: b at x = 5: a

19. Verify each step.

21. Several different lists will work. Here is one example.

 Put $a_n = 1/(n\pi)$ for n = 1, 2, 3, ... so a_n approaches 0 and $\sin(a_n) = \sin(\frac{1}{1/(n\pi)}) = \sin(n\pi) = 0$ for all n.

 Put $bn = \frac{1}{2n\pi + \pi/2}$ for n = 1, 2, 3, ... so bn approaches 0 and
 sin(bn) = sin(2nπ + π/2) = sin(π/2) = 1 for all n.

 Therefore, $\lim_{h \to 0} \sin(1/x)$ does not exist.

Section 1.3

1. Discontinuous at 1, 3, and 4.
3. (a) Discontinuous at $x = 3$. Fails condition (i) there.
 (b) Discontinuous at $x = 2$. Fails condition (i) there.
 (c) Discontinuous where $\cos(x)$ is negative, (e.g., at $x = \pi$). Fails condition (i) there.
 (d) Discontinuous where x^2 is an integer (e.g., at $x = 1$ or $\sqrt{2}$). Fails condition (ii) there.
 (e) Discontinuous where $\sin(x) = 0$ (e.g., at $x = 0, \pm\pi, \pm 2\pi, ...$). Fails condition (i) there.
 (f) Discontinuous at $x = 0$. Fails condition (i) there.
 (g) Discontinuous at $x = 0$. Fails condition (i) there.
 (h) Discontinuous at $x = 3$. Fails condition (i) there.
 (i) Discontinuous at $x = \pi/2$. Fails condition (i) there.

5. (a) $f(x) = 0$ for at least 3 values of x, $0 \le x \le 5$.
 (b) 1 (c) 3 (d) 2 (e) Yes. It does not have to happen, but it is possible.

7. (a) $f(0) = 0, f(3) = 9$ and $0 \le 2 \le 9$. $c = \sqrt{2} \approx 1.414$
 (b) $f(-1) = 1, f(2) = 4$ and $1 \le 3 \le 4$. $c = \sqrt{3} \approx 1.732$
 (c) $f(0) = 0, f(\pi/2) = 1$ and $0 \le 1/2 \le 1$. $c =$ (inverse sine of 1/2) ≈ 0.524
 (d) $f(0) = 0, f(1) = 1$ and $0 \le 1/3 \le 1$. $c = 1/3$
 (e) $f(2) = 2, f(5) = 20$ and $2 \le 4 \le 20$. $c = (1 + \sqrt{17})/2 \approx 2.561$.
 (f) $f(1) = 0, f(10) \approx 2.30$ and $0 \le 2 \le 2.30$. $c =$ (inverse of $\ln(2)$) $= e^2 \approx 7.389$.

9. Neither student is correct. The bisection algorithm converges to the root labeled C.

11. (a) D
 (b) D
 (c) hits B

13. $[-0.9375, -0.875], \approx -0.879$
 $[1.3125, 1.375], \approx 1.347$
 $[2.5, 2.5625], \approx 2.532$

15. $[2.3125, 2.375], \approx 2.32$.

17. $[-0.375, -0.3125], \approx -0.32$.

19. See the three graphs in Fig. 1.3P19.

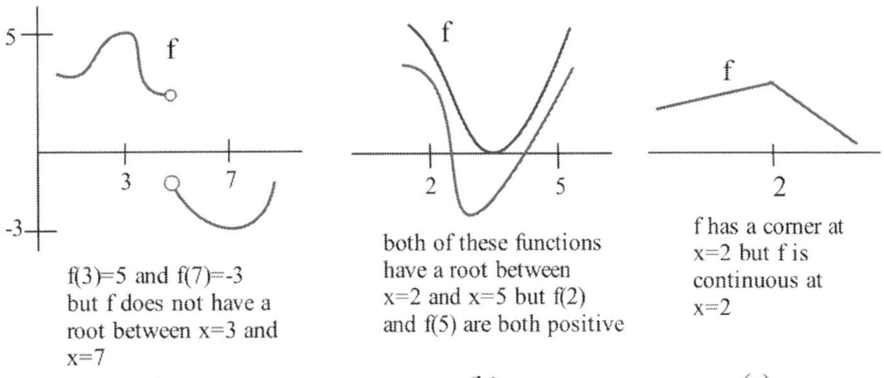
5
-3
f
3
7
f(3)=5 and f(7)=-3
but f does not have a
root between x=3 and
x=7
f
2
5
both of these functions
have a root between
x=2 and x=5 but f(2)
and f(5) are both positive
f
2
f has a corner at
x=2 but f is
continuous at
x=2

(a) (b) (c)

Fig. 1.3P19

21. (a) $A(2.1) - A(2)$ is the area of the region bounded below by the x–axis, above by the graph of f, on the left by the vertical line $x = 2$, and on the right by the vertical line $x = 2.1$.

$$\frac{A(2.1) - A(2)}{0.1} \approx f(2) \text{ or } f(2.1) \text{ so } \frac{A(2.1) - A(2)}{0.1} \approx 1.$$

 (b) $A(4.1) - A(4)$ is the area of the region bounded below by the x–axis, above by the graph of f, on the left by the vertical line $x = 4$, and on the right by the vertical line $x = 4.1$. $\dfrac{A(4.1) - A(4)}{0.1} \approx f(4) \approx 2$.

23. (a) Yes. You supply the justification. (b) Yes (c) Try it.

2.0 INTRODUCTION TO DERIVATIVES

PREVIEW OF CHAPTER 2

The two previous chapters have laid the foundation for the study of calculus. They provided a review of some material you will need, and they started to emphasize the various ways we will need to view and use functions: functions given by graphs, equations, and tables of values.

Chapter 2 will focus on the idea of tangent lines. We will get a definition for the derivative of a function and calculate the derivatives of some functions using this definition. Then we will examine some of the properties of derivatives, see some relatively easy ways to calculate the derivatives, and begin to look at some ways we can use derivatives. Chapter 2 will emphasize what derivatives are, how to calculate them, and some of their applications.

This section begins with a very graphical approach to slopes of tangent lines. It then examines the problem of finding the slopes of the tangent lines for a single function, $y = x^2$, in some detail, and illustrates how these slopes can help us solve fairly sophisticated problems.

Slopes of Tangent Lines: Graphically

Fig. 1 is the graph of a function $y = f(x)$. We can use the information in the graph to fill in the table:

x	y = f(x)	m(x) = the **estimated** SLOPE of the tangent line to y=f(x) at the point (x,y)
0	**0**	**1**
1	**1**	**0**
2	**0**	**– 1**
3	**– 1**	**0**
4	**1**	**1**
5	**2**	**1/2**

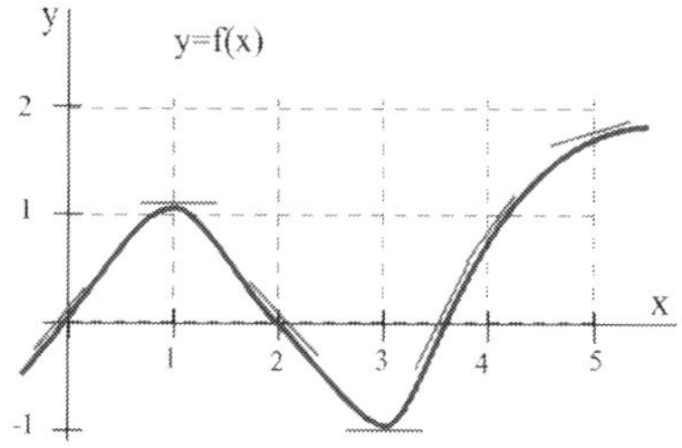

Fig. 1

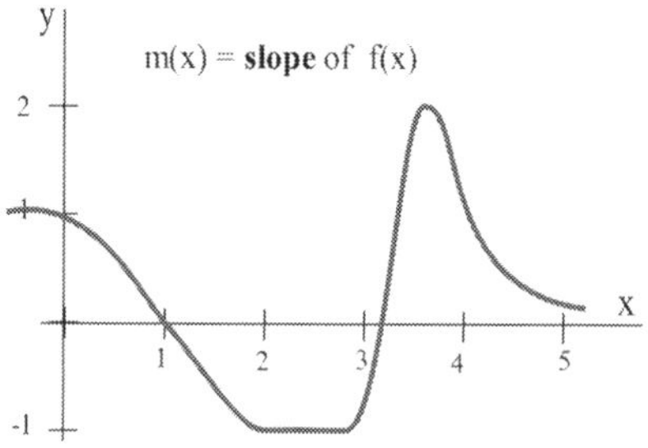

Fig. 2

We can estimate the values of $m(x)$ at some non-integer values of x, $m(.5) \approx 0.5$ and $m(1.3) \approx -0.3$, and even over entire intervals, if $0 < x < 1$, then $m(x)$ is positive.

The values of $m(x)$ definitely depend on the values of x , and $m(x)$ is a function of x. We can use the results in the table to help sketch the graph of $m(x)$ in Fig. 2.

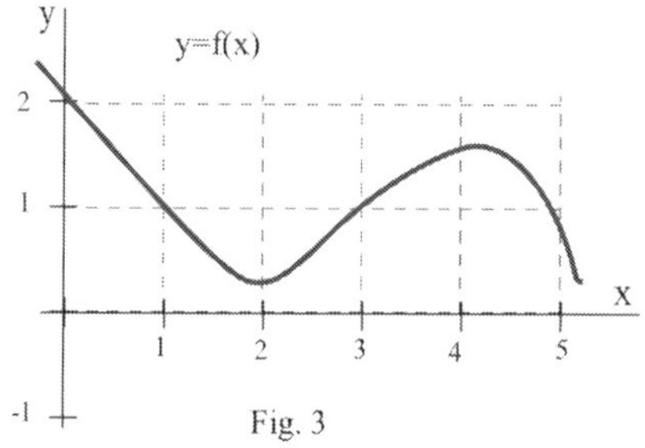

Fig. 3

Practice 1: The graph of $y = f(x)$ is given in Fig. 3. Set up a table of values for x and $m(x)$ (the slope of the line tangent to the graph of y=f(x) at the point (x,y)) and then graph the function $m(x)$.

In some applications, we need to know where the graph of a function f(x) has horizontal tangent lines (slopes = 0). In Fig. 3, the slopes of the tangent lines to graph of $y = f(x)$ are 0 when $x = 2$ or $x \approx 4.5$.

Practice 2: At what values of x does the graph of $y = g(x)$ in Fig. 4 have horizontal tangent lines?

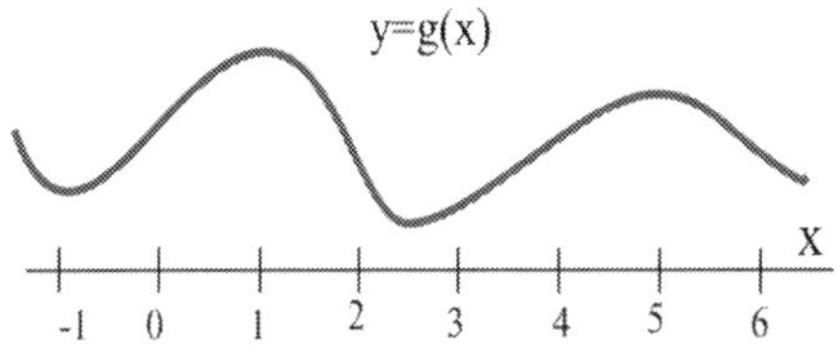

Fig. 4

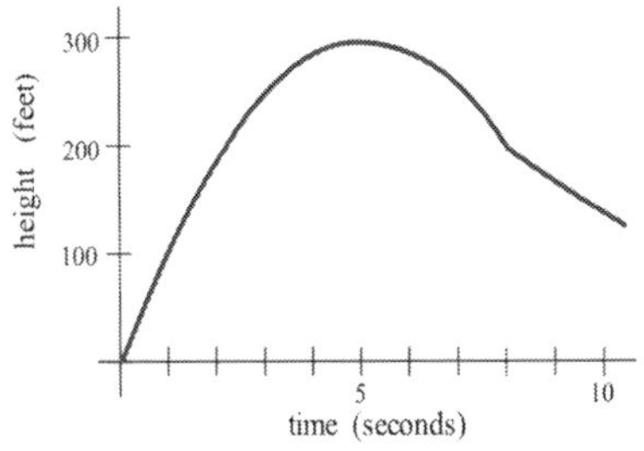

Fig. 5

Example 1: Fig. 5 is the graph of the height of a rocket at time t. Sketch the graph of the **velocity** of the rocket at time t. (Velocity is the **slope of the tangent** to the graph of position or height.)

Solution: The lower graph in Fig. 6 shows the velocity of the rocket.

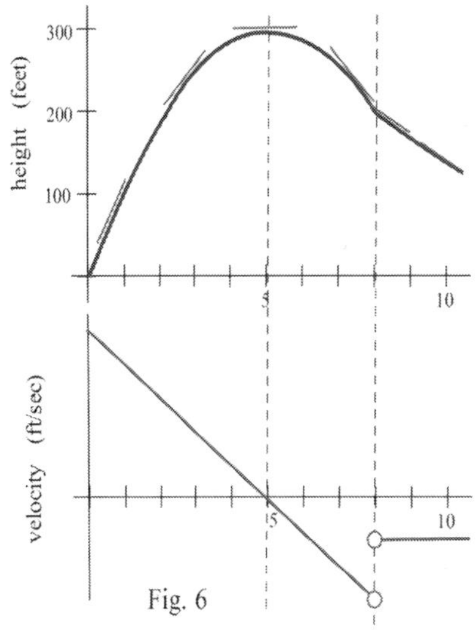

Fig. 6

Practice 3: Fig. 7 shows the temperature during a summer day in Chicago. Sketch the graph of the **rate** at which the temperature is changing. (This is just the graph of the **slopes** of the lines which are tangent to the temperature graph.)

The function m(x) , the slope of the line tangent to the graph of f(x), is called the **derivative of f(x)** . We have used the idea of the slope of the tangent line throughout Chapter 1. In the Section 2.1, we will formally define the derivative of a function and begin to examine some of the properties of the derivative function, but first lets see what we can do when we have a formula for f(x).

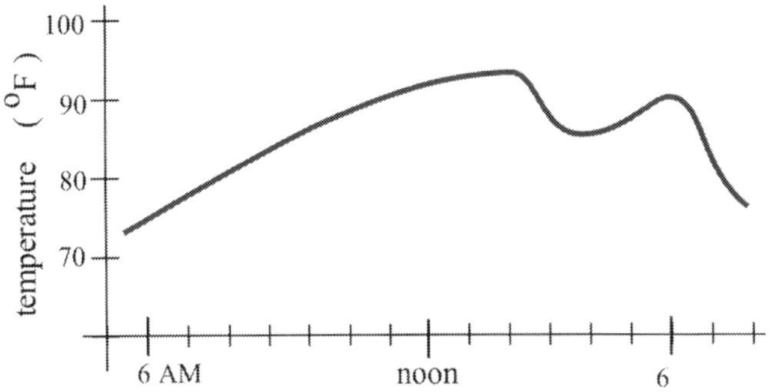

Fig. 7

Tangents to $y = x^2$

When we have a formula for a function, we can determine the slope of the tangent line at a point (x, f(x)) by calculating the slope of the secant line through the points (x, f(x)) and (x+h, f(x+h)),

$m_{sec} = \frac{f(x+h) - f(x)}{(x+h) - (x)}$, and then taking the limit of m_{sec} as h approaches 0 (Fig. 8) :

$$m_{tan} = \lim_{h\to 0} m_{sec} = \lim_{h\to 0} \frac{f(x+h) - f(x)}{(x+h) - (x)} .$$

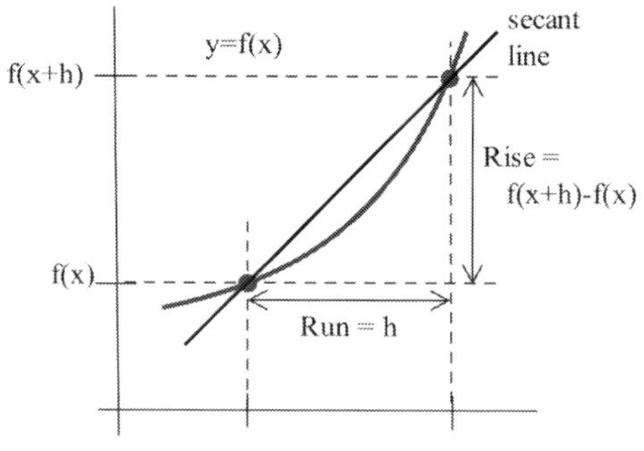

Fig. 8

Example 2: Find the **slope** of the line tangent to the graph of $y = f(x) = x^2$ at the point (2,4). (Fig. 9).

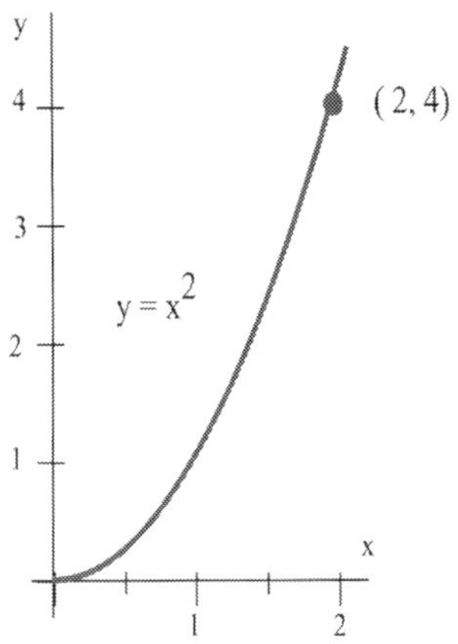

Fig. 9

Solution: In this example x = 2 , so x + h = 2 + h and $f(x + h) = f(2+h) = (2+h)^2$.

The slope of the tangent line at (2,4) is

$$m_{tan} = \lim_{h\to 0} m_{sec} = \lim_{h\to 0} \frac{f(2+h) - f(2)}{(2+h) - (2)}$$

$$= \lim_{h\to 0} \frac{(2+h)^2 - (2)^2}{(2+h) - (2)} = \lim_{h\to 0} \frac{4 + 4h + h^2 - 4}{h}$$

$$= \lim_{h\to 0} \frac{4h + h^2}{h} = \lim_{h\to 0} \frac{h(4+h)}{h} = \lim_{h\to 0} (4+h) = 4 .$$

The tangent line to the graph of $y = x^2$ at the point (2,4) has slope 4 .

We can use the point–slope formula for a line to find the equation of the tangent line:
$y - y_0 = m(x - x_0)$ so $y - 4 = 4(x - 2)$ and $y = 4x - 4$.

Practice 4: Use the method of Example 2 to show that the **slope** of the line tangent to the graph of $y = f(x) = x^2$ at the point (1,1) is $m_{tan} = 2$. Also find the values of m_{tan} at (0,0) and (–1,1).

It is possible to find the slopes of the tangent lines one point at a time, but that is not very efficient. You should have noticed in the Practice 4 that the algebra for each point was very similar, so let's do all the work once for an arbitrary point $(x, f(x)) = (x, x^2)$ and then use the general result for our particular problems. The slope of the line tangent to the graph of $y = f(x) = x^2$ at the arbitrary point (x, x^2) is

$$m_{tan} = \lim_{h\to 0} \frac{f(x+h) - f(x)}{(x+h) - (x)} = \lim_{h\to 0} \frac{(x+h)^2 - (x)^2}{(x+h) - (x)} = \lim_{h\to 0} \frac{x^2 + 2xh + h^2 - x^2}{h}$$

$$= \lim_{h\to 0} \frac{2xh + h^2}{h} = \lim_{h\to 0} \frac{h(2x+h)}{h} = \lim_{h\to 0} (2x+h) = \mathbf{2x} .$$

The **slope** of the line tangent to the graph of $y = f(x) = x^2$ at the point (x, x^2) is $m_{tan} = \mathbf{2x}$. We can use this general result at any value of x without going through all of the calculations again. The slope of the line tangent to $y = f(x) = x^2$ at the point $(4, 16)$ is $m_{tan} = 2(4) = 8$ and the slope at (π, π^2) is $m_{tan} = 2(\pi) = 2\pi$. The value of x determines where we are on the curve (at $y = x^2$) as well as the slope of the tangent line, $m_{tan} = 2x$, at that point. The slope $m_{tan} = 2x$ is a **function** of x and is called the **derivative of $y = x^2$**.

Simply knowing that the slope of the line tangent to the graph of $y = x^2$ is $m_{tan} = 2x$ at a point (x,y) can help us quickly find the equation of the line tangent to the graph of $y = x^2$ at any point and answer a number of difficult–sounding questions.

Example 3: Find the equations of the lines tangent to $y = x^2$ at $(3, 9)$ and (p, p^2).

Solution: At $(3, 9)$, the slope of the tangent line is $2x = 2(3) = 6$, and the equation of the line is

$y - y_0 = m(x - x_0)$ so $y - 9 = 6(x - 3)$ and $y = 6x - 9$.

At (p, p^2), the slope of the tangent line is $2x = 2(p) = 2p$, and the equation of the line is

$y - y_0 = m(x - x_0)$ so $y - p^2 = 2p(x - p)$ and $y = 2px - p^2$.

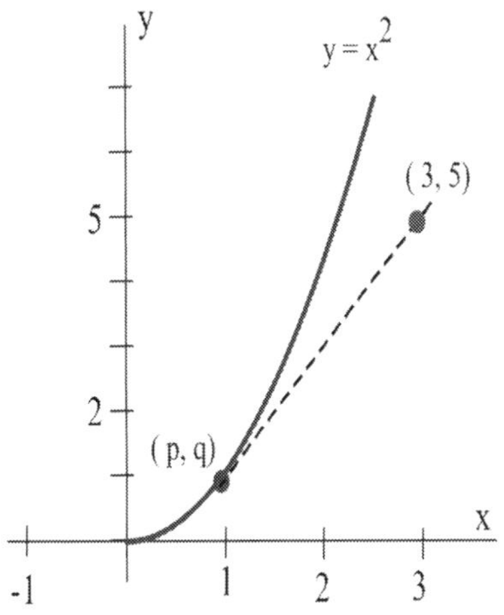

Fig. 10

Example 4: A rocket has been programmed to follow the path $y = x^2$ in space (from left to right along the curve), but an emergency has arisen and the crew must return to their base which is located at coordinates (3,5). At what point on the path $y = x^2$ should the captain turn off the engines so the ship will coast along the tangent to the curve to return to the base? (Fig. 10)

Solution: You might spend a few minutes trying to solve this problem without using the relation $m_{tan} = 2x$, but the problem is much easier if we do use that result.

Lets assume that the captain turns off the engine at the point (p,q) on the curve $y = x^2$, and then try to determine what values p and q must have so that the resulting tangent line to the curve will go through the point (3,5). The point (p,q) is on the curve $y = x^2$, so $q = p^2$, and the equation of the tangent line, found in Example 3, is $y = 2px - p^2$.

To find the value of p so that the tangent line will go through the point (3,5), we can substitute the values $x = 3$ and $y = 5$ into
the equation of the tangent line and solve for p :

$y = 2px - p^2$ so $5 = 2p(3) - p^2$ and $p^2 - 6p + 5 = 0$.

The only solutions of $p^2 - 6p + 5 = (p - 1)(p - 5) = 0$ are $p = 1$ and $p = 5$, so the only possible points are (1,1) and (5,25). You can verify that the tangent lines to $y = x^2$ at (1,1) and (5,25) go through the base at the point (3,5) (Fig. 11). Since the ship is moving from left to right along the curve, the captain should turn off the engines at the point (1,1). Why not at (5,25)?

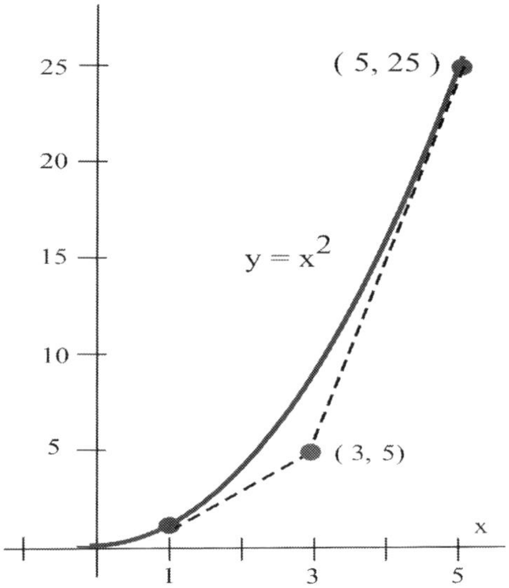

Fig. 11

Practice 5: Verify that if the rocket engines in Example 4 are shut off at (2,4), then the rocket will go through the point (3,8).

PROBLEMS

1. Use the function in Fig. 12 to fill in the table and then graph m(x).

x	y = f(x)	m(x) = the estimated slope of the tangent line to y=f(x) at the point (x,y)
0		
0.5		
1.0		
1.5		
2.0		
2.5		
3.0		
3.5		
4.0		

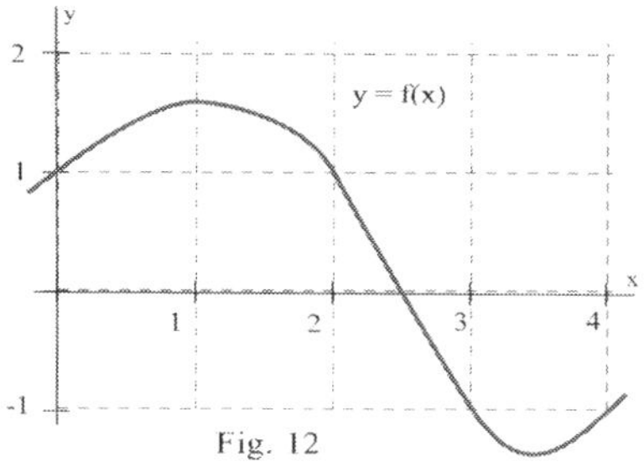

Fig. 12

2. Use the function in Fig. 13 to fill in the table and then graph m(x).

x	y = g(x)	m(x) = the estimated slope of the tangent line to y=g(x) at the point (x,y)
0		
0.5		
1.0		
1.5		
2.0		
2.5		
3.0		
3.5		
4.0		

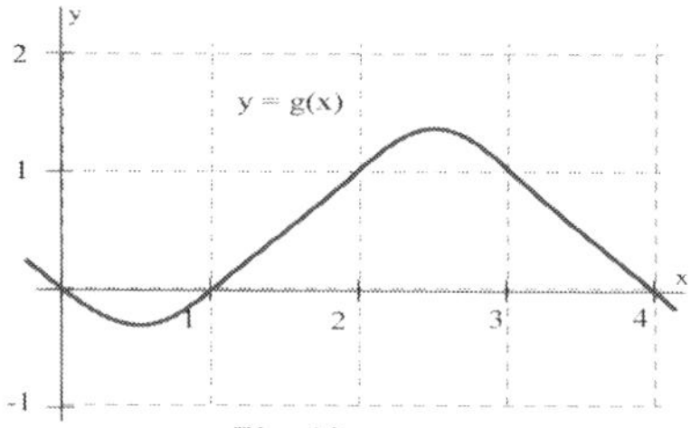

Fig. 13

3. (a) At what values of x does the graph of f in Fig. 14 have a horizontal tangent line?
   (b) At what value(s) of x is the value of f the largest? smallest?
   (c) Sketch the graph of m(x) = the slope of the line tangent to the graph of f at the point (x,y).

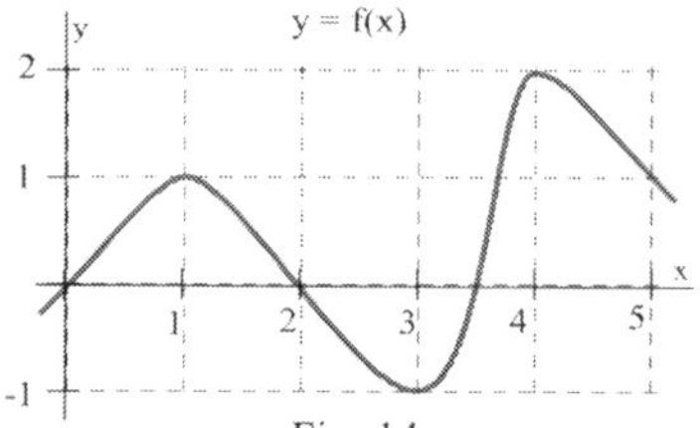

Fig. 14

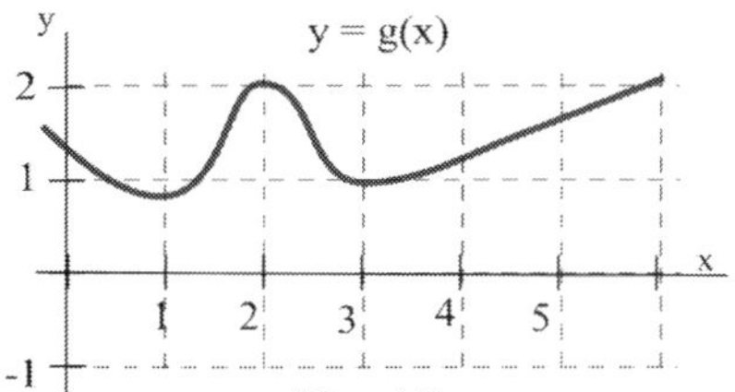

Fig. 15

4. (a) At what values of x does the graph of g in Fig. 15 have a horizontal tangent line?
   (b) At what value(s) of x is the value of g the largest? smallest?
   (c) Sketch the graph of m(x) = the slope of the line tangent to the graph of g at the point (x,y).

5. (a) Sketch the graph of f(x) = sin(x) for $-3 \le x \le 10$.
   (b) Sketch the graph of m(x) = slope of the line tangent to the graph of sin(x) at the point (x, sin(x)).
   (c) Your graph in part (b) should look familiar. What function is it?

6. Match the situation descriptions with the corresponding **time–velocity** graphs in Fig. 16.
   (a) A car quickly leaving from a stop sign.
   (b) A car sedately leaving from a stop sign.
   (c) A student bouncing on a trampoline.
   (d) A ball thrown straight up.
   (e) A student confidently striding across campus to take a calculus test.
   (f) An unprepared student walking across campus to take a calculus test.

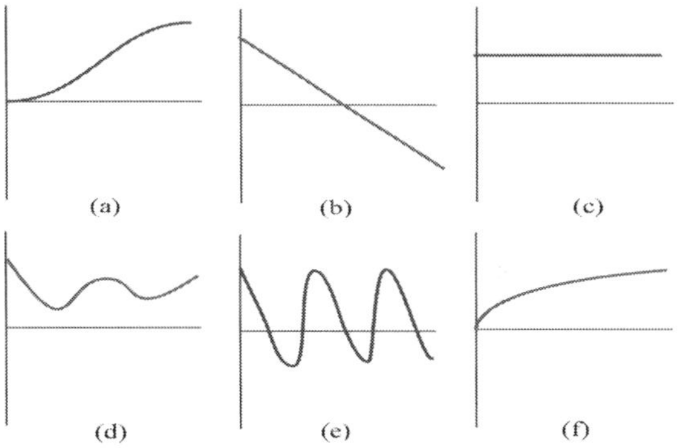

Fig. 16

Problems 7 – 10 assume that a rocket is following the path $y = x^2$, from left to right.

7. At what point should the engine be turned off in order to coast along the tangent line to a base at (5,16)?

8. At what point should the engine be turned off in order to coast along the tangent line to a base at (3,–7)?

9. At what point should the engine be turned off in order to coast along the tangent line to a base at (1,3)?

10. Which points in the plane can not be reached by the rocket? Why not?

For each function f(x) in problems 11 – 16, perform steps (a) – (d):

(a) calculate $m_{sec} = \dfrac{f(x+h) - f(x)}{(x+h) - (x)}$ and simplify (b) determine $m_{tan} = \lim_{h \to 0} m_{sec}$

(c) evaluate m_{tan} at x = 2 , (d) find the equation of the line tangent to the graph of f at (2, f(2))

11. f(x) = 3x – 7 12. f(x) = 2 – 7x 13. f(x) = ax + b where a and b are constants

14. $f(x) = x^2 + 3x$ 15. $f(x) = 8 - 3x^2$ 16. $f(x) = ax^2 + bx + c$ where a, b and c are constants

In problems 17 and 18, use the result that if $f(x) = ax^2 + bx + c$ then $m_{tan} = 2ax + b$.

17. $f(x) = x^2 + 2x$. At which point(s) $(p, f(p))$ does the line tangent to the graph at that point also go through the point $(3, 6)$?

18. (a) If $a \neq 0$, then what is the shape of the graph of $y = f(x) = ax^2 + bx + c$?

(b) At what value(s) of x is the line tangent to the graph of f(x) horizontal?

Section 2.0 **PRACTICE Answers**

Practice 1: Approximate values of m(x) are in the table below. Fig. 17 is a graph of m(x).

x	y = f(x)	m(x) = the **estimated** SLOPE of the tangent line to y=f(x) at the point (x,y)
0	**2**	**–1**
1	**1**	**–1**
2	**1/3**	**0**
3	**1**	**1**
4	**3/2**	**1/2**
5	**1**	**–2**

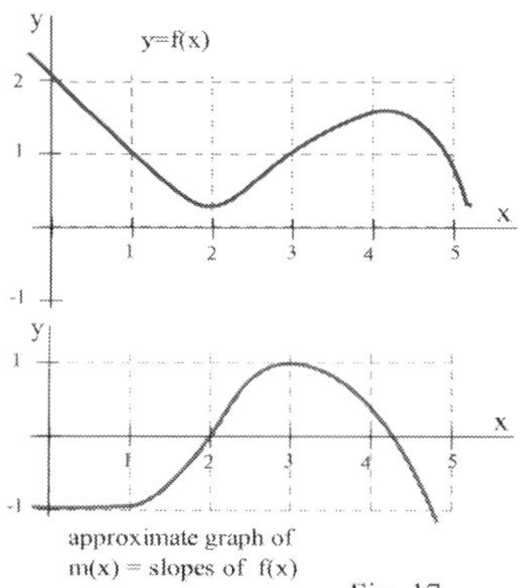

Fig. 17

Practice 2: The tangent lines to the graph of g are horizontal (slope = 0) when $x \approx -1, 1, 2.5$, and 5.

Practice 3: Fig. 18 is a graph of the approximate **rate** of temperature change (slope).

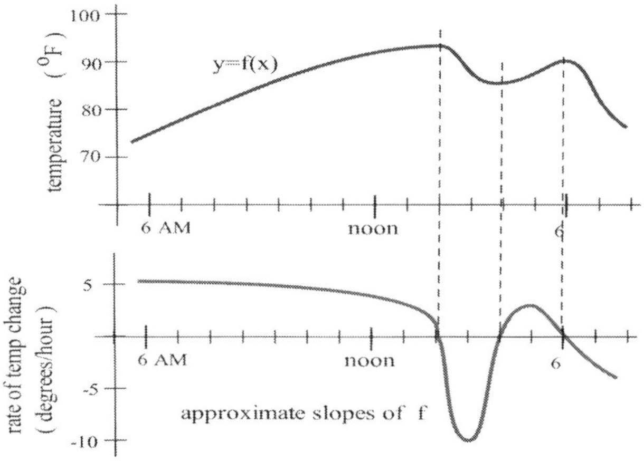

Fig. 18

Practice 4: $y = x^2$.

At (1,1), $m_{tan} = \lim_{h\to 0} \frac{f(1+h)-f(1)}{(1+h)-(1)} = \lim_{h\to 0} \frac{(1+h)^2-(1)^2}{h} = \lim_{h\to 0} \frac{\{1+2h+h^2\}-1}{h}$

$= \lim_{h\to 0} \frac{2h+h^2}{h} = \lim_{h\to 0} \frac{h(1+h)}{h} = \lim_{h\to 0}(2+h) = 2$

At (0,0), $m_{tan} = \lim_{h\to 0} \frac{f(0+h)-f(0)}{(0+h)-(0)} = \lim_{h\to 0} \frac{(0+h)^2-(0)^2}{h} = \lim_{h\to 0} \frac{h^2}{h} = \lim_{h\to 0} h = \mathbf{0}$.

At (–1,1), $m_{tan} = \lim_{h\to 0} \frac{f(-1+h)-f(-1)}{(-1+h)-(-1)} = \lim_{h\to 0} \frac{\{1-2h+h^2\}-1}{h} = \lim_{h\to 0} \frac{-2h+h^2}{h} = \mathbf{-2}$.

Practice 5: From Example 4 we know the slope of the tangent line is $m_{tan} = 2x$ so the slope of the tangent line at (2,4) is $m_{tan} = 2x = 2(2) = 4$. The tangent line has slope 4 and goes through the point (2,4) so the equation of the tangent line (using $y - y_o = m(x - x_o)$) is $y - 4 = 4(x - 2)$ or $y = 4x - 4$. The point (3,8) satisfies the equation $y = 4x - 4$ so the point (3,8) lies on the tangent line.

2.1 THE DEFINITION OF DERIVATIVE

The graphical idea of a **slope of a tangent line** is very useful, but for some uses we need a more algebraic definition of the **derivative of a function.** We will use this definition to calculate the derivatives of several functions and see that the results from the definition agree with our graphical understanding. We will also look at several different interpretations for the derivative, and derive a theorem which will allow us to easily and quickly determine the derivative of any fixed power of x.

In the last section we found the slope of the tangent line to the graph of the function $f(x) = x^2$ at an arbitrary point $(x, f(x))$ by calculating the slope of the secant line through the points $(x, f(x))$ and $(x+h, f(x+h))$,

$$m_{sec} = \frac{f(x+h) - f(x)}{(x+h) - (x)} ,$$

and then by taking the limit of m_{sec} as h approached 0 (Fig. 1). That approach to calculating slopes of tangent lines is the definition of the derivative of a function.

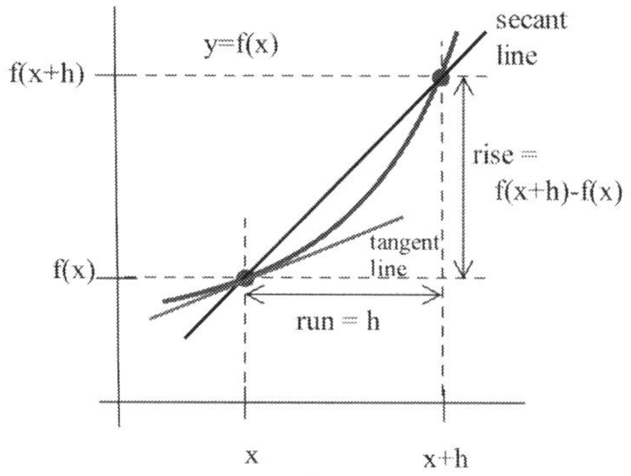

Fig. 1

Definition of the Derivative:

The **derivative** of a function f is a new function, **f '** (pronounced "eff prime"),

whose value at x is $\mathbf{f'(x)} = \lim_{h \to 0} \frac{f(x+h) - f(x)}{h}$ if the limit exists and is finite.

This is <u>the</u> definition of differential calculus, and you must know it and understand what it says. The rest of this chapter and all of Chapter 3 are built on this definition as is much of what appears in later chapters. It is remarkable that such a simple idea (the slope of a tangent line) and such a simple definition (for the derivative f ') will lead to so many important ideas and applications.

Notation: There are three commonly used notations for the **derivative of y = f(x)**:

f '(x) emphasizes that the derivative is a **function** related to f

D(f) emphasizes that we perform an **operation** on f to get the derivative of f

$\frac{\mathbf{df}}{\mathbf{dx}}$ emphasizes that the derivative is the **limit of** $\frac{\Delta f}{\Delta x} = \frac{f(x+h) - f(x)}{h}$.

We will use all three notations so you can get used to working with each of them.

f '(x) represents the slope of the tangent line to the graph of y = f(x) at the point (x, f(x)) or the instantaneous rate of change of the function f at the point (x, f(x)).

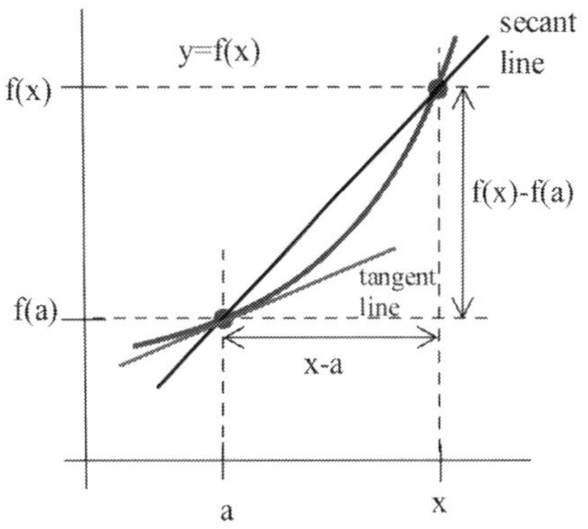

Fig. 2

If, in Fig. 2, we let x be the point a+h, then $h = x - a$. As $h \to 0$, we see that $x \to a$ and

$$\lim_{h\to 0} \frac{f(a+h)-f(a)}{h} = \lim_{x\to a} \frac{f(x)-f(a)}{x-a} \quad \text{so}$$

$$\mathbf{f'(a)} = \lim_{h\to 0} \frac{f(a+h)-f(a)}{h} = \lim_{x\to a} \frac{f(x)-f(a)}{x-a} .$$

We will use whichever of these two forms is more convenient algebraically.

Calculating Some Derivatives Using The Definition

Fortunately, we will soon have some quick and easy ways to calculate most derivatives, but first we will have to use the definition to determine the derivatives of a few basic functions. In Section 2.2 we will use those results and some properties of derivatives to calculate derivatives of combinations of the basic functions. Let's begin by using the graphs and then the definition to find a few derivatives.

Example 1: Graph **y = f(x) = 5** and estimate the **slope** of the tangent line at each point on the graph. Then use the definition of the derivative to calculate the exact slope of the tangent line at each point. Your graphic estimate and the exact result from the definition should agree.

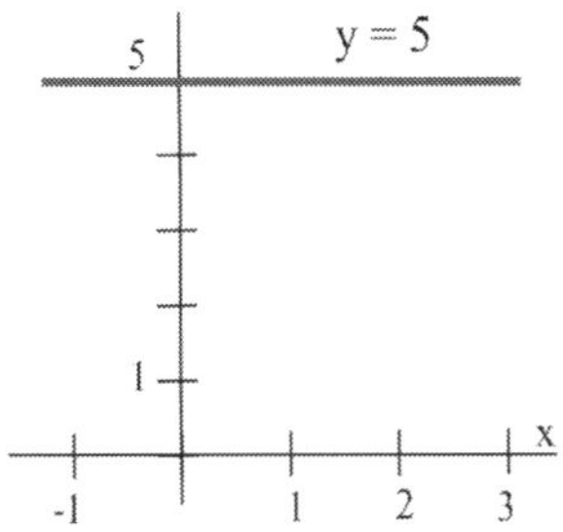

Fig. 3

Solution: The graph of y = f(x) = 5 is a horizontal line (Fig. 3) which has slope 0 so we should expect that its tangent line will also have slope 0.

Using the definition: Since f(x) = 5, then f(x+h) = 5, so

$$\mathbf{D}(\, f(x)\,) \equiv \lim_{h\to 0} \frac{f(x+h)-f(x)}{h} = \lim_{h\to 0} \frac{5-5}{h} = \lim_{h\to 0} \frac{0}{h} = 0 .$$

Using similar steps, it is easy to show that the derivative of any constant function is 0.

> **Theorem:** If f(x) = k, then f '(x) = 0 .

Practice 1: Graph **y = f(x) = 7x** and estimate the slope of the tangent line at each point on the graph. Then use the definition of the derivative to calculate the exact slope of the tangent line at each point.

Example 2: Determine the derivative of **y = f(x) = 5x^3** graphically and using the definition. Find the equation of the line tangent to y = 5x^3 at the point (1,5).

Solution: It appears that the graph of $y = f(x) = 5x^3$ (Fig. 4) is increasing so the slopes of the tangent lines are positive except perhaps at $x = 0$ where the graph seems to flatten out.

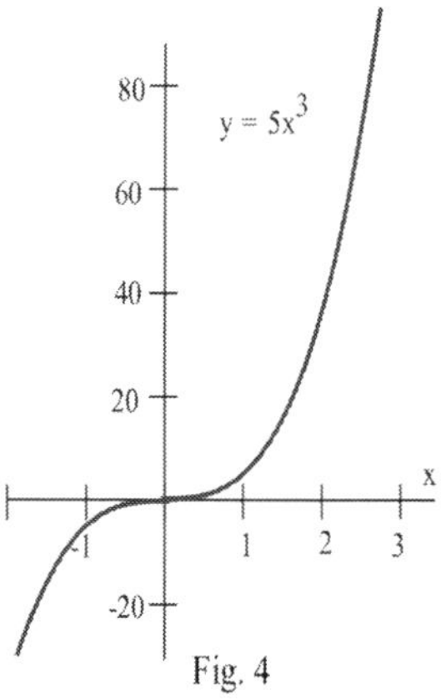

Fig. 4

Using the definition: Since $f(x) = 5x^3$, then $f(x+h) = 5(x+h)^3 = 5(x^3 + 3x^2h + 3xh^2 + h^3)$ so

$$\mathbf{f'(x)} \equiv \lim_{h\to 0} \frac{f(x+h)-f(x)}{h} \qquad \text{the definition}$$

$$= \lim_{h\to 0} \frac{5(x^3+3x^2h+3xh^2+h^3)-5(x^3)}{h} \qquad \text{eliminate } 5x^3 - 5x^3$$

$$= \lim_{h\to 0} \frac{15x^2h+15xh^2+5h^3}{h}$$

$$= \lim_{h\to 0} \frac{h(15x^2+15xh+5h^2)}{h} \qquad \text{divide by } h$$

$$= \lim_{h\to 0} (15x^2+15xh+5h^2) = 15x^2 + 0 + 0 = \mathbf{15x^2}$$

so $\mathbf{D}(\mathbf{5}x^3) = 15x^2$ which is positive except when x=0, and then $15x^2 = 0$.

$f'(x) = 15x^2$ is the slope of the line tangent to the graph of f at the point $(x, f(x))$. At the point (1,5), the slope of the tangent line is $f'(1) = 15(1)^2 = 15$. From the point–slope formula, the equation of the tangent line to f is $y - 5 = 15(x - 1)$ or $y = 15x - 10$.

Practice 2: Use the definition to show that the derivative of $y = x^3$ is $\frac{dy}{dx} = 3x^2$. Find the equation of the line tangent to the graph of $y = x^3$ at the point (2, 8).

If f has a derivative at x, we say that f is **differentiable** at x. If we have a point on the graph of a differentiable function and a slope (the derivative evaluated at the point), it is easy to write the equation of the tangent line.

> **Tangent Line Formula**
>
> If f is differentiable at **a**
>
> then the equation of the tangent line to f at the point (a, f(a)) is $\mathbf{y = f(a) + f'(a)(x - a)}$.

Proof: The tangent line goes through the point (a, f(a)) with slope f '(a) so, using the point–slope formula, $y - f(a) = f'(a)(x - a)$ or $y = f(a) + f'(a)(x - a)$.

Practice 3: The derivatives **D**(x) = 1 , **D**(x^2) = 2x , **D**(x^3) = $3x^2$ exhibit the start of a pattern. Without using the definition of the derivative, what do you think the following derivatives will be? **D**(x^4) , **D**(x^5) , **D**(x^{43}) , **D**($\sqrt{x}$) = **D**($x^{1/2}$) and **D**(x^π) .

(Just make an intelligent "guess" based on the pattern of the previous examples.)

Before going on to the "pattern" for the derivatives of powers of x and the general properties of derivatives, let's try the derivatives of two functions which are not powers of x: sin(x) and | x | .

Theorem: **D**(sin(x)) = cos(x) .

The graph of y = f(x) = sin(x) is well–known (Fig. 5). The graph has horizontal tangent lines (slope = 0) when $x = \pm\frac{\pi}{2}$ and $x = \pm\frac{3\pi}{2}$ and so on. If $0 < x < \frac{\pi}{2}$, then the slopes of the tangent lines to the graph of y = sin(x) are positive. Similarly, if $\frac{\pi}{2} < x < \frac{3\pi}{2}$, then the slopes of the tangent lines are negative. Finally, since the graph of y = sin(x) is periodic, we expect that the derivative of y = sin(x) will also be periodic.

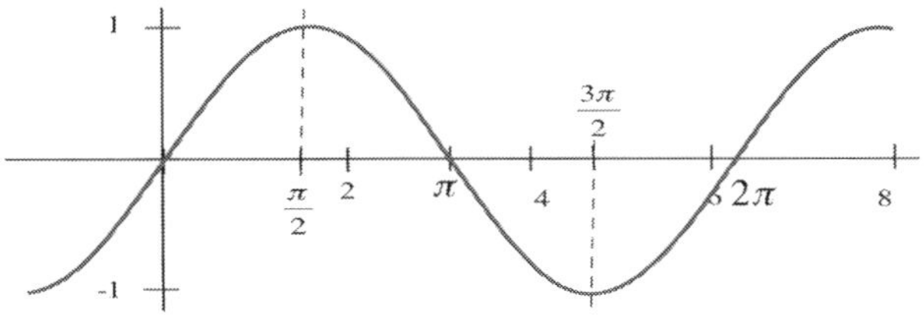

Fig. 5

Proof of the theorem: Since f(x) = sin(x), f(x+h) = sin(x+h) = sin(x)cos(h) + cos(x)sin(h) so

$$\mathbf{f}'(x) \equiv \lim_{h\to 0} \frac{f(x+h)-f(x)}{h}$$

$$= \lim_{h\to 0} \frac{\{\sin(x)\cos(h)+\cos(x)\sin(h)\}-\{\sin(x)\}}{h}$$

this limit looks formidable, but if we just collect the terms containing sin(x) and then those containing cos(x) we get

$$= \lim_{h\to 0} \left\{ \sin(x)\cdot\frac{\cos(h)-1}{h} + \cos(x)\cdot\frac{\sin(h)}{h} \right\}$$

now calculate the limits separately

$$= \left\{\lim_{h\to 0} \sin(x)\right\}\cdot\left\{\lim_{h\to 0} \frac{\cos(h)-1}{h}\right\} + \left\{\lim_{h\to 0} \cos(x)\right\}\cdot\left\{\lim_{h\to 0} \frac{\sin(h)}{h}\right\}$$

the first and third limits do not depend on h, and we calculated the second and fourth limits in Section 1.2

$$= \sin(x)\cdot(0) + \cos(x)\cdot(1) = \mathbf{cos\ (x)}.$$

So **D**(sin(x)) = cos(x) , and the various properties we expected of the derivative of y = sin(x) by examining its graph are true of cos(x).

Practice 4: Use the definition to show that **D**(cos(x)) = – sin(x) . (This is similar to the situation for f(x) = sin(x). You will need the formula cos(x+h) = cos(x)·cos(h) – sin(x)·sin(h). Then collect all the terms containing cos(x) and all the terms with sin(x). At that point you should recognize and be able to evaluate the limits.)

Example 3: For y = |x| find dy/dx .

Solution: The graph of y = f(x) = | x | (Fig. 6) is a "V" with its vertex at the origin. When x > 0, the graph is just y = |x| = x which is a line with slope +1 so we should expect the derivative of |x| to be +1. When x < 0, the graph is y = |x| = –x which is a line with slope –1, so we expect the derivative of |x| to be –1. When x = 0, the graph has a corner, and we should expect the derivative of |x| to be undefined at x = 0.

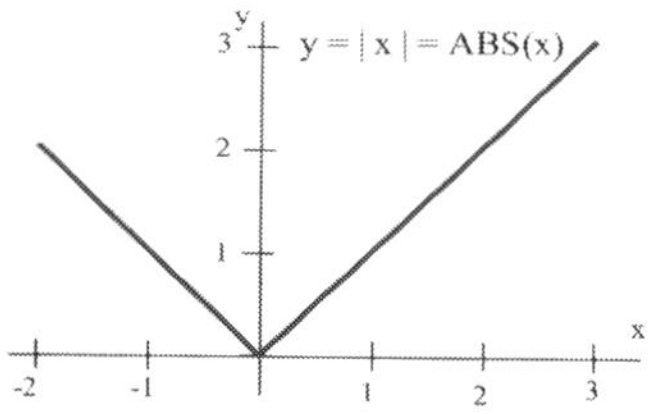

Fig. 6

Using the definition: It is easiest to consider 3 cases in the definition of | x | : x > 0, x < 0 and x = 0.

If x > 0, then, for small values of h, x + h > 0 so **D**f(x) $\equiv \lim_{h\to 0} \frac{|x+h|-|x|}{h} = \lim_{h\to 0} \frac{h}{h} = 1.$

If x < 0, then, for small values of h, we also know that x + h < 0 so **D**f(x) = $\lim_{h\to 0} \frac{-h}{h} = -1.$

When x = 0, the situation is a bit more complicated and

$$\mathbf{D}f(x) \equiv \lim_{h\to 0} \frac{f(x+h)-f(x)}{h} = \lim_{h\to 0} \frac{|0+h|-|0|}{h} = \lim_{h\to 0} \frac{|h|}{h} \text{ which is undefined}$$

since $\lim_{h\to 0^+} \frac{|h|}{h} = +1$ and $\lim_{h\to 0^-} \frac{|h|}{h} = -1$. $\mathbf{D}(\,|x|\,) = \begin{cases} +1 & \text{if } x > 0 \\ \text{undefined} & \text{if } x = 0 \\ -1 & \text{if } x < 0 \end{cases}$.

Practice 5: Graph y = | x – 2 | and y = | 2x | and use the graphs to determine **D**(|x – 2|) and **D**(| 2x |).

INTERPRETATIONS OF THE DERIVATIVE

So far we have emphasized the derivative as the slope of the line tangent to a graph . That interpretation is very visual and useful when examining the graph of a function, and we will continue to use it. Derivatives, however, are used in a wide variety of fields and applications, and some of these fields use other interpretations. The following are a few interpretations of the derivative which are commonly used.

General

Rate of Change $f'(x)$ is the **rate of change** of the function at x. If the units for x are years and the units for f(x) are people, then the units for $\frac{df}{dx}$ are $\frac{\text{people}}{\text{year}}$, a rate of change in population.

Graphical

Slope $f'(x)$ is the **slope of the line tangent to the graph of f at the point (x, f(x)).**

Physical

Velocity If f(x) is the position of an object at time x, then $f'(x)$ is the **velocity** of the object at time x. If the units for x are hours and f(x) is distance measured in miles, then the units for $\mathbf{f}'(x) = \frac{df}{dx}$ are $\frac{\text{miles}}{\text{hour}}$, miles per hour, which is a measure of velocity.

Acceleration If f(x) is the velocity of an object at time x, then $f'(x)$ is the **acceleration** of the object at time x. If the units are for x are hours and f(x) has the units $\frac{\text{miles}}{\text{hour}}$, then the units for the acceleration $\mathbf{f}'(\mathbf{x}) = \frac{df}{dx}$ are $\frac{\text{miles/hour}}{\text{hour}} = \frac{\text{miles}}{\text{hour}^2}$, miles per hour per hour.

Magnification $\mathbf{f}'(x)$ is the **magnification factor** of the function f for points which are close to x. If a and b are two points very close to x, then the distance between f(a) and f(b) will be close to $f'(x)$ times the original distance between a and b: $f(b) - f(a) \approx f'(x)(b - a)$.

Business

Marginal Cost If f(x) is the total cost of x objects, then $\mathbf{f}'(x)$ is the **marginal cost**, at a production level of x. This marginal cost is approximately the additional cost of making one more object once we have already made x objects. If the units for x are bicycles and the units for f(x) are dollars, then the units for $f'(x) = \frac{df}{dx}$ are $\frac{\text{dollars}}{\text{bicycle}}$, the cost per bicycle.

Marginal Profit If f(x) is the total profit from producing and selling x objects, then $\mathbf{f}'(x)$ is the **marginal profit**, the profit to be made from producing and selling one more object. If the units for x are bicycles and the units for f(x) are dollars, then the units for $f'(x) = \frac{df}{dx}$ are $\frac{\text{dollars}}{\text{bicycle}}$, dollars per bicycle, which is the profit per bicycle.

In business contexts, the word "marginal" usually means the derivative or rate of change of some quantity.

One of the strengths of calculus is that it provides a unity and economy of ideas among diverse applications. The vocabulary and problems may be different, but the ideas and even the notations of calculus are still useful.

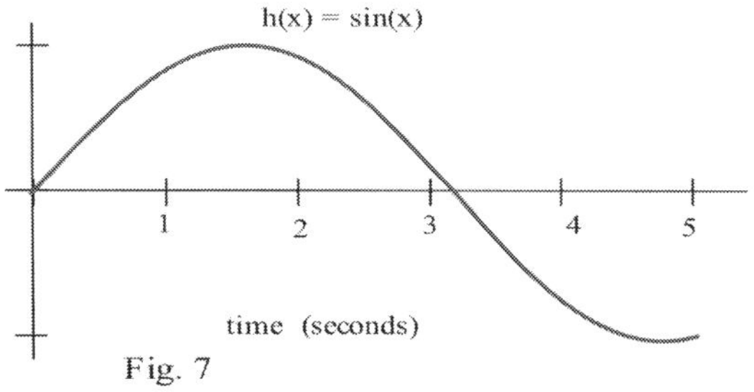

Fig. 7

Example 4: A small cork is bobbing up and down, and at time t seconds it is h(t) = sin(t) feet above the mean water level (Fig. 7). Find the height, velocity and acceleration of the cork when t = 2 seconds. (Include the proper units for each answer.)

Solution: h(t) = sin(t) represents the height of the cork at any time t , so the height of the cork when t = 2 is h(2) = sin(2) ≈ 0.91 feet.

The velocity is the derivative of the position, so $v(t) = \frac{d\ h(t)}{dt} = \frac{d\ \sin(t)}{dt} = \cos(t)$. The derivative of position is the limit of $(\Delta h)/(\Delta t)$, so the units are (feet)/(seconds). After 2 seconds the velocity is v(2) = cos(2) ≈ –0.42 feet per second = –0.42 ft/s .

The acceleration is the derivative of the velocity, so $a(t) = \frac{d\ v(t)}{dt} = \frac{d\ \cos(t)}{dt} = -\sin(t)$. The derivative of velocity is the limit of $(\Delta v)/(\Delta t)$, so the units are (feet/second) / (seconds) or feet/second^2. After 2 seconds the acceleration is a(2) = – sin(2) ≈ –0.91 ft/s^2 .

Practice 6: Find the height, velocity and acceleration of the cork in the previous example after 1 second?

A MOST USEFUL FORMULA: D(x^n)

Functions which include powers of x are very common (every polynomial is a sum of terms which include powers of x), and, fortunately, it is easy to calculate the derivatives of such powers. The "pattern" emerging from the first few examples in this section is, in fact, true for all powers of x. We will only state and prove the "pattern" here for positive integer powers of x, but it is also true for other powers as we will prove later.

Theorem: If n is a positive integer, then D(x^n) = $n \cdot x^{n-1}$.

This theorem is an example of the power of generality and proof in mathematics. Rather than resorting to the definition when we encounter a new power of x (imagine using the definition to calculate the derivative of x^{307}), we can justify the pattern for all positive integer exponents n , and then simply apply the result for whatever exponent we have. We know, from the first examples in this section, that the theorem is true for n= 1, 2 and 3 , but no number of examples would guarantee that the pattern is true for all exponents. We need a proof that what we think is true really is true.

Proof of the theorem: Since $f(x) = x^n$, then $f(x+h) = (x+h)^n$, and in order to simplify $f(x+h) - f(x) = (x+h)^n - x^n$, we will need to expand $(x+h)^n$. However, we really only need to know the first two terms of the expansion and to know that all of the other terms of the expansion contain a power of h of at least 2. The Binomial Theorem from algebra says (for n > 3) that

$(x+h)^n = x^n + n\cdot x^{n-1}h + a\cdot x^{n-2}h^2 + b\cdot x^{n-3}h^3 + \ldots + h^n$ where a and b represent numerical coefficients. (Expand $(x+h)^n$ for at least a few different values of n to convince yourself of this result.)

Then $\mathbf{D}(f(x)) \equiv \lim_{h\to 0} \dfrac{f(x+h)-f(x)}{h} = \lim_{h\to 0} \dfrac{(x+h)^n - x^n}{h}$ then expand to get

$= \lim_{h\to 0} \dfrac{\{x^n + n\cdot x^{n-1}h + a\cdot x^{n-2}h^2 + b\cdot x^{n-3}h^3 + \ \ldots + \ h^n\} - x^n}{h}$ eliminate $x^n - x^n$

$= \lim_{h\to 0} \dfrac{\{n\cdot x^{n-1}h + a\cdot x^{n-2}h^2 + b\cdot x^{n-3}h^3 + \ \ldots + \ h^n\}}{h}$ factor h out of the numerator

$= \lim_{h\to 0} \dfrac{h\cdot\{n\cdot x^{n-1} + a\cdot x^{n-2}h + b\cdot x^{n-3}h^2 + \ \ldots + \ h^{n-1}\}}{h}$ divide by the factor h

$= \lim_{h\to 0} \{ n\cdot x^{n-1} + a\cdot x^{n-2}h + b\cdot x^{n-3}h^2 + \ldots + h^{n-1} \}$ separate the limits

$= n\cdot x^{n-1} + \lim_{h\to 0} \{ a\cdot x^{n-2}\mathbf{h} + b\cdot x^{n-3}\mathbf{h^2} + \ldots + \mathbf{h^{n-1}} \}$ each term has a factor of h, and h→0

$= n\cdot x^{n-1} + \mathbf{0} = \mathbf{n\cdot x^{n-1}}$ so $\mathbf{D}(x^n) = \mathbf{n\cdot x^{n-1}}$.

Practice 7: Use the theorem to calculate $\mathbf{D}(x^5)$, $\dfrac{\mathbf{d}}{\mathbf{dx}}(x^2)$, $\mathbf{D}(x^{100})$, $\dfrac{\mathbf{d}}{\mathbf{dt}}(t^{31})$, and $\mathbf{D}(x^0)$.

We will occasionally use the result of the theorem for the derivatives of **all** constant powers of x even though it has only been proven for positive integer powers, so far. The result for all constant powers of x is proved in Section 2.9

Example 5: Find $\mathbf{D}(1/x)$ and $\dfrac{\mathbf{d}}{\mathbf{dx}}(\sqrt{x})$.

Solution: $\mathbf{D}(\frac{1}{x}) = \mathbf{D}(x^{-1}) = -1x^{(-1)-1} = -1x^{-2} = \dfrac{-1}{x^2}$. $\dfrac{\mathbf{d}}{\mathbf{dx}}(\sqrt{x}) = \mathbf{D}(x^{1/2}) = (1/2)x^{-1/2} = \dfrac{1}{2\sqrt{x}}$.

These results can be obtained by using the definition of the derivative, but the algebra is slightly awkward.

Practice 8: Use the pattern of the theorem to find $\mathbf{D}(x^{3/2})$, $\dfrac{\mathbf{d}}{\mathbf{dx}}(x^{1/3})$, $\mathbf{D}(\frac{1}{\sqrt{x}})$ and $\dfrac{\mathbf{d}}{\mathbf{dt}}(t^{\pi})$.

Example 6: It costs $\sqrt{x}$ hundred dollars to run a training program for x employees.

(a) How much does it cost to train 100 employees? 101 employees? If you already need to train 100 employees, how much additional will it cost to add 1 more employee to those being trained?

(b) For $f(x) = \sqrt{x}$, calculate $f'(x)$ and evaluate f' at $x = 100$. How does $f'(100)$ compare with the last answer in part (a)?

Solution: (a) Put $f(x) = \sqrt{x} = x^{1/2}$ hundred dollars, the cost to train x employees. Then $f(100) = \$1000$ and $f(101) = \$1004.99$, so it costs \$4.99 additional to train the 101st employee.

(b) $f'(x) = \frac{1}{2}\, x^{-1/2} = \frac{1}{2\sqrt{x}}$ so $f'(100) = \frac{1}{2\sqrt{100}} = \frac{1}{20}$ hundred dollars = \$5.00 .

Clearly $f'(100)$ is very close to the actual additional cost of training the 101st employee.

IMPORTANT DEFINITIONS AND RESULTS

Definition of Derivative: $f'(x) \equiv \lim_{h \to 0} \frac{f(x+h) - f(x)}{h}$ if the limit exists and is finite.

Notations For The Derivative: $\mathbf{f'(x),\ Df(x),\ \frac{d\,f(x)}{dx}}$

Tangent Line Equation: The line $\mathbf{y = f(a) + f'(a)\cdot(x - a)}$ is tangent to the graph of f at $(a, f(a))$.

Formulas: **D**(constant) = 0

D(x^n) = $n \cdot x^{n-1}$ (proven for n = positive integer: true for all constants n)

D(sin(x)) = cos(x) and **D**(cos(x)) = –sin(x)

$$\mathbf{D}(|x|) = \begin{cases} +1 & \text{if } x > 0 \\ \text{undefined} & \text{if } x = 0 \\ -1 & \text{if } x < 0 \end{cases} .$$

Interpretations of f '(x):

Slope of a line tangent to a graph

Instantaneous rate of change of a function at a point

Velocity or acceleration

Magnification factor

Marginal change

PROBLEMS

1. Match the graphs of the three functions in Fig. 8 with the graphs of their derivatives.

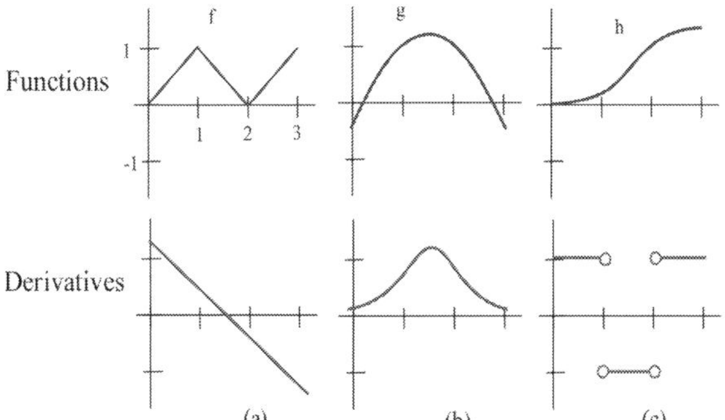

Fig. 8

2. Fig. 9 shows six graphs, three of which are derivatives of the other three. Match the functions with their derivatives.

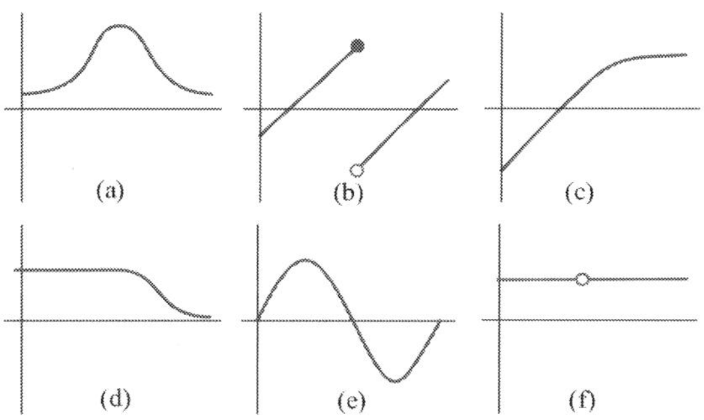

Fig. 9

In problems 3 – 6, find the slope m_{sec} of the secant line through the two given points and then calculate $m_{tan} = \lim_{h \to 0} m_{sec}$.

3. $f(x) = x^2$ (a) $(-2,4), (-2+h, (-2+h)^2)$ (b) $(0.5, 0.25), (0.5+h, (0.5+h)^2)$

4. $f(x) = 3 + x^2$ (a) $(1,4), (1+h, 3+(1+h)^2)$ (b) $(x, 3 + x^2), (x+h, 3 + (x+h)^2)$

5. $f(x) = 7x - x^2$ (a) $(1, 6), (1+h, 7(1+h) - (1+h)^2)$ (b) $(x, 7x - x^2), (x+h, 7(x+h) - (x+h)^2)$

6. $f(x) = x^3 + 4x$ (a) $(1, 5), (1+h, (1+h)^3 + 4(1+h))$ (b) $(x, x^3 + 4x), (x+h, (x+h)^3 + 4(x+h))$

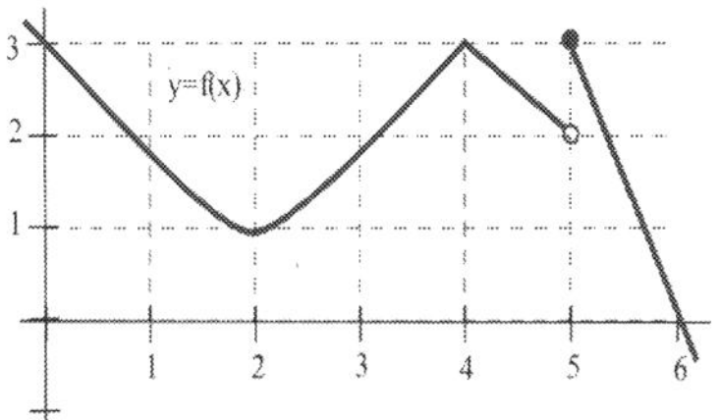

Fig. 10

7. Use the graph in Fig. 10 to estimate the values of these limits. (It helps to recognize what the limit represents.)

(a) $\lim_{h \to 0} \dfrac{f(0+h) - f(0)}{h}$ (b) $\lim_{h \to 0} \dfrac{f(1+h) - f(1)}{h}$ (c) $\lim_{h \to 0} \dfrac{f(2+h) - 1}{h}$

(d) $\lim_{w \to 0} \dfrac{f(3+w) - f(3)}{w}$ (e) $\lim_{h \to 0} \dfrac{f(4+h) - f(4)}{h}$ (f) $\lim_{s \to 0} \dfrac{f(5+s) - f(5)}{s}$

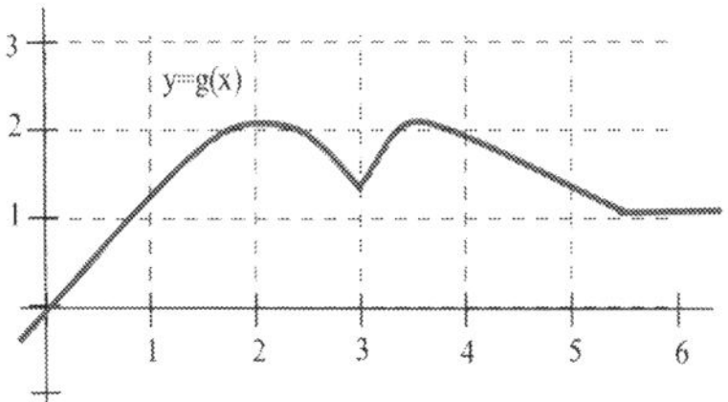

Fig. 11

8. Use the graph in Fig. 11 to estimate the values of these limits.

(a) $\lim_{h \to 0} \dfrac{g(h) - g(0)}{h}$ (b) $\lim_{h \to 0} \dfrac{g(1+h) - g(1)}{h}$ (c) $\lim_{h \to 0} \dfrac{g(2+h) - 2}{h}$

(d) $\lim_{w \to 0} \dfrac{g(3+w) - g(3)}{w}$ (e) $\lim_{h \to 0} \dfrac{g(4+h) - g(4)}{h}$ (f) $\lim_{s \to 0} \dfrac{g(5+s) - g(5)}{s}$

In problems 9 – 12 , use the Definition of the derivative to calculate **f '(x)** and then evaluate f '(3).

9. $f(x) = x^2 + 8$
10. $f(x) = 5x^2 - 2x$
11. $f(x) = 2x^3 - 5x$
12. $f(x) = 7x^3 + x$

13. Graph $f(x) = x^2$, $g(x) = x^2 + 3$ and $h(x) = x^2 - 5$. Calculate the derivatives of f, g, and h.

14. Graph $f(x) = 5x$, $g(x) = 5x + 2$ and $h(x) = 5x - 7$. Calculate the derivatives of f, g, and h.

In problems 15 – 18, find the slopes and equations of the lines tangent to $y = f(x)$ at the given points.

15. $f(x) = x^2 + 8$ at (1,9) and (–2,12)
16. $f(x) = 5x^2 - 2x$ at (2, 16) and (0,0)
17. $f(x) = \sin(x)$ at (π, 0) and (π/2,1)
18. $f(x) = | x + 3 |$ at (0,3) and (–3,0)

19. (a) Find the equation of the line tangent to the graph of $y = x^2 + 1$ at the point (2,5).
(b) Find the equation of the line perpendicular to the graph of $y = x^2 + 1$ at (2,5).
(c) Where is the tangent to the graph of $y = x^2 + 1$ horizontal?
(d) Find the equation of the line tangent to the graph of $y = x^2 + 1$ at the point (p,q).
(e) Find the point(s) (p,q) on the graph of $y = x^2 + 1$ so the tangent line to the curve at (p,q) goes through the point (1,–7).

20. (a) Find the equation of the line tangent to the graph of $y = x^3$ at the point (2,8).
(b) Where, if ever, is the tangent to the graph of $y = x^3$ horizontal?
(c) Find the equation of the line tangent to the graph of $y = x^3$ at the point (p,q).
(d) Find the point(s) (p,q) on the graph of $y = x^3$ so the tangent line to the curve at (p,q) goes through the point (16,0).

21. (a) Find the angle that the tangent line to $y = x^2$ at (1,1) makes with the x–axis.
(b) Find the angle that the tangent line to $y = x^3$ at (1,1) makes with the x–axis.
(c) The curves $y = x^2$ and $y = x^3$ intersect at the point (1,1). Find the angle of intersection of the two curves (actually the angle between their tangent lines) at the point (1,1) .

22. Fig. 12 shows the graph of $y = f(x)$. Sketch the graph of $y = f '(x)$.

23. Fig. 13 shows the graph of the height of an object at time t. Sketch the graph of the object's upward velocity. What are the units for each axis on the velocity graph?

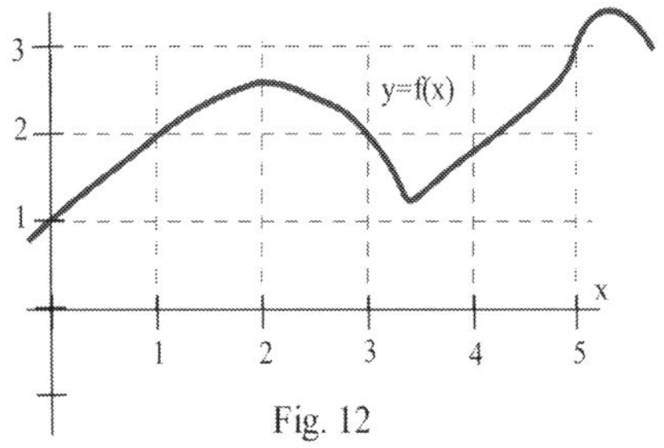

Fig. 12

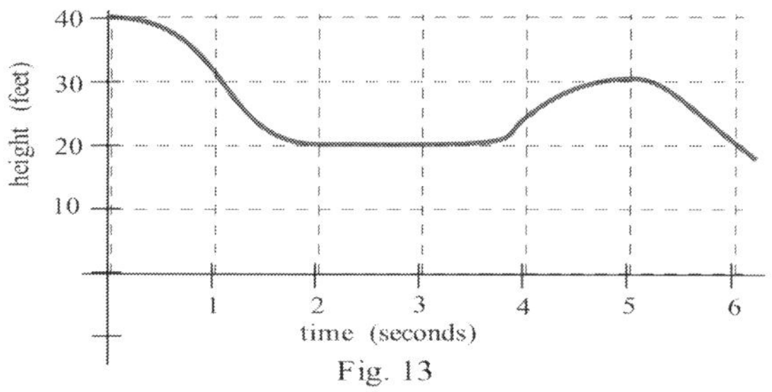

Fig. 13

24. Fill in the table with the appropriate units for $f'(x)$.

units for x	units for f(x)	units for f '(x)
hours	miles	
people	automobiles	
dollars	pancakes	
days	trout	
seconds	miles per second	
seconds	gallons	
study hours	test points	

25. A rock dropped into a deep hole will drop $d(x) = 16x^2$ feet in x seconds.

(a) How far into the hole will the rock be after 4 seconds? 5 seconds?

(b) How fast will it be falling at exactly 4 seconds? 5 seconds? x seconds?

26. It takes $T(x) = x^2$ hours to weave x small rugs. What is the marginal production time to weave a rug? (Be sure to include the units with your answer.)

27. It costs $C(x) = \sqrt{x}$ dollars to produce x golf balls. What is the marginal production cost to make a golf ball? What is the marginal production cost when $x = 25$? when $x = 100$? (Include units.)

28. Define A(x) to be the **area** bounded by the x and y axes, the line $y = 5$, and a vertical line at x (Fig. 14).

(a) Evaluate A(0), A(1), A(2) and A(3).

(b) Find a formula for A(x) for $x \ge 0$: A(x) = ?

(c) Determine $\frac{d\,A(x)}{dx}$. (d) What does $\frac{d\,A(x)}{dx}$ represent?

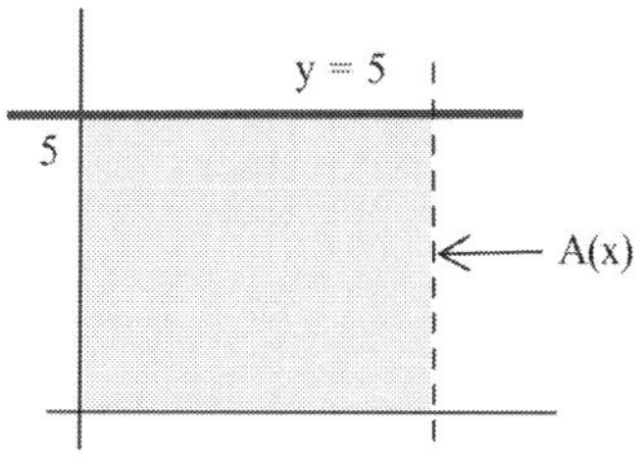

Fig. 14

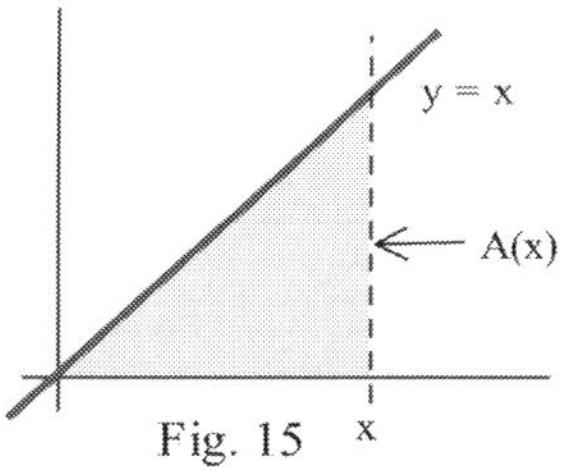

Fig. 15

29. Define A(x) to be the **area** bounded by the x–axis, the line $y = x$, and a vertical line at x (Fig. 15).

(a) Evaluate A(0), A(1), A(2) and A(3).

(b) Find a formula which represents A(x) for all $x \ge 0$: A(x) = ?

(c) Determine $\frac{d\,A(x)}{dx}$. (d) What does $\frac{d\,A(x)}{dx}$ represent?

30. Find (a) $\mathbf{D}(x^{12})$ (b) $\frac{\mathbf{d}}{\mathbf{dx}}(\sqrt[7]{x})$ (c) $\mathbf{D}(\frac{1}{x^3})$ (d) $\frac{\mathbf{d}\,x^e}{\mathbf{dx}}$ (e) $\mathbf{D}(|x-2|)$

31. Find (a) $\mathbf{D}(x^9)$ (b) $\frac{\mathbf{d}\,x^{2/3}}{\mathbf{dx}}$ (c) $\mathbf{D}(\frac{1}{x^4})$ (d) $\mathbf{D}(x^{\pi})$ (e) $\frac{\mathbf{d}\,|x+5|}{\mathbf{dx}}$

In problems 32 – 37, find a function f which has the given derivative. (Each problem has several correct answers, just find one of them.)

32. **f '**(x) = 4x + 3 33. **f '**(x) = $3x^2 + 8x$ 34. **D(** f(x) **)** = $12x^2 - 7$

35. $\frac{\mathbf{d}\,f(t)}{\mathbf{dt}} = 5\cos(t)$ 36. $\frac{\mathbf{d}\,f(x)}{\mathbf{dx}} = 2x - \sin(x)$ 37. **D(** f(x) **)** = $x + x^2$

Section 2.1 **PRACTICE Answers**

Practice 1: The graph of f(x) = 7x is a line through the origin. The slope of the line is 7.

For all x, $m_{tan} = \lim_{h\to 0} \frac{f(x+h) - f(x)}{h} = \lim_{h\to 0} \frac{7(x+h) - 7x}{h} = \lim_{h\to 0} \frac{7h}{h} = \lim_{h\to 0} 7 = \mathbf{7}$.

Practice 2: $f(x) = x^3$ so $f(x+h) = (x+h)^3 = x^3 + 3x^2h + 3xh^2 + h^3$.

$$\frac{dy}{dx} = \lim_{h\to 0} \frac{f(x+h) - f(x)}{h} = \lim_{h\to 0} \frac{\left\{x^3 + 3x^2h + 3xh^2 + h^3\right\} - x^n}{h}$$

$$= \lim_{h\to 0} \frac{3x^2h + 3xh^2 + h^3}{h} = \lim_{h\to 0} (3x^2 + 3xh + h^2) = \mathbf{3x^2}.$$

At the point (2,8), the slope of the tangent line is $3(2)^2 = 12$ so the equation of the tangent line is y – 8 = 12(x – 2) or y = 12x –16.

Practice 3: **D(** x^4 **)** = $4x^3$, **D(** x^5 **)** = $5x^4$, **D(** x^{43} **)** = $43x^{42}$, **D(** $x^{1/2}$ **)** = $\frac{1}{2} x^{-1/2}$, **D(** x^{π} **)** = $\pi x^{\pi-1}$

Practice 4: $\mathbf{D}(\cos(x)) = \lim_{h\to 0} \frac{\cos(x+h) - \cos(x)}{h} = \lim_{h\to 0} \frac{\cos(x)\cos(h) - \sin(x)\sin(h) - \cos(x)}{h}$

$$= \lim_{h\to 0} \cos(x)\frac{\cos(h) - 1}{h} - \sin(x)\frac{\sin(h)}{h} \longrightarrow \cos(x)\cdot(0) - \sin(x)\cdot(1) = \mathbf{-\sin(x)} .$$

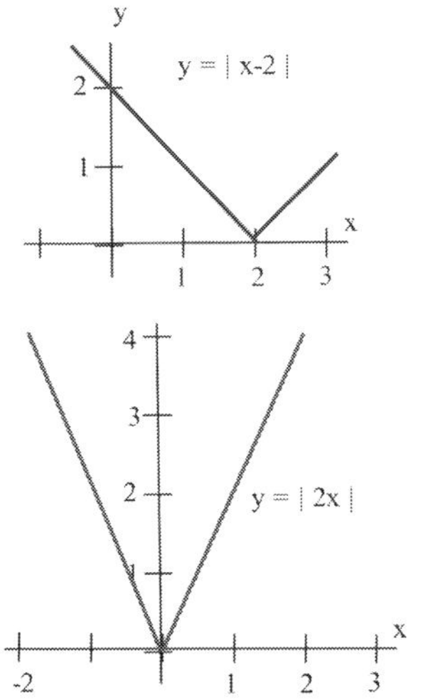

Fig. 16

Practice 5: See Fig. 16 for the graphs of y = | x – 2 | and y = | 2x |.

$$\mathbf{D}(|x-2|) = \begin{cases} +1 & \text{if } x > 2 \\ \text{undefined} & \text{if } x = 2 \\ -1 & \text{if } x < 2 \end{cases}$$

$$\mathbf{D}(|2x|) = \begin{cases} +2 & \text{if } x > 0 \\ \text{undefined} & \text{if } x = 0 \\ -2 & \text{if } x < 0 \end{cases}$$

Practice 6: $h(t) = \sin(t)$ so $h(1) = \sin(1) \approx 0.84$ feet,

$v(t) = \cos(t)$ so $v(1) = \cos(1) \approx 0.54$ feet/second.

$a(t) = -\sin(t)$ so $a(1) = -\sin(1) \approx -0.84$ feet/second^2.

Practice 7: $\mathbf{D}(x^5) = 5x^4$, $\frac{\mathbf{d}\, x^2}{\mathbf{dx}} = 2x^1 = 2x$, $\frac{\mathbf{d}\, x^{100}}{\mathbf{dx}} = 100x^{99}$, $\frac{\mathbf{d}\, t^{31}}{\mathbf{dt}} = 31t^{30}$,

$\mathbf{D}(x^0) = 0x^{-1} = 0$ or $\mathbf{D}(x^0) = \mathbf{D}(1) = 0$.

Practice 8: $\mathbf{D}(x^{3/2}) = \frac{3}{2}\, x^{1/2}$, $\frac{\mathbf{d}\, x^{1/3}}{\mathbf{dx}} = \frac{1}{3}\, x^{-2/3}$, $\mathbf{D}(x^{-1/2}) = \frac{-1}{2}\, x^{-3/2}$, $\frac{\mathbf{d}\, t^{\pi}}{\mathbf{dt}} = \pi\, t^{\pi-1}$.

2.2 DERIVATIVES: PROPERTIES AND FORMULAS

This section begins with a look at which functions have derivatives. Then we'll examine how to calculate derivatives of elementary combinations of basic functions. By knowing the derivatives of some basic functions and just a few differentiation patterns, you will be able to calculate the derivatives of a tremendous variety of functions. This section contains most, but not quite all, of the general differentiation patterns you will ever need.

WHICH FUNCTIONS HAVE DERIVATIVES?

Theorem: If a function is **differentiable** at a point,
then it is **continuous** at that point.

The contrapositive form of this theorem tells about some functions which do not have derivatives:

Contrapositive Form of the Theorem:
If f is **not continuous** at a point,
then f is **not differentiable** at that point.

Proof of the Theorem: We assume that the hypothesis , f is differentiable at the point c , is true so

$\lim\limits_{h\to 0} \dfrac{f(c+h)-f(c)}{h}$ exists and equals f '(c). We want to show that f must necessarily be continuous at c : $\lim\limits_{h\to 0} f(c+h) = f(c)$.

Since f(c + h) can be written as $f(c+h) = f(c) + \left\{\dfrac{f(c+h)-f(c)}{h}\right\}\cdot h$, we have

$$\lim_{h\to 0} f(c+h) = \lim_{h\to 0}\left(f(c) + \left\{\frac{f(c+h)-f(c)}{h}\right\}\cdot h\right)$$

$$= \lim_{h\to 0} f(c) + \lim_{h\to 0}\left\{\frac{f(c+h)-f(c)}{h}\right\}\cdot \lim_{h\to 0}(h) = f(c) + f'(c)\cdot 0 = f(c) .$$

Therefore f is continuous at c .

It is important to clearly understand what is meant by this theorem and what is not meant: If the function is differentiable at a point, then the function is automatically continuous at that point. If the function is continuous at a point, then the function **may** or **may not** have a derivative at that point.
If the function is not continuous at a point, then the function **is not** differentiable at that point.

Example 1: Show that f(x) = [x] = INT(x) is not continuous and not differentiable at 2 (Fig. 1).

Solution: The one–sided limits, $\lim_{x \to 2^+} \text{INT}(x) = 2$ and $\lim_{x \to 2^-} \text{INT}(x) = 1$, have different values so $\lim_{x \to 2} \text{INT}(x)$ does not exist, and INT(x) is not continuous at 2. Since f(x) = INT(x) is not continuous at 2, it is not differentiable there.

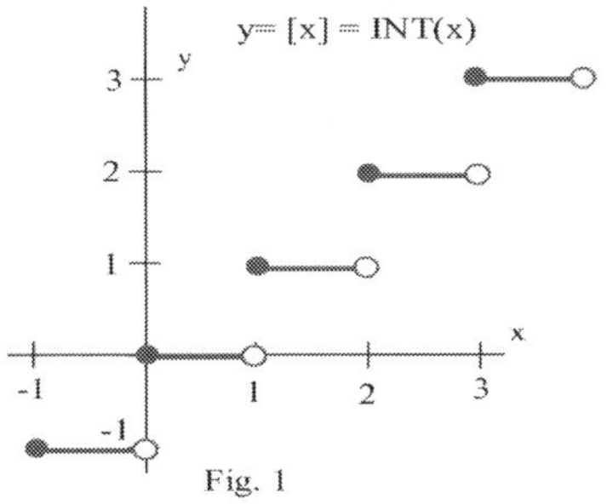

Fig. 1

Lack of continuity is enough to imply lack of differentiability, but the next two examples show that continuity is **not** enough to guarantee differentiability.

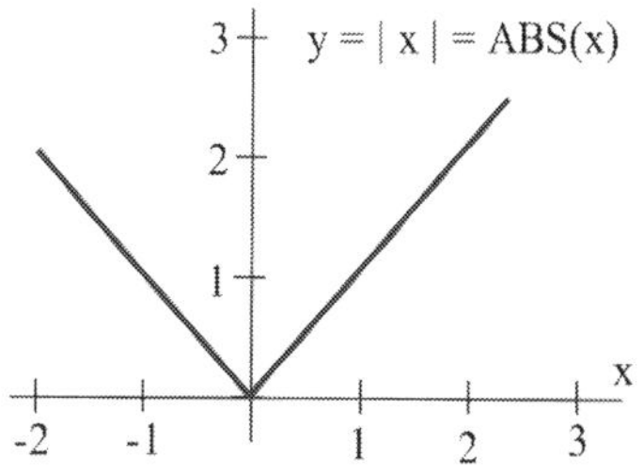

Fig. 2

Example 2: Show that f(x) = | x | is continuous but **not** differentiable at x = 0 (Fig. 2)

Solution: $\lim_{x \to 0} |x| = 0 = |0|$ so f is continuous at 0, but we showed in Section 2.1 that the absolute value function was not differentiable at x = 0.

A function is not differentiable at a cusp or a "corner."

Example 3: Show that $f(x) = \sqrt[3]{x} = x^{1/3}$ is continuous but **not** differentiable at x = 0 (Fig. 3)

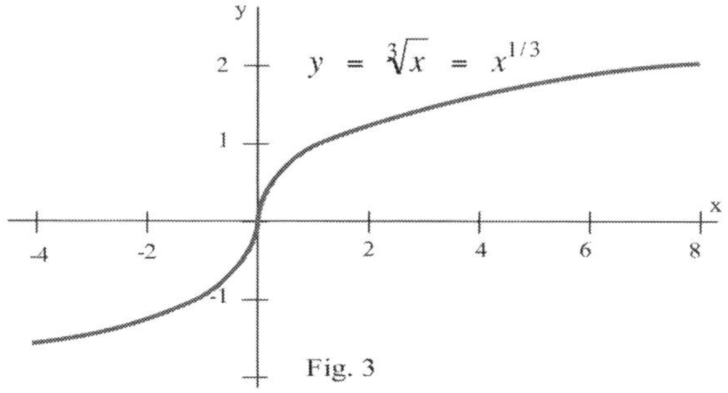

Fig. 3

Solution: $\lim_{x \to 0^-} \sqrt[3]{x} = \lim_{x \to 0^+} \sqrt[3]{x} = 0$ so

$\lim_{x \to 0} \sqrt[3]{x} = 0 = \sqrt{0}$, and f is continuous at 0.

$f'(x) = \frac{1}{3} x^{-(2/3)} = \frac{1}{3x^{2/3}}$ which is undefined at x = 0 so f is not differentiable at 0.

A function is not differentiable where its tangent line is vertical.

Practice 1: At which integer values of x is the graph of f in Fig. 4 continuous? differentiable?

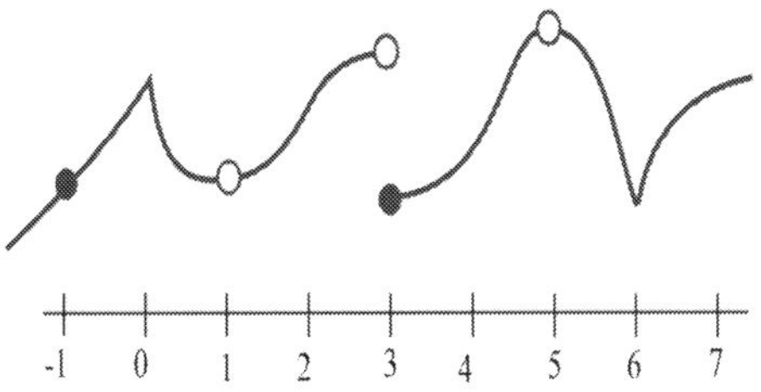
Fig. 4

Graphically, a function is **continuous** if and only if its graph is **connected** and does not have any holes or breaks.

Graphically, a function is **differentiable** if and only if it is continuous and its graph is **smooth** with no corners or vertical tangent lines.

DERIVATIVES OF ELEMENTARY COMBINATIONS OF FUNCTIONS

Example 4: The derivative of $f(x) = x$ is $\mathbf{D}f(x) = 1$, and the derivative of $g(x) = 5$ is $\mathbf{D}g(x) = 0$.

What are the derivatives of their elementary combinations: $3f, f + g, f - g, f\cdot g$ and f/g ?

Solution: $\mathbf{D}(3 f(x)) = \mathbf{D}(3x) = 3 = 3\, \mathbf{D}(f(x))$.

$\mathbf{D}(f(x) + g(x)) = \mathbf{D}(x + 5) = 1 = \mathbf{D}(f(x)) + \mathbf{D}(g(x))$.

$\mathbf{D}(f(x) - g(x)) = \mathbf{D}(x - 5) = 1 = \mathbf{D}(f(x)) - \mathbf{D}(g(x))$.

Unfortunately, the derivatives of $f\cdot g$ and f/g don't follow the same easy patterns:

$\mathbf{D}(f(x) \cdot g(x)) = \mathbf{D}(5x) = 5$ but $\mathbf{D}(f(x)) \cdot \mathbf{D}(g(x)) = (1)\cdot(0) = 0$, and

$\mathbf{D}(f(x) / g(x)) = \mathbf{D}(x / 5) = 1/5$ but $\mathbf{D}(f(x)) / \mathbf{D}(g(x))$ is undefined.

These two very simple functions show that, in general, $\mathbf{D}(f\cdot g) \neq \mathbf{D}(f) \cdot \mathbf{D}(g)$ and $\mathbf{D}(f/g) \neq \mathbf{D}(f) / \mathbf{D}(g)$.

The Main Differentiation Theorem below states the correct patterns for differentiating products and quotients.

Practice 2: For $f(x) = 6x + 8$ and $g(x) = 2$, what are the derivatives of $3f, f+g, f-g, f\cdot g$ and f/g ?

The following theorem says that the simple patterns in the example for constant multiples of functions and sums and differences of functions are true for all differentiable functions. It also includes the correct patterns for derivatives of products and quotients of differentiable functions.

Main Differentiation Theorem: If f and g are differentiable at x , then

(a) **Constant Multiple Rule:** $\mathbf{D}(kf(x)) = k\mathbf{D}(f(x)) = k\mathbf{D}f$

or $(kf(x))' = k f'(x)$

(b) **Sum Rule:** $\mathbf{D}(f(x) + g(x)) = \mathbf{D}(f(x)) + \mathbf{D}(g(x)) = \mathbf{D}f + \mathbf{D}g$

or $(f(x) + g(x))' = f'(x) + g'(x)$

(c) **Difference Rule:** $\mathbf{D}(f(x) - g(x)) = \mathbf{D}(f(x)) - \mathbf{D}(g(x)) = \mathbf{D}f - \mathbf{D}g$

or $(f(x) - g(x))' = f'(x) - g'(x)$

(d) **Product Rule:** $\mathbf{D}(f(x)\cdot g(x)) = f(x) \cdot \mathbf{D}(g(x)) + g(x)\cdot \mathbf{D}(f(x)) = f\,\mathbf{D}g + g\,\mathbf{D}f$

or $(f(x)\cdot g(x))' = f(x)\cdot \mathbf{g'(x)} + g(x)\cdot \mathbf{f'(x)}$

(e) **Quotient Rule:** $$\mathbf{D}\left(\frac{f(x)}{g(x)} \right) = \frac{g(x)\cdot\mathbf{D}(f(x)) - f(x)\cdot\mathbf{D}(g(x))}{(g(x))^2}$$

$$= \frac{g\mathbf{D}f - f\mathbf{D}g}{g^2} \quad \text{or} \quad \frac{g(x)\cdot\mathbf{f'(x)} - f(x)\cdot\mathbf{g'(x)}}{g^2(x)}$$

(provided $g(x) \neq 0$)

The proofs of parts (a), (b), and (c) of this theorem are straightforward, but parts (d) and (e) require some clever algebraic manipulations. Lets look at an example first.

Example 5: Recall that **D**(x^2) = 2x and **D**(sin(x)) = cos(x). Find **D**(3sin(x)) and $\frac{d}{dx}$($5x^2 - 7\sin(x)$) .

Solution: **D**(3sin(x)) is an application of part (a) of the theorem with k = 3 and f(x) = sin(x) so

D(3sin(x)) = 3 **D**(sin(x)) = 3 cos(x).

$\frac{d}{dx}$($5x^2 - 7\sin(x)$) uses part (c) of the theorem with $f(x) = 5x^2$ and g(x) = 7sin(x) so

$$\frac{d}{dx}(5x^2 - 7\sin(x)) = \frac{d}{dx}(5x^2) - \frac{d}{dx}(7\sin(x)) = 5\frac{d}{dx}(x^2) - 7\frac{d}{dx}(\sin(x))$$

$$= 5(\mathbf{2x}) - 7(\mathbf{\cos(x)}) = 10x - 7\cos(x) .$$

Practice 3: Find **D**($x^3 - 5\sin(x)$) and $\frac{d}{dx}$($\sin(x) - 4x^3$) .

Practice 4: Fill in the values in the table for **D**(3f(x)), **D**(2f(x)+g(x)), and **D**(3g(x) – f(x))

x	f(x)	f '(x)	g(x)	g '(x)	**D**(3f(x))	**D**(2f(x) + g(x))	**D**(3g(x) – f(x))
0	3	–2	–4	3			
1	2	–1	1	0			
2	4	2	3	1			

Proof of the Main Derivative Theorem (a) and (c): The only general fact we have about derivatives is the definition as a limit, so our proofs here will have to recast derivatives as limits and then use some results about limits. The proofs are applications of the definition of the derivative and results about limits.

(a) $$\mathbf{D}(kf(x)) \equiv \lim_{h\to 0} \frac{k\cdot f(x+h) - k\cdot f(x)}{h} = \lim_{h\to 0} k\cdot\frac{f(x+h)-f(x)}{h} = k\cdot\lim_{h\to 0}\frac{f(x+h)-f(x)}{h} = k\cdot\mathbf{D}(f(x)).$$

(c) $$\mathbf{D}(f(x) - g(x)) = \lim_{h\to 0}\frac{\{f(x+h)-g(x+h)\} - \{f(x)-g(x)\}}{h}$$

$$= \lim_{h\to 0}\frac{\{f(x+h)-f(x)\} - \{g(x+h)-g(x)\}}{h} = \lim_{h\to 0}\frac{f(x+h)-f(x)}{h} - \lim_{h\to 0}\frac{g(x+h)-g(x)}{h}$$

$$= \mathbf{D}(f(x)) - \mathbf{D}(g(x)) .$$

The proof of part (b) is very similar to these two proofs, and is left for you as the next Practice Problem. The proof for the Product Rule and Quotient Rules will be given later.

Practice 5: Prove part (b) of the theorem, the Sum Rule: **D**(f(x) + g(x)) = **D**(f(x)) + **D**(g(x)) .

Practice 6: Use the Main Differentiation Theorem and the values in the table to fill in the rest of the table.

x	f(x)	f '(x)	g(x)	g '(x)	**D**(f(x)•g(x))	**D**(f(x)/g(x))	**D**(g(x)/f(x))
0	3	–2	–4	3			
1	2	–1	1	0			
2	4	2	3	1			

Example 6: Determine $\mathbf{D}(\ x^2 \cdot \sin(x)\ )$ and $\frac{\mathbf{d}}{\mathbf{dx}}\left(\frac{x^3}{\sin(x)}\right)$.

Solution: (a) We can use the Product Rule with $f(x) = x^2$ and $g(x) = \sin(x)$:

$$\begin{aligned}\mathbf{D}(\ x^2 \cdot \sin(x)\ ) = \mathbf{D}(\ f(x) \cdot g(x)\ ) &= f(x) \cdot \mathbf{D}(\ g(x)\ ) + g(x) \cdot \mathbf{D}(\ f(x)\ ) \\ &= (\ x^2\ ) \cdot \mathbf{D}(\ \sin(x)\ ) + \sin(x) \cdot \mathbf{D}(\ x^2\ ) \\ &= (\ x^2\ ) \cdot (\ \mathbf{\cos(x)}\ ) + \sin(x) \cdot (\ \mathbf{2x}\ ) = x^2 \cdot \cos(x) + 2x \cdot \sin(x)\end{aligned}$$

(b) We can use the Quotient Rule with $f(x) = x^3$ and $g(x) = \sin(x)$:

$$\frac{\mathbf{d}}{\mathbf{dx}}\left(\frac{x^3}{\sin(x)}\right) = \frac{g(x) \cdot \mathbf{D}(\ f(x)\ ) - f(x) \cdot \mathbf{D}(\ g(x)\ )}{g^2(x)} = \frac{\sin(x) \cdot \mathbf{D}(\ x^3\ ) - x^3 \cdot \mathbf{D}(\ \sin(x)\ )}{(\ \sin(x)\ )^2}$$

$$= \frac{\sin(x) \cdot (\mathbf{3x^2}) - x^3 \cdot \mathbf{\cos(x)}}{\sin^2(x)} = \frac{3x^2 \sin(x) - x^3 \cos(x)}{\sin^2(x)}$$

Practice 7: Determine $\mathbf{D}(\ (x^2 + 1)(7x - 3)\ )$, $\frac{\mathbf{d}}{\mathbf{dt}}\left(\frac{3t - 2}{5t + 1}\right)$ and $\mathbf{D}\left(\frac{\cos(x)}{x}\right)$.

Proof of the Product Rule: The proofs of parts (d) and (e) of the theorem are complicated but only involve elementary techniques, used in just the right way. Sometimes we will omit such computational proofs, but the Product and Quotient Rules are fundamental techniques you will need hundreds of times.

By the hypothesis, f and g are differentiable so $\lim_{h \to 0} \frac{f(x+h) - f(x)}{h} = f\,'(x)$

and $\lim_{h \to 0} \frac{g(x+h) - g(x)}{h} = g\,'(x)$.

Also, both f and g are continuous (why?) so $\lim_{h \to 0} f(x+h) = f(x)$ and $\lim_{h \to 0} g(x+h) = g(x)$.

(d) Product Rule: Let $P(x) = f(x) \cdot g(x)$. Then $P(x+h) = f(x+h) \cdot g(x+h)$.

$$\mathbf{D}(\ f(x) \cdot g(x)\ ) = \mathbf{D}(\ P(x)\ ) = \lim_{h \to 0} \frac{P(x+h) - P(x)}{h} = \lim_{h \to 0} \frac{f(x+h)g(x+h) - f(x)g(x)}{h}$$

$$= \lim_{h \to 0} \frac{f(x+h)g(x+h) + \{-f(x)g(x+h) + f(x)g(x+h)\} - f(x)g(x)}{h}$$ adding and subtracting f(x)g(x+h)

$$= \lim_{h \to 0} \frac{f(x+h)g(x+h) - f(x)g(x+h)}{h} + \lim_{h \to 0} \frac{f(x)g(x+h) - f(x)g(x)}{h}$$ regrouping the terms

$$= \lim_{h \to 0} (\ g(x+h)\ ) \cdot \left(\frac{f(x+h) - f(x)}{h}\right) + (\ \mathbf{f(x)}\ ) \cdot \left(\frac{g(x+h) - g(x)}{h}\right)$$ finding common factors

taking limit as $h \to 0$:

$$(\ g(x)\ ) \cdot (\ f\,'(x)\ ) + (\ f(x)\ ) \cdot (\ g\,'(x)\ ) = g\,\mathbf{Df} + f\,\mathbf{Dg}\ .$$

(e) The steps for a proof of the Quotient Rule are shown in Problem 55.

USING THE DIFFERENTIATION RULES

You definitely need to memorize the differentiation rules, but it is vitally important that you also know how to use them. Sometimes it is clear that the function we want to differentiate is a sum or product of two obvious functions, but we commonly need to differentiate functions which involve several operations and functions. Memorizing the differentiation rules is only the first step in learning to use them.

Example 7: Calculate $\mathbf{D}(x^5 + x\cdot\sin(x))$.

Solution: This function is more difficult because it involves both an addition and a multiplication. Which rule(s) should we use, or, more importantly, which rule should we use first?

$\mathbf{D}(x^5 + x\cdot\sin(x)) = \mathbf{D}(x^5) + \mathbf{D}(x\cdot\sin(x))$ applying the Sum Rule and trading one derivative for two easier ones

$= 5x^4 + \{ x\cdot\mathbf{D}(\sin(x)) + \sin(x)\cdot\mathbf{D}(x) \}$ applying the product rule to $\mathbf{D}(x\cdot\sin(x))$

$= 5x^4 + x\cdot\cos(x) + \sin(x)$ this expression has no more derivatives so we are done.

If you were evaluating the function $x^5 + x \sin(x)$ for some particular value of x, you would (1) raise x to the 5th power, (2) calculate $\sin(x)$, (3) multiply $\sin(x)$ by x, and (4) your FINAL evaluation step, SUM the values of x^5 and $x \sin(x)$.

The FINAL step of your evaluation of f indicates the FIRST rule to use to calculate the derivative of f.

Practice 8: Which differentiation rule should you apply FIRST for each of the following:

(a) $x\cdot\cos(x) - x^3\cdot\sin(x)$ (b) $(2x - 3)\cdot\cos(x)$ (c) $2\cos(x) - 7x^2$ (d) $\dfrac{\cos(x) + 3x}{\sqrt{x}}$

Practice 9: Calculate $\mathbf{D}\left(\dfrac{x^2 - 5}{\sin(x)} \right)$ and $\dfrac{d}{dt}\left(\dfrac{t^2 - 5}{t\cdot\sin(t)} \right)$.

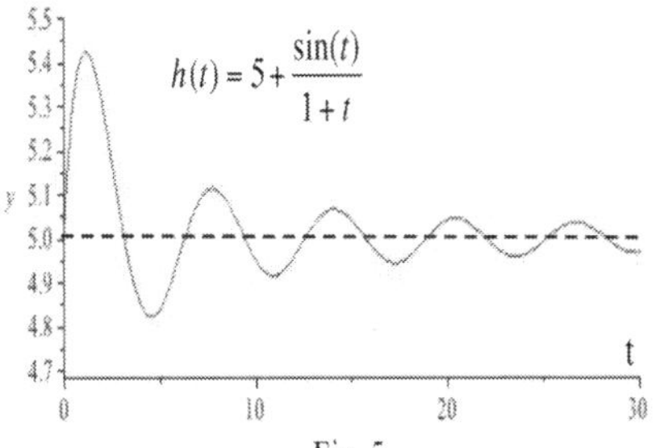

Fig. 5

Example 8: A weight attached to a spring is oscillating up and down. Over a period of time, the motion becomes "damped" because of friction and air resistance (Fig. 5), and its height at time t seconds is $h(t) = 5 + \dfrac{\sin(t)}{1 + t}$ feet.

What are the height and velocity of the weight after 2 seconds?

Solution: The height is

$h(2) = 5 + \dfrac{\sin(2)}{1 + 2} = 5 + \dfrac{.909}{3} = 5.303$ feet above the ground. The velocity is $h'(2)$.

$$h'(t) = 0 + \frac{(1+t)\cdot\mathbf{D}(\sin(t)) - \sin(t)\cdot\mathbf{D}(1+t)}{(1+t)^2} = \frac{(1+t)\cdot\cos(t) - \sin(t)}{(1+t)^2}$$

$$\text{so } h'(2) = \frac{3\cos(2) - \sin(2)}{9} = \frac{-2.158}{9} \approx -0.24 \text{ feet per second .}$$

Practice 10: What are the height and velocity of the weight in the previous example after 5 seconds? What are the height and velocity of the weight be after a "long time" ?

Example 9: Calculate $\mathbf{D}(\, x\cdot\sin(x)\cdot\cos(x)\,)$.

Solution: Clearly we need to use the Product Rule since the only operation in this function is multiplication, but the Product Rule deals with a product of <u>two</u> functions and we have the product of <u>three</u> ; x and sin(x) and cos(x) . However, if we think of our two functions as $f(x) = \mathbf{x\cdot\sin(x)}$ and $g(x) = \mathbf{\cos(x)}$, then we do have the product of two functions and

$$\begin{aligned}\mathbf{D}(\, x\cdot\sin(x)\cdot\cos(x)\,) &= \mathbf{D}(\, f(x)\cdot g(x)\,) = f(x)\cdot\mathbf{D}(\, g(x)\,) + g(x)\cdot\mathbf{D}(\, f(x)\,) \\ &= x\sin(x)\cdot\mathbf{D}(\,\cos(x)\,) + \cos(x)\cdot\mathbf{D}(\, x\sin(x)\,)\end{aligned}$$

We are not done, but we have traded one hard derivative for two easier ones. We know that $\mathbf{D}(\,\cos(x)\,) = -\sin(x)$, and we can use the Product Rule (again) to calculate $\mathbf{D}(\, x\sin(x)\,)$. Then the last line of our calculation becomes

$$= x\sin(x)\cdot(\,-\sin(x)\,) + \cos(x)\cdot\Big\{\, x\,\mathbf{D}(\,\sin(x)\,) + \sin(x)\,\mathbf{D}(\, x\,)\,\Big\}$$

$$= -x\sin^2(x) + \cos(x)\Big\{\, x\,\mathbf{\cos(x)} + \sin(x)\,\mathbf{(1)}\Big\} = -x\sin^2(x) + x\cos^2(x) + \cos(x)\sin(x)\ .$$

EVALUATING A DERIVATIVE AT A POINT

The derivative of a function f is a new **function f '(x)** which gives the slope of the line tangent to the graph of f at each point x. To find the slope of the tangent line at a particular point (c, f(c)) on the graph of f, we should <u>first</u> calculate the derivative f '(x) and <u>then</u> evaluate the **function f '(x)** at the point x = c to get the **number f '(c)**. If you mistakenly evaluate f first, you get a number f(c), and the derivative of a constant is always equal to 0.

Example 10: Determine the slope of the line tangent to $f(x) = 3x + \sin(x)$ at (0, f(0)) and (1, f(1)):

Solution: $f'(x) = \mathbf{D}(\, 3x + \sin(x)\,) = \mathbf{D}(3x) + \mathbf{D}(\,\sin(x)\,) = 3 + \cos(x)$. When $x = 0$, the graph of $y = 3x + \sin(x)$ goes through the point $(\, 0, 3(0)+\sin(0)\,)$ with slope $f'(0) = 3 + \cos(0) = 4$. When $x = 1$, the graph goes through the point $(\, 1, 3(1)+\sin(1)\,) = (1, 3.84)$ with slope $f'(1) = 3 + \cos(1) \approx 3.54$.

Practice 11: Where do $f(x) = x^2 - 10x + 3$ and $g(x) = x^3 - 12x$ have a horizontal tangent lines ?

IMPORTANT RESULTS OF THIS SECTION

Differentiability and Continuity: If a function is differentiable then it must be continuous.

If a function is not continuous then it cannot be differentiable.

A function may be continuous at a point and not differentiable there.

Graphically: CONTINUOUS means **connected.**

DIFFERENTIABLE means continuous, **smooth** and not vertical.

Differentiation Patterns:

$\mathbf{D}(\,kf(x)\,) = k\cdot\mathbf{D}(\,f(x)\,)$

$\mathbf{D}(\,f + g\,) = \mathbf{D}f + \mathbf{D}g$

$\mathbf{D}(\,f - g\,) = \mathbf{D}f - \mathbf{D}g$

$\mathbf{D}(\,f\cdot g\,) = f\cdot\mathbf{D}g + g\cdot\mathbf{D}f$

$$\mathbf{D}(\,f/g\,) = \frac{g\cdot\mathbf{D}f - f\cdot\mathbf{D}g}{g^2}$$

The FINAL STEP used to evaluate f indicates the FIRST RULE to use to differentiate f.

To evaluate a derivative at a point, first differentiate and then evaluate.

PROBLEMS

1. The graph of $y = f(x)$ is given in Fig. 6.
   (a) At which integers is f continuous?
   (b) At which integers is f differentiable?

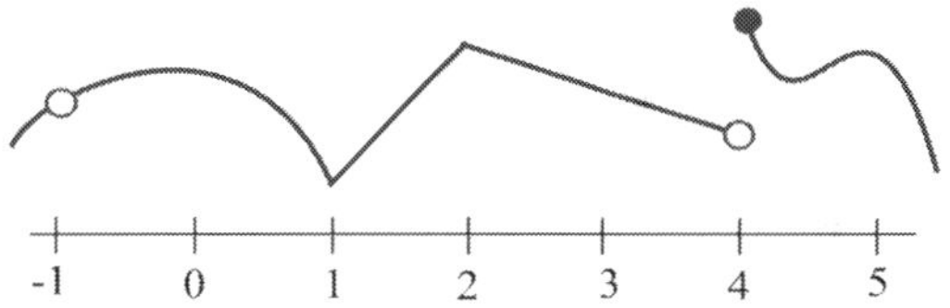

Fig. 6

2. The graph of $y = g(x)$ is given in Fig. 7.
   (a) At which integers is g continuous?
   (b) At which integers is g differentiable?

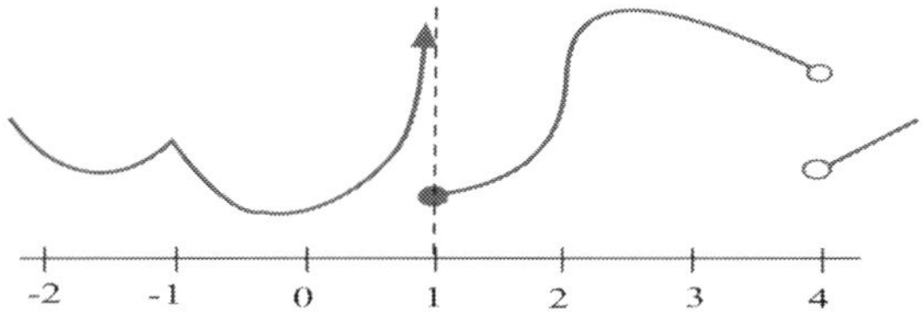

Fig. 7

3. Use the values given in the table to determine the values of $f\cdot g$, $\mathbf{D}(\,f\cdot g\,)$, f/g and $\mathbf{D}(\,f/g\,)$.

x	f(x)	f '(x)	g(x)	g '(x)	f(x)·g(x)	**D**(f(x)·g(x))	f(x)/g(x)	**D**(f(x)/g(x))
0	2	3	1	5				
1	–3	2	5	–2				
2	0	–3	2	4				
3	1	–1	0	3				

4. Use the values given in the table to determine the values of $f\cdot g$, $\mathbf{D}(\,f\cdot g\,)$, f/g and $\mathbf{D}(\,f/g\,)$.

x	f(x)	f '(x)	g(x)	g '(x)	f(x)•g(x)	**D**(f(x)•g(x))	f(x)/g(x)	**D**(f(x)/g(x))
0	4	2	3	–3				
1	0	3	2	1				
2	–2	5	0	–1				
3	–1	–2	–3	4				

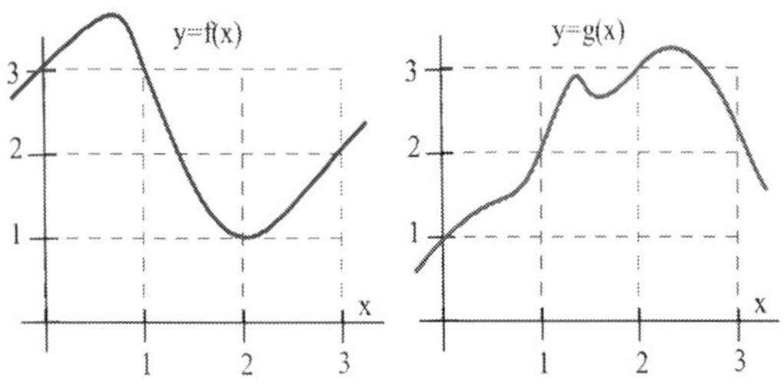

Fig. 8

5. Use the information in Fig. 8 to plot the values of the functions f + g, f•g and f/g and their derivatives at x = 1, 2 and 3 .

6. Use the information in Fig. 8 to plot the values of the functions 2f, f – g and g/f and their derivatives at x = 1, 2 and 3 .

7. Calculate **D**((x – 5)(3x + 7)) by (a) using the product rule and (b) expanding the product and then differentiating. Verify that both methods give the same result.

8. Calculate **D**($\sqrt{x}$ •sin(x)) .

9. Calculate $\frac{d}{dx}\left(\frac{\cos(x)}{x^2}\right)$.

10. Calculate **D**(sin(x) + cos(x)) .

11. Calculate **D**($\sin^2(x)$) and **D**($\cos^2(x)$) .

12. Calculate **D**(sin(x)), $\frac{d}{dx}$(sin(x) + 7) , **D**(sin(x) – 8000) and **D**(sin(x) + k).

13. Find values for the constants a, b and c so that the parabola $f(x) = ax^2 + bx + c$ has f(0) = 0, f '(0) = 0 and f '(10) = 30.

14. If f is a differentiable function,
(a) how are the graphs of y = f(x) and y = f(x) + k related?
(b) how are the derivatives of f(x) and f(x) + k related?

15. If f and g are differentiable functions which always differ by a constant (f(x) – g(x) = k for all x), then what can you conclude about their graphs and their derivatives?

16. If f and g are differentiable functions whose sum is a constant (f(x) + g(x) = k for all x), then what can you conclude about (a) their graphs? (b) their derivatives?

17. If the product of f and g is a constant (f(x)g(x) = k for all x), then how are $\frac{\mathbf{D}(f(x))}{f(x)}$ and $\frac{\mathbf{D}(g(x))}{g(x)}$ related?

18. If the quotient of f and g is a constant (f(x)/g(x) = k for all x), then how are g•f ' and f•g ' related?

In problems 19 – 28, (a) calculate f '(1) and (b) determine when f '(x) = 0.

19. $f(x) = x^2 - 5x + 13$

20. $f(x) = 5x^2 - 40x + 73$

21. $f(x) = 3x - 2\cos(x)$

22. $f(x) = | x + 2 |$

23. $f(x) = x^3 + 9x^2 + 6$

24. $f(x) = x^3 + 3x^2 + 3x - 1$

25. $f(x) = x^3 + 2x^2 + 2x - 1$

26. $f(x) = \dfrac{7x}{x^2 + 4}$

27. $f(x) = x \cdot \sin(x)$ and $0 \le x \le 5$. (You may need to use the Bisection Algorithm or the "trace" option on a calculator to approximate where $f'(x) = 0$.)

28. $f(x) = A x^2 + B x + C$ A, B and C are constants and $A \ne 0$.

29. $f(x) = x^3 + A x^2 + B x + C$ with constants A, B and C. Can you find conditions on the constants A, B and C which will guarantee that the graph of $y = f(x)$ has two distinct "vertices"? (Here a "vertex" means a place where the curve changes from increasing to decreasing or from decreasing to increasing.)

Where are the functions in problems 30 – 37 differentiable?

30. $f(x) = |x| \cos(x)$

31. $f(x) = \dfrac{x - 5}{x + 3}$

32. $f(x) = \tan(x)$

33. $f(x) = \dfrac{x^2 + x}{x^2 - 3x}$

34. $f(x) = | x^2 - 4 |$

35. $f(x) = | x^3 - 1 |$

36. $f(x) = \begin{cases} 0 & \text{if } x < 0 \\ \sin(x) & \text{if } x \ge 0 \end{cases}$

37. $f(x) = \begin{cases} x & \text{if } x < 0 \\ \sin(x) & \text{if } x \ge 0 \end{cases}$

38. For what value(s) of A is $f(x) = \begin{cases} Ax - 4 & \text{if } x < 2 \\ x^2 + x & \text{if } x \ge 2 \end{cases}$ differentiable at $x = 2$?

39. For what values of A and B is $f(x) = \begin{cases} Ax + B & \text{if } x < 1 \\ x^2 + x & \text{if } x \ge 1 \end{cases}$ differentiable at $x = 1$?

40. An arrow shot straight up from ground level with an initial velocity of 128 feet per second will be at height $h(x) = -16x^2 + 128x$ feet at x seconds. (Fig.9)

(a) Determine the velocity of the arrow when $x = 0, 1$ and 2 seconds.

(b) What is the velocity of the arrow, $v(x)$, at any time x?

(c) At what time x will the velocity of the arrow be 0?

(d) What is the greatest height the arrow reaches?

(e) How long will the arrow be aloft?

(f) Use the answer for the velocity in part (b) to determine the acceleration, $a(x) = v'(x)$, at any time x.

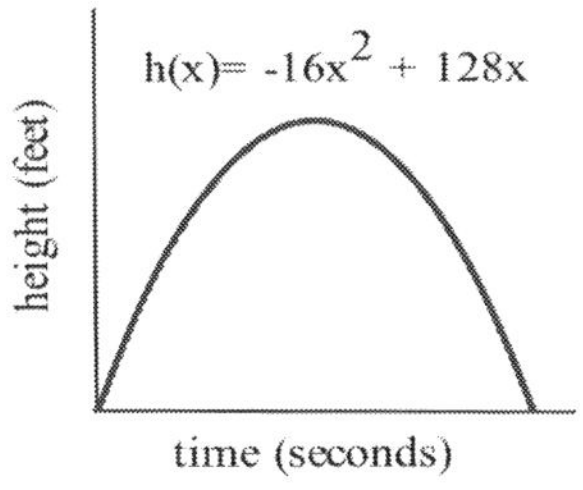

Fig. 9

41. If an arrow is shot straight up from ground level on the moon with an initial velocity of 128 feet per second, its height will be $h(x) = -2.65x^2 + 128x$ feet at x seconds. Do parts (a) – (e) of problem 40 using this new equation for h.

42. In general, if an arrow is shot straight upward with an initial velocity of 128 feet per second from ground level on a planet with a constant gravitational acceleration of g feet per second^2, then its height will be $h(x) = -\frac{g}{2}x^2 + 128x$ feet at x seconds. Answer the questions in problem 40 for arrows shot on Mars and Jupiter (Use the values in Fig. 10).

43. If an object on Earth is propelled upward from ground level with an initial velocity of v_0 feet per second, then its height at x seconds will be $h(x) = -16x^2 + v_0x$.

(a) What will be the object's velocity after x seconds?

(b) What is the greatest height the object will reach?

(c) How long will the object remain aloft?

Object	g (ft/sec^2)	g (cm/sec^2)
Mercury	11.8	358
Venus	20.1	887
Earth	32.2	981
moon	5.3	162
Mars	12.3	374
Jupiter	85.3	2601
Saturn	36.6	1117
Uranus	34.4	1049
Neptune	43.5	1325
Pluto	7.3	221

Source: CRC Handbook of Chemistry and Physics

Fig. 10: Values of g

44. For a 6 foot tall basketball player to dunk the ball, the player must achieve a vertical jump of about 3 feet. Use the information in the previous problems to answer the following questions.

(a) What is the smallest initial vertical velocity the player can have to dunk the ball?

(b) With the initial velocity achieved in part (a), how high would the player jump on the moon?

45. The best high jumpers in the world manage to lift their centers of mass approximately 6.5 feet.

(a) What is the initial vertical velocity these high jumpers attain?

(b) How long are these high jumpers in the air?

(c) With the initial velocity in part (a), how high would they lift their centers of mass on the moon?

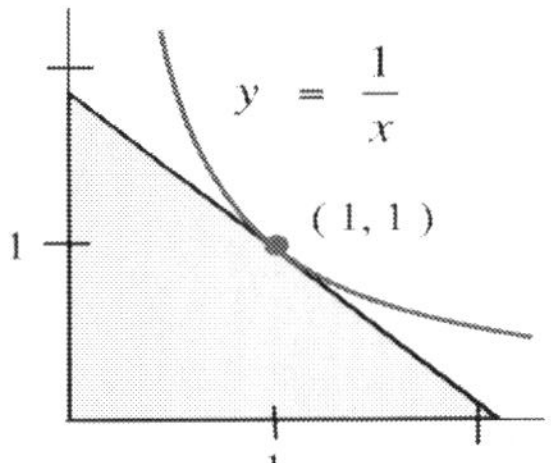

Fig. 11

46. (a) Find the equation of the line L which is tangent to the curve $y = \frac{1}{x}$ at the point (1,1) .

(b) Determine where L intersects the x–axis and the y–axis.

(c) Determine the area of the region in the first quadrant bounded by L, the x–axis and the y–axis. (Fig. 11)

47. (a) Find the equation of the line L which is tangent to the curve $y = \frac{1}{x}$ at the point $(2, \frac{1}{2})$.

(b) Graph $y = 1/x$ and L and determine where L intersects the x–axis and the y–axis.

(c) Determine the area of the region in the first quadrant bounded by L, the x–axis and the y–axis.

48. (a) Find the equation of the line L which is tangent to the curve $y = \frac{1}{x}$ at the point $(p, \frac{1}{p})$, $p \geq 0$.

(b) Determine where L intersects the x–axis and the y–axis.

(c) Determine the area of the region in the first quadrant bounded by L, the x–axis and the y–axis.

(d) How does the area of the triangle in part (c) depend on the initial point $(p, \frac{1}{p})$?

49. Find values for the coefficients a, b and c so that the parabola $f(x) = ax^2 + bx + c$ goes through the point (1,4) and is tangent to the line $y = 9x - 13$ at the point (3, 14).

50. Find values for the coefficients a, b and c so that the parabola $f(x) = ax^2 + bx + c$ goes through the point (0,1) and is tangent to the line $y = 3x - 2$ at the point (2,4).

51. (a) Find a function f so that $\mathbf{D}(f(x)) = 3x^2$.

(b) Find another function g so that $\mathbf{D}(g(x)) = 3x^2$.

(c) Can you find more functions whose derivatives are $3x^2$?

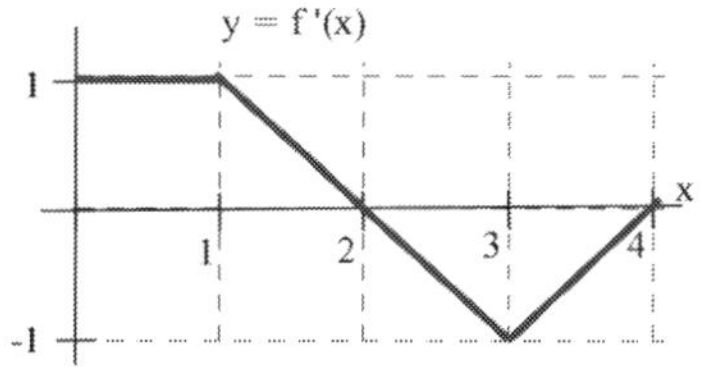

Fig. 12

52. (a) Find a function f so that $\mathbf{f}'(x) = 6x + \cos(x)$.

(b) Find another function g so that $\mathbf{g}'(x) = 6x + \cos(x)$.

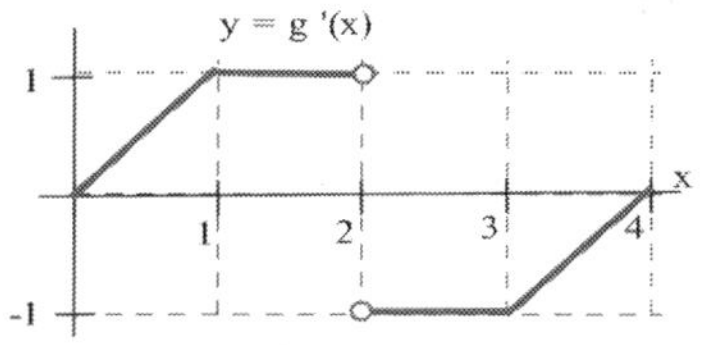

Fig. 13

53. The graph of $y = f'(x)$ is given in Fig. 12.

(a) Assume $f(0) = 0$ and sketch the graph of $y = f(x)$.

(b) Assume $f(0) = 1$ and graph $y = f(x)$.

54. The graph of $y = g'(x)$ is given in Fig. 13. Assume that g is continuous.

(a) Assume $g(0) = 0$ and sketch the graph of $y = g(x)$.

(b) Assume $g(0) = 1$ and graph $y = g(x)$.

55. Assume that f and g are differentiable functions and that $g(x) \neq 0$. State why each step in the following proof of the Quotient Rule is valid.

$$\mathbf{D}\left(\frac{f(x)}{g(x)} \right) = \lim_{h \to 0} \frac{1}{h} \left\{ \frac{f(x+h)}{g(x+h)} - \frac{f(x)}{g(x)} \right\} = \lim_{h \to 0} \frac{1}{h} \left\{ \frac{f(x+h)g(x) - g(x+h)f(x)}{g(x+h)g(x)} \right\}$$

$$= \lim_{h \to 0} \frac{1}{g(x+h)g(x)} \left\{ \frac{f(x+h)g(x) + \left(-f(x)g(x) + f(x)g(x) \right) - g(x+h)f(x)}{h} \right\}$$

$$= \lim_{h \to 0} \frac{1}{g(x+h)g(x)} \left\{ g(x) \frac{f(x+h) - f(x)}{h} - f(x) \frac{g(x+h) - g(x)}{h} \right\}$$

$$= \frac{1}{g^2(x)} \{ g(x) \cdot f'(x) - f(x) \cdot g'(x) \} = \frac{g(x) \cdot f'(x) - f(x) \cdot g'(x)}{g^2(x)} .$$

Section 2.2 **PRACTICE Answers**

Practice 1: f is continuous at x = –1, 0, 2, 4, 6, and 7. f is differentiable at x = –1, 2, 4, and 7.

Practice 2: $f(x) = 6x + 8$ and $g(x) = 2$ so $\mathbf{D}(f(x)) = 6$ and $\mathbf{D}(g(x)) = 0$.

$\mathbf{D}(3f(x)) = 3\mathbf{D}(f(x)) = 3(6) = \mathbf{18}$, $\mathbf{D}(f(x) + g(x)) = \mathbf{D}(f(x)) + \mathbf{D}(g(x)) = 6 + 0 = \mathbf{6}$

$\mathbf{D}(f(x) - g(x)) = \mathbf{D}(f(x)) - \mathbf{D}(g(x)) = 6 - 0 = \mathbf{6}$

$\mathbf{D}(f(x)g(x)) = f(x)\mathbf{D}(g(x)) + g(x)\mathbf{D}(f(x)) = (6x + 8)(0) + (2)(6) = \mathbf{12}$

$$\mathbf{D}(f(x)/g(x)) = \frac{g(x)\mathbf{D}(f(x)) - f(x)\mathbf{D}(g(x))}{(g(x))^2} = \frac{(2)(6) - (6x+8)(0)}{2^2} = \frac{12}{4} = \mathbf{3}$$

Practice 3: $\mathbf{D}(x^3 - 5\sin(x)) = \mathbf{D}(x^3) - 5\mathbf{D}(\sin(x)) = \mathbf{3x^2 - 5\cos(x)}$

$$\frac{\mathbf{d}}{\mathbf{dx}}(\sin(x) - 4x^3) = \frac{\mathbf{d}}{\mathbf{dx}}\sin(x) - 4\frac{\mathbf{d}}{\mathbf{dx}}x^3 = \mathbf{\cos(x) - 12x^2}$$

Practice 4:

x	f(x)	f '(x)	g(x)	g '(x)	**D**(3f(x))	**D**(2f(x) + g(x))	**D**(3g(x) – f(x))
0	3	–2	–4	3	**–6**	**–1**	**11**
1	2	–1	1	0	**–3**	**–2**	**1**
2	4	2	3	1	**6**	**5**	**1**

Practice 5:

$$\mathbf{D}(f(x) + g(x)) = \lim_{h\to 0} \frac{\{f(x+h) + g(x+h)\} - \{f(x) + g(x)\}}{h}$$

$$= \lim_{h\to 0} \frac{f(x+h) - f(x) + g(x+h) - g(x)}{h}$$

$$= \left(\lim_{h\to 0} \frac{f(x+h) - f(x)}{h}\right) + \left(\lim_{h\to 0} \frac{g(x+h) - g(x)}{h}\right) = \mathbf{D}(f(x)) + \mathbf{D}(g(x)) .$$

Practice 6:

x	f(x)	f '(x)	g(x)	g '(x)	**D**(f(x)·g(x))	**D**(f(x)/g(x))	**D**(g(x)/f(x))
0	3	–2	–4	3	3·3+(–4)(–2)=**17**	$\frac{-4(-2)-(3)(3)}{(-4)^2} = \mathbf{\frac{-1}{16}}$	$\frac{(3)(3)-(-4)(-2)}{3^2} = \mathbf{\frac{1}{9}}$
1	2	–1	1	0	2·0+1(–1) = **–1**	$\frac{1(-1)-(2)(0)}{1^2} = \mathbf{-1}$	$\frac{2(0)-1(-1)}{2^2} = \mathbf{\frac{1}{4}}$
2	4	2	3	1	4·1 + 3·2 = **10**	$\frac{3(2)-(4)(1)}{3^2} = \mathbf{\frac{2}{9}}$	$\frac{4(1)-3(2)}{4^2} = \mathbf{\frac{-2}{16}}$

Practice 7:

$\mathbf{D}((x^2+1)(7x-3)) = (x^2+1)\mathbf{D}(7x-3) + (7x-3)\mathbf{D}(x^2+1) = (x^2+1)(\mathbf{7}) + (7x-3)(\mathbf{2x}) = \mathbf{21x^2 - 6x + 7}$

or $\mathbf{D}((x^2+1)(7x-3)) = \mathbf{D}(7x^3 - 3x^2 + 7x) = \mathbf{21x^2 - 6x + 7}$

$$\frac{\mathbf{d}}{\mathbf{dt}}\left(\frac{3t-2}{5t+1}\right) = \frac{(5t+1)\mathbf{D}(3t-2) - (3t-2)\mathbf{D}(5t+1)}{(5t+1)^2} = \frac{(5t+1)(\mathbf{3}) - (3t-2)(\mathbf{5})}{(5t+1)^2} = \frac{13}{(5t+1)^2}$$

$$\mathbf{D}\left(\frac{\cos(x)}{x}\right) = \frac{x\mathbf{D}(\cos(x)) - \cos(x)\mathbf{D}(x)}{(x)^2} = \frac{x(\mathbf{-\sin(x)}) - \cos(x)(\mathbf{1})}{x^2} = \frac{-x\cdot\sin(x) - \cos(x)}{x^2}$$

Practice 8: (a) difference rule (b) product rule (c) difference rule (d) quotient rule

Practice 9:

$$\mathbf{D}\left(\frac{x^2-5}{\sin(x)}\right) = \frac{\sin(x)\mathbf{D}(x^2-5) - (x^2-5)\mathbf{D}(\sin(x))}{(\sin(x))^2} = \frac{\sin(x)(\mathbf{2x}) - (x^2-5)\mathbf{\cos(x)}}{\sin^2(x)}$$

$$\frac{\mathbf{d}}{\mathbf{dt}}\left(\frac{t^2-5}{t\cdot\sin(t)}\right) = \frac{t\cdot\sin(t)\mathbf{D}(t^2-5) - (t^2-5)\mathbf{D}(t\cdot\sin(t))}{(t\cdot\sin(t))^2} = \frac{t\cdot\sin(t)(\mathbf{2t}) - (t^2-5)\{\mathbf{t\cdot\cos(t) + \sin(t)}\}}{t^2\cdot\sin^2(t)}$$

Practice 10: (a) $h(5) = 5 + \frac{\sin(5)}{1+5} \approx \mathbf{4.84\ ft.}$ $v(5) = h'(5) = \frac{(1+5)\cos(5) - \sin(5)}{(1+5)^2} \approx \mathbf{0.074\ ft/sec.}$

"long time": $\lim_{t\to\infty} h(t) = \lim_{t\to\infty} 5 + \frac{\sin(t)}{1+t} = \mathbf{5\ feet}$.

$$\lim_{t\to\infty} h'(t) = \lim_{t\to\infty} \frac{(1+t)\cos(t) - \sin(t)}{(1+t)^2} = \lim_{t\to\infty}\left\{\frac{\cos(t)}{1+t} - \frac{\sin(t)}{(1+t)^2}\right\} = \mathbf{0\ ft/sec.}$$

Practice 11: $f'(x) = 2x - 10$. $f'(x) = 0$ when $2x - 10 = 0$ so when $\mathbf{x = 5}$.

$g'(x) = 3x^2 - 12$. $g'(x) = 0$ when $3x^2 - 12 = 0$ so $x^2 = 4$ and $\mathbf{x = -2, +2}$.

2.3 MORE DIFFERENTIATION PATTERNS

Polynomials are very useful, but they are not the only functions we need. This section uses the ideas of the two previous sections to develop techniques for differentiating **powers of functions**, and to determine the derivatives of some particular functions which occur often in applications, the **trigonometric** and **exponential** functions.

As you focus on learning how to differentiate different types and combinations of functions, it is important to remember what derivatives are and what they measure. Calculators and personal computers are available to calculate derivatives. Part of your job as a professional will be to decide which functions need to be differentiated and how to use the resulting derivatives. You can succeed at that only if you understand what a derivative is and what it measures.

A POWER RULE FOR FUNCTIONS: $\mathbf{D}(f^n(x))$

If we apply the Product Rule to the product of a function with itself, a familiar pattern emerges.

$\mathbf{D}(f^2) = \mathbf{D}(f \cdot f) = f \cdot \mathbf{D}(f) + f \cdot \mathbf{D}(f) = 2f \cdot \mathbf{D}(\mathbf{f})$.

$\mathbf{D}(f^3) = \mathbf{D}(f^2 \cdot f) = f^2 \cdot \mathbf{D}(f) + f \cdot \mathbf{D}(f^2) = f^2 \cdot \mathbf{D}(f) + f \{ 2f \cdot \mathbf{D}(f) \} = f^2 \cdot \mathbf{D}(f) + 2f^2 \cdot \mathbf{D}(f) = 3 f^2 \cdot \mathbf{D}(\mathbf{f})$.

$\mathbf{D}(f^4) = \mathbf{D}(f^3 \cdot f) = f^3 \cdot \mathbf{D}(f) + f \cdot \mathbf{D}(f^3) = f^3 \cdot \mathbf{D}(f) + f\{ 3f^2 \cdot \mathbf{D}(f)\} = f^3 \cdot \mathbf{D}(f) + 3f^3 \cdot \mathbf{D}(f) = 4 f^3 \cdot \mathbf{D}(\mathbf{f})$.

Practice 1: What is the pattern here? What do you think the results will be for $\mathbf{D}(f^5)$ and $\mathbf{D}(f^{13})$?

We could keep differentiating higher and higher powers of f(x) by writing them as products of lower powers of f(x) and using the Product Rule, but the Power Rule For Functions guarantees that the pattern we just saw for the small integer powers also works for all constant powers of functions.

Power Rule For Functions:	If	n is any constant,
	then	$\mathbf{D}(f^n(x)) = n f^{n-1}(x) \cdot \mathbf{D}(\mathbf{f(x)})$.

The Power Rule for Functions is a special case of a more general theorem, the Chain Rule, which we will examine in Section 2.4. The Power Rule For Functions will be proved after the Chain Rule.

Example 1: Use the Power Rule for Functions to find

(a) $\mathbf{D}((x^3 - 5)^2)$ (b) $\frac{\mathbf{d}}{\mathbf{dx}}(\sqrt{2x + 3x^5})$ (c) $\mathbf{D}(\sin^2(x)) = \mathbf{D}((\sin(x))^2)$.

Solution: (a) To match the pattern of the Power Rule for $\mathbf{D}((x^3 - 5)^2)$, let $f(x) = x^3 - 5$ and $n = 2$.

Then $\mathbf{D}((x^3 - 5)^2) = \mathbf{D}(f^n(x)) = n f^{n-1}(x) \cdot \mathbf{D}(f(x))$

$= 2(x^3 - 5)^1 \mathbf{D}(x^3 - 5) = 2(x^3 - 5) (3x^2) = \mathbf{6x^2(x^3 - 5)}$.

(b) To match the pattern for $\frac{\mathbf{d}}{\mathbf{dx}}(\sqrt{2x + 3x^5}) = \frac{\mathbf{d}}{\mathbf{dx}}((2x + 3x^5)^{1/2})$, we can let $f(x) = 2x + 3x^5$ and take $n = 1/2$. Then

$$\frac{\mathbf{d}}{\mathbf{dx}}((2x + 3x^5)^{1/2}) = \frac{\mathbf{d}}{\mathbf{dx}}(f^n(x)) = n\, f^{n-1}(x) \cdot \frac{\mathbf{d}}{\mathbf{dx}}(f(x)) = \frac{1}{2}(2x + 3x^5)^{-1/2} \frac{\mathbf{d}}{\mathbf{dx}}(2x + 3x^5)$$

$$= \frac{\mathbf{1}}{\mathbf{2}}(\mathbf{2x + 3x^5})^{\mathbf{-1/2}} (\mathbf{2 + 15x^4}) = \frac{2 + 15x^4}{2\sqrt{2x + 3x^5}} .$$

(c) To match the pattern for $\mathbf{D}(\sin^2(x))$, Let $f(x) = \sin(x)$ and $n = 2$. Then

$$\mathbf{D}(\sin^2(x)) = \mathbf{D}(f^n(x)) = n\, f^{n-1}(x) \cdot \mathbf{D}(f(x)) = 2\sin^1(x)\, \mathbf{D}(\sin(x)) = \mathbf{2\,\sin(x)\,\cos(x)} .$$

Practice 2: Use the Power Rule for Functions to find
(a) $\frac{\mathbf{d}}{\mathbf{dx}}((2x^5 - \pi)^2)$, (b) $\mathbf{D}(\sqrt{x + 7x^2})$, (c) $\mathbf{D}(\cos^4(x)) = \mathbf{D}((\cos(x))^4)$.

Example 2: Use calculus to show that the line tangent to the circle $x^2 + y^2 = 25$ at the point (3,4) has slope –3/4 .

Solution: The top half of the circle is the graph of $y = f(x) = \sqrt{25 - x^2}$ so $f'(x) = \mathbf{D}((25 - x^2)^{1/2})$

$$= \frac{1}{2} (25 - x^2)^{-1/2}\, \mathbf{D}(25 - x^2) = \frac{-x}{\sqrt{25 - x^2}} \quad \text{and} \quad f'(3) = \frac{-3}{\sqrt{25 - 3^2}} = \frac{-3}{4} .$$

As a check, you can verify that the slope of the radial line through the center of the circle (0,0) and the point (3,4) has slope 4/3 and is perpendicular to the tangent line which has a slope of –3/4.

DERIVATIVES OF TRIGONOMETRIC AND EXPONENTIAL FUNCTIONS

We have some general rules which apply to any elementary combination of differentiable functions, but in order to use the rules we still need to know the derivatives of each of the particular functions. Here we will add to the list of functions whose derivatives we know.

Derivatives of the Trigonometric Functions

We know the derivatives of the sine and cosine functions, and each of the other four trigonometric functions is just a ratio involving sines or cosines. Using the Quotient Rule, we can differentiate the rest of the trigonometric functions.

Theorem:	$\mathbf{D}(\tan(x))$	=	$\sec^2(x)$	$\mathbf{D}(\sec(x))$	=	$\sec(x)\tan(x)$
	$\mathbf{D}(\cot(x))$	=	$-\csc^2(x)$	$\mathbf{D}(\csc(x))$	=	$-\csc(x)\cot(x)$.

Proof: From trigonometry we know $\tan(x) = \frac{\sin(x)}{\cos(x)}$, $\cot(x) = \frac{\cos(x)}{\sin(x)}$, $\sec(x) = \frac{1}{\cos(x)}$, and $\csc(x) = \frac{1}{\sin(x)}$,

and we know **D**(sin(x)) = cos(x) and **D**(cos(x)) = – sin(x) . Using the Quotient Rule,

$$\mathbf{D}(\tan(x)) = \mathbf{D}\left(\frac{\sin(x)}{\cos(x)}\right) = \frac{\cos(x)\cdot\mathbf{D}(\sin(x)) - \sin(x)\cdot\mathbf{D}(\cos(x))}{(\cos(x))^2}$$

$$= \frac{\cos(x)\,\mathbf{\cos(x)} - \sin(x)\{\,\mathbf{-\sin(x)}\,\}}{\cos^2(x)} = \frac{\cos^2(x) + \sin^2(x)}{\cos^2(x)} = \frac{1}{\cos^2(x)} = \mathbf{sec^2(x)}\,.$$

$$\mathbf{D}(\sec(x)) = \mathbf{D}\left(\frac{1}{\cos(x)}\right) = \frac{\cos(x)\,\mathbf{D}(1) - 1\,\mathbf{D}(\cos(x))}{\cos^2(x)}$$

$$= \frac{\cos(x)\,(\,\mathbf{0}\,) - 1\{\,\mathbf{-\sin(x)}\,\}}{\cos^2(x)} = \frac{\sin(x)}{\cos^2(x)} = \frac{\sin(x)}{\cos(x)}\,\frac{1}{\cos(x)} = \mathbf{tan(x)\cdot sec(x)}\,.$$

Instead of the Quotient Rule, we could have used the Power Rule to calculate **D**(sec(x)) = **D**($(\cos(x))^{-1}$) .

Practice 3: Use the Quotient Rule on $f(x) = \cot(x) = \frac{\cos(x)}{\sin(x)}$ to prove that $f'(x) = -\csc^2(x)$.

Practice 4: Prove that **D**(csc(x)) = – csc(x)·cot(x) . The justification of this result is very similar to the justification for **D**(sec(x)).

Practice 5: Find (a) **D**($x^5\cdot\tan(x)$) , (b) $\frac{\mathbf{d}}{\mathbf{dt}}\left(\frac{\sec(t)}{t}\right)$ and (c) **D**($\sqrt{\cot(x) - x}$) .

Derivative of e^x

We can use graphs of exponential functions to estimate the slopes of their tangent lines or we can numerically approximate the slopes.

Example 3: Estimate the derivative of $f(x) = 2^x$ at the point $(0, 2^0) = (0, 1)$ by approximating the slope of the line tangent to $f(x) = 2^x$ at that point.

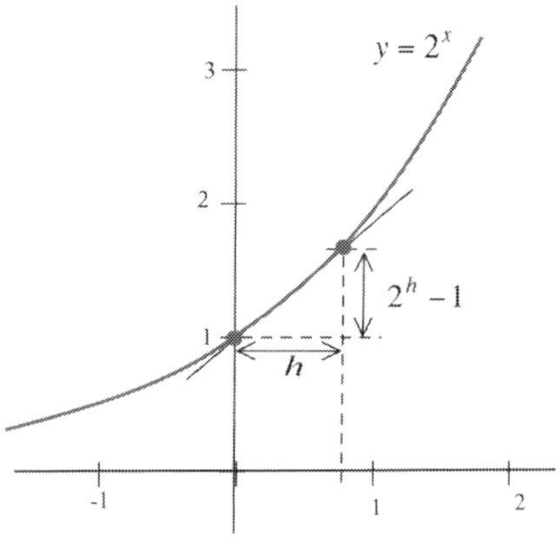

Fig. 1

Solution: We can get estimates from the graph of $f(x) = 2^x$ by carefully graphing $f(x) = 2^x$ for small values of x, sketching secant lines, and then measuring the slopes of the secant lines (Fig. 1).

We can also find the slope numerically by using the definition of the derivative,

$$f'(0) \equiv \lim_{h\to 0} \frac{f(0+h) - f(0)}{h} = \lim_{h\to 0} \frac{2^{0+h} - 2^0}{h} = \lim_{h\to 0} \frac{2^h - 1}{h}$$, and evaluating $\frac{2^h - 1}{h}$

for some very small values of h.

h	$\frac{2^h - 1}{h}$	$\frac{3^h - 1}{h}$	$\frac{e^h - 1}{h}$
0.1	0.717734625		
–0.1	0.669670084		
0.01	0.69555		
–0.01	0.690750451		
0.001	0.6933874		
–0.001	0.69290695		
↓	↓	↓	↓
0	≈ 0.693	≈ 1.099	1

From the table we can see that $f'(0) \approx .693$.

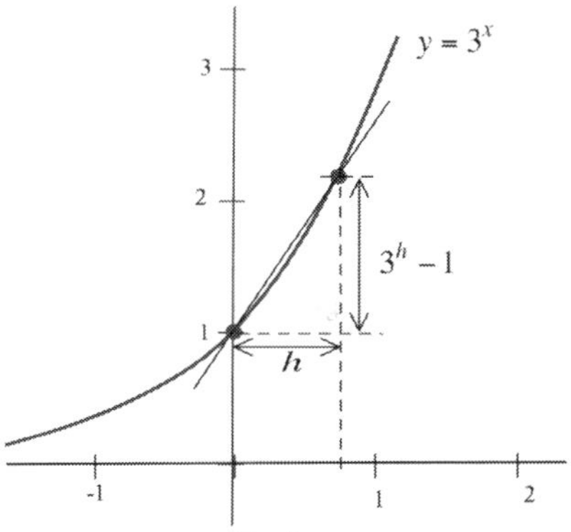

Fig. 2

Practice 6: Fill in the table for $\frac{3^h - 1}{h}$, and show that the slope of the line tangent to $g(x) = 3^X$ at (0,1) is approximately 1.099 . (Fig. 2)

At (0,1), the slope of the tangent to $y = 2^X$ is less than 1 , and the slope of the tangent to $y = 3^X$ is slightly greater than 1. (Fig. 3) There is a number, denoted **e**, between 2 and 3 so that the slope of the tangent to $y = e^X$ is exactly 1: $\lim_{h \to 0} \frac{e^h - 1}{h} = 1$. The number $e \approx 2.71828182845904$.

e is irrational and is very important and common in calculus and applications.

Once we grant that $\lim_{h \to 0} \frac{e^h - 1}{h} = 1$, it is relatively straightforward to calculate **D**(e^X).

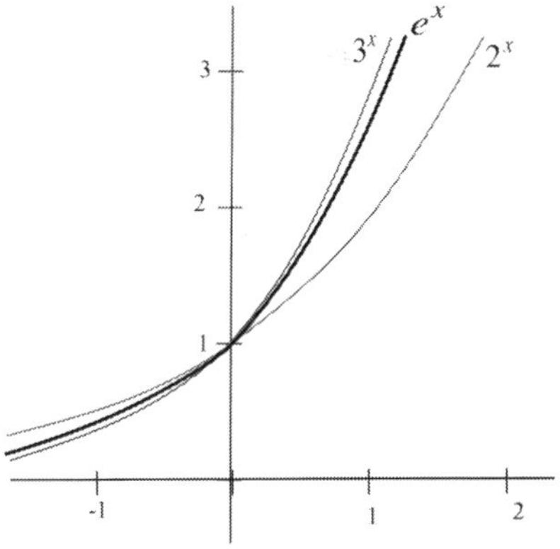

Fig. 3

> **Theorem:** **D**(e^X) = $\mathbf{e^X}$.

Proof: **D**(e^X) $\equiv \lim_{h \to 0} \frac{e^{x+h} - e^x}{h} = \lim_{h \to 0} \frac{e^x \cdot e^h - e^x}{h}$

$$= \lim_{h \to 0} \left(e^x\right) \cdot \left(\frac{e^h - 1}{h}\right)$$

$$= \lim_{h \to 0} \left(e^x\right) \cdot \lim_{h \to 0} \left(\frac{e^h - 1}{h}\right) = (e^X)(1) = \mathbf{e^X} .$$

The function $f(x) = e^X$ is its own derivative: $f'(x) = f(x)$. The **height** of $f(x) = e^X$ at any point and the **slope** of the tangent to $f(x) = e^X$ at that point are the same: as the graph gets higher, its slope gets steeper.

Example 4: Find (a) $\frac{d}{dt}(t \cdot e^t)$, (b) $D(e^x/\sin(x))$ and (c) $D(e^{5x}) = D((e^x)^5)$

Solution: (a) Using the Product Rule with $f(t) = t$ and $g(t) = e^t$,

$$\frac{d}{dt}(t \cdot e^t) = t \cdot D(e^t) + e^t \cdot D(t) = t \cdot e^t + e^t \cdot (1) = t \cdot e^t + e^t = (t+1)e^t .$$

(b) Using the Quotient Rule with $f(x) = e^x$ and $g(x) = \sin(x)$,

$$D\left(\frac{e^x}{\sin(x)}\right) = \frac{\sin(x)\, D(e^x) - e^x\, D(\sin(x))}{\sin^2(x)} = \frac{\sin(x)\, e^x - e^x \cos(x)}{\sin^2(x)} .$$

(c) Using the Power Rule for Functions with $f(x) = e^x$ and $n = 5$,

$$D((e^x)^5) = 5(e^x)^4 \cdot D(e^x) = 5(e^x)^4 \cdot e^x = 5e^{4x}e^x = 5e^{5x} .$$

Practice 7: Find (a) $D(x^3 e^x)$ and (b) $D((e^x)^3)$.

Higher Derivatives: Derivatives of Derivatives

The derivative of a function f is a new function **f '** , and we can calculate the derivative of this new function to get the derivative of the derivative of f, denoted by f '' and called the second derivative of f. For example, if $f(x) = x^5$ then $f'(x) = 5x^4$ and $f''(x) = (f'(x))' = (5x^4)' = 20x^3$.

Definitions:	The first derivative of f is $f'(x)$, the rate of change of f.
	The second derivative of f is $f''(x) = (f'(x))'$, the rate of change of f' .
	The third derivative of f is $f'''(x) = (f''(x))'$, the rate of change of f'' .

For $y = f(x)$, $f'(x) = \frac{dy}{dx}$, $f''(x) = \frac{d}{dx}\left(\frac{dy}{dx}\right) = \frac{d^2y}{dx^2}$, $f'''(x) = \frac{d}{dx}\left(\frac{d^2y}{dx^2}\right) = \frac{d^3y}{dx^3}$, and so on.

Practice 8: Find f ', f '', and f ''' for $f(x) = 3x^7$, $f(x) = \sin(x)$, and $f(x) = x\cos(x)$.

If $f(x)$ represents the position of a particle at time x, then $v(x) = f'(x)$ will represent the velocity (rate of change of the position) of the particle and $a(x) = v'(x) = f''(x)$ will represent the acceleration (the rate of change of the velocity) of the particle.

Example 5: The height (feet) of a particle at time t seconds is $t^3 - 4t^2 + 8t$. Find the height, velocity and acceleration of the particle when $t = 0, 1,$ and 2 seconds.

Solution: $f(t) = t^3 - 4t^2 + 8t$ so $f(0) = 0$ feet, $f(1) = 5$ feet, and $f(2) = 8$ feet.

The velocity is $v(t) = f'(t) = 3t^2 - 8t + 8$ so $v(0) = 8$ ft/s , $v(1) = 3$ ft/s, and $v(2) = 4$ ft/s. At each of these times the velocity is positive and the particle is moving upward, increasing in height.

The acceleration is $a(t) = 6t - 8$ so $a(0) = -8$ ft/s^2 , $a(1) = -2$ ft/s^2 and $a(2) = 4$ ft/s^2 .

We will examine the geometric meaning of the second derivative later.

A REALLY "BENT" FUNCTION

In Section 1.2 we saw that the "holey" function $h(x) = \begin{cases} 2 & \text{if } x \text{ is a } \textbf{rational} \text{ number} \\ 1 & \text{if } x \text{ is an } \textbf{irrational} \text{ number} \end{cases}$

is discontinuous at every value of x, so at every x h(x) is not differentiable. We can create graphs of continuous functions that are not differentiable at several places just by putting corners at those places, but how many corners can a continuous function have? How badly can a continuous function fail to be differentiable?

In the mid–1800s, the German mathematician Karl Weierstrass surprised and even shocked the mathematical world by creating a function which was **continuous everywhere but differentiable nowhere** — a function whose graph was everywhere connected and everywhere bent! He used techniques we have not investigated yet, but we can start to see how such a function could be built.

Start with a function f_1 (Fig. 4) which zigzags between the values +1/2 and –1/2 and has a "corner" at each integer. This starting function f_1 is continuous everywhere and is differentiable everywhere except at the integers. Next create a list of functions $f_2, f_3, f_4, \ldots,$ each of which is a lot shorter but with many more "corners" than the previous ones. For example, we might make f_2 zigzag between the values +1/4 and –1/4 and have "corners" at $\pm 1/2, \pm 3/2, \pm 5/2$, etc., and f_3 zigzag between +1/9 and –1/9 and have "corners" at $\pm 1/3, \pm 2/3, \pm 4/3$, etc. If we add f_1 and f_2, we get a continuous function (since the sum of two continuous functions is continuous) which will have corners at $0, \pm 1/2, \pm 1, \pm 3/2, \ldots$ If we then add f_3 to the previous sum, we get a new continuous function with even more corners. If we continue adding the functions in our list "indefinitely", the final result will be a continuous function which is differentiable nowhere.

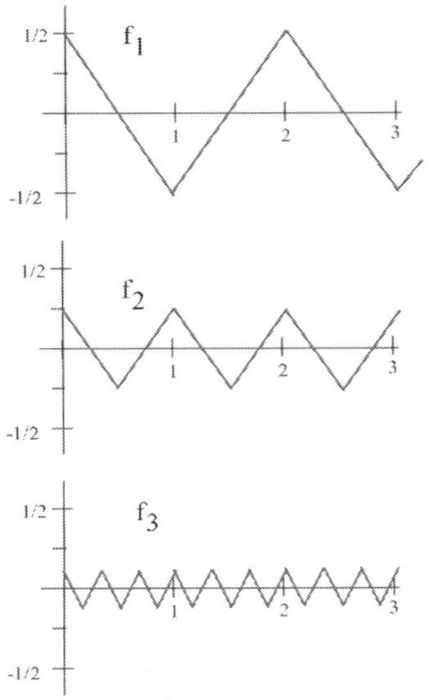

Fig. 4

We haven't developed enough mathematics here to precisely describe what it means to add an infinite number of functions together or to verify that the resulting function is nowhere differentiable, but we will. You can at least start to imagine what a strange, totally "bent" function it must be.

Until Weierstrass created his "everywhere continuous, nowhere differentiable" function, most mathematicians thought a continuous function could only be "bad" in a few places, and Weierstrass' function was (and is) considered "pathological", a great example of how bad something can be. The mathematician Hermite expressed a reaction shared by many when they first encounter Weierstrass' function:

"I turn away with fright and horror from this lamentable evil of functions which do not have derivatives."

IMPORTANT RESULTS

Power Rule For Functions: $\mathbf{D}(f^n(x)) = n \cdot f^{n-1}(x) \cdot \mathbf{D}(\mathbf{f(x)})$

Derivatives of the Trigonometric Functions:

$\mathbf{D}(\sin(x)) = \cos(x)$ $\mathbf{D}(\tan(x)) = \sec^2(x)$ $\mathbf{D}(\sec(x)) = \sec(x)\tan(x)$

$\mathbf{D}(\cos(x)) = -\sin(x)$ $\mathbf{D}(\cot(x)) = -\csc^2(x)$ $\mathbf{D}(\csc(x)) = -\csc(x)\cot(x)$

Derivatives of the Exponential Function: $\mathbf{D}(e^x) = e^x$

PROBLEMS

1. Let $f(1) = 2$ and $f'(1) = 3$. Find the values of $\mathbf{D}(f^2(x))$, $\mathbf{D}(f^5(x))$, and $\mathbf{D}(\sqrt{f(x)})$ at x=1.

2. Let $f(2) = -2$ and $f'(2) = 5$. Find the values of $\mathbf{D}(f^2(x))$, $\mathbf{D}(f^{-3}(x))$, and $\frac{\mathbf{d}}{\mathbf{dx}}(\sqrt{f(x)})$ at x=2.

3. Estimate the values of $f(x)$ and $f'(x)$ in Fig. 5 and determine

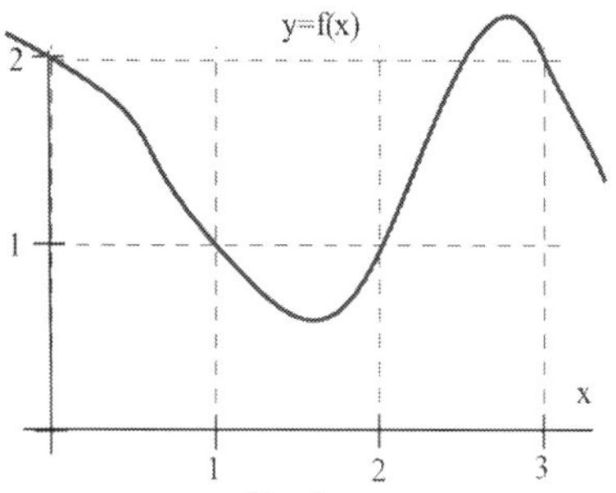

Fig. 5

(a) $\frac{\mathbf{d}}{\mathbf{dx}}(f^2(x))$ at $x = 1$ and 3 (b) $\mathbf{D}(f^3(x))$ at $x = 1$ and 3

(c) $\mathbf{D}(f^5(x))$ at $x = 1$ and 3 .

4. Estimate the values of $f(x)$ and $f'(x)$ in Fig. 5 and determine

(a) $\mathbf{D}(f^2(x))$ at $x = 0$ and 2 (b) $\frac{\mathbf{d}}{\mathbf{dx}}(f^3(x))$ at $x = 0$ and 2

(c) $\frac{\mathbf{d}}{\mathbf{dx}}(f^5(x))$ at $x = 0$ and 2.

In problems 5 – 10 , find the derivative of each function.

5. $f(x) = (2x - 8)^5$ 6. $f(x) = (6x - x^2)^{10}$ 7. $f(x) = x \cdot (3x + 7)^5$

8. $f(x) = (2x + 3)^6 \cdot (x - 2)^4$ 9. $f(x) = \sqrt{x^2 + 6x - 1}$ 10. $f(x) = \dfrac{x - 5}{(x + 3)^4}$

11. A weight attached to a spring is at a height of $h(t) = 3 - 2\sin(t)$ feet above the floor t seconds after it is released. (a) Graph h(t) (b) At what height is the weight when it is released?
(c) How high does the weight ever get above the floor and how close to the floor does it ever get?
(d) Determine the height, velocity and acceleration at time t. (Be sure to include the correct units.)
(e) Why is this an unrealistic model of the motion of a weight on a real spring?

12. A weight attached to a spring is at a height of $h(t) = 3 - \dfrac{2\sin(t)}{1 + 0.1t^2}$ feet above the floor t seconds after it is released. (a) Graph h(t) (b) At what height is the weight when it is released?

(c) Determine the height and velocity at time t.

(d) What happens to the height and the velocity of the weight after a "long time?"

13. The kinetic energy K of an object of mass m and velocity v is $\frac{1}{2} m \cdot v^2$.
(a) Find the kinetic energy of an object with mass m and height $h(t) = 5t$ feet at $t = 1$ and 2 seconds.
(b) Find the kinetic energy of an object with mass m and height $h(t) = t^2$ feet at $t = 1$ and 2 seconds.

14. An object of mass m is attached to a spring and has height $h(t) = 3 + \sin(t)$ feet at time t seconds.
(a) Find the height and kinetic energy of the object when $t = 1, 2,$ and 3 seconds.
(b) Find the rate of change in the kinetic energy of the object when $t = 1, 2,$ and 3 seconds.
(c) Can K ever be negative? Can dK/dt ever be negative? Why?

In problems 15 – 20, find the derivatives **df/dx** .

15. $f(x) = x \cdot \sin(x)$ 16. $f(x) = \sin^5(x)$ 17. $f(x) = e^x - \sec(x)$

18. $f(x) = \sqrt{\cos(x) + 1}$ 19. $f(x) = e^{-x} + \sin(x)$ 20. $f(x) = \sqrt{x^2 - 4x + 3}$

In problems 21 – 26, find the equation of the line tangent to the graph of the function at the given point.

21. $f(x) = (x - 5)^7$ at $(4, -1)$ 22. $f(x) = e^x$ at $(0,1)$ 23. $f(x) = \sqrt{25 - x^2}$ at $(3,4)$

24. $f(x) = \sin^3(x)$ at $(\pi,0)$ 25. $f(x) = (x - a)^5$ at $(a,0)$ 26. $f(x) = x \cdot \cos^5(x)$ at $(0, 0)$

27. Find the equation of the line tangent to $f(x) = e^x$ at the point $(3, e^3)$. Where will this tangent line intersect the x–axis? Where will the tangent line to $f(x) = e^x$ at the point (p, e^p) intersect the x–axis?

In problems 28 – 33, calculate **f '** and **f ''**.

28. $f(x) = 7x^2 + 5x - 3$ 29. $f(x) = \cos(x)$ 30. $f(x) = \sin(x)$

31. $f(x) = x^2 \cdot \sin(x)$ 32. $f(x) = x \cdot \sin(x)$ 33. $f(x) = e^x \cdot \cos(x)$

34. Calculate the first 8 derivatives of $f(x) = \sin(x)$. What is the pattern?
What is the 208th derivative of $\sin(x)$?

35. What will the 2nd derivative of a quadratic polynomial be? The 3rd derivative? The 4th derivative?

36. What will the 3rd derivative of a cubic polynomial be? The 4th derivative?

37. What can you say about the nth and (n+1)st derivatives of a polynomial of degree n?

In problems 38 – 42, you are given f '. Find a function f with the given derivative.

38. $f'(x) = 4x + 2$ 39. $f'(x) = 5e^x$ 40. $f'(x) = 3 \cdot \sin^2(x) \cdot \cos(x)$

41. $f'(x) = 5(1 + e^x)^4 \cdot e^x$ 42. $f'(x) = e^x + \sin(x)$

43. The function $f(x) = \begin{cases} x \cdot \sin(1/x) & \text{if } x \neq 0 \\ 0 & \text{if } x = 0 \end{cases}$ in Fig. 6 is continuous at 0 since $\lim_{h \to 0} f(x) = 0 = f(0)$. Is f differentiable at 0? (Use the definition of f '(0) and consider $\lim_{h \to 0} \dfrac{f(0+h) - f(0)}{h}$.)

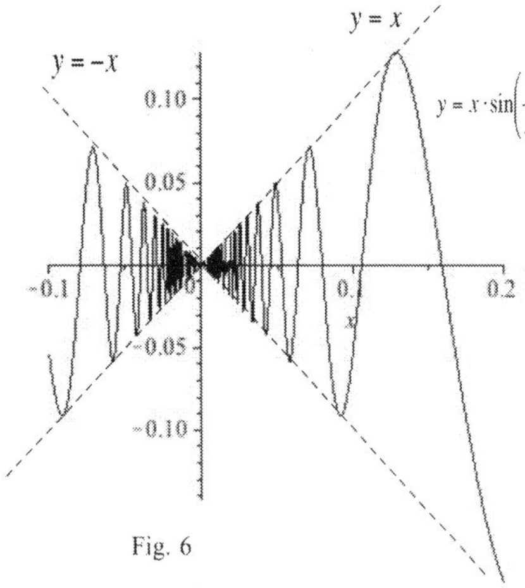

Fig. 6

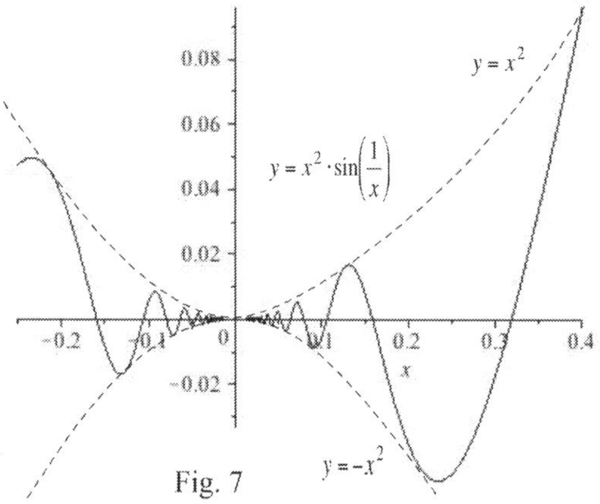

Fig. 7

44. The function $f(x) = \begin{cases} x^2 \cdot \sin(1/x) & \text{if } x \neq 0 \\ 0 & \text{if } x = 0 \end{cases}$ in Fig. 7 is continuous at 0 since $\lim_{h \to 0} f(x) = 0 = f(0)$. Is f differentiable at 0? (Use the definition of f '(0) and consider $\lim_{h \to 0} \dfrac{f(0+h) - f(0)}{h}$.)

The number e appears in a variety of unusual situations. Problems 45 – 48 illustrate a few of them.

45. Use your calculator to examine the values of $(1 + \frac{1}{x})^x$ when x is relatively large, for example, x = 100, 1000, and 10000. Try some other large values for x. If x is large, the value of $(1 + \frac{1}{x})^x$ is close to what number?

46. If you put \$1 into a bank which pays 1% interest per year and compounds the interest x times a year, then after one year you will have earned $(1 + \frac{.01}{x})^x$ dollars in the bank.

(a) How much money will you have after 1 year if the bank calculates the interest once a year?

(b) How much money will you have after 1 year if the bank calculates the interest twice a year?

(c) How much money will you have after 1 year if the bank calculates the interest 365 times a year?

(d) How does your answer in part (c) compare with $e^{.01}$?

47. (a) Calculate the value of the sums $s_1 = 1 + \frac{1}{1!}$, $s_2 = 1 + \frac{1}{1!} + \frac{1}{2!}$, $s_3 = 1 + \frac{1}{1!} + \frac{1}{2!} + \frac{1}{3!}$,

$s_4 = 1 + \frac{1}{1!} + \frac{1}{2!} + \frac{1}{3!} + \frac{1}{4!}$, $s_5 = 1 + \frac{1}{1!} + \frac{1}{2!} + \frac{1}{3!} + \frac{1}{4!} + \frac{1}{5!}$, and

$s_6 = 1 + \frac{1}{1!} + \frac{1}{2!} + \frac{1}{3!} + \frac{1}{4!} + \frac{1}{5!} + \frac{1}{6!}$.

(b) What value do the sums in part (a) seem to be approaching? Calculate s_7 and s_8 .

(n! = product of all positive integers from 1 to n. For example, 2! = 1·2 = 2, 3! = 1·2·3 = 6, 4! = 24.)

48. If it is late at night and you are tired of studying calculus, try the following experiment with a friend. Take the 2 through 10 of hearts from a regular deck of cards and shuffle these 9 cards well. Have your friend do the same with the 2 through 10 of spades. Now compare your cards one at a time. If there is a match, for example you both play a 5, then the game is over and you win. If you make it through the entire 9 cards with no match, then your friend wins. If you play the game **many times**, then the ratio $\frac{\text{total number of games played}}{\text{number of times your friend wins}}$ will be approximately equal to e.

Section 2.3 **PRACTICE Answers**

Practice 1: The pattern is $\mathbf{D}(f^n(x)) = n f^{n-1}(x)\cdot\mathbf{D}(\mathbf{f(x)})$. $\mathbf{D}(f^5) = 5f^4\,\mathbf{D(f)}$ and $\mathbf{D}(f^{13}) = 13f^{12}\mathbf{D(f)}$.

Practice 2: $\frac{d}{dx}(2x^5 - \pi)^2 = 2(2x^5 - \pi)^1\,\mathbf{D}(2x^5 - \pi) = 2(2x^5 - \pi)^1\,(\mathbf{10x^4}) = 40x^9 - 20\pi x^4$.

$\mathbf{D}((x + 7x^2)^{1/2}) = \frac{1}{2}(x + 7x^2)^{-1/2}\,\mathbf{D}(x + 7x^2) = \frac{\mathbf{1 + 14x}}{2\sqrt{x + 7x^2}}$.

$\mathbf{D}((\cos(x))^4) = 4(\cos(x))^3\mathbf{D}(\cos(x)) = 4(\cos(x))^3\,(\mathbf{-\sin(x)}) = \mathbf{-4\cos^3(x)\sin(x)}$.

Practice 3: $\mathbf{D}(\frac{\cos(x)}{\sin(x)}) = \frac{\sin(x)\mathbf{D}(\cos(x)) - \cos(x)\mathbf{D}(\sin(x))}{(\sin(x))^2}$

$= \frac{\sin(x)(\mathbf{-\sin(x)}) - \cos(x)(\mathbf{\cos(x)})}{\sin^2(x)} = \frac{-(\sin^2(x) + \cos^2(x))}{\sin^2(x)} = \frac{-1}{\sin^2(x)} = \mathbf{-\csc^2(x)}$.

Practice 4: $\mathbf{D}(\csc(x)) = \mathbf{D}(\frac{1}{\sin(x)}) = \frac{\sin(x)\mathbf{D}(1) - 1\mathbf{D}(\sin(x))}{\sin^2(x)}$

$= \frac{\sin(x)(\mathbf{0}) - \mathbf{\cos(x)}}{\sin^2(x)} = -\frac{\cos(x)}{\sin(x)}\frac{1}{\sin(x)} = \mathbf{-\cot(x)\csc(x)}$.

Practice 5: $\mathbf{D}(x^5 \cdot \tan(x)) = x^5 \mathbf{D}(\tan(x)) + \tan(x)\mathbf{D}(x^5) = x^5 \mathbf{sec^2(x)} + \tan(x)\mathbf{(5x^4)}$.

$$\frac{\mathbf{d}}{\mathbf{dt}}\left(\frac{\sec(t)}{t}\right) = \frac{t\mathbf{D}(\sec(t)) - \sec(t)\mathbf{D}(t)}{t^2} = \frac{t \cdot \sec(t) \cdot \tan(t) - \sec(t)}{t^2} .$$

$$\mathbf{D}((\cot(x) - x)^{1/2}) = \frac{1}{2}(\cot(x) - x)^{-1/2}\,\mathbf{D}(\cot(x) - x)$$

$$= \frac{1}{2}(\cot(x) - x)^{-1/2}\,(\mathbf{-csc^2(x) - 1}) = \frac{-\csc^2(x) - 1}{2\sqrt{\cot(x) - x}} .$$

Practice 6:

h	$\frac{2^h - 1}{h}$	$\frac{3^h - 1}{h}$	$\frac{e^h - 1}{h}$
0.1	0.717734625	**1.16123174**	1.051709181
–0.1	0.669670084	**1.040415402**	0.9516258196
0.01	0.69555	**1.104669194**	1.005016708
–0.01	0.690750451	**1.092599583**	0.9950166251
0.001	0.6933874	**1.099215984**	1.000500167
–0.001	0.69290695	**1.098009035**	0.9995001666
↓	↓	↓	↓
0	≈ 0.693	≈ 1.099	1

Practice 7: $\mathbf{D}(x^3 e^x) = x^3\,\mathbf{D}(e^x) + e^x\,\mathbf{D}(x^3) = x^3\,(\mathbf{e^x}) + e^x\,(\mathbf{3x^2}) = \mathbf{x^2 \cdot e^x \cdot (x + 3)}$.

$$\mathbf{D}((e^x)^3) = 3(e^x)^2\,\mathbf{D}(e^x) = 3(e^x)^2\,(\mathbf{e^x}) = 3e^{2x} \cdot e^x = \mathbf{3\,e^{3x}} \quad \text{or}$$

$$\mathbf{D}((e^x)^3) = \mathbf{D}(e^{3x}) = e^{3x}\,\mathbf{D}(3x) = \mathbf{3\,e^{3x}} .$$

Practice 8:

$f(x) = 3x^7$	$f(x) = \sin(x)$	$f(x) = x \cdot \cos(x)$
$f'(x) = 21x^6$	$f'(x) = \cos(x)$	$f'(x) = -x \cdot \sin(x) + \cos(x)$
$f''(x) = 126x^5$	$f''(x) = -\sin(x)$	$f''(x) = -x \cdot \cos(x) - 2\sin(x)$
$f'''(x) = 630x^4$	$f'''(x) = -\cos(x)$	$f'''(x) = x \cdot \sin(x) - 3\cos(x)$

2.4 THE CHAIN RULE

The Chain Rule is the most important and most used of the differentiation patterns. It enables us to differentiate **composites** of functions such as $y = \sin(x^2)$. It is a powerful tool for determining the derivatives of some **new functions** such as logarithms and inverse trigonometric functions. And it leads to important **applications** in a variety of fields. You will need the Chain Rule hundreds of times in this course, and practice with it now will save you time and points later. Fortunately, with some practice, the Chain Rule is also easy to use.

We already know how to differentiate the composition of some functions.

Example 1: For $f(x) = 5x - 4$ and $g(x) = 2x + 1$, find $f \circ g(x)$ and $\mathbf{D}(f \circ g(x))$.

Solution: $f \circ g(x) = f(g(x)) = 5(2x+1) - 4 = 10x + 1$, so $\mathbf{D}(f \circ g(x)) = \mathbf{D}(10x + 1) = 10$.

Practice 1: For $f(x) = 5x - 4$ and $g(x) = x^2$, find $f \circ g(x)$, $\mathbf{D}(f \circ g(x))$, $g \circ f(x)$, and $\mathbf{D}(g \circ f(x))$.

Some compositions, however, are still very difficult to differentiate. We know the derivatives of $g(x) = x^2$ and $h(x) = \sin(x)$, and we know how to differentiate some combinations of these functions such as $x^2 + \sin(x)$, $x^2 \cdot \sin(x)$, and even $\sin^2(x)$, but the derivative of the simple composition $f(x) = h \circ g(x) = \sin(x^2)$ is hard — until we know the Chain Rule. To see just how hard, try using the definition of derivative on it.

Example 2:

(a) Suppose amplifier Y doubles the strength of the output signal from amplifier U, and U triples the strength of the original signal x. How does the final signal out of Y compare with the signal x?

original signal x → [amplifier U] → [amplifier Y] → final signal

(b) Suppose y changes twice as fast as u, and u changes three times as fast as x. How does the rate of change of y compare with the rate of change of x?

Solution: In each case we are comparing the result of a composition, and the answer to each question is 6, the product of the two amplifications or rates of change.

In (a), we have that $\frac{\text{signal out of Y}}{\text{signal x}} = \frac{\text{signal out of Y}}{\text{signal out of U}} \cdot \frac{\text{signal out of U}}{\text{signal x}} = (2)(3) = 6.$

In (b), $\frac{\Delta y}{\Delta x} = \frac{\Delta y}{\Delta u} \cdot \frac{\Delta u}{\Delta x} = (2)(3) = 6.$

These examples are simple cases of the Chain Rule for differentiating a composition of functions.

THE CHAIN RULE

Chain Rule (Leibniz notation form)

If y is a differentiable function of u, and u is a differentiable function of x,

then y is a differentiable function of x and $\frac{dy}{dx} = \frac{dy}{du} \cdot \frac{du}{dx}$.

Idea for a proof: $\frac{dy}{dx} = \lim_{\Delta x \to 0} \frac{\Delta y}{\Delta u} \cdot \frac{\Delta u}{\Delta x}$ $\quad$ (*if* $\Delta u \neq 0$)

$$= \left(\lim_{\Delta x \to 0} \frac{\Delta y}{\Delta u} \right) \cdot \left(\lim_{\Delta x \to 0} \frac{\Delta u}{\Delta x} \right)$$

u is continuous, so $\Delta x \to 0$ implies $\Delta u \to 0$

$$= \frac{dy}{du} \cdot \frac{du}{dx}$$

Although this nice short argument gets to the heart of why the Chain Rule works, it is not quite valid. If du/dx $\neq 0$, then it is possible to show that $\Delta u \neq 0$ for all very small values of Δx, and the "idea for a proof" is a real proof. There are, however, functions for which $\Delta u = 0$ for lots of small values of Δx, and these create problems for the previous argument. A justification which is true for ALL cases is much more complicated.

The symbol $\frac{dy}{du}$ is a single symbol (as is $\frac{du}{dx}$) , and we <u>cannot</u> eliminate du from the product $\frac{dy}{du} \cdot \frac{du}{dx}$ in the Chain Rule. It is, however, perfectly fine to use the idea of eliminating du to help you remember the statement of the Chain Rule.

Example 3: $y = \cos(x^2 + 3)$ is $y = \cos(u)$ with $u = x^2 + 3$. Find dy/dx.

Solution: $y = \cos(u)$ so $dy/du = -\sin(u)$. $u = x^2 + 3$ so $du/dx = 2x$. Finally, using the Chain Rule,

$$\frac{dy}{dx} = \frac{dy}{du} \cdot \frac{du}{dx} = -\sin(u) \cdot 2x = -2x \cdot \sin(x^2 + 3) .$$

Practice 2: Find dy/dx for $y = \sin(4x + e^x)$.

There is also a composition of functions form of the Chain Rule. The notation is different, but it means precisely the same as the Leibniz form.

Chain Rule (composition form)

If g is differentiable at x and f is differentiable at g(x),

then the composite $f \circ g$ is differentiable at x , and $(f \circ g)'(x) = \mathbf{D}(f(g(x))) = \mathbf{f}'(g(x)) \cdot \mathbf{g}'(x)$.

You may find it easier to think of the composition form of the Chain Rule in words:

(f(g(x)) ' = "the derivative of the outside function (with respect to the original inside function) times the derivative of the inside function" where f is the outside function and g is the inside function.

Example 4: Differentiate $\sin(x^2)$.

Solution: The function $\sin(x^2)$ is the composition $f\circ g$ of two simple functions: $f(x) = \sin(x)$ and $g(x) = x^2$: $f\circ g\,(x) = f(g(x)) = f(x^2) = \sin(x^2)$ which is the function we want. Both f and g are differentiable functions with derivatives $f'(x) = \cos(x)$ and $g'(x) = 2x$, so, by the Chain Rule,

$$\mathbf{D}(\sin(x^2)) = (f\circ g)'(x) = \mathbf{f'}(g(x))\cdot\mathbf{g'}(x) = \mathbf{cos}(g(x))\cdot\mathbf{2x} = \cos(x^2)\cdot 2x = 2x\cos(x^2).$$

If you tried using the definition of derivative to calculate the derivative of this function at the beginning of the section, you can really appreciate the power of the Chain Rule for differentiating compositions.

Example 5: The table gives values for f , f ' , g and g ' at a number of points. Use these values to determine $(f\circ g)(x)$ and $(f\circ g)'(x)$ at $x = -1$ and 0.

x	f(x)	g(x)	f '(x)	g '(x)	(f∘g)(x)	(f∘g) '(x)
−1	2	3	1	0		
0	−1	1	3	2		
1	1	0	−1	3		
2	3	−1	0	1		
3	0	2	2	−1		

Solution: $(f\circ g)(-1) = f(g(-1)) = f(3) = 0$ and $(f\circ g)(0) = f(g(0)) = f(1) = 1$.

$(f\circ g)'(-1) = \mathbf{f'}(g(-1))\cdot\mathbf{g'}(-1) = \mathbf{f'}(3)\cdot(0) = (2)(0) = 0$ and

$(f\circ g)'(0) = \mathbf{f'}(g(0))\cdot\mathbf{g'}(0) = \mathbf{f'}(1)\cdot(2) = (-1)(2) = -2$.

Practice 3: Fill in the table in Example 5 for $(f\circ g)(x)$ and $(f\circ g)'(x)$ at $x = 1, 2$ and 3.

Neither form of the Chain Rule is inherently superior to the other — use the one you prefer. The Chain Rule will be used hundreds of times in the rest of this book, and it is important that you master its usage. The time you spend now mastering and understanding how to use the Chain Rule will be paid back tenfold in the next several chapters.

Example 6: Determine $\mathbf{D}(e^{\cos(x)})$ using each form of the Chain Rule.

Solution: Using the Leibniz notation: $y = e^u$ and $u = \cos(x)$. $dy/du = e^u$ and $du/dx = -\sin(x)$ so

$dy/dx = (dy/du)\cdot(du/dx) = (e^u)\cdot(-\sin(x)) = -\sin(x)\cdot e^{\cos(x)}$.

The function $e^{\cos(x)}$ is also the composition of $f(x) = e^x$ with $g(x) = \cos(x)$, so

$\mathbf{D}(e^{\cos(x)}) = \mathbf{f}'(g(x))\cdot\mathbf{g}'(x)$ by the Chain Rule

$= \mathbf{e}^{g(x)}\cdot(-\mathbf{sin}(x))$ since $\mathbf{D}(e^x) = e^x$ and $\mathbf{D}(\cos(x)) = -\sin(x)$

$= -\sin(x) \cdot e^{\cos(x)}$.

Practice 4: Calculate $\mathbf{D}(\sin(7x - 1))$, $\frac{\mathbf{d}}{\mathbf{dx}}(\sin(ax + b))$, and $\frac{\mathbf{d}}{\mathbf{dt}}(e^{3t})$.

Practice 5: Use the graph of g in Fig. 1 and the Chain Rule to estimate $\mathbf{D}(\sin(g(x)))$ and $\mathbf{D}(g(\sin(x)))$ at $x = \pi$.

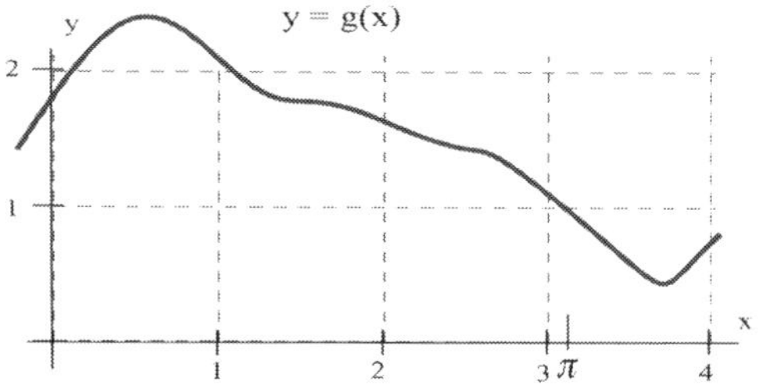

Fig. 1

The Chain Rule is a general differentiation pattern, and it can be used with the other general patterns such as the Product and Quotient Rules.

Example 7: Determine $\mathbf{D}(e^{3x} \cdot \sin(5x + 7))$ and $\frac{\mathbf{d}}{\mathbf{dx}}(\cos(x \cdot e^x))$.

Solution: (a) e^{3x} $\sin(5x + 7)$ is a product of two functions so we need the product rule first:

$\mathbf{D}(e^{3x} \cdot \sin(5x + 7)) = e^{3x} \cdot \mathbf{D}(\sin(5x + 7)) + \sin(5x + 7) \cdot \mathbf{D}(e^{3x})$

$= e^{3x}\, \mathbf{cos}(5x + 7) \cdot \mathbf{5} + \sin(5x + 7)\, \mathbf{e^{3x}} \cdot \mathbf{3} = 5\, e^{3x} \cos(5x + 7) + 3\, e^{3x} \sin(5x + 7)$.

(b) $\cos(x \cdot e^x)$ is a composition of cosine with a product so we need the Chain Rule first:

$\frac{\mathbf{d}}{\mathbf{dx}}(\cos(x \cdot e^x)) = -\mathbf{sin}(x \cdot e^x) \cdot \frac{\mathbf{d}}{\mathbf{dx}}(x \cdot e^x)$

$= -\sin(x e^x)\cdot\{ x \cdot \frac{\mathbf{d}}{\mathbf{dx}}(e^x) + e^x \cdot \frac{\mathbf{d}}{\mathbf{dx}}(x) \} = -\sin(x e^x) \cdot \{ x e^x + e^x \}$.

Sometimes we want to differentiate a composition of more than two functions. We can do so if we proceed in a careful, step–by–step way.

Example 8: Find $\mathbf{D}(\sin(\sqrt{x^3 + 1}))$

Solution: The function $\sin(\sqrt{x^3 + 1})$ can be considered as a composition $f \circ g$ of $f(x) = \sin(x)$ and $g(x) = \sqrt{x^3 + 1}$. Then

$(\sin(\sqrt{x^3 + 1}))' = \mathbf{f}'(g(x))\cdot\mathbf{g}'(\mathbf{x}) = \mathbf{cos}(g(x))\cdot\mathbf{g}'(\mathbf{x}) = \cos(\sqrt{x^3 + 1})\, \mathbf{D}(\sqrt{x^3 + 1})$

For the derivative of $\sqrt{x^3 + 1}$, we can use the Chain Rule again or its special case, the Power Rule:

$$\mathbf{D}(\sqrt{x^3+1}\ ) = \mathbf{D}((x^3+1)^{1/2}) = \frac{1}{2}(x^3+1)^{-1/2}\,\mathbf{D}(x^3+1) = \frac{1}{2}(x^3+1)^{-1/2}\,3x^2 .$$

Finally, $(\sin(\sqrt{x^3+1}\ ))' \ = \cos(\sqrt{x^3+1}\ )\,\mathbf{D}(\sqrt{x^3+1}\ )$

$$= \cos(\sqrt{x^3+1}\ )\ \frac{1}{2}(x^3+1)^{-1/2}(3x^2) = \frac{3x^2\cos(\sqrt{x^3+1}\ )}{2\sqrt{x^3+1}} .$$

This example was more complicated than the earlier ones, but it is just a matter of applying the Chain Rule twice, to a composition of a composition. If you proceed step–by–step and don't get lost in the problem, these multiple applications of the Chain Rule are relatively straightforward.

We can also use the Leibniz form of the Chain Rule for a composition of more than two functions. If $y = \sin(\sqrt{x^3+1}\ )$, then $y = \sin(u)$ with $u = \sqrt{w}$ and $w = x^3+1$. The Leibniz form of the Chain Rule is $\frac{dy}{dx} = \frac{dy}{du}\cdot\frac{du}{dw}\cdot\frac{dw}{dx}$, so $\frac{dy}{dx} = \cos(u)\frac{1}{2\sqrt{w}}\,3x^2 = \cos(\sqrt{x^3+1}\ )\cdot\frac{1}{2\sqrt{x^3+1}}\cdot 3x^2$.

Practice 6: (a) Find $\mathbf{D}(\sin(\cos(5x)))$. (b) For $y = e^{\cos(3x)}$, find dy/dx .

CHAIN RULE AND TABLES OF DERIVATIVES

With the Chain Rule, the derivatives of all sorts of strange and wonderful functions are available. If we know **f '** and **g '**, then we also know the derivatives of their composition: $(f(g(x)))' = \mathbf{f}'(g(x))\,\mathbf{g}'(x)$.

Example 9: Given $\mathbf{D}(\arcsin(x)) = \frac{1}{\sqrt{1-x^2}}$, find $\mathbf{D}(\arcsin(5x))$ and $\frac{\mathbf{d}(\arcsin(e^x))}{dx}$.

Solution: (a) arcsin(5x) is the composition of $f(x) = \arcsin(x)$ with $g(x) = 5x$. We know $\mathbf{g}'(x) = 5$,

and $\mathbf{f}'(x) = \frac{1}{\sqrt{1-x^2}}$ so $\mathbf{f}'(g(x)) = \frac{1}{\sqrt{1-(g(x))^2}} = \frac{1}{\sqrt{1-25x^2}}$.

Then $\mathbf{D}(\arcsin(5x)) = \mathbf{f}'(g(x))\cdot\mathbf{g}'(x) = \frac{1}{\sqrt{1-25x^2}}\cdot(5) = \frac{5}{\sqrt{1-25x^2}}$.

(b) $y = \arcsin(e^x)$ is $y = \arcsin(u)$ with $u = e^x$. We know $dy/du = \frac{1}{\sqrt{1-u^2}}$ and $du/dx = e^x$

so $\frac{dy}{dx} = \frac{dy}{du}\cdot\frac{du}{dx} = \frac{1}{\sqrt{1-u^2}}\cdot e^x = \frac{1}{\sqrt{1-(e^x)^2}}\cdot e^x = \frac{e^x}{\sqrt{1-e^{2x}}}$

In general, $\mathbf{D}(\arcsin(f(x))) = \frac{\mathbf{f}'(\mathbf{x})}{\sqrt{1-(f(x))^2}}$ and $\frac{\mathbf{d}(\arcsin(u))}{dx} = \frac{1}{\sqrt{1-u^2}}\cdot\frac{du}{dx}$.

Practice 7: $\mathbf{D}(\arctan(x)) = \dfrac{1}{1+x^2}$. Find $\mathbf{D}(\arctan(x^3))$ and $\dfrac{\mathbf{d}(\arctan(e^x))}{\mathbf{dx}}$.

Appendix B in the back of this book shows the derivative patterns for a variety of functions. You may not know much about some of the functions, but with the differentiation patterns given and the Chain Rule you should be able to calculate derivatives of compositions. It is just a matter of following the pattern.

Practice 8: Use the patterns $\mathbf{D}(\sinh(x)) = \cosh(x)$ and $\mathbf{D}(\ln(x)) = 1/x$ to determine

(a) $\mathbf{D}(\sinh(5x-7))$ (b) $\dfrac{\mathbf{d}}{\mathbf{dx}}(\ln(3+e^{2x}))$ (c) $\mathbf{D}(\arcsin(1+3x))$.

Example 10: If $\mathbf{D}(F(x)) = e^x \cdot \sin(x)$, find $\mathbf{D}(F(5x))$ and $\dfrac{\mathbf{d}(F(t^3))}{\mathbf{dt}}$

Solution: (a) $\mathbf{D}(F(5x)) = \mathbf{D}(F(g(x))$ with $g(x) = 5x$. $F'(x) = e^x \cdot \sin(x)$ so

$$\mathbf{D}(F(5x)) = F'(g(x)) \cdot g'(x) = e^{g(x)} \cdot \sin(g(x)) \cdot 5 = e^{5x} \cdot \sin(5x) \cdot 5 .$$

(b) $y = F(u)$ with $u = t^3$. $\dfrac{dy}{du} = e^u \cdot \sin(u)$ and $\dfrac{du}{dt} = 3t^2$ so

$$\frac{dy}{dt} = \frac{dy}{du} \cdot \frac{du}{dt} = e^u \cdot \sin(u) \cdot 3t^2 = e^{(t^3)} \cdot \sin(t^3) \cdot 3t^2$$

Proof of the Power Rule For Functions

We started using the Power Rule For Functions in section 2.3. Now we can easily prove it.

Power Rule For Functions: If $y = f^n(x)$ and f is differentiable, then $\dfrac{dy}{dx} = n \cdot f^{n-1}(x) \cdot f'(x)$.

Proof: $y = f^n(x)$ is $y = u^n$ with $u = f(x)$. Then $\dfrac{dy}{du} = n \cdot u^{n-1}$ and $\dfrac{du}{dx} = f'(x)$ so by the Chain Rule, $\dfrac{dy}{dx} = \dfrac{dy}{du} \cdot \dfrac{du}{dx} = n \cdot u^{n-1} \cdot f'(x) = n \cdot f^{n-1}(x) \cdot f'(x)$.

PROBLEMS

In problems 1 – 6, find two functions f and g so that the given function is the composition of f and g.

1. $y = (x^3 - 7x)^5$
2. $y = \sin^4(3x - 8)$
3. $y = \sqrt{(2 + \sin(x))^5}$
4. $y = \dfrac{1}{\sqrt{x^2 + 9}}$
5. $y = |x^2 - 4|$
6. $y = \tan(\sqrt{x})$
7. For each function in problems 1 – 6, write y as a function of u for some u which is a function of x.

Problems 8 and 9 refer to the values given in this table:

x	f(x)	g(x)	f '(x)	g '(x)	(f∘g)(x)	(f∘g)' (x)
–2	2	–1	1	1		
–1	1	2	0	2		
0	–2	1	2	–1		
1	0	–2	–1	2		
2	1	0	1	–1		

8. Use the table of values to determine (f∘g)(x) and (f∘g)' (x) at x = 1 and 2.

9. Use the table of values to determine (f∘g)(x) and (f∘g)' (x) at x = –2, –1 and 0.

10. Use Fig. 2 to estimate the values of g(x), **g '(x)**, (f∘g)(x), **f** '(g(x)), and **(f∘g)** '(x) at x = 1.

11. Use Fig. 2 to estimate the values of g(x), **g '(x)**, (f∘g)(x), **f** '(g(x)), and **(f∘g)** '(x) for x = 2.

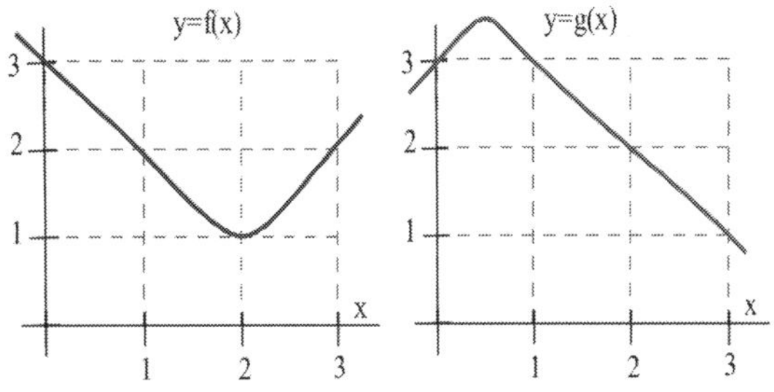

Fig. 2

In problems 12 - 20, differentiate each function.

12. $\mathbf{D}((x^2 + 2x + 3)^{87})$

13. $\mathbf{D}((1 - \frac{3}{x})^4)$

14. $\frac{\mathbf{d}}{\mathbf{dx}} (x + \frac{1}{x})^5$

15. $\mathbf{D}(\frac{5}{\sqrt{2 + \sin(x)}})$

16. $\frac{\mathbf{d}}{\mathbf{dt}}$ $t \cdot \sin(3t + 2)$

17. $\mathbf{D}(x^2 \cdot \sin(x^2 + 3))$

18. $\frac{\mathbf{d}}{\mathbf{dx}}$ $\sin(2x) \cdot \cos(5x + 1)$

19. $\mathbf{D}(\frac{7}{\cos(x^3 - x)})$

20. $\frac{\mathbf{d}}{\mathbf{dt}} \frac{5}{3 + e^t}$

21. $\mathbf{D}(e^x + e^{-x})$

22. $\frac{\mathbf{d}}{\mathbf{dx}} (e^x - e^{-x})$

23. An object attached to a spring is at a height of $h(t) = 3 - \cos(2t)$ feet above the floor t seconds after it is released. (a) At what height was it released?

(b) Determine its height, velocity and acceleration at any time t.

(c) If the object has mass m, determine its kinetic energy $K = \frac{1}{2} mv^2$ and dK/dt at any time t.

24. A manufacturer has determined that an employee with d days of production experience will be able to

produce approximately $P(d) = 3 + 15(1 - e^{-0.2d})$ items per day. Graph P(d).

(a) Approximately how many items will a beginning employee be able to produce each day?

(b) How many items will an experienced employee be able to produce each day?

(c) What is the marginal production rate of an employee with 5 days of experience? (What are the units of your answer, and what does this answer mean?)

25. The air pressure P(h), in pounds per square inch, at an altitude of h feet above sea level is

approximately $P(h) = 14.7\, e^{-0.0000385h}$.

(a) What is the air pressure at sea level? What is the air pressure at an altitude of 30,000 feet?

(b) At what altitude is the air pressure 10 pounds per square inch?

(c) If you are in a balloon which is 2000 feet above the Pacific Ocean and is rising at 500 feet per minute, how fast is the air pressure on the balloon changing?

(d) If the temperature of the gas in the balloon remained constant during this ascent, what would happen to the volume of the balloon?

Find the derivatives in problems 26 – 33 .

26. $\mathbf{D}\left(\frac{(2x+3)^2}{(5x-7)^3} \right)$

27. $\frac{\mathbf{d}}{\mathbf{dz}} \sqrt{1+\cos^2(z)}$

28. $\mathbf{D}(\sin(3x+5))$

29. $\frac{\mathbf{d}}{\mathbf{dx}} \tan(3x+5)$

30. $\frac{\mathbf{d}}{\mathbf{dt}} \cos(7t^2)$

31. $\mathbf{D}(\sin(\sqrt{x+1}))$

32. $\mathbf{D}(\sec(\sqrt{x+1}))$

33. $\frac{\mathbf{d}}{\mathbf{dx}} (e^{\sin(x)})$

In problems 34 – 37 , calculate $\frac{d\,f(x)}{dx}$ and $\frac{d\,x(t)}{dt}$ when $t = 3$ and use these values to determine the value of $\frac{d\,f(x(t))}{dt}$ when $t = 3$.

34. $f(x) = \cos(x)$, $x = t^2 - t + 5$

35. $f(x) = \sqrt{x}$, $x = 2 + \frac{21}{t}$

36. $f(x) = e^x$, $x = \sin(t)$

37. $f(x) = \tan^3(x)$, $x = 8$

In problems 38 – 43, find a function which has the given function as its derivative. (You are given $f'(x)$ in each problem and are asked to find a function $f(x)$.)

38. $f'(x) = (3x+1)^4$

39. $f'(x) = (7x-13)^{10}$

40. $f'(x) = \sqrt{3x-4}$

41. $f'(x) = \sin(2x-3)$

42. $f'(x) = 6e^{3x}$

43. $f'(x) = \cos(x) \cdot e^{\sin(x)}$

If two functions are equal, then their derivatives are also equal. In problems 44 – 47 , differentiate each side of the trigonometric identity to find a new identity.

44. $\sin^2(x) = \frac{1}{2} - \frac{1}{2}\cos(2x)$

45. $\cos(2x) = \cos^2(x) - \sin^2(x)$

46. $\sin(2x) = 2\sin(x)\cos(x)$

47. $\sin(3x) = 3\sin(x) - 4\sin^3(x)$

Derivatives of Families of Functions

So far we have emphasized derivatives of particular functions, but sometimes we want to look at the derivatives of a whole family of functions. In problems 48 –71, the letters A–D represent constants and the given formulas describe families of functions.

For problems 48 – 65, calculate $y' = \frac{dy}{dx}$.

48. $y = Ax^3 - B$
49. $y = Ax^3 + Bx^2 + C$
50. $y = \sin(Ax + B)$
51. $y = \sin(Ax^2 + B)$
52. $y = Ax^3 + \cos(Bx)$
53. $y = \sqrt{A + Bx^2}$
54. $y = \sqrt{A - Bx^2}$
55. $y = A - \cos(Bx)$
56. $y = \cos(Ax + B)$
57. $y = \cos(Ax^2 + B)$
58. $y = A \cdot e^{Bx}$
59. $y = x \cdot e^{Bx}$
60. $y = e^{Ax} + e^{-Ax}$
61. $y = e^{Ax} - e^{-Ax}$
62. $y = \frac{\sin(Ax)}{x}$
63. $y = \frac{Ax}{\sin(Bx)}$
64. $y = \frac{1}{Ax + B}$
65. $y = \frac{Ax + B}{Cx + D}$

In problems 66 – 71, (a) find y', (b) find the value(s) of x so that $y' = 0$, and (c) find y'' . Typically your answer in part (b) will contain As, Bs and (sometimes) Cs.

66. For $y = Ax^2 + Bx + C$, (a) find y', (b) find the value(s) of x so that $y' = 0$, and (c) find y''. (You should recognize the part (b) answer from intermediate algebra. What is it?)

67. $y = Ax(B - x) = ABx - Ax^2$.
68. $y = Ax(B - x^2) = ABx - Ax^3$.
69. $y = Ax^2(B - x) = ABx^2 - Ax^3$.
70. $y = Ax^2 + \frac{B}{x}$.
71. $y = Ax^3 + Bx^2 + C$.

Use the given differentiation patterns to differentiate the composite functions in problems 72 – 83 . We have not derived the derivatives for these functions (yet), but if you are handed the derivative pattern for a function then you should be able to take derivatives of a composition involving that function.

Given: $\mathbf{D}(\arctan(x)) = \frac{1}{1 + x^2}$, $\mathbf{D}(\arcsin(x)) = \frac{1}{\sqrt{1-x^2}}$, $\mathbf{D}(\ln(x)) = \frac{1}{x}$.

72. $\mathbf{D}(\arctan(7x))$
73. $\frac{\mathbf{d}}{\mathbf{dx}}(\arctan(x^2))$
74. $\frac{\mathbf{d}}{\mathbf{dt}}(\arctan(\ln(t)))$
75. $\mathbf{D}(\arctan(e^x))$
76. $\frac{\mathbf{d}}{\mathbf{dw}}(\arcsin(4w))$
77. $\mathbf{D}(\arcsin(x^3))$
78. $\mathbf{D}(\arcsin(\ln(x)))$
79. $\frac{\mathbf{d}}{\mathbf{dt}}(\arcsin(e^t))$
80. $\mathbf{D}(\ln(3x + 1))$
81. $\frac{\mathbf{d}}{\mathbf{dx}}(\ln(\sin(x)))$
82. $\mathbf{D}(\ln(\arctan(x)))$
83. $\frac{\mathbf{d}}{\mathbf{ds}}(\ln(e^s))$

Section 2.4 **PRACTICE Answers**

Practice 1: $f(x) = 5x - 4$ and $g(x) = x^2$ so $f'(x) = 5$ and $g'(x) = 2x$. $f\circ g(x) = f(g(x)) = f(x^2) = 5x^2 - 4$.

$\mathbf{D}(f\circ g(x)) = f'(g(x))\cdot g'(x) = 5\cdot 2x = \mathbf{10x}$ or $\mathbf{D}(f\circ g(x)) = \mathbf{D}(5x^2 - 4) = \mathbf{10x}$.

$g\circ f(x) = g(f(x)) = g(5x - 4) = (5x - 4)^2 = 25x^2 - 40x + 16$.

$\mathbf{D}(g\circ f(x)) = g'(f(x))\cdot f'(x) = 2(5x - 4)\cdot 5 = \mathbf{50x - 40}$ or

$\mathbf{D}(g\circ f(x)) = \mathbf{D}(25x^2 - 40x + 16) = \mathbf{50x - 40}$.

Practice 2: $\frac{\mathbf{d}}{\mathbf{dx}}(\sin(4x + e^x)) = \cos(4x + e^x)\cdot\mathbf{D}(4x + e^x) = \cos(4x + e^x)\cdot(\mathbf{4 + e^x})$

Practice 3: Fill in the table in Example 6 for $(f\circ g)(x)$ and $(f\circ g)'(x)$ at $x = 1, 2$ and 3.

x	f(x)	g(x)	f '(x)	g '(x)	(f∘g)(x)	(f∘g) '(x) = f '(g(x))·g '(x)
–1	2	3	1	0	0	0
0	–1	1	3	2	1	–2
1	1	0	–1	3	**–1**	f '(g(1))·g '(1)=f '(0)·(3)=(3)(3)= **9**
2	3	–1	0	1	**2**	f '(g(2))·g '(2)=f '(–1)·(1)=(1)(1)=**1**
3	0	2	2	–1	**3**	f '(g(3)·g '(3)=f '(2)·(–1)=(0)(–1)=**0**

Practice 4: $\mathbf{D}(\sin(7x - 1)) = \cos(7x - 1)\mathbf{D}(7x - 1) = \mathbf{7}\cdot\cos(7x - 1)$.

$\frac{\mathbf{d}}{\mathbf{dx}} \sin(ax + b) = \cos(ax + b)\, \mathbf{D}(ax + b) = \mathbf{a}\cdot\cos(ax + b)$ $\qquad \frac{\mathbf{d}}{\mathbf{dt}}(e^{3t}) = e^{3t}\frac{\mathbf{d}}{\mathbf{dt}}(3t) = \mathbf{3}\cdot e^{3t}$

Practice 5: $\mathbf{D}(\sin(g(x))) = \cos(g(x))\cdot g'(x)$. At $x = \pi$, $\cos(g(\pi))\cdot g'(\pi) \approx \cos(0.86)\cdot(-1) \approx -0.65$.

$\mathbf{D}(g(\sin(x))) = g'(\sin(x))\cdot\cos(x)$. At $x = \pi$, $g'(\sin(\pi))\cdot\cos(\pi) = g'(0)\cdot(-1) \approx -2$

Practice 6:

$\mathbf{D}(\sin(\cos(5x))) = \cos(\cos(5x))\cdot\mathbf{D}(\cos(5x)) = \cos(\cos(5x))\cdot(-\sin(5x))\cdot\mathbf{D}(5x) = -5\cdot\sin(5x)\cdot\cos(\cos(5x))$

$\frac{\mathbf{d}}{\mathbf{dx}} e^{\cos(3x)} = e^{\cos(3x)}\mathbf{D}(\cos(3x)) = e^{\cos(3x)}(-\sin(3x))\mathbf{D}(3x) = -3\cdot\sin(3x)\cdot e^{\cos(3x)}$.

Practice 7: $\mathbf{D}(\arctan(x^3)) = \frac{1}{1 + (x^3)^2}\,\mathbf{D}(x^3) = \frac{\mathbf{3x^2}}{1 + x^6}$

$\frac{\mathbf{d}}{\mathbf{dx}}(\arctan(e^x)) = \frac{1}{1 + (e^x)^2}\,\mathbf{D}(e^x) = \frac{\mathbf{e^x}}{1 + e^{2x}}$

Practice 8: $\mathbf{D}(\sinh(5x - 7)) = \cosh(5x - 7)\,\mathbf{D}(5x - 7) = \mathbf{5}\cdot\cosh(5x - 7)$

$\frac{\mathbf{d}}{\mathbf{dx}} \ln(3 + e^{2x}) = \frac{1}{3 + e^{2x}}\,\mathbf{D}(3 + e^{2x}) = \frac{\mathbf{2e^{2x}}}{3 + e^{2x}}$

$\mathbf{D}(\arcsin(1 + 3x)) = \frac{1}{\sqrt{1 - (1 + 3x)^2}}\,\mathbf{D}(1 + 3x) = \frac{\mathbf{3}}{\sqrt{1 - (1 + 3x)^2}}$

2.5 SOME APPLICATIONS OF THE CHAIN RULE

The Chain Rule will help us determine the derivatives of logarithms and exponential functions a^x. We will also use it to answer some applied questions and to find slopes of graphs given by parametric equations.

DERIVATIVES OF LOGARITHMS

$$\mathbf{D}(\ln(x)) = \frac{1}{x} \quad \text{and} \quad \mathbf{D}(\ln(g(x))) = \frac{\mathbf{g'(x)}}{g(x)} .$$

Proof: We know that the natural logarithm $\ln(x)$ is the logarithm with base e, and $e^{\ln(x)} = x$ for $x > 0$.

We also know that $\mathbf{D}(e^x) = e^x$, so using the Chain Rule we have $\mathbf{D}(e^{f(x)}) = e^{f(x)} \mathbf{f'(x)}$.

Differentiating each side of the equation $e^{\ln(x)} = x$, we get that

$\mathbf{D}(e^{\ln(x)}) = \mathbf{D}(x)$	use $\mathbf{D}(e^{f(x)}) = e^{f(x)} \cdot \mathbf{f'(x)}$ with $f(x) = \ln(x)$
$e^{\ln(x)} \cdot \mathbf{D}(\ln(x)) = 1$	replace $e^{\ln(x)}$ with x
$x \cdot \mathbf{D}(\ln(x)) = 1$	and solve for $\mathbf{D}(\ln(x))$ to get $\mathbf{D}(\ln(x)) = \frac{1}{x}$.

The function $\ln(g(x))$ is the composition of $f(x) = \ln(x)$ with $g(x)$, so by the Chain Rule,

$$\mathbf{D}(\ln(g(x))) = \mathbf{D}(f(g(x))) = \mathbf{f'}(g(x)) \cdot \mathbf{g}'(x) = \frac{1}{g(x)} \cdot \mathbf{g}'(x) = \frac{\mathbf{g}'(x)}{g(x)} .$$

Example 1: Find $\mathbf{D}(\ln(\sin(x)))$ and $\mathbf{D}(\ln(x^2 + 3))$.

Solution: (a) Using the pattern $\mathbf{D}(\ln(g(x)) = \frac{\mathbf{g}'(x)}{g(x)}$ with $g(x) = \sin(x)$, then

$$\mathbf{D}(\ln(\sin(x))) = \frac{\mathbf{g}'(x)}{g(x)} = \frac{\mathbf{D}(\sin(x))}{\sin(x)} = \frac{\cos(x)}{\sin(x)} = \cot(x).$$

(b) Using the pattern with $g(x) = x^2 + 3$, we have $\mathbf{D}(\ln(x^2 + 3)) = \frac{\mathbf{g}'(x)}{g(x)} = \frac{2x}{x^2 + 3}$.

We can use the Change of Base Formula from algebra to rewrite any logarithm as a natural logarithm, and then we can differentiate the resulting natural logarithm.

Change of Base Formula for logarithms: $\log_a x = \frac{\log_b x}{\log_b a}$ for all positive a, b and x.

Example 2: Use the Change of Base formula and your calculator to find $\log_\pi 7$ and $\log_2 8$.

Solution: $\log_\pi 7 = \frac{\ln 7}{\ln \pi} \approx \frac{1.946}{1.145} \approx 1.700$. (Check that $\pi^{1.7} \approx 7$) $\log_2 8 = \frac{\ln 8}{\ln 2} = 3$.

Practice 1: Find the values of $\log_9 20$, $\log_3 20$ and $\log_\pi e$.

Putting b = e in the Change of Base Formula, $\log_a x = \frac{\log_e x}{\log_e a} = \frac{\ln x}{\ln a}$, so any logarithm can be written as a natural logarithm divided by a constant. Then any logarithm is easy to differentiate.

$$\mathbf{D}(\log_a(x)) = \frac{1}{x \ln(a)} \quad \text{and} \quad \mathbf{D}(\log_a(f(x))) = \frac{\mathbf{f'(x)}}{f(x)} \cdot \frac{1}{\ln(a)}$$

Proof: $\mathbf{D}(\log_a(x)) = \mathbf{D}(\frac{\ln x}{\ln a}) = \frac{1}{\ln(a)} \cdot \mathbf{D}(\ln x) = \frac{1}{\ln(a)} \cdot \frac{1}{x} = \frac{1}{x \ln(a)}$.

The second differentiation formula follows from the Chain Rule.

Practice 2: Calculate $\mathbf{D}(\log_{10}(\sin(x)))$ and $\mathbf{D}(\log_{\pi}(e^x))$.

The number e might seem like an "unnatural" base for a natural logarithm, but of all the logarithms to different bases, the logarithm with base e has the nicest and easiest derivative. The natural logarithm is even related to the distribution of prime numbers. In 1896, the mathematicians Hadamard and Valle–Poussin proved the following conjecture of Gauss: (The Prime Number Theorem) For large values of x,
$\{\text{number of primes less than } x\} \approx \frac{x}{\ln(x)}$.

DERIVATIVE OF a^x

Once we know the derivative of e^x and the Chain Rule, it is relatively easy to determine the derivative of a^x for any $a > 0$.

$$\mathbf{D}(a^x) = a^x \cdot \ln a \quad \text{for } a > 0.$$

Proof: If $a > 0$, then $a^x > 0$ and $a^x = e^{\ln(a^x)} = e^{x \cdot \ln a}$.

$$\mathbf{D}(a^x) = \mathbf{D}(e^{\ln(a^x)}) = \mathbf{D}(e^{x \cdot \ln a}) = e^{x \cdot \ln a} \cdot \mathbf{D}(x \cdot \ln a) = a^x \cdot \mathbf{\ln a} .$$

Example 3: Calculate $\mathbf{D}(7^x)$ and $\frac{\mathbf{d}}{\mathbf{dt}}(2^{\sin(t)})$

Solution: (a) $\mathbf{D}(7^x) = 7^x \ln 7 \approx (1.95)\, 7^x$.

(b) We can write $y = 2^{\sin(t)}$ as $y = 2^u$ with $u = \sin(t)$. Using the Chain Rule,

$$\frac{dy}{dt} = \frac{dy}{du} \cdot \frac{du}{dt} = 2^u \cdot \ln(2) \cdot \cos(t) = 2^{\sin(t)} \cdot \ln(2) \cdot \cos(t) .$$

Practice 3: Calculate $\mathbf{D}(\sin(2^x))$ and $\frac{\mathbf{d}}{\mathbf{dt}}(3^{(t^2)})$.

SOME APPLIED PROBLEMS

Now we can examine applications which involve more complicated functions.

Example 4: A ball at the end of a rubber band (Fig. 1) is oscillating up and down, and its height (in feet) above the floor at time t seconds is h(t) = 5 + 2 sin(t/2) . (t is in radians)

(a) How fast is the ball travelling after 2 seconds? after 4 seconds? after 60 seconds?

(b) Is the ball moving up or down after 2 seconds? after 4 seconds? after 60 seconds?

(c) Is the vertical velocity of the ball ever 0 ?

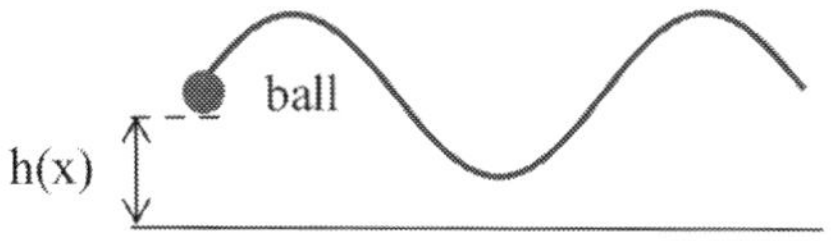

Fig. 1

Solution: (a) v(t) = **D**(h(t)) = **D**(5 + 2 sin(t/2))

= 2 cos(t/2) **D**(t/2) = cos(t/2) feet/second so

v(2) = cos(2/2) ≈ 0.540 ft/s , v(4) = cos(4/2) ≈ –0.416 ft/s ,

and v(60) = cos(60/2) ≈ 0.154 ft/s.

(b) The ball is moving upward when t = 2 and 60 seconds, downward when t = 4.

(c) v(t) = cos(t/2) and cos(t/2) = 0 when t = π ± n·2π (n = 1, 2, . . .) .

Example 5: If 2400 people now have a disease, and the number of people with the disease appears to double every 3 years, then the number of people expected to have the disease in t years is $y = 2400 \cdot 2^{t/3}$.

(a) How many people are expected to have the disease in 2 years?

(b) When are 50,000 people expected to have the disease?

(c) How fast is the number of people with the disease expected to grow now and 2 years from now?

Solution: (a) In 2 years, $y = 2400 \cdot 2^{2/3} \approx 3{,}810$ people.

(b) We know y = 50,000 , and we need to solve $50{,}000 = 2400 \cdot 2^{t/3}$ for t. Taking logarithms of each side of the equation, $\ln(50{,}000) = \ln(2400 \cdot 2^{t/3}) = \ln(2400) + (t/3) \cdot \ln(2)$ so 10.819 = 7.783 + .231t and t ≈ 13.14 years. We expect 50,000 people to have the disease about 13.14 years from now.

(c) This is asking for dy/dt when t = 0 and 2 years. $\frac{dy}{dt} = \frac{d(2400 \cdot 2^{t/3})}{dt} = 2400 \cdot 2^{t/3} \cdot \ln(2) \cdot (1/3)$ $\approx 554.5 \cdot 2^{t/3}$. Now, at t = 0, the rate of growth of the disease is approximately $554.5 \cdot 2^0 \approx 554.5$ people/year. In 2 years the rate of growth will be approximately $554.5 \cdot 2^{2/3} \approx 880$ people/year.

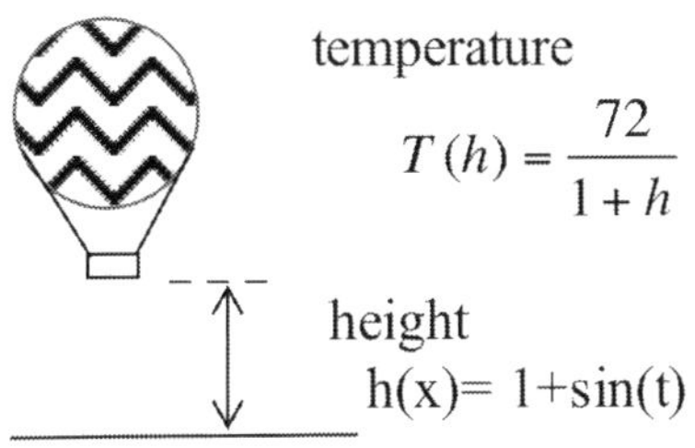

Fig. 2

Example 6: You are riding in a balloon, and at time t (in minutes) you are h(t) = t + sin(t) feet high. If the temperature at an elevation h is $T(h) = \frac{72}{1+h}$ degrees Fahrenheit, then how fast is your temperature changing when t = 5 minutes? (Fig. 2)

Solution: As t changes, your elevation will change, and, as your elevation changes, so will your temperature. It is not difficult to write the temperature as a function of time, and then we could calculate $\frac{d\,T(t)}{dt} = T\,'(t)$ and evaluate $T\,'(5)$, or we could use the Chain Rule:

$$\frac{\mathbf{d}\,T(t)}{\mathbf{dt}} = \frac{\mathbf{d}\,T(\,h(t)\,)}{\mathbf{d}\,h(t)} \cdot \frac{d\,h(t)}{dt} = \frac{\mathbf{d}\,T(\,h\,)}{\mathbf{d}\,h} \cdot \frac{\mathbf{d}\,h(t)}{\mathbf{dt}} = \frac{-72}{(1+h)^2} \cdot (\,1 + \cos(\,t\,)\,)\,.$$

When $t = 5$, then $h(\,t\,) = 5 + \sin(\,5\,) \approx 4.04$ so $T\,'(5) \approx \frac{-72}{(1+4.04)^2} \cdot (1 + .284) \approx -3.64\,^{\mathrm{o}}$/minute.

Practice 4: Write the temperature T in the previous example as a function of the variable t alone and then differentiate T to determine the value of dT/dt when t = 5 minutes.

Example 7: A scientist has determined that, under optimum conditions, an initial population of 40 bacteria will grow "exponentially" to $f(t) = 40 \cdot e^{t/5}$ bacteria after t hours.

(a) Graph $y = f(t)$ for $0 \le t \le 15$. Calculate f(0), f(5), f(10).

(b) How fast is the population increasing at time t? (Find f '(t).)

(c) Show that the rate of population increase, f '(t), is proportional to the population, f(t), at any time t. (Show $f\,'(t) = K \cdot f(t)$ for some constant K.)

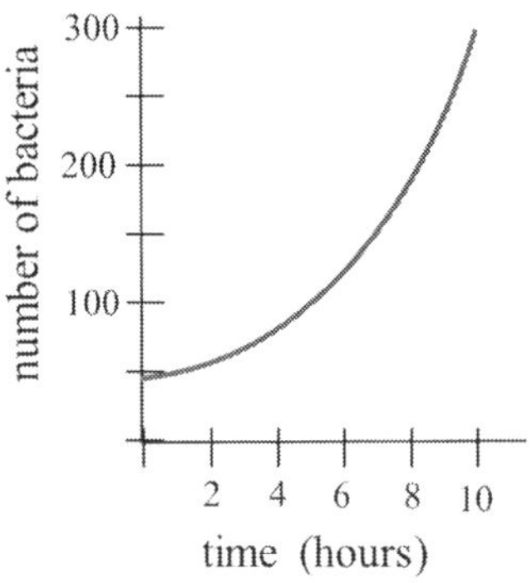

Fig. 3

Solution: (a) The graph of $y = f(t)$ is given in Fig. 3.
$f(0) = 40 \cdot e^{0/5} = 40$ bacteria. $f(5) = 40 \cdot e^{5/5} \approx 109$ bacteria, and $f(10) = 40 \cdot e^{10/5} \approx 296$ bacteria.

(b) $f\,'(t) = \frac{\mathbf{d}}{\mathbf{dt}}(\,f(t)\,) = \frac{\mathbf{d}}{\mathbf{dt}}(\,40 \cdot e^{t/5}\,) = 40 \cdot e^{t/5} \frac{\mathbf{d}}{\mathbf{dt}}(\,t/5\,)$
$= 40 \cdot e^{t/5}\,(\,\mathbf{1/5}\,) = 8 \cdot e^{t/5}$ bacteria/hour.

(c) $f\,'(t) = 8 \cdot e^{t/5} = \frac{1}{5} \cdot (\,40 \cdot e^{t/5}\,) = \frac{1}{5}\,f(t)$ so $f\,'(t) = K \cdot f(t)$ with $K = 1/5$.

PARAMETRIC EQUATIONS

Suppose a robot has been programmed to move in the xy–plane so at time t its x coordinate will be sin(t) and its y coordinate will be t^2. Both x and y are functions of the independent parameter t, x(t) = sin(t) and $y(t) = t^2$, and the path of the robot (Fig. 4) can be found by plotting (x,y) = (x(t), y(t)) for lots of values of t.

t	x(t) = sin(t)	$y(t) = t^2$	plot point at
0	0	0	(0, 0)
.5	.48	.25	(.48, .25)
1.0	.84	1	(.84, 1)
1.5	1.00	2.25	(1, 2.25)
2.0	.91	4	(.91, 4)

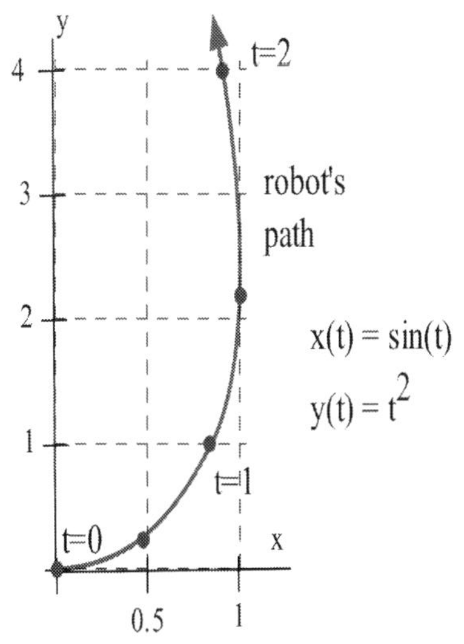

Fig. 4

Typically we know x(t) and y(t) and need to find dy/dx, the slope of the tangent line to the graph of (x(t), y(t)). The Chain Rule says that

$\frac{dy}{dt} = \frac{dy}{dx} \cdot \frac{dx}{dt}$, so , algebraically solving for $\frac{dy}{dx}$, we get $\frac{dy}{dx} = \frac{dy/dt}{dx/dt}$.

If we can calculate dy/dt and dx/dt, the derivatives of y and x with respect to the parameter t, then we can determine dy/dx, the rate of change of y with respect to x.

> If $x = x(t)$ and $y = y(t)$ are differentiable with respect to t , and $\frac{dx}{dt} \neq 0$,
>
> then $\frac{dy}{dx} = \frac{dy/dt}{dx/dt}$.

Example 8: Find the slope of the tangent line to the graph of $(x,y) = (\sin(t), t^2)$ when $t = 2$?

Solution: $dx/dt = \cos(t)$ and $dy/dt = 2t$. When $t = 2$, the object is at the point ($\sin(\mathbf{2})$, $\mathbf{2}^2$) ≈ (.91 , 4)

and the slope of the tangent line to the graph is $\frac{dy}{dx} = \frac{dy/dt}{dx/dt} = \frac{2t}{\cos(t)} = \frac{2\cdot\mathbf{2}}{\cos(\mathbf{2})} \approx \frac{4}{-.42} \approx -9.61$.

Practice 5: Graph $(x,y) = (3\cos(t), 2\sin(t))$ and find the slope of the tangent line when $t = \pi/2$.

When we calculated $\frac{dy}{dx}$, the slope of the tangent line to the graph of (x(t), y(t)), we used the derivatives $\frac{dx}{dt}$ and $\frac{dy}{dt}$, and each of these derivatives also has a geometric meaning:

$\frac{dx}{dt}$ measures the rate of change of x(t) with respect to t -- it tells us whether the x-coordinate is increasing or decreasing as the t-variable increases.

$\frac{dy}{dt}$ measures the rate of change of y(t) with respect to t.

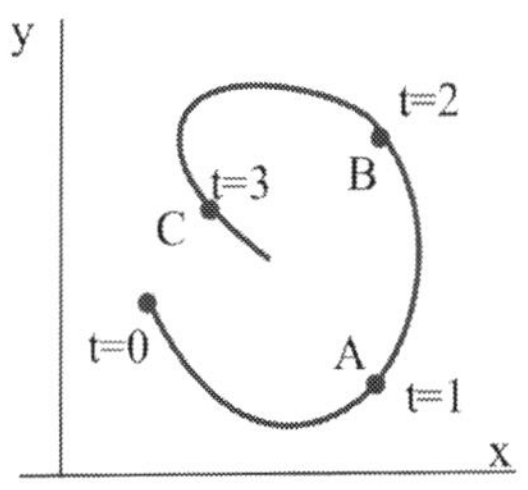

Fig. 5

Example 9: For the parametric graph in Fig. 5, tell whether $\frac{dx}{dt}$, $\frac{dy}{dt}$ *and* $\frac{dy}{dx}$ is positive or negative when t=2.

Solution: As we move through the point B (where t=2) in the direction of increasing values of t, we are moving to the left so x(t) is decreasing and $\frac{dx}{dt}$ is negative.

Similarly, the values of y(t) are increasing so $\frac{dy}{dt}$ is positive. Finally, the slope of the tangent line, $\frac{dy}{dx}$, is negative.

(As check on the sign of $\frac{dy}{dx}$ we can also use the result $\frac{dy}{dx} = \frac{dy/dt}{dx/dt} = \frac{positive}{negative} = negative$.)

Practice 6: For the parametric graph in the previous example, tell whether $\frac{dx}{dt}$, $\frac{dy}{dt}$ and $\frac{dy}{dx}$ is positive or negative when t=1 and when t=3.

Speed

If we know the position of an object at every time, then we can determine its speed. The formula for speed comes from the distance formula and looks a lot like it, but with derivatives.

> If $x = x(t)$ and $y = y(t)$ give the location of an object at time t and are differentiable functions of t,
>
> then the **speed** of the object is $\sqrt{\left(\frac{dx}{dt}\right)^2 + \left(\frac{dy}{dt}\right)^2}$.

Proof: The speed of an object is the limit, as $\Delta t \to 0$, of $\frac{\text{change in position}}{\text{change in time}}$. (Fig. 6)

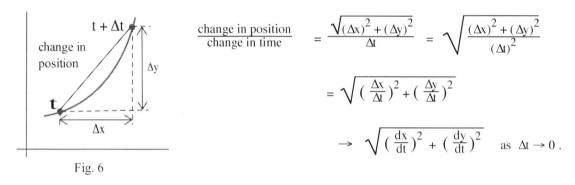

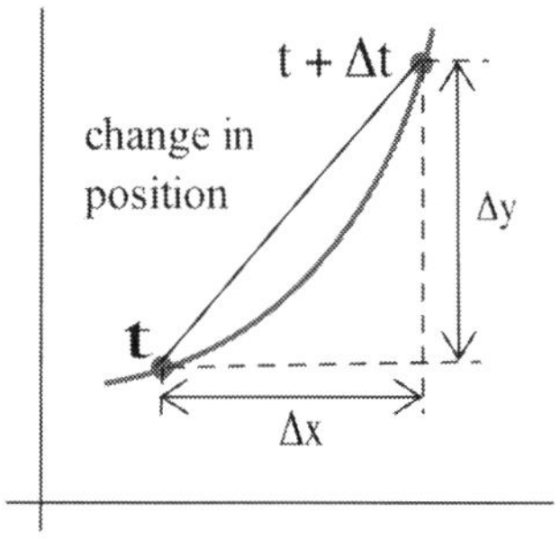

Fig. 6

$$\frac{\text{change in position}}{\text{change in time}} = \frac{\sqrt{(\Delta x)^2 + (\Delta y)^2}}{\Delta t} = \sqrt{\frac{(\Delta x)^2 + (\Delta y)^2}{(\Delta t)^2}}$$

$$= \sqrt{\left(\frac{\Delta x}{\Delta t}\right)^2 + \left(\frac{\Delta y}{\Delta t}\right)^2}$$

$$\to \sqrt{\left(\frac{dx}{dt}\right)^2 + \left(\frac{dy}{dt}\right)^2} \quad \text{as } \Delta t \to 0 .$$

Exercise 10: Find the speed of the object whose location at time t is $(x,y) = (\sin(t), t^2)$ when $t = 0$ and $t = 1$.

Solution: $dx/dt = \cos(t)$ and $dy/dt = 2t$ so speed $= \sqrt{(\cos(t))^2 + (2t)^2} = \sqrt{\cos^2(t) + 4t^2}$.

When $t = 0$, speed $= \sqrt{\cos^2(\mathbf{0}) + 4(\mathbf{0})^2} = \sqrt{1 + 0} = 1$. When $t = 1$,

speed $= \sqrt{\cos^2(\mathbf{1}) + 4(\mathbf{1})^2} = \sqrt{0.29 + 4} \approx 2.07$.

Practice 7: Show that an object whose location at time t is $(x,y) = (3\sin(t), 3\cos(t))$ has a **constant** speed. (This object is moving on a circular path.)

Practice 8: Is the object whose location at time t is $(x,y) = (3\cos(t), 2\sin(t))$ travelling faster at the top of the ellipse (at $t = \pi/2$) or at the right edge of the ellipse (at $t = 0$)?

PROBLEMS

Differentiate the functions in problems 1 – 19.

1. $\ln(5x)$ 2. $\ln(x^2)$ 3. $\ln(x^k)$ 4. $\ln(x^x) = x \cdot \ln(x)$ 5. $\ln(\cos(x))$

6. $\cos(\ln(x))$ 7. $\log_2 5x$ 8. $\log_2 kx$ 9. $\ln(\sin(x))$ 10. $\ln(kx)$

11. $\log_2(\sin(x))$ 12. $\ln(e^x)$ 13. $\log_5 5^x$ 14. $\ln(e^{f(x)})$ 15. $x \cdot \ln(3x)$

16. $e^x \cdot \ln(x)$ 17. $\dfrac{\ln(x)}{x}$ 18. $\sqrt{x + \ln(3x)}$ 19. $\ln(\sqrt{5x-3})$

20. $\dfrac{d}{dt} \ln(\cos(t))$ 21. $\dfrac{d}{dw} \cos(\ln(w))$ 22. $\dfrac{d}{dx} \ln(ax + b)$

23. $\dfrac{d}{dt} \ln(\sqrt{t+1})$ 24. $\mathbf{D}(3^x)$ 25. $\mathbf{D}(5^{\sin(x)})$

26. $\mathbf{D}(x \cdot \ln(x) - x)$ 27. $\dfrac{d}{dx} \ln(\sec(x) + \tan(x))$

28. Find the slope of the line tangent to $f(x) = \ln(x)$ at the point $(e, 1)$. Find the slope of the line tangent to $g(x) = e^x$ at the point $(1, e)$. How are the slopes of f and g at these points related?

29. Find a point P on the graph of $f(x) = \ln(x)$ so the tangent line to f at P goes through the origin.

30. You are moving from left to right along the graph of $y = \ln(x)$ (Fig. 7).
    (a) If the x–coordinate of your location at time t seconds is $x(t) = 3t + 2$, then how fast is your elevation increasing?
    (b) If the x–coordinate of your location at time t seconds is $x(t) = e^t$, then how fast is your elevation increasing?

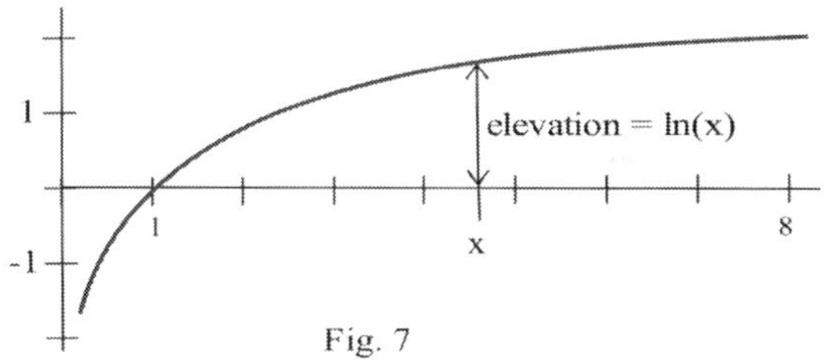

Fig. 7

31. Rumor. The percent of a population, $p(t)$, who have heard a rumor by time t is often modeled $p(t) = \dfrac{100}{1 + Ae^{-t}} = 100(1 + Ae^{-t})^{-1}$ for some positive constant A. Calculate how fast the rumor is spreading, $\dfrac{d\,p(t)}{dt}$.

32. Radioactive decay. If we start with A atoms of a radioactive material which has a "half–life" (the amount of time for half of the material to decay) of 500 years, then the number of radioactive atoms left after t years is $r(t) = A \cdot e^{-Kt}$ where $K = \dfrac{\ln(2)}{500}$. Calculate $r'(t)$ and show that $r'(t)$ is proportional to $r(t)$ ($r'(t) = b \cdot r(t)$ for some constant b).

In problems 33 – 41 , find a function with the given derivative.

33. $f'(x) = \frac{8}{x}$ 34. $h'(x) = \frac{3}{3x+5}$ 35. $f'(x) = \frac{\cos(x)}{3+\sin(x)}$

36. $g'(x) = \frac{x}{1+x^2}$ 37. $g'(x) = 3e^{5x}$ 38. $h'(x) = e^2$

39. $f'(x) = 2x \cdot e^{(x^2)}$ 40. $g'(x) = \cos(x) \cdot e^{\sin(x)}$ 41. $h'(x) = \frac{\cos(x)}{\sin(x)}$

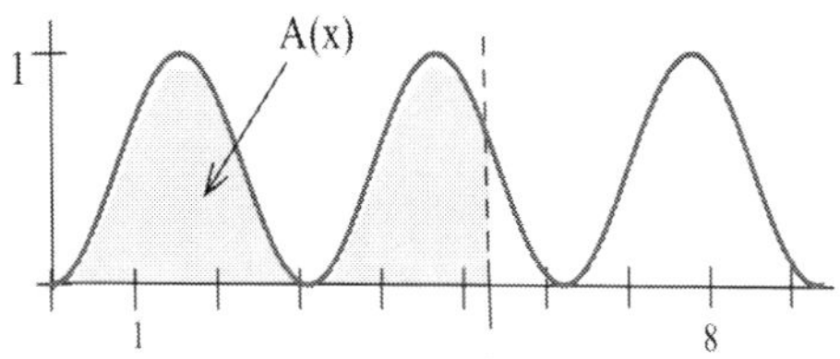

Fig. 8

42. Define A(x) to be the **area** bounded between the x–axis, the graph of f(x), and a vertical line at x (Fig. 8). The area under each "hump" of f is 2 square inches.

(a) Graph A(x) for $0 \le x \le 9$.

(b) Graph A '(x) for $0 \le x \le 9$.

Problems 43 – 48 involve parametric equations.

43. At time t minutes, robot A is at (t, 2t + 1) and robot B is at ($t^2, 2t^2 + 1$).

(a) Where is each robot when t=0 and t = 1?

(b) Sketch the path each robot follows during the first minute.

(c) Find the slope of the tangent line, dy/dx , to the path of each robot at t = 1 minute.

(d) Find the speed of each robot at t = 1 minute.

(e) Discuss the motion of a robot which follows the path (sin(t), 2sin(t) + 1) for 20 minutes.

44. $x(t) = t + 1$, $y(t) = t^2$. (a) Graph (x(t), y(t)) for $-1 \le t \le 4$.

(b) Find dx/dt, dy/dt, the tangent slope dy/dx , and speed when t = 1 and t = 4.

45. For the parametric graph in Fig. 9, determine whether $\frac{dx}{dt}$, $\frac{dy}{dt}$ *and* $\frac{dy}{dx}$ are positive, negative or zero when t = 1 and t = 3.

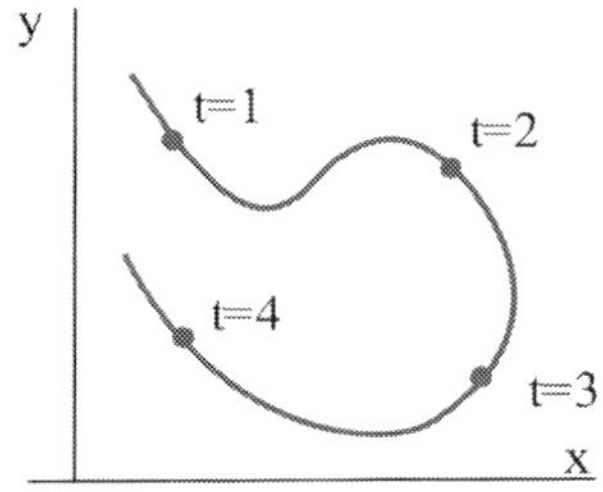

Fig. 9

46. For the parametric graph in Fig. 10, determine whether $\frac{dx}{dt}$, $\frac{dy}{dt}$ *and* $\frac{dy}{dx}$ are positive, negative or zero when t = 1 and t = 3.

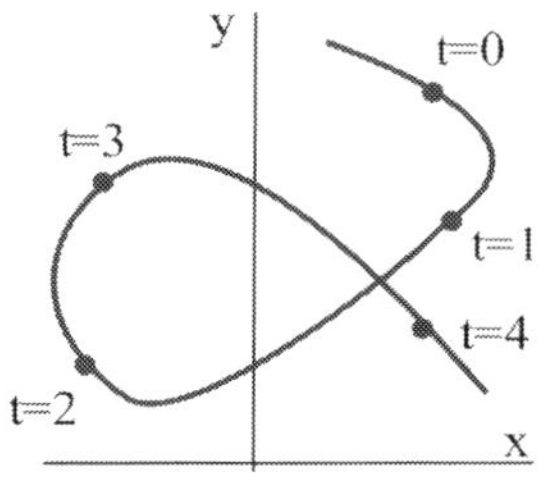

Fig. 10

47. $x(t) = R\cdot(t - \sin(t))$, $y(t) = R\cdot(1 - \cos(t))$. (a) Graph $(x(t), y(t))$ for $0 \le t \le 4\pi$.

(b) Find dx/dt, dy/dt, the tangent slope dy/dx , and speed when $t = \pi/2$ and π.

(The graph of $(x(t), y(t))$ is called a **cycloid** and is the path of a light attached to the edge of a rolling wheel with radius R.)

48. Describe the motion of two particles whose locations at time t are $(\cos(t), \sin(t))$ and $(\cos(t), -\sin(t))$.

49. Describe the path of a robot whose location at time t is

(a) $(3\cdot\cos(t), 5\cdot\sin(t))$ (b) $(A\cdot\cos(t), B\cdot\sin(t))$

(c) Give the parametric equations so the robot will move along the same path as in part (a) but in the opposite direction.

50. After t seconds, a projectile hurled with initial velocity v and angle θ will be at $x(t) = v\cdot\cos(\theta)\cdot t$ feet and $y(t) = v\cdot\sin(\theta)\cdot t - 16t^2$ feet. (Fig. 11) (This formula neglects air resistance.)

(a) For an initial velocity of 80 feet/second and an angle of $\pi/4$, find $t > 0$ so $y(t) = 0$. What does this value for t represent physically? Evaluate x(t).

(b) For v and θ in part (a), calculate dy/dx. Find t so dy/dx $= 0$ at t, and evaluate x(t). What does x(t) represent physically?

(c) What initial velocity is needed so a ball hit at an angle of $\pi/4 \approx 0.7854$ will go over a 40 foot high fence 350 feet away?

(d) What initial velocity is needed so a ball hit at an angle of 0.7 will go over a 40 foot high fence 350 feet away?

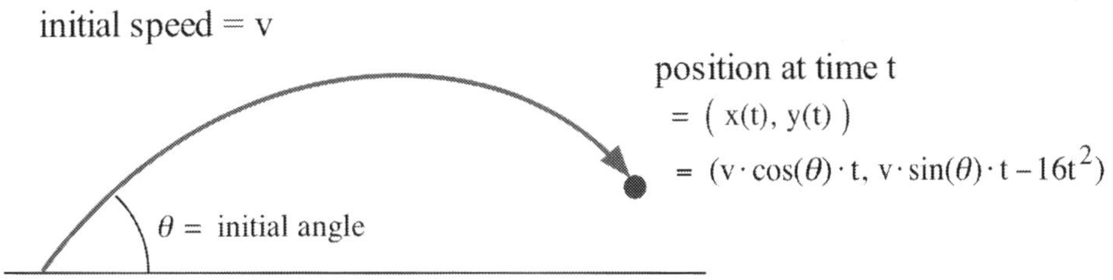

Fig. 11

Section 2.5 **PRACTICE Answers**

Practice 1: $\log_9 20 = \dfrac{\log(20)}{\log(9)} \approx \mathbf{1.3634165} \approx \dfrac{\ln(20)}{\ln(9)}$, $\log_3 20 = \dfrac{\log(20)}{\log(3)} \approx \mathbf{2.726833} \approx \dfrac{\ln(20)}{\ln(3)}$

$\log_\pi e = \dfrac{\log(e)}{\log(\pi)} \approx \mathbf{0.8735685} \approx \dfrac{\ln(e)}{\ln(\pi)} = \dfrac{1}{\ln(\pi)}$

Practice 2: $\mathbf{D}(\log_{10}(\sin(x))) = \dfrac{1}{\sin(x)\cdot\ln(10)}\,\mathbf{D}(\sin(x)) = \dfrac{\mathbf{\cos(x)}}{\sin(x)\cdot\ln(10)}$

$\mathbf{D}(\log_\pi(e^x)) = \dfrac{1}{e^x\cdot\ln(\pi)}\,\mathbf{D}(e^x) = \dfrac{\mathbf{e^x}}{e^x\cdot\ln(\pi)} = \dfrac{1}{\ln(\pi)}$

Practice 3: $\mathbf{D}(\sin(2^x)) = \cos(2^x)\,\mathbf{D}(2^x) = \cos(2^x)\cdot\mathbf{2^x\cdot\ln(2)}$

$\dfrac{\mathbf{d}}{\mathbf{dt}}\, 3^{(t^2)} = 3^{(t^2)}\ln(3)\,\mathbf{D}(t^2) = 3^{(t^2)}\ln(3)\cdot\mathbf{2t}$

Practice 4: $T = \dfrac{72}{1+h} = \dfrac{72}{1+t+\sin(t)}$.

$\dfrac{\mathbf{dT}}{\mathbf{dt}} = \dfrac{(1+t+\sin(t))\cdot\mathbf{D}(72) - 72\cdot\mathbf{D}(1+t+\sin(t))}{(1+t+\sin(t))^2} = \dfrac{-72(1+\cos(t))}{(1+t+\sin(t))^2}$

When $t = 5$, $\dfrac{\mathbf{dT}}{\mathbf{dt}} = \dfrac{-72(1+\cos(5))}{(1+t+\sin(5))^2} \approx \mathbf{-3.63695}$.

Practice 5: $x(t) = 3\cos(t)$ so $dx/dt = -3\sin(t)$. $y(t) = 2\sin(t)$

so $dy/dt = 2\cos(t)$. $\dfrac{dy}{dx} = \dfrac{dy/dt}{dx/dt} = \dfrac{2\cos(t)}{-3\sin(t)}$.

When $t = \pi/2$, $\dfrac{dy}{dx} = \dfrac{2\cos(\pi/2)}{-3\sin(\pi/2)} = \dfrac{2\cdot 0}{-3\cdot 1} = \mathbf{0}$. (See Fig. 12)

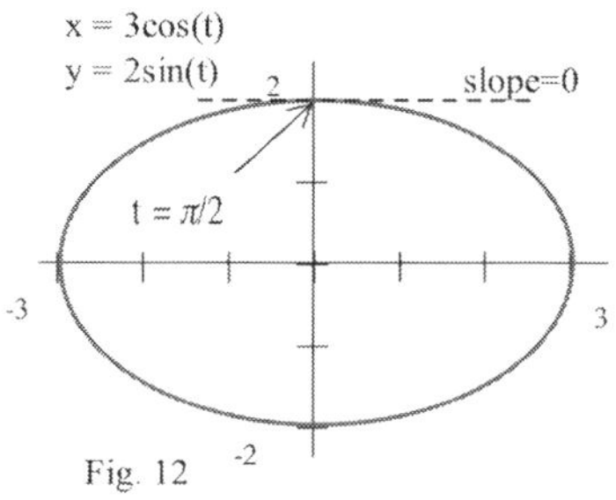

Fig. 12

Practice 6: When x=1: pos., pos., pos. When x=3: pos., neg., neg.

Practice 7: $x(t) = 3\sin(t)$ and $y(t) = 3\cos(t)$ so $dx/dt = 3\cos(t)$ and $dy/dt = -3\sin(t)$. Then

speed $= \sqrt{(dx/dt)^2 + (dy/dt)^2} = \sqrt{(3\cos(t))^2 + (-3\sin(t))^2}$

$= \sqrt{9\cdot\cos^2(t) + 9\cdot\sin^2(t)} = \sqrt{9} = \mathbf{3}$ **, a constant**.

Practice 8: $x(t) = 3\cos(t)$ and $y(t) = 2\sin(t)$ so $dx/dt = -3\sin(t)$ and $dy/dt = 2\cos(t)$. Then

speed $= \sqrt{(dx/dt)^2 + (dy/dt)^2} = \sqrt{(-3\sin(t))^2 + (2\cos(t))^2} = \sqrt{9\sin^2(t) + 4\cos^2(t)}$.

When $t = 0$, the speed is $\sqrt{9\cdot(0)^2 + 4\cdot(1)^2} = 2$.

When $t = \pi/2$, the speed is $\sqrt{9\cdot(1)^2 + 4\cdot(0)^2} = \mathbf{3}$ **(faster)**.

2.6 RELATED RATES: An Application of Derivatives

In the next several sections we'll look at more uses of derivatives. Probably no single application will be of interest or use to everyone, but at least some of them should be useful to you. Applications also reinforce what you have been practicing; they require that you recall what a derivative means and use the techniques covered in the last several sections. Most people gain a deeper understanding and appreciation of a tool as they use it, and differentiation is both a powerful concept and a useful tool.

The Derivative As A Rate of Change

In Section 2.1, several interpretations were given for the derivative of a function. Here we will examine how the "rate of change of a function" interpretation can be used. If several variables or quantities are related to each other and some of the variables are changing at a known rate, then we can use derivatives to determine how rapidly the other variables must be changing.

Example 1: Suppose we know that the radius of a circle is increasing at a rate of 10 feet each second (Fig. 1), and we want to know how fast the area of the circle is increasing when the radius is 5 feet. What can we do?

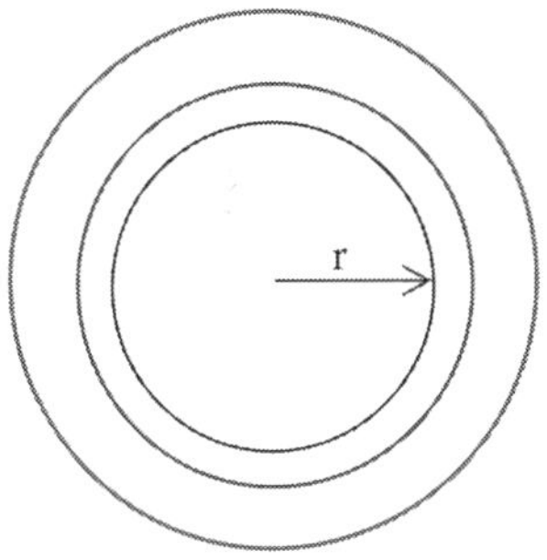

Fig. 1

Solution: We could get an approximate answer by calculating the area of the circle when the radius is 5 feet ($A = \pi r^2 = \pi(5 \text{ feet})^2 \approx 78.6 \text{ feet}^2$) and 1 second later when the radius is 10 feet larger than before ($A = \pi r^2 = \pi(15 \text{ feet})^2 \approx 706.9 \text{ feet}^2$) and then finding $\Delta\text{Area}/\Delta\text{time} = (706.9 \text{ ft}^2 - 78.6 \text{ ft}^2)/(1 \text{ sec}) = 628.3 \text{ ft}^2/\text{sec}$. This approximate answer represents the average change in area during the 1 second period when the radius increased from 5 feet to 15 feet. It is the slope of the secant line through the points P and Q in Fig. 2, and it is clearly not a very good approximation of the instantaneous rate of change of the area, the slope of the tangent line at the point P.

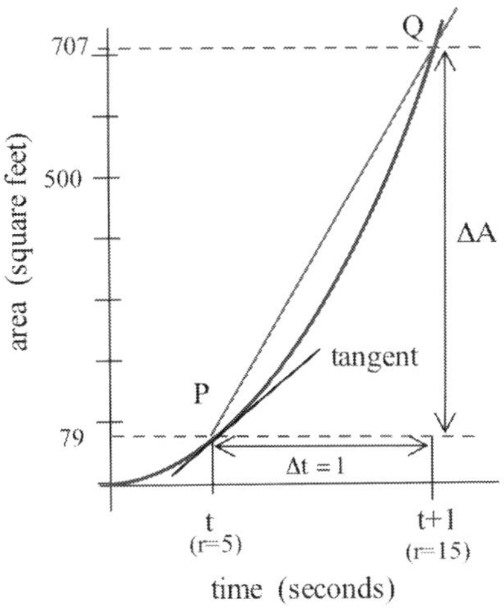

Fig. 2

We could get a better approximation by calculating $\Delta A/\Delta t$ over a shorter time interval, say $\Delta t = .1$ seconds. Then the original area was 78.6 ft^2, the new area is $A = \pi(6 \text{ feet})^2 \approx 113.1 \text{ feet}^2$ (Why is the new radius 6 feet?) so $\Delta A/\Delta t = (113.1 \text{ ft}^2 - 78.6 \text{ ft}^2)/(.1 \text{ sec}) = 345 \text{ ft}^2/\text{sec}$. This is the slope of the secant line through the point P and R in Fig. 3, and it is a much better approximation of the slope of the tangent line at P, but it is still only an approximation. Using derivatives, we can get an **exact** answer without doing very much work.

We know that the two variables in this problem, the radius r and the area A, are related to each other by the formula $A = \pi r^2$, and we know that both r and A are changing over time so each of them is a function of an additional variable t = time. We will continue to write the radius and area variables as r and A, but it is important to remember that each of them is really a **function** of t, $r = r(t)$ and $A = A(t)$. The statement that "the radius is increasing at a rate of 10 feet each second" can be translated into a mathematical statement about the rate of change, the derivative of r with respect to time: $\frac{dr}{dt} = 10$ ft/sec . The question about the rate of change of the area is a question about $\frac{dA}{dt}$. Collecting all of this information, we have

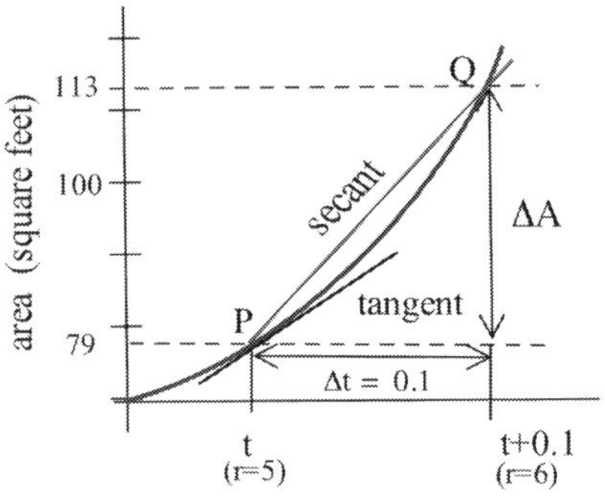

Fig. 3

Variables: r(t) = radius at time t , A(t) = area at time t

We Know: r = 5 feet and $\frac{d\,r(t)}{dt} = 10$ ft/sec.

We Want: $\frac{d\,A(t)}{dt}$ when r = 5

Connecting Equation: $A = \pi r^2$ or $A(t) = \pi r^2(t)$.

Finally, we are ready to find $\frac{dA}{dt}$ — we just need to differentiate each side of the equation $A = \pi r^2$ with respect to the independent variable t.

$\frac{dA}{dt} = \frac{d(\pi r^2)}{dt} = \pi \frac{dr^2}{dt} = \pi\, 2r \,\mathbf{\frac{dr}{dt}}$. The last piece, $\mathbf{\frac{dr}{dt}}$, appears in the derivative because r is a **function** of t and we must use the differentiation rule for a function to a power (or the Chain Rule):

$\frac{d}{dt} f^n(t) = nf^{n-1}(t)\cdot \mathbf{\frac{df(t)}{dt}}$.

We know from the problem that $\frac{dr}{dt} = 10$ ft/sec so $\frac{dA}{dt} = \pi\, 2r\, \mathbf{\frac{dr}{dt}} = \pi 2r(10 \text{ ft/s}) = 20\pi r$ ft/s. This answer tells us that the rate of increase of the area of the circle, $\frac{dA}{dt}$, depends on the value of the radius r as well as on the value of $\frac{dr}{dt}$. Since r = 5 feet, the area of the circle will be increasing at a rate of

$$\frac{dA}{dt} = 20\pi \mathbf{r} \text{ ft/s} = 20\pi(\mathbf{5\ feet}) \text{ ft/s} = 100\pi \text{ ft}^2/\text{s} \approx 314.2 \text{ square feet per second.}$$

The key steps in finding the exact rate of change of the area of the circle were to:

- write the known information in a mathematical form, expressing rates of change as derivatives (r = 5 feet and dr/dt = 10 ft/sec)
- write the question in a mathematical form (dA/dt = ?)
- find an equation connecting or relating the variables ($A = \pi r^2$)
- differentiate both sides of the equation relating the variables, remembering that the variables are **functions** of t ($dA/dt = 2\pi r\, dr/dt$)
- put all of the known values into the equation in the previous step and solve for the desired part in the resulting equation ($dA/dt = 2\pi(5\text{ ft})(10\text{ ft/sec}) = 314.2\text{ ft}^2\text{/sec}$)

Example 2: Divers lives depend on understanding situations involving related rates. In water, the pressure at a depth of x feet is approximately $P(x) = 15(1 + \frac{x}{33})$ pounds per square inch (compared to approximately 15 pounds per square inch at sea level = P(0)). Volume is inversely proportional to the pressure , v = k/p , so doubling the pressure will result in half the original volume. Remember that volume is a function of the pressure: v = v(p).

(a) Suppose a diver's lungs, at a depth of 66 feet, contained 1 cubic foot of air , and the diver ascended to the surface without releasing any air, what would happen?

(b) If a diver started at a depth of 66 feet and ascended at a rate of 2 feet per second, how fast would the pressure be changing?

(Dives deeper than 50 feet also involve a risk of the "bends" or decompression sickness if the ascent is too rapid. Tables are available which show the safe rates of ascent from different depths.)

Solution: (a) The diver would risk getting ruptured lungs. The 1 cubic foot of air at a depth of 66 feet would be at a pressure of $P(\mathbf{66}) = 15(1 + \frac{\mathbf{66}}{33}) = 45$ pounds per square inch (psi). Since the pressure at sea level , P(0) = 15 psi , is only 1/3 as great, each cubic foot of air would expand to 3 cubic feet, and the diver's lungs would be in danger. Divers are taught to release air as they ascend to avoid this danger.

(b) The diver is ascending at a rate of 2 feet/second so the rate of change of the diver's depth x(t) is $\frac{\mathbf{d}}{\mathbf{dt}}x(t) = -2$ ft/s. The pressure , $P = 15(1 + \frac{x}{33}) = 15 + \frac{15}{33}x$, is a function of x (or x(t)) so $\frac{\mathbf{d}}{\mathbf{dt}}P = \frac{\mathbf{d}P}{\mathbf{dx}} \cdot \frac{\mathbf{d}x}{\mathbf{dt}} = (\frac{15}{33}\text{ psi/ft})\cdot(-2\text{ ft/sec}) = -\frac{30}{33}\text{ psi/sec} \approx -0.91$ psi/sec.

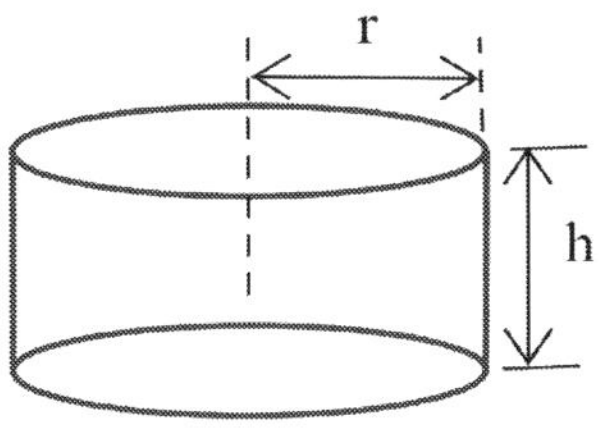

Fig. 4

Example 3: The height of a cylinder is increasing at 7 meters per second and the radius is increasing at 3 meters per second. How fast is the volume changing when the cylinder is 5 meters high and has a radius of 6 meters? (Fig. 4)

Solution: First we need to translate our known information into a mathematical format. The height and radius are given: h = height = 5 m and r = radius = 6 m. We are also told how fast h and r are changing: dh/dt = 7 m/s and dr/dt = 3 m/s. Finally, we are asked to find dV/dt, and we should expect the units of dV/dt to be the same as $\Delta V/\Delta t$ which are m^3/s.

> **Variables**: h(t) = height at time t , r(t) = radius at t , V(t) = volume at t.
>
> **Know:** h = 5 m , $\frac{d\,h(t)}{dt} = 7$ m/s , r = 6 m , $\frac{d\,r(t)}{dt} = 3$ m/s .
>
> **Want:** $\frac{d\,V(t)}{dt}$

We also need an equation which relates the variables h, r and V (all of which are **functions** of time t) to each other:

> **Connecting Equation:** $V = \pi r^2 h$ or $V(t) = \pi r^2(t)\, h(t)$

Then, differentiating each side of this equation with respect to t (remembering that h, r and V are **functions**), we have

$$\frac{dV}{dt} = \frac{d(\pi r^2 h)}{dt} = \pi \frac{d(r^2 h)}{dt} = \pi\left\{ r^2 \frac{dh}{dt} + h \frac{dr^2}{dt} \right\} \qquad \text{by the Product Rule}$$

$$= \pi\left\{ r^2 \frac{dh}{dt} + h(2r)\frac{dr}{dt} \right\} \qquad \text{by the Power Rule for functions.}$$

The rest is just substituting values and doing some arithmetic:

$$= \pi\left\{ (6\text{ m})^2 (7\text{ m/s}) + (5\text{ m})2(6\text{ m})(3\text{ m/s}) \right\}$$
$$= \pi\{ 252\text{ m}^3/\text{s} + 180\text{ m}^3/\text{s} \}$$
$$= 432\pi\text{ m}^3/\text{s} \approx 1357.2\text{ m}^3/\text{s} .$$

The volume of the cylinder is increasing at a rate of 1357.2 cubic meters per second. (It is always encouraging when the units of our answer are the ones we expect.)

Practice 1: How fast is the **surface area** of the cylinder changing in the previous example? (Assume that h, r, dh/dt, and dr/dt have the same values as in the example and use Fig. 5 to help you determine an equation relating the surface area of the cylinder to the variables h and r. The cylinder has a top and bottom.)

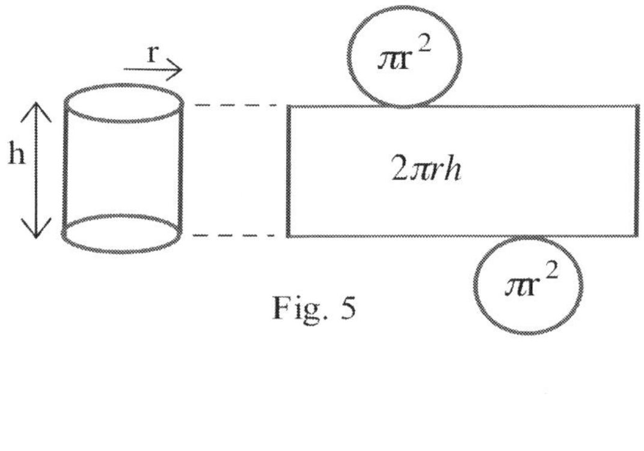

Fig. 5

Practice 2: How fast is the volume of the cylinder in the previous example changing if the radius is **decreasing** at a rate of 3 meters per second? (The height, radius and rate of change of the height are the same as in the previous example: 5 m, 6 m and 7 m/s respectively.)

Usually, the most difficult part of Related Rate problems is to find an equation which relates or connects all of the variables. In the previous problems, the relating equations required a knowledge of geometry and formulas for areas and volumes (or knowing where to find them). Other Related Rates problems may require other information about similar triangles, the Pythagorean formula, or trigonometry — it depends on the problem.

It is a good idea, a very good idea, to draw a picture of the physical situation whenever possible. It is also a good idea, particularly if the problem is very important (your next raise depends on getting the right answer), to calculate at least one approximate answer as a check of your exact answer.

Example 4: Water is flowing into a conical tank at a rate of $5 \text{ m}^3/\text{s}$. If the radius of the top of the cone is 2 m (see Fig. 6), the height is 7 m, and the depth of the water is 4 m, then how fast is the water level rising?

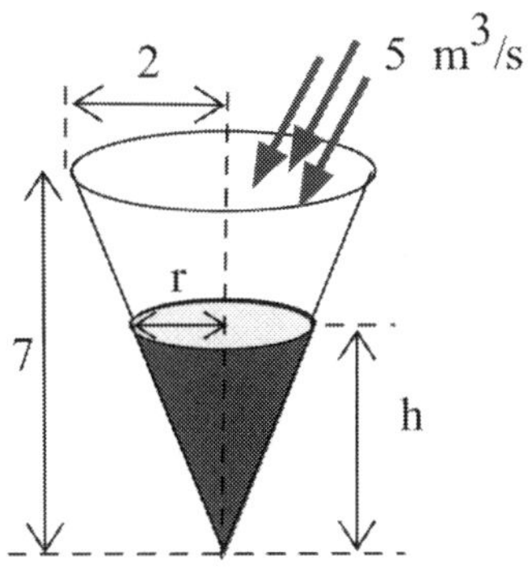

Fig. 6

Solution: Lets define our variables to be h = height (or depth) of the water in the cone and V = the volume of the water in the cone. Both h and V are changing, and both of them are functions of time t. We are told in the problem that h = 4 m and $dV/dt = 5 \text{ m}^3/\text{s}$, and we are asked to find dh/dt. We expect that the units of dh/dt will be the same as $\Delta h/\Delta t$ which are meters/second.

Variables: h(t) = height at t , r(t) = radius of the top surface of the water at t, V(t) = volume of water at time t

Know: h = 4 m , $\frac{d\,V(t)}{dt} = 5 \text{ m}^3/\text{s}$

Want: $\frac{d\,h(t)}{dt}$

Unfortunately, the equation for the volume of a cone, $V = \frac{1}{3}\pi r^2 h$, also involves an additional variable r, the radius of the cone at the top of the water . This is a situation in which the picture can be a great help by suggesting that we have a pair of similar triangles so r/h = (top radius)/(total height) = (2 m)/(7 m) = 2/7 and $r = \frac{2}{7}h$. Then we can rewrite the volume of the cone of water, $V = \frac{1}{3}\pi r^2 h$, as a function of the single variable h:

Connecting Equation: $V = \frac{1}{3}\pi r^2 h = \frac{1}{3}\pi(\frac{2}{7}h)^2 h = \frac{4}{147}\pi h^3$.

The rest of the solution is straightforward.

$$\frac{dV}{dt} = \frac{d(\frac{4}{147}\pi h^3)}{dt} = \frac{4}{147}\pi\frac{d(h^3)}{dt} = \frac{4}{147}\pi\, 3h^2\frac{dh}{dt} \quad \text{remember, h is a function of t}$$

$$= \frac{4}{147}\pi\, 3(4\text{ m})^2\frac{dh}{dt} \approx (4.10\text{ m}^2)\frac{dh}{dt}\ .$$

We know that $\frac{dV}{dt} = 5\text{ m}^3/\text{s}$ and $\frac{dV}{dt} = (4.10\text{ m}^2)\frac{dh}{dt}$ so it is easy to solve for

$$\frac{dh}{dt} = \frac{dV/dt}{(4.10\text{ m}^2)} = \frac{5\text{ m}^3/\text{s}}{4.10\text{ m}^2} \approx 1.22\text{ m/s}\ .$$

This example was a little more difficult than the others because we needed to use similar triangles to get an equation relating V to h and because we eventually needed to do a little arithmetic to solve for dh/dt.

Practice 3: A rainbow trout has taken the fly at the end of a 60 foot line, and the line is being reeled in at a rate of 30 feet per minute. If the tip of the rod is 10 feet above the water and the trout is at the surface of the water, how fast is the trout being pulled toward the angler? (Suggestion: Draw a picture and use the Pythagorean formula.)

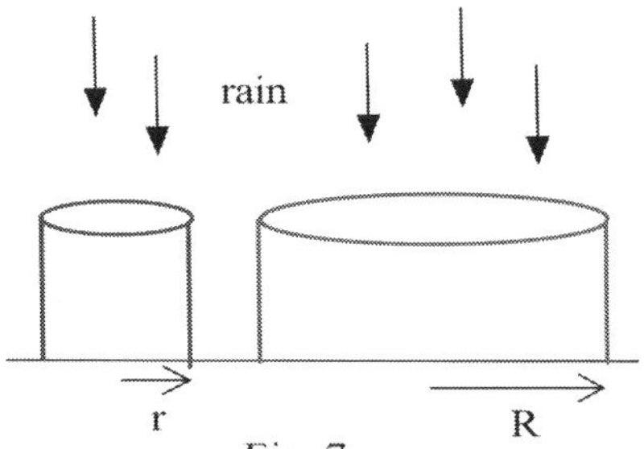

Fig. 7

Example 5: When rain is falling vertically, the amount (volume) of rain collected in a cylinder is proportional to the area of the **opening** of the cylinder. If you place a narrow cylindrical glass and a wide cylindrical glass out in the rain (Fig. 7),

(a) which glass will collect water faster, and

(b) in which glass will the water level rise faster?

Solution: Let's assume that the smaller glass has a radius of r and the larger has a radius of R, R > r, so the areas of their openings are $\pi\cdot r^2$ and $\pi\cdot R^2$ respectively.

(a) The smaller glass will collect water at the rate $\frac{dv}{dt} = K\cdot\pi\cdot r^2$, and the larger at the rate $\frac{dV}{dt} = K\cdot\pi\cdot R^2$ so $\frac{dV}{dt} > \frac{dv}{dt}$, and the larger glass will collect water faster than the smaller glass.

(b) The volume of water in each glass is a function of the radius of the glass and the height of the water in the glass: $v = \pi r^2 h$ and $V = \pi R^2 H$ where h and H are the heights of the water levels in the smaller and larger glasses, respectively. The heights h and H vary with t (are functions of t) so

$$\frac{dv}{dt} = \frac{d(\pi r^2 h)}{dt} = \pi r^2\frac{dh}{dt} \text{ and } \frac{dh}{dt} = \frac{dv/dt}{\pi r^2} = \frac{K\pi r^2}{\pi r^2} = K \quad (\text{we got } dv/dt = K\pi r^2 \text{ in part (a)}).$$

Similarly, $\frac{dV}{dt} = \frac{d(\pi R^2 H)}{dt} = \pi R^2 \frac{dH}{dt}$ so $\frac{dH}{dt} = \frac{dV/dt}{\pi R^2} = \frac{K\pi R^2}{\pi R^2} = K$.

Then $\frac{dh}{dt} = K = \frac{dH}{dt}$ so the water level in each glass is rising at the same rate. In a one minute period, the larger glass will collect more rain, but the larger glass also requires more rain to raise its water level by each inch. How do you think the volumes and water levels would change if we placed a small glass and a large plastic box side by side in the rain?

PROBLEMS

1. An expandable sphere is being filled with liquid at a constant rate from a tap (imagine a water balloon connected to a faucet). When the radius of the sphere is 3 inches, the radius is increasing at 2 inches per minute. How fast is the liquid coming out of the tap? ($V = \frac{4}{3}\pi r^3$)

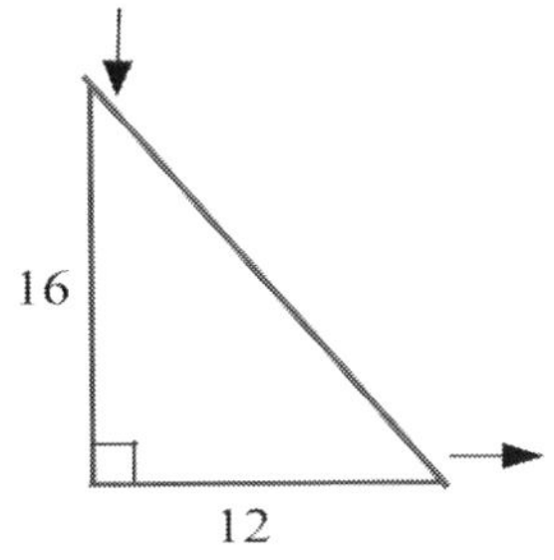

Fig. 8

2. The 12 inch base of a right triangle is growing at 3 inches per hour, and the 16 inch height is shrinking at 3 inches per hour (Fig. 8)
   (a) Is the area increasing or decreasing?
   (b) Is the perimeter increasing or decreasing?
   (c) Is the hypotenuse increasing or decreasing?

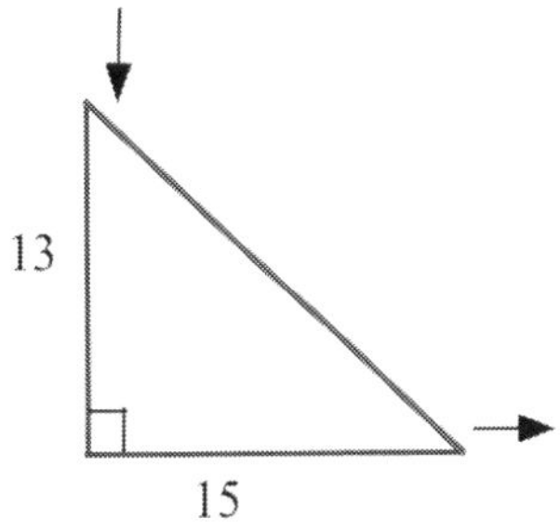

Fig. 9

3. One hour later the right triangle in Problem 2 is 15 inches long and 13 inches high (Fig. 9), and the base and height are changing at the same rate as in Problem 2.
   (a) Is the area increasing or decreasing now?
   (b) Is the hypotenuse increasing or decreasing now?
   (c) Is the perimeter increasing or decreasing now?

4. A young woman and her boyfriend plan to elope, but she must rescue him from his mother who has locked him in his room. The young woman has placed a 20 foot long ladder against his house and is knocking on his window when his mother begins pulling the bottom of the ladder away from the house at a rate of 3 feet per second (Fig. 10). How fast is the top of the ladder (and the young couple) falling when the bottom of the ladder is
   (a) 12 feet from the bottom of the wall?
   (b) 16 feet from the bottom of the wall?
   (c) 19 feet from the bottom of the wall?

Fig. 10

5. The length of a 12 foot by 8 foot rectangle is increasing at a rate of 3 feet per second and the width is decreasing at 2 feet per second (Fig. 11).

(a) How fast is the perimeter changing?

(b) How fast is the area changing?

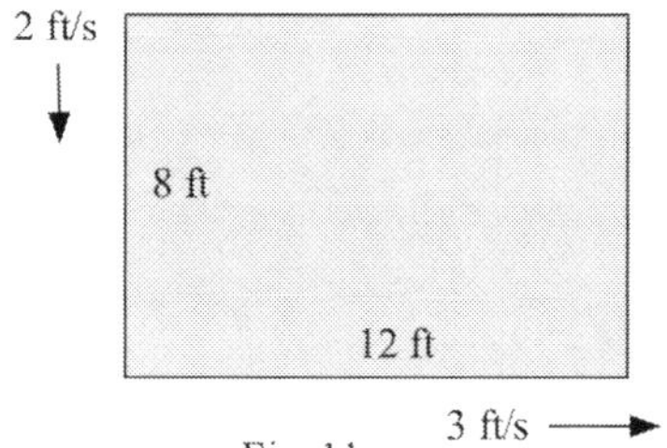

Fig. 11

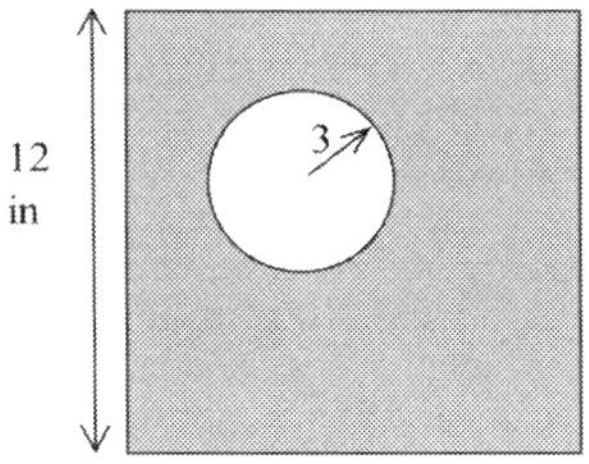

Fig. 12

6. A circle of radius 3 inches is inside a square with 12 inch sides (Fig. 12). How fast is the area between the circle and square changing if the radius is increasing at 4 inches per minute and the sides are increasing at 2 inches per minute?

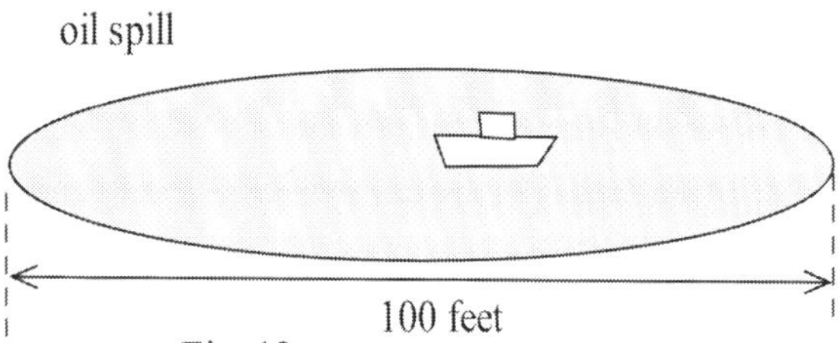

Fig. 13

7. An oil tanker in Puget Sound has sprung a leak, and a circular oil slick is forming (Fig. 13). The oil slick is 4 inches thick everywhere, is 100 feet in diameter, and the diameter is increasing at 12 feet per hour. Your job, as the Coast Guard commander or the tanker's captain, is to determine how fast the oil is leaking from the tanker.

8. A mathematical species of slug has a semicircular cross section and is always 5 times as long as it is high (Fig. 14). When the slug is 5 inches long, it is growing at .2 inches per week.

(a) How fast is its volume increasing?

(b) How fast is the area of its "foot" (the part of the slug in contact with the ground) increasing?

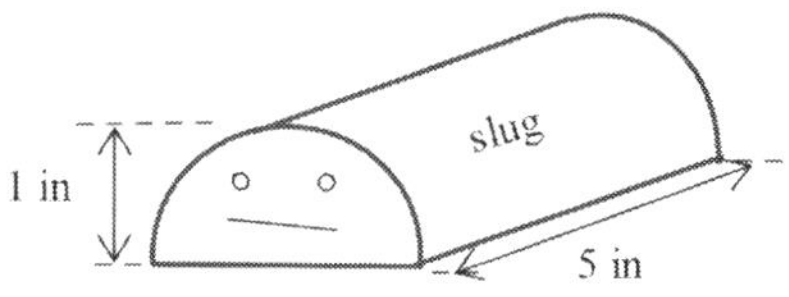

Fig. 14

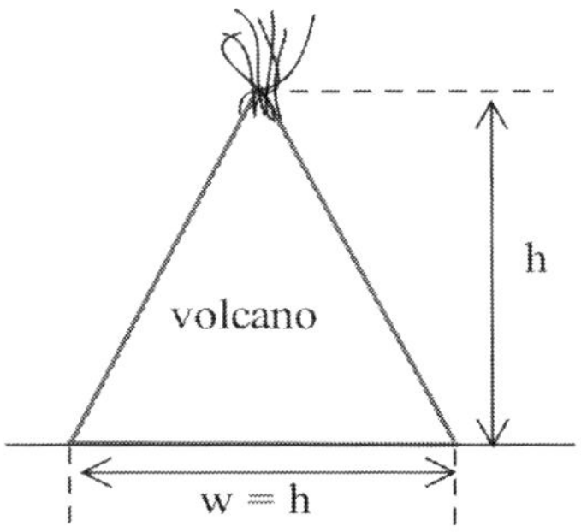

Fig. 15

9. Lava flowing from a hole at the top of a hill is forming a conical mountain whose height is always the same as the width of its base (Fig. 15). If the mountain is increasing in height at 2 feet per hour when it is 500 feet high, how fast is the lava flowing (how fast is the volume of the mountain increasing)? ($V = \frac{1}{3}\pi r^2 h$)

10. A six foot tall person is walking away from a 14 foot tall lamp post at 3 feet per second (Fig. 16). When the person is 10 feet from the lamp post,
   (a) How fast is the length of the person's shadow changing?
   (b) How fast is the tip of the shadow moving away from the lamp post?

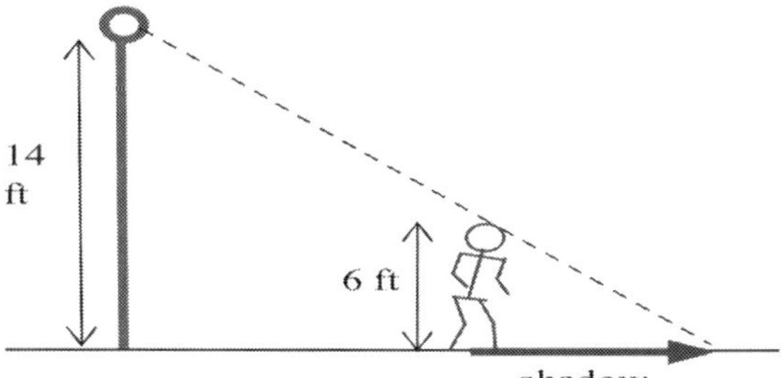

Fig. 16

11. Answer parts (a) and (b) in Problem 10 for when the person is 20 feet from the lamp post.

12. Water is being poured at a rate of 15 cubic feet per minute into a conical reservoir which is 20 feet deep and has a top radius of 10 feet (Fig. 17).
   (a) How long will it take to fill the empty reservoir?
   (b) How fast is the water level rising when the water is 4 feet deep?
   (c) How fast is the water level rising when the water is 16 feet deep?

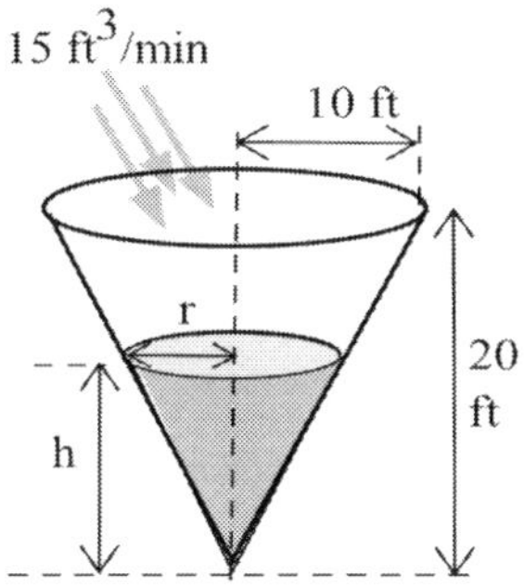

Fig. 17

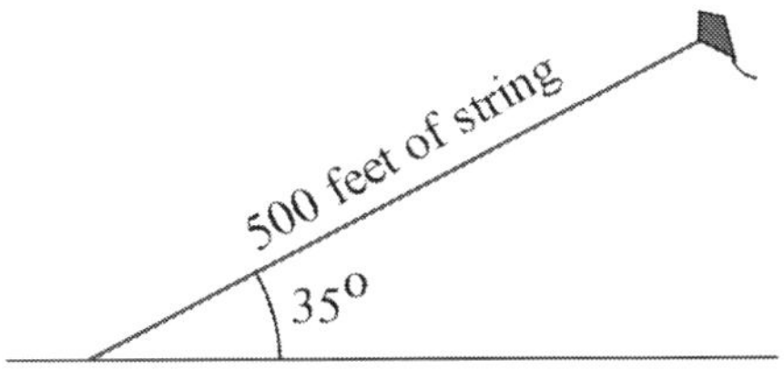

Fig. 18

13. The string of a kite is perfectly taut and always makes an angle of 35° above horizontal (Fig. 18).
   (a) If the kite flyer has let out 500 feet of string, how high is the kite?
   (b) If the string is let out at a rate of 10 feet per second, how fast is the kite's height increasing?

14. A small tracking telescope is viewing a hot air balloon rise from a point 1000 meters away from a point directly under the balloon (Fig. 19).
   (a) When the viewing angle is 20°, it is increasing at a rate of 3° per minute. How high is the balloon, and how fast is it rising?
   (b) When the viewing angle is 80°, it is increasing at a rate of 2° per minute. How high is the balloon, and how fast is it rising?

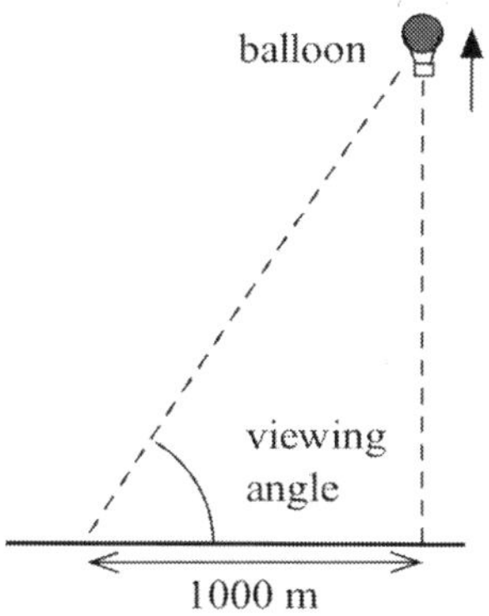

Fig. 19

15. The 8 foot diameter of a spherical gas bubble is increasing at 2 feet per hour, and the 12 foot long edges of a cube containing the bubble are increasing at 3 feet per hour. Is the volume contained between the spherical bubble and the cube increasing or decreasing? At what rate?

16. In general, the strength S of an animal is proportional to the cross–sectional area of its muscles, and this area is proportional to the square of its height H, so the strength $S = aH^2$. Similarly, the weight W of the animal is proportional to the cube of its height, so $W = bH^3$. Finally, the relative strength R of an animal is the ratio of its strength to its weight. As the animal grows, show that its strength and weight increase, but that the relative strength decreases.

17. The snow in a hemispherical pile melts at a rate proportional to its exposed surface area (the surface area of the hemisphere). Show that the height of the snow pile is decreasing at a constant rate.

18. If the rate at which water vapor condenses onto a spherical raindrop is proportional to the surface area of the raindrop, show that the radius of the raindrop will increase at a constant rate.

19. Define A(x) to be the **area** bounded by the x and y axes, the horizontal line $y = 5$, and a vertical line at x (Fig. 20).

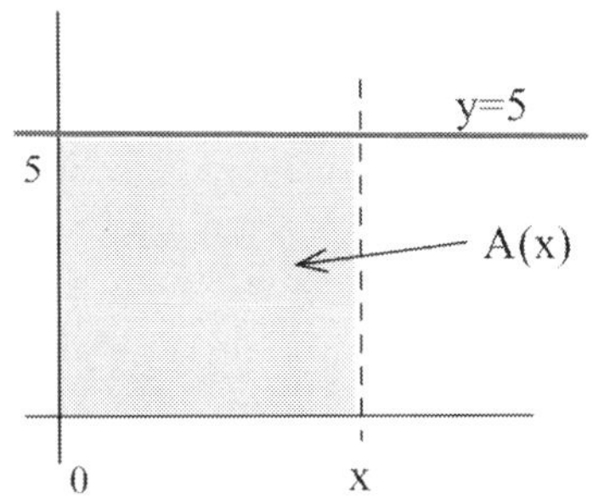

Fig. 20

(a) Find a formula for A as a function of x.

(b) Determine $\frac{d\,A(x)}{dx}$ when $x = 1, 2, 4$ and 9.

(c) Suppose x is a function of time, $x(t) = t^2$, and find a formula for A as a function of t.

(d) Determine $\frac{d\,A}{dt}$ when $t = 1, 2,$ and 3.

(e) Suppose $x(t) = 2 + \sin(t)$. Find a formula for A(t) and determine $\frac{d\,A}{dt}$.

20. The point P is going around the circle $x^2 + y^2 = 1$ twice a minute (Fig. 21). How fast is the distance between the point P and the point (4,3) changing

(a) when P = (1,0)? (b) when P = (0,1)? (c) when P = (.8, .6)?

(Suggestion: Write x and y as parametric functions of time t.)

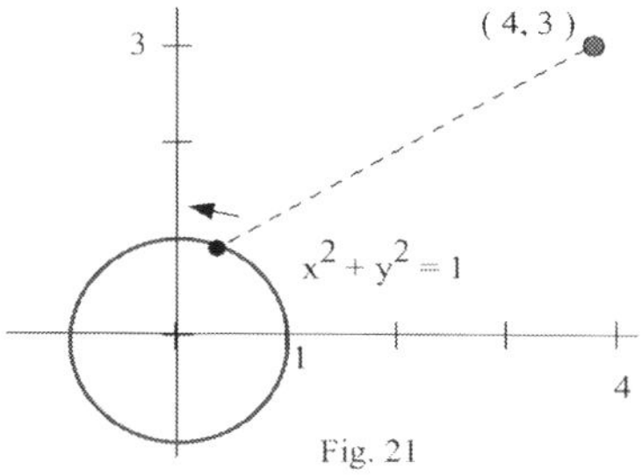

Fig. 21

21. You are walking along a sidewalk toward a 40 foot wide sign which is adjacent to the sidewalk and perpendicular to it (Fig. 22).

(a) If your viewing angle θ is 10°, then how far are you from the nearest corner of the sign?

(b) If your viewing angle is 10° and you are walking at 25 feet per minute, then how fast is your viewing angle changing?

(c) If your viewing angle is 10° and is increasing at 2° per minute, then how fast are you walking?

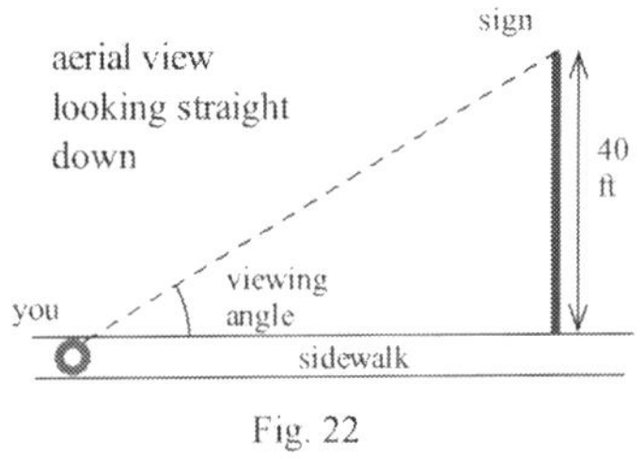

Fig. 22

Section 2.6 **PRACTICE Answers**

Practice 1: The surface area of the cylinder is $SA = 2\pi rh + 2\pi r^2$. From the Example, we know that $h' = 7$ m/s and $r' = 3$ m/s, and we want to know how fast the surface area is changing when $h = 5$ m and $r = 6$ m.

$$\frac{d\,SA}{dt} = 2\pi r \cdot h' + 2\pi \cdot r' \cdot h + 2\pi \cdot 2r \cdot r'$$

$$= 2\pi(6\text{ m})(7\text{ m/s}) + 2\pi(3\text{ m/s})(5\text{ m}) + 2\pi(2\cdot 6\text{ m})(3\text{ m/s}) = 186\pi\ \text{m}^2/\text{s}$$

$\approx$ **584.34 square meters per second**. (Note that the units represent a rate of change of area.)

Practice 2: The volume of the cylinder is V = (area of the bottom)(height) = $\pi r^2 h$. We are told that $r' = -3$ m/s, and that $h = 5$ m, $r = 6$ m, and $h' = 7$ m/s.

$$\frac{dV}{dt} = \pi r^2 \cdot h' + \pi \cdot 2r \cdot r' \cdot h = \pi(6\text{ m})^2(7\text{ m/s}) + \pi(2\cdot 6\text{ m})(-3\text{ m/s})(5\text{ m}) = 72\pi\ \text{m}^3/\text{s}$$

$\approx$ **226.19 cubic meters per second**. (Note that the units represent a rate of change of volume.)

Practice 3: Fig. 23 represents the situation described in this problem. We are told that $L' = -30$ ft/min. The variable F represents the distance of the fish from the angler, and we are asked to find F ', the rate of change of F when $L = 60$ ft.

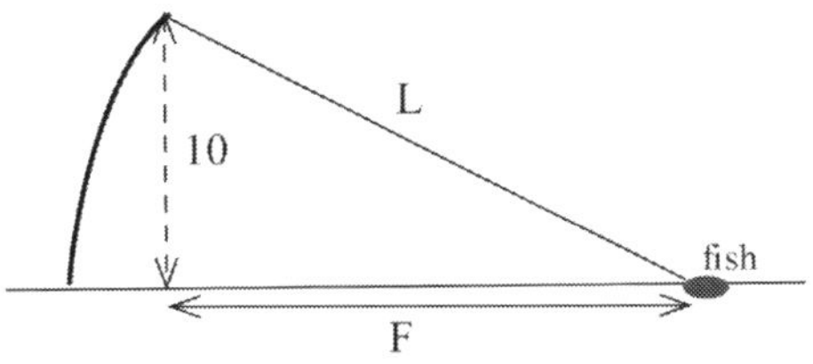

Fig. 23

Fortunately, the problem contains a right triangle so there is a formula (the Pythagorean formula) connecting F and L: $F^2 + 10^2 = L^2$ so

$$F = \sqrt{L^2 - 100}\ .$$

Then $F' = \frac{1}{2}(L^2 - 100)^{-1/2}\,\frac{d(L^2 - 100)}{dt} = \frac{2L\cdot L'}{2\sqrt{L^2 - 100}}$.

When $L = 60$ feet, $F' = \dfrac{2(60\text{ ft})(-30\text{ ft/min})}{2\sqrt{(60\text{ ft})^2 - (10\text{ ft})^2}} \approx \dfrac{-3600\text{ ft}^2/\text{min}}{118.32\text{ ft}} \approx$ **–30.43 ft/min.**

We could also find F ' implicitly: $F^2 = L^2 - 100$ so, differentiating each side,

$$2F\cdot F' = 2L\cdot L' \text{ and } F' = \frac{L\cdot L'}{F}\ .$$

Then we could use the given values for L and L ' and value of F (found using the Pythagorean formula) to evaluate F '.

2.7 NEWTON'S METHOD FOR FINDING ROOTS

Newton's method is a process which can find roots of functions whose graphs cross or just kiss the x–axis. Although this method is a bit harder to apply than the Bisection Algorithm, it often finds roots that the Bisection Algorithm misses, and it usually finds them faster.

Off On A Tangent

The basic idea of Newton's Method is remarkably simple and graphic (Fig. 1):

> at a point (x, f(x)) on the graph of f, the tangent line to the graph of f "points toward" a root of f, a place where the graph touches the x–axis.

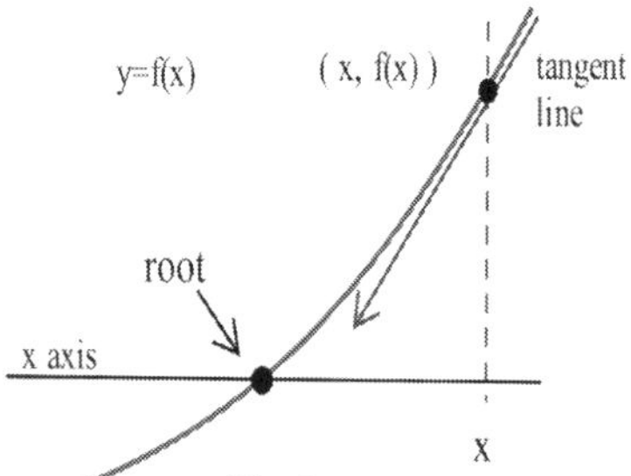

Fig. 1

If we want to find a root of f , all we need to do is pick a starting value x_0, go up or down to the point $(x_0 , f(x_0))$ on the graph of f, build a tangent line there, and follow the tangent line to where it crosses the x–axis, say at x_1. If x_1 is a root of f , then we are done. If x_1 is not a root of f, then x_1 is usually closer to the root than x_0 was, and we can repeat the process, using x_1 as our new starting point. Newton's method is an **iterative procedure**, that is, the output from one application of the method becomes the starting point for the next application.

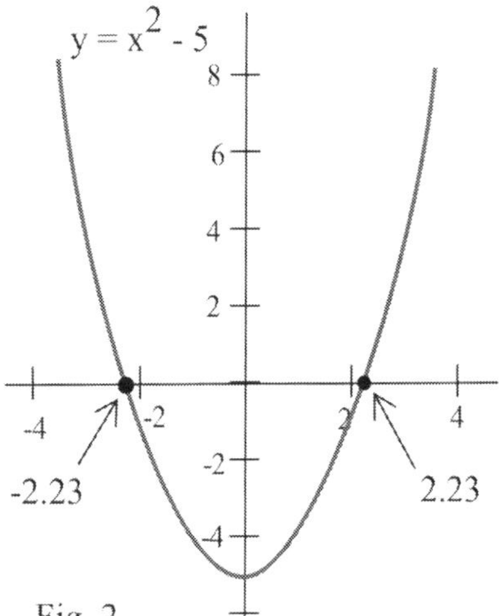

Fig. 2

Let's start with a differentiable function $f(x) = x^2 – 5$, (Fig. 2) whose roots we already know, $x = \pm\sqrt{5} \approx \pm 2.236067977$, and illustrate how Newton's method works. First we pick some value for x_0 , say $x_0 = 4$ for this example, and move to the point $(x_0 , f(x_0)) = (4 , 11)$ on the graph of f.

At (4 , 11) , the graph of f "points to" a location on the x–axis which is closer to the root of f (Fig. 3). We can calculate this location on the x–axis by finding the equation of the line tangent to the graph of f at the point (4 , 11) and then finding where this tangent line intersects the x–axis:

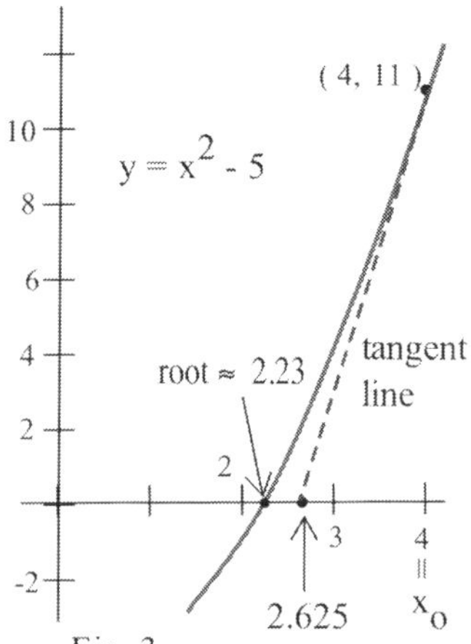

Fig. 3

At the point (4 , 11), the line tangent to f has slope $m = f '(4) = 2(4) = 8$, so the equation of the tangent line is $y – 11 = 8(x – 4)$. Setting $y = 0$, we can find where the tangent line crosses the x–axis:

$$0 – 11 = 8(x – 4) \text{ , so } x = 4 – \frac{11}{8} = \frac{21}{8} = 2.625 .$$

Call this new value x_1 : $x_1 = 2.625$.

The point $x_1 = 2.625$ is closer to the actual root , but it certainly does not equal the actual root. If Newton's method stopped after one step with the estimate of 2.625, it would not be very useful. Instead, we can use this new value for x , $x_1 = 2.625$, to repeat the procedure (Fig. 4):

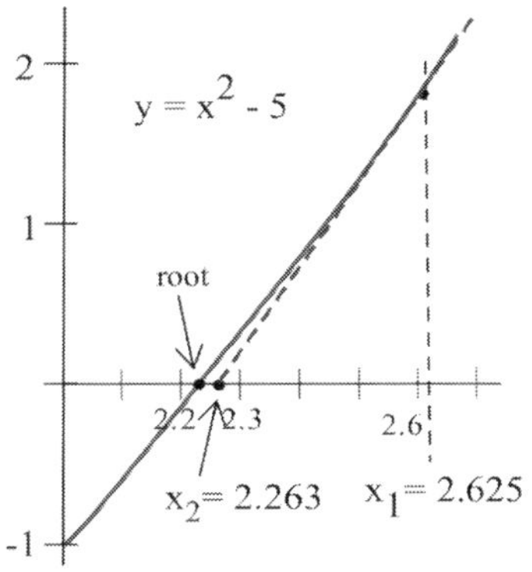

Fig. 4

(i) move to the point $(x_1, f(x_1)) = (2.625 , 1.890625)$ on the graph,

(ii) find the equation of the tangent line at the point $(x_1, f(x_1))$:

$$y-1.890625 = 5.25(x- 2.625)$$

(iii) find the new value where the tangent line intersects the x–axis and call it x_2. ($x_2 = 2.262880952$)

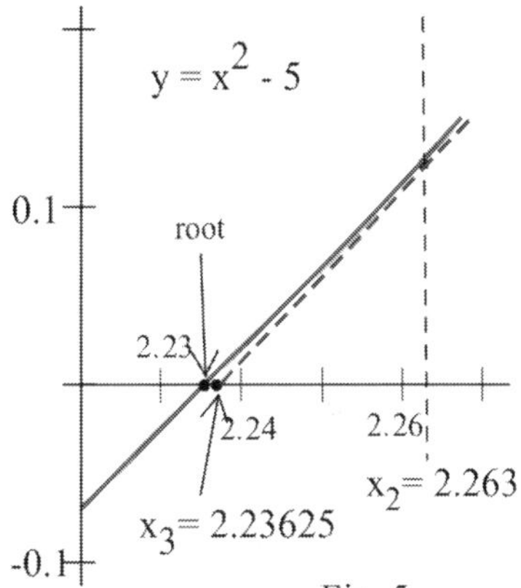

Fig. 5

When we continue repeating this process, (Fig. 5) using each new estimate for the root of $f(x) = x^2 - 5$ as the beginning point for calculating the next estimate, we get:

Beginning estimate:	$x_0 = 4$	(0 correct digits)
after 1 iteration:	$x_1 = \underline{2}.625$	(1 correct digit)
after 2 iterations:	$x_2 = \underline{2.2}62880952$	(2 correct digits)
after 3 iterations:	$x_3 = \underline{2.236}251252$	(4 correct digits)
after 4 iterations:	$x_4 = \underline{2.2360679}85$	(8 correct digits)

It only took 4 iterations to get an approximation of $\sqrt{5}$ which is within 0.000000008 of the exact value. One more iteration gives an approximation x_5 which has 16 correct digits. If we start with $x_0 = -2$ (or any negative number), then the values of x_n approach $-\sqrt{5} \approx -2.23606$.

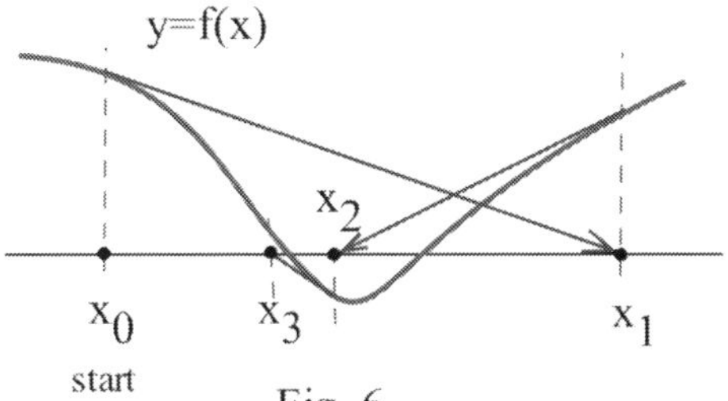

Fig. 6

Fig. 6 shows the process for Newton's Method, starting with x_0 and graphically finding the locations on the x–axis of x_1, x_2, and x_3 .

Practice 1: Find where the tangent line to $f(x) = x^3 + 3x - 1$ at (1, 3) intersects the x–axis.

Practice 2: A starting point and a graph of f are given in Fig. 7. Label the approximate locations of the next two points on the x–axis which are found by Newton's method.

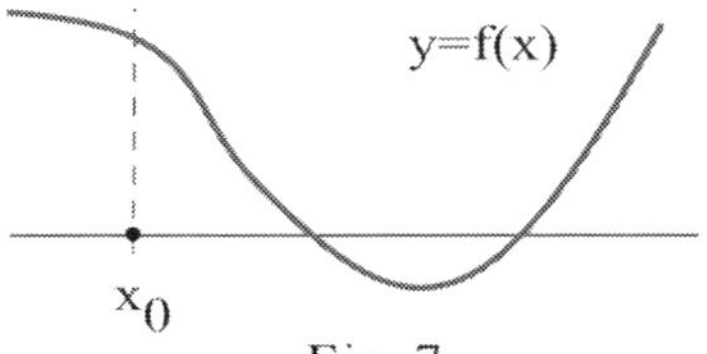

Fig. 7

The Algorithm for Newton's Method

Rather than deal with each particular function and starting point, let's find a pattern for a general function f. For the starting point x_0, the slope of the tangent line at the point $(x_0 , f(x_0))$ is $f'(x_0)$ so the equation of the tangent line is $y - f(x_0) = f'(x_0)(x - x_0)$. This line intersects the x–axis when $y = 0$, so $0 - f(x_0) = f'(x_0)(x - x_0)$ and $x_1 = x = x_0 - \frac{f(x_0)}{f'(x_0)}$. Starting with x_1 and repeating this process we have $x_2 = x_1 - \frac{f(x_1)}{f'(x_1)}$; starting with x_2, we get $x_3 = x_2 - \frac{f(x_2)}{f'(x_2)}$; and so on.

In general, if we start with x_n, the line tangent to the graph of f at the point $(x_n , f(x_n))$ intersects the x–axis at the point $x_{n+1} = x_n - \frac{f(x_n)}{f'(x_n)}$, our new estimate for the root of f.

Algorithm for Newton's Method:

(1) Pick a starting value x_0 (preferably close to a root of f).

(2) For each estimate x_n, calculate a new estimate $x_{n+1} = x_n - \frac{f(x_n)}{f'(x_n)}$.

(3) Repeat step (2) until the estimates are "close enough" to a root or until the method "fails".

When the algorithm for Newton's method is used with $f(x) = x^2 - 5$, the function at the beginning of this section, we have $f'(x) = 2x$ so

$$x_{n+1} = x_n - \frac{f(x_n)}{f'(x_n)} = x_n - \frac{x_n^2 - 5}{2x_n} = \frac{2x_n^2 - (x_n^2 - 5)}{2x_n}$$

$$= \frac{x_n^2 + 5}{2x_n} = \frac{1}{2}\left\{ x_n + \frac{5}{x_n} \right\}.$$

The new approximation, x_{n+1}, is the average of the previous approximation, x_n, and 5 divided by the previous approximation, $5/x_n$. Problem 16 asks you to show that this pattern, called Heron's method, approximates the square root of any positive number. Just replace the "5" with the number whose square root you want.

Example 1: Use Newton's method to approximate the root(s) of $f(x) = 2x + x\sin(x+3) - 5$.

Solution: $f'(x) = 2 + x\cos(x+3) + \sin(x+3)$ so

$$x_{n+1} = x_n - \frac{f(x_n)}{f'(x_n)} = x_n - \frac{2x_n + x_n \sin(x_n+3) - 5}{2 + x_n \cos(x_n+3) + \sin(x_n+3)}$$

The graph of f(x) for $-4 \le x \le 6$ (Fig. 8) indicates only one root of f, and that root is near $x = 3$ so pick $x_0 = 3$. Then Newton's method yields the values $x_0 = 3$, $x_1 = \underline{2.964}84457$, $x_2 = \underline{2.964462}77$, $x_3 = \underline{2.96446273}$ (the underlined digits agree with the exact root).

If we had picked $x_0 = 4$, Newton's method would have required 4 iterations to get 9 digits of accuracy. If $x_0 = 5$, then 7 iterations are needed to get 9 digits of accuracy. If we pick $x_0 = 5.1$, then the values of x_n are not close to the actual root after even 100 iterations, $x_{100} \approx -49.183$. Picking a good value for x_0 can result in values of x_n which get close to the root quickly. Picking a poor value for x_0 can result in x_n values which take longer to get close to the root or which don't approach the root at all.

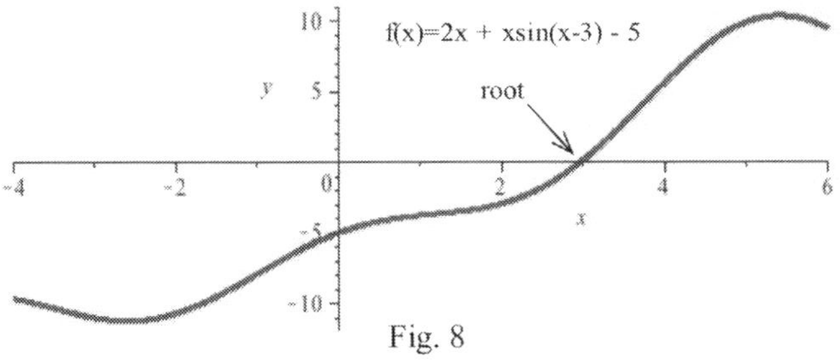

Fig. 8

Note: An examination of the graph of the function can help you pick a "good" x_0.

Practice 3: Put $x_0 = 3$ and use Newton's method to find the first two iterates, x_1 and x_2, for the function $f(x) = x^3 - 3x^2 + x - 1$.

Example 2: The function in Fig. 9 has roots at $x = 3$ and $x = 7$. If we pick $x_0 = 1$ and apply Newton's method, which root do the iterates, the x_n, approach?

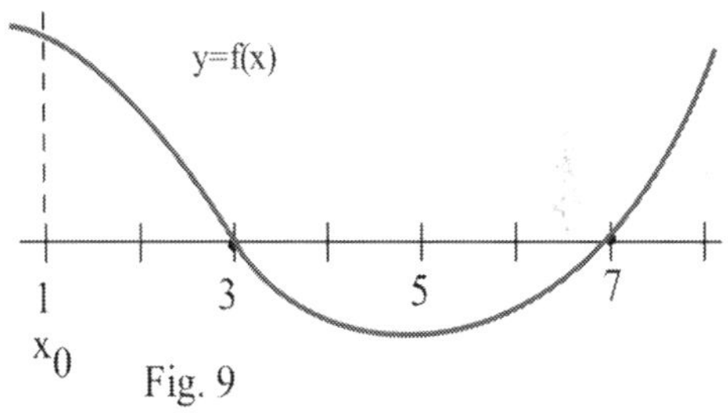

Fig. 9

Solution: The iterates of $x_0 = 1$ are labeled in Fig. 10. They are approaching the root at 7.

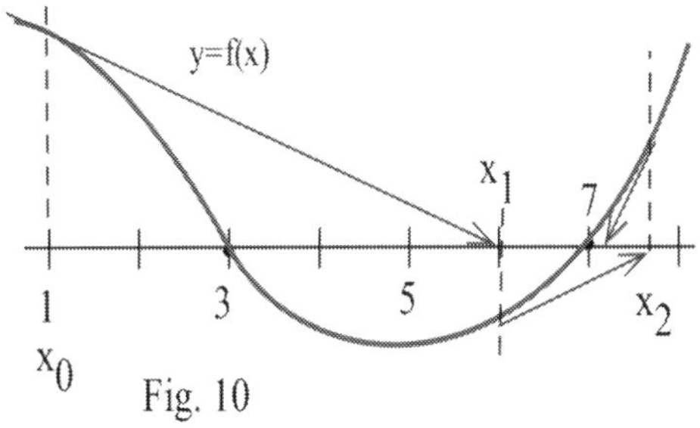

Fig. 10

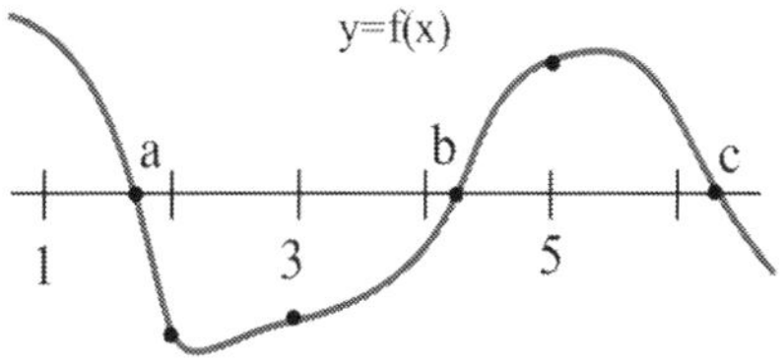

Fig. 11

Practice 4: For the function in Fig. 11, which root do the iterates of Newton's method approach if (a) $x_0 = 2$? (b) $x_0 = 3$? (c) $x_0 = 5$?

Iteration

We have been emphasizing the geometric nature of Newton's method, but Newton's method is also an example of **iterating a function**. If $N(x) = x - \frac{f(x)}{f'(x)}$, the "pattern" in the algorithm, then

$$x_1 = x_0 - \frac{f(x_0)}{f'(x_0)} = N(x_0),$$

$$x_2 = x_1 - \frac{f(x_1)}{f'(x_1)} = N(x_1) = N(N(x_0)) = N \circ N(x_0),$$

$$x_3 = N(x_2) = N \circ N \circ N(x_0), \text{ and , in general,}$$

$$x_n = N(x_{n-1}) = n^{th} \text{ iteration of } N \text{ starting with } x_0.$$

At each step, we are using the output from the function N as the next input into N .

What Can Go Wrong?

When Newton's method works, it usually works very well and the values of the x_n approach a root of f very quickly, often doubling the number of correct digits with each iteration. There are, however, several things which can go wrong.

One obvious problem with Newton's method is that $f'(x_n)$ can be 0. Then we are trying to divide by 0 and x_{n+1} is undefined. Geometrically, if $f'(x_n) = 0$, then the tangent line to the graph of f at x_n is horizontal and does not intersect the x–axis at one point (Fig. 12). If $f'(x_n) = 0$, just pick another starting value x_0 and begin again. In practice, a second or third choice of x_0 usually succeeds.

There are two other less obvious difficulties that are not as easy to overcome — the values of the iterates x_n may become locked into an infinitely repeating loop (Fig. 13), or they may actually move farther away from a root (Fig. 14).

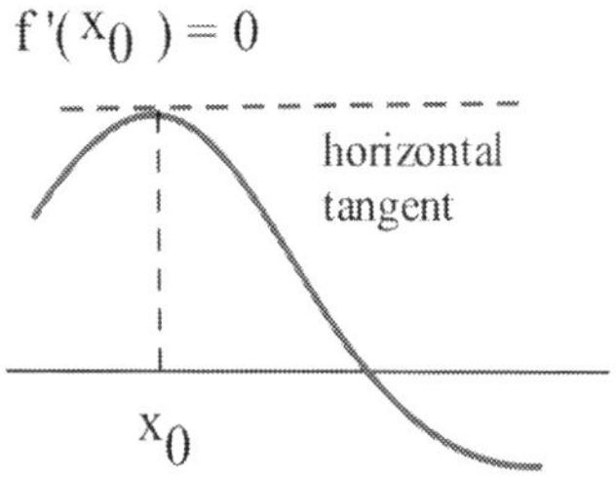

Fig. 12

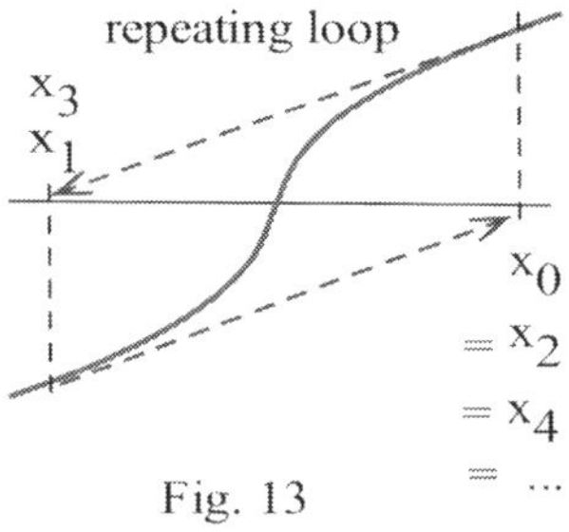

Fig. 13

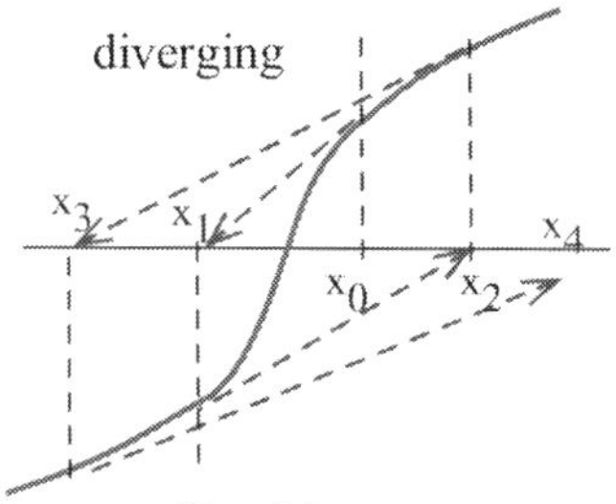

Fig. 14

Example 3: Put $x_0 = 1$ and use Newton's method to find the first two iterates, x_1 and x_2 , for the function $f(x) = x^3 - 3x^2 + x - 1$.

Solution: This is the same function as in the previous Practice problem, but we are using a different starting value for x_0 . $f'(x) = 3x^2 - 6x + 1$ so

$$x_1 = x_0 - \frac{f(x_0)}{f'(x_0)} = 1 - \frac{f(1)}{f'(1)} = 1 - \frac{-2}{-2} = 0 \quad \text{and} \quad x_2 = x_1 - \frac{f(x_1)}{f'(x_1)} = 0 - \frac{f(0)}{f'(0)} = 0 - \frac{-1}{1} = 1$$

which is the same as x_0 , so $x_3 = x_1 = 0$ and $x_4 = x_2 = 1$. The values of x_n alternate between 1 and 0 and do not approach a root.

Newton's method behaves badly at only a few starting points for this particular function. For most starting points Newton's method converges to the root of this function.

There are some functions which defeat Newton's method for **almost every** starting point.

Practice 5: For $f(x) = \sqrt[3]{x} = x^{1/3}$ and $x_0 = 1$, verify that $x_1 = -2, x_2 = 4, x_3 = -8$. Also try $x_0 = -3$, and verify that the same pattern holds: $x_{n+1} = -2x_n$. Graph f and explain why the Newton's method iterates get farther and farther away from the root at 0.

Newton's method is powerful and quick and very easy to program on a calculator or computer. It usually works so well that many people routinely use it as the first method they apply. If Newton's method fails for their particular function, they simply try some other method.

Chaotic Behavior and Newton's Method

An algorithm leads to **chaotic behavior** if two starting points which are close together generate iterates which are sometimes far apart and sometimes close together: $| a_0 - b_0 |$ is small but $| a_n - b_n |$ is large for lots (infinitely many) of values of n and $| a_n - b_n |$ is small for lots of values of n.

The iterates of the next simple algorithm exhibit chaotic behavior.

A Simple Chaotic Algorithm: Starting with any number between 0 and 1, double the number and keep the fractional part of the result: x_1 is the fractional part of $2x_0$, x_2 is the fractional part of $2x_1$, and in general, $x_{n+1} = 2x_n - [\, 2x_n \,] = 2x_n - \text{INT}(\, 2x_n \,)$.

If $x_0 = 0.33$, then the iterates of the algorithm are 0.66, 0.32 (= fractional part of 2·0.66), 0.64, 0.28, 0.56, . . .

The iterates for two other starting values close to .33 are given below as well as the iterates of 0.470 and 0.471 :

start = x_0	**0.32**	**0.33**	**0.34**	**0.470**	**0.471**
x_1	0.64	0.66	0.68	0.940	0.942
x_2	0.28	0.32	0.36	0.880	0.884
x_3	0.56	0.64	0.72	0.760	0.768
x_4	0.12	0.28	0.44	0.520	0.536
x_5	0.24	0.56	0.88	0.040	0.072
x_6	0.48	0.12	0.76	0.080	0.144
x_7	0.96	0.24	0.56	0.160	0.288
x_8	0.92	0.48	0.12	0.320	0.576
x_9	0.84	0.96	0.24	0.640	0.152

There are starting values as close together as we want whose iterates are far apart infinitely often.

Many physical, biological, and business phenomena exhibit chaotic behavior. Atoms can start out within inches of each other and several weeks later be hundreds of miles apart. The idea that small initial differences can lead to dramatically diverse outcomes is sometimes called the "Butterfly Effect" from the title of a talk ("Predictability: Does the Flap of a Butterfly's Wings in Brazil Set Off a Tornado in Texas?") given by Edward Lorenz, one of the first people to investigate chaos. The "butterfly effect" has important implications about the possibility, or rather the impossibility, of accurate long–range weather forecasting. Chaotic behavior is also an important aspect of studying turbulent air and water flows, the incidence and spread of diseases, and even the fluctuating behavior of the stock market.

Newton's method often exhibits chaotic behavior, and, since it is a relatively easy to study, is often used as a model to study the properties of chaotic behavior. If we use Newton's method to approximate the roots of $f(x) = x^3 - x$ (with roots $0, +1$ and -1), then starting points which are very close together can have iterates which converge to different roots. The iterates of .4472 and .4473 converge to the roots 0 and +1, respectively. The iterates of the middle point .44725 converge to the root -1, and the iterates of another nearby point, $\sqrt{1/5} \approx .44721$, simply cycle between $-\sqrt{1/5}$ and $+\sqrt{1/5}$ and do not converge at all.

Practice 6: Find the first 4 Newton's method iterates of $x_0 = .997$ and $x_0 = 1.02$ for $f(x) = x^2 + 1$. Try two other starting values very close to 1 (but not equal to 1) and find their first 4 iterates. Use the graph of $f(x) = x^2 + 1$ to explain how starting points so close together can quickly have iterates so far apart.

PROBLEMS

1. The graph of $y = f(x)$ is given in Fig. 15. Estimate the locations of x_1 and x_2 when Newton's method is applied to f with the given starting value x_0.

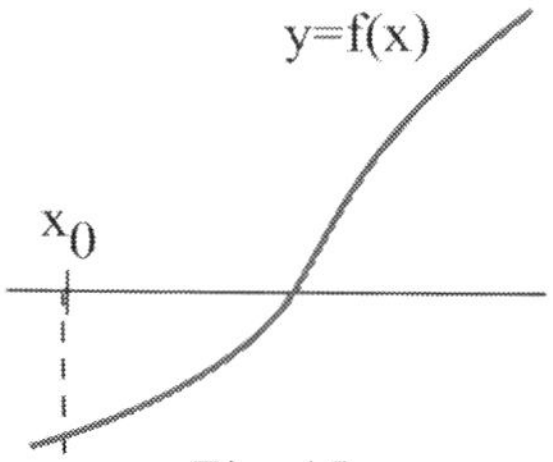

Fig. 15

2. The graph of $y = g(x)$ is given in Fig. 16. Estimate the locations of x_1 and x_2 when Newton's starting value x_0.

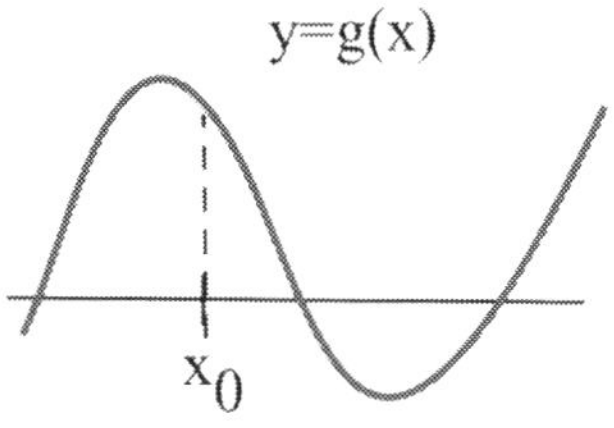

Fig. 16

3. The function in Fig. 17 has several roots. Which root do the iterates of Newton's method converge to if we start with $x_0 = 1$? $x_0 = 5$?

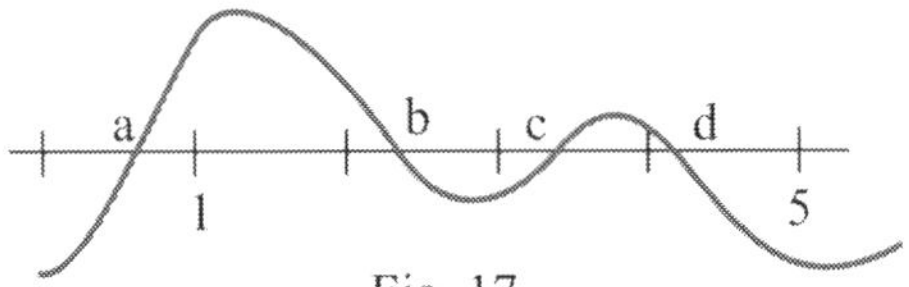

Fig. 17

4. The function in Fig. 18 has several roots. Which root do the iterates of Newton's method converge to if we start with $x_0 = 2$? $x_0 = 6$?

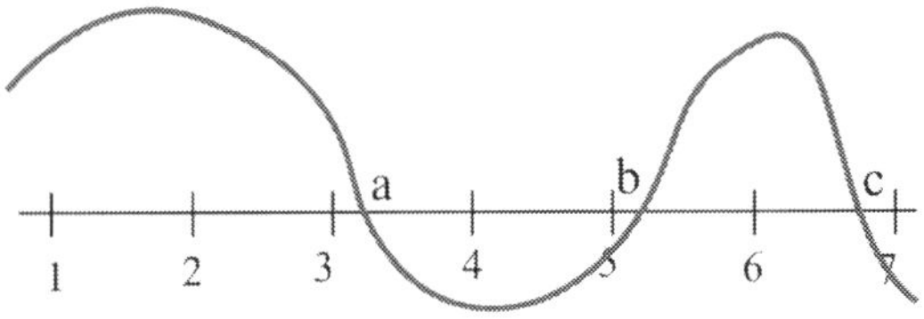

Fig. 18

5. What happens to the iterates if we apply Newton's method to the function in Fig. 19 and start with $x_0 = 1$? $x_0 = 5$?

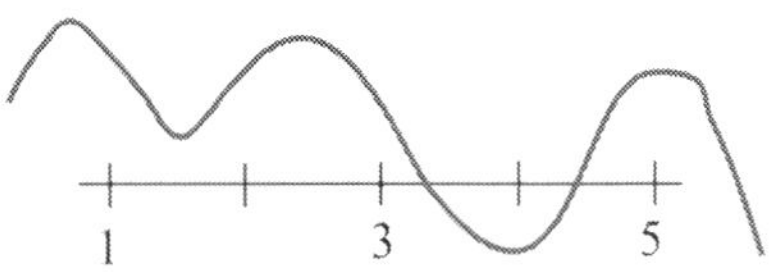

Fig. 19

6. What happens if we apply Newton's method to a function f and start with x_0 = a root of f?

7. What happens if we apply Newton's method to a function f and start with x_0 = a maximum of f?

In problems 8 and 9, a function and a value for x_0 are given. Apply Newton's method to find x_1 and x_2.

8. $f(x) = x^3 + x - 1$ and $x_0 = 1$

9. $f(x) = x^4 - x^3 - 5$ and $x_0 = 2$

In problems 10 and 11, use Newton's method to find a root or solution, accurate to 2 decimal places, of the given functions using the given starting points.

10. $f(x) = x^3 - 7$ and $x_0 = 2$

11. $f(x) = x - \cos(x)$ and $x_0 = 0.7$

In problems 12 – 15, use Newton's method to find **all** roots or solutions, accurate to 2 decimal places. It is helpful to examine a graph of the function to determine a "good" starting value x_0.

12. $2 + x = e^x$

13. $\frac{x}{x+3} = x^2 - 2$

14. $x = \sin(x)$

15. $x = \sqrt[5]{3}$

16. Show that if Newton's method is applied to $f(x) = x^2 - A$ to approximate the square root of A, then $x_{n+1} = \frac{1}{2}(x_n + \frac{A}{x_n})$ so the new estimate of the square root is the average of the previous estimate and A divided by the previous estimate. This method of approximating square roots is called Heron's method.

17. Use Newton's method to devise an algorithm for approximating the cube root of a number A.

18. Use Newton's method to devise an algorithm for approximating the nth root of a number A.

Problems 19 – 22 involve chaotic behavior.

19. The iterates of numbers using the Simple Chaotic Algorithm have a number of properties.
    (a) Verify that the iterates of $x = 0$ are all equal to 0.
    (b) Verify that if $x = 1/2, 1/4, 1/8$, and, in general, $1/2^n$, then the nth iterate of x is 0 (and so are all of the iterates beyond the nth iterate.)

20. When Newton's method is applied to $f(x) = x^2 + 1$, most starting values for x_0 lead to chaotic x_n. Find a value for x_0 so the iterates alternate: $x_1 = -x_0$ and $x_2 = -x_1 = x_0$.

21. $f(x) = \begin{cases} 2x & \text{if } 0 \le x < 1/2 \\ 2 - 2x & \text{if } 1/2 \le x \le 1 \end{cases}$ is called a "stretch and fold" function.
    (a) Describe what f does to the points in the interval [0, 1].
    (b) Examine and describe the behavior of the iterates of 2/3, 2/5, 2/7, and 2/9.
    (c) Examine and describe the behavior of the iterates of .10, .105, and .11 .
    (d) Do the iterates of f lead to chaotic behavior?

22. (This problem requires a computer or programmable calculator.)

For each of the following functions start with x = .5 .

(a) If $f(x) = 2x(1-x)$, then what happens to the iterates of f after many iterations? ("many" = 50 is fine.)

(b) If $f(x) = 3.3x(1-x)$, then what happens to the iterates of f after many iterations?

(c) If $f(x) = 3.83x(1-x)$, then what happens to the iterates of f after many iterations?

(d) What do you think happens to the iterates of $f(x) = 3.7x(1-x)$? What actually does happen.

(e) Repeat parts (a)–(d) with some other starting values for x strictly between 0 and 1 ($0<x<1$). Does the starting value seem to effect the eventual behavior of the iterates?

(The behavior of the iterates of f depends in a strange way on the numerical value of the leading coefficient. The behavior exhibited in part (d) is an example of "chaos".)

Section 2.7 **PRACTICE Answers**

Practice 1: $f(x) = x^3 + 3x + 1$ so $f'(x) = 3x^2 + 3$ and the slope of the tangent line at the point (1,3) is $f'(1) = 6$. Using the point–slope form for the equation of a line, the equation of the tangent line is $y - 3 = 6(x - 1)$ or $y = 6x - 3$.

The y–coordinate of a point on the x–axis is 0 so we need to put $y = 0$ and solve the linear equation for x: $0 = 6x - 3$ so $x = 1/2$.

The line tangent to the graph of $f(x) = x^3 + 3x + 1$ at the point (1,3) intersects the x–axis at the point **(1/2, 0)**.

Practice 2: The approximate locations of x_1 and x_2 are shown in Fig. 20.

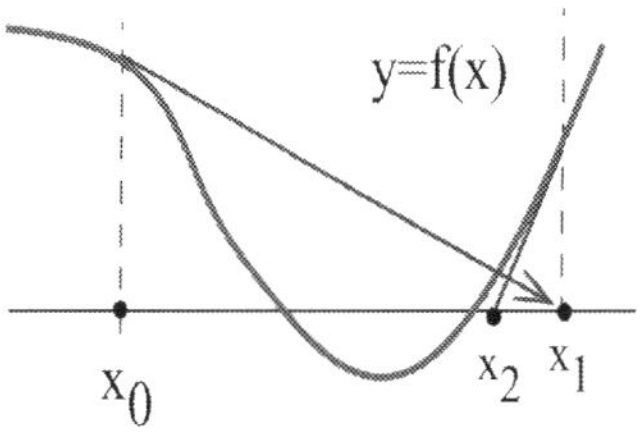

Fig. 20

Practice 3: $f(x) = x^3 - 3x^2 + x - 1$ so $f'(x) = 3x^2 - 6x + 1$. $x_0 = 3$.

$$x_1 = x_0 - \frac{f(x_0)}{f'(x_0)} = 3 - \frac{f(3)}{f'(3)} = 3 - \frac{2}{10} = \mathbf{2.8}.$$

$$x_2 = x_1 - \frac{f(x_1)}{f'(x_1)} = 2.8 - \frac{f(2.8)}{f'(2.8)} = 2.8 - \frac{0.232}{7.72} \approx \mathbf{2.769948187}$$

$$x_3 = x_2 - \frac{f(x_2)}{f'(x_2)} \approx 2.769292663 .$$

Practice 4: Fig. 21 shows the first iteration of Newton's Method for $x_0 = 2, 3,$ and 5.

If $x_0 = 2$, the iterates approach the root at a.

If $x_0 = 3$, the iterates approach the root at c.

If $x_0 = 5$, the iterates approach the root at a.

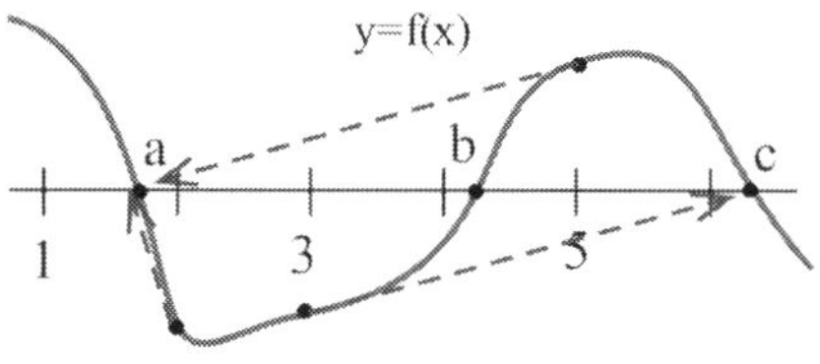

Fig. 21

Practice 5: $f(x) = x^{1/3}$ so $f'(x) = \frac{1}{3} x^{-2/3}$.

If $x_0 = \mathbf{1}$, then $x_1 = 1 - \frac{f(1)}{f'(1)} = 1 - \frac{1}{1/3} = 1 - 3 = \mathbf{-2}$,

$$x_2 = -2 - \frac{f(-2)}{f'(-2)} = -2 - \frac{(-2)^{1/3}}{\frac{1}{3}(-2)^{-2/3}} = -2 - \frac{-2}{1/3} = \mathbf{4} ,$$

$$x_3 = 4 - \frac{f(4)}{f'(4)} = 4 - \frac{(4)^{1/3}}{\frac{1}{3}(4)^{-2/3}} = 4 - \frac{4}{1/3} = \mathbf{-8}$$, and so on.

If $x_0 = \mathbf{-3}$, then $x_1 = -3 - \frac{f(-3)}{f'(-3)} = -3 - \frac{(-3)^{1/3}}{\frac{1}{3}(-3)^{-2/3}} = -3 + 9 = \mathbf{6}$,

$$x_2 = 6 - \frac{f(6)}{f'(6)} = 6 - \frac{6^{1/3}}{\frac{1}{3}6^{-2/3}} = 6 - \frac{6}{1/3} = \mathbf{-12} .$$

The graph of the cube root $f(x) = x^{1/3}$ has a shape similar to Fig. 14, and the behavior of the iterates is similar to the pattern in that figure. Unless $x_0 = 0$ (the only root of f) the iterates alternate in sign and double in magnitude with each iteration: they get progressively farther from the root with each iteration.

Practice 6: If $x_0 = 0.997$, then $x_1 \approx -0.003$, $x_2 \approx 166.4$, $x_3 \approx 83.2$, $x_4 \approx 41.6$.

If $x_0 = 1.02$, then $x_1 \approx 0.0198$, $x_2 \approx -25.2376$, $x_3 \approx -12.6$, $x_4 \approx -6.26$.

2.8 Linear Approximation and Differentials

Newton's method used tangent lines to "point toward" a root of the function. In this section we examine and use another geometric characteristic of tangent lines:

If	f is differentiable at a and x is close to a,
then	the tangent line L(x) is close to f(x). (Fig. 1)

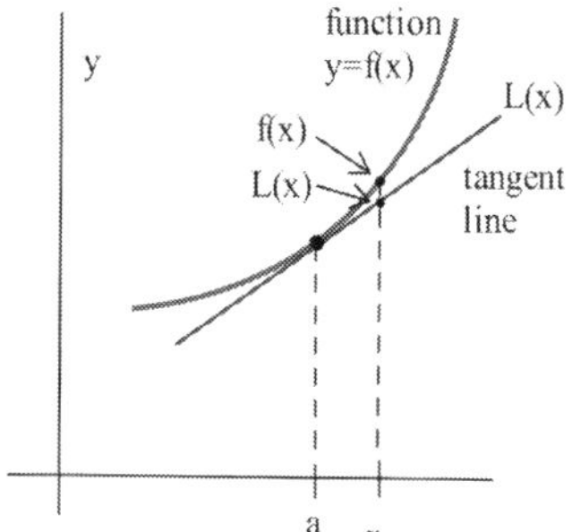

Fig. 1

This idea is used to approximate the values of some commonly used functions and to predict the "error" or uncertainty in a final calculation if we know the "error" or uncertainty in our original data. Finally, we define and give some examples of a related concept called the **differential** of a function.

Linear Approximation

Since this section uses tangent lines frequently, it is worthwhile to recall how we find the equation of the line tangent to f at a point x = a. The line tangent to f at x = a goes through the point (a, f(a)) and has slope f '(a), so, using the point–slope form $y - y_0 = m(x - x_0)$ for linear equations, we have

$$y - f(a) = f'(a)\cdot(x - a) \quad\text{and}\quad y = f(a) + f'(a)\cdot(x - a) .$$

This final result is the equation of the line tangent to f at x = a.

If	f is differentiable at x = a,
then	the equation of the line L tangent to f at x = a is
	$\mathbf{L(x) = f(a) + f'(a)\cdot(x - a)}$.

Example 1: Find the equation of the line L(x) which is tangent to the graph of $f(x) = \sqrt{x}$ at the point (9,3). Evaluate L(9.1) and L(8.88) to approximate $\sqrt{9.1}$ and $\sqrt{8.88}$.

Solution: $f(x) = \sqrt{x} = x^{1/2}$ and $f'(x) = \frac{1}{2}x^{-1/2} = \frac{1}{2\sqrt{x}}$ so $f(9) = 3$

and $f'(9) = \frac{1}{2\sqrt{9}} = \frac{1}{6}$. Then

$L(x) = f(9) + f'(9)\cdot(x - 9) = 3 + \frac{1}{6}(x - 9)$. If x is close to 9, then the value of L(x) is a good approximation of the value of $\sqrt{x}$ (Fig. 2) .

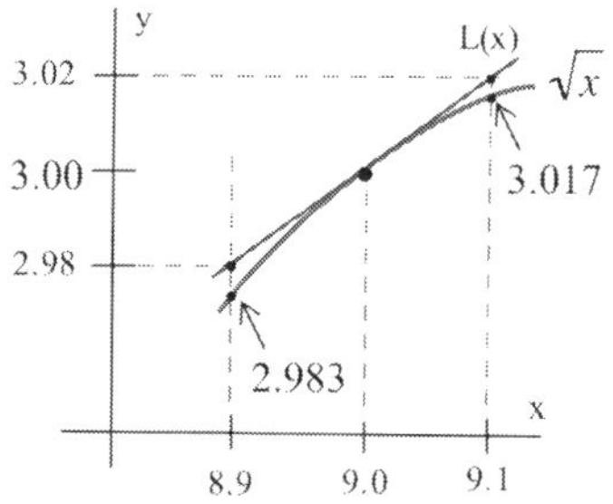

Fig. 2

The number 9.1 is close to 9 so

$\sqrt{9.1} = f(9.1) \approx L(9.1) = 3 + \frac{1}{6}(9.1 - 9) = 3.016666.$

Similarly,

$\sqrt{8.88} = f(8.88) \approx L(8.88) = 3 + \frac{1}{6}(8.88 - 9) = 2.98$. In fact, $\sqrt{9.1} \approx 3.016621$, so our estimate, using L(9.1), is within 0.000045 of the exact answer. $\sqrt{8.88} \approx 2.979933$ (accurate to 6 decimal places) and our estimate is within 0.00007 of the exact answer.

In each example, we got a good estimate of a square root with very little work. The graph indicates the tangent line L(x) is slightly above f(x), and each estimate is slightly larger than the exact value.

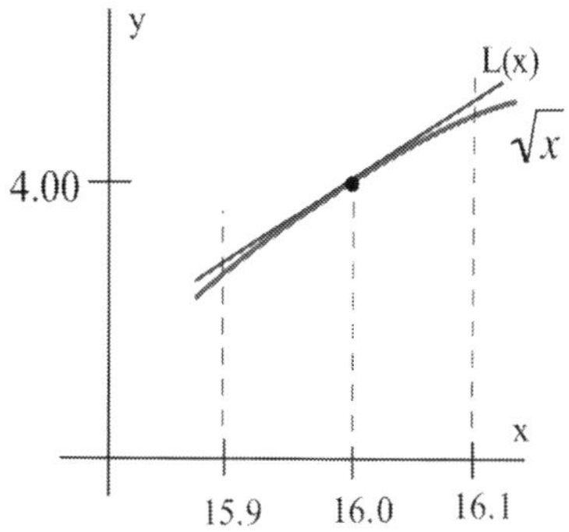

Fig. 3

Practice 1: Find the equation of the line L(x) tangent to the graph of $f(x) = \sqrt{x}$ at the point (16,4) (Fig. 3). Evaluate L(16.1) and L(15.92) to approximate $\sqrt{16.1}$ and $\sqrt{15.92}$. Are your approximations using L larger or smaller than the exact values of the square roots?

Practice 2: Find the equation of the line L(x) tangent to the graph of $f(x) = x^3$ at the point (1,1) and use L(x) to approximate $(1.02)^3$ and $(0.97)^3$. Do you think your approximations using L are larger or smaller than the exact values?

The process we have used to approximate square roots and cubics can be used to approximate any differentiable function, and the main result about the linear approximation follows from the two statements in the boxes. Putting these two statements together, we have the process for Linear Approximation.

Linear Approximation Process: (Fig. 4)

If f is differentiable at a and **x is close to a,**

then (geometrically) the graph of the tangent line L(x) is close to the graph of f(x), and

(algebraically) the values of the tangent line function $L(x) = f(a) + f'(a)\cdot(x - a)$ approximate the values of f(x):

$$\mathbf{f(x) \approx L(x) = f(a) + f'(a)\cdot(x - a)} .$$

Sometimes we replace "x – a" with "Δx" in the last equation, and the statement becomes $f(x) \approx f(a) + f'(a)\cdot\Delta x$.

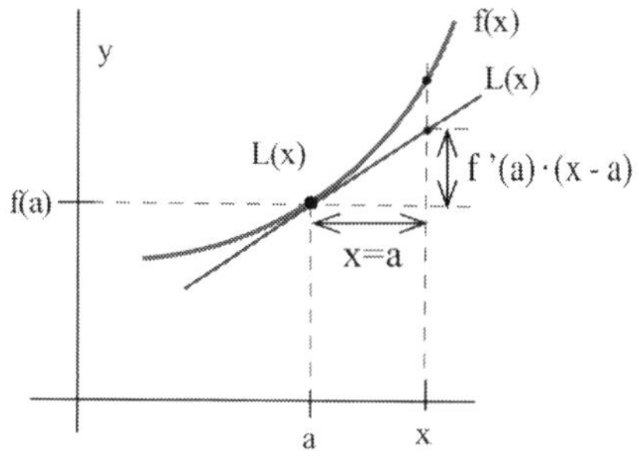

Fig. 4

Example 2: Use the linear approximation process to approximate the value of $e^{0.1}$.

Solution: $f(x) = e^x$ so $f'(x) = e^x$. We need to pick a value **a** near $x = 0.1$ for which we know the exact value of $f(\mathbf{a})$ and $f'(\mathbf{a})$, and $\mathbf{a} = 0$ is the obvious choice. Then

$$e^{0.1} = f(0.1) \approx L(0.1) = f(0) + f'(0)\cdot(0.1 - 0)$$

$$= e^0 + e^0\cdot(0.1) = 1 + 1\cdot(0.1) = 1.1 .$$

This approximation is within 0.0052 of the exact value of $e^{0.1}$.

Practice 3: Approximate the value of $(1.06)^4$, the amount \$1 becomes after 4 years in a bank which pays 6% interest compounded annually. (Take $f(x) = x^4$ and $a = 1$.)

Practice 4: Use the linear approximation process and the values in the table to estimate the value of f when $x = 1.1, 1.23$ and 1.38.

x	f(x)	f '(x)
1	0.7854	0.5
1.2	0.8761	0.4098
1.4	0.9505	0.3378

We can also approximate **functions** as well as numbers.

Example 3: Find a linear approximation formula $L(x)$ for $\sqrt{1+x}$ when x is small. Use your result to approximate $\sqrt{1.1}$ and $\sqrt{.96}$.

Solution: $f(x) = \sqrt{1+x} = (1+x)^{1/2}$ so $f'(x) = \frac{1}{2}(1+x)^{-1/2} = \frac{1}{2\sqrt{1+x}}$. Since "x is small", we know that x is close to 0, and we can pick $a = 0$. Then $f(a) = f(0) = 1$ and $f'(a) = f'(0) = \frac{1}{2}$ so

$$\sqrt{1+x} \approx L(x) = f(0) + f'(0)\cdot(x - 0) = 1 + \frac{1}{2}x = 1 + \frac{x}{2} .$$

If x is small, then $\mathbf{\sqrt{1+x} \approx 1 + \frac{x}{2}}$. $\sqrt{1.1} = \sqrt{1 + 0.1} \approx 1 + \frac{0.1}{2} = 1.05$ and $\sqrt{0.96} = \sqrt{1 + (-.04)} \approx 1 + \frac{-.04}{2} = 0.98$. Use your calculator to determine by how much each estimate differs from the true value.

Applications of Linear Approximation to Measurement "Error"

Most scientific experiments involve using instruments to take measurements, but the instruments are not perfect, and the measurements we get from them are only accurate up to a certain level. If we know the level of accuracy of our instruments and measurements, we can use the idea of linear approximation to estimate the level of accuracy of results we calculate from our measurements.

If we measure the side x of a square to be 8 inches, then, of course, we would calculate its area to be $A(x) = x^2 = 64$ square inches. Suppose, as is reasonable in a real measurement, that our measuring instrument could only measure or be read to the nearest 0.05 inches. Then our measurement of 8 inches would really mean some number between $8-0.05 = 7.95$ inches and $8+0.05 = 8.05$ inches, and the true area of the square would be between $A(7.95) = 63.2025$ and $A(8.05) = 64.8025$ square inches. Our possible "error" or "uncertainty", because of the limitations of the instrument, could be as much as $64.8025 - 64 = 0.8025$ square inches so we could report the area of the square to be 64 ± 0.8025 square inches. We can also use the linear approximation method to estimate the "error" or uncertainty of the area. (For a function as simple as the area of a square, this linear approximation method really isn't needed, but it is used to illustrate the idea.)

For a square with side x, the area is $A(x) = x^2$ and $A'(x) = 2x$. If Δx represents the "error" or uncertainty of our measurement of the side, then, using the linear approximation technique for $A(x)$, $A(x) \approx A(a) + A'(a) \cdot \Delta x$ so the uncertainty of our calculated area is $A(x) - A(a) \approx A'(x) \cdot \Delta x$. In this example, $a = 8$ inches and $\Delta x = 0.05$ inches so

$$A(8.05) \approx A(8) + A'(8) \cdot (0.05) = 64 + 2(8) \cdot (0.05) = 64.8 \text{ square inches,}$$

and the uncertainty in our calculated area is approximately

$$A(8 + 0.05) - A(8) \approx A'(8) \cdot \Delta x = 2(8 \text{ inches})(0.05 \text{ inches}) = 0.8 \text{ square inches.}$$

This process can be summarized as:

Linear Approximation Error:

If the value of the x–variable is measured to be $x = a$ with an "error" of Δx units,

then Δf, the "error" in estimating $f(x)$, is $\boldsymbol{\Delta f = f(x) - f(a) \approx f'(a) \cdot \Delta x}$.

Practice 5: If we measure the side of a cube to be 4 cm with an uncertainty of 0.1 cm, what is the volume of the cube and the uncertainty of our calculation of the volume? (Use linear approximation.)

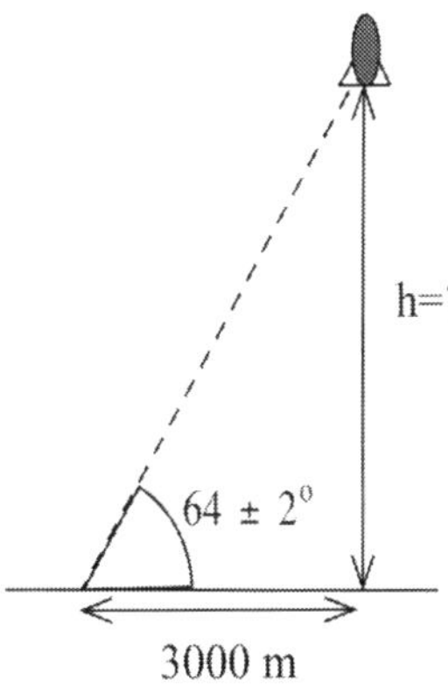

Fig. 5

Example 4: We are using a tracking telescope to follow a small rocket. Suppose we are 3000 meters from the launch point of the rocket, and, 2 seconds after the launch, we measure the angle of the inclination of the rocket to be 64° with a possible "error" of 2° (Fig. 5). How high is the rocket and what is the possible error in this calculated height?

Solution: Our measured angle is $x = 1.1170$ radians and $\Delta x = 0.0349$ radians (all of our trigonometric work is in radians), and the height of the rocket at an angle x is $f(x) = 3000 \cdot \tan(x)$ so $f(1.1170) \approx 6151$ m. Our uncertainty in the height is

$$\Delta f(x) \approx f'(x) \cdot \Delta x \approx 3000 \cdot \sec^2(x) \cdot \Delta x = 3000 \sec^2(1.1170) \cdot (0.0349) = 545 \text{ m.}$$

If our measured angle of 64° can be in error by as much as 2°, then our calculated height of 6151 m can be in error by as much as 545 m. The height is 6151 ± 545 meters.

Practice 6: Suppose we measured the angle of inclination of in the previous Example to be $43^{o} \pm 1^{o}$. Estimate the height of the rocket in the form "height ± error ."

In some scientific and engineering applications, the calculated **result** must be within some given specification. You might need to determine how accurate the initial measurements must be in order to guarantee the final calculation is within the specification. Added precision usually costs time and money, so it is important to choose a measuring instrument which is good enough for the job but not too good or too expensive.

Example 5: Your company produces ball bearings (spheres) with a volume of 10 cm^3, and the volume must be accurate to within 0.1 cm^3. What radius should the bearings have and what error can you tolerate in the radius measurement to meet the accuracy specification for the volume? ($V = \frac{4}{3} \pi r^3$)

Solution: Since we want $V = 10$, we can solve $10 = \frac{4}{3} \pi r^3$ for r to get $r = 1.3365$ cm.

$V(r) = \frac{4}{3} \pi r^3$ and $V'(r) = 4\pi r^2$ so $\Delta V \approx V'(r) \cdot \Delta r$. In this case we have been given that $\Delta V = 0.1 \text{ cm}^3$, and we have calculated $r = 1.3365$ cm so $0.1 \text{ cm}^3 = V'(1.3365 \text{ cm}) \cdot \Delta r = (22.45 \text{ cm}^2) \cdot \Delta r$.

Solving for Δr, we get $\Delta r \approx 0.0045$ cm. To meet the specifications for the allowable error in the volume, we must allow no more than 0.0045 cm variation in the radius. If we measure the diameter of the sphere rather than the radius, then we want $d = 2r = 2(1.3365 \pm 0.0045) = 2.673 \pm 0.009$ cm .

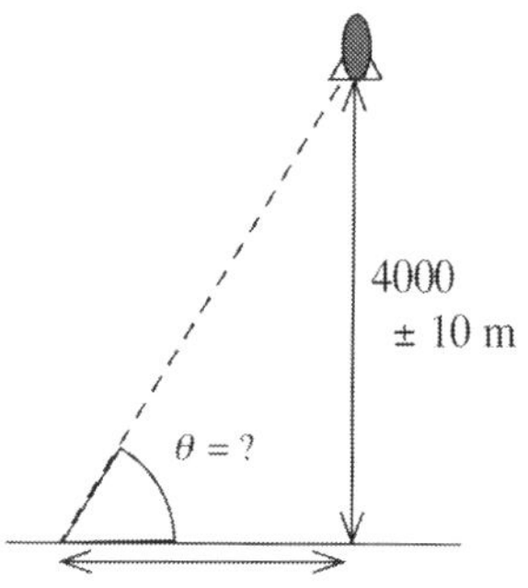

Fig. 6

Practice 7: You want to determine the height of the rocket to within 10 meters when it is 4000 meters high (Fig. 6). How accurate must your angle measurement be? (Do your calculations in radians.)

Relative Error and Percentage Error

The "error" we have been examining is called the **absolute error** to distinguish it from two other commonly used terms, the **relative error** and the **percentage error** which compare the absolute error with the magnitude of the number being measured. An "error" of 6 inches in measuring the circumference of the earth would be extremely small, but a 6 inch error in measuring your head for a hat would result in a very bad fit.

Definitions: The **Relative Error** of f is $\frac{\text{error of f}}{\text{value of f}} = \frac{\Delta f}{f}$

The **Percentage Error** of f is $\frac{\Delta f}{f} \cdot (100)$.

Example 6: If the relative error in the calculation of the area of a circle must be less than 0.4 , then what relative error can we tolerate in the measurement of the radius?

Solution: $A(r) = \pi r^2$ so $A'(r) = 2\pi r$ and $\Delta A \approx A'(r)\,\Delta r = 2\pi r\,\Delta r$. The Relative Error of A is

$\frac{\Delta A}{A} \approx \frac{2\pi r\,\Delta r}{\pi r^2} = 2\frac{\Delta r}{r}$. We can guarantee that the Relative Error of A , $\frac{\Delta A}{A}$, is less than 0.4

if the Relative Error of r , $\frac{\Delta r}{r} = \frac{1}{2}\frac{\Delta A}{A}$, is less than $\frac{1}{2}(0.4) = 0.2$.

Practice 8: If you can measure the side of a cube with a percentage error less than 3%, then what will the percentage error for your calculation of the surface area of the cube be?

The Differential of f

In Fig. 7, the change in value of the function f near the point (x, f(x)) is $\Delta f = f(x + \Delta x) - f(x)$ and the change along the tangent line is $f'(x)\cdot\Delta x$. If Δx is small, then we have used the approximation that $\Delta f \approx f'(x)\cdot\Delta x$. This leads to the definition of a new quantity, **df**, called the differential of f.

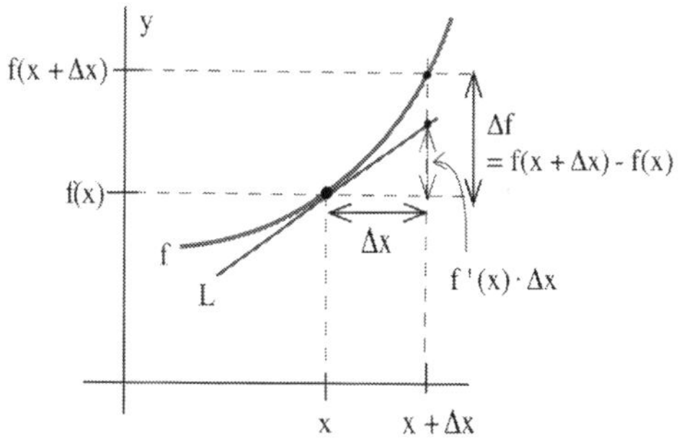

Fig. 7

Definition: The **differential of f** is $\mathbf{df \equiv f'(x)\cdot dx}$ where **dx** is any real number.

The differential of f represents the change in f , as x changes from x to x + dx, along the tangent line to the graph of f at the point (x, f(x)) . If we take dx to be the number Δx, then the differential is an approximation of Δf: $\Delta f \approx f'(x)\cdot\Delta x = f'(x)\cdot dx = df$.

Example 7: Determine the differential df of each of $f(x) = x^3 - 7x$, $g(x) = \sin(x)$, and $h(r) = \pi r^2$.

Solution: $df = f'(x)\cdot dx = (3x^2 - 7)\,dx$, $dg = g'(x)\cdot dx = \cos(x)\,dx$, and $dh = h'(r)\,dr = 2\pi r\,dr$.

Practice 9: Determine the differentials of $f(x) = \ln(x)$, $u = \sqrt{1 - 3x}$, and $r = 3\cos(\theta)$.

We will do little with differentials for a while, but are used extensively in integral calculus.

The Linear Approximation "Error" $|f(x) - L(x)|$

An approximation is most valuable if we also have some measure of the size of the "error", the distance between the approximate value and the value being approximated,. Typically, we will not know the exact value of the error (why not?), but it is useful to know that the error must be less than some number. For example, if one scale gives the weight of a gold pendant as 10.64 grams with an error less than .3 grams (10.64 ± .3 grams) and another scale gives the weight of the same pendant as 10.53 grams with an error less than .02 grams (10.53 ± .02 grams), then we can have more faith in the second approximate weight because of the smaller "error" guarantee. Before finding a guarantee on the size of the error of the linear approximation process, we will check how well the linear approximation process approximates some functions we can compute exactly. Then we will prove one bound on the possible error and state a somewhat stronger bound.

Example 8: Let $f(x) = x^2$. Evaluate $f(2+\Delta x)$, $L(2+\Delta x)$ and $|f(2+\Delta x) - L(2+\Delta x)|$ for $\Delta x = 0.1, 0.05, 0.01, 0.001$ and for a general value of Δx.

Solution: $f(2+\Delta x) = (2+\Delta x)^2 = 2^2 + 4\Delta x + (\Delta x)^2$ and $L(2+\Delta x) = f(2) + f'(2)\cdot\Delta x = 2^2 + 4\cdot\Delta x$. Then

Δx	$f(2+\Delta x)$	$L(2+\Delta x)$	$\lvert f(2+\Delta x) - L(2+\Delta x)\rvert$
0.1	4.41	4.4	0.01
0.05	4.2025	4.2	0.0025
0.01	4.0401	4.04	0.0001
0.001	4.004001	4.004	0.000001

Cutting the value of Δx in half makes the error 1/4 as large. Cutting Δx to 1/10 as large makes the error 1/100 as large. In general, $|f(2+\Delta x) - L(2+\Delta x)| = |(2^2 + 4\cdot\Delta x + (\Delta x)^2) - (2^2 + 4\cdot\Delta x)| = (\Delta x)^2$.

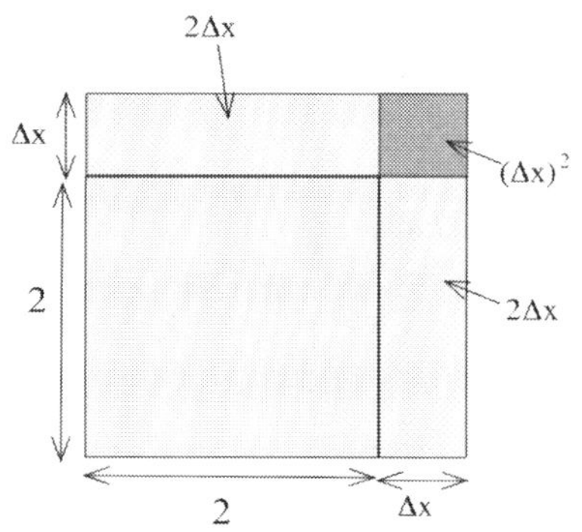

Fig. 8

This function and error also have a nice geometric interpretation (Fig. 8): $f(x) = x^2$ is the area of a square of side x so $f(2+\Delta x)$ is the area of a square of side $2+\Delta x$, and that area is the sum of the pieces with areas 2^2, $2\cdot\Delta x$, $2\cdot\Delta x$, and $(\Delta x)^2$. The linear approximation $L(2+\Delta x) = 2^2 + 4\cdot\Delta x$ to the area of the square includes the 3 largest pieces 2^2, $2\cdot\Delta x$ and $2\cdot\Delta x$, but it omits the small square with area $(\Delta x)^2$ so the approximation is in error by the amount $(\Delta x)^2$.

Practice 10: Let $f(x) = x^3$. Evaluate $f(4+\Delta x)$, $L(4+\Delta x)$ and $|f(4+\Delta x) - L(4+\Delta x)|$ for $\Delta x = 0.1, 0.05, 0.01, 0.001$ and for a general value of Δx. Use Fig. 9 to give a geometric interpretation of $f(4+\Delta x)$, $L(4+\Delta x)$ and $|f(4+\Delta x) - L(4+\Delta x)|$.

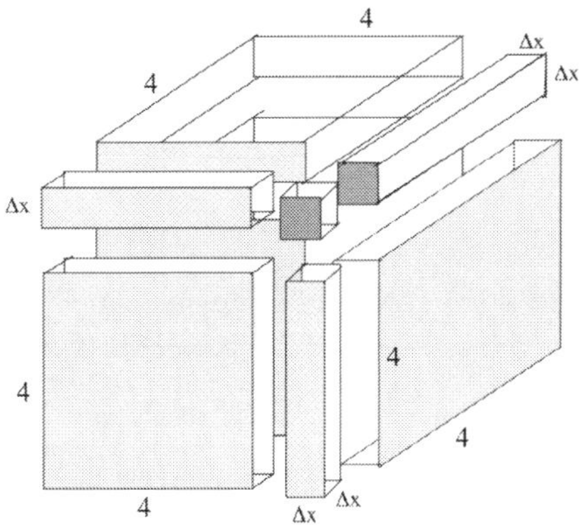

Fig. 9

In both the example and practice problem, the error $|f(a+\Delta x) - L(a+\Delta x)|$ turned out to be very small, proportional to $(\Delta x)^2$, when Δx was small. In general, the error approaches 0 as Δx approaches 0.

Theorem : If $f(x)$ is differentiable at a and $L(a+\Delta x) = f(a) + f'(a)\cdot\Delta x$,

then $\lim_{\Delta x \to 0} |f(a+\Delta x) - L(a+\Delta x)| = 0$ and

$$\lim_{\Delta x \to 0} \frac{|f(a+\Delta x) - L(a+\Delta x)|}{\Delta x} = 0$$

Proof: $|f(a+\Delta x) - L(a+\Delta x)| = |f(a+\Delta x) - f(a) - f'(a)\cdot\Delta x| = \left\{ \frac{f(a+\Delta x) - f(a)}{\Delta x} - f'(a) \right\}\cdot \Delta x$. But f is differentiable at x=a so $\lim_{\Delta x \to 0} \frac{f(a+\Delta x) - f(a)}{\Delta x} = f'(a)$ and $\lim_{\Delta x \to 0} \left\{ \frac{f(a+\Delta x) - f(a)}{\Delta x} - f'(a) \right\} = 0$.

Then $\lim_{\Delta x \to 0} |f(a+\Delta x) - L(a+\Delta x)| = \lim_{\Delta x \to 0} \left\{ \frac{f(a+\Delta x) - f(a)}{\Delta x} - f'(a) \right\} \cdot \lim_{\Delta x \to 0} \Delta x = 0 \cdot 0 = 0$

Not only does the difference $f(a+\Delta x) - L(a+\Delta x)$ approach 0, but this difference approaches 0 so fast that we can divide it by Δx, another quantity approaching 0, and the quotient still approaches 0.

In the next chapter we can prove that the error of the linear approximation process is proportional to $(\Delta x)^2$. For now we just state the result.

Theorem: If f is differentiable at a and $|f''(x)| \le M$ for all x between a and $a+\Delta x$

then $|\text{"error"}| = |f(a+\Delta x) - L(a+\Delta x)| \le \frac{1}{2} M\cdot(\Delta x)^2$.

Problems

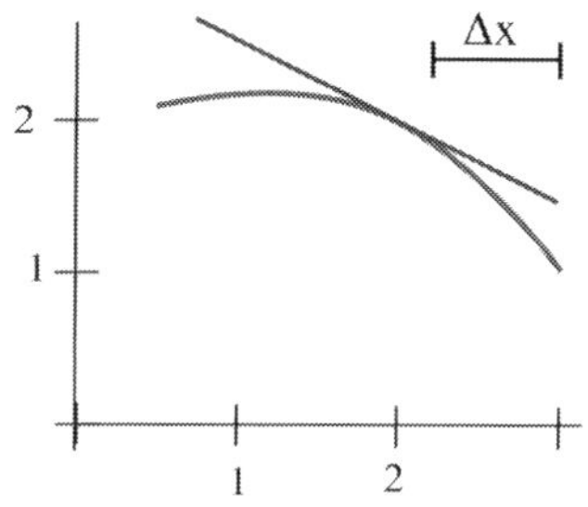

Fig. 10

1. Fig. 10 shows the tangent line to a function g at the point (2,2) and a line segment Δx units long. On the figure, label the locations of
   (a) $2 + \Delta x$ on the x–axis, (b) the point $(2 + \Delta x, g(2 + \Delta x))$, and
   (c) the point $(2 + \Delta x, g(2) + g'(2)\cdot\Delta x)$.
   (d) How large is the "error", $(g(2) + g'(2)\cdot\Delta x) - (g(2+\Delta x))$?

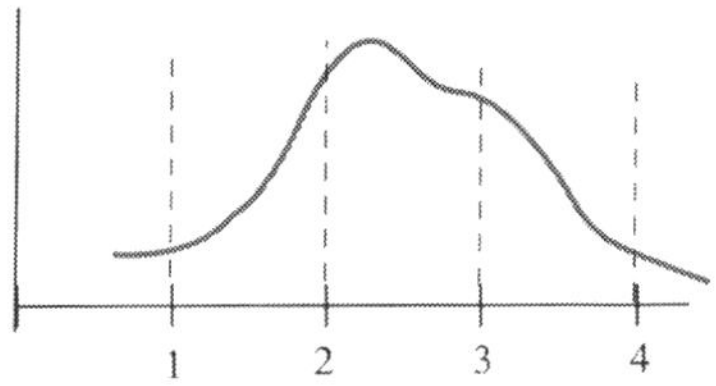

Fig. 11

2. In Fig. 11 is the linear approximation $L(a + \Delta x)$ larger or smaller than the value of $f(a + \Delta x)$ when
   (a) $a = 1$ and $\Delta x = 0.2$? (b) $a = 2$ and $\Delta x = -0.1$? (c) $a = 3$ and $\Delta x = 0.1$?
   (d) $a = 4$ and $\Delta x = 0.2$? and (e) $a = 4$ and $\Delta x = -0.2$?

In problems 3 and 4, find the equation of the tangent line L to the given function f at the given point $(a, f(a))$. Use the value $L(a + \Delta x)$ to approximate the value of $f(a + \Delta x)$.

3. (a) $f(x) = \sqrt{x}$, $a = 4$, $\Delta x = 0.2$ (b) $f(x) = \sqrt{x}$, $a = 81$, $\Delta x = -1$ (c) $f(x) = \sin(x)$, $a = 0$, $\Delta x = 0.3$

4. (a) $f(x) = \ln(x)$, $a = 1$, $\Delta x = 0.3$ (b) $f(x) = e^x$, $a = 0$, $\Delta x = 0.1$ (c) $f(x) = x^5$, $a = 1$, $\Delta x = 0.03$

5. Show that $(1 + x)^n \approx 1 + nx$ if x is "close to" 0. (Suggestion: Put $f(x) = (1+x)^n$, $a = 0$, and $\Delta x = x$.)

In problems 6 and 7, use the Linear Approximation Process to derive each approximation formula for x **"close to" 0**.

6. (a) $(1 - x)^n \approx 1 - nx$ (b) $\sin(x) \approx x$ (c) $e^x \approx 1 + x$

7. (a) $\ln(1 + x) \approx x$ (b) $\cos(x) \approx 1$ (c) $\tan(x) \approx x$ (d) $\sin(\pi/2 + x) \approx 1$

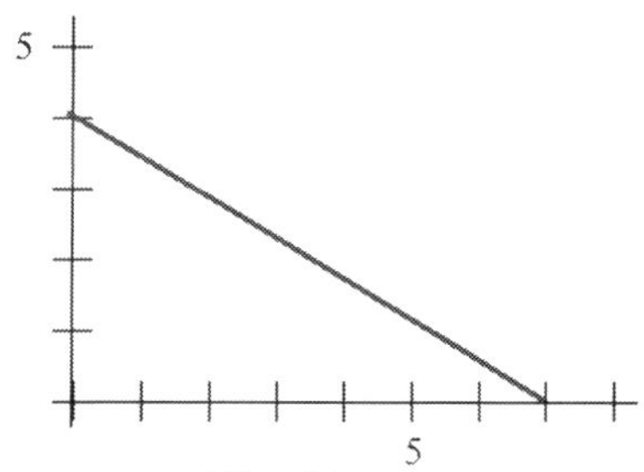

Fig. 12

8. The height of a triangle is exactly 4 inches, and the base is measured to be 7 ± 0.5 inches (Fig. 12). Shade a part of the figure which represents the "error" in the calculation of the area of the triangle.

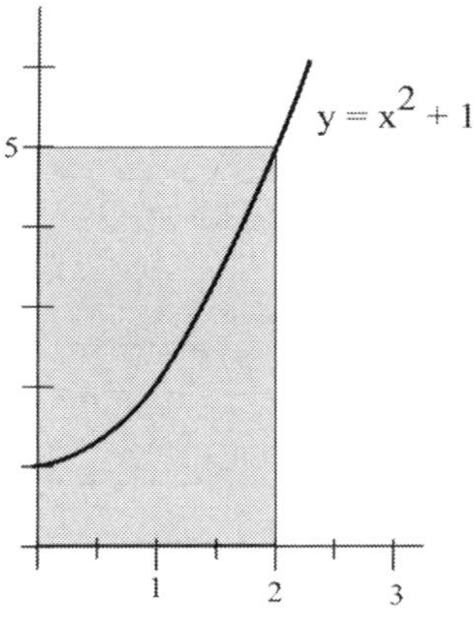

Fig. 13

9. A rectangle has one side on the x–axis, one side on the y–axis, and a corner on the graph of $y = x^2 + 1$ (Fig. 13).
   (a) Use Linear Approximation of the area formula to estimate the increase in the area of the rectangle if the base grows from 2 to 2.3 inches.
   (b) Calculate exactly the increase in the area of the rectangle as the base grows from 2 to 2.3 inches.

10. You can measure the diameter of a circle to within 0.3 cm.
    (a) How large is the "error" in the calculated area of a circle with a measured diameter of 7.4 cm?
    (b) How large is the "error" in the calculated area of a circle with a measured diameter of 13.6 cm?
    (c) How large is the percentage error in the calculated area of a circle with a measured diameter of d?

11. You are minting gold coins which must have a volume of $47.3 \pm 0.1 \text{ cm}^3$. If you can manufacture the coins to be exactly 2 cm high, how much variation can you allow for the radius?

12. If F is the fraction of carbon–14 remaining in a plant sample Y years after it died, then $Y = \frac{5700}{\ln(0.5)} \cdot \ln(F)$.
    (a) Estimate the age of a plant sample in which 83±2 % (0.83 ± 0.02) of the carbon–14 remains.
    (b) Estimate the age of a plant sample in which 13±2 % (0.13 ± 0.02) of the carbon–14 remains.

13. Your company is making dice (cubes) and the specifications require that their volume be $87 \pm 2 \text{ cm}^3$. How long should each side be and how much variance can a side have in order to meet the specifications?

14. If the specifications require a cube with a surface area of $43 \pm 0.2 \text{ cm}^2$, how long should each side be and how much variance can a side have in order to meet the specifications?

15. The period P, in seconds, for a pendulum to make one complete swing and return to the release point is $P = 2\pi\sqrt{L/g}$ where L is the length of the pendulum in feet and g is 32 feet/sec^2.
    (a) If L = 2 feet, what is the period of the pendulum?
    (b) If P = 1 second, how long is the pendulum?
    (c) Estimate the change in P if L increases from 2 feet to 2.1 feet.
    (d) The length of a 24 foot pendulum is increasing 2 inches per hour. Is the period getting longer or shorter? How fast is the period changing?

16. A ball thrown at an angle θ with an initial velocity v will land $\frac{v^2}{g} \cdot \sin(2\theta)$ feet from the thrower.

(a) How far away will the ball land if $\theta = \pi/4$ and v = 80 feet/second?

(b) Which will result in a greater change in the distance: a 5% error in the angle θ or a 5% error in the initial velocity v?

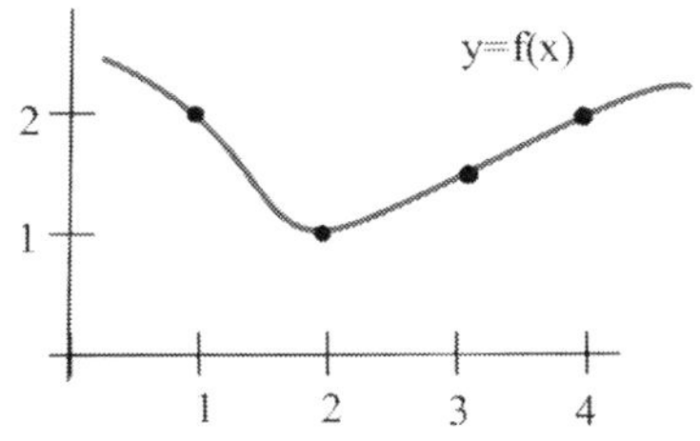

Fig. 14

17. For the function in Fig. 14, estimate the value of **df** when

(a) x = 2 and dx = 1 (b) x = 4 and dx = –1

(c) x = 3 and dx = 2

18. For the function in Fig. 15, estimate the value of **df** when

(a) x = 1 and dx = 2 (b) x = 2 and dx = –1

(c) x = 3 and dx = 1

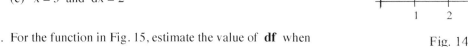

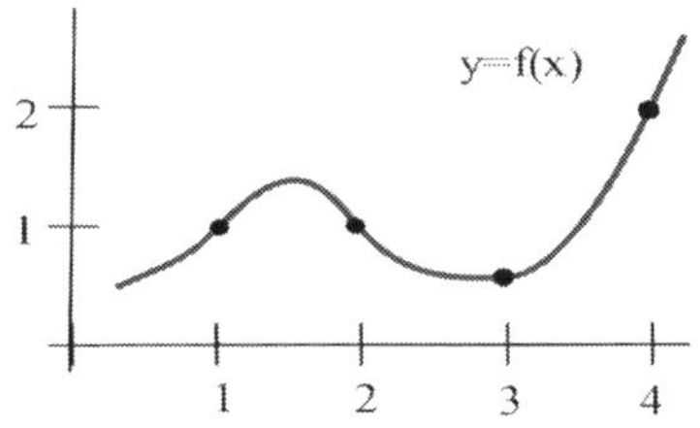

Fig. 15

19. Calculate the differentials **df** of the following functions:

(a) $f(x) = x^2 - 3x$ (b) $f(x) = e^x$

(c) $f(x) = \sin(5x)$ (d) $f(x) = x^3 + 2x$ with x = 1 and dx = 0.2

(e) $f(x) = \ln(x)$ with x = e and dx = – 0.1 (f) $f(x) = \sqrt{2x+5}$ with x = 22 and dx = 3 .

Section 2.8 PRACTICE Answers

Practice 1: $f(x) = x^{1/2}$ so $f'(x) = \frac{1}{2\sqrt{x}}$. At the point (16,4) on the graph of f, the slope of the tangent line is $f'(16) = \frac{1}{2\sqrt{16}} = \frac{1}{8}$. The equation of the tangent line is

$y - 4 = \frac{1}{8}(x - 16)$ or $y = \frac{1}{8}x + 2$: $L(x) = \frac{1}{8}x + 2$. Then

$\sqrt{16.1} \approx L(16.1) = \frac{1}{8}(16.1) + 2 =$ **4.0125** and $\sqrt{15.92} \approx L(15.92) = \frac{1}{8}(15.92) + 2 =$ **3.99**

Practice 2: $f(x) = x^3$ so $f'(x) = 3x^2$. At (1,1), the slope of the tangent line is $f'(1) = 3$. The equation of the tangent line is $y - 1 = 3(x - 1)$ or $y = 3x - 2$: $L(x) = 3x - 2$. Then $(1.02)^3 \approx L(1.02) = 3(1.02) - 2 =$ **1.06** and $(0.97)^3 \approx L(0.97) = 3(0.97) - 2 =$ **0.91** .

Practice 3: $f(x) = x^4$ so $f'(x) = 4x^3$. a = 1 and $\Delta x = 0.06$.

$(1.06)^4 = f(1.06) \approx L(1.06) = f(1) + f'(1)\cdot(0.06) = 1^4 + 4(1)^3\cdot(0.06) =$ **1.24** .

Practice 4: $f(1.1) \approx f(1) + f'(1)\cdot(0.1) = 0.7854 + (0.5)\cdot(0.1) =$ **0.8354** .

$f(1.23) \approx f(1.2) + f'(1.2)\cdot(0.03) = 0.8761 + (0.4098)\cdot(0.03) =$ **0.888394** .

$f(1.38) \approx f(1.4) + f'(1.4)\cdot(-0.02) = 0.9505 + (0.3378)\cdot(-0.02) =$ **0.943744** .

Practice 5: $x = 4$ cm and $\Delta x = 0.1$ cm. $f(x) = x^3$ so $f'(x) = 3x^2$ and $f(4) = 4^3 = 64 \text{ cm}^3$. Then "error" $\Delta f \approx f'(x)\Delta x = 3x^2 \cdot \Delta x$. When $x = 4$ and $\Delta x = 0.1$, "error" $\Delta f \approx 3(4)^2(0.1) = \mathbf{4.8 \text{ cm}^3}$.

Practice 6: $43 \pm 1^{\circ}$ is 0.75049 ± 0.01745 radians. $f(x) = 3000 \cdot \tan(x)$ so

$f(0.75049) = 3000 \cdot \tan(0.75049) \approx \mathbf{2797.5 \text{ m}}$. $f'(x) = 3000 \sec^2(x)$ so

$\Delta f(x) \approx f'(x) \cdot \Delta x = 3000 \cdot \sec^2(x) \cdot \Delta x = 3000 \cdot \sec^2(0.75049) \cdot (0.01745) = \mathbf{97.9 \text{ m}}$.

The height of the rocket is **2797.5 ± 97.9 m** .

Practice 7: $f(\theta) = 2000 \cdot \tan(\theta)$ so $f'(\theta) = 2000 \cdot \sec^2(\theta)$. We know $4000 = 2000 \cdot \tan(\theta)$ so $\tan(\theta) = 2$ and $\theta \approx 1.10715$ (radians). $f'(\theta) = 2000 \cdot \sec^2(\theta)$ so $f'(1.10715) = 2000 \cdot \sec^2(1.10715) \approx 10{,}000$. Finally, "error" $\Delta f \approx f'(\theta) \cdot \Delta\theta$ so $10 \approx 10{,}000 \cdot \Delta\theta$ and $\Delta\theta \approx 10/10{,}000 = \mathbf{0.001}$ (radians) $\approx 0.057^{\circ}$.

Practice 8: $A(r) = 6r^2$ so $A'(r) = 12r$ and $\Delta A \approx A'(r) \cdot \Delta r = 12r \cdot \Delta r$. We are also told that $\Delta r/r < 0.03$.

Percentage error is $\frac{\Delta A}{A} \cdot 100 = \frac{12r \cdot \Delta r}{6r^2} \cdot 100 = \frac{2 \cdot \Delta r}{r} \cdot 100 < 200 \cdot (0.03) = \mathbf{6}$.

Practice 9: $f(x) = \ln(x)$ $\quad df = f'(x) \cdot dx = \mathbf{\frac{1}{x}\, dx}$

$u = \sqrt{1 - 3x}$ $\quad du = \frac{du}{dx} \cdot dx = \mathbf{\frac{-3}{2\sqrt{1 - 3x}} \cdot dx}$

$r = 3\cos(\theta)$ $dr = \frac{dr}{d\theta} \cdot d\theta = \mathbf{-3 \sin(\theta) \cdot d\theta}$

Practice 10: $f(x) = x^3$, $f'(x) = 3x^2$, and $L(4 + \Delta x) = f(4) + f'(4)\Delta x = 4^3 + 3(4)^2\Delta x = 64 + 48 \cdot \Delta x$.

Δx	$f(4 + \Delta x)$	$L(4 + \Delta x)$	$\lvert f(4+\Delta x) - L(4+\Delta x) \rvert$
0.1	68.921	68.8	0.121
0.05	66.430125	66.4	0.030125
0.01	64.481201	64.48	0.001201
0.001	64.048012	64.048	0.000012

$f(4+\Delta x)$ is the actual volume of the cube with side length $4+\Delta x$.

$L(4+\Delta x)$ is the volume of the cube with side length 4 ($v = 64$) plus the volume of the 3 "slabs" ($v = 3 \cdot 4^2 \cdot \Delta x$)

$\lvert f(4+\Delta x) - L(4+\Delta x) \rvert$ is the volume of the "leftover" pieces from L: the 3 "rods" ($v = 3 \cdot 4 \cdot (\Delta x)^2$) and the tiny cube ($v = (\Delta x)^3$).

2.9 IMPLICIT and LOGARITHMIC DIFFERENTIATION

This short section presents two final differentiation techniques. These two techniques are more specialized than the ones we have already seen and they are used on a smaller class of functions. For some functions, however, one of these may be the only method that works. The idea of each method is straightforward, but actually using each of them requires that you proceed carefully and practice.

Implicit Differentiation

In our work up until now, the functions we needed to differentiate were either given **explicitly,** such as $y = f(x) = x^2 + \sin(x)$, or it was possible to get an explicit formula for them, such as solving $y^3 - 3x^2 = 5$ to get $y = \sqrt[3]{5 + 3x^2}$. Sometimes, however, we will have an equation relating x and y which is either difficult or impossible to solve explicitly for y, such as $y^2 + 2y = \sin(x) + 4$ or $y + \sin(y) = x^3 - x$. In any case, we can still find $y' = f'(x)$ by using implicit differentiation.

The key idea behind implicit differentiation is to **assume that y is a function of x** even if we cannot explicitly solve for y. This assumption does not require any work, but we need to be very careful to treat y as a function when we differentiate and to use the Chain Rule or the Power Rule for Functions.

Example 1: Assume that y is a function of x.

Calculate (a) $\mathbf{D}(y^3)$, (b) $\frac{\mathbf{d}}{\mathbf{dx}}(x^3y^2)$, and (c) $(\sin(y))'$

Solution: (a) We need the Power Rule for Functions since **y is a function of x**:

$$\mathbf{D}(y^3) = 3y^2 \cdot \mathbf{D}(y) = 3y^2 \cdot \mathbf{y}'.$$

(b) We need to use the product rule and the Chain Rule:

$$\frac{\mathbf{d}}{\mathbf{dx}}(x^3y^2) = x^3 \cdot \frac{\mathbf{d}}{\mathbf{dx}}(y^2) + y^2 \cdot \frac{\mathbf{d}}{\mathbf{dx}}(x^3) = x^3\, \mathbf{2y}\frac{\mathbf{dy}}{\mathbf{dx}} + y^2 \cdot \mathbf{3x^2} = 2x^3y\frac{\mathbf{dy}}{\mathbf{dx}} + 3x^2y^2.$$

(c) We just need to know that $\mathbf{D}(\sin(x)) = \cos(x)$ and then use the Chain Rule:

$$(\sin(y))' = \cos(y) \cdot \mathbf{y}'.$$

Practice 1: Assume that y is a function of x. Calculate (a) $\mathbf{D}(x^2 + y^2)$ and (b) $\frac{\mathbf{d}}{\mathbf{dx}}(\sin(2 + 3y))$.

> **IMPLICIT DIFFERENTIATION:**
>
> To determine **y '**, differentiate each side of the defining equation, **treating y as a function of x**, and then algebraically solve for y '.

Example 2: Find the slope of the tangent line to the circle $x^2 + y^2 = 25$ at the point (3,4) with and without implicit differentiation.

Solution:

Explicitly: We can solve the equation of the circle for $y = +\sqrt{25 - x^2}$ or $y = -\sqrt{25 - x^2}$.

Since the point (3,4) is on the top half of the circle (Fig. 1), $y = +\sqrt{25 - x^2}$ and

$$\mathbf{D}(\, y\,) = \mathbf{D}(\, +\sqrt{25 - x^2}\,) = \frac{1}{2}\,(\,25 - x^2\,)^{-1/2}\,\mathbf{D}(\,25 - x^2\,) = \frac{-x}{\sqrt{25 - x^2}}$$

Replacing x with 3, we have $\mathbf{y'} = \dfrac{-3}{\sqrt{25 - 3^2}} = -3/4$.

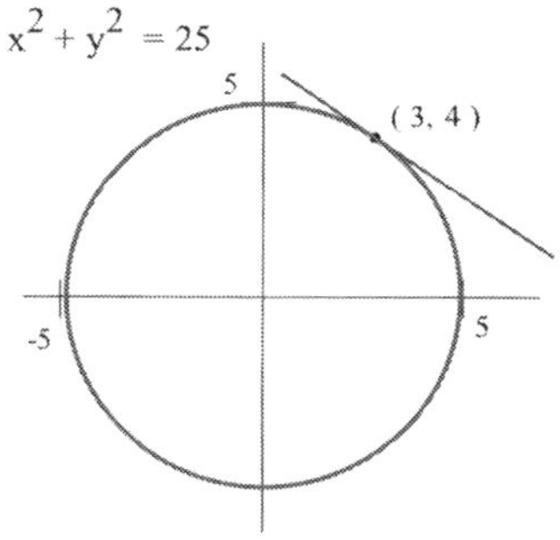

Fig. 1

Implicitly: We differentiate each side of the equation $x^2 + y^2 = 25$ and then solve for $\mathbf{y'}$. $\mathbf{D}(\,x^2 + y^2\,) = \mathbf{D}(\,25\,)$ so $2x + 2y{\cdot}\mathbf{y'} = 0$.

Solving for $\mathbf{y'}$, we have $\mathbf{y'} = -\dfrac{2x}{2y} = -x/y$, and, at the point (3,4),

$\mathbf{y'} = -3/4$, the same answer we found explicitly.

Practice 2: Find the slope of the tangent line to $y^3 - 3x^2 = 15$ at the point (2,3) with and without implicit differentiation.

In the previous example and practice problem, it was easy to explicitly solve for y , and then we could differentiate y to get y '. Because we could explicitly solve for y , we had a choice of methods for calculating y '. Sometimes, however, we can not explicitly solve for y , and the only way of determining y ' is implicit differentiation.

Example 3: Determine y ' at (0,2) for $y^2 + 2y = \sin(x) + 8$.

Solution: Assuming that y is a function of x and differentiating each side of the equation, we get

$$\mathbf{D}(\,y^2 + 2y\,) = \mathbf{D}(\,\sin(x) + 8\,) \text{ so } 2y\,\mathbf{y'} + 2\,\mathbf{y'} = \cos(x) \text{ and } (2y + 2)\,\mathbf{y'} = \cos(x).$$

$$\text{Then } \mathbf{y'} = \frac{\cos(x)}{2y + 2} \text{ so , at the point } (0,2),\ \mathbf{y'} = \frac{\cos(0)}{2(\mathbf{2}) + 2} = 1/6\ .$$

Practice 3: Determine y ' at (1,0) for $y + \sin(y) = x^3 - x$.

In practice, the equations may be rather complicated, but if you proceed carefully and step–by–step, implicit differentiation is not difficult. Just remember that **y must be treated as a function** so every time you differentiate a term containing a y you should get something which has a **y'** . The algebra needed to solve for y' is always easy -- if you differentiated correctly the resulting equation will be a linear equation in the variable y'.

Example 4: Find the equation of the tangent line L to the "tilted' parabola in Fig. 1 at the point (1,2).

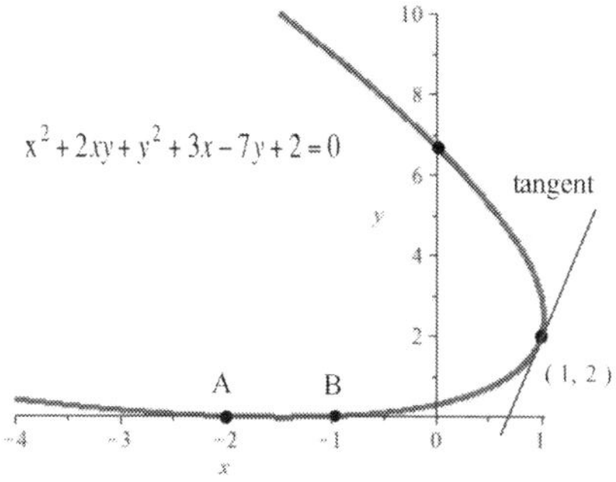

Fig. 2

Solution: The line goes through the point (1, 2) so we need to find the slope there. Differentiating each side of the equation, we get

$\mathbf{D}(\ x^2 + 2xy + y^2 + 3x - 7y + 2\ ) = \mathbf{D}(\ 0\ )$ so

$2x + 2x\,\mathbf{y}\,' + 2y + 2y\,\mathbf{y}\,' + 3 - 7\mathbf{y}\,' = 0$ and

$(2x + 2y - 7)\,\mathbf{y}\,' = -2x - 2y - 3$.

Solving for $\mathbf{y}\,'$, $\mathbf{y}\,' = \dfrac{-2x - 2y - 3}{2x + 2y - 7}$, so the slope at (1,2) is $m = \mathbf{y}\,' = \dfrac{-2 - 4 - 3}{2 + 4 - 7} = 9$.

Finally, the equation of the line is $y - 2 = 9(x - 1)$ so $y = 9x - 7$.

Practice 4: Find the points where the graph in Fig. 2 crosses the y–axis, and find the slopes of the tangent lines at those points.

Implicit differentiation is an alternate method for differentiating equations which can be solved explicitly for the function we want, and it is the only method for finding the derivative of a function which we cannot describe explicitly.

Logarithmic Differentiation

In section 2.5 we saw that $\mathbf{D}(\ \ln(\ f(x)\ )\ ) = \dfrac{f\,'(x)}{f(x)}$. If we simply multiply each side by f(x) , we have $\mathbf{f\,'(x)} = f(x) \cdot \mathbf{D}(\ \ln(\ f(x)\ )\ )$. When the logarithm of a function is simpler than the function itself, it is often easier to differentiate the logarithm of f than to differentiate f itself.

Logarithmic Differentiation: $\mathbf{f\,'(x)} = f(x) \cdot \mathbf{D}(\ \ln(\ f(x)\ )\ )$ **.**

The derivative of f is f times the derivative of the natural logarithm of f. Usually it is easiest to proceed in three steps:

(i) calculate ln(f(x)) and simplify,

(ii) calculate **D**(ln(f(x))) and simplify, and

(iii) multiply the result in step (ii) by f(x).

Let's examine what happens when we use this process on an "easy" function, $f(x) = x^2$, and a "hard" one, $f(x) = 2^x$. Certainly we don't need to use logarithmic differentiation to find the derivative of $f(x) = x^2$, but sometimes it is instructive to try a new algorithm on a familiar function. Logarithmic differentiation is the easiest way to find the derivative of $f(x) = 2^x$.

$f(x) = x^2$

(i) $\ln(f(x)) = \ln(x^2) = 2\cdot\ln(x)$

(ii) $\mathbf{D}(\ln(f(x))) = \mathbf{D}(2\cdot\ln(x)) = \frac{2}{x}$

(iii) $f'(x) = f(x)\cdot \mathbf{D}(\ln(f(x))) = x^2 \cdot \frac{2}{x} = \mathbf{2x}$

$f(x) = 2^x$

(i) $\ln(f(x)) = \ln(2^x) = x \cdot\ln(2)$

(ii) $\mathbf{D}(\ln(f(x))) = \mathbf{D}(x \cdot\ln(2)) = \ln(2)$

(iii) $f'(x) = f(x)\cdot \mathbf{D}(\ln(f(x))) = \mathbf{2^x\cdot\ln(2)}$

Example 5: Use the pattern $\mathbf{f'(x)} = f(x) \cdot\mathbf{D}(\ln(f(x)))$ to find the derivative of $f(x) = (3x+7)^5\cdot\sin(2x)$.

Solution: (i) $\ln(f(x)) = \ln((3x+7)^5\cdot\sin(2x)) = 5\cdot\ln(3x+7) + \ln(\sin(2x))$ so

(ii) $\mathbf{D}(\ln(f(x))) = \mathbf{D}(5\cdot\ln(3x+7) + \ln(\sin(2x))) = 5\frac{3}{3x+7} + \frac{2\cos(2x)}{\sin(2x)}$.

Then (iii) $\mathbf{f'(x)} = f(x)\cdot\mathbf{D}(\ln(f(x))) = (3x+7)^5\cdot\sin(2x) \left(\frac{15}{3x+7} + \frac{2\cos(2x)}{\sin(2x)} \right)$

$= 15\,(3x+7)^4 \sin(2x) + 2\,(3x+7)^5\cos(2x)$,

the same result we would obtain using the product rule.

Practice 5: Use logarithmic differentiation to find the derivative of $f(x) = (2x+1)^3 \cdot(3x^2 - 4)^7 \cdot(x+7)^4$.

We could have differentiated the functions in the example and practice problem without logarithmic differentiation. There are, however, functions for which logarithmic differentiation is the only method we can use. We know how to differentiate x to a constant power, $\mathbf{D}(x^n) = n \cdot x^{n-1}$, and a constant to the variable power, $\mathbf{D}(c^x) = c^x \cdot\ln(c)$, but the function $f(x) = x^x$ has both a variable base and a variable power so neither differentiation rule applies to x^x. We need to use logarithmic differentiation.

Example 6: Find $\mathbf{D}(x^x)$ $(x > 0)$.

Solution: (i) $\text{Ln}(f(x) = \ln(x^x) = x\cdot\ln(x)$

(ii) $\mathbf{D}(\ln(f(x))) = \mathbf{D}(x \cdot\ln(x)) = x \cdot\mathbf{D}(\ln(x)) + \ln(x) \cdot\mathbf{D}(x) = x (\frac{1}{x}) + \ln(x)\,(1) = 1 + \ln(x)$.

Then (iii) $\mathbf{D}(x^x) = \mathbf{f'(x)} = f(x) \cdot\mathbf{D}(\ln(f(x))) = x^x \cdot\left(1 + \ln(x) \right)$.

Practice 6: Find $\mathbf{D}(x^{\sin(x)})$ $(x > 0)$.

Logarithmic differentiation is an alternate method for differentiating some functions such as products and quotients, and it is the only method we've seen for differentiating some other functions such as variable bases to variable exponents.

PROBLEMS

In problems 1 – 10 find dy/dx in two ways: (a) by differentiating implicitly and (b) by explicitly solving for y and then differentiating. Then find the value of dy/dx at the given point using your results from both the implicit and the explicit differentiation.

1. $x^2 + y^2 = 100$, point (6, 8)

2. $x^2 + 5y^2 = 45$, point (5, 2)

3. $x^2 - 3xy + 7y = 5$, point (2,1)

4. $\sqrt{x} + \sqrt{y} = 5$, point (4,9)

5. $\frac{x^2}{9} + \frac{y^2}{16} = 1$, point (0,4)

6. $\frac{x^2}{9} + \frac{y^2}{16} = 1$, point (3,0)

7. $\ln(y) + 3x - 7 = 0$, point (2,e)

8. $x^2 - y^2 = 16$, point (5,3)

9. $x^2 - y^2 = 16$, point (5,–3)

10. $y^2 + 7x^3 - 3x = 8$, point (1,2)

11. Find the slopes of the lines tangent to the graph in Fig. 3 at the points (3,1), (3,3), and (4,2) .

12. Find the slopes of the lines tangent to the graph in Fig. 3 where the graph crosses the y–axis.

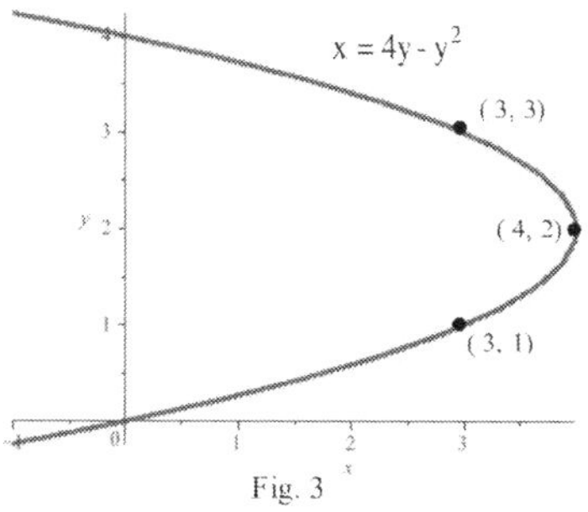

Fig. 3

13. Find the slopes of the lines tangent to the graph in Fig. 4 at the points ((5,0), (5,6), and (–4,3).

14. Find the slopes of the lines tangent to the graph in Fig. 4 where the graph crosses the y–axis.

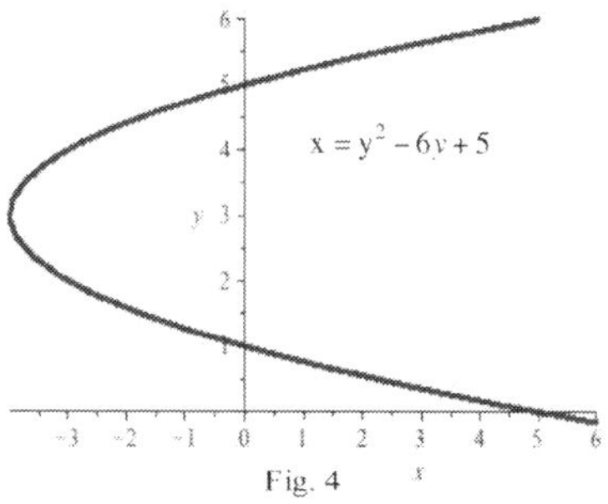

Fig. 4

In problems 15 – 22 , find dy/dx using **implicit differentiation** and then find the slope of the line tangent to the graph of the equation at the given point.

15. $y^3 - 5y = 5x^2 + 7$, point (1,3)

16. $y^2 - 5xy + x^2 + 21 = 0$, point (2,5)

17. $y^2 + \sin(y) = 2x - 6$, point (3,0)

18. $y + 2x^2y^3 = 4x + 7$, point (3,1)

19. $e^y + \sin(y) = x^2 - 3$, point (2,0)

20. $(x^2 + y^2 + 1)^2 - 4x^2 = 81$, point $(0, 2\sqrt{2})$

21. $x^{2/3} + y^{2/3} = 5$, point (8,1)

22. $x + \cos(xy) = y + 3$, point (2,0)

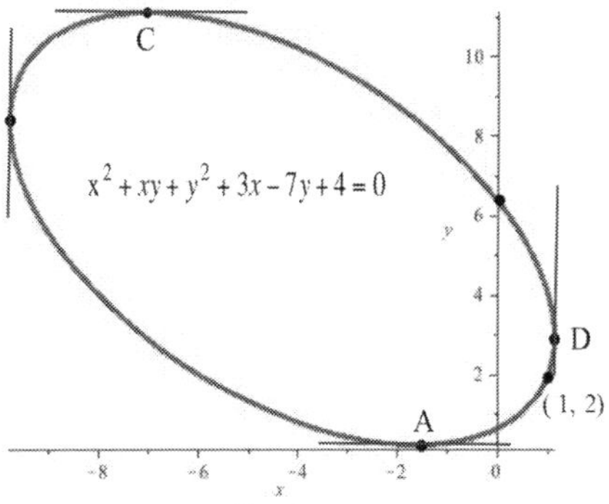

Fig. 5

23. Find the slope of the line tangent to the ellipse in Fig. 5 at the point (1, 2).

24. Find the slopes of the tangent lines at the points where the ellipse in Fig. 5 crosses the y–axis.

25. Find y ' for $y = Ax^2 + Bx + C$ and for $x = Ay^2 + By + C$.

26. Find y ' for $y = Ax^3 + B$ and for $x = Ay^3 + B$.

27. Find y ' for $Ax^2 + Bxy + Cy^2 + Dx + Ey + F = 0$.

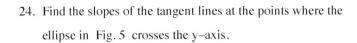

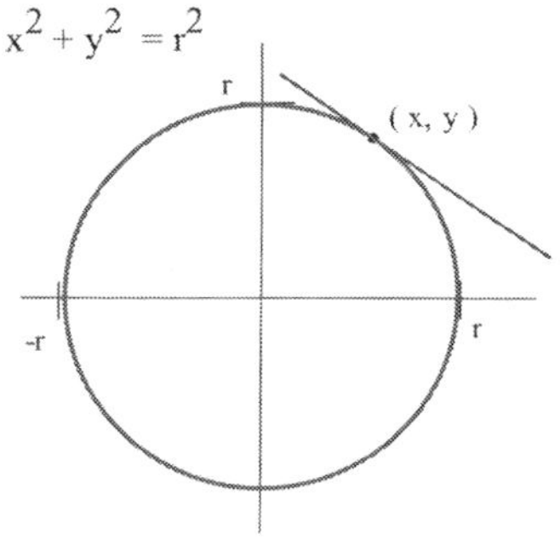

Fig. 6

28. In chapter 1 we assumed that the tangent line to a circle at a point was perpendicular to the radial line through the point and the center of the circle. Use implicit differentiation to prove that the line tangent to the circle $x^2 + y^2 = r^2$ (Fig. 6) at (x,y) is perpendicular to the line through (0,0) and (x,y).

29. Find the coordinates of point A where the tangent line to the ellipse in Fig. 5 is horizontal.

30. Find the coordinates of point B where the tangent line to the ellipse in Fig. 5 is vertical.

31. Find the coordinates of points C and D on the ellipse in Fig. 5..

In problems 32 – 40 find dy/dx in two ways: (a) by using the "usual" differentiation patterns and (b) by using logarithmic differentiation.

32. $y = x \cdot \sin(3x)$

33. $y = (x^2 + 5)^7 \cdot (x^3 - 1)^4$

34. $y = \dfrac{\sin(3x - 1)}{x + 7}$

35. $y = x^5 \cdot (3x + 2)^4$

36. $y = 7^x$

37. $y = e^{\sin(x)}$

38. $y = \cos^7(2x + 5)$

39. $y = \sqrt{25 - x^2}$

40. $y = \dfrac{x \cdot \cos(x)}{x^2 + 1}$

In problems 41 – 46 , use logarithmic differentiation to find dy/dx .

41. $y = x^{\cos(x)}$

42. $y = (\cos(x))^x$

43. $y = x^4 \cdot (x - 2)^7 \cdot \sin(3x)$

44. $y = \dfrac{\sqrt{x + 10}}{(2x + 3)^3 \cdot (5x - 1)^7}$

45. $y = (3 + \sin(x))^x$

46. $y = \sqrt{\dfrac{x^2 + 1}{x^2 - 1}}$

In problems 47 – 50, use the values in each table to calculate the values of the derivative in the last column.

47. Use Table 1. 48. Use Table 2 49. Use Table 3. 50. Use Table 4.

Table 1

x	f(x)	ln(f(x))	**D**(ln(f(x)))	f '(x)
1	1	0	1.2	
2	9	2.2	1.8	
3	64	4.2	2.1	

Table 2

x	g(x)	ln(g(x))	**D**(ln(g(x)))	g '(x)
0	5	1.6	0.6	
1	10	2.3	0.7	
2	20	3.0	0.8	

Table 3

x	f(x)	ln(f(x))	**D**(ln(f(x)))	f '(x)
1	5	1.6	–1	
2	2	0.7	0	
3	7	1.9	2	

Table 4

x	g(x)	ln(g(x))	**D**(ln(g(x)))	g '(x)
2	1.4	0.3	1.2	
3	3.3	1.2	0.6	
7	13.6	2.6	0.2	

Problems 51 – 55 illustrate how logarithmic differentiation can be used to verify some differentiation patterns we already know (51 and 52) and to derive some new patterns (53 – 55). Assume that all of the functions are differentiable and that the function combinations are defined.

51. Use logarithmic differentiation on $f \cdot g$ to rederive the product rule: $\mathbf{D}(f \cdot g) = f \cdot g' + g \cdot f'$.

52. Use logarithmic differentiation on f/g to rederive the quotient rule: $\mathbf{D}(f/g) = \dfrac{g \cdot f' - f \cdot g'}{g^2}$.

53. Use logarithmic differentiation on $f \cdot g \cdot h$ to derive a product rule for three functions: $\mathbf{D}(f \cdot g \cdot h)$.

54. Use logarithmic differentiation on the exponential function a^x to determine its derivative: $\mathbf{D}(a^x)$.

55. Use logarithmic differentiation to determine a pattern for the derivative of f^g: $\mathbf{D}(f^g)$.

Section 2.9 **PRACTICE Answers**

Practice 1: $\mathbf{D}(x^2 + y^2) = 2x + 2y \cdot \mathbf{y'}$

$\dfrac{d}{dx}(\sin(2 + 3y)) = \cos(2 + 3y) \cdot \mathbf{D}(2 + 3y) = \cos(2 + 3y) \cdot 3\mathbf{y'}$

Practice 2: Explicitly: $y = (3x^2 + 15)^{1/3}$ so $y' = \frac{1}{3}(3x^2 + 15)^{-2/3} \cdot \mathbf{D}(3x^2 + 15) = \frac{1}{3}(3x^2 + 15)^{-2/3} \cdot (6x)$.

When $(x,y) = (2,3)$, $y' = \frac{1}{3}(3(2)^2 + 15)^{-2/3} \cdot (6 \cdot 2) = 4 (27)^{-2/3} = \mathbf{\frac{4}{9}}$.

Implicitly: $\mathbf{D}(y^3 - 3x^2) = \mathbf{D}(15)$ so $3y^2 \cdot \mathbf{y'} - 6x = 0$ and $y' = \dfrac{2x}{y^2}$.

When $(x,y) = (2,3)$, $y' = \dfrac{2 \cdot (2)}{(3)^2} = \mathbf{\frac{4}{9}}$.

Practice 3: $y + \sin(y) = x^3 - x$

$\mathbf{D}(y + \sin(y)) = \mathbf{D}(x^3 - x)$ differentiating each side

$\mathbf{y'} + \cos(y)\cdot\mathbf{y'} = 3x^2 - 1$

$\mathbf{y'}\cdot(1 + \cos(y)) = 3x^2 - 1$

$$\mathbf{y'} = \frac{3x^2 - 1}{1 + \cos(y)}$$

Then when $(x,y) = (1,0)$, $\mathbf{y'} = \dfrac{3(1)^2 - 1}{1 + \cos(0)} = \mathbf{1}$.

Practice 4: To find where the parabola crosses the y–axis, we can set $x = 0$ and solve for the values of y. Replacing x with 0 in $x^2 + 2xy + y^2 + 3x - 7y + 2 = 0$, we have $y^2 - 7y + 2 = 0$ so $y = \dfrac{7 \pm \sqrt{(-7)^2 - 4(1)(2)}}{2(1)} = \dfrac{7 \pm \sqrt{41}}{2} \approx 0.3$ and 6.7. The parabola crosses the y–axis approximately at the points $(0, 0.3)$ and $(0, 6.7)$.

From Example 4, we know that $y' = \dfrac{-2x - 2y - 3}{2x + 2y - 7}$, so

at the point $(0, 0.3)$, the slope is approximately $\dfrac{0 - 0.6 - 3}{0 + 0.6 - 7} \approx 0.56$, and

at the point $(0, 6.7)$, the slope is approximately $\dfrac{0 - 13.4 - 3}{0 + 13.4 - 7} \approx -2.56$.

Practice 5: $f'(x) = f(x)\cdot\mathbf{D}(\ln(f(x)))$ and $f(x) = (2x + 1)^3 (3x^2 - 4)^7 (x + 7)^4$

(i) $\ln(f(x)) = 3\cdot\ln(2x + 1) + 7\cdot\ln(3x^2 - 4) + 4\cdot\ln(x + 7)$.

(ii) $\mathbf{D}(\ln(f(x))) = \dfrac{3}{2x + 1}(2) + \dfrac{7}{3x^2 - 4}(6x) + \dfrac{4}{x + 7}(1)$

(iii) $f'(x) = f(x)\cdot\mathbf{D}(\ln(f(x))) = (2x + 1)^3 (3x^2 - 4)^7 (x + 7)^4 \cdot \left\{ \dfrac{2\cdot 3}{2x + 1} + \dfrac{7\cdot 6x}{3x^2 - 4} + \dfrac{4}{x + 7} \right\}$.

Practice 6: $f'(x) = f(x)\cdot\mathbf{D}(\ln(f(x)))$ and $f(x) = x^{\sin(x)}$ so

(i) $\ln(f(x)) = \ln(x^{\sin(x)}) = \sin(x)\cdot\ln(x)$

(ii) $\mathbf{D}(\ln(f(x))) = \mathbf{D}(\sin(x)\cdot\ln(x)) = \sin(x)\cdot\mathbf{D}(\ln(x)) + \ln(x)\cdot\mathbf{D}(\sin(x)) = \sin(x)\cdot\dfrac{1}{x} + \ln(x)\cdot\cos(x)$

(iii) $f'(x) = f(x)\cdot\mathbf{D}(\ln(f(x))) = x^{\sin(x)} \cdot \left\{ \sin(x)\cdot\dfrac{1}{x} + \ln(x)\cdot\cos(x) \right\}$

Chapter Two

Section 2.0

1.

x	y = f(x)	m(x) = the estimated slope of the tangent line to y=f(x) at the point (x,y)
0	1	1
0.5	1.4	1/2
1.0	1.6	0
1.5	1.4	–1/2
2.0	1	–2
2.5	0	–2
3.0	–1	–2
3.5	–1.3	0
4.0	–1	1

3. (a) At $x = 1, 3$, and 4. (b) f is largest at $x = 4$. f is smallest at $x = 3$.

5. (a) Graph (b) Graph (c) $m(x) = \cos(x)$

7. The solution is similar to the method used in Example 4. Assume we turn off the engine at the point (p,q) on the curve $y = x^2$, and then find values of p and q so the tangent line to $y = x^2$ at the point (p,q) goes through the given point (5,16). (p,q) is on $y = x^2$ so $q = p^2$. The equation of the tangent line to $y = x^2$ at (p, p^2) is $y = 2px - p^2$ so, substituting $x = 5$ and $y = 16$, we have $16 = 2p(5) - p^2$. Solving $p^2 - 10p + 16 = 0$ we get $p = 2$ or $p = 8$. The solution we want (moving left to right along the curve) is $p = 2, q = p^2 = 4$. ($p = 8, q = 64$ would be the solution if we were moving right to left.)

9. Impossible. The point (1, 3) is "inside" the parabola.

11. (a) $m_{sec} = \dfrac{f(x+h) - f(x)}{(x+h) - (x)} = \dfrac{\{ 3(x + h) - 7 \} - \{ 3x - 7 \}}{(x + h) - x} = \dfrac{3h}{h} = 3.$
(b) $m_{tan} = \lim\limits_{h\to 0} m_{sec} = \lim\limits_{h\to 0} 3 = 3.$ (c) At $x = 2$, $m_{tan} = 3$.
(d) $f(2) = -1$ so the tangent line is $y - (-1) = 3(x - 2)$ or $y = 3x - 7$.

13. (a) $m_{sec} = \dfrac{f(x+h) - f(x)}{(x+h) - (x)} = \dfrac{\{ a(x + h) + b \} - \{ ax + b\}}{(x + h) - x} = \dfrac{ah}{h} = a.$
(b) $m_{tan} = \lim\limits_{h\to 0} m_{sec} = \lim\limits_{h\to 0} a = a$ (c) At $x = 2$, $m_{tan} = a$.
(d) $f(2) = 2a + b$ so the tangent line is $y - (2a + b) = a(x - 2)$ or $y = ax + b$..

15. (a) $m_{sec} = \dfrac{f(x+h) - f(x)}{(x+h) - (x)} = \dfrac{\{ 8 - 3(x + h)^2 \} - \{ 8 - 3x^2 \}}{(x + h) - x} = \dfrac{-6xh - 3h^2}{h} = -6x - 3h.$
(b) $m_{tan} = \lim\limits_{h\to 0} m_{sec} = \lim\limits_{h\to 0} -6x - 3h = -6x$ (c) At $x = 2$, $m_{tan} = -6(2) = -12$.
(d) $f(2) = -4$ so the tangent line is $y - (-4) = -12(x - 2)$ or $y = -12x + 20$...

17. $a = 1, b = 2, c = 0$, so $m_{tan} = (2)(1)(x) + 2 = 2x + 2$. The problem is to find p for which

$$6 - (p^2 + 2p) = (2p + 2)(3 - p).$$

This reduces to $p^2 - 6p = 0$ so $p = 0$ or 6 and the required points are (0, 0) and (6, 48).

Section 2.1

1. (a) derivative of g (b) derivative of h (c) derivative of f
3. (a) $m_{sec} = h - 4$, $m_{tan} = \lim_{h\to 0} m_{sec} = -4$. (b) $m_{sec} = h + 1$, $m_{tan} = \lim_{h\to 0} m_{sec} = 1$.
5. (a) $m_{sec} = 5 - h$, $m_{tan} = \lim_{h\to 0} m_{sec} = 5$. (b) $m_{sec} = 7 - 2x - h$, $m_{tan} = \lim_{h\to 0} m_{sec} = 7 - 2x$.
7. (a) –1 (b) –1 (c) 0 (d) +1 (e) DNE (f) DNE
9. $f'(x) = \lim_{h\to 0} \frac{f(x+h)-f(x)}{h} = \lim_{h\to 0} \frac{\{(x+h)^2+8\}-\{x^2+8\}}{h} = \lim_{h\to 0} \frac{2xh+h^2}{h} = \lim_{h\to 0} 2x+h = 2x$. $f'(3)=6$.
11. $f'(x) = \lim_{h\to 0} \frac{\{2(x+h)^3-5(x+h)\}-\{2x^3-5x\}}{h} = \lim_{h\to 0} \frac{6x^2h+6xh^2+2h^3-5h}{h} = 6x^2-5$. $f'(3) = 49$.
13. For any constant C, if $f(x) = x^2 + C$, then

$$f'(x) = = \lim_{h\to 0} \frac{f(x+h)-f(x)}{h} = \lim_{h\to 0} \frac{\{(x+h)^2+C\}-\{x^2+C\}}{h} = \lim_{h\to 0} \frac{2xh+h^2}{h} = \lim_{h\to 0} 2x+h = 2x.$$

The graphs of $f(x) = x^2$, $g(x) = x^2 + 3$ and $h(x) = x^2 - 5$ are "parallel" parabolas: g is f shifted up 3 units, and h is f shifted down 5 units.

15. $f'(x) = 2x$. Then $f'(1) = 2$ and the equation of the tangent line at (1,9) is $y - 9 = 2(x - 1)$ or $y = 2x + 7$.
$f'(-2) = -4$ and the equation of the tangent line at (–2,12) is $y - 12 = -4(x + 2)$ or $y = -4x + 4$.

17. $f'(x) = \cos(x)$. Then $f'(\pi) = \cos(\pi) = -1$ and the equation of the tangent line at $(\pi,0)$ is $y - 0 = -1(x - \pi)$ or $y = -x + \pi$.
$f'(\pi/2) = \cos(\pi/2) = 0$ and the equation of the tangent line at $(\pi/2,1)$ is $y - 1 = 0(x - \pi/2)$ or $y = 1$.

19. (a) $y - 5 = 4(x - 2)$ or $y = 4x - 3$ (b) $x + 4y = 22$ or $y = -0.25x + 5.5$
(c) $f'(x) = 2x$ so the tangent line is horizontal when $x = 0$: at the point (0,1).
(d) $f'(p) = 2p$ (the slope of the tangent line) so $y - q = 2p(x - p)$ or $y = 2px + (q - 2p^2)$.
Since $q = p^2 + 1$, the equation of the tangent line becomes $y = 2px + (p^2 + 1 - 2p^2) = 2px - p^2 + 1$.
(e) We need p such that $-7 = 2p(1) - p^2 + 1$ or $p^2 - 2p - 8 = 0$. Then $p = -2, 4$. There are two points with the property we want: (–2, 5) and (4, 17).
21. (a) $y' = 2x$, so when $x = 1$, $y' = 2$. Angle = arctan(2) ≈ 1.107 radians ≈ 63°.
(b) $y' = 3x^2$, so when $x = 1$, $y' = 3$. Angle = arctan(3) ≈ 1.249 radians ≈ 72°.
(c) Angle ≈ 1.249 – 1.107 radians = 0.142 radians (or angle = $72^o - 63^o = 9^o$)

23. Graph. On the graph of upward velocity, the units on the horizontal axis are "seconds" and the units on the vertical axis are "feet per second."

25. (a) d(4) = 256 ft. d(5) = 400 ft. (b) $d'(x) = 32x$ $d'(4) = 128$ ft/sec $d'(5) = 160$ ft/sec.
27. $C(x) = \sqrt{x}$ dollars to produce x golf balls.

Marginal production cost is $C'(x) = \frac{1}{2\sqrt{x}}$ dollars per golf ball.

$C'(25) = \frac{1}{2\sqrt{25}} = \frac{1}{10}$ dollars per golf ball. $C'(100) = \frac{1}{2\sqrt{100}} = \frac{1}{20}$ dollars per golf ball.

29. (a) A(0) = 0, A(1) = 1/2, A(2) = 2 and A(3) = 9/2. (b) $A(x) = x^2/2$ ($x \ge 0$). (c) $\frac{d\,A(x)}{dx} = x$.
(d) $\frac{d\,A(x)}{dx}$ represents the rate at which A(x) is increasing, the rate at which area is accumulating.
31. (a) $9x^8$ (b) $\frac{2}{3x^{1/3}}$ (c) $\frac{-4}{x^5}$ (d) $\pi x^{\pi-1}$ (e) 1 if $x > -5$ and –1 if $x < -5$

33. $f(x) = x^3 + 4x^2$ (plus any constant) 35. $f(x) = 5\cdot\sin(t)$ 37. $f(x) = \frac{1}{2}x^2 + \frac{1}{3}x^3$

Section 2.2

1. (a) Cont. at 0, 1, 2, 3, 5 (b) Diff. at 0, 3, 5

3.

x	f(x)	f '(x)	g(x)	g '(x)	f(x)•g(x)	**D**(f(x)•g(x))	f(x)/g(x)	**D**(f(x)/g(x))
0	2	3	1	5	**2**	**13**	**2**	**–7**
1	–3	2	5	–2	**–15**	**16**	**–3/5**	**4/25**
2	0	–3	2	4	**0**	**–6**	**0**	**–3/2**
3	1	–1	0	3	**0**	**3**	**undef**	**undef**

5.

x	f(x)	f '(x)	g(x)	g '(x)	f(x)+g(x)	f(x)•g(x)	f(x)/g(x)	**D**(f(x)+g(x))	**D**(f(x)•g(x))	**D**(f(x)/g(x))
1	3	–2	2	2	5	6	3/2	0	2	–10/4
2	1	0	3	1/2	4	3	1/3	1/2	1/2	–1/18
3	2	1	2	–1	4	4	1	0	0	1

7. (a) $\mathbf{D}(\,(x-5)(3x+7)\,) = (x-5)3 + (3x+7)1 = 6x - 8$

(b) $\mathbf{D}(\,3x^2 - 8x - 35\,) = 6x - 8$, the same result as in (a)

9. $$\frac{d}{dx}\,\frac{\cos(x)}{x^2} = \frac{x^2(\,-\sin(x)\,) - (\,\cos(x)\,)(\,2x\,)}{(\,x^2\,)^2} = -\,\frac{x\sin(x) + 2\cos(x)}{x^3}$$

11. $\mathbf{D}(\,\sin^2(x)\,) = \sin(x)\cos(x) + \sin(x)\cos(x) = 2\sin(x)\cos(x)$,

$\mathbf{D}(\,\cos^2(x)\,) = \cos(x)(\,-\sin(x)\,) + \cos(x)(\,-\sin(x)\,) = -2\sin(x)\cos(x)$

13. $f(x) = ax^2 + bx + c$ so $f(0) = c$. Then $f(0) = 0$ implies that $c = 0$.
$f'(x) = 2ax + b$ so $f'(0) = b$ and $f'(0) = 0$ implies that $b = 0$.
Finally, $f'(10) = 20a + b = 20a$ so $f'(10) = 30$ implies that $20a = 30$ and $a = 3/2$.

$$f(x) = \frac{3}{2}x^2 + 0x + 0 \text{ has } f(0) = 0, f'(0) = 0, \text{ and } f'(10) = 30.$$

15. Their graphs are vertical shifts of each other, and their derivatives are equal.

17. $f(x)g(x) = k$ so $\mathbf{D}(\,f(x)g(x)\,) = \mathbf{D}(\,k\,) = 0$ and $f(x)g'(x) + g(x)f'(x) = 0$.
If $f(x) \neq 0$ and $g(x) \neq 0$, then $\dfrac{f'(x)}{f(x)} = -\dfrac{g'(x)}{g(x)}$.

19. $f'(x) = 2x - 5$ so $f'(1) = -3$. $f'(x) = 0$ if $x = 5/2$.

21. $f'(x) = 3 + 2\sin(x)$ so $f'(1) = 3 + 2\sin(1) \approx 4.68$. $f'(x)$ never equals 0 since $\sin(x)$ never equals $-3/2$.

23. $f'(x) = 3x^2 + 18x = 3x(x+6)$ so $f'(1) = 21$. $f'(x) = 0$ if $x = 0$ or -6.

25. $f'(x) = 3x^2 + 4x + 2$ so $f'(1) = 9$. $f'(x) = 0$ for no values of x (the discriminant $4^2 - 4(3)(2) < 0$).

27. $f'(x) = x\cdot\cos(x) + \sin(x)$ so $f'(1) = 1\cdot\cos(1) + \sin(1) \approx 1.38$. The graph of $f'(x)$ crosses the x–axis infinitely often. The root of f' at $x = 0$ is easy to see (and verify). Other roots of f', such as near $x = 2.03$ and 4.91 and -2.03, can be found numerically using the Bisection algorithm or graphically using the "zoom" or "trace" features on some calculators.

29. $f'(x) = 3x^2 + 2Ax + B$. The graph of $y = f(x)$ has two distinct "vertices" if $f'(x) = 0$ for two distinct values of x. This occurs if the discriminant of $3x^2 + 2Ax + B$ is greater than 0: $(2A)^2 - 4(3)(B) > 0$.

31. Everywhere except at $x = -3$.

33. Everywhere except at $x = 0$ and 3.

35. Everywhere except at $x = 1$.

37. Everywhere. The only possible difficulty is at $x = 0$, and the definition of the derivative gives $f'(0) = 1$. The derivatives of the "two pieces" of f match at $x = 0$ to give a differentiable function there.

39. Continuity of f at $x = 1$ requires $A + B = 2$. The "left derivative" of f at $x = 1$ is $\mathbf{D}(Ax + B) = A$ and the "right derivative" of f at $x = 1$ is 3 (if $x > 1$ then $\mathbf{D}(x^2 + x) = 2x + 1$) so to achieve differentiability $A = 3$ and $B = 2 - A = -1$.

41. $h(x) = 128x - 2.65x^2$ ft.
(a) $h'(x) = 128 - 5.3x$ so $h'(0) = 128$ ft/sec, $h'(1) = 122.7$ ft/sec, and $h'(2) = 117.4$ ft/sec.
(b) $v(x) = h'(x) = 128 - 5.3x$ ft/sec.
(c) $v(x) = 0$ when $x = 128/5.3 \approx 24.15$ sec.
(d) $h(24.15) \approx 1{,}545.66$ ft.
(e) about 48.3 seconds: 24.15 up and 24.15 down

43. $h(x) = v_o x - 16x^2$ ft.
(a) $h'(x) = v_o - 32x$ ft/sec
(b) Max height when $x = v_o/32$: max height $= h(v_o/32) = v_o(v_o/32) - 16(v_o/32)^2 = (v_o)^2/64$ feet.
(c) Time aloft $= 2(v_o/32) = v_o/16$ seconds.

45. (a) $(v_o)^2/64 = 6.5$, so $v_o = 8\sqrt{6.5} \approx 20.396$ ft/sec.
(b) $2(v_o/32) = \dfrac{8\sqrt{6.5}}{16} \approx 1.27$ seconds.
(c) Max height $= \dfrac{(v_o)^2}{2g} = \dfrac{(8\sqrt{6.5})^2}{2(5.3)} = \dfrac{416}{10.6} \approx 39.25$ feet.

47. (a) $y' = -\dfrac{1}{x^2}$; $y'(2) = -1/4$ so $y - 1/2 = (-1/4)(x - 2)$ or $y = (-1/4)x + 1$.
(b) x–intercept at $x = 4$, y–intercept at $y = 1$
(c) $A = \frac{1}{2}(\text{base})(\text{height}) = \frac{1}{2}(4)(1) = 2$.

49. Since (1,4) and (3,14) are on the parabola, we need $a + b + c = 4$ and $9a + 3b + c = 14$. Then, subtracting the first equation from the second, $8a + 2b = 10$. $f'(x) = 2ax + b$ so $f'(3) = 6a + b = 9$, the slope of $y = 9x - 13$. Now solve the system $8a + 2b = 10$ and $6a + b = 9$ to get $a = 2$ and $b = -3$. Then use $a + b + c = 4$ to get $c = 5$. $a = 2, b = -3, c = 5$.

51. (a) $f(x) = x^3$ (b) $g(x) = x^3 + 1$
(c) If $h(x) = x^3 + C$ for any constant C, then $\mathbf{D}(h(x)) = 3x^2$.

53. (a) For $0 \le x \le 1$, $f'(x) = 1$ so $f(x) = x + C$. Since $f(0) = 0$, we know $C = 0$ and $f(x) = x$.
For $1 \le x \le 3$, $f'(x) = 2 - x$ so $f(x) = 2x - \frac{1}{2}x^2 + K$. Since $f(1) = 1$, we know
$K = -1/2$ and $f(x) = 2x - \frac{1}{2}x^2 - \frac{1}{2}$.
For $3 \le x \le 4$, $f'(x) = x - 4$ so $f(x) = \frac{1}{2}x^2 - 4x + L$. Since $f(3) = 1$, we know $L = 17/2$
and $f(x) = \frac{1}{2}x^2 - 4x + \frac{17}{2}$.
(b) This graph is a vertical shift, up 1 unit, of the graph in part (a).

Section 2.3

1. $\mathbf{D}(f^2(x)) = 2\cdot f^1(x)\cdot f'(x)$. At $x = 1$, $\mathbf{D}(f^2(x)) = 2(2)(3) = 12$.
$\mathbf{D}(f^5(x)) = 5\cdot f^4(x)\cdot f'(x)$. At $x = 1$, $\mathbf{D}(f^5(x)) = 5(2)^4(3) = 240$.
$\mathbf{D}(f^{1/2}(x)) = (1/2)\cdot f^{-1/2}(x)\cdot f'(x)$. At $x = 1$, $\mathbf{D}(f^{1/2}(x)) = (1/2)(2)^{-1/2}(3) = \frac{3}{2\sqrt{2}} = \frac{3\sqrt{2}}{4}$.

3.

x	f(x)	f '(x)	$\mathbf{D}(f^2(x))$	$\mathbf{D}(f^3(x))$	$\mathbf{D}(f^5(x))$
1	1	–1	–2	–3	–5
3	2	–3	–12	–36	–240

5. $f'(x) = 5\cdot(2x - 8)^4\cdot(2)$

7. $f'(x) = x\cdot 5\cdot(3x + 7)^4\cdot 3 + 1\cdot(3x + 7)^5 = (3x + 7)^4 \{ 15x + (3x + 7) \} = (3x + 7)^4\cdot(18x + 7)$

9. $f'(x) = \frac{1}{2}(x^2 + 6x - 1)^{-1/2}\cdot(2x + 6) = \frac{x + 3}{\sqrt{x^2 + 6x - 1}}$

11. (a) graph $h(t) = 3 - 2\sin(t)$ (b) When $t = 0$, $h(0) = 3$ feet.
(c) Highest = 5 feet above the floor. Lowest = 1 foot above the floor.
(d) $h(t) = 3 - 2\sin(t)$ feet, $v(t) = h'(t) = -2\cos(t)$ ft/sec, and $a(t) = v'(t) = 2\sin(t)$ ft/sec^2 .
(e) This spring oscillates forever. The motion of a real spring would "damp out" due to friction.

13. $K = \frac{1}{2}mv^2$ (a) If $h(t) = 5t$, then $v(t) = h'(t) = 5$. Then $K(1) = K(2) = \frac{1}{2}m(5)^2 = 12.5m$.
(b) If $h(t) = t^2$, then $v(t) = h'(t) = 2t$ so $v(1) = 2$ and $v(2) = 4$. Then $K(1) = \frac{1}{2}m(2)^2 = 2m$
and $K(2) = \frac{1}{2}m(4)^2 = 8m$.

15. $\frac{df}{dx} = x\cdot\mathbf{D}(\sin(x)) + \sin(x)\cdot\mathbf{D}(x) = x\cdot\cos(x) + \sin(x)$

17. $f'(x) = e^x - \sec(x)\cdot\tan(x)$

19. $f'(x) = -e^{-x} + \cos(x)$

21. $f'(x) = 7(x - 5)^6(1)$ so $f'(4) = 7(-1)^6(1) = 7$. Then $y - (-1) = 7(x - 4)$ and $y = 7x - 29$.

23. $f'(x) = \frac{1}{2}(25 - x^2)^{-1/2}(-2x) = \frac{-x}{\sqrt{25 - x^2}}$ so $f'(3) = \frac{-3}{4}$. Then $y - 4 = -\frac{3}{4}(x - 3)$ or $3x + 4y = 25$.

25. $f'(x) = 5(x - a)^4(1)$ so $f'(a) = 5(a - a)^4(1) = 0$. Then $y - 0 = 0(x - a)$ or $y = 0$.

27. $f'(x) = e^x$ so $f'(3) = e^3$. Then $y - e^3 = e^3 (x - 3)$.
x–intercept (y=0): $0 - e^3 = e^3 (x - 3)$ so $-1 = x - 3$ and $x = 2$.
(y–intercept (x=0): $y - e^3 = e^3 (0 - 3)$ so $y = -3e^3 + e^3 = -2e^3$.)

At (p, e^p), $f'(p) = e^p$ so $y - e^p = e^p (x - p)$.
x–intercept (y=0): $0 - e^p = e^p (x - p)$ so $-1 = x - p$ and $x = p - 1$.

29. $f'(x) = -\sin(x)$, $f''(x) = -\cos(x)$

31. $f'(x) = x^2 \cos(x) + 2x \sin(x)$,
$f''(x) = -x^2 \sin(x) + 2x \cos(x) + 2x \cos(x) + 2 \sin(x) = -x^2 \sin(x) + 4x \cos(x) + 2\sin(x)$

33. $f'(x) = e^x \cdot \cos(x) - e^x \cdot \sin(x)$, $f''(x) = -2\, e^x \cdot \sin(x)$

35. $q' =$ linear, $q'' =$ constant, $q''' = q^{(4)} = q^{(5)} = \ldots = 0$

37. $p^{(n)} =$ constant, $p^{(n+1)} = 0$ 39. $f(x) = 5e^x$ 41. $f(x) = (1 + e^x)^5$

43. No. $\displaystyle \lim_{h\to 0} \frac{f(0+h) - f(0)}{h} = \lim_{h\to 0} \frac{(0+h)\cdot \sin\left(\frac{1}{0+h}\right) - 0}{h} = \lim_{h\to 0} \sin\left(\frac{1}{h}\right)$

which does not exist.

(To see that this last limit does not exist, graph $\sin(1/h)$ for $-1 \le h \le 1$, or evaluate $\sin(1/h)$ for lots of small values of h, e.g., h 0.1, 0.01, 0.001, ...)

45. $(1 + \frac{1}{x})^x \approx 2.718 \ldots = e$ when x is large.

47. (a) $s_2 = 2.5$, $s_3 \approx 2.67$, $s_4 \approx 2.708$, $s_5 \approx 2.716$, $s_6 \approx 2.718$, $s_7 \approx 2.71825$, $s_8 \approx 2.718178$

(b) They are approaching e .

Section 2.4

1. If $f(x) = x^5$ and $g(x) = x^3 - 7x$, then $f \circ g(x) = (x^3 - 7x)^5$

3. If $f(x) = x^{5/2}$ and $g(x) = 2 + \sin(x)$, then $f \circ g(x) =$

$\sqrt{(2 + \sin(x))^5}$ (The pair $f(x) = \sqrt{x}$ and $g(x) = (2 + \sin(x))^5$ also work.)

5. If $f(x) = |x|$ and $g(x) = x^2 - 4$, then $f \circ g(x) = |x^2 - 4|$

7. (1) $y = u^5$, $u = x^3 - 7x$ (2) $y = u^4$, $u = \sin(3x - 8)$

(3) $y = u^{5/2}$, $u = 2 + \sin(x)$ (4) $y = 1/\sqrt{u}$, $u = x^2 + 9$

(5) $y = |u|$, $u = x^2 - 4$ (6) $y = \tan(u)$, $u = \sqrt{x}$

8. & 9.

x	f(x)	g(x)	f '(x)	g '(x)	(f∘g)(x)	(f∘g)' (x)
–2	2	–1	1	1	**1**	**0**
–1	1	2	0	2	**1**	**2**
0	–2	1	2	–1	**0**	**1**
1	0	–2	–1	2	**2**	**2**
2	1	0	1	–1	**–2**	**–2**

11. $g(2) \approx 2$, $g'(2) \approx -1$, $(f \circ g)(2) = f(g(2)) \approx f(2) \approx 1$

$f'(g(2)) \approx f'(2) \approx 0$, $(f \circ g)'(2) = f'(g(2)) \cdot g'(2) \approx 0$

13. $\mathbf{D}\left((1 - \frac{3}{x})^4 \right) = 4(1 - \frac{3}{x})^3 (\frac{3}{x^2})$

15. $\mathbf{D}\left(\frac{5}{\sqrt{2 + \sin(x)}} \right) = 5(\frac{-1}{2})(2 + \sin(x))^{-3/2} \cos(x) = \frac{-5\cos(x)}{2(2 + \sin(x))^{3/2}}$

17. $\mathbf{D}(x^2 \cdot \sin(x^2 + 3)) = x^2\{\cos(x^2 + 3)\}(2x) + \{\sin(x^2 + 3)\}(2x) = 2x\{x^2 \cdot \cos(x^2 + 3) + \sin(x^2 + 3)\}$

19. $\mathbf{D}\left(\frac{7}{\cos(x^3 - x)} \right) = \mathbf{D}(7 \sec(x^3 - x)) = 7(3x^2 - 1)\cdot\sec(x^3 - x)\cdot\tan(x^3 - x)$

21. $\mathbf{D}(e^x + e^{-x}) = e^x - e^{-x}$

23. $h(t) = 3 - \cos(2t)$ feet. (a) $h(0) = 2$ feet above the floor.

(b) $h(t) = 3 - \cos(2t)$ feet, $v(t) = h'(t) = 2\sin(2t)$ ft/sec , $a(t) = v'(t) = 4\cos(2t)$ ft/sec^2

(c) $K = \frac{1}{2} mv^2 = \frac{1}{2} m(2\sin(2t))^2 = 2m\cdot\sin^2(2t)$, $dK/dt = 8m\cdot\sin(2t)\cdot\cos(2t)$

25. $P(h) = 14.7 e^{-0.0000385h}$. (a) $P(0) = 14.7$ psi (pounds per square inch), $P(30{,}000) \approx 4.63$ psi

(b) $10 = 14.7 e^{-0.0000385h}$ so $\frac{10}{14.7} = e^{-0.0000385h}$ and $h = \frac{1}{-0.0000385} \ln(\frac{10}{14.7}) \approx 10{,}007$ ft.

(c) $dP/dh = 14.7(-0.0000385) e^{-0.0000385h}$ psi/ft

At $h = 2{,}000$ feet, $dP/dh = 14.7(-0.0000385) e^{-0.0000385(2{,}000)}$ psi/ft ≈ -0.000524 psi/ft.
Finally, $dP/dt = 500(-0.000524) \approx -0.262$ psi/minute

(d) If the temperature is constant, then (pressure)(volume) is a constant (from physics!) so a decrease in pressure results in an increase in volume.

27. $\frac{\mathbf{d}}{\mathbf{dz}} \sqrt{1 + \cos^2(z)} = \frac{2\cos(z)\{ -\sin(z) \}}{2\sqrt{1 + \cos^2(z)}} = \frac{-\sin(2z)}{2\sqrt{1 + \cos^2(z)}}$

29. $\frac{\mathbf{d}}{\mathbf{dx}} \tan(3x + 5) = 3\cdot\sec^2(3x + 5)$

31. $\mathbf{D}(\sin(\sqrt{x + 1})) = \{ \cos(\sqrt{x + 1}) \frac{1}{2\sqrt{x + 1}}$

33. $\frac{\mathbf{d}}{\mathbf{dx}} (e^{\sin(x)}) = e^{\sin(x)} \cdot\cos(x)$

35. $f(x) = \sqrt{x}$ so $\frac{d f(x)}{dx} = \frac{1}{2\sqrt{x}}$. $x(t) = 2 + \frac{21}{t}$ so $\frac{d x(t)}{dt} = -\frac{21}{t^2}$.

At $t = 3$, $x = 9$ and $\frac{d x(t)}{dt} = -\frac{21}{9} = -\frac{7}{3}$ so $\frac{d (f_\infty x)}{dt} = (\frac{1}{2\sqrt{9}})(-\frac{7}{3}) = -\frac{7}{18}$.

37. $f(x) = \tan^3(x)$ so $\frac{d f(x)}{dx} = 3\cdot\tan^2(x)\cdot\sec^2(x)$. $x(t) = 8$ so $\frac{d x(t)}{dt} = 0$.

At $t = 3$, $x = 8$ and $\frac{d x(t)}{dt} = 0$ so $\frac{d (f_\infty x)}{dt} = 0$.

39. $f(x) = \frac{1}{77} (7x - 13)^{11}$ 41. $f(x) = -\frac{1}{2} \cos(2x - 3)$ 43. $f(x) = e^{\sin(x)}$

45. Then $-2\sin(2x) = 2\cos(x)\{-\sin(x)\} - 2\sin(x)\cdot\cos(x)$ or $\sin(2x) = 2\sin(x)\cdot\cos(x)$.

47. $3\cos(3x) = 3\cos(x) - 12\sin^2(x)\cdot\cos(x)$

so $\cos(3x) = \cos(x)\{ 1 - 4\sin^2(x) \} = \cos(x)\{ 1 - 4 + 4\cos^2(x) \} = 4\cos^3(x) - 3\cos(x)$

49. $y' = 3Ax^2 + 2Bx$ 51. $y' = 2Ax\cdot\cos(Ax^2 + B)$ 53. $y' = \frac{Bx}{\sqrt{A + Bx^2}}$

55. $y' = B\cdot\sin(Bx)$ 57. $y' = -2Ax\cdot\sin(Ax^2 + B)$ 59. $y' = x(B\cdot e^{Bx}) + e^{Bx} = (Bx+1)\cdot e^{Bx}$

61. $y' = A \cdot e^{Ax} + A \cdot e^{-Ax}$ 63. $y' = \dfrac{A \cdot \sin(Bx) - Ax \cdot B \cdot \cos(Bx)}{\sin^2(Bx)}$

65. $y' = \dfrac{(Cx+D)A - (Ax+B)C}{(Cx+D)^2} = \dfrac{AD - BC}{(Cx+D)^2}$

67. (a) $y' = AB - 2Ax$, (b) $x = \dfrac{AB}{2A} = \dfrac{B}{2}$, (c) $y'' = -2A$.

69. (a) $y' = 2ABx - 3Ax^2 = Ax \cdot (2B - 3x)$, (b) $x = 0, 2B/3$, (c) $y'' = 2AB - 6Ax$.

71. (a) $y' = 3Ax^2 + 2Bx = x \cdot (3Ax + 2B)$, (b) $x = 0, \dfrac{-2B}{3A}$, (c) $y'' = 6Ax + 2B$.

73. $\dfrac{\mathbf{d}}{\mathbf{dx}}(\arctan(x^2)) = \dfrac{2x}{1+x^4}$

75. $\mathbf{D}(\arctan(e^x)) = \dfrac{1}{1+(e^x)^2} \cdot e^x = \dfrac{e^x}{1+e^{2x}}$

77. $\mathbf{D}(\arcsin(x^3)) = \dfrac{3x^2}{\sqrt{1-x^6}}$

79. $\dfrac{\mathbf{d}}{\mathbf{dt}}(\arcsin(e^t)) = \dfrac{1}{\sqrt{1-(e^t)^2}} \cdot e^t = \dfrac{e^t}{\sqrt{1-e^{2t}}}$

81. $\dfrac{\mathbf{d}}{\mathbf{dx}}(\ln(\sin(x))) = \dfrac{1}{\sin(x)} \cos(x) = \cot(x)$

83. $\dfrac{\mathbf{d}}{\mathbf{ds}}(\ln(e^s)) = \dfrac{1}{e^s} \cdot e^s = 1$, or $\dfrac{\mathbf{d}}{\mathbf{ds}}(\ln(e^s)) = \dfrac{\mathbf{d}}{\mathbf{ds}}(s) = 1$

Section 2.5

1. $\mathbf{D}(\ln(5x)) = \dfrac{1}{5x} 5 = \dfrac{1}{x}$

3. $\mathbf{D}(\ln(x^k)) = \dfrac{1}{x^k} k x^{k-1} = \dfrac{k}{x}$

5. $\mathbf{D}(\ln(\cos(x))) = \dfrac{1}{\cos(x)}(-\sin(x)) = -\tan(x)$

7. $\mathbf{D}(\log_2(5x)) = \dfrac{1}{5x \ln(2)}(5) = \dfrac{1}{x \ln(2)}$

9. $\mathbf{D}(\ln(\sin(x))) = \dfrac{1}{\sin(x)}(\cos(x)) = \cot(x)$

11. $\mathbf{D}(\log_2(\sin(x))) = \dfrac{1}{\sin(x)} \dfrac{1}{\ln(2)}(\cos(x)) = \dfrac{\cot(x)}{\ln(2)}$

13. $\mathbf{D}(\log_5(5^x)) = \mathbf{D}(x) = 1$

15. $\mathbf{D}(x \ln(3x)) = x \cdot \dfrac{1}{3x} \cdot 3 + \ln(3x) = 1 + \ln(3x)$

17. $\mathbf{D}\left(\dfrac{\ln(x)}{x}\right) = \dfrac{x \cdot \frac{1}{x} - \ln(x) \cdot 1}{x^2} = \dfrac{1 - \ln(x)}{x^2}$

19. $\mathbf{D}(\ln((5x-3)^{1/2})) = \dfrac{1}{(5x-3)^{1/2}} \cdot \mathbf{D}((5x-3)^{1/2}) = \dfrac{1}{(5x-3)^{1/2}} \cdot \dfrac{1}{2} \cdot (5x-3)^{-1/2} \cdot \mathbf{D}(5x-3) = \dfrac{5}{2} \cdot \dfrac{1}{5x-3}$.

21. $\dfrac{d}{dw}(\cos(\ln(w))) = \{-\sin(\ln(w))\} \dfrac{1}{w} = \dfrac{-\sin(\ln(w))}{w}$

23. $\dfrac{d}{dt}(\ln(\sqrt{t+1})) = \dfrac{1}{2(t+1)}$

25. $\mathbf{D}(5^{\sin(x)}) = 5^{\sin(x)} \ln(5) \cos(x)$

27. $\dfrac{d}{dx} \ln(\sec(x) + \tan(x)) = \dfrac{1}{\sec(x) + \tan(x)}(\sec(x)\tan(x) + \sec^2(x)) = \sec(x)$

29. $f(x) = \ln(x)$, $f'(x) = \frac{1}{x}$. Let $P = (p, \ln(p))$. Then we must satisfy $y - \ln(p) = \frac{1}{p}(x - p)$ with $x = 0$ and $y = 0$: $-\ln(p) = -1$ so $p = e$ and $P = (e, 1)$.

31. $p(t) = 100(1 + Ae^{-t})^{-1}$. $\frac{d}{dt}\ p(t) = 100(-1)(1 + Ae^{-t})^{-2}\ (Ae^{-t}(-1)) = \frac{100\,Ae^{-t}}{(1 + Ae^{-t})^2}$.

33. $f(x) = 8\ln(x)$ + any constant

35. $f(x) = \ln(3 + \sin(x))$ + any constant

37. $g(x) = \frac{3}{5}\ e^{5x}$ + any constant

39. $f(x) = e^{x^2}$ + any constant

41. $h(x) = \ln(\sin(x))$ + any constant

43. A: $(t, 2t + 1)$, B: $(t^2, 2t^2 + 1)$

(a) When $t = 0$, A is at (0,1) and B is at (0,1). When $t = 1$, A is at (1,3), B is at (1,3)

(b) graph

(c) $dy/dx = 2$ for each, since $y = 2x + 1$.

(d) A: $dx/dt = 1$, $dy/dt = 2$ so speed $= \sqrt{1^2 + 2^2} = \sqrt{5}$

B: $dx/dt = 2t$, $dy/dt = 4t$ so speed $= \sqrt{(2t)^2 + (4t)^2} = 2\sqrt{5}\ t$. At t=1, B's speed is $2\sqrt{5}$.

(e) This robot moves on the same path $y = 2x + 1$, but it moves to the right and up for about 1.57 minutes, reverses its direction and returns to its starting point, then continues left and down for another 1.57 minutes, reverses, and continues to oscillate.

45. When t=1, dx/dt = +, dy/dt= –, dy/dx= – . WHen t=3, dx/dt = – , dy/dt= – , dy/dx= + .

47. $x(t) = R(t - \sin(t))$
$y(t) = R(1 - \cos(t))$

(a) graph

(b) $dx/dt = R(1 - \cos(t))$, $dy/dt = R\sin(t)$, so $\frac{dy}{dx} = \frac{\sin(t)}{1 - \cos(t)}$.

When $t = \pi/2$, then $dx/dt = R$, $dy/dt = R$ so $dy/dx = 1$ and speed $= \sqrt{R^2 + R^2} = R\sqrt{2}$

When $t = \pi$, $dx/dt = 2R$, $dy/dt = 0$ so $dy/dx = 0$ and speed $= \sqrt{(2R)^2 + 0} = 2R$.

49. (a) The ellipse $(\frac{x}{3})^2 + (\frac{y}{5})^2 = 1$.

(b) The ellipse $(\frac{x}{A})^2 + (\frac{y}{B})^2 = 1$ if $A \neq 0$ and $B \neq 0$. (c) $(3\cdot\cos(t), -5\cdot\sin(t))$ works.

If $A = 0$, the motion is oscillatory along the x–axis.
If $B = 0$, the motion is oscillatory along the y–axis.

Section 2.6

1. $V = \frac{4}{3}\pi r^3$ ($r = r(t)$) so $\frac{dV}{dt} = 4\pi r^2 \frac{dr}{dt}$.

When $r = 3$ in., $\frac{dr}{dt} = 2$ in/min, so $\frac{dV}{dtt} = 4\pi(3\text{ in})^2(2\text{ in/min}) = 72\pi\text{ in}^3\text{/min} \approx 226.19\text{ in}^3\text{/min}$.

3. $b = 15$ in., $h = 13$ in., $\frac{db}{dt} = 3$ in/hr, $\frac{dh}{dt} = -3$ in/hr.

(a) $A = \frac{1}{2}\ bh$ so $\frac{dA}{dt} = \frac{1}{2}\{ b\frac{dh}{dt} + h\frac{db}{dt} \} = \frac{1}{2}\{ (15\text{ in})(-3\text{ in/hr}) + (13\text{ in})(3\text{ in/hr}) \} < 0$
so A is decreasing.

(b) Hypotenuse $C = \sqrt{b^2 + h^2}$ so $\frac{dC}{dt} = \frac{b\frac{db}{dt} + h\frac{dh}{dt}}{\sqrt{b^2 + h^2}} = \frac{15(3) + 13(-3)}{\sqrt{15^2 + 13^2}} > 0$ so C is increasing.

(c) Perimeter $P = b + h + C$ so $\frac{dP}{dt} = \frac{db}{dt} + \frac{dh}{dt} + \frac{dC}{dt} = (3) + (-3) + \frac{6}{\sqrt{394}} > 0$ so P is increasing.

5. (a) $P = 2x + 2y$ so $\frac{dP}{dt} = 2\frac{dx}{dt} + 2\frac{dy}{dt} = 2(3 \text{ ft/sec}) + 2(-2 \text{ ft/sec}) = 2 \text{ ft/sec}$.

(b) $A = xy$ so $\frac{dA}{dt} = x\frac{dy}{dt} + y\frac{dx}{dt} = (12 \text{ ft})(-2 \text{ ft/sec}) + (8 \text{ ft})(3 \text{ ft/sec}) = 0 \text{ ft}^2\text{/sec}$.

7. $V = \pi r^2 h = \pi r^2 (1/3)$ so $\frac{dV}{dt} = \frac{2\pi}{3} r \frac{dr}{dt}$.

When $r = 50$ ft. and $\frac{dr}{dt} = 6$ ft/hr, then $\frac{dV}{dt} = \frac{2\pi}{3}(50 \text{ ft})(6 \text{ ft/hr}) = 200\pi \text{ ft}^3\text{/hr} \approx 628.32 \text{ ft}^3\text{/hr}$.

9. $w(t) = h(t)$ for all t so $\frac{dw}{dt} = \frac{dh}{dt}$. $V = \frac{1}{3}\pi r^2 h$ and $r = w/2 = h/2$ so $V = \frac{1}{3}\pi(h/2)^2 h = \frac{1}{12}\pi h^3$

and $\frac{dV}{dt} = \frac{1}{4}\pi h^2 \frac{dh}{dt}$. When $h = 500$ ft and $\frac{dh}{dt} = 2$ ft/hr, then $\frac{dV}{dt} = \frac{1}{4}\pi(500)^2(2) = 125{,}000\pi \text{ ft}^3\text{/hr}$.

11. Let x be the distance from the lamp post to the person, and L be the length of the shadow, both in feet. By similar triangles, $\frac{L}{6} = \frac{x}{8}$ so $L = \frac{3}{4}x$. $\frac{dx}{dt} = 3$ ft/sec.

(a) $\frac{dL}{dt} = \frac{3}{4}\frac{dx}{dt} = \frac{3}{4}(3 \text{ ft/sec}) = 2.25$ ft/sec. (b) $\frac{d}{dt}(x + L) = \frac{dx}{dt} + \frac{dL}{dt} = 5.25$ ft/sec.

(The value of x does not enter into the calculations.)

13. (a) $\sin(35^o) = \frac{h}{500}$ so $h = 500 \cdot \sin(35^o) \approx 287$ ft.

(b) L = length of the string so $h = L \cdot \sin(35^o)$ and $\frac{dh}{dt} = \sin(35^o)\frac{dL}{dt} = \sin(35^o)(10 \text{ ft/sec}) \approx 5.7$ ft/sec.

15. $V = s^3 - \frac{4}{3}\pi r^3$. $r = \frac{1}{2}(\text{diameter}) = 4$ ft, $\frac{dr}{dt} = 1$ ft/hr, $s = 12$ ft, $\frac{ds}{dt} = 3$ ft/hr.

$\frac{dV}{dt} = 3s^2\frac{ds}{dt} - 4\pi r^2\frac{dr}{dt} = 3(12 \text{ ft})^2(3 \text{ ft/hr}) - 4\pi(4 \text{ ft})^2(1 \text{ ft/hr}) \approx 1094.94 \text{ ft}^3\text{/hr}$. The volume is **increasing** at about $1094.94 \text{ ft}^3\text{/hr}$.

17. Given: $\frac{dV}{dt} = k \cdot 2\pi r^2$ with k constant. We also have $V = \frac{2}{3}\pi r^3$ so $\frac{dV}{dt} = 2\pi r^2 \frac{dr}{dt}$.

Therefore, $k \cdot 2\pi r^2 = 2\pi r^2 \frac{dr}{dt}$ so $\frac{dr}{dt} = k$. The radius r is changing at a constant rate.

19. (a) $A = 5x$ (b) $\frac{dA}{dx} = 5$ for all $x > 0$. (c) $A = 5t^2$

(d) $\frac{dA}{dt} = 10t$. When $t = 1$, $\frac{dA}{dt} = 10$; when $t = 2$, $\frac{dA}{dt} = 20$; when $t = 3$, $\frac{dA}{dt} = 30$.

(e) $A = 10 + 5 \cdot \sin(t)$. $\frac{dA}{dt} = 5 \cdot \cos(t)$.

21. (a) $\tan(10^o) = \frac{40}{x}$ so $x = \frac{40}{\tan(10^o)} \approx 226.9$ ft.

For parts (b) and (c) we need to work in radians since our formulas for the derivatives of the trigonometric functions assume that the angles are measured in radians: $360^o \approx 2\pi$ radians so $10^o \approx 0.1745$ radians and $2^o \approx 0.0349$ radians,

(b) $x = \frac{40}{\tan(\theta)} = 40\cot(\theta)$ so $\frac{dx}{dt} = -40\csc^2(\theta)\frac{d\theta}{dt}$ and

$$\frac{d\theta}{dt} = \frac{\sin^2(\theta)\frac{dx}{dt}}{-40} = \frac{\sin^2(0.1745)(-25)}{-40} \approx \frac{(0.1736)^2(-25)}{-40} \approx 0.0188 \text{ radians/min} \approx 1.079^o\text{/min}.$$

(c) $\frac{dx}{dt} -40\csc^2(\theta)\frac{d\theta}{dt} = -40\frac{1}{\sin^2(\theta)}\frac{d\theta}{dt} \approx -40\frac{1}{(0.1736)^2}(0.0349) \approx -46.3$ ft/min. (The " – " indicates the distance to the sign is decreasing: you are approaching the sign.) Your speed is 46.3 ft/min.

Section 2.7

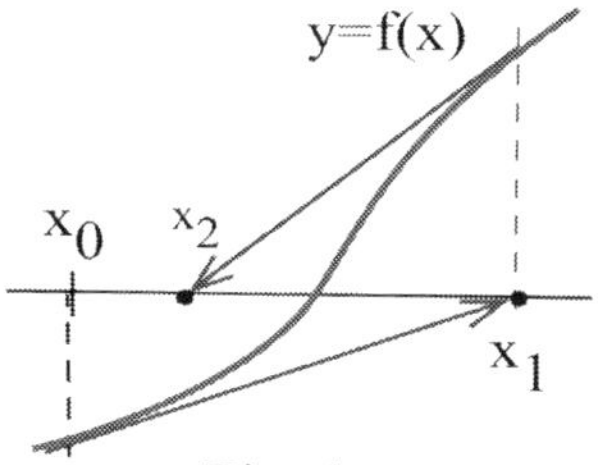

Fig. 1

1. See Fig. 15

3. $x_o = 1$: a. $x_o = 5$: b.

5. $x_o = 1$: 1, 2, 1, 2, 1, ...
   $x_o = 5$: x_1 is undefined since $f'(5) = 0$.

7. If f is differentiable, then $f'(x_o) = 0$ and $x_1 = x_o - \frac{f(x_o)}{f'(x_o)}$ is undefined.

9. $f(x) = x^4 - x^3 - 5$, $x_o = 2$. $f'(x) = 4x^3 - 3x^2$.

   Then $x_1 = 2 - \frac{3}{20} = \frac{37}{20} = 1.85$ and $x_2 = 1.85 - \frac{1.85^4 - 1.85^3 - 5}{4(1.85)^3 - 3(1.85)^2} \approx 1.824641$.

11. $f(x) = x - \cos(x)$, $f'(x) = 1 + \sin(x)$, $x_o = 0.7$. Then $x_1 = 0.7394364978$, $x_2 = 0.7390851605$, and root ≈ 0.74 .

13. $\frac{x}{x+3} = x^2 - 2$ so we can use $f(x) = x^2 - 2 - \frac{x}{x+3}$. If $x_o = -4$, then the iterates $x_n \to -3.3615$. If $x_o = -2$, then $x_n \to -1.1674$. If $x_o = 2$, then the iterates $x_n \to 1.5289$.

15. $x^5 - 3 = 0$ and $x_o = 1$. Then $x_n \to 1.2457$.

17. $f(x) = x^3 - A$ so $f'(x) = 3x^2$.

    Then $x_{n+1} = x_n - \frac{(x_n)^3 - A}{3(x_n)^2} = x_n - \frac{x_n}{3} + \frac{A}{3(x_n)^2} = \frac{1}{3}\left\{ 2x_n + \frac{A}{(x_n)^2} \right\}$.

19. (a) $2(0) - \text{INT}(2(0)) = 0 - 0 = 0$.

    (b) $2(1/2) - \text{INT}(2(1/2)) = 1 - 1 = 0$.
    $2(1/4) - \text{INT}(2(1/4)) = 1/2 - 0 = 1/2 \to 0$.
    $2(1/8) - \text{INT}(2(1/8)) = 1/4 - 0 = 1/4 \to 1/2 \to 0$.
    $2(1/2^n) - \text{INT}(2(1/2^n)) = 1/2^{n-1} - 0 = 1/2^{n-1} \to 1/2^{n-2} \to \ldots \to 1/4 \to 1/2 \to 0$.

21. (a) If $0 \le x \le 1/2$, then f stretches x to twice its value, 2x .
    If $1/2 < x \le 1$, then f stretches x to twice its value (2x) and "folds" the part above the value 1 (2x – 1) to below 1: $1 - (2x - 1) = 2 - 2x$.

    (b) f(2/3) = 2/3 .
    f(2/5) = 4/5, f (4/5) = 2/5 , and the values continues to cycle.
    f(2/7) = 4/7, f(4/7) = 6/7, f(6/7) = 2/7, and the values continues to cycle.
    f(2/9) = 4/9, f(4/9) = 8/9, f(8/9) = 2/9, and the values continues to cycle.

    (c) 0.1, 0.2, **0.4**, 0.8, **0.4**, 0.8, and the pair of values 0.4 and 0.8 continue to cycle.
    0.105, 0.210, 0.42, 0.84, **0.32**, 0.64, 0.72, 0.56, 0.88, 0.24, 0.48, 0.96, 0.08, 0.16, **0.32** , ...
    0.11, 0.22, 0.44, **0.88**, 0.24, 0.48, 0.96, 0.08, 0.16, 0.32, 0.64, 0.72, 0.56, **0.88**, ...

    (d) Probably so.

Section 2.8

1. See Fig. 1.

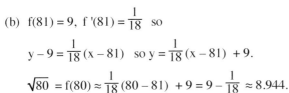

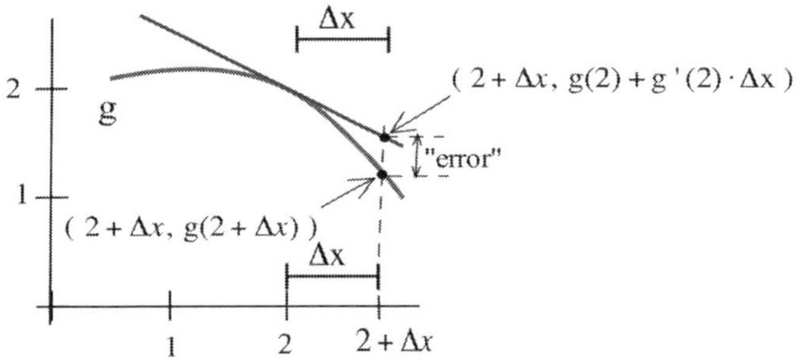

Fig. 1

3. (a) $f(4) = 2, f'(4) = \frac{1}{2\sqrt{4}} = \frac{1}{4}$.

Then $y - 2 = \frac{1}{4}(x - 4)$ so $y = \frac{1}{4}x + 1$.

$\sqrt{4.2} = f(4.2) \approx \frac{1}{4}(4.2) + 1 = 2.05$.

(b) $f(81) = 9$, $f'(81) = \frac{1}{18}$ so

$y - 9 = \frac{1}{18}(x - 81)$ so $y = \frac{1}{18}(x - 81) + 9$.

$\sqrt{80} = f(80) \approx \frac{1}{18}(80 - 81) + 9 = 9 - \frac{1}{18} \approx 8.944$.

(c) $f(0) = 0, f'(0) = 1$ so $y - 0 = 1(x - 0)$ and $y = x$. Then $\sin(0.3) = f(0.3) \approx 0.3$.

4. (a) $f(1) = 0, f'(1) = 1$ so $y - 0 = 1(x - 1)$. Then $\ln(1.3) = f(1.3) \approx 1.3 - 1 = 0.3$.

(b) $f(0) = 1, f'(0) = 1$ so $y - 1 = 1(x - 0)$ and $y = 1 + x$. Then $e^{0.1} = f(0.1) \approx 1.1$.

(c) $f(1) = 1, f'(1) = 5$ so $y - 1 = 5(x - 1)$ or $y = 1 + 5(x-1)$. Then $(1.03)^5 \approx 1 + 5(0.03) = 1.15$.

5. $f(x) = (1 + x)^n$, $f'(x) = n(1 + x)^{n-1}$, $f(0) = 1$ and $f'(0) = n$. Then $y - 1 = n(x - 0)$ or $y = 1 + nx$.

Therefore, $(1 + x)^n \approx 1 + nx$ (when x is close to 0).

6. (a) $f(x) = (1 - x)^n$, $f'(x) = -n(1 - x)^{n-1}$, and $f'(0) = -n$. Then $y - 1 = -n(x - 0)$ or $y = 1 - nx$.

Therefore, $(1 - x)^n \approx 1 - nx$ (when x is close to 0).

(b) $f(x) = \sin(x)$, $f'(x) = \cos(x)$, and $f'(0) = 1$. Then $y - 0 = 1(x - 0)$ or $y = x$.

Therefore, $\sin(x) \approx x$ for x close to 0.

(c) $f(x) = e^x$, $f'(x) = e^x$, and $f'(0) = 1$. Then $y - 1 = 1(x - 0)$ or $y = x + 1$. $e^x \approx x + 1$.

7. (a) $f(x) = \ln(1 + x)$, $f'(x) = \frac{1}{1+x}$, and $f'(0) = 1$. Then $y - 0 = 1(x - 0)$ so $y = x$ and $\ln(1 + x) \approx x$.

(b) $f(x) = \cos(x)$, $f'(x) = -\sin(x)$, and $f'(0) = 0$. Then $y - 1 = 0(x - 0)$ so $y = 1$ and $\cos(x) \approx 1$.

(c) $f(x) = \tan(x)$, $f'(x) = \sec^2(x)$, and $f'(0) = 1$. Then $y - 0 = 1(x - 0)$ so $y = x$ and $\tan(x) \approx x$.

(d) $f(x) = \sin(\frac{\pi}{2} + x)$, $f'(x) = \cos(\frac{\pi}{2} + x)$, and $f'(0) = \cos(\frac{\pi}{2} + 0) = 0$. Then $y - 1 = 0(x - 0)$ so $y = 1$

and $\sin(\frac{\pi}{2} + x) \approx 1$.

9. (a) Area $A(x) = (base)(height) = x(x^2 + 1) = x^3 + x$. Then $A'(x) = 3x^2 + 1$ so $A'(2) = 13$

Then $\Delta A \approx A'(2) \cdot \Delta x = (13)(2.3 - 2) = 3.9$

(b) (base)(height) $= (2.3)((2.3)^2 + 1) = 14.467$. Then the actual difference $= 14.467 - (2)(2^2+1) = 4.467$.

11. $V = \pi r^2 h = 2\pi r^2$ and $\Delta V = 2\pi 2r\ \Delta r = 4\pi r\ \Delta r$. Since $V = 2\pi r^2 = 47.3$, we have $r = \sqrt{47.3/(2\pi)} \approx 2.7437$ cm. We know $\Delta V = \pm 0.1$ so, using $\Delta V = 4\pi r\ \Delta r$, we have $\pm 0.1 = 4\pi(2.7437)\Delta r$ and $\Delta r = \dfrac{\pm 0.1}{4\pi(2.7437)} \approx \pm\ 0.0029$ cm. The required tolerance is $\pm\ 0.0029$ cm. (C.W. comment: "A coin 2 cm high is one hell of a coin! My eyeball estimate is that 47.3 cm^3 of gold weighs around 2 pounds.")

13. $V = x^3$. When $V = 87$, $x \approx 4.431$ cm. $\Delta V = 3x^2\ \Delta x$ so $\Delta x = \dfrac{\Delta V}{3x^2} \approx \dfrac{2}{3(4.431)^2} = 0.034$ cm.

15. $P = 2\pi\sqrt{\dfrac{L}{g}}$ with $g = 32\ ft/sec^2$. (a) $P = 2\pi\sqrt{\dfrac{2}{32}} = \dfrac{\pi}{2} \approx 1.57$ seconds.

(b) $1 = 2\pi\sqrt{\dfrac{L}{32}}$ so $L = \dfrac{8}{\pi^2} \approx 0.81$ feet.

(c) $dP = \dfrac{2\pi}{\sqrt{32}}\ \dfrac{1}{2\sqrt{2}}\ dL = \dfrac{2\pi}{4\sqrt{2}}\ \dfrac{1}{2\sqrt{2}}\ (0.1) \approx 0.039$ seconds.

(d) 2 in/hr = 1/6 ft/hr = $\dfrac{1}{21600}$ ft/sec. $\dfrac{dP}{dt} = \dfrac{2\pi}{4\sqrt{2}}\ \dfrac{1}{2\sqrt{24}}\ \dfrac{1}{21600} \approx 5.25 \times 10^{-6}$.

17. (a) $df = f\,'(2)\ dx \approx (0)(1) = 0$ (b) $df = f\,'(4)\ dx \approx (0.3)(-1) = -0.3$ (c) $df = f\,'(3)\ dx \approx (0.5)(2) = 1$

19. (a) $f(x) = x^2 - 3x$. $f\,'(x) = 2x - 3$. $df = (2x - 3)\ dx$. (b) $f(x) = e^x$. $f\,'(x) = e^x$. $df = e^x\ dx$.

(c) $f(x) = \sin(5x)$. $f\,'(x) = 5\cos(5x)$. $df = 5\cos(5x)\ dx$.

(d) $f(x) = x^3 + 2x$. $f\,'(x) = 3x^2 + 2$. $df = (3x^2 + 2)\ dx$. When $x = 1$ and $dx = 0.2$, $df = (3\cdot 1^2 + 2)\ (0.2) = 1$.

(e) $f\,'(x) = 1/x$. $df = \dfrac{1}{x}\ dx$. When $x = e$ and $dx = -0.1$, $df = = \dfrac{1}{e}(-0.1) = -\dfrac{1}{10e}$.

(f) $f(x) = \sqrt{2x + 5}$. $f\,'(x) = \dfrac{1}{\sqrt{2x + 5}}$. $df = \dfrac{1}{\sqrt{2x + 5}}\ dx$. When $x = 22$ and $dx = 3$, $df = \dfrac{1}{\sqrt{49}}\ (3) = \dfrac{3}{7}$.

Section 2.9

1. (a) $x^2 + y^2 = 100$ so $2x + 2y\mathbf{y}' = 0$ and $\mathbf{y}' = -x/y$. At (6,8), $y' = -6/8 = -3/4$.

(b) $y = \sqrt{100 - x^2}$ so $y' = \dfrac{-x}{\sqrt{100 - x^2}}$. At (6,8), $y' = \dfrac{-6}{\sqrt{100 - 36}} = \dfrac{-6}{8} = -\dfrac{3}{4}$.

3. (a) $x^2 - 3xy + 7y = 5$ so $2x - 3(y + x\mathbf{y}') + 7\mathbf{y}' = 0$ and $\mathbf{y}' = \dfrac{3y - 2x}{7 - 3x}$. At (2,1), $y' = \dfrac{3-4}{7-6} = -1$.

(b) $y = \dfrac{5 - x^2}{7 - 3x}$ so $y' = \dfrac{(7 - 3x)(-2x) - (5 - x^2)(-3)}{(7 - 3x)^2}$. At (2,1), $y' = \dfrac{(1)(-4) - (1)(-3)}{(1)^2} = -1$.

5. (a) $\dfrac{x^2}{9} + \dfrac{y^2}{16} = 1$ so $\dfrac{2x}{9} + \dfrac{2y}{16}\ \mathbf{y}' = 0$ and $\mathbf{y}' = -\dfrac{16}{9}\dfrac{x}{y}$. At (0,4), $y' = 0$.

(b) $y = 4\sqrt{1 - \dfrac{x^2}{9}} = \dfrac{4}{3}\sqrt{9 - x^2}$ so $y' = \dfrac{4}{3}\ \dfrac{-x}{\sqrt{9 - x^2}}$. At (0,4), $y' = 0$.

7. (a) $\ln(y) + 3x - 7 = 0$ so $\frac{1}{y}\,\mathbf{y}' + 3 = 0$ and $\mathbf{y}' = -3y$. At $(2,e)$, $y' = -3e$.

(b) $y = e^{7-3x}$ so $y' = -3\,e^{7-3x}$. At $(2,e)$, $y' = -3\,e^{7-6} = -3e$.

9. (a) $x^2 - y^2 = 16$ so $2x - 2y\mathbf{y}' = 0$ and $\mathbf{y}' = x/y$. At $(5,-3)$, $y' = -5/3$.

(b) The point $(5,-3)$ is on the bottom half of the circle so

$$y = -\sqrt{x^2 - 16} \text{ . Then } y' = -\frac{x}{\sqrt{x^2 - 16}} \text{ . At } (5,-3),\ y' = -\frac{5}{\sqrt{25 - 16}} = -\frac{5}{3} \text{ .}$$

11. $x = 4y - y^2$ so, differentiating each side, $1 = 4\mathbf{y}' - 2y\,\mathbf{y}'$ and $\mathbf{y}' = \frac{1}{4 - 2y}$.

At $(3,1)$, $y' = \frac{1}{4 - 2(1)} = \frac{1}{2}$. At $(3,3)$, $y' = \frac{1}{4 - 2(3)} = -\frac{1}{2}$.

At $(4,2)$, $y' = \frac{1}{4 - 2(2)}$ is undefined (the tangent line is vertical).

13. $x = y^2 - 6y + 5$. Differentiating each side, $1 = 2y\,\mathbf{y}' - 6\,\mathbf{y}'$ and $\mathbf{y}' = \frac{1}{2y - 6}$.

At $(5,0)$, $y' = \frac{1}{2(0) - 6} = -\frac{1}{6}$. At $(5,6)$, $y' = \frac{1}{2(6) - 6} = \frac{1}{6}$.

At $(-4,3)$, $y' = \frac{1}{2(3) - 6}$ is undefined (vertical tangent line).

15. $3y^2\mathbf{y}' - 5\mathbf{y}' = 10x$ so $\mathbf{y}' = \frac{10x}{3y^2 - 5}$. At $(1,3)$, $m = 10/22 = 5/11$.

17. $y^2 + \sin(y) = 2x - 6$ so $2y\mathbf{y}' + \cos(y)\mathbf{y}' = 2$ and $\mathbf{y}' = \frac{2}{2y + \cos(y)}$. At $(3,0)$, $m = \frac{2}{0+1} = 2$.

19. $e^y + \sin(y) = x^2 - 3$ so $e^y\mathbf{y}' + \cos(y)\mathbf{y}' = 2x$ and $\mathbf{y}' = \frac{2x}{e^y + \cos(y)}$. At $(2,0)$, $m = \frac{4}{1+1} = 2$.

21. $x^{2/3} + y^{2/3} = 5$ so $\frac{2}{3}\,x^{-1/3} + \frac{2}{3}\,y^{-1/3}\mathbf{y}' = 0$ and $\mathbf{y}' = -(x/y)^{-1/3}$. At $(8,1)$, $m = -(8/1)^{-1/3} = -\frac{1}{2}$.

23. Using implicit differentiation, $2x + x\mathbf{y}' + y + 2y\mathbf{y}' + 3 - 7\mathbf{y}' = 0$ so $\mathbf{y}' = \mathbf{\frac{-2x - y - 3}{x + 2y - 7}}$.

At $(1,2)$, $y' = \frac{-2(1) - (2) - 3}{(1) + 2(2) - 7} = \frac{7}{2}$.

25. $y = Ax^2 + Bx + C$ so $y' = 2Ax + B$. (explicitly)

$x = Ay^2 + By + C$ so (implicitly) $1 = 2Ay\mathbf{y}' + B\mathbf{y}'$ and $\mathbf{y}' = \frac{1}{2Ay + B}$.

27. $Ax^2 + Bxy + Cy^2 + Dx + Ey + F = 0$.

Then $2Ax + Bx\mathbf{y}' + By + 2Cy\mathbf{y}' + D + E\mathbf{y}' = 0$ so $\mathbf{y}' = \frac{-2Ax - By - D}{Bx + 2Cy + E}$.

28. $x^2 + y^2 = r^2$ so $2x + 2yy' = 0$ and $y' = -\frac{x}{y}$. The slope of the tangent line is $-x/y$. The slope of the line through the points $(0,0)$ and (x,y) is y/x, so the slopes of the lines are negative reciprocals of each other and the lines are perpendicular.

29. From problem 23, $y' = \frac{-2x - y - 3}{x + 2y - 7}$ so $y' = 0$ when $-2x - y - 3 = 0$ and $y = -2x - 3$.

Substituting $y = -2x - 3$ into the original equation, we have

$x^2 + x(-2x - 3) + (-2x - 3)^2 + 3x - 7(-2x - 3) + 4 = 0$ so

$3x^2 + 26x + 34 = 0$ and $x = \frac{-26 \pm \sqrt{26^2 - 4(3)(34)}}{2(3)} = \frac{-26 \pm \sqrt{268}}{6} \approx -1.605$ and -7.062.

If $x \approx -1.605$ (point A), then $y = -2x - 3 \approx -2(-1.605) - 3 = 0.21$. Point A is **(–1.605, 0.21)**.

If $x \approx -7.062$ (point C), then $y = -2x - 3 \approx -2(-7.062) - 3 = 11.124$. Point C is **(–7.062, 11.124)**.

31. From the solution to problem 29, point C is (–7.062, 11.124).

At D, $y' = \frac{-2x - y - 3}{x + 2y - 7}$ is undefined so $x + 2y - 7 = 0$ and $x = 7 - 2y$.

Substituting $x = 7 - 2y$ into the original equation, we have

$(7-2y)^2 + (7-2y)y + y^2 + 3(7-2y) - 7y + 4 = 0$ so (simplifying) $3y^2 - 34y + 74 = 0$. Then

$y = \frac{34 \pm \sqrt{34^2 - 4(3)(74)}}{2(3)} = \frac{34 \pm \sqrt{268}}{6} \approx 8.398$ and 2.938.

If $y \approx 2.938$ (point D), then $x = 7 - 2y \approx 7 - 2(2.938) = 1.124$. Point D is **(1.124, 2.938)**.

Point B is (–9.79, 8.395).

33. (a) $y = (x^2 + 5)^7 (x^3 - 1)^4$.

$y' = (x^2 + 5)^7 (4)(x^3 - 1)^3 (3x^2) + (7)(x^2 + 5)^6 (2x)(x^3 - 1)^4$

$= (x^2 + 5)^6 (x^3 - 1)^3 (2x)\{6x^3 + 30x + 7x^3 - 7\} = (x^2 + 5)^6 (x^3 - 1)^3 (2x)\{13x^3 + 30x - 7\}$.

(b) $\ln(y) = 7\ln(x^2 + 5) + 4\ln(x^3 - 1)$

$\frac{y'}{y} = \frac{14x}{x^2 + 5} + \frac{12x^2}{x^3 - 1}$ so

$y' = y\left\{ \frac{14x}{x^2 + 5} + \frac{12x^2}{x^3 - 1} \right\} = (x^2 + 5)^7 (x^3 - 1)^4 \left\{ \frac{14x}{x^2 + 5} + \frac{12x^2}{x^3 - 1} \right\}$ and this is the same as in part (a). (Really it is.)

35. (a) $y = x^5 (3x + 2)^4$. $y' = x^5\, \mathbf{D}((3x + 2)^4) + (3x + 2)^4 \mathbf{D}(x^5) = x^5 \cdot 4(3x + 2)^3 (3) + (3x + 2)^4 \cdot (5x^4)$

(b) $\ln(y) = 5\ln(x) + 4\ln(3x + 2)$.

$\frac{y'}{y} = \frac{5}{x} + \frac{12}{3x + 2}$ so $y' = y\left\{ \frac{5}{x} + \frac{12}{3x + 2} \right\} = \{ x^5 (3x + 2)^4 \}\left\{ \frac{5}{x} + \frac{12}{3x + 2} \right\}$

and this is the same as in part (a).

37. (a) $y = e^{\sin(x)}$ so $y' = e^{\sin(x)} \cos(x)$.

(b) $\ln(y) = \sin(x)$ so $\frac{y'}{y} = \cos(x)$ and $y' = y \cos(x) = e^{\sin(x)} \cos(x)$.

39. (a) $y = \sqrt{25 - x^2}$ so $y' = \frac{-x}{\sqrt{25 - x^2}}$.

(b) $\ln(y) = \frac{1}{2}\ln(25 - x^2)$ so $\frac{y'}{y} = \frac{1}{2}\frac{-2x}{25 - x^2} = \frac{-x}{25 - x^2}$. Then

$$y' = y\frac{-x}{25 - x^2} = \sqrt{25 - x^2}\,\frac{-x}{25 - x^2} = \frac{-x}{\sqrt{25 - x^2}} .$$

41. $y = x^{\cos(x)}$ so $\ln(y) = \cos(x)\cdot\ln(x)$ and $\frac{y'}{y} = \cos(x)\cdot\frac{1}{x} - \ln(x)\cdot\sin(x)$.

Then $y' = y\,\{ \cos(x)\cdot\frac{1}{x} - \ln(x)\cdot\sin(x) \} = x^{\cos(x)}\{ \cos(x)\cdot\frac{1}{x} - \ln(x)\cdot\sin(x) \}$.

43. $y = x^4(x - 2)^7\sin(3x)$ so $\ln(y) = 4\ln(x) + 7\ln(x - 2) + \ln(\sin(3x))$ and $\frac{y'}{y} = \frac{4}{x} + \frac{7}{x - 2} + \frac{3\cos(3x)}{\sin(3x)}$.

Then $y' = y\,\{ \frac{4}{x} + \frac{7}{x - 2} + \frac{3\cos(3x)}{\sin(3x)} \} = x^4(x - 2)^7\sin(3x)\,\{ \frac{4}{x} + \frac{7}{x - 2} + \frac{3\cos(3x)}{\sin(3x)} \}$.

45. $\ln(y) = x\cdot\ln(3 + \sin(x))$ so $\frac{y'}{y} = \frac{x\cos(x)}{3 + \sin(x)} + \ln(3 + \sin(x))$. Then

$y' = (3 + \sin(x))^x\,\{ \frac{x\cos(x)}{3 + \sin(x)} + \ln(3 + \sin(x)) \}$.

47.

x	f '(x)
1	1(1.2) = 1.2
2	9(1.8) = 16.2
3	64(2.1) = 134.4

49.

x	f '(x)
1	5(–1) = –5
2	2(0) = 0
3	7(2) = 14

51. $\ln(f\cdot g) = \ln(f) + \ln(g)$ so $\frac{D(f\cdot g)}{f\cdot g} = \frac{f'}{f} + \frac{g'}{g}$. Then $D(f\cdot g) = (f\cdot g)\{ \frac{f'}{f} + \frac{g'}{g} \} = f'\cdot g + g'\cdot f$.

52. $\ln(f/g) = \ln(f) - \ln(g)$ so $\frac{D(f/g)}{f/g} = \frac{f'}{f} - \frac{g'}{g}$. Then

$D(f/g) = \frac{f}{g}\{ \frac{f'}{f} - \frac{g'}{g} \} = \frac{f'}{g} - \frac{f\cdot g'}{g^2} = \frac{f\cdot g' - g\cdot f'}{g^2}$.

53. $\ln(f\cdot g\cdot h) = \ln(f) + \ln(g) + \ln(h)$ so $\frac{D(f\cdot g\cdot h)}{f\cdot g\cdot h} = \frac{f'}{f} + \frac{g'}{g} + \frac{h'}{h}$.

Then $D(f\cdot g\cdot h) = (f\cdot g\cdot h)\{ \frac{f'}{f} + \frac{g'}{g} + \frac{h'}{h} \} = f'\cdot g\cdot h + f\cdot g'\cdot h + f\cdot g\cdot h'$.

54. $\ln(a^x) = x\ln(a)$ so $\frac{D(a^x)}{a^x} = \ln(a)$ and $D(a^x) = a^x\ln(a)$.

55. On your own.

3.1 FINDING MAXIMUMS AND MINIMUMS

In theory and applications, we often want to maximize or minimize some quantity. An engineer may want to maximize the speed of a new computer or minimize the heat produced by an appliance. A manufacturer may want to maximize profits and market share or minimize waste. A student may want to maximize a grade in calculus or minimize the hours of study needed to earn a particular grade.

Also, many natural objects follow minimum or maximum principles, so if we want to model some natural phenomena we may need to maximize or minimize. A light ray travels along a "minimum time" path. The shape and surface texture of some animals tend to minimize or maximize heat loss. Systems reach equilibrium when their potential energy is minimized. A basic tenet of evolution is that a genetic characteristic which maximizes the reproductive success of an individual will become more common in a species.

Calculus provides tools for analyzing functions and their behavior and for finding maximums and minimums.

Methods for Finding Maximums and Minimums

We can try to find where a function f is largest or smallest by evaluating f at lots of values of x, a method which is not very efficient and may not find the exact place where f achieves its extreme value. However, if we try hundreds or thousands of values for x , then we can often find a value of f which is close to the maximum or minimum. In general, this type of exhaustive search is only practical if you have a computer do the work.

The graph of a function is a visual way of examining lots of values of f, and it is a good method, particularly if you have a computer to do the work for you. However, it is inefficient, and we still may not find the exact location of the maximum or minimum.

Calculus provides ways of drastically narrowing the number of points we need to examine to find the exact locations of maximums and minimums. Instead of examining f at thousands of values of x, calculus can often guarantee that the maximum or minimum must occur at one of 3 or 4 values of x, a substantial improvement in efficiency.

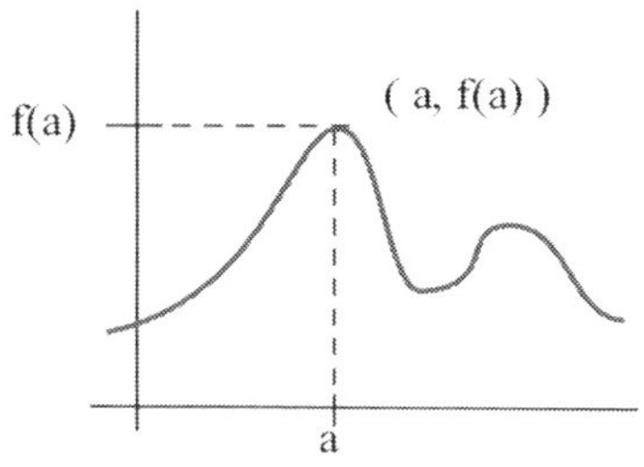

Fig. 1

A Little Terminology

Before we examine how calculus can help us find maximums and minimums, we need to define the concepts we will develop and use.

Definitions: f has a **maximum** or **global maximum** at a if $f(a) \geq f(x)$ for all x in the domain of f.
The maximum value of f **is f(a)**, and this maximum value of f **occurs at a.**
The **maximum point** on the graph of f is **(a , f(a)).** (Fig. 1)

Definition: f has a **local** or **relative maximum** at a if $f(a) \geq f(x)$ for all x near a or in some open interval which contains a.

Global and local minimums are defined similarly by replacing the ≥ with ≤ in the previous definitions.

Definition: f has a **global extreme** at a if f(a) is a **global maximum or minimum.**
f has a **local extreme** at a if f(a) is a **local maximum or minimum.**

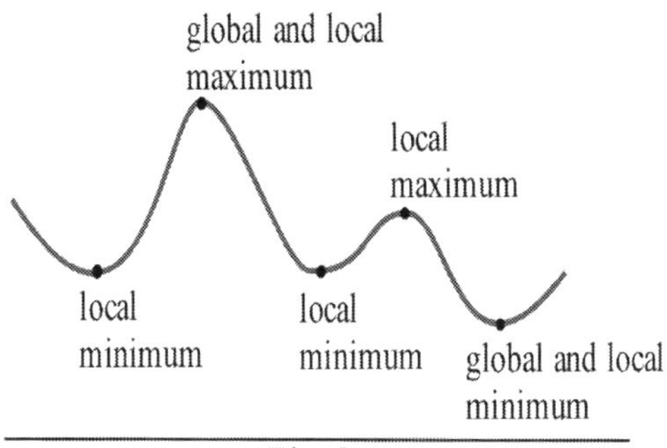

Fig. 2

The local and global extremes of the function in Fig. 2 are labeled. You should notice that every global extreme is also a local extreme, but there are local extremes which are not global extremes. If h(x) is the height of the earth above sea level at the location x, then the global maximum of h is h(summit of Mt. Everest) = 29,028 feet. The local maximum of h for the United States is h(summit of Mt. McKinley) = 20,320 feet. The local minimum of h for the United States is h(Death Valley) = – 282 feet.

Practice 1: The table shows the annual calculus enrollments at a large university. Which years had relative maximum or minimum calculus enrollments? What were the global maximum and minimum enrollments in calculus?

year	1980	81	82	83	84	85	86	87	88	89	90
enrollment	1257	1324	1378	1336	1389	1450	1523	1582	1567	1545	1571

Finding Maximums and Minimums of a Function

One way to narrow our search for a maximum value of a function f is to eliminate those values of x which, for some reason, cannot possibly make f maximum.

Theorem : If $f'(a) > 0$ or $f'(a) < 0$,
then f(a) is not a local maximum or minimum. (Fig. 3)

Proof: Assume that $f'(a) > 0$. By definition, $f'(a) = \lim_{\Delta x \to 0} \frac{f(a+\Delta x) - f(a)}{\Delta x}$,

so $\lim_{\Delta x \to 0} \frac{f(a+\Delta x) - f(a)}{\Delta x} > 0$ and the right and left limits are both positive:

$$\lim_{\Delta x \to 0^+} \frac{f(a+\Delta x) - f(a)}{\Delta x} > 0 \text{ and } \lim_{\Delta x \to 0^-} \frac{f(a+\Delta x) - f(a)}{\Delta x} > 0.$$

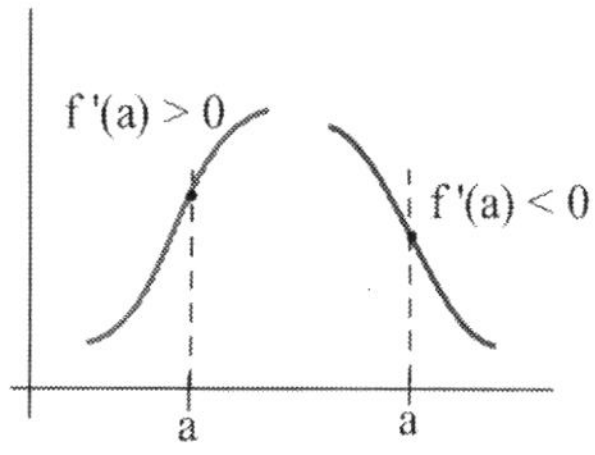

Fig. 3

Since the right limit, $\Delta x \to 0^+$, is positive, there are values of $\Delta x > 0$ so $\frac{f(a+\Delta x)-f(a)}{\Delta x} > 0$.

Multiplying each side of this last inequality by the positive Δx, we have $f(a+\Delta x) - f(a) > 0$ and $f(a+\Delta x) > f(a)$ so $f(a)$ is not a maximum.

Since the left limit, $\Delta x \to 0^-$, is positive, there are values of $\Delta x < 0$ so $\frac{f(a+\Delta x)-f(a)}{\Delta x} > 0$.

Multiplying each side of the last inequality by the negative Δx, we have that $f(a+\Delta x) - f(a) < 0$ and $f(a+\Delta x) < f(a)$ so $f(a)$ is not a minimum.

The proof for the "$f'(a) < 0$" case is similar.

When we evaluate the derivative of a function f at a point $x = a$, there are only four possible outcomes: $f'(a) > 0$, $f'(a) < 0$, $f'(a) = 0$ or $f'(a)$ is undefined. If we are looking for extreme values of f, then we can eliminate those points at which f' is positive or negative, and only two possibilities remain: $f'(a) = 0$ or $f'(a)$ is undefined.

Theorem : If f is defined on an open interval, and $f(a)$ is a local extreme of f, then either $f'(a) = 0$ or f is not differentiable at a.

Example 1: Find the local extremes of $f(x) = x^3 - 6x^2 + 9x + 2$ for all values of x.

Solution: An extreme value of f can occur only where $f'(x) = 0$ or where f is not differentiable.

$f'(x) = 3x^2 - 12x + 9 = 3(x^2 - 4x + 3) = 3(x-1)(x-3)$ so $f'(x) = 0$ only at $x = 1$ and $x = 3$. f' is a polynomial, so f is differentiable for all x.

The only possible locations of local extremes of f are at $x = 1$ and $x = 3$. We don't know yet whether $f(1)$ or $f(3)$ is a local extreme of f, but we can be certain that no other point is a local extreme. The graph of f (Fig. 4) shows that $(1, f(1)) = (1, 6)$ is a local maximum and $(3, f(3)) = (3, 2)$ is a local minimum. This function does not have a global maximum or minimum.

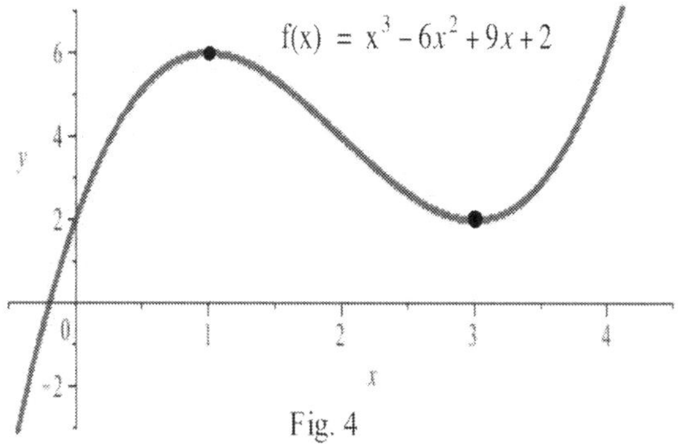

Fig. 4

Practice 2: Find the local extremes of $f(x) = x^2 + 4x - 5$ and $g(x) = 2x^3 - 12x^2 + 7$.

It is important to recognize that the conditions "$f'(a) = 0$" or "f not differentiable at a" do **not** guarantee that $f(a)$ is a local maximum or minimum. They only say that $f(a)$ **might be** a local extreme or that $f(a)$ is a **candidate** for being a local extreme.

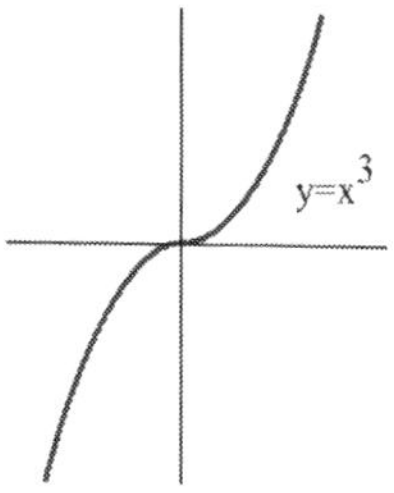

Fig. 5

Example 2: Find all local extremes of $f(x) = x^3$.

Solution: $f(x) = x^3$ is differentiable for all x, and $f'(x) = 3x^2$. The only place where $f'(x) = 0$ is at $x = 0$, so the only candidate is the point (0,0). But if $x > 0$ then $f(x) = x^3 > 0 = f(0)$, so f(0) is not a local maximum. Similarly, if $x < 0$ then $f(x) = x^3 < 0 = f(0)$ so f(0) is not a local minimum. The point (0,0) is the only candidate to be a local extreme of f, and this candidate did not turn out to be a local extreme of f. The function $f(x) = x^3$ does not have any local extremes. (Fig. 5)

If $f'(a) = 0$ or f is not differentiable at a
then the point (a , f(a)) is a <u>candidate</u> to be a local extreme and **<u>may</u> or <u>may not</u>** be a local extreme.

Practice 3: Sketch the graph of a differentiable function f which satisfies the conditions:

(i) $f(1) = 5, f(3) = 1, f(4) = 3$ and $f(6) = 7$, (ii) $f'(1) = 0, f'(3) = 0, f'(4) = 0$ and $f'(6) = 0$,

(iii) the only local maximums of f are at (1,5) and (6,7), and the only local minimum is at (3,1).

Is f(a) a Maximum or Minimum or Neither?

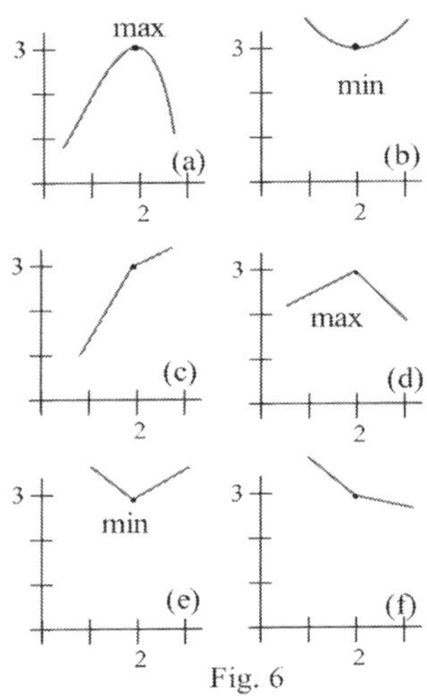

Fig. 6

Once we have found the candidates (a, f(a)) for extreme points of f, we still have the problem of determining whether the point is a maximum, a minimum or neither.

One method is to graph (or have your calculator graph) the function near a, and then draw your conclusion from the graph. All of the graphs in Fig. 6 have $f(2) = 3$, and, on each of the graphs, $f'(2)$ either equals 0 or is undefined. It is clear from the graphs that the point (2,3) is a local maximum in (a) and (d), (2,3) is a local minimum in (b) and (e), and (2,3) is not a local extreme in (c) and (f).

In sections 3.3 and 3.4, we will investigate how information about the first and second derivatives of f can help determine whether the candidate (a, f(a)) is a maximum, a minimum, or neither.

Endpoint Extremes

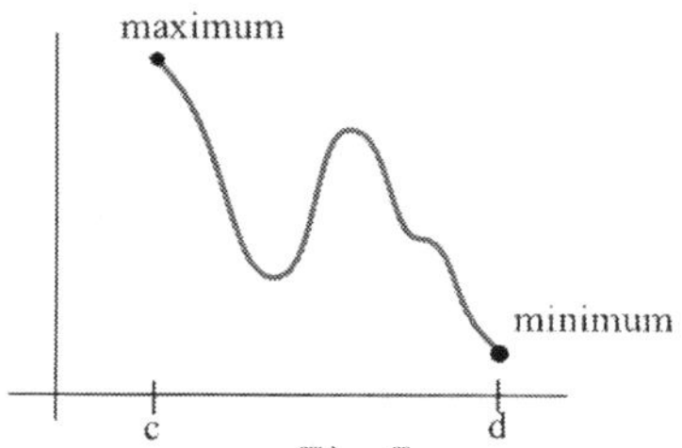

Fig. 7

So far we have been discussing finding extreme values of functions over the entire real number line or on an open interval, but, in practice, we may need to find the extreme of a function over some closed interval $[c, d]$. If the extreme value of f occurs at $x = a$ between c and d, $c < a < d$, then the previous reasoning and results still apply: either $f'(a) = 0$ or f is not differentiable at a.

On a closed interval, however, there is one more possibility: an extreme can occur at an **endpoint** of the closed interval (Fig. 7), at $x = c$ or $x = d$.

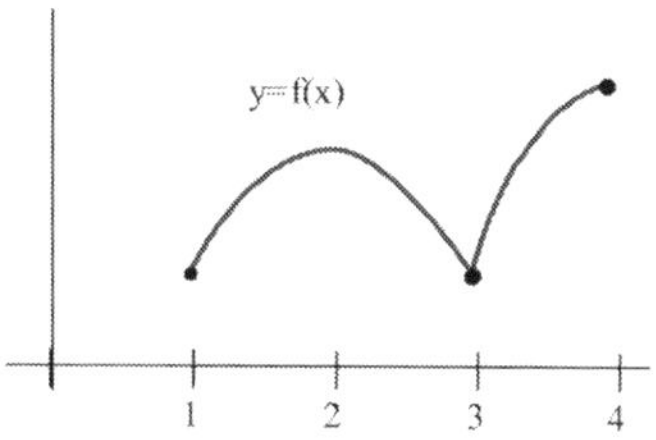

Fig. 8

Practice 4: List all of the local extremes $(a, f(a))$ of the function in Fig. 8 on the interval $[1,4]$ and state whether (i) $f'(a) = 0$ or (ii) f is not differentiable at a or (iii) a is an endpoint.

Example 3: Find the extreme values of $f(x) = x^3 - 3x^2 - 9x + 5$ for $-2 \le x \le 6$.

Solution: $f'(x) = 3x^2 - 6x - 9 = 3(x + 1)(x - 3)$. We need to find where (i) $f'(x) = 0$, (ii) f is not differentiable, and (iii) the endpoints.

(i) $f'(x) = 3(x + 1)(x - 3) = 0$ when $x = \mathbf{-1}$ and $x = \mathbf{3}$.

(ii) f is a polynomial so it is differentiable everywhere.

(iii) The endpoints of the interval are $x = \mathbf{-2}$ and $x = \mathbf{6}$.

Altogether we have four points in the interval to examine, and any extreme values of f can only occur when x is one of those four points: $f(-2) = 3$, $f(-1) = 10$, $f(3) = -22$, and $f(6) = 59$. The minimum of f on $[-2, 6]$ is -22 when $x = 3$, and the maximum of f on $[-2, 6]$ is 59 when $x = 6$.

Sometimes the function we need to maximize or minimize is more complicated, but the same methods work.

Example 4: Find the extreme values of $f(x) = \frac{1}{3}\sqrt{64 + x^2} + \frac{1}{5}(10 - x)$ for $0 \le x \le 10$.

Solution: This function comes from an application we will examine in section 3.5. The only possible locations of extremes are where $f'(x) = 0$ or $f'(x)$ is undefined or where x is an endpoint of the interval $[0, 10]$.

$$f'(x) = \mathbf{D}\left(\frac{1}{3}(64 + x^2)^{1/2} + \frac{1}{5}(10 - x)\right) = \frac{1}{3}\frac{1}{2}(64 + x^2)^{-1/2}(2x) - \frac{1}{5} = \frac{x}{3\sqrt{64 + x^2}} - \frac{1}{5}.$$

To determine where $f'(x) = 0$, we need to set the derivative equal to 0 and solve for x.

If $f'(x) = \dfrac{x}{3\sqrt{64 + x^2}} - \dfrac{1}{5} = 0$ then $\dfrac{x}{3\sqrt{64 + x^2}} = \dfrac{1}{5}$ so $\dfrac{x^2}{576 + 9x^2} = \dfrac{1}{25}$.

Then $16x^2 = 576$ so $x = \pm 6$, and the only point <u>in the interval</u> $[0,10]$ where $f'(x) = 0$ is at $x = 6$. Putting $x = 6$ into the original equation for f gives $f(6) \approx 4.13$.

We can evaluate the formula for $f'(x)$ for any value of x, so the derivative is always defined. Finally, the interval $[0, 10]$ has two endpoints, $x = 0$ and $x = 10$. $f(0) \approx 4.67$ and $f(10) \approx 4.27$.

The maximum of f on [0,10] must occur at one of the points (0, 4.67), (6, 4.13) and (10, 4.27), and the minimum must occur at one of these three points.

The maximum value of f is 4.67 at $x = 0$, and the minimum value of f is 4.13 at $x = 6$. The graph of f is shown in Fig. 9.

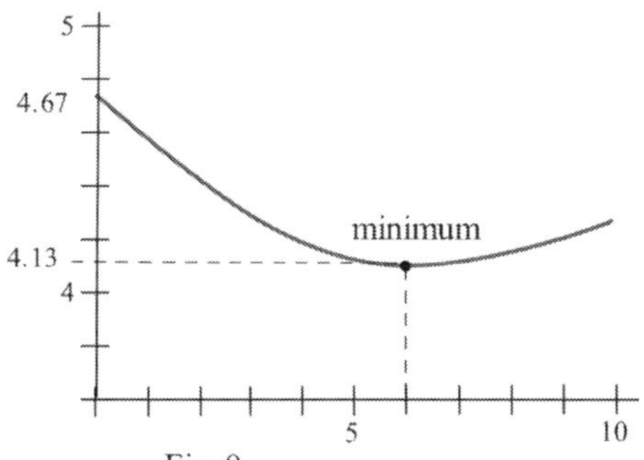

Fig. 9

Practice 5: Find the extreme values of

$$f(x) = \frac{1}{3}\sqrt{64 + x^2} + \frac{1}{5}(10 - x) \quad \text{for } \mathbf{0 \le x \le 5}.$$

Critical Numbers

The points at which a function might have an extreme value are called **critical numbers**.

Definitions: A **critical number** for a function f is a value $x = \mathbf{a}$ in the domain of f so

(i) $f'(\mathbf{a}) = 0$,

or (ii) f is not differentiable at **a**,

or (iii) **a** is an endpoint

If we are trying to find the extreme values of f on an open interval $c < x < d$ or on the entire number line, then there will not be any endpoints so there will not be any endpoint critical numbers to worry about.

We can now give a very succinct description of where to look for extreme values of a function:

An extreme value of f can only occur at a critical number.

The critical numbers only give the **possible** locations of extremes, and some critical numbers are not the locations of extremes. The critical numbers are the **candidates** for the locations of maximums and minimums (Fig. 10). Section 3.5 is devoted entirely to translating and solving maximum and minimum problems.

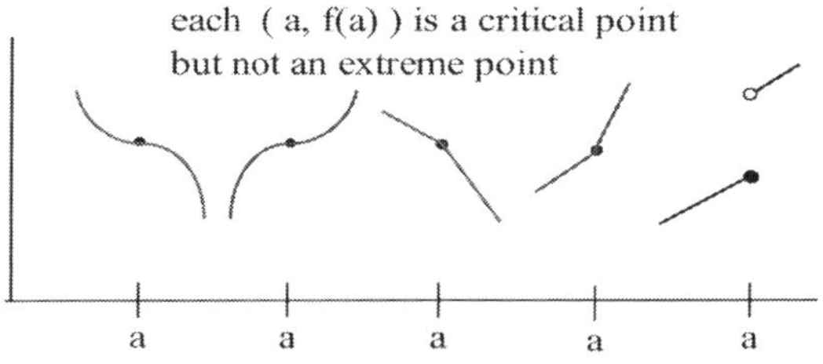

Fig. 10

Which Functions Have Extremes?

So far we have concentrated on finding the extreme values of functions, but some functions don't have extreme values. Example 2 showed that $f(x) = x^3$ did not have a maximum or minimum.

Example 5: Find the extreme values of $f(x) = x$.

Solution: Since $f'(x) = 1 > 0$ for all x, the first theorem in this section guarantees that f has no extreme values. The function $f(x) = x$ does not have a maximum or minimum on the real number line.

The difficulty with the previous function was that the domain was so large that we could always make the function larger or smaller than any given value. The next example shows that we can encounter the same difficulty even on a small interval.

Example 6: Show that $f(x) = \frac{1}{x}$ (Fig. 11) does not have a maximum or minimum on the interval $(0,1)$.

Solution: f is continuous for all $x \neq 0$ so f is continuous on the interval $(0,1)$.

For $0 < x < 1$, $f(x) = \frac{1}{x} > 0$ (Fig. 11). For any number a strictly between 0 and 1, we can show that $f(a)$ is neither a maximum nor a minimum of f on $(0,1)$. Pick b to be any number between 0 and a, $0 < b < a$. Then

$f(b) = \frac{1}{b} > \frac{1}{a} = f(a)$, so $f(a)$ is not a maximum. Similarly, pick c to be any number between a and 1, $a < c < 1$. Then $f(a) = \frac{1}{a} > \frac{1}{c} = f(c)$, so $f(a)$ is not a minimum. The interval $(0,1)$ is not large, but f still does not have an extreme value in $(0,1)$.

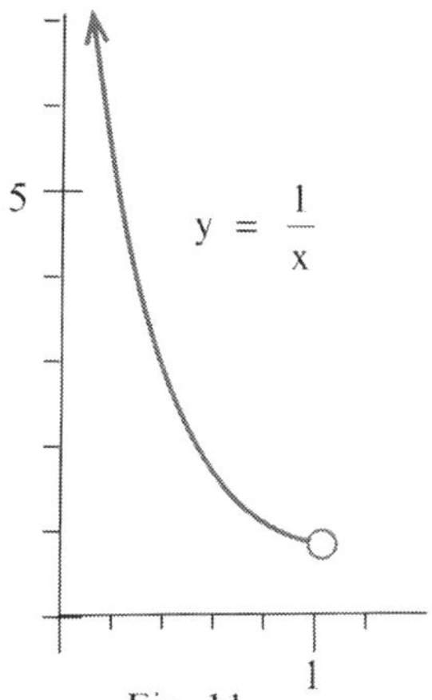

Fig. 11

The Extreme Value Theorem gives conditions so that a function is guaranteed to have a maximum and a minimum.

> **Extreme Value Theorem:** If f is **continuous** on a **closed** interval $[a,b]$,
> then f attains both a maximum and minimum on $[a,b]$.

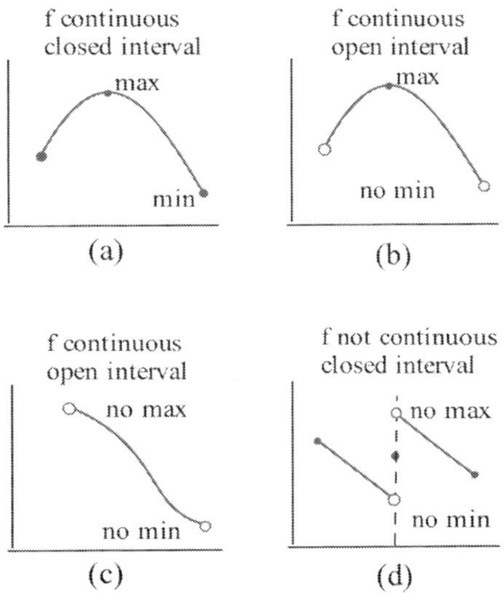

Fig. 12

The proof of this theorem is difficult and is omitted. Fig. 12 illustrates some of the possibilities for continuous and discontinuous functions on open and closed intervals. The Extreme Value Theorem guarantees that certain functions (continuous) on certain intervals (closed) must have maximums and minimums. Other functions on other intervals **may** or **may not** have maximums and minimums.

PROBLEMS

1. Label all of the local maximums and minimums of the function in Fig. 13. Also label all of the critical points.

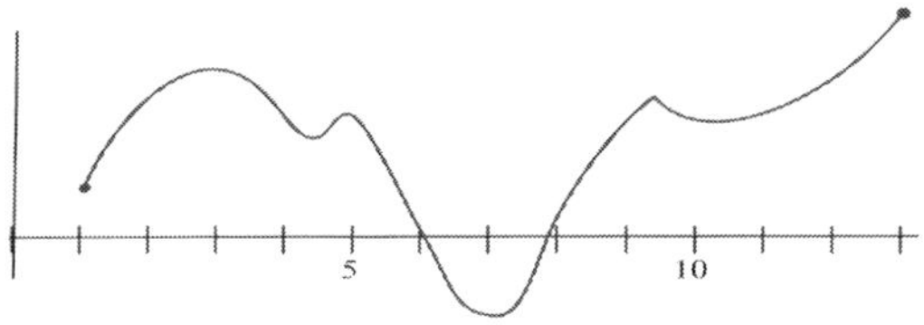

Fig. 13

2. Label all of the local maximums and minimums of the function in Fig. 14. Also label all of the critical points.

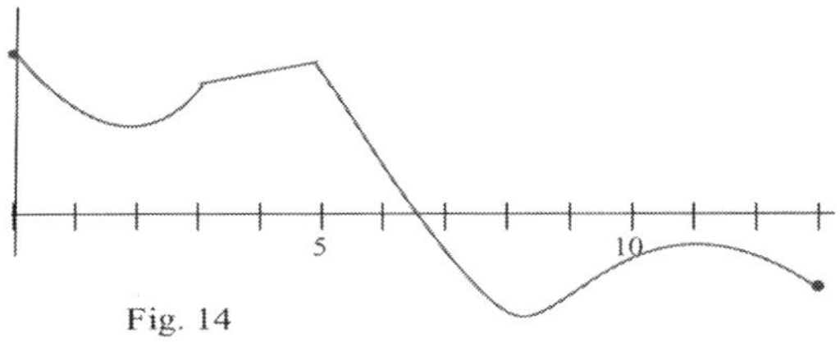

Fig. 14

In problems 3 – 14, find all of the critical points and local maximums and minimums of each function.

3. $f(x) = x^2 + 8x + 7$

4. $f(x) = 2x^2 - 12x + 7$

5. $f(x) = \sin(x)$

6. $f(x) = x^3 - 6x^2 + 5$

7. $f(x) = (x - 1)^2 (x - 3)$

8. $f(x) = \ln(x^2 - 6x + 11)$

9. $f(x) = 2x^3 - 96x + 42$

10. $f(x) = 5x - 2$

11. $f(x) = 5x + \cos(2x+1)$

12. $f(x) = \sqrt[3]{x}$

13. $f(x) = e^{-(x - 2)^2}$

14. $f(x) = | x + 5 |$

15. Sketch the graph of a continuous function f so that
   (a) $f(1) = 3$, $f'(1) = 0$, and the point (1,3) is a relative maximum of f.
   (b) $f(2) = 1$, $f'(2) = 0$, and the point (2,1) is a relative minimum of f.
   (c) $f(3) = 5$, f is not differentiable at 3, and the point (3,5) is a relative maximum of f.
   (d) $f(4) = 7$, f is not differentiable at 4, and the point (4,7) is a relative minimum of f.
   (e) $f(5) = 4$, $f'(5) = 0$, and the point (5,4) is not a relative minimum or maximum of f.
   (f) $f(6) = 3$, f is not differentiable at 6, and the point (6,3) is not a relative minimum or maximum of f.

In problems 16 – 25, find all critical points and local extremes of each function on the given intervals.

16. $f(x) = x^2 - 6x + 5$ on the entire real number line.

17. $f(x) = x^2 - 6x + 5$ on $[-2, 5]$.

18. $f(x) = 2 - x^3$ on the entire real number line.

19. $f(x) = 2 - x^3$ on $[-2, 1]$.

20. $f(x) = x^3 - 3x + 5$ on the entire real number line.

21. $f(x) = x^3 - 3x + 5$ on $[-2, 1]$.

22. $f(x) = x^5 - 5x^4 + 5x^3 + 7$ on the entire real number line.

23. $f(x) = x^5 - 5x^4 + 5x^3 + 7$ on $[0, 2]$.

24. $f(x) = \dfrac{1}{x^2 + 1}$ on the entire real number line.

25. $f(x) = \dfrac{1}{x^2 + 1}$ on $[1, 3]$.

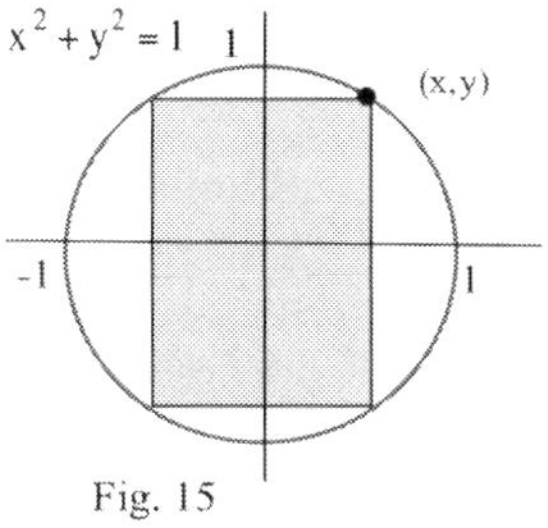

Fig. 15

26. (a) Find two numbers whose sum is 22 and whose product is as large as possible. (Suggestion: call the numbers x and $22 - x$).
(b) Find two numbers whose sum is $A > 0$ and whose product is as large as possible.

27. Find the coordinates of the point in the first quadrant on the circle $x^2 + y^2 = 1$ so that the rectangle formed in Fig. 15 has the largest possible area. (Suggestion: the coordinates of a point on the circle are $(x, \sqrt{1 - x^2})$.)

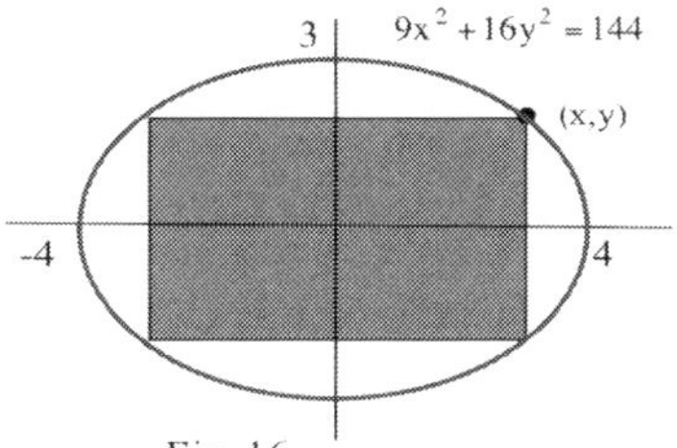

Fig. 16

28. Find the coordinates of the point in the first quadrant on the ellipse $9x^2 + 16y^2 = 144$ so that the rectangle formed in Fig. 16 has the largest possible area. The smallest possible area.

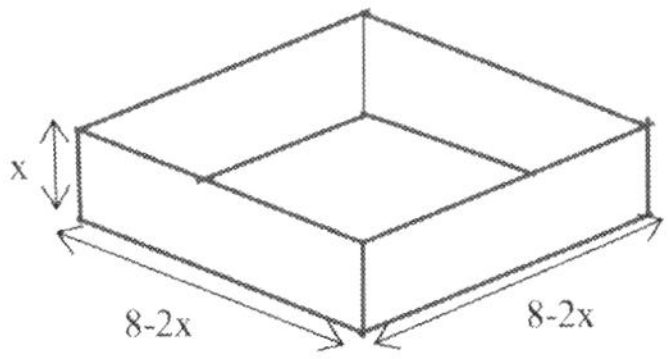

Fig. 17

29. Find the value for x so the box in Fig. 17 has the largest possible volume? The smallest volume?

30. Find the radius and height of the cylinder that has the largest volume ($V = \pi \cdot r^2 \cdot h$) if the sum of the radius and height is 9.

31. Suppose you are working with a polynomial of degree 3 on a closed interval.
(a) What is the largest number of critical points the function can have on the interval?
(b) What is the smallest number of critical points it can have?
(c) What are the patterns for the most and fewest critical points a polynomial of degree n on a closed interval can have?

32. Suppose you have a polynomial of degree 3 divided by a polynomial of degree 2 on a closed interval.
(a) What is the largest number of critical points the function can have on the interval?
(b) What is the smallest number of critical points it can have?

33. Suppose $f(1) = 5$ and $f'(1) = 0$. What can we conclude about the point (1,5) if
(a) $f'(x) < 0$ for $x < 1$, and $f'(x) > 0$ for $x > 1$? (b) $f'(x) < 0$ for $x < 1$, and $f'(x) < 0$ for $x > 1$?
(c) $f'(x) > 0$ for $x < 1$, and $f'(x) < 0$ for $x > 1$? (d) $f'(x) > 0$ for $x < 1$, and $f'(x) > 0$ for $x > 1$?

34. Suppose $f(2) = 3$ and f is continuous but not differentiable at $x = 2$. What can we conclude about the point (2,3) if

(a) $f'(x) < 0$ for $x < 2$, and $f'(x) > 0$ for $x > 2$? (b) $f'(x) < 0$ for $x < 2$, and $f'(x) < 0$ for $x > 2$?

(c) $f'(x) > 0$ for $x < 2$, and $f'(x) < 0$ for $x > 2$? (d) $f'(x) > 0$ for $x < 2$, and $f'(x) > 0$ for $x > 2$?

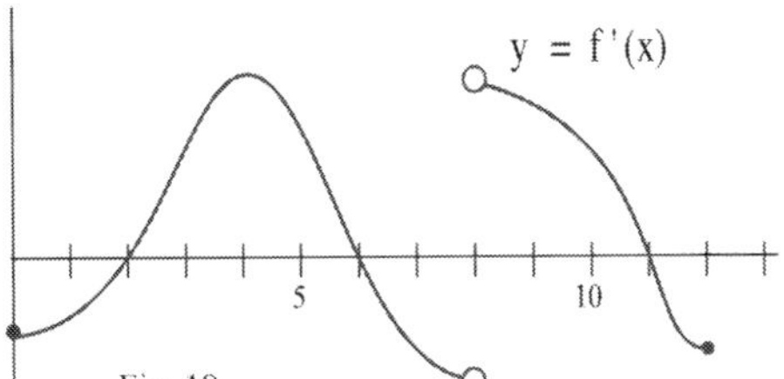

Fig. 18

35. f is a continuous function, and Fig. 18 shows the graph of f ' .

(a) Which values of x are critical points?

(b) At which values of x is f a local maximum?

(c) At which values of x is f a local minimum?

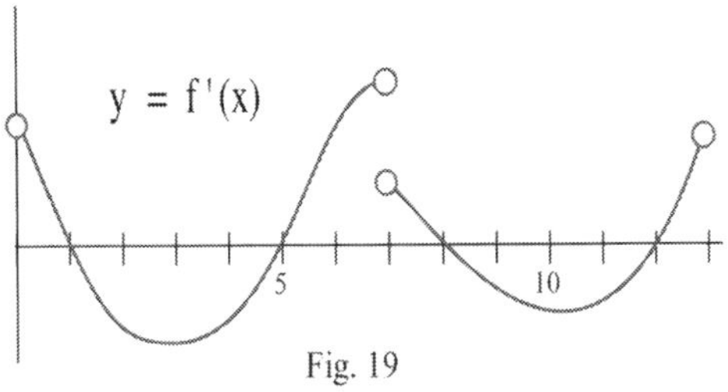

Fig. 19

36. f is a continuous function, and Fig. 19 shows the graph of f ' .

(a) Which values of x are critical points?

(b) At which values of x is f a local maximum?

(c) At which values of x is f a local minimum?

37. State the contrapositive form of the Extreme Value Theorem.

38. Imagine the graph of $f(x) = 1 - x$. Does f have a **maximum** value for x in the interval I?

(a) $I = [0, 2]$ (b) $I = [0, 2)$ (c) $I = (0, 2]$ (d) $I = (0,2)$ (e) $I = (1, \pi]$

39. Imagine the graph of $f(x) = 1 - x$. Does f have a **minimum** value for x in the interval I?

(a) $I = [0, 2]$ (b) $I = [0, 2)$ (c) $I = (0, 2]$ (d) $I = (0,2)$ (e) $I = (1, \pi]$

40. Imagine the graph of $f(x) = x^2$. Does f have a **maximum** value for x in the interval I?

(a) $I = [-2, 3]$ (b) $I = [-2, 3)$ (c) $I = (-2, 3]$ (d) $I = [-2, 1)$ (e) $I = (-2, 1]$

41. Imagine the graph of $f(x) = x^2$. Does f have a **minimum** value for x in the following intervals?

(a) $I = [-2, 3]$ (b) $I = [-2, 3)$ (c) $I = (-2, 3]$

(d) $I = [-2, 1)$ (e) $I = (-2, 1]$

42. Define A(x) to be the **area** bounded between the x–axis, the graph of f, and a vertical line at x (Fig. 20).

(a) At what value of x is A(x) minimum?

(b) At what value of x is A(x) maximum?

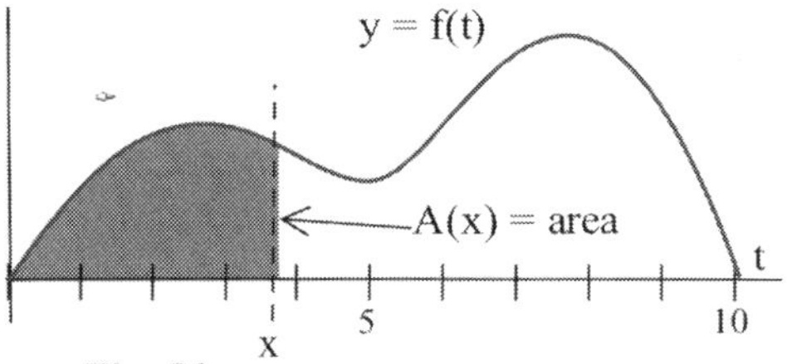

Fig. 20

43. Define S(x) to be the **slope** of the line through the points (0,0) and (x, f(x)) in Fig. 21.

(a) At what value of x is S(x) minimum?

(b) At what value of x is S(x) maximum?

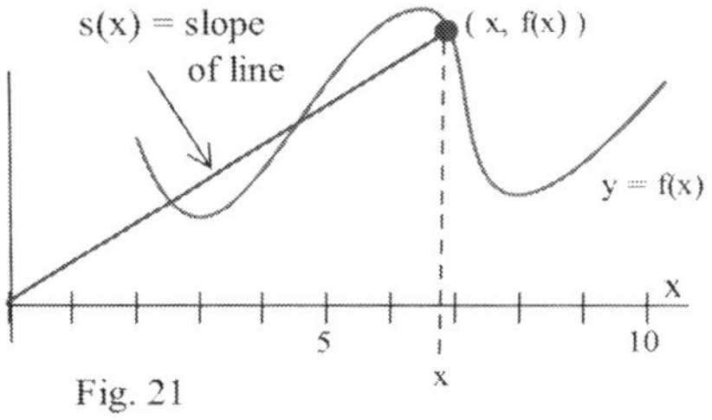

Fig. 21

Section 3.1 **PRACTICE Answers**

Practice 1: The enrollments were relative maximums in '82, '87, and '90.

The global maximum was in '87. The enrollments were relative minimums in '80, '83, and '89. The global minimum occurred in '80.

Practice 2: $f(x) = x^2 + 4x - 5$ is a polynomial so f is differentiable for all x, and $f'(x) = 2x + 4$. $f'(x) = 0$ when $x = -2$ so the only candidate for a local extreme is $x = -2$. Since the graph of f is a parabola opening up, the point $(-2, f(-2)) = (-2, -9)$ is a local minimum.

$g(x) = 2x^3 - 12x^2 + 7$ is a polynomial so g is differentiable for all x, and $g'(x) = 6x^2 - 24x = 6x(x - 4)$. $g'(x) = 0$ when $x = 0, 4$ so the only candidates for a local extreme are $x = 0$ and $x = 4$. The graph of g (Fig. 22) shows that g has a local maximum at (0,7) and a local minimum at (4, –57).

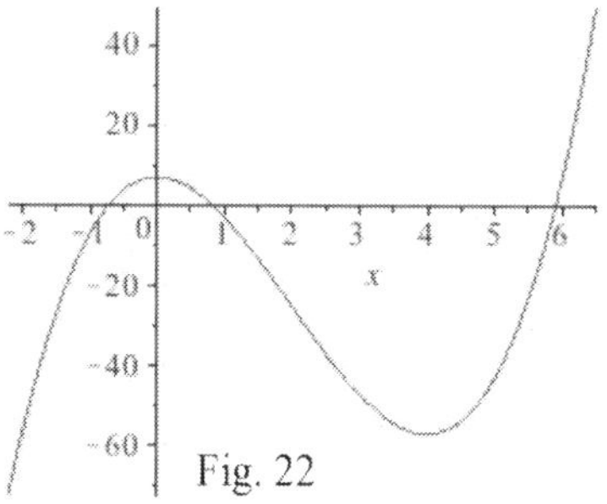

Fig. 22

Practice 3: see Fig. 23

x	f(x)	f '(x)	max/min
1	5	0	local max
2			
3	1	0	local min
4	3	0	neither
5			
6	7	0	local max

Practice 4: (1, f(1)) is a local minimum. x = 1 is an endpoint.

(2, f(2)) is a local maximum. $f'(2) = 0$.

(3, f(3)) is a local minimum. f is not differentiable at x = 3.

(4, f(4)) is a local maximum. x = 4 is an endpoint.

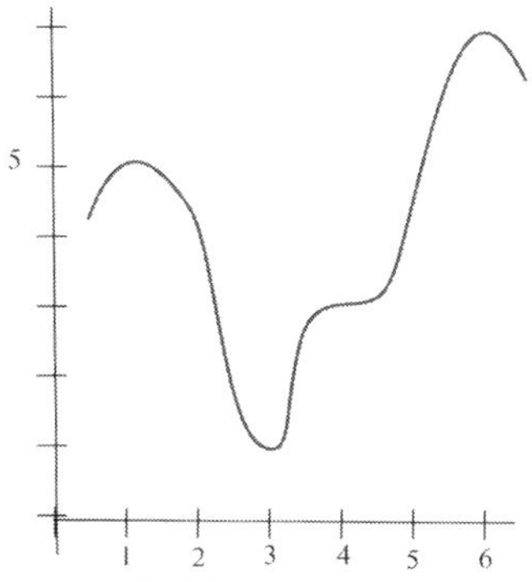

Fig. 23

Practice 5: This is the same function that was used in Example 4, but in this Practice problem the interval is [0, 5] instead of [0, 10] in the Example. See the Example for the calculations.

Critical points: endpoints **x = 0** and **x = 5**.

f is differentiable for all $0 < x < 5$: none.

$f'(x) = 0$: none in **[0, 5]**.

$f(0) \approx 4.67$ is the maximum of f on [0, 5]. $f(5) \approx 4.14$ is the minimum of f on [0, 5].

3.2 The Mean Value Theorem and Its Consequences

If you averaged 30 miles per hour during a trip, then at some instant during the trip you were traveling exactly 30 miles per hour.

That relatively obvious statement is the Mean Value Theorem as it applies to a particular trip. It may seem strange that such a simple statement would be important or useful to anyone, but the Mean Value Theorem is important and some of its consequences are very useful for people in a variety of areas. Many of the results in the rest of this chapter depend on the Mean Value Theorem, and one of the corollaries of the Mean Value Theorem will be used every time we calculate an "integral" in later chapters. A truly delightful aspect of mathematics is that an idea as simple and obvious as the Mean Value Theorem can be so powerful.

Before we state and prove the Mean Value Theorem and examine some of its consequences, we will consider a simplified version called Rolle's Theorem.

Rolle's Theorem

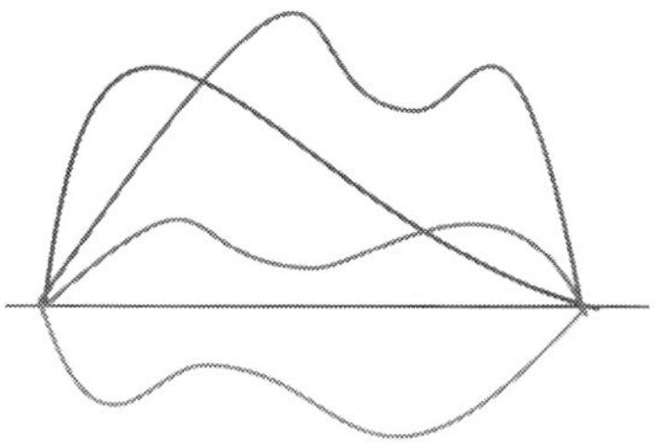
Fig. 1

Suppose we pick any two points on the x–axis and think about all of the differentiable functions which go through those two points (Fig. 1). Since our functions are differentiable, they must be continuous and their graphs can not have any holes or breaks. Also, since these functions are differentiable, their derivatives are defined everywhere between our two points and their graphs can not have any "corners" or vertical tangents. The graphs of the functions in Fig. 1 can still have all sorts of shapes, and it may seem unlikely that they have any common properties other than the ones we have stated, but Michel Rolle (1652–1719) found one. He noticed that every one of these functions has one or more points where the tangent line is horizontal (Fig. 2), and this result is named after him.

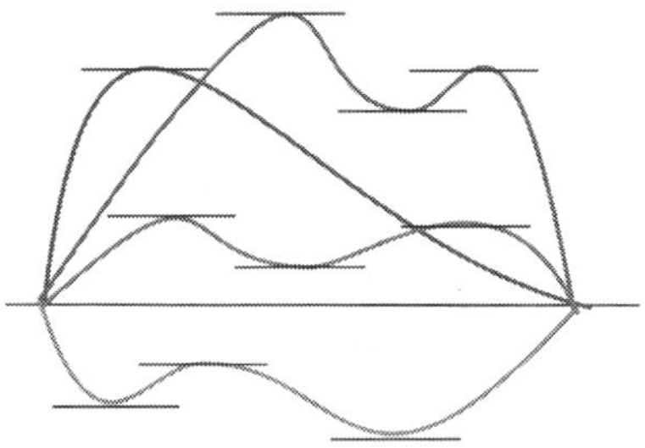
Fig. 2

> **Rolle's Theorem:**
>
> If $f(a) = f(b)$, and $f(x)$ is continuous for $a \le x \le b$ and differentiable for $a < x < b$,
>
> then there is at least one number c, between a and b, so that $f'(c) = 0$.

Proof: We consider two cases: when $f(x) = f(a)$ for all x in (a,b) and when $f(x) \ne f(a)$ for some x in (a,b).

Case I, $f(x) = f(a)$ for all x in (a,b) : If $f(x) = f(a)$ for all x between a and b, then f is a horizontal line segment and $f'(c) = 0$ for all values of c strictly between a and b.

Case II, $f(x) \ne f(a)$ for some x in (a,b): Since f is continuous on the closed interval [a,b], we know from the Extreme Value Theorem that f must have a maximum value in the closed interval [a,b] and a minimum value in the interval.

If f(x) > f(a) for some value of x in [a,b], then the maximum of f must occur at some value c strictly between a and b, a < c < b. (Why can't the maximum be at a or b?) Since f(c) is a local maximum of f, then c is a critical number of f and f '(c) = 0 or f '(c) is undefined. But f is differentiable at all x between a and b, so the only possibility left is that f '(c) = 0.

If f(x) < f(a) for some value of x in [a,b], then f has a minimum at some value x = c strictly between a and b, and f '(c) = 0.

In either case, there is at least one value of c between a and b so that f '(c) = 0.

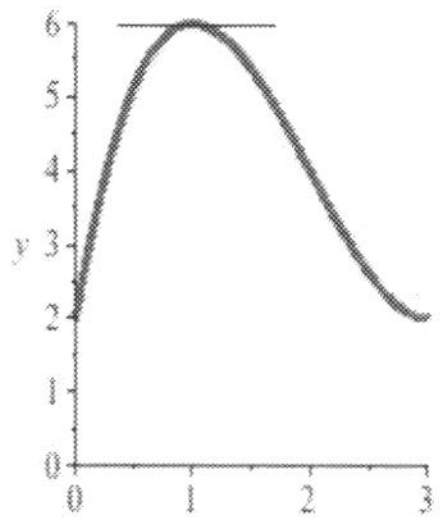

Fig. 3

Example 1: Show that $f(x) = x^3 - 6x^2 + 9x + 2$ satisfies the hypotheses of Rolle's Theorem on the interval [0, 3] and find the value of c which the theorem says exists.

Solution: f is a polynomial so it is continuous and differentiable everywhere. f(0) = 2 and f(3) = 2. $f'(x) = 3x^2 - 12x + 9 = 3(x-1)(x-3)$ so f '(x) = 0 at 1 and 3. The value c = 1 is between 0 and 3. Fig. 3 shows the graph of f.

Practice 1: Find the value(s) of c for Rolle's Theorem for the functions in Fig. 4.

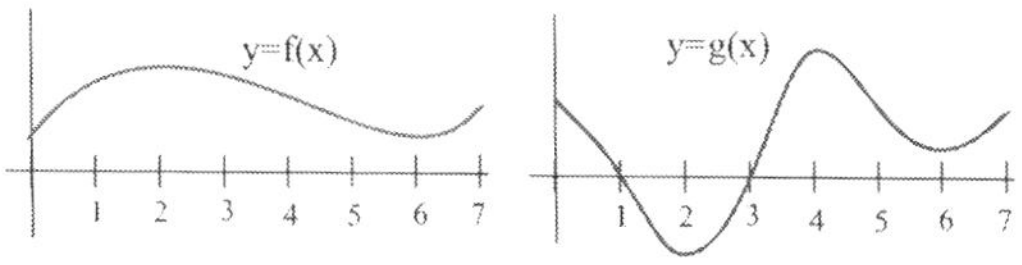

Fig. 4

The Mean Value Theorem

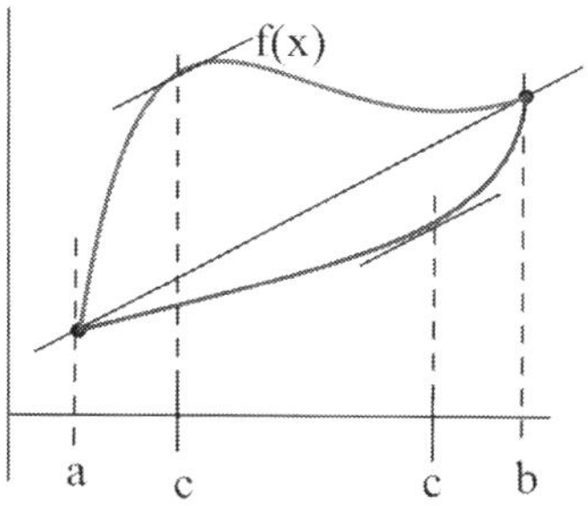

Fig. 5

Geometrically, the Mean Value Theorem is a "tilted" version of Rolle's Theorem (Fig. 5). In each theorem we conclude that there is a number c so that the slope of the tangent line to f at x = c is the same as the slope of the line connecting the two ends of the graph of f on the interval [a,b]. In Rolle's Theorem, the two ends of the graph of f are at the same height, f(a) = f(b), so the slope of the line connecting the ends is zero. In the Mean Value Theorem, the two ends of the graph of f do not have to be at the same height so the line through the two ends does not have to have a slope of zero.

Mean Value Theorem:

If f(x) is continuous for $a \le x \le b$ and differentiable for $a < x < b$,

then there is at least one number c, between a and b, so the tangent line at c is parallel to the secant line through the points (a, f(a)) and (b,f(b)): $f'(c) = \dfrac{f(b) - f(a)}{b - a}$.

Proof: The proof of the Mean Value Theorem uses a tactic common in mathematics: introduce a new function which satisfies the hypotheses of some theorem we already know and then use the conclusion of that previously proven theorem. For the Mean Value Theorem we introduce a new function, $h(x)$, which satisfies the hypotheses of Rolle's Theorem. Then we can be certain that the conclusion of Rolle's Theorem is true for $h(x)$, and the Mean Value Theorem for f follows from the conclusion of Rolle's Theorem for h.

First, let $g(x)$ be the straight line through the ends $(a, f(a))$ and $(b, f(b))$ of the graph of f. The function g goes through the point $(a, f(a))$ so $g(a) = f(a)$. Similarly, $g(b) = f(b)$. The slope of the linear function g is $\frac{f(b) - f(a)}{b - a}$ so $g'(x) = \frac{f(b) - f(a)}{b - a}$ for all x between a and b, and g is continuous and differentiable. (The formula for g is $g(x) = f(a) + m(x - a)$ with $m = (f(b) - f(a))/(b - a)$.)

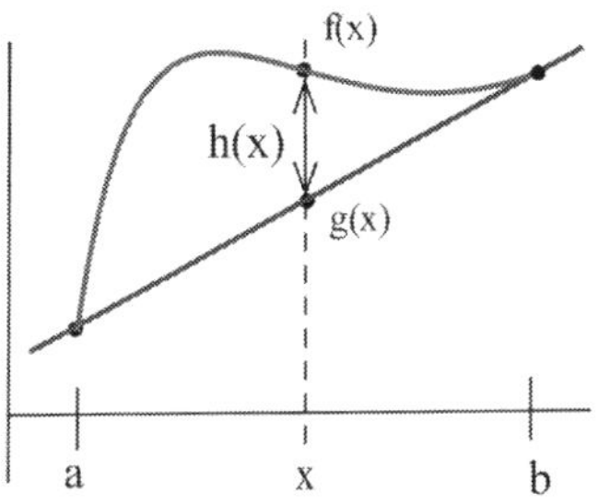

Fig. 6

Define $h(x) = f(x) - g(x)$ for $a \le x \le b$ (Fig. 6). The function h satisfies the hypotheses of Rolle's theorem:

$h(a) = f(a) - g(a) = 0$ and $h(b) = f(b) - g(b) = 0$,

$h(x)$ is continuous for $a \le x \le b$ since both f and g are continuous there, and

$h(x)$ is differentiable for $a < x < b$ since both f and g are differentiable there,

so the conclusion of Rolle's Theorem applies to h:

there is a c, between a and b, so that $h'(c) = 0$.

The derivative of $h(x) = f(x) - g(x)$ is $h'(x) = f'(x) - g'(x)$ so we know that there is a number c, between a and b, with $h'(c) = 0$. But $0 = h'(c) = f'(c) - g'(c)$ so $f'(c) = g'(c) = \frac{f(b) - f(a)}{b - a}$.

Graphically, the Mean Value Theorem says that there is at least one point c where the slope of the tangent line, $f'(c)$, equals the slope of the line through the end points of the graph segment, $(a, f(a))$ and $(b, f(b))$. Fig. 7 shows the locations of the parallel tangent lines for several functions and intervals.

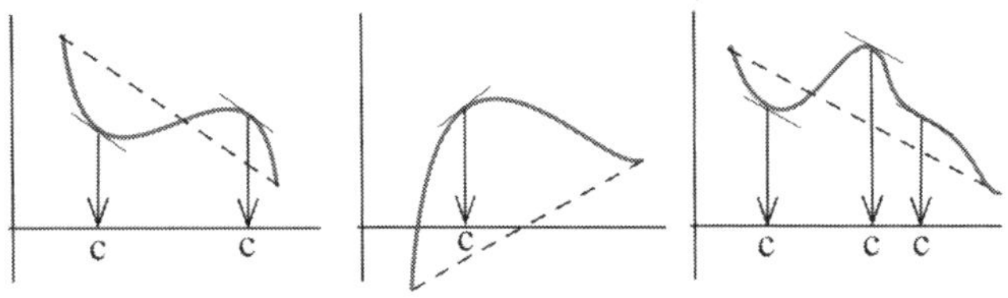

Fig. 7

The Mean Value Theorem also has a very natural interpretation if $f(x)$ represents the position of an object at time x: $f'(x)$ represents the velocity of the object at the **instant** x, and

$\frac{f(b) - f(a)}{b - a} = \frac{\text{change in position}}{\text{change in time}}$ represents the **average** (mean) velocity of the object during the time interval from time a to time b. The Mean Value Theorem says that there is a time c, between a and b, when the instantaneous velocity, $f'(c)$, is equal to the average velocity for the entire trip, $\frac{f(b) - f(a)}{b - a}$. If your average velocity during a trip is 30 miles per hour, then at some instant during the trip you were traveling exactly 30 miles per hour.

Practice 2: For $f(x) = 5x^2 - 4x + 3$ on the interval $[1,3]$, calculate $m = \frac{f(b) - f(a)}{b - a}$ and find the value of c so that $f'(c) = m$.

Some Consequences of the Mean Value Theorem

If the Mean Value Theorem was just an isolated result about the existence of a particular point c, it would not be very important or useful. However, the Mean Value Theorem is the basis of several results about the behavior of functions over entire intervals, and it is these consequences which give it an important place in calculus for both theoretical and applied uses.

The next two corollaries are just the first of many results which follow from the Mean Value Theorem.

We already know, from the Main Differentiation Theorem, that the derivative of a constant function $f(x) = k$ is always 0, but can a nonconstant function have a derivative which is always 0? The first corollary says no.

Corollary 1: If $f'(x) = 0$ for all x in an interval I, then $f(x) = K$, a constant, for all x in I.

Proof: Assume $f'(x) = 0$ for all x in an interval I, and pick any two points a and b $(a \neq b)$ in the interval. Then, by the Mean Value Theorem, there is a number c between a and b so that $f'(c) = \frac{f(b) - f(a)}{b - a}$. By our assumption, $f'(x) = 0$ for all x in I so we know that $0 = f'(c) = \frac{f(b) - f(a)}{b - a}$ and we can conclude that $f(b) - f(a) = 0$ and $f(b) = f(a)$. But a and b were any two points in I, so the value of $f(x)$ is the same for any two values of x in I, and f is a constant function on the interval I.

We already know that if two functions are parallel (differ by a constant), then their derivatives are equal, but can two nonparallel functions have the same derivative? The second corollary says no.

Corollary 2: If $f'(x) = g'(x)$ for all x in an interval I,
then $f(x) - g(x) = K$, a constant, for all x in I,
so the graphs of f and g are "parallel" on the interval I.

Proof: This corollary involves two functions instead of just one, but we can imitate the proof of the Mean Value Theorem and introduce a new function $h(x) = f(x) - g(x)$. The function h is differentiable, and $h'(x) = f'(x) - g'(x) = 0$ for all x in I, so, by Corollary 1, $h(x)$ is a constant function and $K = h(x) = f(x) - g(x)$ for all x in the interval. Then $f(x) = g(x) + K$.

We will use Corollary 2 hundreds of times in Chapters 4 and 5 when we work with "integrals". Typically you will be given the derivative of a function, f '(x) , and asked to find **all** functions f which have that derivative. Corollary 2 tells us that if we can find **one** function f which has the derivative we want, then the only other functions which have the same derivative are f(x) + K : once you find one function with the right derivative, you have essentially found all of them.

Example 2: (a) Find **all** functions whose derivatives equal 2x .

(b) Find a function g(x) with g '(x) = 2x and g(3) = 5.

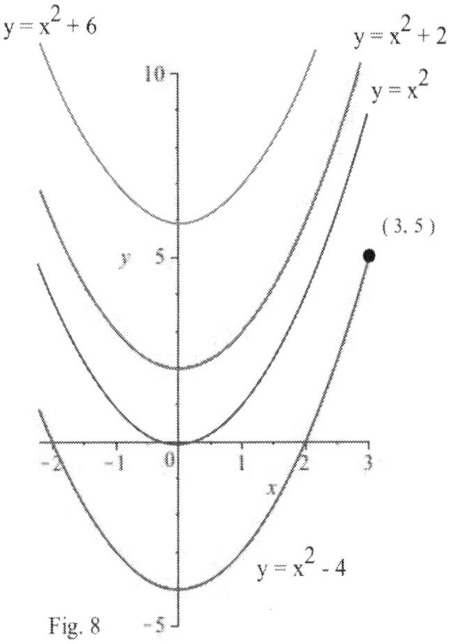

Fig. 8

Solution: (a) We can recognize that if $f(x) = x^2$ then $f'(x) = 2x$ so one function with the derivative we want is $f(x) = x^2$. Corollary 2 guarantees that every function g whose derivative is 2x has the form $g(x) = f(x) + K = x^2 + K$. The only functions with derivative 2x have the form $x^2 + K$.

(b) Since $g'(x) = 2x$, we know that g must have the form $g(x) = x^2 + K$, but this is a whole "family" of functions (Fig. 8), and we want to find one member of the family . We know that g(3) = 5 so we want to find the member of the family which goes through the point (3,5). All we need to do is replace the g(x) with 5 and the x with 3 in the formula $g(x) = x^2 + K$, and then solve for the value of K:

$5 = g(3) = (3)^2 + K$ so $K = -4$.

The function we want is $g(x) = x^2 - 4$.

Practice 3: Restate Corollary 2 as a statement about the positions and velocities of two cars.

PROBLEMS

1. In Fig. 9, find the location of the number(s) "c" which Rolle's Theorem promises (guarantees).

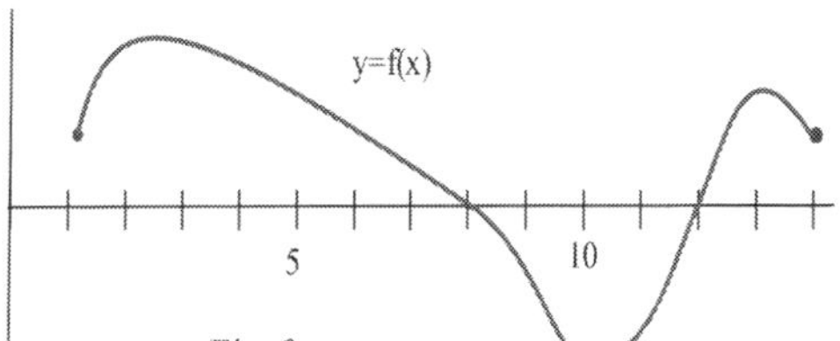

Fig. 9

For problems 2 – 4, verify that the hypotheses of Rolle's Theorem are satisfied for each of the functions on the given intervals, and find the value of the number(s) "c" which Rolle's Theorem promises exists.

2. (a) $f(x) = x^2$ on $[-2, 2]$ (b) $f(x) = x^2 - 5x + 8$ on $[0, 5]$
3. (a) $f(x) = \sin(x)$ on $[0, \pi]$ (b) $f(x) = \sin(x)$ on $[\pi, 5\pi]$
4. (a) $f(x) = x^3 - x + 3$ on $[-1, 1]$ (b) $f(x) = x \cdot \cos(x)$ on $[0, \pi/2]$

5. Suppose you toss a ball straight up and catch it when it comes down. If h(t) is the height of the ball at time t, then what does Rolle's Theorem say about the velocity of the ball? Why is it easier to catch a ball which someone on the ground tosses up to you on a balcony, than for you to be on the ground and catch a ball which someone on a balcony tosses down to you?

6. If $f(x) = 1/x^2$, then $f(-1) = 1$ and $f(1) = 1$ but $f'(x) = -2/x^3$ is never equal to 0. Why doesn't this function violate Rolle's Theorem?

7. If $f(x) = |x|$, then $f(-1) = 1$ and $f(1) = 1$ but $f'(x)$ is never equal to 0. Why doesn't this function violate Rolle's Theorem?

8. If $f(x) = x^2$, then $f'(x) = 2x$ is never 0 on the interval $[1, 3]$. Why doesn't this function violate Rolle's Theorem?

9. If I take off in an airplane, fly around for awhile and land at the same place I took off from, then my starting and stopping heights are the same but the airplane is always moving. Doesn't this violate Rolle's theorem which says there is an instant when my velocity is 0?

10. Prove the following corollary of Rolle's Theorem: If P(x) is a polynomial, then between any two roots of P there is a root of P'.

11. Use the corollary in problem 10 to justify the conclusion that the **only** root of $f(x) = x^3 + 5x - 18$ is 2. (Suggestion: What could you conclude about f' if f had another root?)

12. In Fig. 10, find the location(s) of the "c" which the Mean Value Theorem promises (guarantees).

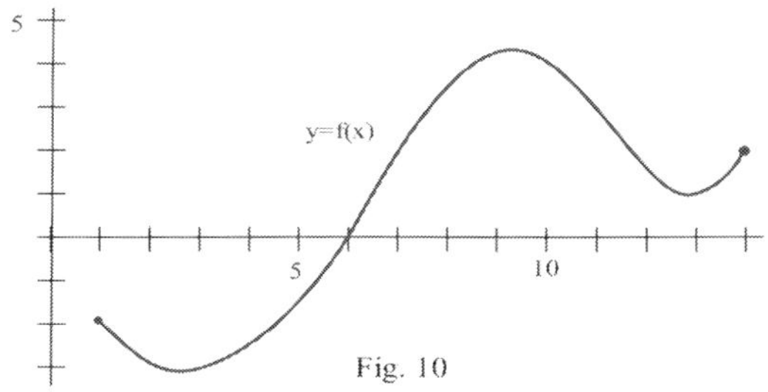

Fig. 10

In problems 13–15, verify that the hypotheses of the Mean Value Theorem are satisfied for each of the functions on the given intervals, and find the value of a number(s) "c" which Mean Value Theorem guarantees.

13. (a) $f(x) = x^2$ on $[0, 2]$ (b) $f(x) = x^2 - 5x + 8$ on $[1, 5]$

14. (a) $f(x) = \sin(x)$ on $[0, \pi/2]$ (b) $f(x) = x^3$ on $[-1, 3]$

15. (a) $f(x) = 5 - \sqrt{x}$ on $[1, 9]$ (b) $f(x) = 2x + 1$ on $[1, 7]$

16. For the quadratic functions in parts (a) and (b) of problem 13, the number c turned out to be the midpoint of the interval, $c = (a + b)/2$.

(a) For $f(x) = 3x^2 + x - 7$ on $[1, 3]$, show that $f'(2) = \dfrac{f(3) - f(1)}{3 - 1}$.

(b) For $f(x) = x^2 - 5x + 3$ on $[2, 5]$, show that $f'(7/2) = \dfrac{f(5) - f(2)}{5 - 2}$.

(c) For $f(x) = Ax^2 + Bx + C$ on $[a, b]$, show that $f'\left(\dfrac{a+b}{2}\right) = \dfrac{f(b) - f(a)}{b - a}$.

17. If $f(x) = | x |$, then $f(-1) = 1$ and $f(3) = 3$ but $f'(x)$ is never equal to $\frac{f(3) - f(-1)}{3 - (-1)} = \frac{1}{2}$. Why doesn't this function violate the Mean Value Theorem?

In problems 18 and 19, you are a traffic court judge. In each case, a speeding ticket has been given and you need to decide if the ticket is appropriate.

18. The toll taker says, "Your Honor, based on the elapsed time from when the car entered the toll road until the car stopped at my booth, I know the average speed of the car was 83 miles per hour. I did not actually see the car speeding, but I know it was and I gave the driver a speeding ticket."

19. The driver in the next case heard the toll taker and says, "Your Honor, my average velocity on that portion of the toll road was only 17 miles per hour, so I could not have been speeding. I don't deserve a ticket."

20. Find three different functions f, g and h so that $f'(x) = g'(x) = h'(x) = \cos(x)$.

21. Find a function f so that $f'(x) = 3x^2 + 2x + 5$ and $f(1) = 10$.

22. Find a function g so that $g'(x) = x^2 + 3$ and $g(0) = 2$.

23. Find values for A and B so that the graph of the parabola $f(x) = Ax^2 + B$ is
   (a) tangent to the line $y = 4x + 5$ at the point $(1,9)$
   (b) tangent to the line $y = 7 - 2x$ at the point $(2,3)$
   (c) tangent to the parabola $y = x^2 + 3x - 2$ at the point $(0,2)$

24. Sketch the graphs of several members of the "family" of functions whose derivatives always equal 3. Give a formula which defines every function in this family.

25. Sketch the graphs of several members of the "family" of functions whose derivatives always equal $3x^2$. Give a formula which defines every function in this family.

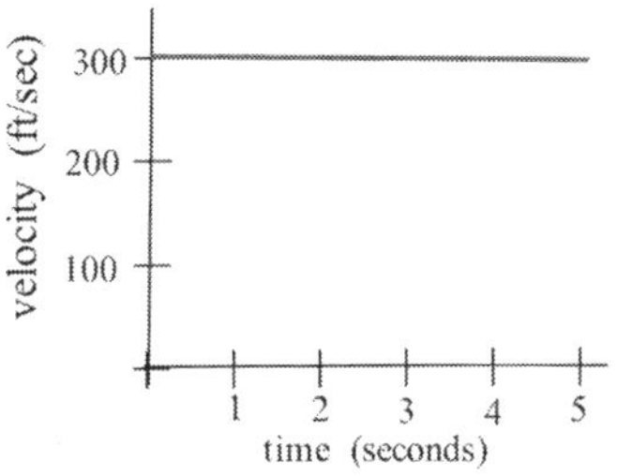

Fig. 11

26. At t seconds after takeoff, the upward velocity of a helicopter was $v(t) = 2t - 7$ feet/second. Two seconds after takeoff, the helicopter was 8 feet above sea level. Find a formula for the height of the helicopter at every time t.

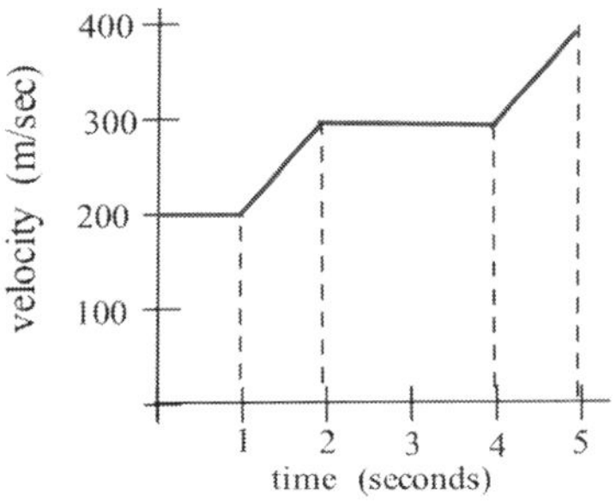

Fig. 12

27. Assume that a rocket is fired from the ground and has the upward velocity shown in Fig. 11. Estimate the height of the rocket when t = 1, 2, and 5 seconds.

28. Fig. 12 shows the upward velocity of a rocket. Use the information in the graph to estimate the height of the rocket when t = 1, 2, and 5 seconds.

29. Use the following information to determine an equation for f(x): $f''(x) = 6$, $f'(0) = 4$, and $f(0) = -5$.

30. Use the following information to determine an equation for g(x): $g''(x) = 12x$, $g'(1) = 9$, and $g(2) = 30$.

31. Define A(x) to be the **area** bounded by the x–axis, the line $y = 3$, and a vertical line at x (Fig. 13).

(a) Find a formula for A(x)? (b) Determine $A'(x)$

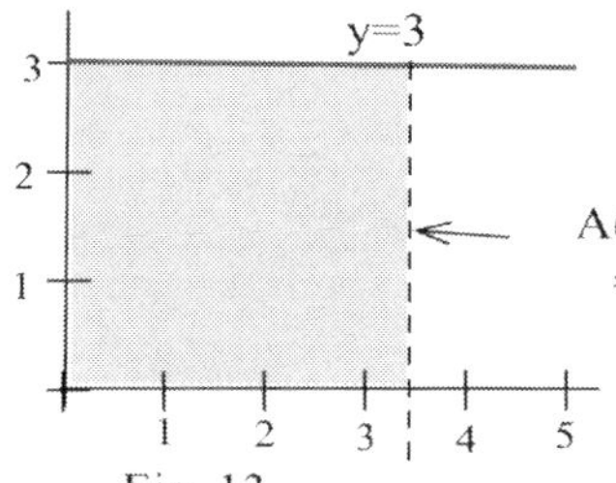

Fig. 13

32. Define A(x) to be the **area** bounded by the x–axis, the line $y = 2x$, and a vertical line at x (Fig. 14).

(a) Find a formula for A(x)? (b) Determine $A'(x)$

33. Define A(x) to be the **area** bounded by the x–axis, the line $y = 2x + 1$, and a vertical line at x (Fig. 15).

(a) Find a formula for A(x)? (b) Determine $A'(x)$

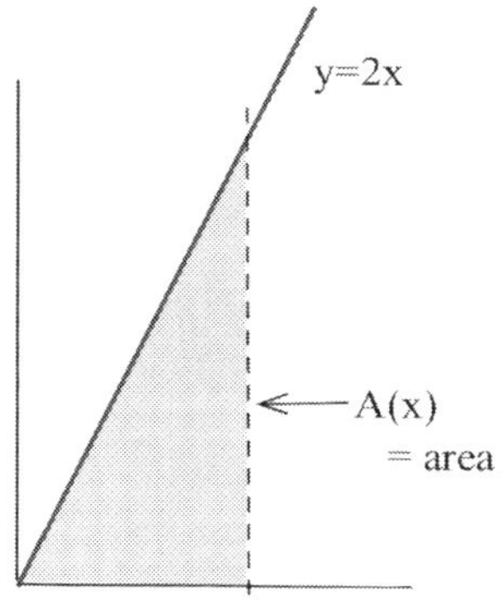

Fig. 14

In problems 34 – 36, we have a list of numbers $a_1, a_2, a_3, a_4, \ldots$, and the consecutive differences between numbers in the list are $a_2 - a_1, a_3 - a_2, a_4 - a_3, \ldots$

34. If $a_1 = 5$ and the difference between consecutive numbers in the list is always 0, what can you conclude about the numbers in the list?

35. If $a_1 = 5$ and the difference between consecutive numbers in the list is always 3, find a formula for a_n?

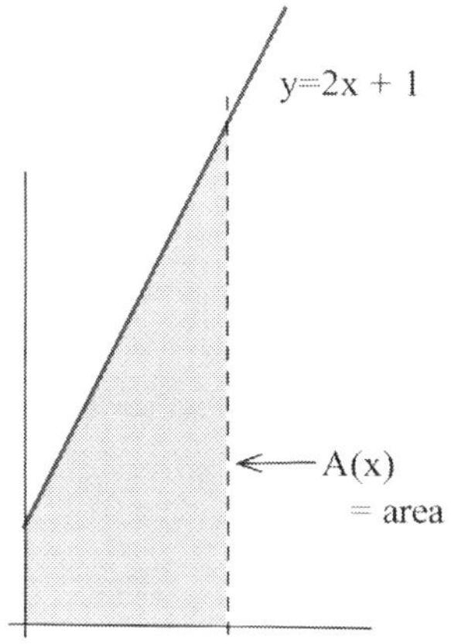

Fig. 15

36. Suppose the "a" list starts 3, 4, 7, 8, 6, 10, 13, . . ., and there is a "b" list which has the same differences between consecutive numbers as the "a" list.

(a) If $b_1 = 5$, find the next six numbers in the "b" list. How is b_n related to a_n?

(b) If $b_1 = 2$, find the next six numbers in the "b" list. How is b_n related to a_n?

(c) If $b_1 = B$, find the next six numbers in the "b" list. How is b_n related to a_n?

Section 3.2 **PRACTICE Answers**

Practice 1: $f'(x) = 0$ when $x = 2$ and 6 so $c = 2$ and $c= 6$.

$g'(x) = 0$ when $x = 2, 4$, and 6 so $c = 2, c = 4$, and $c = 6$.

Practice 2: $f(x) = 5x^2 - 4x + 3$ on $[1, 3]$. $f(1) = 4$ and $f(3) = 36$ so

$$m = \frac{f(b) - f(a)}{b - a} = \frac{36 - 4}{3 - 1} = 16.$$

$f'(x) = 10x - 4$ so $f'(c) = 10c - 4 = 16$ if $10c = 20$ and $\mathbf{c = 2}$.

The graph of f and the location of c are shown in Fig. 16.

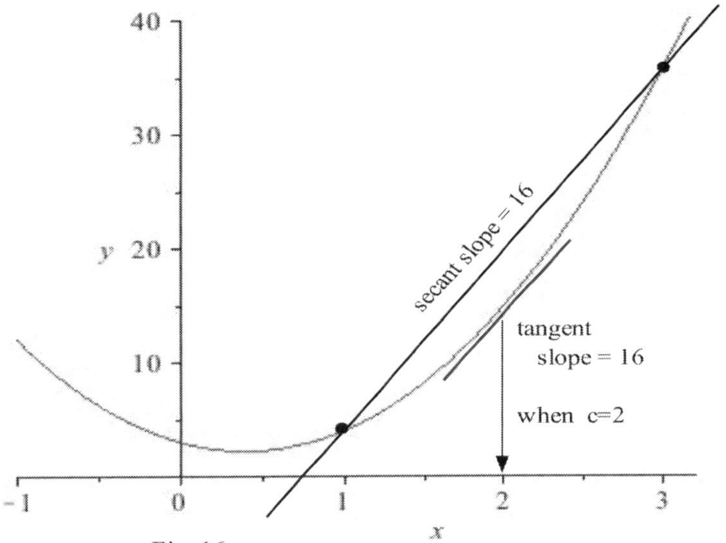

Fig. 16

Practice 3:

If two cars have the same velocities during an interval of time ($f'(t) = g'(t)$ for t in I)

then the cars are always a constant distance apart during that time interval.

(Note: The "same velocity" means **same speed** and **same direction**. If two cars are traveling at the same speed but in different directions, then the distance between them changes and is not constant.)

3.3 THE FIRST DERIVATIVE AND THE SHAPE OF f

This section examines some of the interplay between the shape of the graph of f and the behavior of f '. If we have a graph of f , we will see what we can conclude about the values of f '. If we know values of f ', we will see what we can conclude about the graph of f .

Definitions: The function f is **increasing on (a,b)** if $a < x_1 < x_2 < b$ implies $f(x_1) < f(x_2)$.
The function f is **decreasing on (a,b)** if $a < x_1 < x_2 < b$ implies $f(x_1) > f(x_2)$.

f is **monotonic on (a,b)** if f is increasing on (a,b) or if f is decreasing on (a,b).

Graphically, f is **increasing** (decreasing) if, as we move from left to right along the graph of f, the height of the graph **increases** (decreases).

These same ideas make sense if we consider h(t) to be the height (in feet) of a rocket at time t seconds. We naturally say that the rocket is rising or that its height is increasing if the height h(t) increases over a period of time, as t increases.

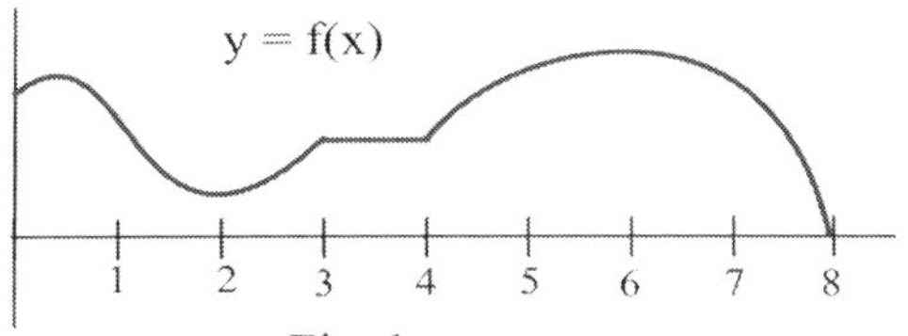

Fig. 1

Example 1: List the intervals on which the function given in Fig. 1 is increasing or decreasing.

Solution: f is increasing on the intervals [0,1] , [2,3] and [4,6]. f is decreasing on [1,2] and [6,8]. On the interval [3,4] the function is not increasing or decreasing, it is constant. It is also valid to say that f is increasing on the intervals [0.3, 0.8] and (0.2, 0.5) as well as many others, but we usually talk about the longest intervals on which f is monotonic.

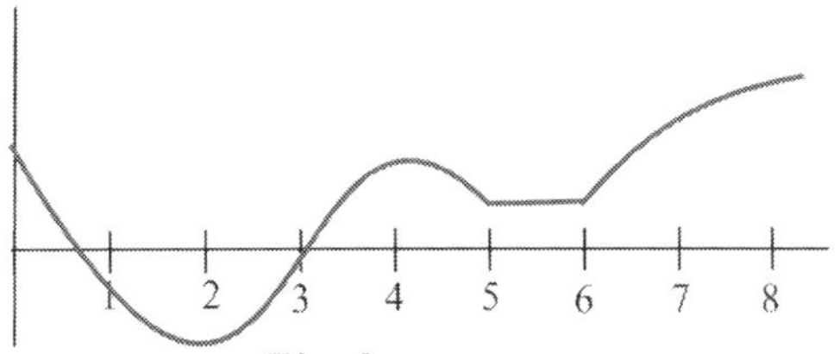

Fig. 2

Practice 1: List the intervals on which the function given in Fig. 2 is increasing or decreasing.

If we have an accurate graph of a function, then it is relatively easy to determine where f is monotonic, but if the function is defined by an equation, then a little more work is required. The next two theorems relate the values of the derivative of f to the monotonicity of f. The first theorem says that if we know where f is monotonic, then we also know something about the values of f '. The second theorem says that if we know about the values of f ' then we can draw conclusions about where f is monotonic.

First Shape Theorem

For a function f which is differentiable on an interval (a,b);

(i) if f is increasing on (a,b), then $f'(x) \geq 0$ for all x in (a,b)

(ii) if f is decreasing on (a,b), then $f'(x) \leq 0$ for all x in (a,b)

(iii) if f is constant on (a,b), then $f'(x) = 0$ for all x in (a,b).

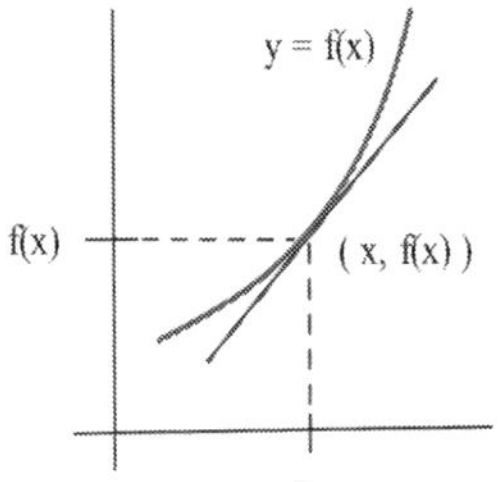

Fig. 3

Proof: Most people find a picture such as Fig. 3 to be a convincing justification of this theorem: if the graph of f increases near a point (x, f(x)), then the tangent line is also increasing, and the slope of the tangent line is positive (or perhaps zero at a few places). A more precise proof, however, requires that we use the definitions of the derivative of f and of "increasing".

(i) Assume that f is increasing on (a,b). We know that f is differentiable, so if x is any number in the interval (a,b) then

$$f'(x) = \lim_{h\to 0} \frac{f(x+h) - f(x)}{h}$$, and this limit exists and is a finite value.

If h is any small enough **positive** number so that x + h is also in the interval (a,b), then $x < x + h$ and $f(x) < f(x + h)$. We know that the numerator, $f(x + h) - f(x)$, and the denominator, h, are both positive so the limiting value, $f'(x)$, must be positive or zero: $f'(x) \geq 0$.

(ii) Assume that f is decreasing on (a,b): The proof of this part is very similar to part (i). If $x < x + h$, then $f(x) > f(x + h)$ since f is decreasing on (a,b). Then the numerator of the limit, $f(x + h) - f(x)$, will be negative and the denominator, h, will still be positive, so the limiting value, $f'(x)$, must be negative or zero: $f'(x) \leq 0$.

(iii) The derivative of a constant is zero, so if f is constant on (a,b) then $f'(x) = 0$ for all x in (a,b).

The previous theorem is easy to understand, but you need to pay attention to exactly what it says and what it does not say. It is possible for a differentiable function which is increasing on an interval to have horizontal tangent lines at some places in the interval (Fig 4). It is also possible for a continuous function which is increasing on an interval to have an undefined derivative at some places in the interval (Fig. 4). Finally, it is possible for a function which is increasing on an interval to fail to be continuous at some places in the interval (Fig. 5).

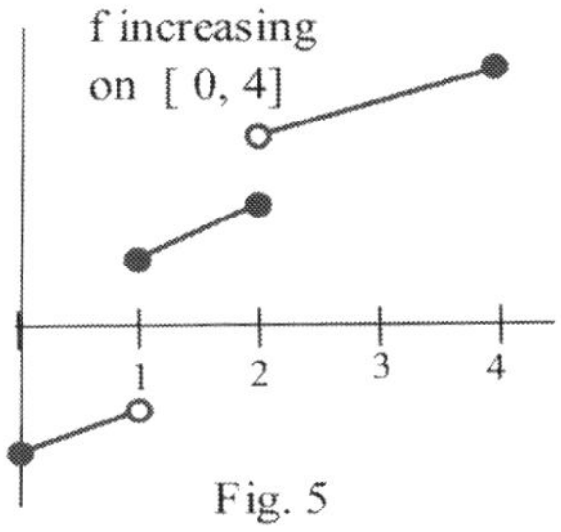

Fig. 5

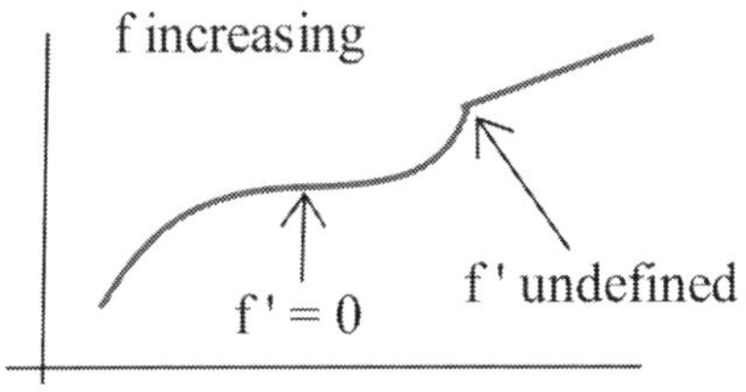

Fig. 4

The First Shape Theorem has a natural interpretation in terms of the height h(t) and upward velocity h '(t) of a helicopter at time t. If the height of the helicopter is increasing (h(t) is increasing) , then the helicopter has a positive or zero upward velocity: h '(t) ≥ 0. If the height of the helicopter is not changing, then its upward velocity is 0: h '(t) = 0.

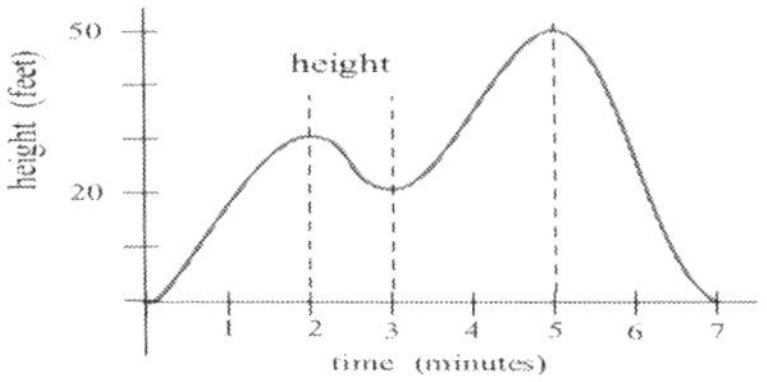

Fig. 6

Example 2: Fig. 6 shows the height of a helicopter during a period of time. Sketch the graph of the upward velocity of the helicopter, dh/dt.

Solution: The graph of v(t) = dh/dt is shown in Fig. 7. Notice that the h(t) has a local maximum when t = 2 and t = 5, and v(2) = 0 and v(5) = 0. Similarly, h(t) has a local minimum when t = 3, and v(3) = 0. When h is increasing, v is positive. When h is decreasing, v is negative.

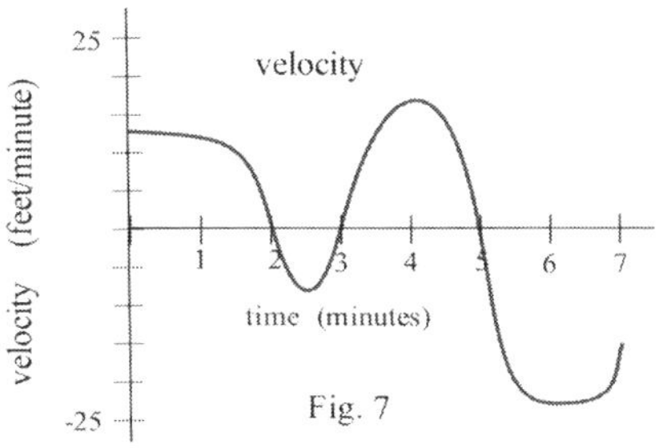

Fig. 7

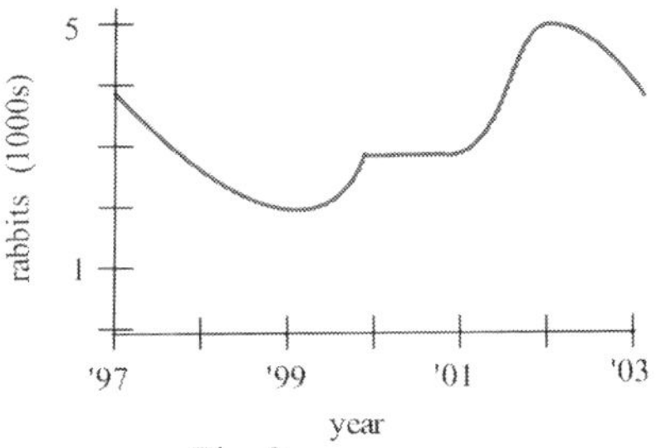

Fig. 8

Practice 2: Fig. 8 shows the population of rabbits on an island during 6 years. Sketch the graph of the rate of population change, dR/dt, during those years.

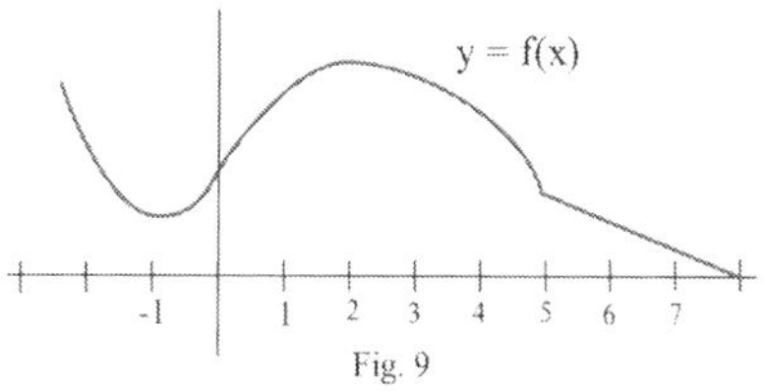

Fig. 9

Example 3: The graph of f is shown in Fig. 9 . Sketch the graph of f ' .

Solution: It is a good idea to look first for the points where f '(x) = 0 or where f is not differentiable, the critical points of f. These locations are usually easy to spot, and they naturally break the problem into several smaller pieces. The only numbers at which f '(x) = 0 are x = –1 and x = 2, so the only places the graph of f '(x) will cross the x–axis

are at x = –1 and x = 2 , and we can plot the point (–1,0) and (2,0) on the graph of f '. The only place that f is not differentiable is at the "corner" above x = 5, so the graph of f ' will not have a point for x = 5. The rest of the graph of f is relatively easy:

if x < –1 then	f(x) is decreasing	so f '(x) is negative,
if –1 < x < 2 then	f(x) is increasing	so f '(x) is positive,
if 2 < x < 5 then	f(x) is decreasing	so f '(x) is negative, and
if 5 < x then	f(x) is decreasing	so f '(x) is negative.

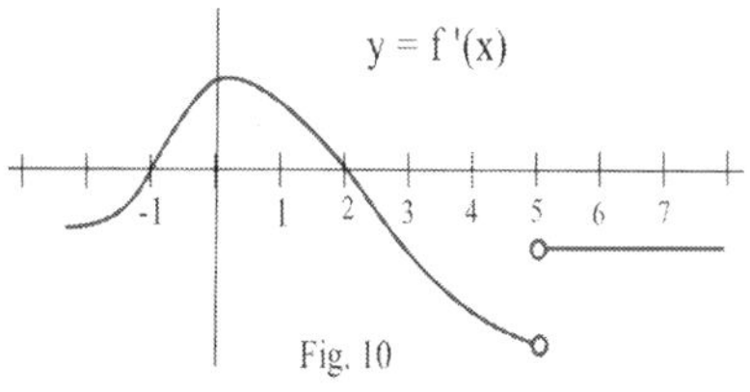

Fig. 10

The graph of f ' is shown in Fig. 10. f(x) is continuous at x =5, but f is not differentiable at x = 5, as is indicated by the "hole" in the graph.

Practice 3: The graph of f is shown in Fig. 11. Sketch the graph of f '. (The graph of f has a "corner" at x = 5.)

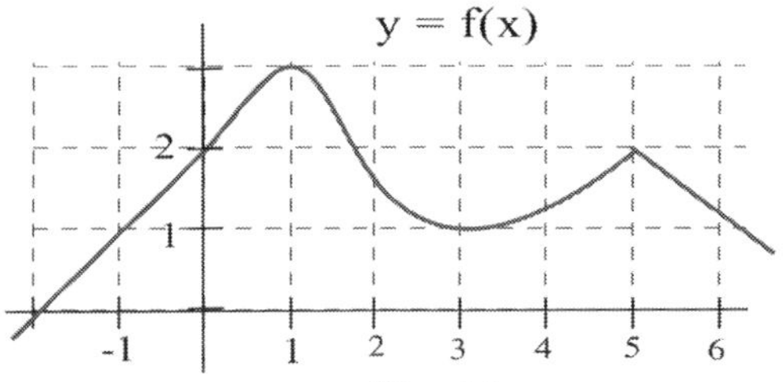

Fig. 11

The next theorem is almost the converse of the First Shape Theorem and explains the relationship between the values of the derivative and the graph of a function from a different perspective. It says that if we know something about the values of f ', then we can draw some conclusions about the shape of the graph of f.

> **Second Shape Theorem**
>
> For a function f which is differentiable on an interval I;
>
> (i) if $f'(x) > 0$ for all x in the interval I, then f is increasing on I,
>
> (ii) if $f'(x) < 0$ for all x in the interval I, then f is decreasing on I,
>
> (iii) if $f'(x) = 0$ for all x in the interval I, then f is constant on I.

Proof: This theorem follows directly from the Mean Value Theorem, and part (c) is just a restatement of the First Corollary of the Mean Value Theorem.

(a) Assume that $f'(x) > 0$ for all x in I and pick any points a and b in I with $a < b$. Then, by the Mean Value Theorem, there is a point c between a and b so that $\frac{f(b) - f(a)}{b - a} = f'(c) > 0$, and we can conclude that $f(b) - f(a) > 0$ and $f(b) > f(a)$. Since $a < b$ implies that $f(a) < f(b)$, we know that f is increasing on I.

(b) Assume that $f'(x) < 0$ for all x in I and pick any points a and b in I with $a < b$. Then there is a point c between a and b so that $\frac{f(b) - f(a)}{b - a} = f'(c) < 0$, and we can conclude that $f(b) - f(a) = (b-a) f'(c) < 0$ so $f(b) < f(a)$. Since $a < b$ implies that $f(a) > f(b)$, we know f is decreasing on I.

Practice 4: Rewrite the Second Shape Theorem as a statement about the height h(t) and upward velocity h '(t) of a helicopter at time t seconds.

The value of the function at a number x tells us the height of the graph of f above or below the point x on the x–axis. The value of f ' at a number x tells us whether the graph of f is increasing or decreasing (or neither) as the graph passes through the point (x, f(x)) on the graph of f. If f(x) is positive, it is possible for f '(x) to be positive, negative, zero or undefined: the value of f(x) has absolutely nothing to do with the value of f '. Fig. 12 illustrates some of the combinations of values for f and f '.

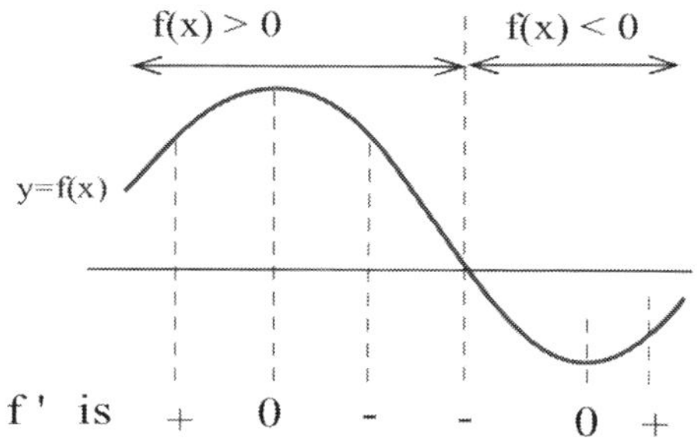

Fig. 12

Practice 5: Graph a continuous function which satisfies the conditions on f and f ' given below:

x	–2	–1	0	1	2	3
f(x)	1	–1	–2	–1	0	2
f '(x)	–1	0	1	2	–1	1

The Second Shape Theorem is particularly useful if we need to graph a function f which is defined by an equation. Between any two consecutive critical numbers of f, the graph of f is monotonic (why?). If we can find all of the critical numbers of f, then the domain of f will be naturally broken into a number of pieces on which f will be monotonic.

Example 4: Use information about the values of f ' to help graph $f(x) = x^3 - 6x^2 + 9x + 1$.

Solution: $f'(x) = 3x^2 - 12x + 9 = \mathbf{3(x-1)(x-3)}$ so $f'(x) = 0$ only when $x = 1$ or $x = 3$. f ' is a polynomial so it is always defined. The only critical numbers for f are $x = 1$ and $x = 3$, and they divide the real number line into three pieces on which f is monotonic: $(-\infty, 1)$, $(1,3)$ and $(3, \infty)$.

If $x < 1$, then $f'(x) = 3$(negative number)(negative number) > 0 so f is increasing.

If $1 < x < 3$, then $f'(x) = 3$(positive number)(negative number) < 0 so f is decreasing.

If $3 < x$, then $f'(x) = 3$(positive number)(positive number) > 0 so f is increasing.

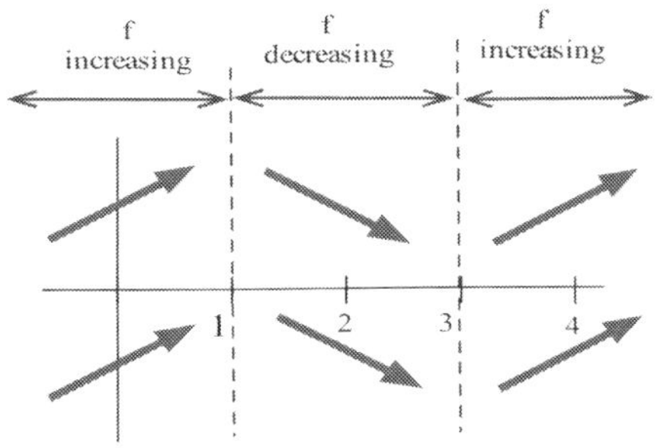

Fig. 13

Even though we don't know the value of f anywhere yet, we do know a lot about the shape of the graph of f : as we move from left to right along the x–axis, the graph of f increases until $x = 1$, then the graph decreases until $x = 3$, and then the graph increases again (Fig. 13) The graph of f makes "turns" when $x = 1$ and $x = 3$.

To plot the graph of f, we still need to evaluate f at a few values of x, but only at a very few values. $f(1) = 5$, and $(1,5)$ is a local maximum of f. $f(3) = 1$, and $(3,1)$ is a local minimum of f. The graph of f is shown in Fig. 14.

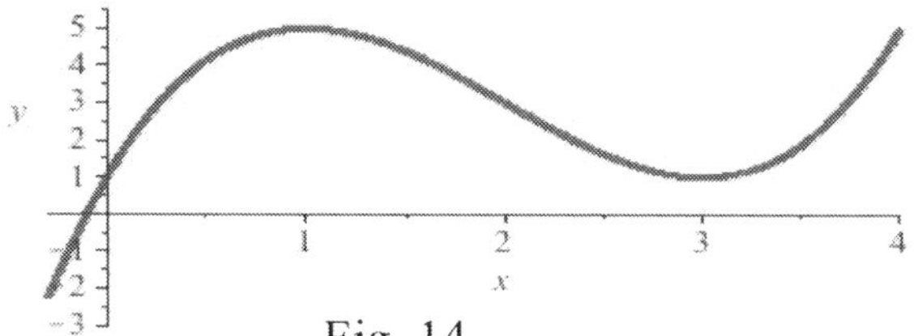

Fig. 14

Practice 6: Use information about the values of f ' to help graph $f(x) = x^3 - 3x^2 - 24x + 5$.

Example 5: Use the graph of f ' in Fig. 15 to sketch the shape of the graph of f. Why isn't the graph of f ' enough to completely determine the graph of f ?

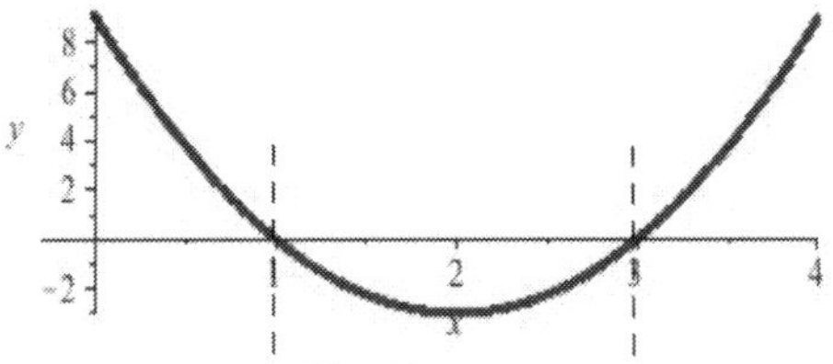
Fig. 15

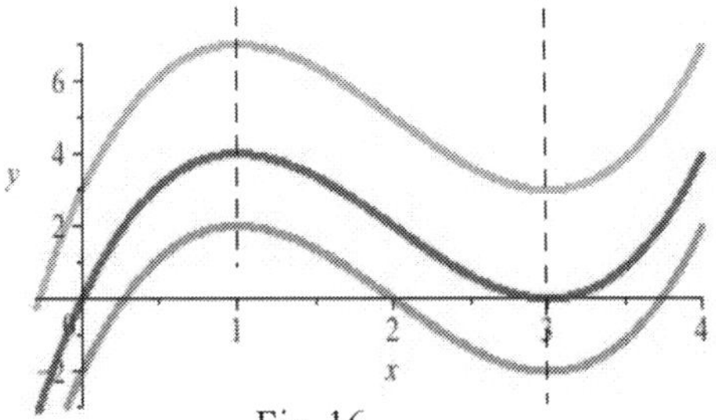
Fig. 16

Solution: Several functions which have the derivative we want are given in Fig. 16 , and each of them is a correct answer. By the Second Corollary to the Mean Value Theorem, we know there is a whole family of parallel functions which have the derivative we want, and each of these functions is a correct answer. If we had additional information about the function such as a point it went through, then only one member of the family would satisfy the extra condition and that function would be the only correct answer.

Practice 7: Use the graph of g ' in Fig. 17 to sketch the shape of a graph of g.

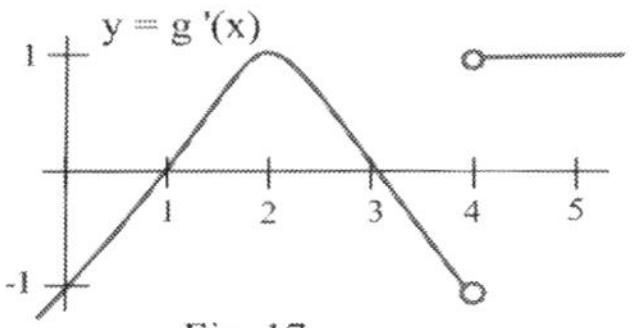

Fig. 17

Practice 8: A weather balloon is released from the ground and sends back its upward velocity measurements (Fig. 18). Sketch a graph of the height of the balloon. When was the balloon highest?

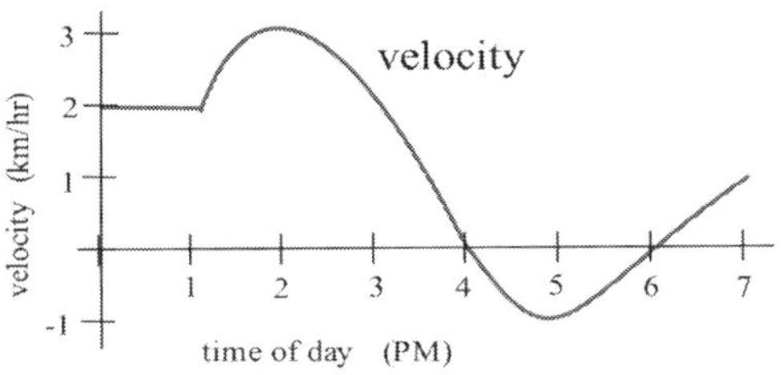

Fig. 18

Using the Derivative to Test for Extremes

The first derivative of a function tells about the general shape of the function, and we can use that shape information to determine if an extreme point is a maximum or minimum or neither.

First Derivative Test for Local Extremes

Let f be a continuous function with $f'(a) = 0$ or $f'(a)$ is undefined.

(i) If f '(left of a) > 0 and f '(right of a) < 0, then (a,f(a)) is a local maximum (Fig. 19a)

(ii) If f '(left of a) < 0 and f '(right of a) > 0, then (a,f(a)) is a local minimum (Fig. 19b)

(iii) If f '(left of a) > 0 and f '(right of a) > 0, then (a,f(a)) is **not** a local extreme (Fig. 19c)

(iv) If f '(left of a) < 0 and f '(right of a) < 0, then (a,f(a)) is **not** a local extreme (Fig. 19d)

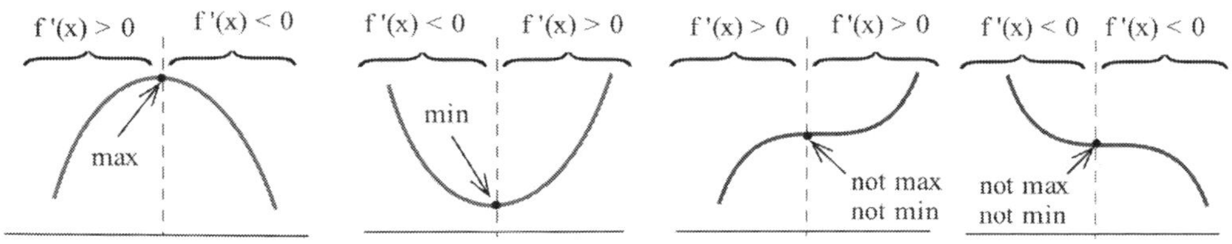

Fig. 19

Practice 9: Find all extremes of $f(x) = 3x^2 - 12x + 7$ and use the First Derivative Test to determine if they are maximums, minimums or neither.

A variant of the First Derivative Test can also be used to determine whether an endpoint gives a maximum or minimum for a function.

PROBLEMS

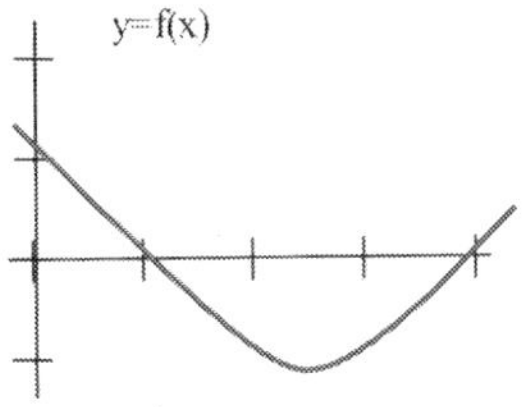

Fig. 20

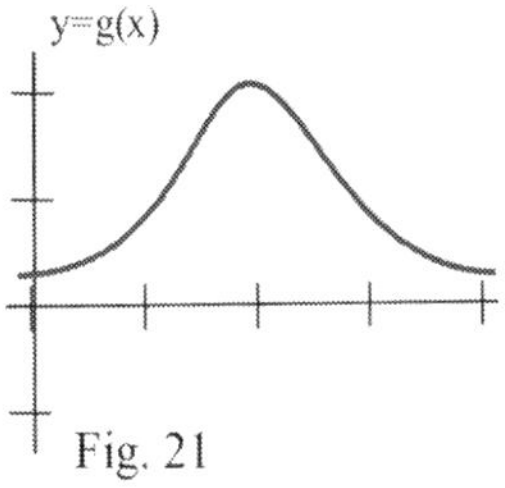

Fig. 21

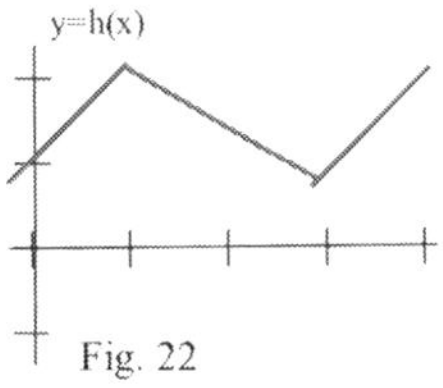

Fig. 22

In problems 1–3, sketch the graph of the derivative of each function.

1. Use Fig. 20.
2. Use Fig. 21.
3. Use Fig. 22

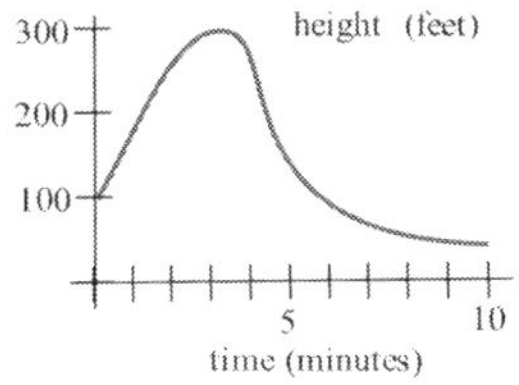

Fig. 23

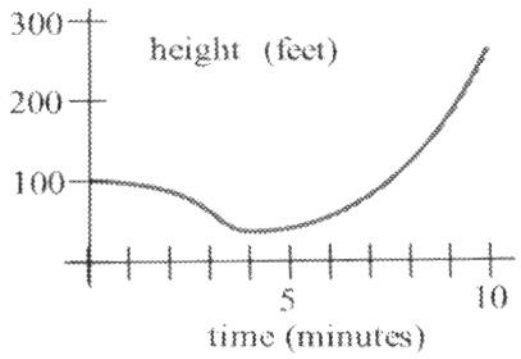

Fig. 24

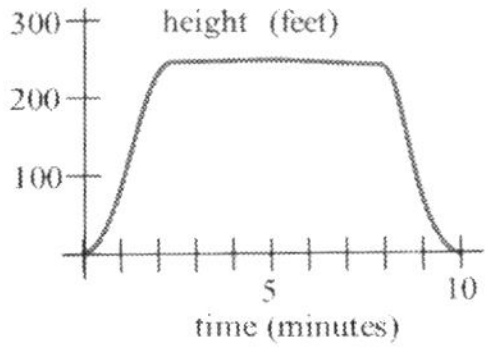

Fig. 25

In problems 4–6, the graph of the height of a helicopter is shown. Sketch the graph of the upward velocity of the helicopter.

4. Use Fig. 23.
5. Use Fig. 24.
6. Use Fig. 25.

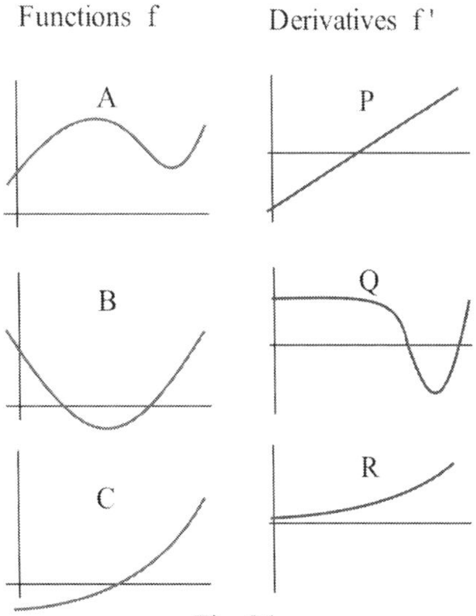

Fig. 26

7. In Fig. 26, match the graphs of the functions with those of their derivatives.
8. In Fig. 27, match the graphs showing the heights of rockets with those showing their velocities.
9. Use the Second Shape Theorem to show that $f(x) = \ln(x)$ is monotonic increasing on $(0, \infty)$.
10. Use the Second Shape Theorem to show that $g(x) = e^x$ is increasing on the entire real number line.

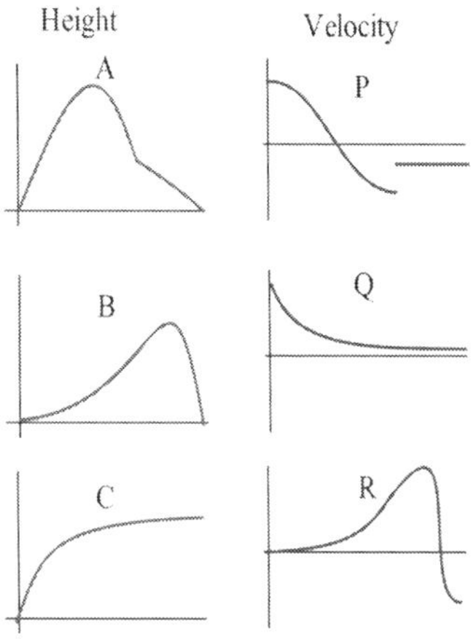

Fig. 27

11. A student is working with a complicated function f and has shown that the derivative of f is always positive. A minute later the student also claims that $f(x) = 2$ when $x = 1$ and when $x = \pi$. Without checking the student's work, how can you be certain that it contains an error?

12. Fig. 28 shows the graph of the **derivative** of a continuous function f.
    (a) List the critical numbers of f.
    (b) For what values of x does f have a local maximum?
    (c) For what values of x does f have a local minimum?

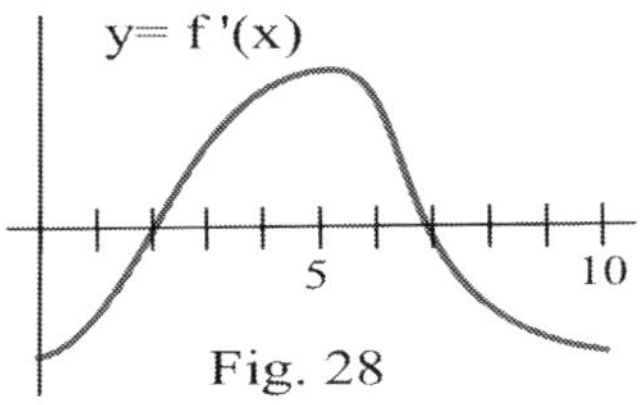

Fig. 28

13. Fig. 29 shows the graph of the **derivative** of a continuous function g.
    (a) List the critical numbers of g.
    (b) For what values of x does g have a local maximum?
    (c) For what values of x does g have a local minimum?

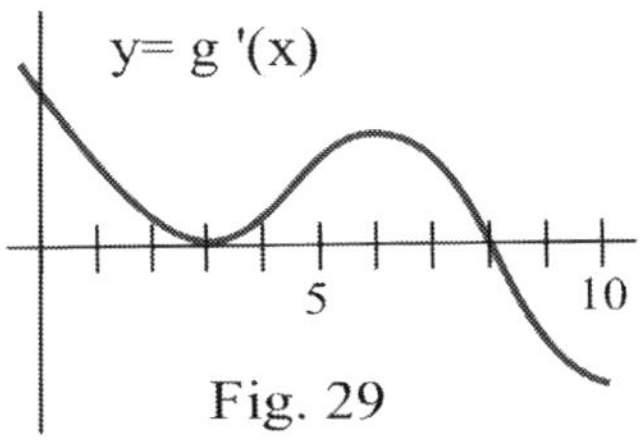

Fig. 29

In problems 14–16, the graphs of the upward velocities of several helicopters are shown. Use each graph to determine when each helicopter was at a relatively maximum and minimum height.

14. Use Fig. 30. 15. Use Fig. 31. 16. Use Fig. 32.

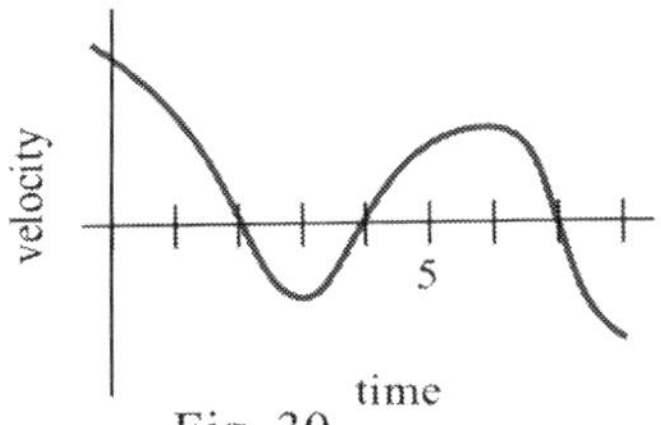

Fig. 30

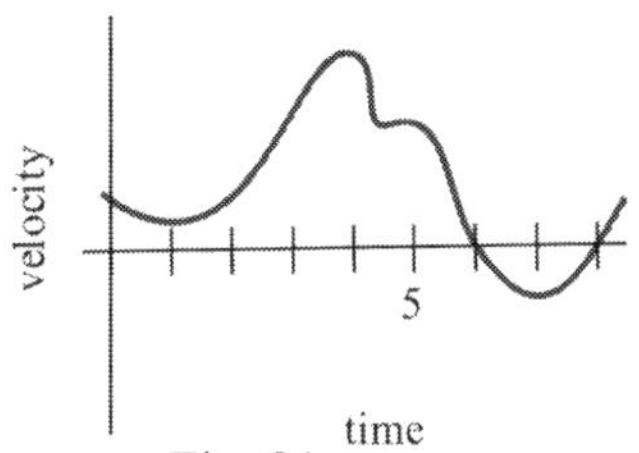

Fig. 31

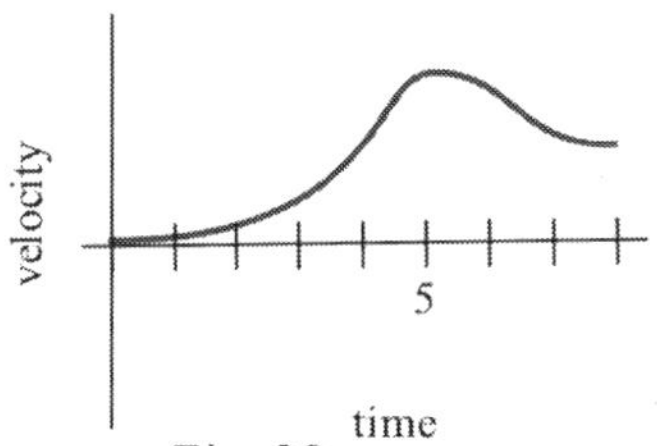

Fig. 32

In problems 17 – 22 , use information from the derivative of each function to help you graph the function. Find all local maximums and minimums of each function.

17. $f(x) = x^3 - 3x^2 - 9x - 5$ 18. $g(x) = 2x^3 - 15x^2 + 6$ 19. $h(x) = x^4 - 8x^2 + 3$

20. $s(t) = t + \sin(t)$ 21. $r(t) = \dfrac{2}{t^2 + 1}$ 22. $f(x) = \dfrac{x^2 + 3}{x}$

23. $f(x) = 2x + \cos(x)$ so $f(0) = 1$. Without graphing the function, you can be certain that f has how many **positive** roots? (zero, one, two, more than two)

24. $g(x) = 2x - \cos(x)$ so $g(0) = -1$. Without graphing the function, you can be certain that g has how many **positive** roots? (zero, one, two, more than two)

25. $h(x) = x^3 + 9x - 10$ has a root at $x = 1$. Without graphing h, show that h has no other roots.

26. Sketch the graphs of monotonic decreasing functions which have exactly (a) zero roots, (b) one root, and (c) two roots.

27. Each of the following statements is false. Give (or sketch) a counterexample for each statement.
    (a) If f is increasing on an interval I, then $f'(x) > 0$ for all x in I.
    (b) If f is increasing and differentiable on I, then $f'(x) > 0$ for all x in I.
    (c) If cars A and B always have the same speed, then they will always be the same distance apart.

28. (a) Give the equations of several different functions f which all have the same derivative f '(x) = 2.

(b) Give the equation of the function f with derivative f '(x) = 2 which also satisfies f(1) = 5.

(c) Give the equation of the function g with g '(x) = 2 , and the graph of g goes through (2, 1).

29. (a) Give the equations of several different functions h which all have the same derivative h '(x) = 2x.

(b) Give the equation of the function f with derivative f '(x) = 2x which also satisfies f(3) = 20.

(c) Give the equation of the function g with g '(x) = 2x , and the graph of g goes through (2, 7).

30. Sketch functions with the given properties to help you determine whether each statement is True or False.

(a) If f '(7) > 0 and f '(x) > 0 for all x near 7, then f(7) is a local maximum of f on [1,7].

(b) If g '(7) < 0 and g '(x) < 0 for all x near 7, then g(7) is a local minimum of g on [1,7].

(c) If h '(1) > 0 and h '(x) > 0 for all x near 1, then h(1) is a local minimum of h on [1,7].

(d) If r '(1) < 0 and r '(x) < 0 for all x near 1, then r(1) is a local maximum of r on [1,7].

(e) If s '(7) = 0, then s(7) is a local maximum of s on [1,7].

Section 3.3 **PRACTICE Answers**

Practice 1: g is increasing on [2, 4] and [6, 8].

g is decreasing on [0, 2] and [4, 5].

g is constant on [5, 6].

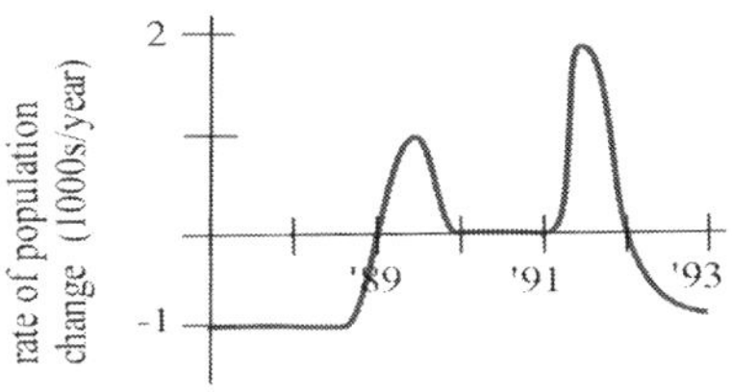

Fig. 33

Practice 2: The graph in Fig. 33 shows the rate of population change, dR/dt.

Practice 3: The graph of f ' is shown in Fig. 34. Notice how the graph of f ' is 0 where f has a maximum and minimum.

Practice 4: The Second Shape Theorem for helicopters:

(i) If the upward velocity h ' is positive during time interval I then the height h is increasing during time interval I.

(ii) If the upward velocity h ' is negative during time interval I then the height h is decreasing during time interval I.

(iii) If the upward velocity h ' is zero during time interval I then the height h is constant during time interval I.

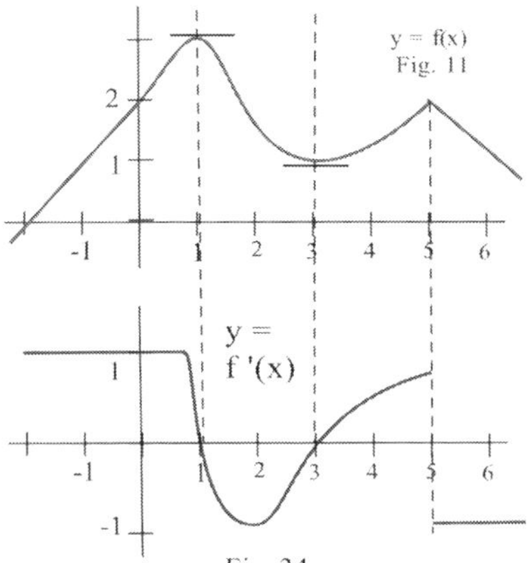

Fig. 34

Practice 5: A graph satisfying the conditions in the table is shown in Fig. 35.

x	–2	–1	0	1	2	3
f(x)	1	–1	–2	–1	0	2
f '(x)	–1	0	1	2	–1	1

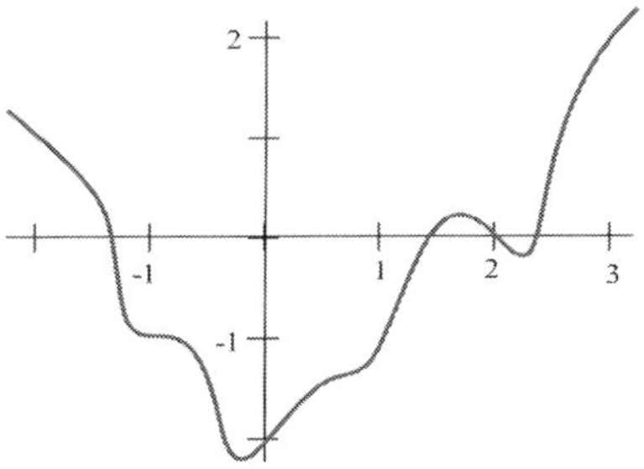

Fig. 35

Practice 6: $f(x) = x^3 - 3x^2 - 24x + 5$. $f'(x) = 3x^2 - 6x - 24 = 3(x - 4)(x + 2)$.

$f'(x) = 0$ if $x = -2, 4$.

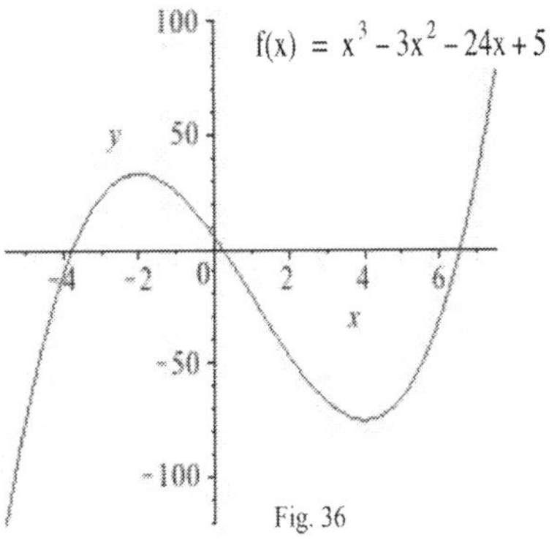

Fig. 36

If $x < -2$,

then $f'(x) = 3(x - 4)(x + 2) = 3(\text{negative})(\text{negative}) > 0$ so f is increasing.

If $-2 < x < 4$,

then $f'(x) = 3(x - 4)(x + 2) = 3(\text{negative})(\text{positive}) < 0$ so f is decreasing.

If $x > 4$,

then $f'(x) = 3(x - 4)(x + 2) = 3(\text{positive})(\text{positive}) > 0$ so f is increasing.

f has a relative maximum at $x = -2$ and a relative minimum at $x = 4$.

The graph of f is shown in Fig. 36.

Practice 7: Fig. 37 shows several possible graphs for g. Each of the g graphs has the correct shape to give the graph of g '. Notice that the graphs of g are "parallel," differ by a constant.

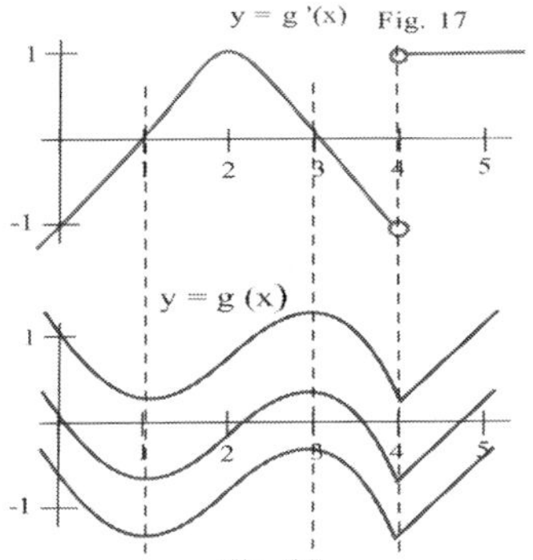

Fig. 37

Practice 8: Fig. 38 shows the height graph for the balloon. The balloon was highest at 4 pm and had a local minimum at 6pm.

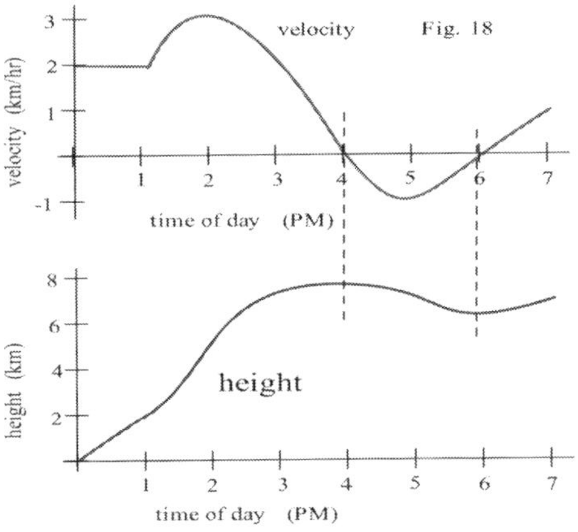

Fig. 38

Practice 9:

$f(x) = 3x^2 - 12x + 7$ so $f'(x) = 6x - 12$. $f'(x) = 0$ if $x = 2$.

If $x < 2$, then $f'(x) < 0$ and f is decreasing.

If $x > 2$, then $f'(x) > 0$ and f is increasing.

From this we can conclude that f has a minimum when $x = 2$ and has a shape similar to Fig. 19(b).

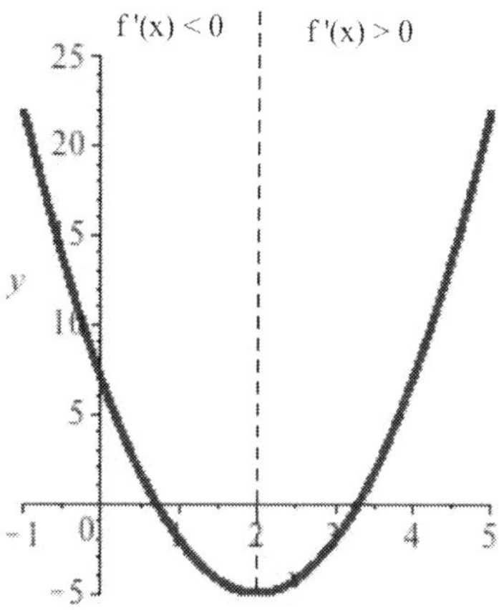

Fig. 39

We could also notice that the graph of the quadratic $f(x) = 3x^2 - 12x + 7$ is an upward opening parabola. The graph of f is shown in Fig. 39.

3.4 SECOND DERIVATIVE AND THE SHAPE OF f

The first derivative of a function gives information about the shape of the function, so the second derivative of a function gives information about the shape of the first derivative and about the shape of the function. In this section we investigate how to use the second derivative and the shape of the first derivative to reach conclusions about the shape of the function. The first derivative tells us whether the graph of f is increasing or decreasing. The second derivative will tell us about the "concavity" of f, whether f is curving upward or downward.

Concavity

Graphically, a function is concave up if its graph is curved with the opening upward (Fig. 1a). Similarly, a function is concave down if its graph opens downward (Fig. 1b). The concavity of a function can be important in applied problems and can even affect billion–dollar decisions.

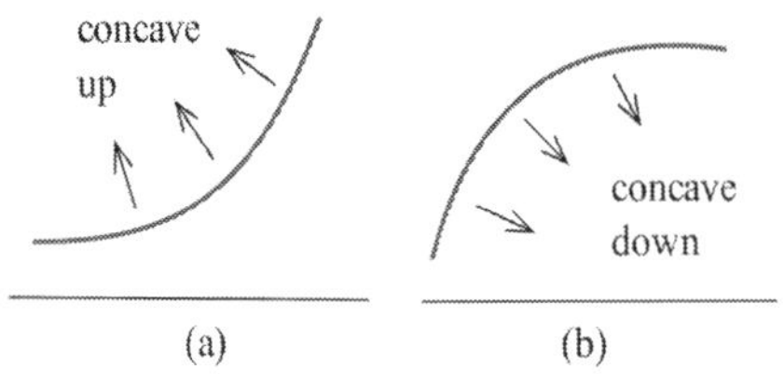

Fig. 1

An Epidemic: Suppose an epidemic has started, and you, as a member of congress, must decide whether the current methods are effectively fighting the spread of the disease or whether more drastic measures and more money are needed. In Fig. 2, f(x) is the number of people who have the disease at time x, and two different situations are shown. In both (a) and (b), the number of people with the disease, f(now), and the rate at which new people are getting sick, f '(now), are the same. The difference in the two situations is the concavity of f, and that difference in concavity might have a big effect on your decision. In (a), f is concave down at "now", and it appears that the current methods are starting to bring the epidemic under control. In (b), f is concave up, and it appears that the epidemic is still out of control.

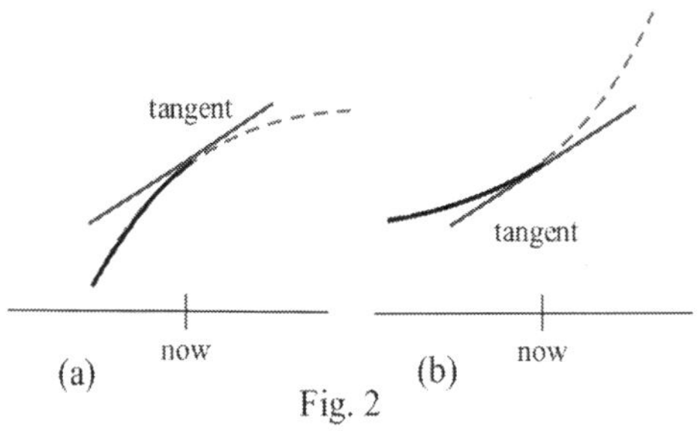

Fig. 2

Usually it is easy to determine the concavity of a function by examining its graph, but we also need a definition which does not require that we have a graph of the function, a definition we can apply to a function described by a formula without having to graph the function.

Definition: Let f be a differentiable function.

f is **concave up** at a if the graph of f is above the tangent line L to f for all x close to a (but not equal to a) : $\mathbf{f(x) > L(x)} = f(a) + f'(a)(x - a)$.

f is **concave down** at a if the graph of f is below the tangent line L to f for all x close to a (but not equal to a) : $\mathbf{f(x) < L(x)} = f(a) + f'(a)(x - a)$.

Fig. 3 shows the concavity of a function at several points. The next theorem gives an easily applied test for the concavity of a function given by a formula.

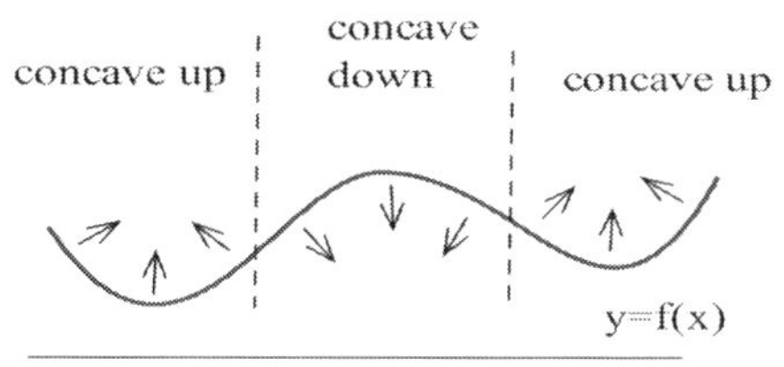

Fig. 3

The Second Derivative Condition for Concavity

(a) If $f''(x) > 0$ on an interval I, then $f'(x)$ is increasing on I and f is concave up on I.

(b) If $f''(x) < 0$ on an interval I, then $f'(x)$ is decreasing on I and f is concave down on I.

(c) If $f''(a) = 0$, then $f(x)$ may be concave up or concave down or neither at a.

Proof: (a) Assume that $f''(x) > 0$ for all x in I, and let a be any point in I. We want to show that f is concave up at a so we need to prove that the graph of f (Fig. 4) is above the tangent line to f at a: $f(x) > L(x) = f(a) + f'(a)(x-a)$ for x close to a.

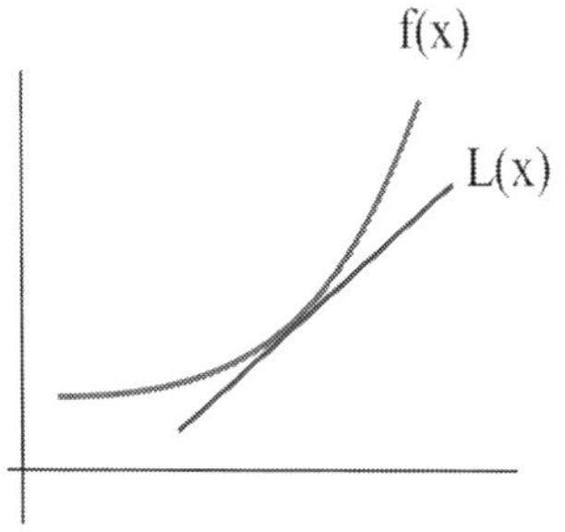

Fig. 4

Assume that x is in I, and apply the Mean Value Theorem to f on the interval from a to x. Then there is a number c between a and x so that

$$f'(c) = \frac{f(x) - f(a)}{x - a} \quad \text{and} \quad f(x) = f(a) + f'(c)(x-a).$$

Since $f'' > 0$ between a and x, we know from the Second Shape Theorem that f' is increasing between a and x. We will consider two cases: $x > a$ and $x < a$.

If $x > a$, then $x-a > 0$ and c is in the interval $[a,x]$ so $a < c$. Since f' is increasing, $a < c$ implies that $f'(a) < f'(c)$. Multiplying each side of the inequality $f'(a) < f'(c)$ by the positive amount $x-a$, we get that $f'(a)(x-a) < f'(c)(x-a)$. Adding $f(a)$ to each side of this last inequality, we have $L(x) = f(a) + f'(a)(x-a) < f(a) + f'(c)(x-a) = f(x)$.

If $x < a$, then $x-a < 0$ and c is in the interval $[x,a]$ so $c < a$. Since f' is increasing, $c < a$ implies that $f'(c) < f'(a)$. Multiplying each side of the inequality $f'(c) < f'(a)$ by the **negative** amount $x-a$, we get that $f'(c)(x-a) > f'(a)(x-a)$ and $f(x) = f(a) + f'(c)(x-a) > f(a) + f'(a)(x-a) = L(x)$.

In each case we get that the function $f(x)$ is above the tangent line $L(x)$. The proof of (b) is similar.

(c) Let $f(x) = x^4$, $g(x) = -x^4$, and $h(x) = x^3$ (Fig. 5). The second derivative of each of these functions is zero at $a = 0$, and at (0,0) they all have the same tangent line: $L(x) = 0$, the x–axis. However, at (0,0) they all have different concavity: f is concave up, g is concave down, and h is neither concave up nor down.

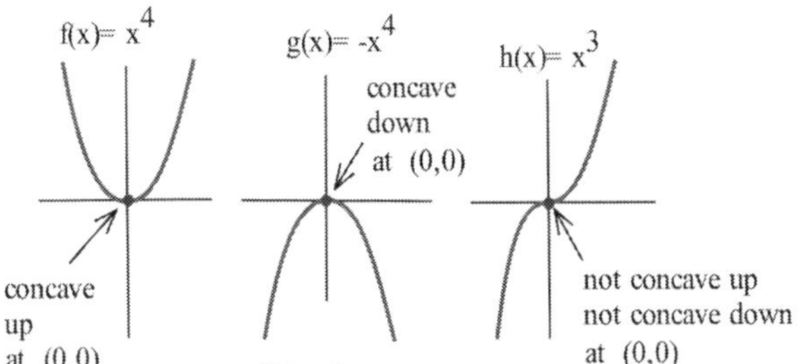

Fig. 5

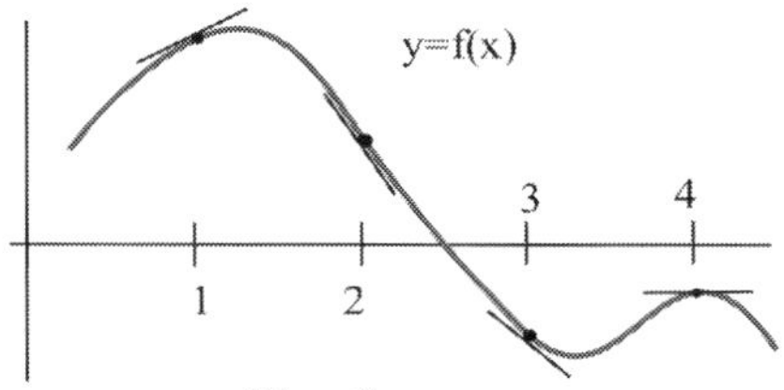

Fig. 6

Practice 1: Use the graph of f in Fig. 6 to finish filling in the table with "+" for positive, "–" for negative" , or "0".

x	f(x)	f '(x)	f "(x)	Concavity (up or down)
1	+	+	–	down
2	+			
3	–			
4				

Feeling the Second Derivative

Earlier we saw that if a function f(t) represents the position of a car at time t, then f '(t) is the velocity and f "(t) is the acceleration of the car at the instant t.

If we are driving along a straight, smooth road, then what we feel is the acceleration of the car:

a large positive acceleration feels like a "push" **toward the back** of the car,

a large negative acceleration (a deceleration) feels like a "push" **toward the front** of the car, and

an acceleration of 0 for a period of time means the velocity is constant and we do not feel pushed in either direction.

On a moving vehicle it is possible to measure these "pushes", the acceleration, and from that information to determine the velocity of the vehicle, and from the velocity information to determine the position. Inertial guidance systems in airplanes use this tactic: they measure front–back, left–right and up–down acceleration several times a second and then calculate the position of the plane. They also use computers to keep track of time and the rotation of the earth under the plane. After all, in 6 hours the earth has made a quarter of a revolution, and Dallas has rotated more than 5000 miles!

Example 1: The upward acceleration of a rocket was $a(t) = 30\ m/s^2$ for the first 6 seconds of flight, $0 \le t \le 6$. The velocity of the rocket at t=0 was 0 m/s and the height of the rocket above the ground at t=0 was 25 m. Find a formula for the height of the rocket at time t and determine the height at t = 6 seconds.

Solution: $v'(t) = a(t) = 30$ so $v(t) = 30t + K$ for some constant K. We also know $v(\mathbf{0}) = 0$ so $30(\mathbf{0}) + K = 0$ and $K = 0$. Therefore, $v(t) = 30t$.

Similarly, $h'(t) = v(t) = 30t$ so $h(t) = 15t^2 + C$ for some constant C. We know that $h(\mathbf{0}) = 25$ so $15(\mathbf{0})^2 + C = 25$ and $C = 25$. Then $h(t) = 15t^2 + 25$. $h(6) = 15(6)^2 + 25 = 565$ m.

f '' and Extreme Values of f

The concavity of a function can also help us determine whether a critical point is a maximum or minimum or neither. For example, if a point is at the bottom of a concave up function (Fig. 7), then the point is a minimum.

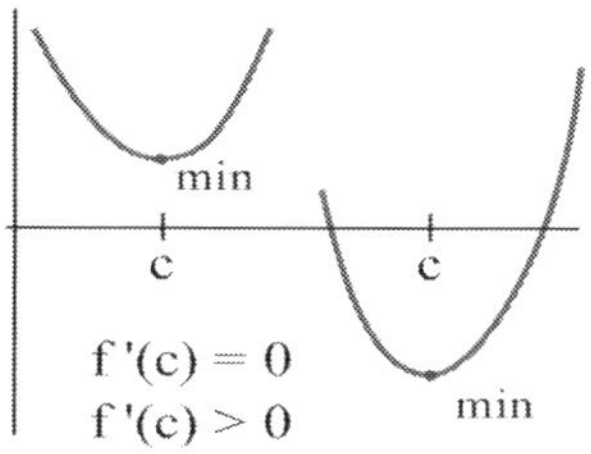

Fig. 7

The Second Derivative Test for Extremes:

(a) If $f'(c) = 0$ and $f''(c) < 0$ then f is concave down and has a local maximum at $x = c$.

(b) If $f'(c) = 0$ and $f''(c) > 0$ then f is concave up and has a local minimum at $x = c$.

(c) If $f'(c) = 0$ and $f''(c) = 0$ then f may have a local maximum, a minimum or neither at $x = c$.

Proof: (a) Assume that $f'(c) = 0$. If $f''(c) < 0$ then f is concave down at $x = c$ so the graph of f will be below the tangent line $L(x)$ for values of x near c. The tangent line, however, is given by $L(x) = f(c) + f'(c)(x - c) = f(c) + 0(x - c) = f(c)$, so if x is close to c then $f(x) < L(x) = f(c)$ and f has a local maximum at $x = c$. The proof of (b) for a local minimum of f is similar.

(c) If $f'(c) = 0$ and $f''(c) = 0$, then we cannot immediately conclude anything about local maximums or minimums of f at $x = c$. The functions $f(x) = x^4$, $g(x) = -x^4$, and $h(x) = x^3$ all have their first and second derivatives equal to zero at $x = 0$, but f has a local minimum at 0, g has a local maximum at 0, and h has neither a local maximum nor a local minimum at $x = 0$.

The Second Derivative Test for Extremes is very useful when f '' is easy to calculate and evaluate. Sometimes, however, the First Derivative Test or simply a graph of the function is an easier way to determine if we have a local maximum or a local minimum — it depends on the function and on which tools you have available to help you.

Practice 2: $f(x) = 2x^3 - 15x^2 + 24x - 7$ has critical numbers $x = 1$ and 4. Use the Second Derivative Test for Extremes to determine whether $f(1)$ and $f(4)$ are maximums or minimums or neither.

Inflection Points

Definition: An **inflection point** is a point on the graph of a function where the concavity of the function changes, from concave up to down or from concave down to up.

Practice 3: Which of the labelled points in Fig. 8 are inflection points?

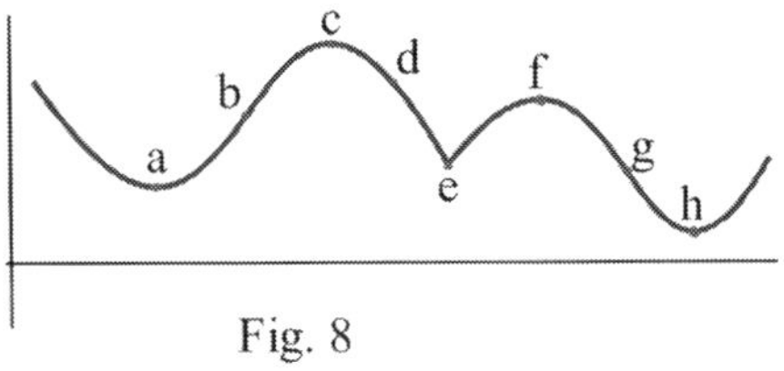

Fig. 8

To find the inflection points of a function we can use the second derivative of the function. If $f''(x) > 0$, then the graph of f is concave up at the point $(x, f(x))$ so $(x, f(x))$ is not an inflection point. Similarly, if $f''(x) < 0$, then the graph of f is concave down at the point $(x,f(x))$ and the point is not an inflection point. The only points left which can possibly be inflection points are the places where $f''(x)$ is 0 or undefined (f' is not differentiable). To find the inflection points of a function we only need to check the points where $f''(x)$ is 0 or undefined. If $f''(c) = 0$ or is undefined, then the point $(c,f(c))$ **may** or **may not** be an inflection point — we would need to check the concavity of f on each side of $x = c$. The functions in the next example illustrate what can happen.

Example 2: Let $f(x) = x^3$, $g(x) = x^4$ and $h(x) = x^{1/3}$ (Fig. 9). For which of these functions is the point $(0,0)$ an inflection point?

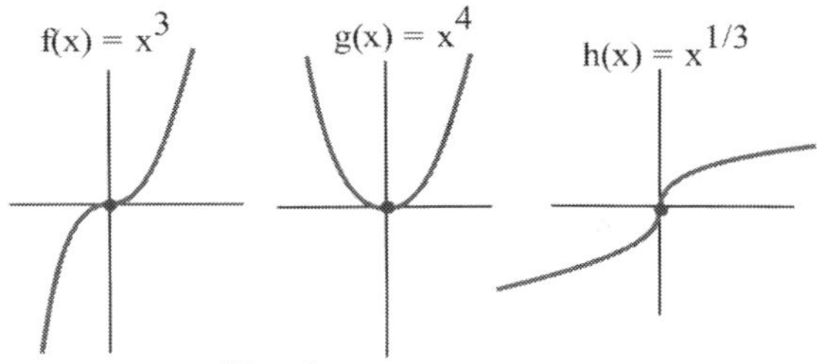

Fig. 9

Solution: Graphically, it is clear that the concavity of $f(x) = x^3$ and $h(x) = x^{1/3}$ changes at $(0,0)$, so $(0,0)$ is an inflection point for f and h. The function $g(x) = x^4$ is concave up everywhere so $(0,0)$ is not an inflection point of g.

If $f(x) = x^3$, then $f'(x) = 3x^2$ and $f''(x) = 6x$. The only point at which $f''(x) = 0$ or is undefined (f' is not differentiable) is at $x = 0$. If $x < 0$, then $f''(x) < 0$ so f is concave down. If $x > 0$, then $f''(x) > 0$ so f is concave up. At $x = 0$ the concavity changes so the point $(0,f(0)) = (0,0)$ is an inflection point of x^3.

If $g(x) = x^4$, then $g'(x) = 4x^3$ and $g''(x) = 12x^2$. The only point at which $g''(x) = 0$ or is undefined is at $x = 0$. If $x < 0$, then $g''(x) > 0$ so g is concave up. If $x > 0$, then $g''(x) > 0$ so g is also concave up. At $x = 0$ the concavity **does not change** so the point $(0, g(0)) = (0,0)$ is **not an inflection point** of x^4.

If $h(x) = x^{1/3}$, then $h'(x) = \frac{1}{3} x^{-2/3}$ and $h''(x) = -\frac{2}{9} x^{-5/3}$. h" is not defined if $x = 0$, but h "(negative number) > 0 and h "(positive number) < 0 so h changes concavity at (0,0) and (0,0) is an inflection point of h.

Practice 4: Find the inflection points of $f(x) = x^4 - 12x^3 + 30x^2 + 5x - 7$.

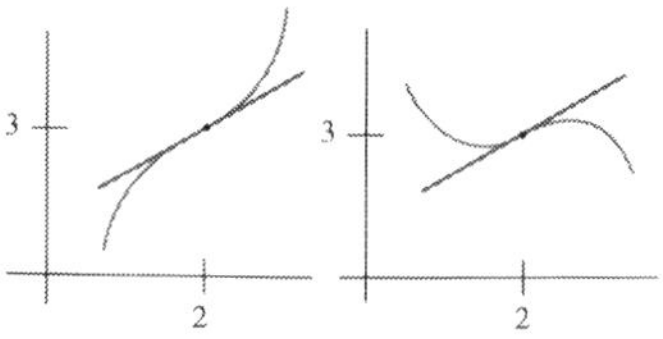

Fig. 10

Example 3: Sketch graph of a function with $f(2) = 3$, $f'(2) = 1$, and an inflection point at (2,3) .Solution: Two solutions are given in Fig. 10.

PROBLEMS

In problems 1 and 2, each quotation is a statement about a quantitity of something changing over time. Let f(t) represent the quantity at time t. For each quotation, tell what f represents and whether the first and second derivatives of f are positive or negative.

1. (a) "Unemployment rose again, but the rate of increase is smaller than last month."
   (b) "Our profits declined again, but at a slower rate than last month."
   (c) "The population is still rising and at a faster rate than last year."

2. (a) "The child's temperature is still rising, but slower than it was a few hours ago."
   (b) "The number of whales is decreasing, but at a slower rate than last year."
   (c) "The number of people with the flu is rising and at a faster rate than last month."

3. Sketch the graphs of functions which are defined and concave up everywhere and which have
   (a) no roots. (b) exactly 1 root. (c) exactly 2 roots. (d) exactly 3 roots.

4. On which intervals is the function in Fig. 11
   (a) concave up? (b) concave down?

5. On which intervals is the function in Fig. 12
   (a) concave up? (b) concave down?

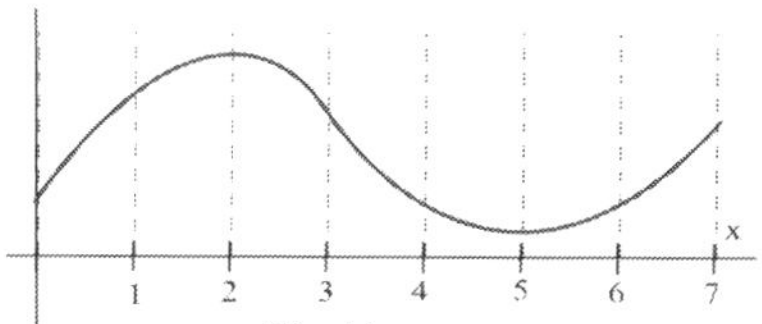

Fig. 11

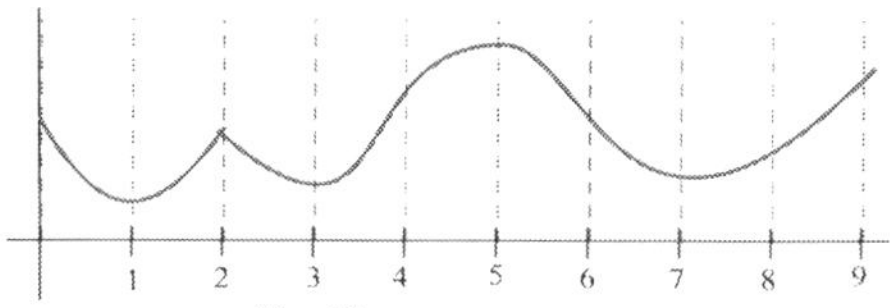

Fig. 12

In problems 6 – 10, a function and values of x so that $f'(x) = 0$ are given. Use the Second Derivative Test to determine whether each point (x, f(x)) is a local maximum, a local minimum or neither

6. $f(x) = 2x^3 - 15x^2 + 6$, $x = 0, 5$.

7. $g(x) = x^3 - 3x^2 - 9x + 7$, $x = -1, 3$.

8. $h(x) = x^4 - 8x^2 - 2$, $x = -2, 0, 2$.

9. $f(x) = \sin^5(x)$, $x = \pi/2, \pi, 3\pi/2$

10. $f(x) = x \cdot \ln(x)$, $x = 1/e$.

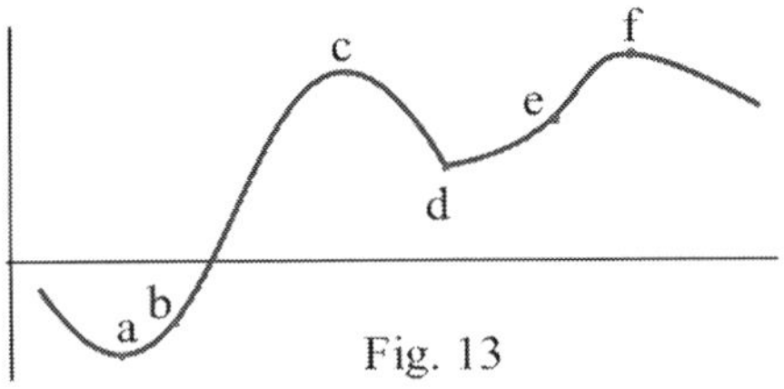

Fig. 13

11. At which labeled values of x in Fig. 13 is the point $(x, f(x))$ an inflection point?

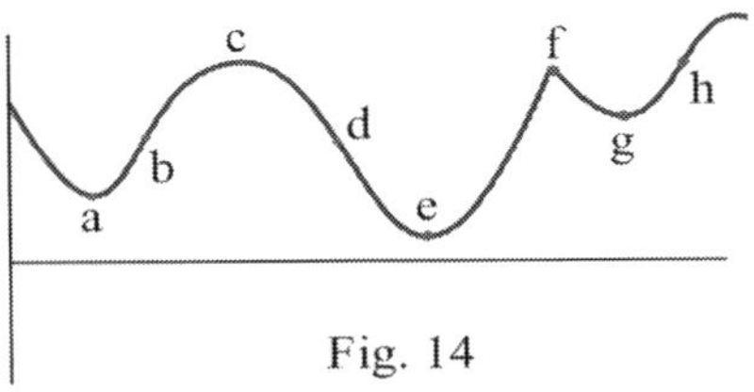

Fig. 14

12. At which labeled values of x in Fig. 14 is the point $(x, g(x))$ an inflection point?

13. How many inflection points can a

(a) quadratic polynomial have? (b) cubic polynomial have?

(c) polynomial of degree n have?

14. Fill in the table with "+", "–", or "0" for the function in Fig. 15.

x	f(x)	f '(x)	f "(x)
0			
1			
2			
3			

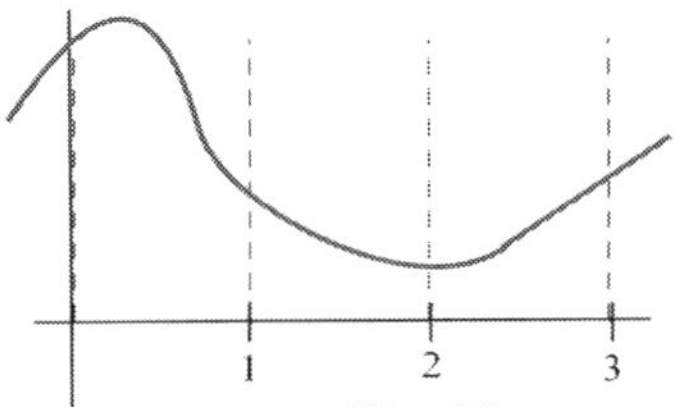

Fig. 15

15. Fill in the table with "+", "–", or "0" for the function in Fig. 16

x	g(x)	g '(x)	g "(x)
0			
1			
2			
3			

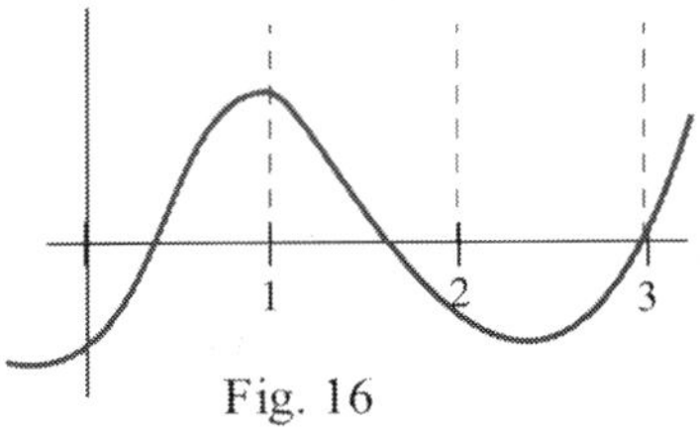

Fig. 16

16. Sketch functions f for x–values near 1 so $f(1) = 2$ and

(a) $f'(1) = +$, $f''(1) = +$ (b) $f'(1) = +$, $f''(1) = -$

(c) $f'(1) = -$, $f''(1) = +$

(d) $f'(1) = +$, $f''(1) = 0$, $f''(1^-) = -$, $f''(1^+) = +$ (e) $f'(1) = +$, $f''(1) = 0$, $f''(1^-) = +$, $f''(1^+) = -$

17. Some people like to think of a concave up graph as one which will "hold water" and of a concave down graph as one which will "spill water." That description is accurate for a concave down graph, but it can fail for a concave up graph. Sketch the graph of a function which is concave up on an interval, but which will not "hold water".

18. The function $f(x) = \frac{1}{\sqrt{2\pi}} e^{-(x-c)^2/(2b^2)}$ is called the Gaussian distribution, and its graph is a bell–shaped curve (Fig. 17) that occurs commonly in statistics.

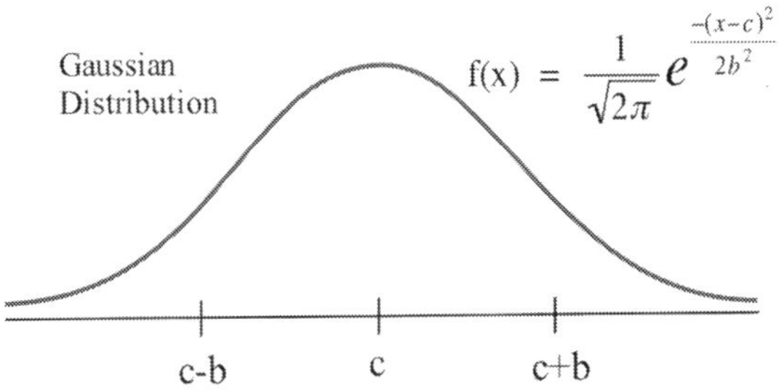

Fig. 17

(i) Show that f has a maximum at $x = c$. (The value c is called the mean of this distribution.)

(ii) Show that f has inflection points where $x = c + b$ and $x = c - b$. (The value b is called the standard deviation of this distribution.)

Section 3.4 **PRACTICE Answers**

Practice 1: See Fig. 6.

x	f(x)	f '(x)	f "(x)	Concavity (up or down)
1	+	+	–	down
2	+	**–**	**–**	**down**
3	–	**–**	**+**	**up**
4	**–**	**0**	**–**	**down**

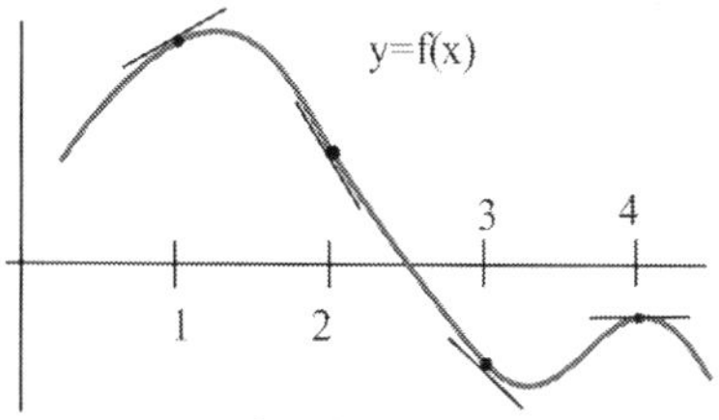

Fig. 6

Practice 2: $f(x) = 2x^3 - 15x^2 + 24x - 7$.

$f'(x) = 6x^2 - 30x + 24$ which is defined for all x.

$f'(x) = 0$ if $x = 1, 4$ (critical values).

$f''(x) = 12x - 30$.

$f''(1) = -18$ so f is concave down at the critical value $x = 1$ so $(1, f(1)) = (1,4)$ is a rel. max.

$f''(4) = +18$ so f is concave up at the critical value $x = 4$ so $(4, f(4)) = (4, -23)$ is a rel. min.

Fig. 18 shows the graph of f.

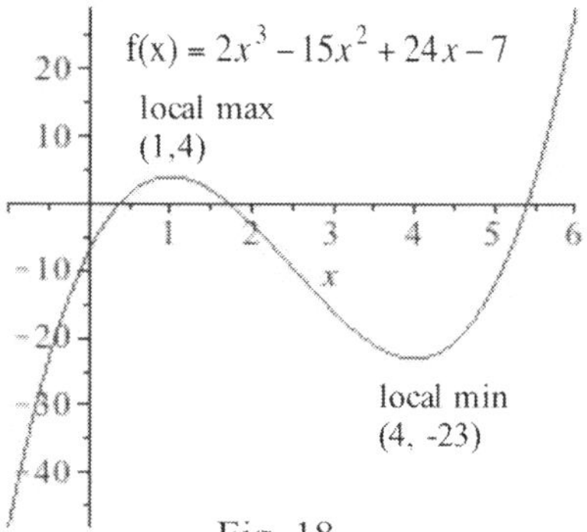

Fig. 18

Practice 3: The points labeled (b) and (g) in Fig. 8 are inflection points.

Practice 4: $f(x) = x^4 - 12x^3 + 30x^2 + 5x - 7$. $f'(x) = 4x^3 - 36x^2 + 60x + 5$.

$f''(x) = 12x^2 - 72x + 60 = 12(x^2 - 6x + 5) = 12(x - 1)(x - 5)$.

The only candidates to be Inflection Points are $x = 1$ and $x = 5$.

If $x < 1$, then $f''(x) = 12(x - 1)(x - 5) = 12($ neg $)($ neg $)$ is positive.

If $1 < x < 5$, then

$f''(x) = 12(x - 1)(x - 5) = 12($ pos $)($ neg $)$ is negative.

If $5 < x$, then $f''(x) = 12(x - 1)(x - 5) = 12($ pos $)($ pos $)$ is positive.

f changes concavity at $x = 1$ and $x = 5$ so

x = 1 and x = 5 are Inflection Points.

Fig. 19 shows the graph of f.

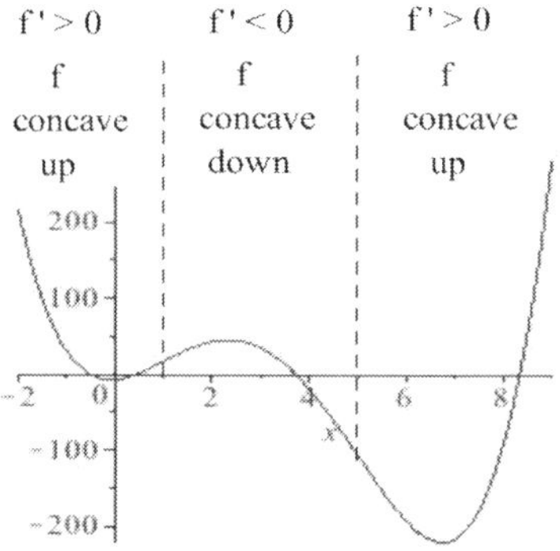

$f(x) = x^4 - 12x^3 + 30x^2 + 5x - 7$

Fig. 19

3.5 APPLIED MAXIMUM AND MINIMUM PROBLEMS

We have used derivatives to help find the maximums and minimums of some functions given by equations, but it is very unlikely that someone will simply hand you a function and ask you to find its extreme values. More typically, someone will describe a problem and ask your help in maximizing or minimizing something: "What is the largest volume package which the post office will take?"; "What is the quickest way to get from here to there?"; or "What is the least expensive way to accomplish some task?" Usually these problems have some restrictions or constraints on what is allowed, and sometimes neither the problem nor the constraints are clearly stated.

Before we can use calculus or other mathematical techniques to solve the max/min problem, we need to **understand** what is really being asked. We need to **translate** the problem into a mathematical form which we can **solve**, and we need to **check** our mathematical solution to see if it is really a solution of the original problem. Often, the hardest parts of the problem are understanding the problem and translating it into a mathematical form.

In this section we examine some problems which require understanding, translation, solution, and checking. Most of these problems are not as complicated as those a working scientist, engineer or economist needs to solve, but they represent a step in developing the required skills.

Example 1: The company you own has a large supply of 8 inch by 15 inch rectangular pieces of tin, and you decide to make them into boxes by cutting a square from each corner and folding up the sides (Fig. 1). For example, if you cut a 1 inch square from each corner the resulting 6 inch by 13 inch by 1 inch box has a volume of 78 cubic inches. The amount of money you get for a box depends on how much the box holds, so you want to make boxes with the largest possible volumes. How large a square should you cut from each corner?

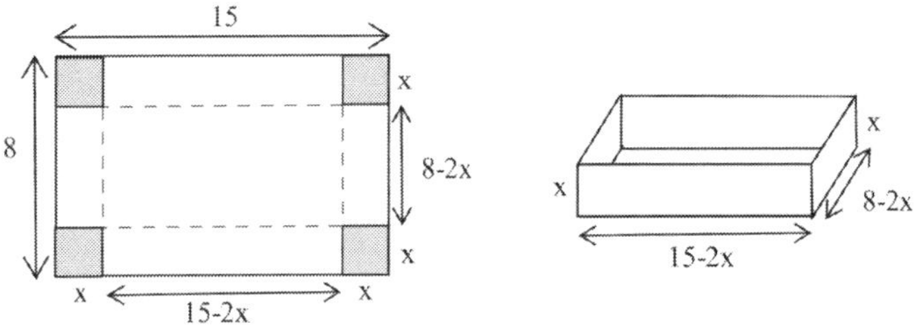

Fig. 1

Solution: First we need to understand the problem, and a diagram can be very helpful. Then we need to translate it into a mathematical problem:

* **identify the variables**,
* **label the variable and constant parts of the diagram**, and
* **represent the quantity to be maximized as a function.**

If we label the side of the square as x inches, then the box is x inches high, $8 - 2x$ inches wide, and $15 - 2x$ inches long, so the volume is (length)·(width)·(height) $= (15 - 2x)\cdot(8 - 2x)\cdot(x)$ $= 4x^3 - 46x^2 + 120x$ cubic inches. Now we have a mathematical problem, maximize $V(x) = 4x^3 - 46x^2 + 120x$, and we can use the calculus techniques from the previous sections.

$V'(x) = 12x^2 - 92x + 120$, and we need to find the critical points. (i) We can find where $V'(x) = 0$ by factoring or using the quadratic formula: $V'(x) = 12x^2 - 92x + 120 = 4(3x - 5)(x - 6) = 0$ if $x = 5/3$ or $x = 6$, so $x = 5/3$ and $x = 6$ are critical points of V. (ii) $V'(x)$ is a polynomial so it is always defined and there are no critical points from an undefined derivative. (iii) What are the endpoints for x in this problem? A square cannot have a negative length so $x \geq 0$. We cannot remove more than half of the width, so $8 - 2x \geq 0$ and $x \leq 4$. Together, these two inequalities say that $0 \leq x \leq 4$, so the endpoints are $x = 0$ and $x = 4$. (Note that the value $x = 6$ is not in this interval, so $x = 6$ does not maximize the volume and we do not consider it further.)

The maximum volume must occur at one of the critical points $x = 0, 5/3$, or 4: $V(0) = 0$, $V(5/3) = 2450/27 \approx 90.74$ cubic inches, and $V(4) = 0$. The maximum volume of the box occurs when a 5/3 inch by 5/3 inch square is removed from each corner, and resulting box is 5/3 inches high, $8 - 2(5/3) = 14/3$ inches wide, and $15 - 2(5/3) = 35/3$ inches long.

Practice 1: If you start with 7 inch by 15 inch pieces of tin, what size square should you remove from each corner so the box will have as large a volume as possible?
(Hint: $12x^2 - 88x + 105 = (2x - 3)(6x - 35)$)

We were fortunate in the previous example and practice problem because the functions we created to describe the volume were functions of only one variable. In some problems, the function we get will have more than one variable, and we will need to use additional information to change our function into a function of one variable. Typically the constraints will contain the additional information we need.

Example 2: We want to fence a rectangular area in our backyard for a garden. One side of the garden is along the edge of the yard which is already fenced, so we only need to build a new fence along the other 3 sides of the rectangle (Fig. 2). If we have 80 feet of fencing available, what dimensions should the garden have in order to enclose the largest possible area?

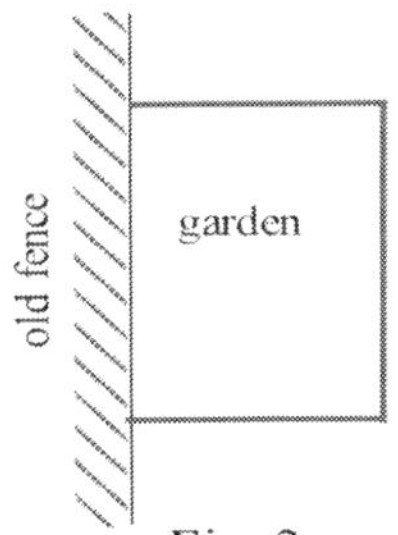

Fig. 2

Solution: The first step is to understand the problem, and a **diagram** or picture of the situation often helps. Next, we need to **identify the variables**: in this case, the length, call it x, and width, call it y, of the garden. Fig. 3 shows the **labeled diagram** so now we can write a formula for the function which we want to maximize:

Maximize A = area of the rectangle = (length)·(width) = $x \cdot y$.

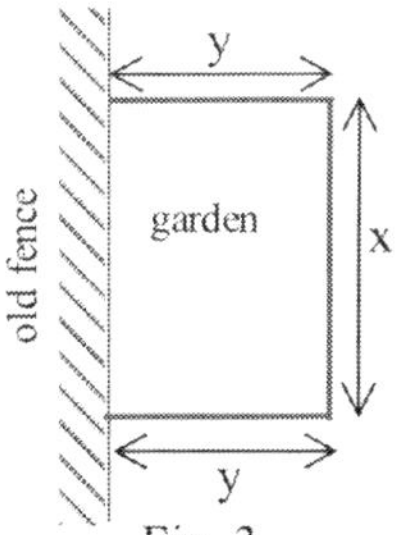

Fig. 3

Unfortunately, our function A has two variables, x and y , so we need to find a relationship between them (an equation containing both x and y) which we can solve for one of x or y. The constraint in this problem says that "we have 80 feet of fencing available" so $x + 2y = 80$ and $y = 40 - (x/2)$. Then

$A = x \cdot y = x \cdot (40 - (x/2)) = 40x - \frac{x^2}{2}$, a function of one variable. We want to maximize A.

$A' = 40 - x$. The only time $A' = 0$ is when $x = 40$, so $x = 40$ so there is only one critical point of type (i). A is differentiable for all x so there are no critical numbers of the type (ii). Finally, $0 \le x \le 80$ (why?) so the only critical points of type (iii) are when $x = 0$ and $x = 80$. The only critical points of A are when $x = 0, 40$, and 80, and the maximum area occurs at one of them:

at the critical number $x = 0$, $A = 40(0) - \frac{(0)^2}{2} = 0$ square feet

at the critical number $x = 40$, $A = 40(40) - \frac{(40)^2}{2} = 800 \text{ ft}^2$

at the critical number $x = 80$, $A = 40(80) - \frac{(80)^2}{2} = 0 \text{ ft}^2$

so the largest rectangular garden has an area of 800 square feet and dimensions $x = 40$ feet by $y = 40 - (x/2) = 40 - (40/2) = 20$ feet.

Practice 2: Suppose you decide to fence the rectangular garden in the corner of your yard. Then two sides of the garden are bounded by the yard fence which is already there, so you only need to use the 80 feet of fencing to enclose the other two sides. What are the dimensions of the new garden of largest area? What are the dimensions of the rectangular garden of largest area in the corner of the yard if you have F feet of new fencing available?

Example 3: You need to reach home as quickly as possible, but you are in a rowboat 4 miles from shore and your home is 2 miles up the coast (Fig. 4). If you can row at 3 miles per hour and walk at 5 miles per hour, toward which point on the shore should you row? Toward which point should you row if your home is 7 miles up the coast?

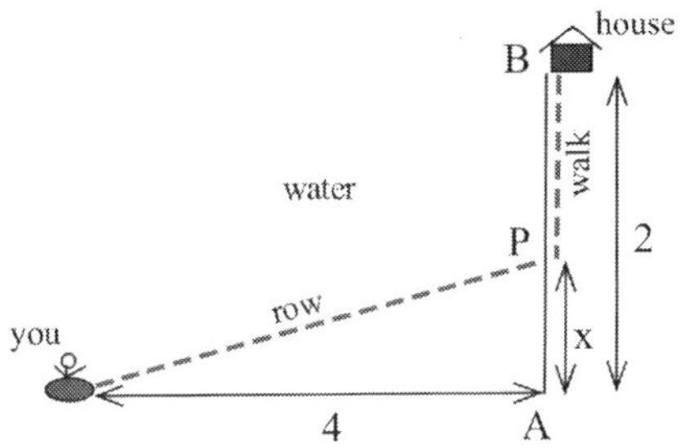

Fig. 4

Solution: Fig. 4 shows a labeled diagram with the variable x representing the distance from point A, the nearest shore point, to point P, the point you row toward. Then the total time, rowing and walking, is

T = total time = (rowing time from boat to P) + (walking time from P to B)

= (distance from boat to P)/(rate from boat to P) + (distance from P to B)/(rate from P to B)

$= \sqrt{x^2 + 4^2}\,/3 + (2 - x)/5 = \frac{\sqrt{x^2 + 16}}{3} + \frac{2 - x}{5}$.

It is not reasonable to row to a point below A and then walk home, so $x \ge 0$. Similarly, we can conclude that $x \le 2$, so our interval is $0 \le x \le 2$ and the endpoints are $x = 0$ and $x = 2$.

To find the other critical numbers of T between $x = 0$ and $x = 2$, we need the derivative of T.

$$T'(x) = \frac{1}{3}\cdot\frac{1}{2}(x^2+16)^{-1/2}(2x) - \frac{1}{5} = \frac{x}{3\sqrt{x^2+16}} - \frac{1}{5}.$$

To find where $T'(x)$ is zero, set $T'(x) = 0$ and solve:

$$T'(x) = \frac{x}{3\sqrt{x^2+16}} - \frac{1}{5} = 0 \text{ so } \frac{x}{3\sqrt{x^2+16}} = \frac{1}{5} \text{ and}$$

$5x = 3\sqrt{x^2+16}$ so $25x^2 = 9x^2 + 144$ and $x = \pm 3$. Neither of these numbers, however, is in our interval $0 \le x \le 2$ so neither of them gives a minimum time.

T is differentiable for all values of x, so there are no critical numbers of type (ii).

The only critical numbers for T on this interval are $x = \mathbf{0}$ and $x = \mathbf{2}$: $T(\mathbf{0}) = \frac{\sqrt{\mathbf{0}+16}}{3} + \frac{2-\mathbf{0}}{5} = \frac{4}{3} + \frac{2}{5} \approx 1.73$ hours and $T(\mathbf{2}) = \frac{\sqrt{\mathbf{2}^2+16}}{3} + \frac{2-\mathbf{2}}{5} = \frac{\sqrt{20}}{3} + 0 \approx 1.49$ hours. The quickest route is when P is 2 miles down the coast. You should row directly toward home.

If your home is 7 miles down the coast, then the interval for x is $0 \le x \le 7$ which has the endpoints $x = 0$ and $x = 7$. Our function for the travel time is $T(x) = \frac{\sqrt{x^2+16}}{3} + \frac{7-x}{5}$ and $T'(x) = \frac{x}{3\sqrt{x^2+16}} - \frac{1}{5}$ so the only point in our interval where $T(x)' = 0$ is at $x = 3$.

The only critical numbers for T in the interval are $x = \mathbf{0}$, $x = \mathbf{3}$, and $x = \mathbf{7}$:

$$T(\mathbf{0}) = \frac{\sqrt{\mathbf{0}^2+16}}{3} + \frac{7-\mathbf{0}}{5} = \frac{4}{3} + \frac{7}{5} \approx 2.73 \text{ hours}$$

$$T(\mathbf{7}) = \frac{\sqrt{\mathbf{7}^2+16}}{3} + \frac{7-\mathbf{7}}{5} = \frac{\sqrt{65}}{3} + 0 \approx 2.68 \text{ hours}$$

$$T(\mathbf{3}) = \frac{\sqrt{\mathbf{3}^2+16}}{3} + \frac{7-\mathbf{3}}{5} = \frac{5}{3} + \frac{4}{5} \approx 2.47 \text{ hours.}$$

The quickest way home is to aim for a point P which is 3 miles down the coast, row directly to P, and then walk along the coast to home.

One challenge of max/min problems is that they may require geometry or trigonometry or other mathematical facts and relationships.

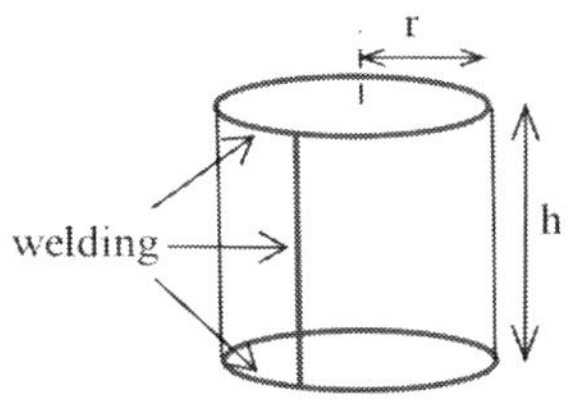

Fig. 5

Example 4: Find the height and radius of the least expensive closed cylinder which has a volume of 1000 cubic inches. Assume that the materials are free, but that it costs 80¢ per inch to weld the top and bottom onto the cylinder and to weld the seam up the side of the cylinder (Fig. 5).

Solution: If we let r be the radius of the cylinder and h be its height, then the volume $V = \pi r^2 h = 1000$. The function we want to minimize is cost, and

C = total welding cost = (top seam cost) + (bottom seam cost) + (side seam cost)

= (top seam length)·(80¢/inch) + (bottom seam length)·(80¢/in) + (side seam length)·(80¢/in)

$= (2\pi r)\cdot(80) + (2\pi r)\cdot(80) + (h)\cdot(80) = 320\pi r + 80\,h.$

Unfortunately, our function C is a function of two variables, r and h, but we can use the information in the constraint, $V = \pi r^2 h = 1000$, to solve for h and then substitute this h into the formula for C: $1000 = \pi r^2 h$ so $h = \frac{1000}{\pi r^2}$ and then $C = 320\pi r + 80\,\mathbf{h} = 320\pi r + 80(\frac{\mathbf{1000}}{\boldsymbol{\pi r^2}})$, a function of one variable. $C' = 320\pi - \frac{160000}{\pi r^3}$, and C is a minimum when $C' = 0$: at

$$r = \sqrt[3]{\frac{500}{\pi^2}} \approx 3.7 \text{ inches and } h = \frac{1000}{\pi r^2} \approx \frac{1000}{\pi (3.7)^2} \approx 23.3 \text{ inches.}$$

Practice 3: Find the height and radius of the least expensive closed cylinder which has a volume of 1000 cubic inches. Assume that the only cost for this cylinder is the cost of the materials: the material for the top and bottom costs 5¢ per square inch, and the material for the sides costs 3¢ per square inch (Fig. 6).

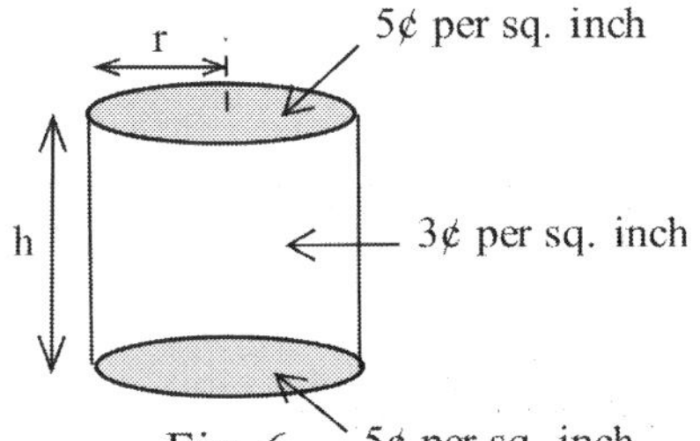

Fig. 6

Example 5: Find the dimensions of the least expensive rectangular box which is three times as long as it is wide and which holds 100 cubic centimeters of water. The material for the bottom costs 7¢ per cm^2, the sides cost 5¢ per cm^2 and the top costs 2¢ per cm^2.

Solution: Label the box so w = width, l = length, and h = height. Then our cost function C is

C = (bottom cost) + (cost of front and back) + (cost of ends) + (top cost)

= (bottom area)·(7¢) + (front and back area)·(5¢) + (ends area)·(5¢) + (top area)·(2¢)

$= (wl)\cdot(7) + (2lh)\cdot(5) + (2wh)\cdot(5) + (wl)\cdot(2) = 7wl + 10lh + 10wh + 2wl = 9(wl) + 10(lh) + 10(wh).$

Unfortunately, C is a function of 3 variables, w, l, and h, but we can use the other information in the constraints to eliminate some of the variables:

the box is "three times as long as it is wide" so $l = 3w$ and

$C = 9w\mathbf{l} + 10\mathbf{l}h + 10wh = 9w(\mathbf{3w}) + 10(\mathbf{3w})h + 10wh = 27w^2 + 40wh.$

We also know that the volume V is 100 in^3 and $V = lwh = 3w^2h$ (since $l = 3w$), so $h = \frac{100}{3w^2}$.

Then $C = 27w^2 + 40w\mathbf{h} = 27w^2 + 40w(\frac{100}{3w^2}) = 27w^2 + \frac{4000}{3w}$, a function of one variable.

$C' = 54w - \frac{4000}{3w^2}$, and C is minimized when $w = \sqrt[3]{\frac{4000}{162}} \approx 2.91$ inches ($l = 3w \approx 8.73$ inches,

and $h = \frac{100}{3w^2} \approx 3.94$ inches). The minimum cost is approximately \$6.87 .

Problems described in words are usually more difficult to solve because we first need to understand and "translate" the problem into a mathematical problem, and, unfortunately, those skills only seem to come with practice. With practice, however, you will start to recognize patterns for understanding, translating, and solving these problems, and you will develop the skills you need. So read carefully, draw pictures, think hard, and do the best you can.

Problems

1. (a) You have 200 feet of fencing to enclose a rectangular vegetable garden. What should the dimensions of your garden be in order to enclose the largest area?
   (b) Show that if you have P feet of fencing available, the garden of greatest area is a square.
   (c) What are the dimensions of the largest rectangular garden you can enclose with P feet of fencing if one edge of the garden borders a straight river and does not need to be fenced?
   (d) Just thinking — calculus will not help with this one: What do you think is the shape of the largest garden which can be enclosed with P feet of fencing if we do not require the garden to be rectangular? What do you think is the shape of the largest garden which can be enclosed with P feet of fencing if one edge of the garden borders a river and does not need to be fenced?

2. (a) You have 200 feet of fencing available to construct a rectangular pen with a fence divider down the middle (see Fig. 7). What dimensions of the pen enclose the largest total area?
   (b) If you need 2 dividers, what dimensions of the pen enclose the largest area?
   (c) What are the dimensions in parts (a) and (b) if one edge of the pen borders on a river and does not require any fencing?

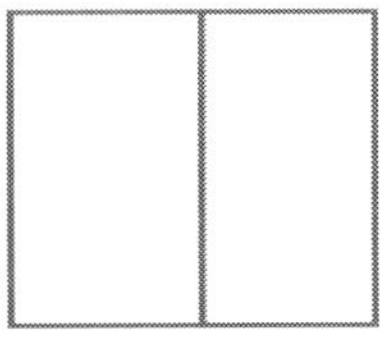

Fig. 7

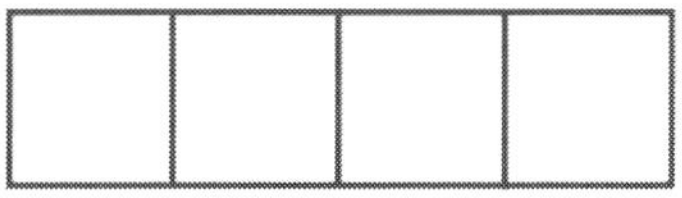

Fig. 8

3. You have 120 feet of fencing to construct a pen with 4 equal sized stalls.
   (a) If the pen is rectangular and shaped like the Fig. 8, what are the dimensions of the pen of largest area and what is that area?
   (b) The square pen in Fig. 9 uses 120 feet of fencing and encloses a larger area (400 square feet) than the best design in part (a). Design a pen which uses only 120 feet of fencing and has 4 equal sized stalls but which encloses even more than 400 square feet. (Suggestion: don't use rectangles and squares.)

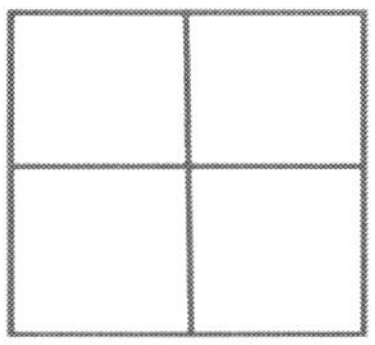

Fig. 9

4. (a) You have a 10 inch by 15 inch piece of tin which you plan to form into a box (without a top) by cutting a square from each corner and folding up the sides (see Fig. 10). How much should you cut from each corner so the resulting box has the greatest volume?

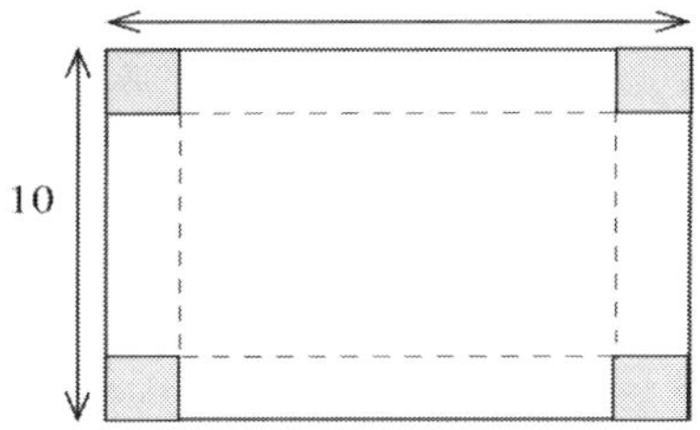

Fig. 10

(b) If the piece of tin is A inches by B inches, how much should you cut from each corner so the resulting box has the greatest volume?

5. You have a 10 inch by 10 inch piece of cardboard which you plan to cut and fold as shown in Fig. 11 to form a box with a top. Find the dimensions of the box which has the largest volume.

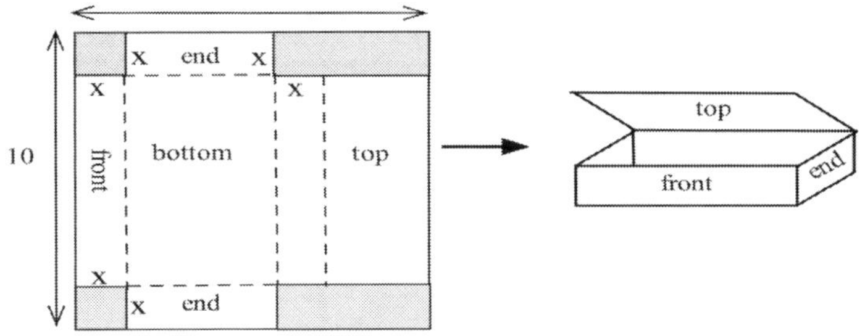

Fig. 11

6. (a) You have been asked to bid on the construction of a square-bottomed box with no top which will hold 100 cubic inches of water. If the bottom and sides are made from the same material, what are the dimensions of the box which uses the least material? (Assume that no material is wasted.)

(b) Suppose the box in part (a) uses different materials for the bottom and the sides. If the bottom material costs 5¢ per square inch and the side material costs 3¢ per square inch, what are the dimensions of the least expensive box which will hold 100 cubic inches of water?

(This is a "classic" problem which has many variations. We could require that the box be twice as long as it is wide, or that the box have a top, or that the ends cost a different amount than the front and back, or even that it costs some amount of money to weld each inch of edge. You should be able to set up the cost equations for these variations.)

7. (a) Determine the dimensions of the least expensive cylindrical can which will hold 100 cubic inches if the materials cost 2¢, 5¢ and 3¢ respectively for the top, bottom and sides.

(b) How do the dimensions of the least expensive can change if the bottom material costs more than 5¢ per square inch?

8. You have 100 feet of fencing to build a pen in the shape of a circular sector, the "pie slice" in Fig. 12. The area of such a sector is rs/2.

(a) What value of r maximizes the enclosed area?

(b) What is the central angle when the area is maximized?

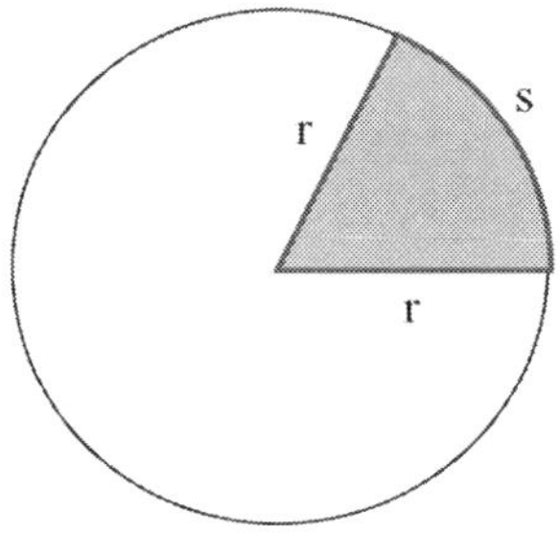

Fig. 12

9. You are a lifeguard standing at the edge of the water when you notice a swimmer in trouble (Fig. 13). Assuming you can run about 8 meters per second and swim about 2 m/s, how far along the shore should you run before diving into the water in order to reach the swimmer as quickly as possible?

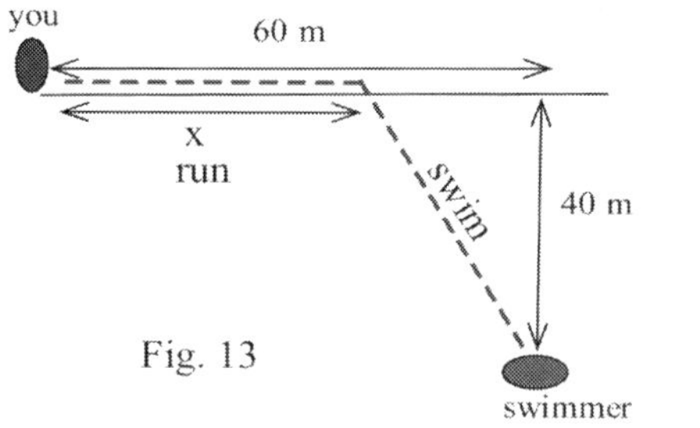

Fig. 13

10. (a) You have been asked to determine the least expensive route for a telephone cable which connects Andersonville with Beantown (see Fig. 14). If it costs $5000 per mile to lay the cable on land and $8000 per mile to lay the cable across the river and the cost of the cable is negligible, find the least expensive route.

(b) What is the least expensive route if the cable costs $7000 per mile plus the cost to lay it.

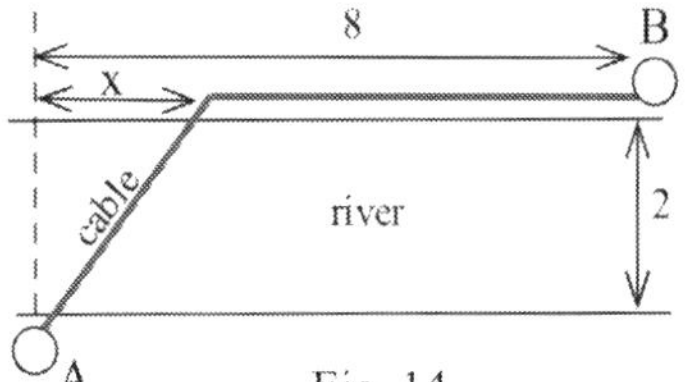

Fig. 14

11. You have been asked to determine where a water works should be built along a river between Chesterville and Denton (see Fig. 15) to minimize the total cost of the pipe to the towns.

(a) Assume that the same size (and cost) pipe is used to each town. (This part can be done quickly without using calculus.)

(b) Assume that the pipe to Chesterville costs $3000 per mile and to Denton it costs $7000 per mile.

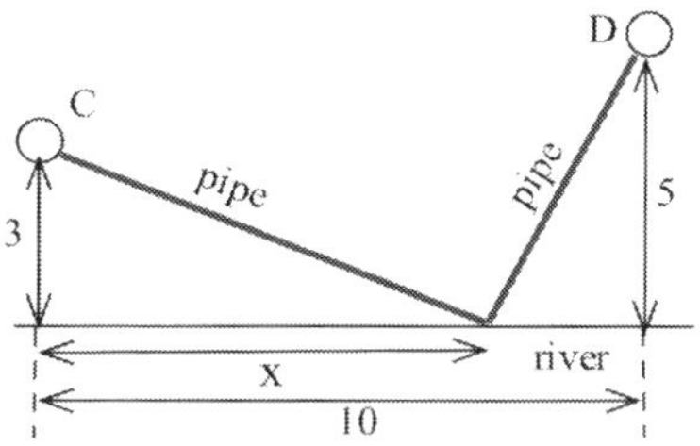

Fig. 15

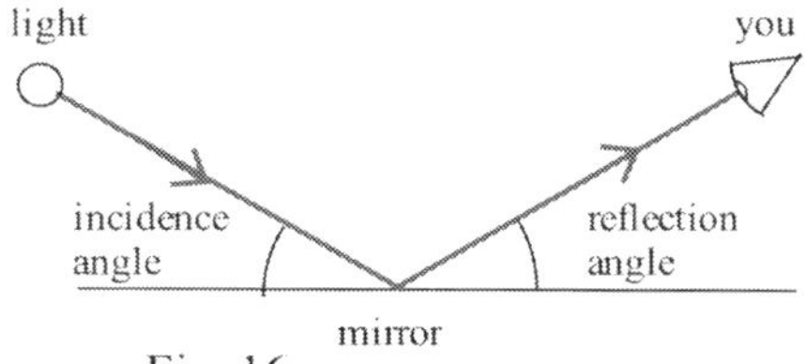

Fig. 16

12. Light from a bulb at A is reflected off a flat mirror to your eye at point B (see Fig. 16). If the time (and length of the path) from A to the mirror and then to your eye is a minimum, show that the angle of incidence equals the angle of reflection. (Hint: This is similar to the previous problem.)

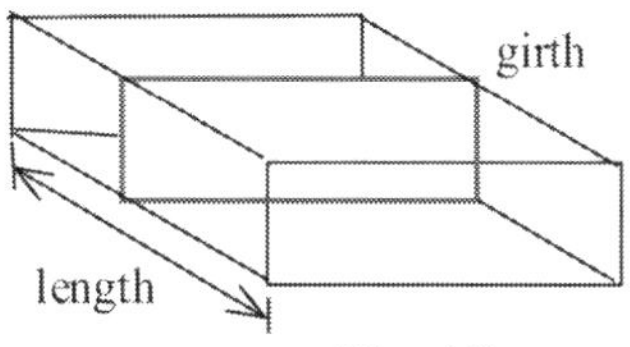

Fig. 17

13. U.S. postal regulations state that the sum of the length and girth (distance around) of a package must be no more than 108 inches. (Fig. 17)

(a) Find the dimensions of the acceptable box with a square end which has the largest volume.

(b) Find the dimensions of the acceptable box which has the largest volume if its end is a rectangle twice as long as it is wide.

(c) Find the dimensions of the acceptable box with a circular end which has the largest volume.

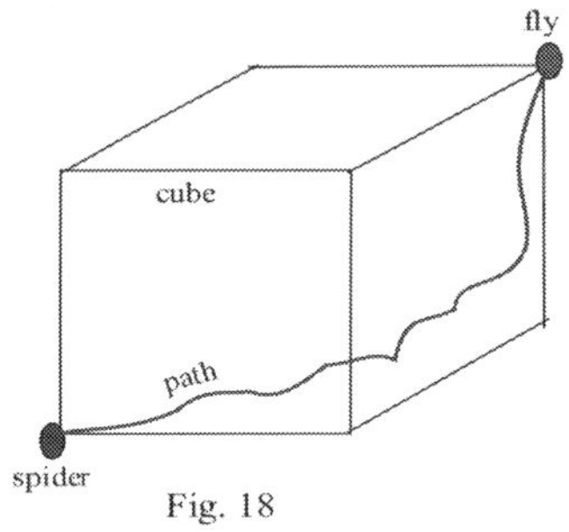

Fig. 18

14. Just thinking — you don't need calculus for this problem: A spider and a fly are located on opposite corners of a cube (see Fig. 18). What is the shortest path along the surface of the cube from the spider to the fly?

15. Two sides of a triangle are 7 and 10 inches respectively. What is the length of the third side so the area of the triangle will be greatest? (This problem can be done without using calculus. How? If you do use calculus, consider the angle θ between the two sides.)

16. Find the shortest distance from the point (2,0) to the curve

(a) $y = 3x - 1$ (b) $y = x^2$ (c) $x^2 + y^2 = 1$ (d) $y = \sin(x)$

17. Find the dimensions of the rectangle with the largest area if the base must be on the x-axis and its other two corners are on the graph of

(a) $y = 16 - x^2$ on $[-4, 4]$ (b) $x^2 + y^2 = 1$ on $[-1, 1]$

(c) $|x| + |y| = 1$ on $[-1, 1]$ (d) $y = \cos(x)$ on $[-\pi/2, \pi/2]$

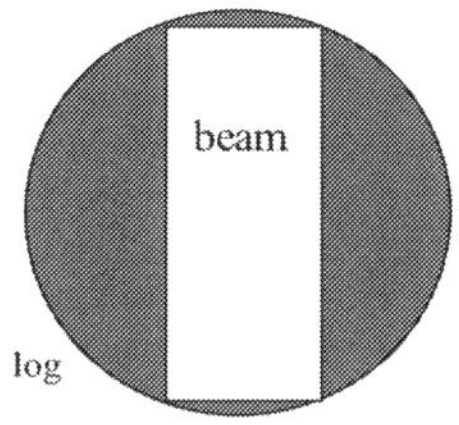

Fig. 19

18. The strength of a wooden beam is proportional to the product of its width and the square of its height (Fig. 19).

(a) What are the dimensions of the strongest beam that can be cut from a log with diameter 12 inches?

(b) What are the dimensions of the strongest beam that can be cut from a log with diameter d inches?

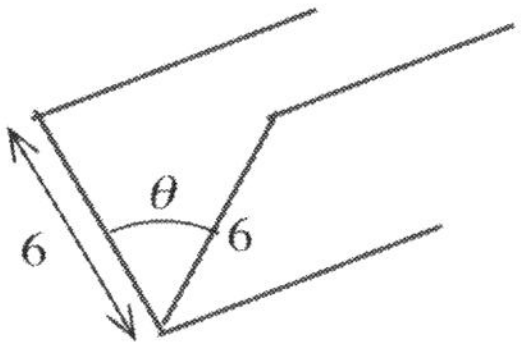

Fig. 20

19. You have a long piece of 12 inch wide metal which you are going to fold along the center line to form a V–shaped gutter (Fig. 20). What angle θ will give the gutter which holds the most water (the largest cross–sectional area)?

20. You have a long piece of 8 inch wide metal which you are going to make into a gutter by bending up 3 inches on each side (Fig. 21). What angle θ will give the gutter which holds the most water (with the largest cross–sectional area)?

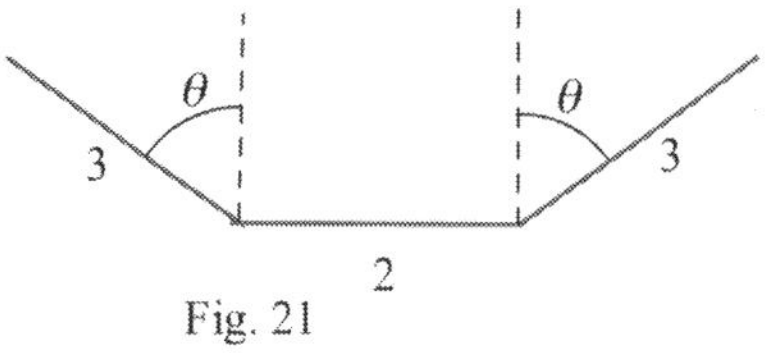

Fig. 21

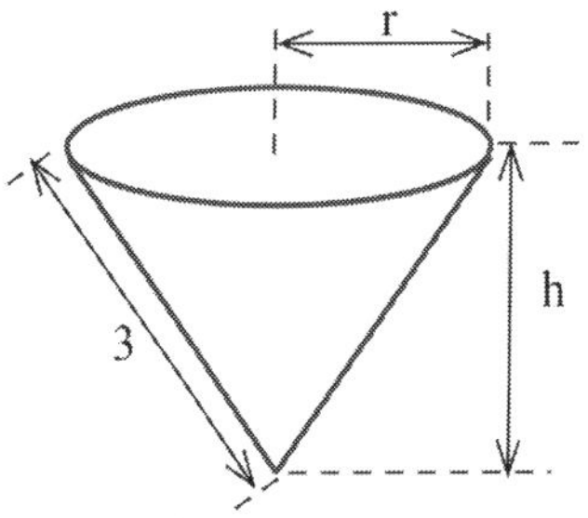

Fig. 22

21. You have a 6 inch diameter circle of paper which you want to form into a drinking cup by removing a pie–shaped wedge and forming the remaining paper into a cone (Fig. 22). Find the height and top radius of the cone so the volume of the cup is as large as possible.

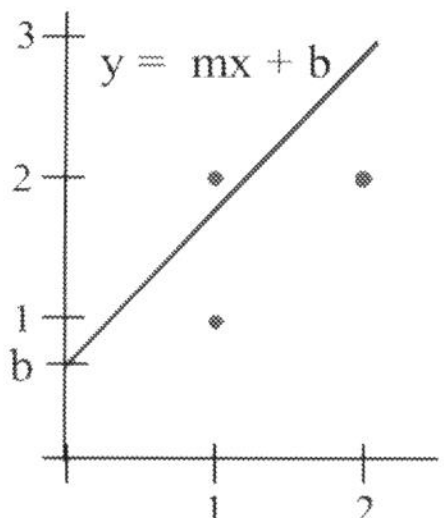

Fig. 23

22. (a) What value of b minimizes the sum of the squares of the vertical distances of the line $y=2x+b$ from the points (1, 1), (1, 2) and (2, 2)? (Fig. 23)

(b) What slope m minimizes the sum of the squares of the vertical distances of the line $y=mx$ from the points (1, 1), (1, 2) and (2, 2)?

(c) What slope m minimizes the sum of the squares of the vertical distances of the line $y = mx$ from the points (2, 1), (4, 3), (–2, –2), and (–4, –2)?

23. You own a small airplane which holds a maximum of 20 passengers. It costs you \$100 per flight from St. Thomas to St. Croix for gas and wages plus an additional \$6 per passenger for the extra gas required by the extra weight. The charge per passenger is \$30 each if 10 people charter your plane (10 is the minimum number you will fly), and this charge is reduced by \$1 per passenger for each passenger over 10 who goes (that is, if 11 go they each pay \$29, if 12 go they each pay \$28, etc.). What number of passengers on a flight will maximize your profits?

24. Prove: If f and g are differentiable functions and if the vertical distance between f and g is greatest at $x = c$, then $f'(c) = g'(c)$ and the tangent lines to f and g are parallel when $x = c$. (Fig. 24)

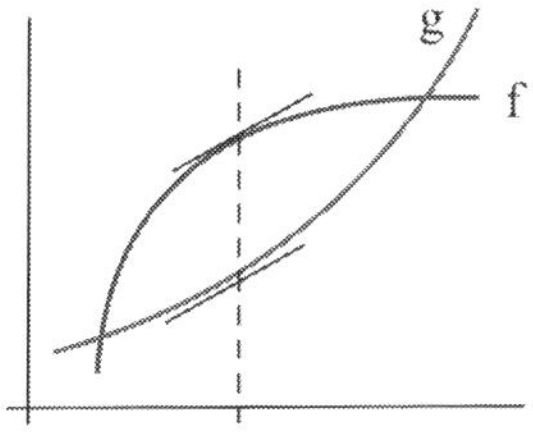

Fig. 24

25. Profit is revenue minus expenses. Assume that revenue and expenses are differentiable functions and show that when profit is maximized, then marginal revenue (dR/dx) equals marginal expense (dE/dx).

26. D. Simonton claims that the "productivity levels" of people in different fields can be described as a function of their "career age" t by $p(t) = e^{-at} - e^{-bt}$ where a and b are constants which depend on the field of work, and career age is approximately 20 less than the actual age of the individual.

(a) Based on this model, at what ages do mathematicians (a=.03, b=.05), geologists (a=.02, b=.04), and historians (a=.02, b=.03) reach their maximum productivity?

(b) Simonton says "With a little calculus we can show that the curve (p(t)) maximizes at $t = \frac{1}{b-a} \ln(\frac{b}{a})$." Use calculus to show that Simonton is correct.

Note: Models of this type have uses for describing the behavior of groups, but it is dangerous and usually invalid to apply group descriptions or comparisons to **individuals** in the group. (Scientific Genius, by Dean Simonton, Cambridge University Press, 1988, pp. 69 – 73)

27. After the table was wiped and the potato chips dried off, the question remained: "Just how far could a can of cola be tipped before it fell over?"

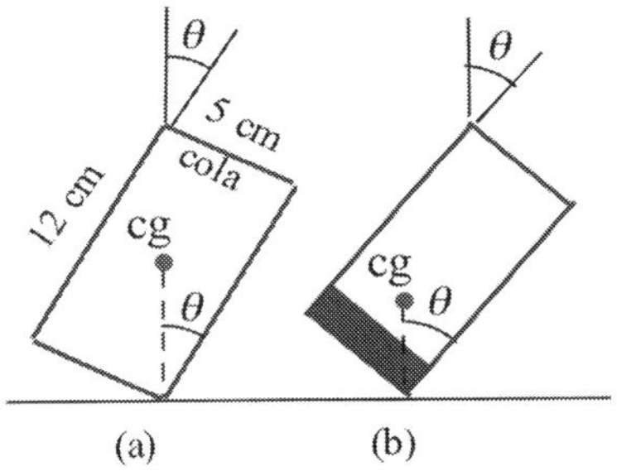

Fig. 25

(i) For a full can or an empty can the answer was easy: the center of gravity (cg) of the can is at the middle of the can, half as high as the height of the can, and we can tilt the can until the cg is directly above the bottom rim. (Fig. 25a) Find θ.

(ii) For a partly filled can more thinking was needed. Some ideas you will see in chapter 5 let us calculate that the cg of a can containing x cm of cola is $C(x) = \frac{360 + 9.6x^2}{60 + 19.2x}$ cm above the bottom of the can. Find the height **x** of cola in the can which will make the cg as low as possible.

(iii) Assuming that the cola is frozen solid (so the top of the cola stays parallel to the bottom of the can), how far can we tilt a can containing **x** cm of cola. (Fig. 25b)

(iv) If the can contained **x** cm of liquid cola, could we tilt it more or less far than the frozen cola before it would fall over?

28. Just thinking — calculus will not help with this one:

(a) Four towns are located at the corners of a square (see Fig. 26). What is the shortest length of road we can construct so that it is possible to travel along the road from any town to any other town?

(b) What is the shortest connecting path for 5 towns located on the corners of a pentagon?

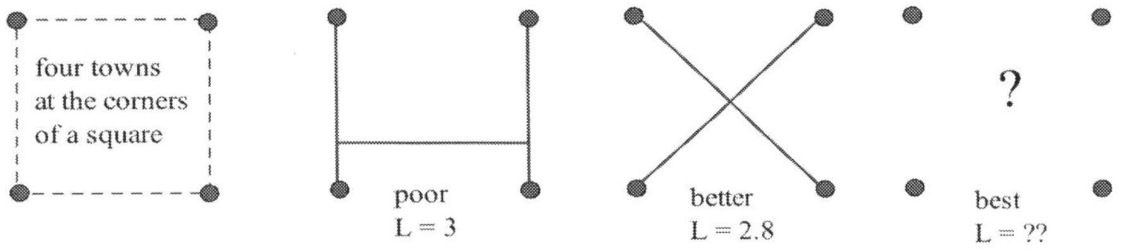

Fig. 26

(The problem of finding the shortest path connecting several points in the plane is called the "Steiner problem." It is important for designing computer chips and telephone networks to be as efficient as possible.)

Generalized Max/Min Problems

The previous max/min problems were all numerical problems: the amount of fencing in problem 2 was 200 feet, the sides of the piece of tin in problem 4 were 10 and 15, and the parabola in problem 17a was $y = 16 - x^2$. In doing those problems you might have noticed some patterns among the numbers in the problem and the numbers in your answers, and you might have wondered if the pattern was an accident of the numbers or if it there really was a pattern at work. Rather than trying several numerical examples to see if the "pattern" holds, mathematicians, engineers, scientists and others sometimes resort to generalizing the problem. We free the problem from the particular numbers by replacing the numbers with letters, and then we solve the generalized problem. In this way, relationships between the values in the problem and those in the solution can become more obvious. Solutions to these generalized problems are also useful if you want to program a computer or calculator to quickly provide numerical answers.

29. (a) Find the dimensions of the rectangle with the greatest area that can be built so the base of the rectangle is on the x–axis between 0 and 1 ($0 \le x \le 1$) and one corner of the rectangle is on the curve $y = x^2$ (Fig. 27a). What is the area of this rectangle?
    (b) Generalize the problem in part (a) for the parabola $y = Cx^2$ with $C > 0$ and $0 \le x \le 1$ (Fig. 27b).
    (c) Generalize for the parabola $y = Cx^2$ with $C > 0$ and $0 \le x \le B$ (Fig. 27c).

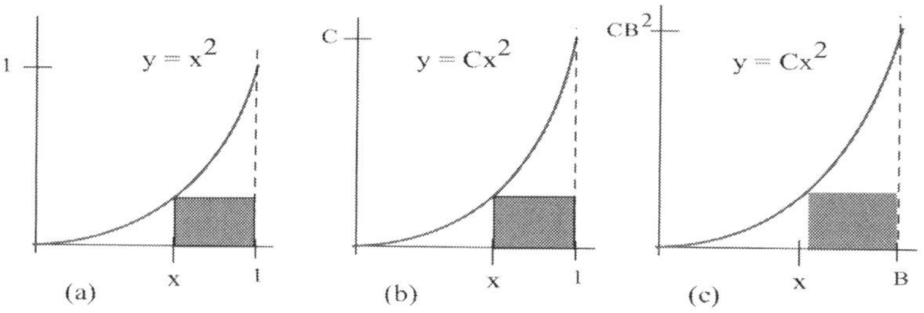

Fig. 27

30. (a) Find the dimensions of the rectangle with the greatest area that can be built so the base of the rectangle is on the x–axis between 0 and 1 ($0 \le x \le 1$) and one corner of the rectangle is on the curve $y = x^3$. What is the area of this rectangle?
    (b) Generalize the problem in part (a) for the curve $y = Cx^3$ with $C > 0$ and $0 \le x \le 1$.
    (c) Generalize for the curve $y = Cx^3$ with $C > 0$ and $0 \le x \le B$.
    (d) Generalize for the curve $y = Cx^n$ with $C > 0$, n a positive integer, and $0 \le x \le B$.

31. (a) The base of a right triangle is 50 and the height is 20 (Fig. 28a). Find the dimensions and area of the rectangle with the greatest area that can be enclosed in the triangle if the base of the rectangle must lie on the base of the triangle.
    (b) The base of a right triangle is B and the height is H (Fig. 28b). Find the dimensions and area of the rectangle with the greatest area that can be enclosed in the triangle if the base of the rectangle must lie on the base of the triangle.

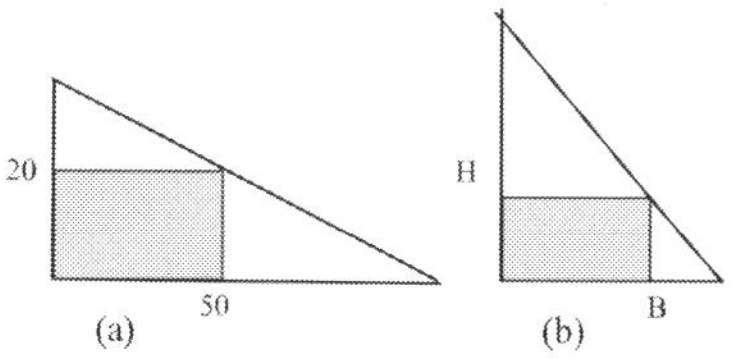

Fig. 28

    (c) State your general conclusion from part (b) in words.

32. (a) You have T dollars to buy fence to enclose a rectangular plot of land (Fig. 29). The fence for the top and bottom costs \$5 per foot and for the sides it costs \$3 per foot. Find the dimensions of the plot with the largest area. For this largest plot, how much money was used for the top and bottom, and for the sides?

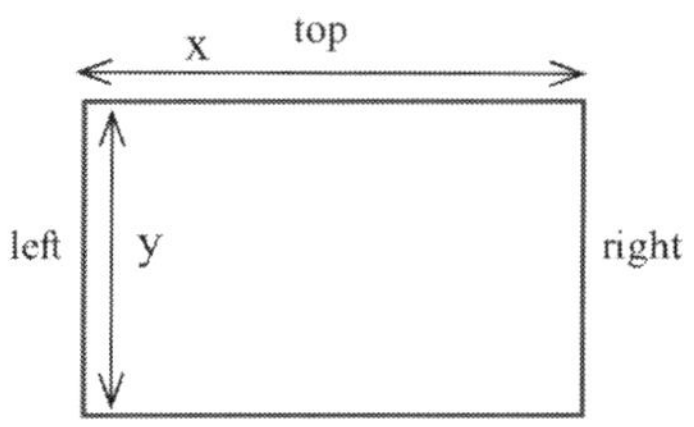

Fig. 29

(b) You have T dollars to buy fence to enclose a rectangular plot of land. The fence for the top and bottom costs \$A per foot and for the sides it costs \$B per foot. Find the dimensions of the plot with the largest area. For this largest plot, how much money was used for the top and bottom (together), and for the sides (together)?

(c) You have T dollars to buy fence to enclose a rectangular plot of land. The fence costs \$A per foot for the top, \$B/foot for the bottom, \$C/ft for the left side and \$D/ft for the right side. Find the dimensions of the plot with the largest area. For this largest plot, how much money was used for the top and bottom (together), and for the sides (together)?

33. Determine the dimensions of the least expensive cylindrical can which will hold V cubic inches if the top material costs \$A per square inch, the bottom material costs \$B/in^2, and the side material costs \$C/in^2.

34. Find the location of C in Fig. 30 so the sum of the distances from A to C and from C to B is a minimum.

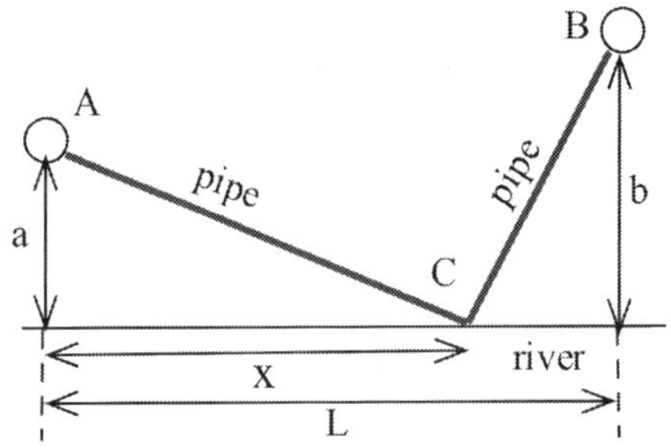

Fig. 30

Section 3.5 **PRACTICE Answers**

Practice 1: $V(x) = x(15 - 2x)(7 - 2x) = 4x^3 - 44x^2 + 105x$.

$V'(x) = 12x^2 - 88x + 105 = (2x - 3)(6x - 35)$ which is defined for all x so the only critical numbers are the endpoints $x = 0$ and $x = 7/2$ and the places where V ' equals 0, at $x = 3/2$ and $x = 35/6$ (but 35/6 is not in the interval $[0, 7/2]$ so it is not practical for this applied problem).

The maximum volume must occur when $x = 0$, $x = 3/2$, or $x = 7/2$):

$V(0) = 0\cdot(15 - 2\cdot0)\cdot(7 - 2\cdot0) = 0$

$V(\frac{3}{2}) = \frac{3}{2}\cdot(15 - 2\cdot\frac{3}{2})\cdot(7 - 2\cdot\frac{3}{2})$

$= \frac{3}{2}(12)(4) = 72$ max.

$V(\frac{7}{2}) = \frac{7}{2}\cdot(15 - 2\cdot\frac{7}{2})\cdot(7 - 2\cdot\frac{7}{2})$

$= \frac{7}{2}(8)(0) = 0$

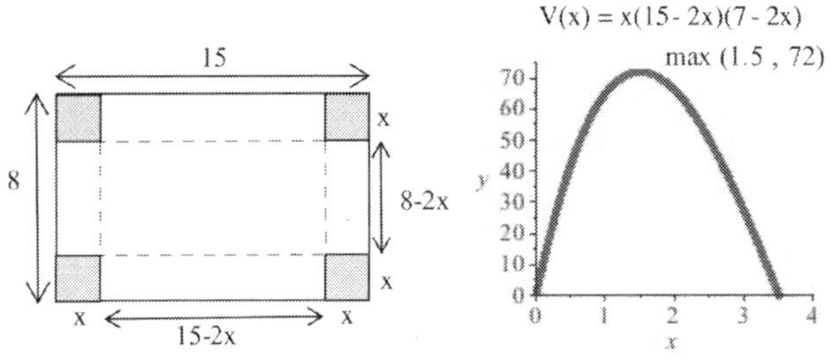

Fig. 31

Fig. 31 shows the graph of V(x).

Practice 2: (a) We have 80 feet of fencing. (See Fig. 32). Our assignment is to maximize the area of the garden: $A = x \cdot y$ (two variables). Fortunately we have the constraint that $x + y = 80$ so $y = 80 - x$, and our assignment reduces to maximizing a function of one variable:

maximize $A = x \cdot y = x \cdot (80 - x) = 80x - x^2$.

$A' = 80 - 2x$ so $A' = 0$ when $x = 40$.

$A'' = -2$ so A is concave down, and A has a maximum at $x = 40$.

The maximum area is $A = 40 \cdot 40 = 1600$ square feet when $x = 40$ ft. and $y = 40$ ft. The maximum area garden is a square.

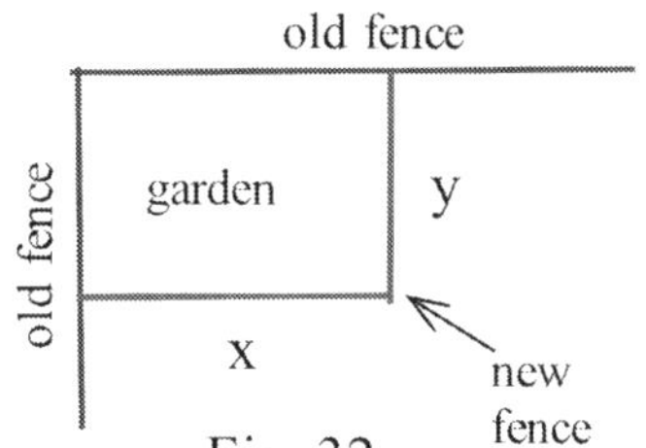

Fig. 32

(b) This is very similar to part (a) except we have F feet of fencing instead of 80 feet.

$x + y = F$ so $y = F - x$, and we want to maximize $A = xy = x(F - x) = Fx - x^2$.

$A' = F - 2x$ so $A' = 0$ when $x = F/2$ and $y = F/2$. The maximum area is $A = F^2/4$ square feet and that occurs when the garden is a square and half of the new fence is used on each of the two new sides.

Practice 3: Cost C = 5(area of top) + 3(area of sides) + 5(area of bottom) $= 5(\pi r^2) + 3(2\pi rh) + 5(\pi r^2)$ so our assignment is to minimize $C = 10\pi r^2 + 6\pi rh$, a function of two variables r and h.

Fortunately we also have the constraint that volume $= 1000 \text{ in}^3 = \pi r^2 h$ so $h = \dfrac{1000}{\pi r^2}$. Then

$C = 10\pi r^2 + 6\pi r\left(\dfrac{1000}{\pi r^2}\right) = 10\pi r^2 + \dfrac{6000}{r}$ so $C' = 20\pi r - \dfrac{6000}{r^2}$. $C' = 0$ if $20\pi r - \dfrac{6000}{r^2} = 0$

so $20\pi r^3 = 6000$ and $r = \left(\dfrac{6000}{20\pi}\right)^{1/3} \approx 4.57$ in. Then $h = \dfrac{1000}{\pi r^2} \approx \dfrac{1000}{\pi(4.57)^2} \approx 15.24$ in.

($C'' = 20\pi + \dfrac{12000}{r^3} > 0$ for all $r > 0$ so C is concave up and we have found a minimum of C.)

3.6 ASYMPTOTIC BEHAVIOR OF FUNCTIONS

When you turn on an automobile or a light bulb many things happen, and some of them are uniquely part of the start up of the system. These "transient" things occur only during start up, and then the system settles down to its steady–state operation. The start up behavior of systems can be very important, but sometimes we want to investigate the steady–state or long term behavior of the system: how is the system behaving "after a long time?" In this section we consider ways of investigating and describing the long term behavior of functions and the systems they may model: how is a function behaving "when x (or –x) is arbitrarily large?"

Limits As X Becomes Arbitrarily Large ("Approaches Infinity")

The same type of questions we considered about a function f as x approached a finite number can also be asked about f as x "becomes arbitrarily large," "increases without bound," and is eventually larger than any fixed number.

Example 1: What happens to the values of $f(x) = \frac{5x}{2x+3}$ (Fig. 1) and $g(x) = \frac{\sin(7x+1)}{3x}$ as x becomes arbitrarily large, as x increases without bound?

Solution: One approach is numerical: evaluate f(x) and g(x) for some "large" values of x and see if there is a pattern to the values of f(x) and g(x). Fig. 1 shows the values of f(x) and g(x) for several large values of x. When x is very large, it appears that the values of f(x) are close to 2.5 = 5/2 and the values of g(x) are close to 0. In fact, we can guarantee that the values of f(x) are as close to 5/2 as someone wants by taking x to be "big enough." The values of $f(x) = \frac{5x}{2x+3}$ may or may not ever equal 5/2 (they never do), but if x is "large," then f(x) is "close to" 5/2. Similarly, we can guarantee that the values of g(x) are as close to 0 as someone wants by taking x to be "big enough." The graphs of f and g are shown in Fig. 2 for "large" values of x.

x	$\frac{5x}{2x+3}$	$\frac{\sin(7x+1)}{3x}$
10	2.17	0.03170
100	2.463	–0.00137
1000	2.4962	0.00033
10,000	2.4996	0.000001

Fig. 1

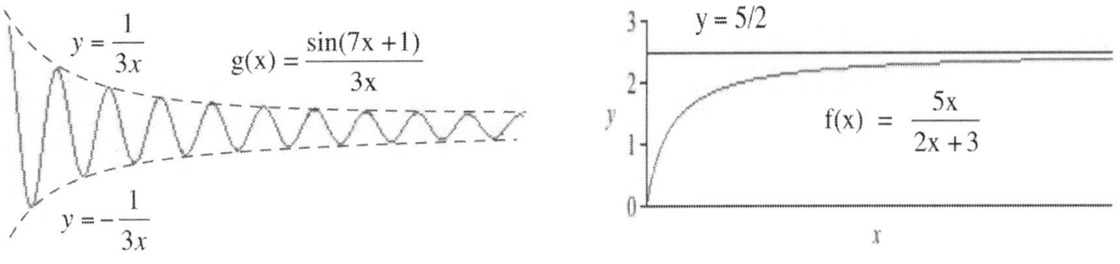

Fig. 2

Practice 1: What happens to the values of $f(x) = \frac{3x+4}{x-2}$ and $g(x) = \frac{\cos(5x)}{2x+7}$ as x becomes arbitrarily large?

The answers for Example 1 can be written as limit statements:

"As x becomes arbitrarily large, the values of $\frac{5x}{2x+3}$ approach $\frac{5}{2}$ " can be written " $\lim_{x\to\infty} \frac{5x}{2x+3} = \frac{5}{2}$ " and

"the values of $\frac{\sin(7x+1)}{3x}$ approach 0." can be written " $\lim_{x\to\infty} \frac{\sin(7x+1)}{3x} = 0$."

The symbol " $\lim_{x\to\infty}$ " is read "the limit as x approaches infinity" and means "the limit as x becomes arbitrarily large" or as x increases without bound. (During this discussion and throughout this book, we do not treat "infinity" or "∞", as a number, but only as a useful notation. "Infinity" is not part of the real number system, and we use the common notation "$x\to\infty$" and the phrase "x approaches infinity" only to mean that "x becomes arbitrarily large." The notation "$x \to -\infty$," read as "x approaches negative infinity," means that the values of $-x$ become arbitrarily large.)

Practice 2: Write your answers to Practice 1 using the limit notation.

The $\lim_{x\to\infty} f(x)$ asks about the behavior of f(x) as the values of x get larger and larger without any bound, and one way to determine this behavior is to look at the values of f(x) at some values of x which are "large". If the values of the function get arbitrarily close to a single number as x gets larger and larger, then we will say that number is the limit of the function as x approaches infinity. A definition of the limit as "$x\to\infty$" is given at the end of this section.

x	$\frac{6x+7}{3-2x}$	$\frac{\sin(3x)}{x}$
10		
200		
5000		
20,000		

Fig. 3

Practice 3: Fill in the table in Fig. 3 for $f(x) = \frac{6x+7}{3-2x}$ and $g(x) = \frac{\sin(3x)}{x}$, and then use those values to estimate $\lim_{x\to\infty} f(x)$ and $\lim_{x\to\infty} g(x)$.

Example 2: How large does x need to be to guarantee that $f(x) = \frac{1}{x} < 0.1$? 0.001? $< E$ (assume $E>0$)?

Solution: If $x > 10$, then $\frac{1}{x} < \frac{1}{10} = 0.1$ (Fig. 4). If $x > 1000$, then $\frac{1}{x} < \frac{1}{1000} = 0.001$.

In general, if E is any positive number, then we can guarantee that $| f(x) | < E$ by picking only values of $x > \frac{1}{E} > 0$: if $x > \frac{1}{E}$, then $\frac{1}{x} < E$.

From this we can conclude that $\lim_{x \to \infty} \frac{1}{x} = 0$.

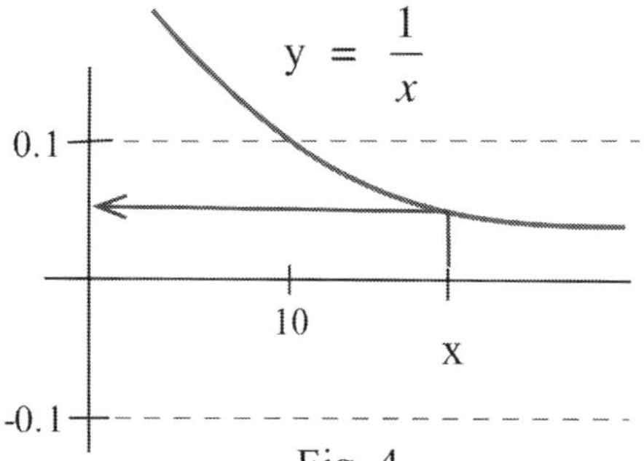

Fig. 4

Practice 4: How large does x need to be to guarantee that $f(x) = \frac{1}{x^2} < 0.1$? 0.001? $< E$ (assume $E>0$)?

Evaluate $\lim_{x \to \infty} \frac{1}{x^2}$.

The Main Limit Theorem (Section 1.2) about limits of combinations of functions is true if the limits as "$x \to a$" are replaced with limits as "$x \to \infty$", but we will not prove those results.

Polynomials occur commonly, and we often need the limit, as $x \to \infty$, of ratios of polynomials or functions containing powers of x. In those situations the following technique is often helpful:

(i) factor the highest power of x in the denominator from both the numerator and the denominator, and

(ii) cancel the common factor from the numerator and denominator.

The limit of the new denominator is a constant, so the limit of the resulting ratio is easier to determine.

Example 3: Determine $\lim_{x \to \infty} \frac{7x^2 + 3x - 4}{3x^2 - 5}$ and $\lim_{x \to \infty} \frac{9x + 2}{3x^2 - 5x + 1}$.

Solutions: $\lim_{x \to \infty} \frac{7x^2 + 3x - 4}{3x^2 - 5} = \lim_{x \to \infty} \frac{x^2(7 + 3/x - 4/x^2)}{x^2(3 - 5/x^2)}$ factoring out x^2

$= \lim_{x \to \infty} \frac{7 + 3/x - 4/x^2}{3 - 5/x^2} = \frac{7}{3}$ canceling x^2 and noting $\frac{3}{x}, \frac{4}{x^2}, \frac{5}{x^2} \to 0$.

Similarly, $\lim_{x \to \infty} \frac{9x + 2}{3x^2 - 5x + 1} = \lim_{x \to \infty} \frac{x^2(9/x + 2/x^2)}{x^2(3 - 5/x + 1/x^2)}$

$= \lim_{x \to \infty} \frac{9/x + 2/x^2}{3 - 5/x + 1/x^2} = \frac{0}{3} = 0$.

If we have a difficult limit, as $x \to \infty$, it is often useful to algebraically manipulate the function into the form of a ratio and then use the previous technique.

If the values of the function oscillate and do not approach a single number as x becomes arbitrarily large, then the function does not have a limit as x approaches infinity: the limit Does Not Exist.

Example 4: Evaluate $\lim_{x\to\infty} \sin(x)$ and $\lim_{x\to\infty} x - [x]$.

Solution: $f(x) = \sin(x)$ and $g(x) = x - [x]$ do not have limits as $x \to \infty$. As x grows without bound, the values of $f(x) = \sin(x)$ oscillate between -1 and $+1$ (Fig. 5), and these values of $\sin(x)$ do not approach a single number. Similarly, $g(x) = x - [x]$ continues to take on values between 0 and 1, and these values are not approaching a single number.

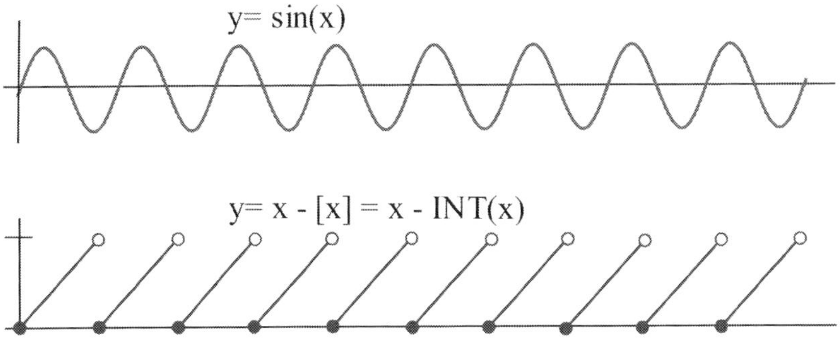

Fig. 5

Using Calculators To Help Find Limits as "$x \to \infty$" or "$x \to -\infty$"

Calculators only store a limited number of digits of a number, and this is a severe limitation when we are dealing with extremely large numbers.

Example: The value of $f(x) = (x + 1) - x$ is clearly equal to 1 for all values of x, and your calculator will give the right answer if you use it to evaluate f(4) or f(5). Now use it to evaluate f for a big value of x, say $x = 10^{40}$. $f(10^{40}) = (10^{40} + 1) - 10^{40} = 1$, but most calculators do not store 40 digits of a number, and they will respond that $f(10^{40}) = 0$ which is **wrong**. In this example the calculator's error is obvious, but the same type of errors can occur in less obvious ways when very large numbers are used on calculators.

You need to be careful with and somewhat suspicious of the answers your calculator gives you.

Calculators can still be helpful for examining some limits as "$x \to \infty$" and "$x \to -\infty$" as long as we do not place too much faith in their responses.

Even if you have forgotten some of the properties of natural logarithm function ln(x) and the cube root function $\sqrt[3]{x}$, a little experimentation on your calculator can help you determine that $\lim_{x\to\infty} \frac{\ln(x)}{\sqrt[3]{x}} = 0$.

The Limit is Infinite

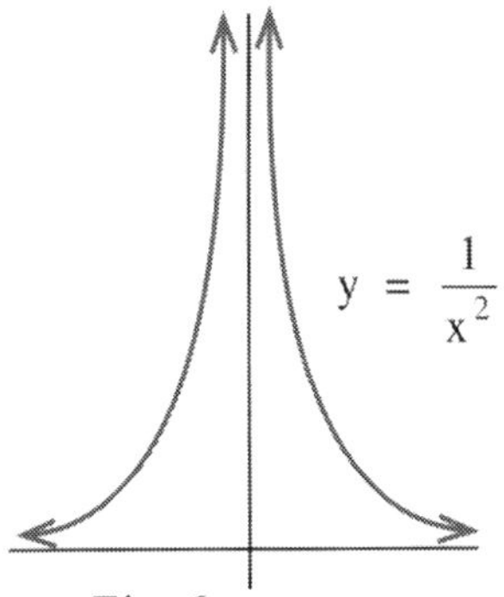

Fig. 6

The function $f(x) = \frac{1}{x^2}$ is undefined at $x = 0$, but we can still ask about the behavior of $f(x)$ for values of x "close to" 0. Fig. 6 indicates that if x is very small, close to 0, then $f(x)$ is very large. As the values of x get closer to 0, the values of $f(x)$ grow larger and can be made as large as we want by picking x to be close enough to 0. Even though the values of f are not approaching any number, we use the "infinity" notation to indicate that the values of f are growing without bound, and write

$$\lim_{x \to 0} = \infty .$$

The values of $\frac{1}{x^2}$ do not equal "infinity:" $\lim_{x \to 0} = \infty$ means that the values of $\frac{1}{x^2}$ can be made arbitrarily large by picking values of x very close to 0.

The limit, as $x \to 0$, of $\frac{1}{x}$ is slightly more complicated. If x is close to 0, then the value of $f(x) = 1/x$ can be a large positive number or a large negative number, depending on the sign of x.

The function $f(x) = 1/x$ does not have a (two–sided) limit as x approaches 0, but we can still ask about one–sided limits:

$$\lim_{x \to 0^+} \frac{1}{x} = \infty \quad \text{and} \quad \lim_{x \to 0^-} \frac{1}{x} = -\infty .$$

Example 5: Determine $\lim_{x \to 3^+} \frac{x-5}{x-3}$ and $\lim_{x \to 3^-} \frac{x-5}{x-3}$.

Solution: (a) As $x \to 3^+$, then $x - 5 \to -2$ and $x - 3 \to 0$. Since the denominator is approaching 0 we cannot use the Main Limit Theorem, and we need to examine the functions more carefully. If $x \to 3^+$, then $x > 3$ so $x - 3 > 0$. If x is close to 3 and slightly larger than 3, then the ratio of $x - 5$ to $x - 3$ is the ratio $\frac{\text{a number close to } -2}{\text{small positive number}}$ = large negative number. As $x > 3$ gets closer to 3, $\frac{x-5}{x-3}$ is $\frac{\text{a number closer to } -2}{\text{positive and closer to } 0}$ = larger negative number. By taking $x > 3$ closer to 3, the denominator gets closer to 0 but is always positive, so the ratio gets arbitrarily large and negative: $\lim_{x \to 3^+} \frac{x-5}{x-3} = -\infty$.

(b) As $x \to 3^-$, then $x - 5 \to -2$ and $x - 3$ gets arbitrarily close to 0, and $x - 3$ is negative. The value of the ratio $\frac{x-5}{x-3}$ is $\frac{\text{a number close to } -2}{\text{arbitrarily small negative number}}$ = arbitrarily large positive number: $\lim_{x \to 3^-} \frac{x-5}{x-3} = +\infty$.

Practice 5: Determine $\lim_{x \to 2^+} \frac{7}{2-x}$, $\lim_{x \to 2^+} \frac{3x}{2x-4}$, $\lim_{x \to 2^+} \frac{3x^2-6x}{x-2}$

Horizontal Asymptotes

The limits of f , as "$x \to \infty$" and "$x \to -\infty$," give us information about horizontal asymptotes of f.

Definition: The line $y = K$ is a **horizontal asymptote** of f if $\lim_{x\to\infty} f(x) = K$ or $\lim_{x\to-\infty} f(x) = K$.

Example 6: Find any horizontal asymptotes of $f(x) = \frac{2x + \sin(x)}{x}$.

Solution: $\lim_{x\to\infty} \frac{2x+\sin(x)}{x} = \lim_{x\to\infty} \frac{2x}{x} + \frac{\sin(x)}{x} = 2 + 0 = 2$ so the line $\mathbf{y = 2}$ is a horizontal asymptote of f. The limit, as "$x \to -\infty$," is also 2 so $y = 2$ is the only horizontal asymptote of f. The graphs of f and $y = 2$ are given in Fig. 7 . A function may or may not cross its asymptote.

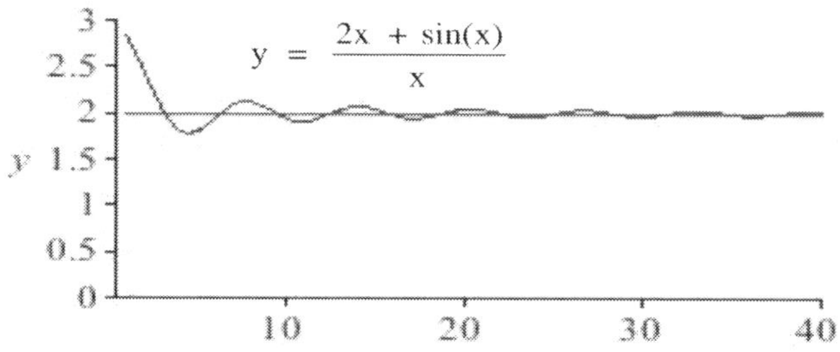

Fig. 7

Vertical Asymptotes

Definition: The vertical line $x = a$ is a **vertical asymptote** of the graph of f if either or both of the one–sided limits, as $x \to a^-$ or $x \to a^+$, of f is infinite.

If our function f is the ratio of a polynomial P(x) and a polynomial Q(x), $f(x) = \frac{P(x)}{Q(x)}$, then the only **candidates** for vertical asymptotes are the values of x where $Q(x) = 0$. However, the fact that $Q(a) = 0$ is **not** enough to guarantee that the line $x = a$ is a vertical asymptote of f ; we also need to evaluate P(a). If $Q(a) = 0$ and $P(a) \neq 0$, then the line $x = a$ is a vertical asymptote of f. If $Q(a) = 0$ and $P(a) = 0$, then the line $x = a$ may or may not be a vertical asymptote.

Example 7: Find the vertical asymptotes of $f(x) = \frac{x^2 - x - 6}{x^2 - x}$ and $g(x) = \frac{x^2 - 3x}{x^2 - x}$.

Solution: $f(x) = \dfrac{x^2 - x - 6}{x^2 - x} = \dfrac{(x-3)(x+2)}{x(x-1)}$ so the only values which make the denominator 0 are

$x = 0$ and $x = 1$, and these are the only candidates to be vertical asymptotes.

$\lim\limits_{x \to 0^+} f(x) = +\infty$ and $\lim\limits_{x \to 1^+} f(x) = -\infty$ so $x = 0$ and $x = 1$ are both vertical asymptotes of f.

$g(x) = \dfrac{x^2 - 3x}{x^2 - x} = \dfrac{x(x-3)}{x(x-1)}$ so the only candidates to be vertical asymptotes are $x = 0$ and $x = 1$.

$\lim\limits_{x \to 1^+} g(x) = \lim\limits_{x \to 1^+} \dfrac{x(x-3)}{x(x-1)} = \lim\limits_{x \to 1^+} \dfrac{x-3}{x-1} = -\infty$ so $x = 1$ is a vertical asymptote of g.

$\lim\limits_{x \to 0} g(x) = \lim\limits_{x \to 0} \dfrac{x(x-3)}{x(x-1)} = \lim\limits_{x \to 0} \dfrac{x-3}{x-1} = 3 \neq \infty$ so $x = 0$ is **not** a vertical asymptote.

Practice 6: Find the vertical asymptotes of $f(x) = \dfrac{x^2 + x}{x^2 + x - 2}$ and $g(x) = \dfrac{x^2 - 1}{x - 1}$.

Other Asymptotes as "$x \to \infty$" and "$x \to -\infty$"

If the limit of f(x) as "$x \to \infty$" or "$x \to -\infty$" is a constant K, then the graph of f gets close to the horizontal line $y = K$, and we said that $y = K$ was a horizontal asymptote of f. Some functions, however, approach other lines which are not horizontal.

Example 8: Find all asymptotes of $f(x) = \dfrac{x^2 + 2x + 1}{x} = x + 2 + \dfrac{1}{x}$.

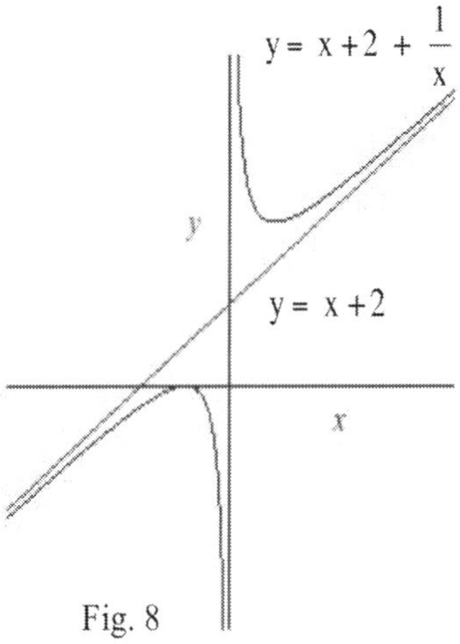

Fig. 8

Solution: If x is a large positive number or a large negative number,

then $\frac{1}{x}$ is very close to 0, and the graph of f(x) is very close to the line $y = x + 2$ (Fig. 8). The line $\mathbf{y = x + 2}$ is an asymptote of the graph of f.

If x is a large positive number, then $1/x$ is positive, and the graph of f is slightly above the graph of $y = x + 2$. If x is a large negative number, then $1/x$ is negative, and the graph of f will be slightly below the graph of $y = x + 2$. The $1/x$ piece of f never equals 0 so the graph of f never crosses or touches the graph of the asymptote $y = x + 2$.

The graph of f also has a vertical asymptote at $x = 0$ since $\lim\limits_{x \to 0^+} f(x) = \infty$ and $\lim\limits_{x \to 0^-} f(x) = -\infty$.

Practice 7: Find all asymptotes of $g(x) = \dfrac{2x^2 - x - 1}{x + 1} = 2x - 3 + \dfrac{2}{x + 1}$.

Some functions even have **nonlinear asymptotes**, asymptotes which are not straight lines. The graphs of these functions approach some nonlinear function when the values of x are arbitrarily large.

Example 9: Find all asymptotes of $f(x) = \dfrac{x^4 + 3x^3 + x^2 + 4x + 5}{x^2 + 1} = x^2 + 3x + \dfrac{x + 5}{x^2 + 1}$.

Solution: When x is very large, positive or negative, then $\dfrac{x + 5}{x^2 + 1}$ is very close to 0, and the graph of f is very close to the graph of $g(x) = x^2 + 3x$. The function $\mathbf{g(x) = x^2 + 3x}$ is a nonlinear asymptote of f. The denominator of f is never 0, and f has no vertical asymptotes.

Practice 8: Find all asymptotes of $f(x) = \dfrac{x^3 + 2\sin(x)}{x} = x^2 + \dfrac{2\sin(x)}{x}$.

If f(x) can be written as a sum of two other functions, $f(x) = g(x) + r(x)$, with $\lim_{x \to \pm\infty} r(x) = 0$, then the graph of f is asymptotic to the graph of g , and g is an asymptote of f.

> Suppose $f(x) = g(x) + r(x)$ and $\lim_{x \to \pm\infty} r(x) = 0$:
>
> if $g(x) = K$, then f has a horizontal asymptote $y = K$;
>
> if $g(x) = ax + b$, then f has a linear asymptote $y = ax + b$; or
>
> if $g(x) =$ a nonlinear function, then f has a nonlinear asymptote $y = g(x)$.

Definition of $\lim_{x \to \infty} f(x) = K$

The following definition states precisely what is meant by the phrase "we can guarantee that the values of f(x) are arbitrarily close to K by using sufficiently large values of x."

Definition: $\lim_{x \to \infty} f(x) = K$

means

for every given $\varepsilon > 0$, there is a number N so that

if x is larger than N

then f(x) is within ε units of K.

(equivalently; $| f(x) - K | < \varepsilon$ whenever $x > N$.)

Example 10: Show that $\lim_{x \to \infty} \frac{x}{2x+1} = \frac{1}{2}$.

Solution: Typically we need to do two things. First we need to find a value of N, usually depending on ε. Then we need to show that the value of N we found satisfies the conditions of the definition.

(i) Assume that $| f(x) - K |$ is less than ε and solve for $x > 0$.

If $\varepsilon > \left| \frac{x}{2x+1} - \frac{1}{2} \right| = \left| \frac{2x-(2x+1)}{2(2x+1)} \right| = \left| \frac{-1}{4x+2} \right| = \frac{1}{4x+2}$, then $4x + 2 > \frac{1}{\varepsilon}$ and $x > \frac{1}{4}\left(\frac{1}{\varepsilon} - 2\right)$. For any $\varepsilon > 0$, take $N = \frac{1}{4}\left(\frac{1}{\varepsilon} - 2\right)$.

(ii) For any $\varepsilon > 0$, take $N = \frac{1}{4}\left(\frac{1}{\varepsilon} - 2\right)$. (Now we can just reverse the order of the steps in part (i).)

If $x > 0$ and $x > N = \frac{1}{4}\left(\frac{1}{\varepsilon} - 2\right)$,

then $4x + 2 > \frac{1}{\varepsilon}$ so $\varepsilon > \frac{1}{4x+2} = \left| \frac{x}{2x+1} - \frac{1}{2} \right| = | f(x) - K |$.

We have shown that "for every given ε, there is an N" that satisfies the definition.

PROBLEMS

1. Fig. 9 shows f(x) and g(x) for $0 \le x \le 5$. Let $h(x) = \frac{f(x)}{g(x)}$.
   (a) At what value of x does h(x) have a root?
   (b) Determine the limits of h(x) as $x \to 1^+, x \to 1^-, x \to 3^+$, and $x \to 3^-$.
   (c) Where does h(x) have a vertical asymptote?

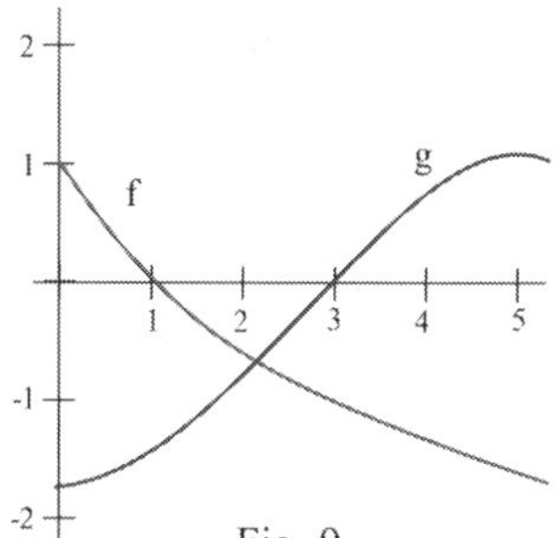

Fig. 9

2. Fig. 10 shows f(x) and g(x) for $0 \le x \le 5$. Let $h(x) = \frac{f(x)}{g(x)}$.
   (a) At what value(s) of x does h(x) have a root?
   (b) Where does h(x) have vertical asymptotes?

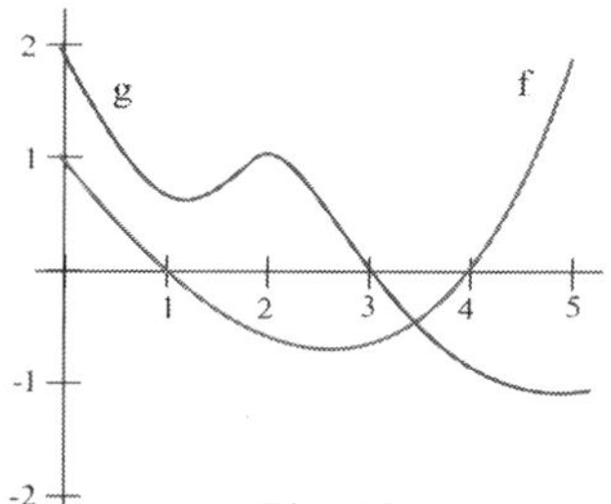

Fig. 10

3. Fig. 11 shows f(x) and g(x) for $0 \le x \le 5$. Let $h(x) = \frac{f(x)}{g(x)}$, and determine the limits of h(x) as $x \to 2^+, x \to 2^-, x \to 4^+$, and $x \to 4^-$.

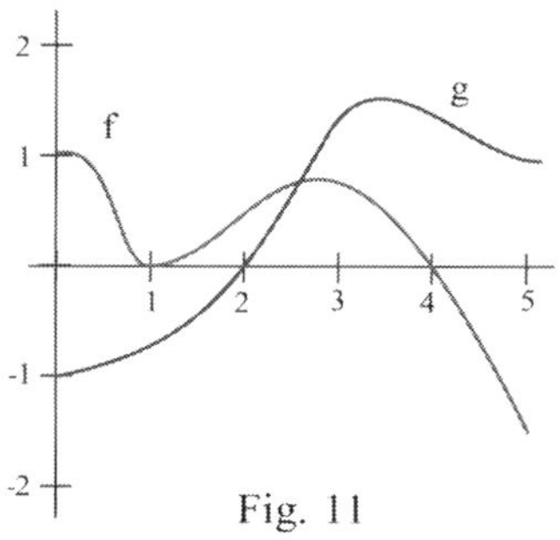

Fig. 11

For problems 4 – 24, calculate the limit of each expression as "**x→ ∞**."

4. $\frac{6}{x+2}$

5. $\frac{28}{3x-5}$

6. $\frac{7x+12}{3x-2}$

7. $\frac{4-3x}{x+8}$

8. $\frac{5\sin(2x)}{2x}$

9. $\frac{\cos(3x)}{5x-1}$

10. $\dfrac{2x - 3\sin(x)}{5x - 1}$ 11. $\dfrac{4 + x\cdot\sin(x)}{2x - 3}$ 12. $\dfrac{x^2 - 5x + 2}{x^2 + 8x - 4}$

13. $\dfrac{2x^2 - 9}{3x^2 + 10x}$ 14. $\dfrac{\sqrt{x + 5}}{\sqrt{4x - 2}}$ 15. $\dfrac{5x^2 - 7x + 2}{2x^3 + 4x}$

16. $\dfrac{x + \sin(x)}{x - \sin(x)}$ 17. $\dfrac{7x^2 + x\cdot\sin(x)}{3 - x^2 + \sin(7x^2)}$ 18. $\dfrac{7x^{143} + 734x - 2}{x^{150} - 99x^{83} + 25}$

19. $\dfrac{\sqrt{9x^2 + 16}}{2 + \sqrt{x^3 + 1}}$ 20. $\sin\left(\dfrac{3x + 5}{2x - 1}\right)$ 21. $\cos\left(\dfrac{7x + 4}{x^2 + x + 1}\right)$

22. $\ln\left(\dfrac{3x^2 + 5x}{x^2 - 4}\right)$ 23. $\ln(x + 8) - \ln(x - 5)$ 24. $\ln(3x + 8) - \ln(2x + 1)$

25. Salt water with a concentration of 0.2 pounds of salt per gallon flows into a large tank that initially contains 50 gallons of pure water.

(a) If the flow rate of salt water into the tank is 4 gallons per minute, what is the volume V(t) of water and the amount A(t) of salt in the tank t minutes after the flow begins?

(b) Show that the salt concentration C(t) at time t is $C(t) = \dfrac{.8t}{4t + 50}$.

(c) What happens to the concentration C(t) after a "long" time?

(d) Redo parts (a) – (c) for a large tank which initially contains 200 gallons of pure water.

26. Under certain laboratory conditions, an agar plate contains $B(t) = 100(2 - \dfrac{1}{e^t}) = 100(2 - e^{-t})$ bacteria t hours after the start of the experiment.

(a) How many bacteria are on the plate at the start of the experiment (t = 0)?

(b) Show that the population is always increasing. (Show $B'(t) > 0$ for all $t > 0$.)

(c) What happens to the population B(t) after a "long" time?

(d) Redo parts (a) – (c) for $B(t) = A(2 - \dfrac{1}{e^t}) = A(2 - e^{-t})$.

For problems 27 – 41 , calculate the limits.

27. $\lim\limits_{x\to 0} \dfrac{x+5}{x^2}$ 28. $\lim\limits_{x\to 3} \dfrac{x-1}{(x-3)^2}$ 29. $\lim\limits_{x\to 5} \dfrac{x-7}{(x-5)^2}$

30. $\lim\limits_{x\to 2^+} \dfrac{x-1}{x-2}$ 31. $\lim\limits_{x\to 2^-} \dfrac{x-1}{x-2}$ 32. $\lim\limits_{x\to 2} \dfrac{x-1}{x-2}$

33. $\lim\limits_{x\to 4^+} \dfrac{x+3}{4-x}$ 34. $\lim\limits_{x\to 1^-} \dfrac{x^2+5}{1-x}$ 35. $\lim\limits_{x\to 3^+} \dfrac{x^2-4}{x^2-2x-3}$

36. $\lim\limits_{x\to 2} \dfrac{x^2-x-2}{x^2-4}$ 37. $\lim\limits_{x\to 0} \dfrac{x-2}{1-\cos(x)}$ 38. $\lim\limits_{x\to\infty} \dfrac{x^3+7x-4}{x^2+11x}$

39. $\lim\limits_{x\to 5} \dfrac{\sin(x-5)}{x-5}$ 40. $\lim\limits_{x\to 0} \dfrac{x+1}{\sin^2(x)}$ 41. $\lim\limits_{x\to 0^+} \dfrac{1+\cos(x)}{1-e^x}$

In problems 42 – 50, write the **equation** of each asymptote for each function and state whether it is a vertical or horizontal asymptote.

42. $f(x) = \dfrac{x+2}{x-1}$ 43. $f(x) = \dfrac{x-3}{x^2}$ 44. $f(x) = \dfrac{x-1}{x^2-x}$

45. $f(x) = \dfrac{x+5}{x^2-4x+3}$ 46. $f(x) = \dfrac{x+\sin(x)}{3x-3}$ 47. $f(x) = \dfrac{x^2-4}{x^2+1}$

48. $f(x) = \dfrac{\cos(x)}{x^2}$ 49. $f(x) = 2 + \dfrac{3-x}{x-1}$ 50. $f(x) = \dfrac{x\cdot\sin(x)}{x^2-x}$

In problems 51 – 59, write the **equation** of each asymptote for each function.

51. $f(x) = \dfrac{2x^2+x+5}{x}$ 52. $f(x) = \dfrac{x^2+x}{x+1}$ 53. $f(x) = \dfrac{1}{x-2} + \sin(x)$

54. $f(x) = x + \dfrac{x}{x^2+1}$ 55. $f(x) = x^2 + \dfrac{x}{x^2+1}$ 56. $f(x) = x^2 + \dfrac{x}{x+1}$

57. $f(x) = \dfrac{x\cdot\cos(x)}{x-3}$ 58. $f(x) = \dfrac{x^3-x^2+2x-1}{x-1}$ 59. $f(x) = \sqrt{\dfrac{x^2+3x+2}{x+3}}$

Section 3.6 **PRACTICE Answers**

Practice 1: As x becomes arbitrarily large, the values of f(x) approach 3 and the values of g(x) approach 0.

Practice 2: $\lim\limits_{x\to\infty} \dfrac{3x+4}{x-2} = 3$ and $\lim\limits_{x\to\infty} \dfrac{\cos(5x)}{2x+7} = 0$

Practice 3: The completed table is shown in Fig. 12.

Practice 4: If $x > \sqrt{10} \approx 3.162$, then $f(x) = \dfrac{1}{x^2} < 0.1$.

If $x > \sqrt{1000} \approx 31.62$, then $f(x) = \dfrac{1}{x^2} < 0.001$.

If $x > \sqrt{1/E}$, then $f(x) = \dfrac{1}{x^2} < E$.

x	$\dfrac{6x+7}{3-2x}$	$\dfrac{\sin(3x)}{x}$
10	**–3.94117647**	**–0.09880311**
200	**–3.04030227**	**0.00220912**
5000	**–3.00160048**	**0.00017869**
20,000	**–3.00040003**	**0.00004787**
	↓ **–3**	↓ **0**

Fig. 12

Practice 5: (a) $\lim_{x \to 2^+} \frac{7}{2-x} = -\infty$.

As $x \to 2^+$ the values $2 - x \to 0$, and $x > 2$ so $2 - x < 0$: $2 - x$ takes small negative values.

Then the values of $\frac{7}{2-x} = \frac{7}{\text{small negative values}}$ are large negative values so we represent the limit as "$-\infty$."

(b) $\lim_{x \to 2^+} \frac{3x}{2x-4} = +\infty$.

As $x \to 2^+$ the values of $2x - 4 \to 0$, and $x > 2$ so $2x - 4 > 0$: $2x - 4$ takes small positive values. As $x \to 2^+$ the values of $3x \to +6$.

Then the values of $\frac{3x}{2x-4} = \frac{\text{values near } +6}{\text{small positive values}}$ are large positive values so we represent the limit as "$+\infty$."

(c) $\lim_{x \to 2^+} \frac{3x^2 - 6x}{x-2} = 6$.

As $x \to 2^+$, the values of $3x^2 - 6x \to 0$ and $x - 2 \to 0$ so we need to do more work. The numerator can be factored $3x^2 - 6x = 3x(x - 2)$ and then the rational function can be reduced (since $x \to 2$ we know $x \neq 2$):

$$\lim_{x \to 2^+} \frac{3x^2 - 6x}{x-2} = \lim_{x \to 2^+} \frac{3x(x-2)}{x-2} = \lim_{x \to 2^+} 3x = 6 .$$

Practice 6: (a) $f(x) = \frac{x^2 + x}{x^2 + x - 2} = \frac{x(x+1)}{(x-1)(x+2)}$.

f has vertical asymptotes at $x = 1$ and $x = -2$.

(b) $g(x) = \frac{x^2 - 1}{x - 1} = \frac{(x+1)(x-1)}{x-1}$.

The value $x = 1$ is not in the domain of g. If $x \neq 1$, then $g(x) = x + 1$.

g has a "hole" when $x = 1$ and no vertical asymptotes.

Practice 7: $g(x) = 2x - 3 + \frac{2}{x + 1}$.

g has a vertical asymptote at $x = -1$.

g has no horizontal asymptotes.

$\lim_{x \to \infty} \frac{2}{x + 1} = 0$ so g has the linear asymptote $y = 2x - 3$.

Practice 8: $f(x) = x^2 + \frac{2 \cdot \sin(x)}{x}$.

f is not defined at $x = 0$, so f has a vertical asymptote or a "hole" when $x = 0$.

$\lim_{x \to 0} x^2 + \frac{2 \cdot \sin(x)}{x} = 0 + 2 = 2$ so f has a "hole" when $x = 0$.

$\lim_{x \to \infty} \frac{2 \cdot \sin(x)}{x} = 0$ so f has the nonlinear asymptote $y = x^2$.

Appendix: MAPLE, infinite limits and limits as "$x \to \infty$"

command	output	comment
limit(1/x, x = 0);	undefined	Maple uses the convention "x=0" even though what we really mean is "x approaches 0"
limit(1/x, x=0, right);	∞	"x=0, right" means $0 \le x$
limit(1/x. x=0, left);	$-\infty$	"x=0, left" means $x \ge 0$
limit(sin(x)/x, x = infinity);	0	
limit(2*x, x= infinity);	∞	
limit(5-3*x, x= infinity);	$-\infty$	
limit((1=cos(x))/(1-exp(x)). x = 0, right);	$-\infty$	

3.7 L'HÔPITAL'S RULE

When we began taking limits of slopes of secant lines, $m_{sec} = \frac{f(x+h) - f(x)}{h}$ as $h \to 0$, we frequently encountered one difficulty: both the numerator and the denominator approached 0. And since the denominator approached 0, we could not apply the Main Limit Theorem. In each case, however, we managed to get past this "0/0" difficulty by using algebra or geometry or trigonometry, but there was no common approach or pattern. The algebraic steps we used to evaluate

$\lim_{h\to 0} \frac{(2+h)^2 - 4}{h}$ seem quite different from the trigonometric steps needed for $\lim_{h\to 0} \frac{\sin(2+h) - \sin(2)}{h}$.

In this section we consider a single technique, called l'Hôpital's Rule (pronounced Low–Pee–Tall), which enables us to quickly and easily evaluate limits of the form "0/0" as well as several other difficult forms.

A Linear Example

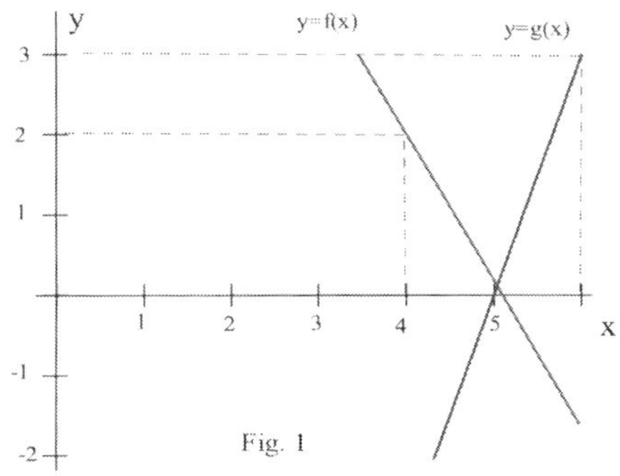

Fig. 1

Two linear functions are given in Fig. 1, and we need to find $\lim_{x\to 5} \frac{f(x)}{g(x)}$. Unfortunately, $\lim_{x\to 5} f(x) = 0$ and $\lim_{x\to 5} g(x) = 0$ so we cannot apply the Main Limit Theorem. However, we know f and g are linear, we can calculate their slopes from Fig. 1, and we know that they both go through the point (5,0) so we can find their equations: $f(x) = \mathbf{-2}(x - 5)$ and $g(x) = \mathbf{3}(x - 5)$.

Now the limit is easier: $\lim_{x\to 5} \frac{f(x)}{g(x)} = \lim_{x\to 5} \frac{-2(x-5)}{3(x-5)} = \frac{\mathbf{-2}}{\mathbf{3}} = \frac{\text{slope of f}}{\text{slope of g}}$.

In fact, this pattern works for any two linear functions:

If f and g are linear functions with slopes **m** and **n** $\neq 0$ and a common root at $x = a$, ($f(x) - f(a) = \mathbf{m}(x - a)$ and $g(x) - g(a) = \mathbf{n}(x - a)$ so $f(x) = \mathbf{m}(x - a)$ and $g(x) = \mathbf{n}(x - a)$)

then $\lim_{x\to a} \frac{f(x)}{g(x)} = \lim_{x\to a} \frac{m(x-a)}{n(x-a)} = \frac{\mathbf{m}}{\mathbf{n}} = \frac{\text{slope of f}}{\text{slope of g}}$.

The really powerful result, discovered by John Bernoulli and named for the Marquis de l'Hôpital who published it in his calculus book, is that the same pattern is true for differentiable functions even if they are not linear.

L'Hôpital's Rule ("0/0" form)

If f and g are differentiable at x = a ,

and f(a) = 0, g(a) = 0, and g '(a) ≠ 0 ,

then $\lim_{x \to a} \frac{f(x)}{g(x)} = \frac{f'(a)}{g'(a)} = \frac{\text{slope of f at a}}{\text{slope of g at a}}$.

Idea for a proof: Even though f and g may not be linear functions, they are differentiable so at the point x = a they are "almost linear" in the sense that they are well approximated by their tangent lines at that point (Fig. 2): since f(a) = g(a) = 0

f(x) ≈ f(a) + f '(a)(x–a) = f '(a)(x–a) and g(x) ≈ g(a) + g '(a)(x–a) = g '(a)(x–a).

Then $\lim_{x \to a} \frac{f(x)}{g(x)} \approx \lim_{x \to a} \frac{f'(a)(x-a)}{g'(a)(x-a)} = \frac{f'(a)}{g'(a)}$.

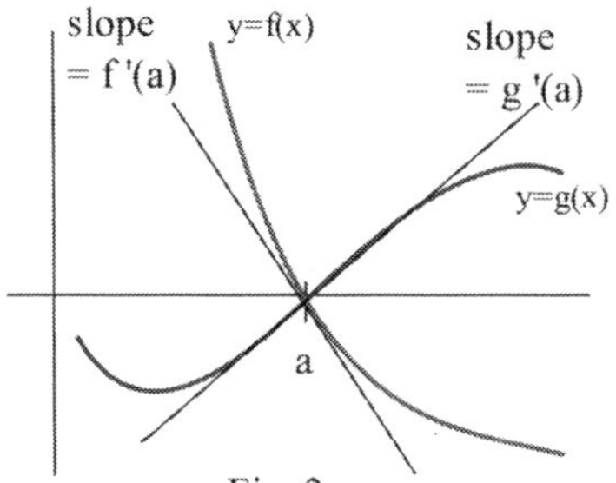

Fig. 2

(Unfortunately, we have ignored a couple subtle difficulties such as g(x) or g '(x) possibly being 0 when x is close to a. A proof of l'Hôpital's Rule is difficult and is not included.)

Example 1: Use l'Hôpital's Rule to determine $\lim_{x \to 0} \frac{x^2 + \sin(5x)}{3x}$ and $\lim_{x \to 1} \frac{\ln(x)}{e^x - e}$

Solution: (a) We could evaluate this limit without l'Hôpital's Rule but let's use it. We can match the pattern of l'Hôpital's Rule by letting a = 0, f(x) = x^2 + sin(5x) and g(x) = 3x. Then f(0) = 0, g(0) = 0, and f and g are differentiable with f '(x) = 2x + 5cos(5x) and g '(x) = 3 so

$$\lim_{x \to 0} \frac{x^2 + \sin(5x)}{3x} = \frac{f'(\mathbf{0})}{g'(\mathbf{0})} = \frac{2\cdot\mathbf{0} + 5\cos(5\cdot\mathbf{0})}{3} = \frac{5}{3} .$$

(b) Let a = 1, f(x) = ln(x) and g(x) = e^x – e. Then f(1) = 0, g(1) = 0, f and g are differentiable for x near 1 (x ≠ 0), and f '(x) = 1/x and g '(x) = e^x . Then

$$\lim_{x \to 1} \frac{\ln(x)}{e^x - e} = \frac{f'(\mathbf{1})}{g'(\mathbf{1})} = \frac{1/\mathbf{1}}{e^{\mathbf{1}}} = \frac{1}{e} .$$

Practice 1: Use l'Hôpital's Rule to find $\lim_{x \to 0} \frac{1 - \cos(5x)}{3x}$ and $\lim_{x \to 2} \frac{x^2 + x - 6}{x^2 + 2x - 8}$.

Strong Version of L'Hôpital's Rule

L'Hôpital's Rule can be strengthened to include the case when g '(a) = 0 and the indeterminate form "∞/∞", the case when both f and g increase without any bound.

L'Hôpital's Rule (Strong " 0/0 " and " ∞/∞ " forms)

If f and g are differentiable on an open interval I which contains the point a, g '(x) ≠ 0 on I except possibly at a, and

$$\lim_{x\to a} \frac{f(x)}{g(x)} = \text{"}\frac{0}{0}\text{"} \text{ or } \text{"}\frac{\infty}{\infty}\text{"}$$

then $\displaystyle\lim_{x\to a} \frac{f(x)}{g(x)} = \lim_{x\to a} \frac{f'(x)}{g'(x)}$ provided the limit on the right exists.

("a" can represent a finite number or "∞.")

Example 2: Evaluate $\displaystyle\lim_{x\to\infty} \frac{e^{7x}}{5x}$.

Solution: As "x →∞", both f(x) = e^{7x} and g(x) = 5x increase without bound so we have an "∞/∞" indeterminate form and can use the Strong Version l'Hôpital's Rule:

$$\lim_{x\to\infty} \frac{e^{7x}}{5x} = \lim_{x\to\infty} \frac{7e^{7x}}{5} = \infty .$$

The limit of f '/g ' may also be an indeterminate form, and then we can apply l'Hôpital's Rule to the ratio f '/g '. We can continue using l'Hôpital's Rule at each stage as long as we have an indeterminate quotient.

Example 3: $\displaystyle\lim_{x\to 0} \frac{x^3}{x-\sin(x)}$

Solution: As x → 0, f(x) = x^3 → 0 and g(x) = x – sin(x) → 0 so

$$\lim_{x\to 0} \frac{x^3}{x-\sin(x)} = \lim_{x\to 0} \frac{3x^2}{1-\cos(x)} \to \text{"}\frac{0}{0}\text{"}$$ so we can use l'Hôpital's Rule again

$$= \lim_{x\to 0} \frac{6x}{\sin(x)} \to \text{"}\frac{0}{0}\text{"}$$ and again

$$= \lim_{x\to 0} \frac{6}{\cos(x)} = \frac{6}{1} = 6 .$$

Practice 2: Use l'Hôpital's Rule to find $\displaystyle\lim_{x\to\infty} \frac{x^2+e^x}{x^3+8x}$.

Which Function Grows Faster?

Sometimes we want to compare the asymptotic behavior of two systems or functions for large values of x, and l'Hôpital's Rule can be a useful tool. For example, if we have two different algorithms for sorting names, and each algorithm takes longer and longer to sort larger collections of names, we may want to know which algorithm will accomplish the task more efficiently for really large collections of names.

Example 4: Algorithm A requires $n \cdot \ln(n)$ steps to sort n names and algorithm B requires $n^{1.5}$ steps. Which algorithm will be better for sorting very large collections of names?

Solution: We can compare the ratio of the number of steps each algorithm requires, $\frac{n \cdot \ln(n)}{n^{1.5}}$, and then take the limit of this ratio as n grows arbitrarily large: $\lim_{n\to\infty} \frac{n \cdot \ln(n)}{n^{1.5}}$. If this limit is infinite, we say that $n \cdot \ln(n)$ "grows faster" than $n^{1.5}$. If the limit is 0, we say that $n^{1.5}$ grows faster than $n \cdot \ln(n)$. Since $n \cdot \ln(n)$ and $n^{1.5}$ both grow arbitrarily large when n is large, we can algebraically simplify the ratio to $\frac{\ln(n)}{n^{0.5}}$ and then use L'Hopital's Rule:

$$\lim_{n\to\infty} \frac{\ln(n)}{n^{0.5}} = \lim_{n\to\infty} \frac{1/n}{0.5n^{-0.5}} = \lim_{n\to\infty} \frac{2}{\sqrt{n}} = 0 .$$

$n^{1.5}$ grows faster than $n \cdot \ln(n)$ so algorithm A requires fewer steps for really large sorts.

Practice 3: Algorithm A requires e^n operations to find the shortest path connecting n towns, algorithm B requires $100 \cdot \ln(n)$ operations for the same task, and algorithm C requires n^5 operations. Which algorithm is best for finding the shortest path connecting a very large number of towns? Worst?

Other "Indeterminate Forms"

"0/0" is called an indeterminate form because knowing that f approaches 0 and g approaches 0 is **not** enough to determine the limit of f/g, even if it has a limit. The ratio of a "small" number divided by a "small" number can be almost anything as the three simple "0/0" examples show:

$\lim_{x\to 0} 3x/x = 3$, $\quad \lim_{x\to 0} x^2/x = 0$, and $\quad \lim_{x\to 0} 5x/x^3 = \infty$.

Similarly, "∞/∞" is an indeterminate form because knowing that f and g both grow arbitrarily large is **not** enough to determine the value limit of f/g or if the limit exists:

$\lim_{x\to\infty} 3x/x = 3$, $\quad \lim_{x\to\infty} x^2/x = \infty$, and $\quad \lim_{x\to\infty} 5x/x^3 = 0$.

Besides the indeterminate quotient forms "0/0" and "∞/∞" there are several other "indeterminate forms." In each case, the resulting limit depends not only on each function's limit but also on how quickly each function approaches its limit.

Product: If f approaches 0, and g grows arbitrarily large,
then the product f · g has the indeterminant form "0 · ∞."

Exponent: If f and g both approach 0,
then the function f^g has the indeterminant form "0^0."

If f approaches 1, and g grows arbitrarily large,
then the function f^g has the indeterminant form "1^∞."

If f grows arbitrarily large, and g approaches 0, the function f^g has the indeterminant form "∞^0."

Difference: If f and g both grow arbitrarily large, the function f – g has the indeterminant form "∞ – ∞."

Unfortunately, l'Hôpital's Rule can only be used directly with an indeterminate quotient ("0/0" or "∞/∞'), but these other forms can be algebraically manipulated into quotients, and then l'Hôpital's Rule can be applied to the resulting quotient.

Example 5: Evaluate $\lim_{x\to 0^+} x \cdot \ln(x)$ (" $0\cdot(-\infty)$ " form)

Solution: This limit involves an indeterminate product, and we need a quotient in order to apply l'Hôpital's Rule. We can rewrite the product $x \cdot \ln(x)$ as the quotient $\frac{\ln(x)}{1/x}$, and then

$$\lim_{x\to 0^+} x\ln(x) = \lim_{x\to 0^+} \frac{\ln(x)}{1/x} \to \frac{\infty}{\infty}$$ so apply l'Hôpital's Rule

$$= \lim_{x\to 0^+} \frac{1/x}{-1/x^2} = \lim_{x\to 0^+} -x = 0 .$$

> A product $f \cdot g$ with the indeterminant form "$0\cdot\infty$" can be rewritten as a quotient, $\frac{f}{1/g}$ or $\frac{g}{1/f}$, and then l'Hôpital's Rule can be used.

Example 6: Evaluate $\lim_{x\to 0^+} x^x$ (" 0^0 " form)

Solution: An indeterminate exponent can be converted to a product by recalling a property of exponential and logarithm functions: for any positive number a, $a = e^{\ln(a)}$ so $f^g = e^{\ln(f^g)} = e^{g\ln(f)}$.

$$\lim_{x\to 0^+} x^x = \lim_{x\to 0^+} e^{\ln(x^x)} = \lim_{x\to 0^+} e^{x\cdot\ln(x)}$$ and this last limit involves an indeterminate

product $x \cdot \ln(x) \to 0\cdot(-\infty)$ which we converted to a quotient and evaluated to be 0 in Example 5.
Our final answer is then $e^{\mathbf{0}} = \mathbf{1}$: $\lim_{x\to 0^+} x^x = \lim_{x\to 0^+} e^{\ln(x^x)} = \lim_{x\to 0^+} e^{x\cdot\ln(x)} = e^{\mathbf{0}} = \mathbf{1}$.

> An indeterminate form involving exponents, f^g with the form "0^0," "1^∞," or "∞^0," can be converted to an indeterminate product by recognizing that $f^g = e^{g \cdot \ln(f)}$ and then determining the limit of $g \cdot \ln(f)$. The final result is $e^{(\text{limit of } g \cdot \ln(f))}$.

Example 7: Evaluate $\lim_{x\to\infty} \left(1+\frac{a}{x}\right)^x$ ("1^∞" form)

Solution: $\lim_{x\to\infty} \left(1+\frac{a}{x}\right)^x = \lim_{x\to\infty} e^{x \cdot \ln(1 + a/x)}$ so we need $\lim_{x\to\infty} x \cdot \ln(1 + \frac{a}{x})$.

$\lim_{x\to\infty} x \cdot \ln(1 + \frac{a}{x}) \to$ "$\infty \cdot 0$" an indeterminate product so rewrite it as a quotient

$$= \lim_{x\to\infty} \frac{\ln\left(1+\frac{a}{x}\right)}{1/x} \to \text{"}\frac{0}{0}\text{"}$$ an indeterminate quotient so use l'Hôpital's Rule

$$= \lim_{x\to\infty} \frac{\left(\dfrac{-a/x^2}{1+\frac{a}{x}}\right)}{\left(-\frac{1}{x^2}\right)} = \lim_{x\to\infty} \frac{a}{1+\frac{a}{x}} \to \frac{a}{1} = \mathbf{a}.$$

Finally, $\lim_{x\to\infty} \left(1+\frac{a}{x}\right)^x = \lim_{x\to\infty} e^{x \cdot \ln(1 + a/x)} = e^{\mathbf{a}}$.

PROBLEMS

Determine the limits in problems 1 – 15.

1. $\lim_{x\to 1} \frac{x^3-1}{x^2-1}$
2. $\lim_{x\to 2} \frac{x^4-16}{x^5-32}$
3. $\lim_{x\to 0} \frac{\ln(1+3x)}{5x}$
4. $\lim_{x\to\infty} \frac{e^x}{x^3}$
5. $\lim_{x\to 0} \frac{x \cdot e^x}{1-e^x}$
6. $\lim_{x\to 0} \frac{\cos(a+x)-\cos(a)}{x}$
7. $\lim_{x\to\infty} \frac{\ln(x)}{x}$
8. $\lim_{x\to\infty} \frac{\ln(x)}{\sqrt{x}}$
9. $\lim_{x\to\infty} \frac{\ln(x)}{x^p}$ (p is any positive number)
10. $\lim_{x\to 0} \frac{e^{3x}-e^{2x}}{4x}$
11. $\lim_{x\to 0} \frac{1-\cos(3x)}{x^2}$
12. $\lim_{x\to 0} \frac{1-\cos(2x)}{x}$
13. $\lim_{x\to a} \frac{x^m-a^m}{x^n-a^n}$
14. $\lim_{x\to 0} \frac{2^x-1}{x}$
15. $\lim_{x\to 0} \frac{1-\cos(x)}{x \cdot \cos(x)}$
16. Find a value for p so $\lim_{x\to\infty} \frac{3x}{px+7} = 2$.

17. Find a value for p so $\lim_{x \to 0} \frac{e^{px} - 1}{3x} = 5$.

18. $\lim_{x \to \infty} \frac{\sqrt{3x+5}}{\sqrt{2x-1}}$ has the indeterminate form "∞/∞". Why doesn't l'Hôpital's Rule work with this limit? (Hint: Apply l'Hôpital's Rule twice and see what happens.) Evaluate the limit without using l'Hôpital's Rule.

19. (a) Evaluate $\lim_{x \to \infty} \frac{e^x}{x}$, $\lim_{x \to \infty} \frac{e^x}{x^2}$, $\lim_{x \to \infty} \frac{e^x}{x^5}$.

(b) An algorithm is "exponential" if it requires $a \cdot e^{bn}$ steps (a and b are positive constants). An algorithm is "polynomial" if it requires $c \cdot n^d$ steps (c and d are positive constants). Show that polynomial algorithms require fewer steps than exponential algorithms for large problems.

20. The problem $\lim_{x \to 0} \frac{x^2}{3x^2 + x}$ appeared on a test.

One student determined the limit was an indeterminate "0/0" form, and applied l'Hôpital's Rule to get:

$$\text{"}\lim_{x \to 0} \frac{x^2}{3x^2 + x} = \lim_{x \to 0} \frac{2x}{6x+1} = \lim_{x \to 0} \frac{2}{6} = \frac{1}{3}.\text{"}$$

Another student also determined the limit was an indeterminate "0/0" form and wrote,

$$\text{"}\lim_{x \to 0} \frac{x^2}{3x^2 + x} = \lim_{x \to 0} \frac{2x}{6x+1} = \frac{0}{0+1} = 0.\text{"}$$

Which student is correct? Why?

Determine the limits in problems 21 – 29.

21. $\lim_{x \to 0^+} \sin(x) \cdot \ln(x)$

22. $\lim_{x \to \infty} x^3 \cdot e^{-x}$

23. $\lim_{x \to 0^+} \sqrt{x} \cdot \ln(x)$

24. $\lim_{x \to 0^+} x^{\sin(x)}$

25. $\lim_{x \to \infty} (1 - 3/x^2)^x$

26. $\lim_{x \to \infty} (1 - \cos(3x))^x$

27. $\lim_{x \to 0} \left(\frac{1}{x} - \frac{1}{\sin(x)} \right)$

28. $\lim_{x \to \infty} (x - \ln(x))$

29. $\lim_{x \to \infty} \left(\frac{x+5}{x} \right)^{1/x}$

30. $\lim_{x \to \infty} (1 + 3/x)^{2/x}$

Section 3.7 **PRACTICE Answers**

Practice 1: (a) $\lim_{x \to 0} \frac{1-\cos(5x)}{3x}$. The numerator and denominator are both differentiable and both equal 0 when x = 0, so we can apply l'Hôpital's Rule:

$$\lim_{x \to 0} \frac{1-\cos(5x)}{3x} = \lim_{x \to 0} \frac{5\cdot\sin(5x)}{3} \to \frac{0}{3} = \mathbf{0}.$$

(b) $\lim_{x \to 2} \frac{x^2+x-6}{x^2+2x-8}$. The numerator and denominator are both differentiable functions and they both equal 0 when x = 0, so we can apply l'Hôpital's Rule:

$$\lim_{x \to 2} \frac{x^2+x-6}{x^2+2x-8} = \lim_{x \to 2} \frac{2x+1}{2x+2} = \mathbf{\frac{5}{6}}.$$

Practice 2: $\lim_{x \to \infty} \frac{x^2+e^x}{x^3+8x}$. The numerator and denominator are both differentiable and both become arbitrarily large as x becomes large, so we can apply l'Hôpital's Rule:

$$\lim_{x \to \infty} \frac{x^2+e^x}{x^3+8x} = \lim_{x \to \infty} \frac{2x+e^x}{3x^2+8} \to \text{"}\frac{\infty}{\infty}\text{"}.$$ Using l'Hôpital's Rule again:

$$\lim_{x \to \infty} \frac{2x+e^x}{3x^2+8} = \lim_{x \to \infty} \frac{2+e^x}{6x} \to \text{"}\frac{\infty}{\infty}\text{"}$$ and again:

$$\lim_{x \to \infty} \frac{2+e^x}{6x} = \lim_{x \to \infty} \frac{e^x}{6} \to \infty.$$

Practice 3: Comparing A with e^n operations to B with 100·ln(n) operations.

$$\lim_{n \to \infty} \frac{e^n}{100\cdot\ln(n)} \to \text{"}\frac{\infty}{\infty}\text{"}$$ so use L'Hopital's Rule:

$$\lim_{n \to \infty} \frac{e^n}{100\cdot\ln(n)} = \lim_{n \to \infty} \frac{e^n}{100/n} = \lim_{n \to \infty} \frac{n\cdot e^n}{100} = \infty$$ so B requires fewer operations than A.

Comparing B with 100·ln(n) operations to C with n^5 operations.

$$\lim_{n \to \infty} \frac{100\cdot\ln(n)}{n^5} \to \frac{\infty}{\infty}. \quad \lim_{n \to \infty} \frac{100\cdot\ln(n)}{n^5} = \lim_{n \to \infty} \frac{100/n}{5n^4} = \lim_{n \to \infty} \frac{100}{5n^5} = 0$$

so B requires fewer operations than C. B requires the fewest operations of the three algorithms.

Comparing A with e^n operations to C with n^5 operations. Using l'Hôpital's Rule several times:

$$\lim_{n \to \infty} \frac{e^n}{n^5} = \lim_{n \to \infty} \frac{e^n}{5n^4} = \lim_{n \to \infty} \frac{e^n}{20n^3} = \lim_{n \to \infty} \frac{e^n}{60n^2} = \lim_{n \to \infty} \frac{e^n}{120n} = \lim_{n \to \infty} \frac{e^n}{120} = \infty$$

so A requires more operations than C. A requires the most operations of the three algorithms.

Chapter Three

Section 3.1

1. Local maximums at x = 3, x = 5, x = 9, and x = 13. Global maximum at x = 13.
   Local minimums at x = 1, x = 4.5, x = 7, and x = 10.5 . Global minimum at x = 7 .

3. $f(x) = x^2 + 8x + 7$ so $f'(x) = 2x + 8$ which is defined for all values of x. $f'(x) = 0$ when $x = -4$ so $x = -4$ is a critical number. There are no endpoints.
   The only critical number is $x = -4$, and the only critical point is $(-4, f(-4)) = (-4, -9)$ which is the global (and local) minimum.

5. $f(x) = \sin(x)$ so $f'(x) = \cos(x)$ which is defined for all values of x. $f'(x) = 0$ when $x = \frac{\pi}{2} + n\pi$ so the values $x = \frac{\pi}{2} + n\pi$ are critical numbers. There are no endpoints.
   $f(x) = \sin(x)$ has local and global maximums at $x = \frac{\pi}{2} + 2n\pi$, and global and local minimums at $x = \frac{3\pi}{2} + 2n\pi$.

7. $f(x) = (x-1)^2(x-3)$ so $f'(x) = (x-1)^2 + 2(x-1)(x-3) = (x-1)(3x-7)$ which is defined for all values of x. $f'(x) = 0$ when $x = 1$ and $x = 7/3$ so $x = 1$ and $x = 7/3$ are critical numbers. There are no endpoints.
   The only critical points are $(1, 0)$ which is a local maximum and $(7/3, -32/27)$ which is a local minimum. When the interval is the entire real number line, this function does not have a global maximum or global minimum.

9. $f(x) = 2x^3 - 96x + 42$ so $f'(x) = 6x^2 - 96$ which is defined for all values of x.
   $f'(x) = 6(x+4)(x-4) = 0$ when $x = -4$ and $x = 4$ so $x = -4$ and $x = 4$ are critical numbers. There are no endpoints. The only critical points are $(-4, 298)$ which is a local maximum and $(4, -214)$ which is a local minimum. When the interval is the entire real number line, this function does not have a global maximum or global minimum.

11. $f(x) = 5x + \cos(2x+1)$ so $f'(x) = 5 - 2\sin(2x+1)$ which is defined for all values of x. $f'(x)$ is always positve (why?) so $f'(x)$ is never equal to 0. There are no endpoints. The function $f(x) = 5x + \cos(2x+1)$ is always increasing and has no critical numbers, no critical points, no local or global maximums or minimums.

13. $f(x) = e^{-(x-2)^2}$ so $f'(x) = -2(x-2)\,e^{-(x-2)^2}$ which is defined for all values of x. $f'(x) = 0$ when $x = 2$ so $x = 2$ is a critical number. There are no endpoints. The only critical point is $(2, 1)$ which is a local and global maximum. When the interval is the entire real number line, this function does not have a local or global minimum.

15. See Fig. 3.1P15

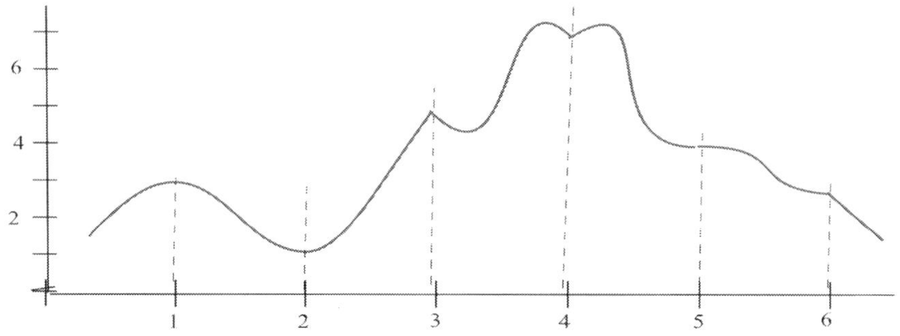

Fig. 3.1P15

17. $f(x) = x^2 - 6x + 5$ on $[-2, 5]$ so $f'(x) = 2x - 6$ which is defined for all values of x. $f'(x) = 0$ when $x = 3$ so $x = 3$ is a critical number. The endpoints are $x = -2$ and $x = 5$ which are also critical numbers. The critical points are $(3, -4)$ which is the local and global minimum, $(-2, 21)$ which is a local and global maximum, and $(5, 0)$ which is a local maximum.

19. $f(x) = 2 - x^3$ on $[-2, 1]$ so $f'(x) = -3x^2$ which is defined for all values of x. $f'(x) = 0$ when $x = 0$ so $x = 0$ is a critical number. The endpoints are $x = -2$ and $x = 1$ which are also critical numbers. The critical points are $(-2, 10)$ which is a local and global maximum, $(0,2)$ which is not a local or global maximum or minimum, and $(1,1)$ which is a local and global minimum.

21. $f(x) = x^3 - 3x + 5$ on $[-2, 1]$ so $f'(x) = 3x^2 - 3 = 3(x-1)(x+1)$ which is defined for all values of x. $f'(x) = 0$ when $x = -1$ and $x = +1$ so these are critical numbers. The endpoints $x = -2$ and $x = 1$ are also critical numbers. The critical points are $(-2, 3)$ which is a local and global minimum on $[-2, 1]$, the point $(-1, 7)$ which is a local and global maximum on $[-2, 1]$, and the point $(1, 3)$ which is a local and global minimum on $[-2, 1]$.

23. $f(x) = x^5 - 5x^4 + 5x^3 + 7$ so $f'(x) = 5x^4 - 20x^3 + 15x^2 = 5x^2(x^2 - 4x + 3) = 5x^2(x-3)(x-1)$ which is defined for all values of x. $f'(x) = 0$ when $x = 0$ and $x = 1$ in the interval $[0, 2]$ so each of these values is a critical number. The endpoints $x = 0$ and $x = 2$ are also critical numbers. The critical points are $(0, 7)$ which is a local minimum, $(1, 8)$ which is a local and global maximum, and $(2, -1)$ which is a local and global minimum. ($f'(3) = 0$ too, but $x = 3$ is not in the interval $[0, 2]$.)

25. $f(x) = \dfrac{1}{x^2+1}$ so $f'(x) = \dfrac{-2x}{(x^2+1)^2}$ which is defined for all values of x. $f'(x) = 0$ when $x = 0$ but $x = 0$ is not in the interval $[1, 3]$ so $x = 0$ is a not a critcal number. The endpoints $x = 1$ and $x = 3$ are critical numbers. The critical points are $(1, 1/2)$ which is a local and global maximum, and $(3, 1/10)$ which is a local and global minimum.

27. $A(x) = 4x\sqrt{1-x^2}$ $(0<x<1)$.

$$A'(x) = 4\left[\frac{-x^2}{\sqrt{1-x^2}} + \sqrt{1-x^2}\right] = 4\frac{1-2x^2}{\sqrt{1-x^2}} \begin{cases} >0 & \text{if } 0 < x < 1/\sqrt{2} \\ <0 & \text{if } 1/\sqrt{2} < x < 1 \end{cases}$$

A maximum is attained when $x = 1/\sqrt{2}$: $A(1/\sqrt{2}) = 4\frac{1}{\sqrt{2}}\sqrt{1-\frac{1}{2}} = 4\frac{1}{\sqrt{2}}\frac{1}{\sqrt{2}} = 2.$

29. $V = x(8-2x)^2$ for $0 < x < 4$.

$V' = x(2)(8-2x)(-2) + (8-2x)^2 = (8-2x)(-4x+8-2x) = (8-2x)(8-6x) = 4(4-x)(4-3x)$ so

$$V' \begin{cases} <0 & \text{if } 4/3 < x \\ >0 & \text{if } 0 < x < 4/3. \end{cases}$$

$V(4/3) = \frac{4}{3}(8 - \frac{8}{3})^2 = \frac{4}{3}(\frac{16}{3})^2 = \frac{1024}{27} \approx 37.926$ cubic units is the largest volume.

Smallest volume is 0 which occurs when $x = 0$ and $x = 4$.

31. (a) 4. The endpoints and two values of x for which $f'(x) = 0$.
(b) 2. The endpoints.
(c) At most $n + 1$. The 2 endpoints and the $n - 1$ interior points x for which $f'(x) = 0$. At least 2. The 2 endpoints.

33. (a) local minimum at (1,5) (b) no extrema at (1,5)
(c) local maximum at (1,5) (d) no extrema at (1,5)

35. (a) 0, 2, 6, 8, 11, 12 (b) 0, 6, 11 (c) 2, 8, 12

37. If f does not attain a maximum on [a,b] or f does not attain a mimimum on [a,b], then f must have a discontinuity on [a,b].

39 (a) yes, –1 (b) no (c) yes, –1 (d) no (e) yes, $1 - \pi$

41. (a) yes, 0 (b) yes, 0 (c) yes, 0 (d) yes, 0 (e) yes, 0

43. (a) S(x) is minimum when $x \approx 8$. (b) S(x) is maximum when x = 2.

Section 3.2

1. $c \approx 3$, 10, and 13. 3. (a) $c = \pi/2$ (b) $c = 3\pi/2, 5\pi/2, 7\pi/2, 9\pi/2$

5. Rolle's Theorem asserts that the velocity h '(t) will equal 0 at some point between the time the ball is tossed and the time it comes back down. The ball is not moving as fast when it reaches the balcony from below.

7. The function does not violate Rolle's Thm. because the function does not satisfy the hypotheses of the theorem: f is not differentiable at 0, a point in the interval $-1 < x < 1$.

9. No. The velocity is not the same as the rate of change of altitude, since altitude is only one of the components of position. Rolle's Theorem only says there was a time when my altitude was not changing.

11. Since $f'(x) = 3x^2 + 5$, $f'(x) = 0$ has no real roots. If $f(x) = 0$ for a value of x other than 2, then by the corollary from Problem 8, we would have an immediate contradiction.

13. (a) $f(0) = 0, f(2) = 4, f'(c) = 2c. \ \frac{4-0}{2-0} = 2c$ implies that $c = 1$.

(b) $f(1) = 4, f(5) = 8, f'(c) = 2c - 5. \ \frac{8-4}{5-1} = 2c - 5$ implies that $c = 3$.

14. (a) $f(0) = 0, f(\pi/2) = 1, f'(c) = \cos(c). \ \frac{1-0}{\pi/2 - 0} = \cos(c)$ implies $c = \arccos(2/\pi) \approx 0.88$.

(b) $f(-1) = -1, f(3) = 27, f'(c) = 3c^2. \ \frac{27+1}{3+1} = 3c^2$ implies $c^2 = 7/3$ so $c = \sqrt{7/3}$ (since $c > -1$)

15 (a) $f(1) = 4, f(9) = 2, \ f'(c) = \frac{-1}{2\sqrt{c}} . \ \frac{2-4}{1-9} = \frac{-1}{2\sqrt{c}}$ implies $\frac{-1}{4} = \frac{-1}{2\sqrt{c}}$ so $c = 4$.

(b) $f(1) = 3, f(7) = 15, \ f'(c) = 2. \ \frac{15-3}{7-1} = 2$ so any c between 1 and 7 will do.

17. The hypotheses are not all satisfied since f '(x) does not exist at x = 0 which is between –1 and 3.

19. Guilty. All we know is that f '(c) = 17 at some point, but this does not prove that the motorist "could not have been speeding."

21. $f(x) = x^3 + x^2 + 5x + c$. $f(1) = 7 + c = 10$ when $c = 3$. Therefore, $f(x) = x^3 + x^2 + 5x + 3$.

23. (a) $f'(x) = 2Ax$. We need $A(1)^2 + B = 9$ and $2A(1) = 4$ so $A = 2$ and $B = 7$ and $f(x) = 2x^2 + 7$.

(b) $A(2)^2 + B = 3$ and $2A(2) = -2$ so $A = -1/2$ and $B = 5$ and $f(x) = \frac{-1}{2} x^2 + 5$.

(c) $A(0)^2 + B = 2$ and $\mathbf{2A(0) = 3}$. There is no such A. The point (0,2) is not on the parabola $y = x^2 + 3x - 2$.

25. $f(x) = x^3 + C$, a family of "parallel" curves for different values of C.

27. v(t) = 300. Assuming the rocket left the ground at t = 0, we have y(1) = 300 ft, y(2) = 600 ft, y(5) = 1500 ft.

29. $f''(x) = 6, f'(0) = 4, f(0) = -5$. $f(x) = 3x^2 + 4x - 5$.

31. (a) $A(x) = 3x$ (b) $A'(x) = 3$. 33. (a) $A(x) = x^2 + x$ (b) $A'(x) = 2x + 1$.

35. $a_1 = 5, a_2 = a_1 + 3 = 5 + 3 = 8, a_3 = a_2 + 3 = (5 + 3) + 3 = 11$, $a_4 = a_3 + 3 = (5+3) + 3 + 3 = 14$.
In general, $a_n = 5 + 3(n - 1) = 2 + 3n$.

Section 3.3

1. See Fig. 3.3P1.

3. See Fig. 3.3P3.

4. See Fig. 3.3P4.

5. See Fig. 3.3P5.

6. See Fig. 3.3P6.

7. A–Q, B–P, C–R

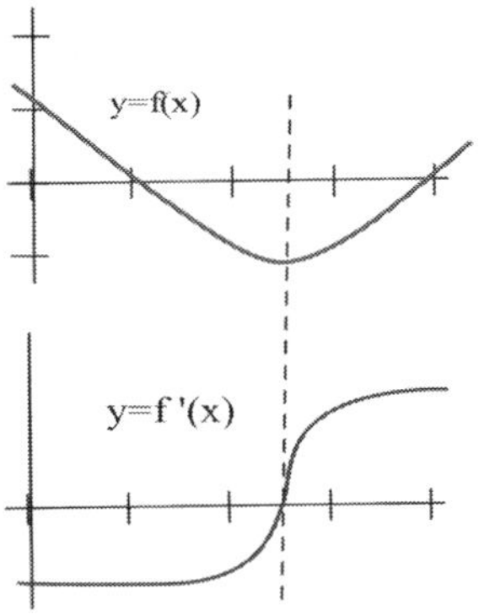

Fig 3.3P1

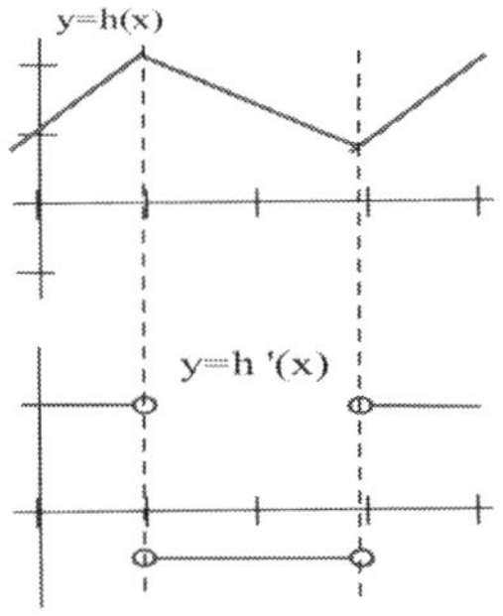

Fig. 3.3P3

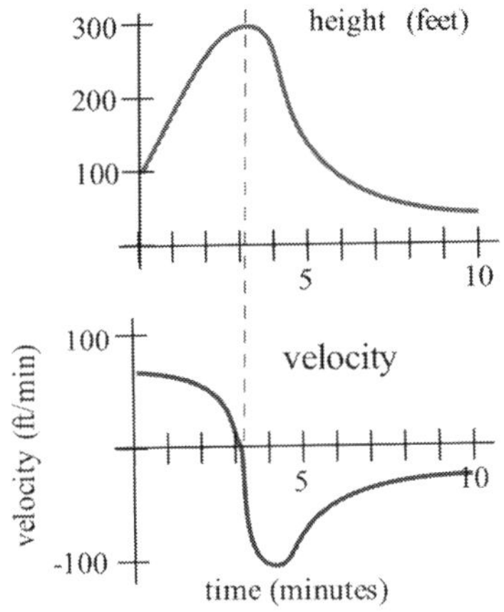

Fig. 3.3P4

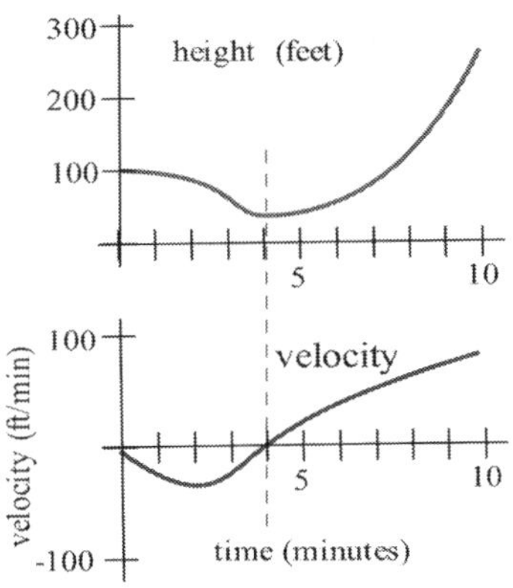

Fig. 3.3P5

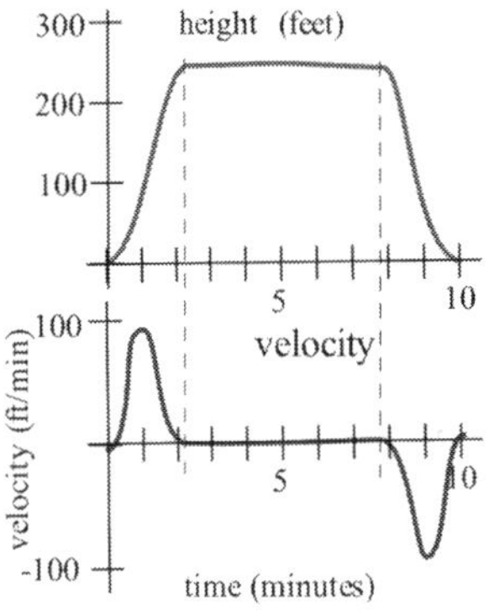

Fig. 3.3P6

9. $f'(x) = \frac{1}{x} > 0$ for $x > 0$ so $f(x) = \ln(x)$ is increasing on $(0, \infty)$.

11. If f is increasing then $f(1) < f(\pi)$ so $f(1)$ and $f(\pi)$ cannot both equal 2.

13. (a) $x = 3$, $x = 8$ (b) maximum at $x = 8$ (c) none (or only at right endpoint)

14. Relative maximum height at $x = 2$ and $x = 7$. Relative minimum height at $x = 4$.

15. Relative maximum height at $x = 6$. Relative minimum height at $x = 8$.

17. $f(x) = x^3 - 3x^2 - 9x - 5$ has a relative minimum at $(3, -32)$ and a relative maximum at $(-1, 0)$.

19. $h(x) = x^4 - 8x^2 + 3$ has a relative maximum at $(0, 3)$ and relative minimums at $(2, -13)$ and $(-2, -13)$.

21. $r(t) = 2(t^2 + 1)^{-1}$ has a relative maximum at $(0, 2)$ and no relative minimums.

23. No positive roots. $f(x) = 2x + \cos(x)$ is continuous. $f(0) = 1 > 0$. Since $f'(x) = 2 - \sin(x) > 0$ for all x, f is increasing and never decreases back to the x–axis (a root).

24. One positive root. $g(x) = 2x - \cos(x)$ is continuous. $g(0) = -1 < 0$ and $g(1) = 2 - \cos(1) > 0$ so by the Intermediate Value Theorem g has a root between 0 and 1. Since $g'(x) = 2 + \sin(x) > 0$ for all x, g is increasing and can have only that 1 root.

25. $h(x) = x^3 + 9x - 10$ and $h(1) = 0$. $h'(x) = 3x^2 + 9 = 3(x^2 + 3) > 0$ for all x so h is always increasing and can cross the x–axis at most at one place. Since the graph of h crosses the x–axis at x = 1, that is the only root of h.

27.

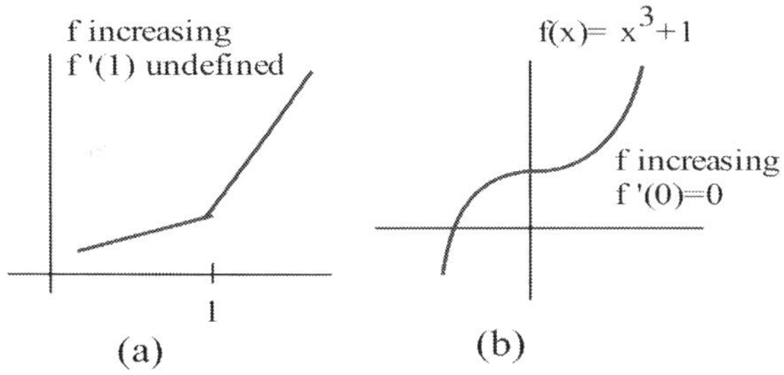

If they travel at the same positive speed in different directions, then their distance apart is not constant.

(c)

Fig. 3.3P27

29. (a) $h(x) = x^2, x^2 + 1, x^2 - 7$, or, in general, $x^2 + C$ for any constant C.
(b) $f(x) = x^2 + C$ for some value C and $20 = f(3) = 3^2 + C$ so $C = 20 - 9 = 11$. $f(x) = x^2 + 11$.
(c) $g(x) = x^2 + C$ for some value C and $7 = g(2) = 2^2 + C$ so $C = 7 - 4 = 3$. $g(x) = x^2 + 3$.

Section 3.4

1. (a) f(t) = number of workers unemployed at time t. $f'(t) > 0$ and $f''(t) < 0$
(b) f(t) = profit at time t. $f'(t) < 0$ and $f''(t) > 0$.
(c) f(t) = population at time t. ($f'(t) > 0$ and $f''(t) > 0$.

3. See Fig. 3.4P3.

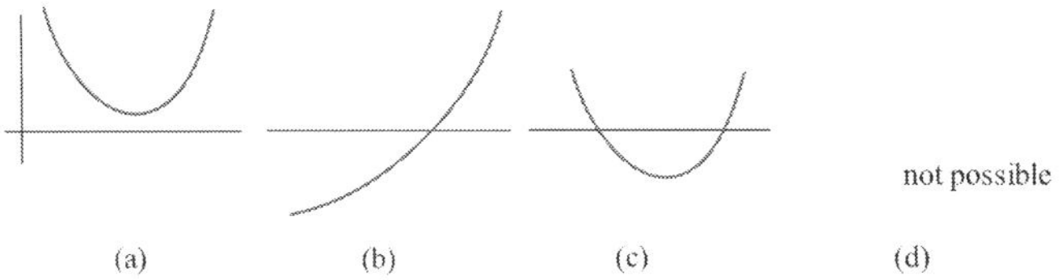

Fig. 3.4P3

5. (a) Concave up on (0, 2), (2, 3+), (6, 9). Concave down on (3+, 6). (A small technical note: we have defined concavity only at points where the function is differentiable, so we exclude the endpoints and points where the function is not differentiable from the intervals of concave up and concave down.)

7. $g(x) = x^3 - 3x^2 - 9x + 7$. $g''(x) = 6x - 6$. $g''(-1) < 0$ so $(-1, 12)$ is a local maximum. $g''(3) > 0$ so $(3, -20)$ is a local minimum.

9. $f(x) = \sin^5(x)$. $f''(x) = 5\{ -\sin^5(x) + 4\sin^3(x)\cdot\cos^2(x) \}$. $f''(\pi/2) < 0$ so $(\pi/2, 1)$ is a local maximum. $f''(3\pi/2) > 0$ so $(3\pi/2, -1)$ is a local minimum.
$f''(\pi) = 0$ and f changes concavity at $x = \pi$ so $(\pi, 0)$ is an inflection point.

11. d and e. 13. (a) 0 (b) at most 1 (c) at most n – 2.

15.

x	g(x)	g '(x)	g "(x)
0	–	+	+
1	+	0	–
2	–	–	+
3	0	+	+

17. See Fig. 3.4P17.

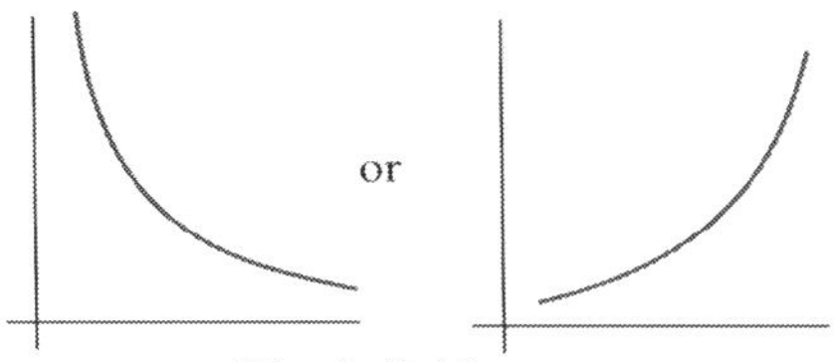

Fig. 3.4P17

Section 3.5

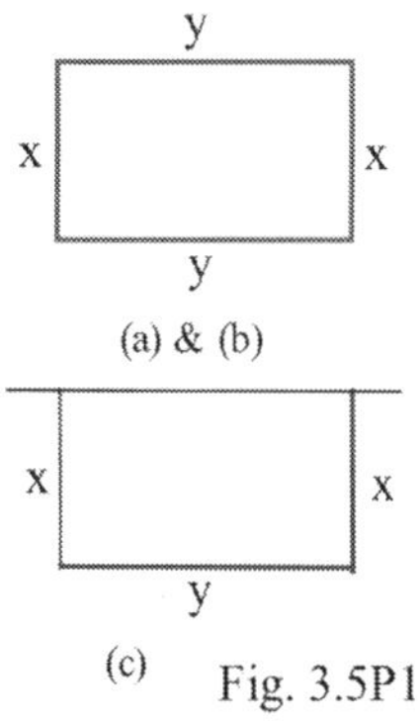

Fig. 3.5P1

1. (a) $2x + 2y = 200$ so $y = 100 - x$. Maximize $A = x \cdot y = x \cdot (100 - x) = 100x - x^2$.
$A' = 100 - 2x$ and $A' = 0$ when $x = 50$ ($y = 100 - x = 50$). $A'' = -2 < 0$ so $x = 50$ yields the maximum enclosed area. When $x = 50$, $A = 50(100 - 50) = 2500$ square feet.

   (b) $2x + 2y = P$ so $y = P/2 - x$. Maximize $A = x \cdot y = x \cdot (P/2 - x) = (P/2)x - x^2$.
$A' = P/2 - 2x$ and $A' = 0$ when $x = P/4$ (then $y = P/2 - x = P/4$). $A'' = -2 < 0$ so $x = P/4$ yields the maximum enclosed area.
This garden is a $P/4$ by $P/4$ square.

   (c) $2x + y = P$ so $y = P - 2x$. Maximize $A = xy = x(P - 2x) = Px - 2x^2$.
$A' = P - 4x$ and $A' = 0$ when $x = P/4$ (then $y = P - 2x = P/2$).

   (d) A circle. A semicircle.

3. (a) $120 = 2x + 5y$ so $y = 24 - \frac{2}{5}x$. Maximize $A = xy = x(24 - \frac{2}{5}x) = 24x - \frac{2}{5}x^2$.
$A' = 24 - \frac{4}{5}x$ and $A' = 0$ when $x = 30$ (then $y = 12$). $A'' = -4/5 < 0$ so $x = 30$ yields the maximum enclosed area. Area is (30 ft)(12 ft) = 360 square feet.

(a)
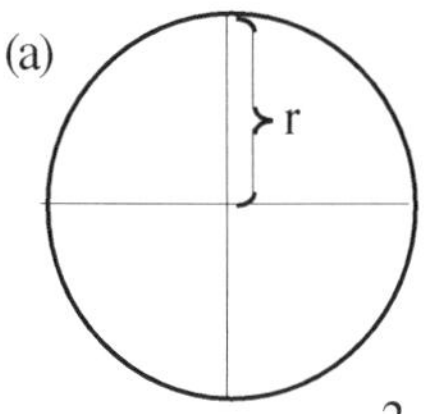

area ≈ 427.8 ft^2

   (b) A circular pen divided into 4 equal stalls by two diameters shown in diagram (a) does a better job than a square with 400 square feet. If the radius is r, then $4r + 2\pi r = 120$ so $r = 120/(4 + 2\pi) \approx 11.67$.
The resulting enclosed area is $A = \pi r^2 \approx \pi(11.67)^2 \approx 427.8$ sq. ft.

   The pen shown in diagram (b) does even better. If each semicircle has radius r, then the figure uses $4\sqrt{2}\, r + 4\pi r = 120$ feet of fence so $r = 120/(4\sqrt{2} + 4\pi) \approx 6.585$. The resulting enclosed area is
A = (square) + (four semicircles) $= (2r)^2 + 4(\frac{1}{2}\pi r^2) \approx 445.90$ sq. ft.

(b)
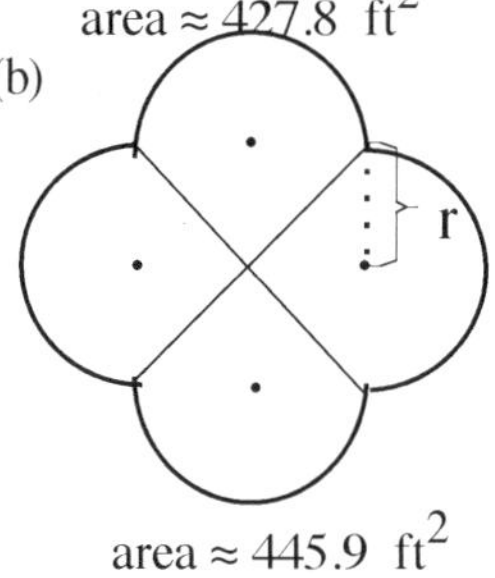

area ≈ 445.9 ft^2

5. $2x + 2y = 10$ so $y = 5 - x$.
Maximize $V = xy(10 - 2x) = x(5 - x)(10 - 2x) = 50x - 20x^2 + 2x^3$.
$V' = 50 - 40x + 6x^2 = 2(3x - 5)(x - 5)$ and $V' = 0$ when $x = 5$ and $x = 5/3$. When $x = 5$, then $V = 0$, clearly not a maximum, so $x = 5/3$. The dimensions of the box with the largest volume are 5/3, 10/3, and 20/3.

7. (a) $V = \pi r^2 h = 100$ so $h = \dfrac{100}{\pi r^2}$.
Minimize C = 2(top area) + 5(bottom area) + 3(side area)
$= 2(\pi r^2) + 5(\pi r^2) + 3(2\pi rh) = 7\pi r^2 + 6\pi r \left(\dfrac{100}{\pi r^2} \right) = 7\pi r^2 + \dfrac{600}{r}$.
$C' = 14\pi r - \dfrac{600}{r^2}$ and $C' = 0$ when $r = \sqrt[3]{600/(14\pi)} \approx 2.39$ (then $h = \dfrac{100}{\pi r^2} \approx 5.57$).

   (b) Let k = top + bottom rate = 2¢ + the bottom rate > 2¢ + 5¢ = 7¢. Minimize $C = k\pi r^2 + \dfrac{600}{r}$.
$C' = 2k\pi r - \dfrac{600}{r^2}$ and $C' = 0$ when $r = \sqrt[3]{600/(2k\pi)}$. If $k = 8$, then $r \approx 2.29$.
If $k = 9$, then $r \approx 2.20$. If $k = 10$, then $r \approx 2.12$. As the cost of the bottom material increases, the radius of the least expensive cylindrical can decreases: the least expensive can becomes narrower and taller

9. Time = distance/rate. Run distance = x ($0 \le x \le 60$ Why?) so run time = x/8.

Swim distance = $\sqrt{40^2 + (60-x)^2}$ so swim time = $\frac{1}{2}\sqrt{40^2 + (60-x)^2}$ and the total time is

$$T = \frac{x}{8} + \frac{1}{2}\sqrt{40^2 + (60-x)^2} .$$

$$T' = \frac{1}{8} + \frac{1}{2}\frac{1}{2}(40^2 + (60-x)^2)^{-1/2} \cdot 2 \cdot (60-x) \cdot (-1) = \frac{1}{8} - \frac{60-x}{2\sqrt{40^2 + (60-x)^2}} .$$

$T' = 0$ when $x = 60 \pm \frac{40}{\sqrt{15}}$. The value $x = 60 + \frac{40}{\sqrt{15}} > 60$ so the least total time occurs when

$x = 60 - \frac{40}{\sqrt{15}} \approx 49.7$ meters. In this situation, the lifeguard should run about 5/6 of the way along the beach before going into the water.

11. (a) Consider a similar problem with a new town D* located at the "mirror image" of D across the river (Fig. 3.5P11a). If the water works is built at any location W along the river, then the distances are the same from W to D and to D*: dist(W,D) = dist(W,D*).
Then dist(C,W) + dist(W,D) = dist(C,W) + dist(W,D*). The shortest distance from C to D* is a straight line (Fig. 3.5P11b), and this straight line gives similar triangles with equal side ratios: $\frac{x}{3} = \frac{10-x}{5}$ so $x = 15/4 = 3.75$ miles. A consequence of this "mirror image" view of the problem is that "at the best location W the angle of incidence α equals the angle of reflection β "

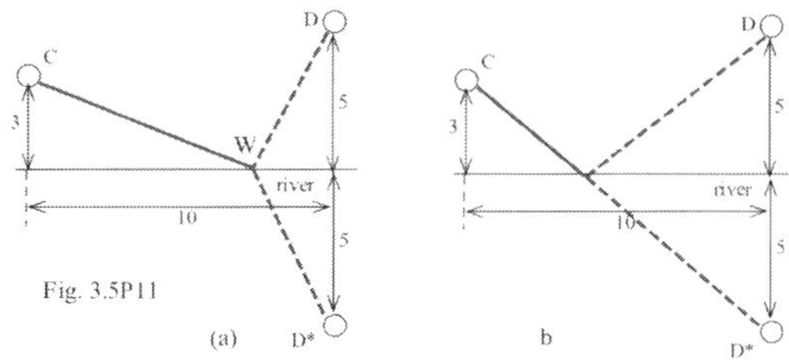

Fig. 3.5P11

(b) Minimize C = 3000dist(C,W) + 7000dist(W,D) = $3000\sqrt{x^2+9} + 7000\sqrt{(10-x)^2+25}$.

$$C' = \frac{3000x}{\sqrt{x^2+9}} + \frac{-7000(10-x)}{\sqrt{(10-x)^2+25}} \text{ so}$$

$C' = 0$ when $\frac{3x}{\sqrt{x^2+9}} = \frac{7(10-x)}{\sqrt{(10-x)^2+25}}$ and $x \approx 7.82$ miles.

As it becomes relatively more expensive to build the pipe from a point W on the river to D, the cheapest route tends to shorten the distance from W to D.

13. (a) Let x be the length of one edge of the square end. Then $V = x^2(108-4x) = 108x^2 - 4x^3$.
$V' = 216x - 12x^2 = 6x(18-x)$ so $V' = 0$ when $x = 0$ or $x = 18$. The dimensions of the greatest volume acceptable box with a square end are 18 by 18 by 36 inches: $V = 11{,}664 \text{ in}^3$.

(b) Let x be the length of the shorter edge of the end. Then $V = 2x^2(108-6x) = 216x^2 - 12x^3$.
$V' = 432x - 36x^2 = 36x(12-x)$ so $V' = 0$ when $x = 0$ or $x = 12$. The dimensions of the largest box acceptable box with this shape are 12 by 24 by 36 inches: $V = 10{,}368 \text{ in}^3$.

(c) Let x be the radius of the circular end. Then $V = \pi x^2(108 - 2\pi x) = 108\pi x^2 - 2\pi^2 x^3$.
$V' = 216\pi x - 6\pi^2 x^2 = 6\pi x(36 - \pi x)$ so $V' = 0$ when $x = 0$ or $x = 36/\pi \approx 11.46$ inches. The dimensions of the largest box acceptable box with a circular end are a radius of $36/\pi \approx 11.46$ and a length of 36 inches: $V \approx 14{,}851 \text{ in}^3$.

15. Without calculus: The area of the triangle is $\frac{1}{2}$(base)(height) $= \frac{1}{2}$(7)(height) and the height is maximum when the angle between the sides is a right angle.
Using calculus: Let θ be the angle between the sides. Then the area of the triangle is
$A = \frac{1}{2}$(base)(height) $= \frac{1}{2}$(7)(height) $= \frac{1}{2}(7)(10 \sin\theta) = 35 \sin\theta$. $A' = 35 \cos\theta$ so $A' = 0$ when $\theta = \pi/2$, and the triangle is a right triangle with sides 7 and 10.
Using either approach, the maximum area of the triangle is $\frac{1}{2}(7)(10) = 35$ square inches, and the other side is the hypotenuse with length $\sqrt{7^2 + 10^2} = \sqrt{149} \approx 12.2$ inches.

17. (a) $A = 2x(16 - x^2) = 32x - 2x^3$. Then $A' = 32 - 6x^2$ so $A' = 0$ when $x = \sqrt{32/6} \approx 2.31$.
The dimensions are $2\sqrt{32/6} \approx 4.62$ and $16 - (\sqrt{32/6})^2 = 64/6 \approx 10.67$.
(b) $A = 2x(\sqrt{1 - x^2})$. Then $A' = 2\left(\sqrt{1 - x^2} - \frac{x^2}{\sqrt{1 - x^2}}\right)$ so $A' = 0$ when $x = 1/\sqrt{2} \approx 0.707$. The dimensions are $2(1/\sqrt{2}) \approx 1.414$ and $1/\sqrt{2} \approx 0.707$.
(c) The graph of $|x| + |y| = 1$ is a "diamond" (a square) with corners at (1,0), (0,1), (–1,0) and (0,–1).
For $0 \le x \le 1$, $A = 2x \cdot 2(1 - x)$ so $A = 4x - 4x^2$. Then $A' = 4 - 8x$ and $A' = 0$ when $x = 1/2$. $A'' = -8$ so we have a local max. The dimensions are $2(1/2) = 1$ and $2(1 - 1/2) = 1$.
(d) $A = 2x \cos(x)$ $(0 \le x \le \pi/2)$. Then $A' = 2\cos(x) - 2x \cdot \sin(x)$ so $A' = 0$ when $x \approx 0.86$. The dimensions are $2(0.86) = 1.72$ and $\cos(0.86) \approx 0.65$.

19. $A = 6 \cdot \sin(\theta/2) \cdot 6 \cdot \cos(\theta/2) = 36 \cdot \frac{1}{2} \sin(\theta) = 18 \sin(\theta)$ and this is a maximum when $\theta = \pi/2$. Then the maximum area is $A = 18 \sin(\pi/2) = 18$ square inches. (This problem is similar to problem 15.)

21. $V = \frac{1}{3}\pi r^2 h$ and $h = \sqrt{9 - r^2}$ so $V = \frac{1}{3}\pi r^2 \sqrt{9 - r^2} = \frac{\pi}{3}\sqrt{9r^4 - r^6}$. Then
$V' = \frac{\pi}{6} \frac{36r^3 - 6r^5}{\sqrt{9r^4 - r^6}}$, and $V' = 0$ when $36r^3 = 6r^5$ so $r = \sqrt{6} \approx 2.45$ inches and
$h = \sqrt{9 - r^2} = \sqrt{3} \approx 1.73$ inches.

23. Let $n \ge 10$ be the number of passengers. The income is $I = n(30 - (n-10)) = 40n - n^2$. The cost is $C = 100 + 6n$ so the profit is $P = \text{Income} - \text{Cost} = (40n - n^2) - (100 + 6n) = 34n - n^2 - 100$.
$P' = 34 - 2n$ and $P' = 0$ when $n = 17$. 17 passengers on the flight maximize your profit.
(This is an example of treating a naturally discrete variable, the number of passengers, as a continuous variable.)

25. Apply the result of problem 24 with $R = f$ and $E = g$.

27. (i) Let D = diameter of the base of the can, and let H = the height of the can.
Then $\theta = \arctan\left(\frac{\text{radius of can}}{\text{height of cg}}\right) = \arctan\left(\frac{D/2}{H/2}\right)$.
For this can, D = 5 cm and H = 12 cm (sorry this should be in the statement of the problem) so
$\theta = \arctan(2.5/6) = \arctan(0.42) \approx 0.395$ which is about 22.6°. The can can be tilted about 22.6° before it falls over.

(ii) $C(x) = \frac{360 + 9.6x^2}{60 + 19.2x}$ so $C'(x) = \frac{(60 + 19.2x)(19.2x) - (360 + 9.6x^2)(19.2)}{(0 + 19.2x)^2}$. $C'(x) = 0$ when

$(19.2)(9.6x^2 + 60x - 360) = 0$ so $x = 3.75$: the height of the cola is $h = 3.75$ cm.

(iii) $C(3.75) = 3.75$ (The center of gravity is exactly at the top edge of the cola. It turns out that when the cg of a can and liquid system is as low as possible then the cg is at the top edge of the liquid.) Then $\theta = \arctan(\text{radius}/(\text{height of cg})) = \arctan(2.5/3.75) = \arctan(0.667) \approx 0.588$ which is about 33.7^O. In this situation, the can can be tilted about 33.7^O before it falls over.

(iv) Less.

28. (a) & (b) See Fig. 3.5P28. In the solutions to these "shortest path" problems, the roads all meet at 120° angles.

Fig. 3.5P28

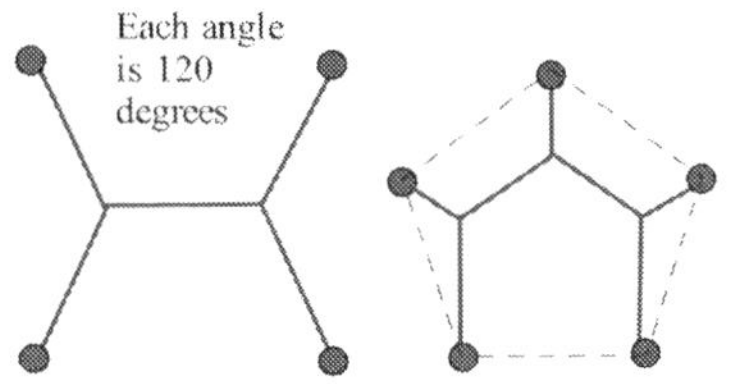
Each angle
is 120
degrees

29. (a) A=(base)(height)= $(1-x)(x^2) = x^2 - x^3$ *for* $0 \le x \le 1$. $A' = 2x - 3x^2 = 0$ *if* $x = 2/3$.

(Clearly the endpoints x=0 and x=1 will not give the largest area.) *Then* $A = \left(\frac{1}{3}\right)\left(\frac{2}{3}\right)^2 = \frac{4}{27}$.

(b) A=(base)(height)= $(1-x)(Cx^2) = Cx^2 - Cx^3$ for $0 \le x \le 1$.

$A' = 2Cx - 3Cx^2 = 0$ if $x = 2/3$. Then $A = \left(\frac{1}{3}\right)(C)\left(\frac{2}{3}\right)^2 = \frac{4C}{27}$.

(c) A=(base)(height)= $(B-x)(Cx^2) = BCx^2 - Cx^3$ for $0 \le x \le 1$.

$A' = 2BCx - 3Cx^2 = 0$ if $x = \frac{2}{3}B$. Then $A = \left(\frac{B}{3}\right)(C)\left(\frac{2B}{3}\right)^2 = \frac{4}{27}B^3C$.

31. (a) $y = 20 - \frac{20}{50}x$. $A = (\text{base})\text{height} = xy = x(20 - \frac{2}{5}x) = 20x - \frac{2}{5}x^2$.

$A' = 20 - \frac{4}{5}x = 0$ when $x = 25$. Then $y = 10$ and Area = 250.

(b) $y = H - \frac{H}{B}x$. $A = (\text{base})\text{height} = x(H - \frac{H}{B}x) = Hx - \frac{H}{B}x^2$.

$A' = H - \frac{2H}{B}x = 0$ when $x = \frac{B}{2}$. Then $y = \frac{H}{2}$ and Area $= \frac{BH}{4}$.

33. $F = \text{cost} = (\text{top cost}) + (\text{bottom}) + (\text{sides}) = (\pi r^2)A + (\pi r^2)B + (2\pi rh)C$

But we know $V = \pi r^2 h$ so $h = \frac{V}{\pi r^2}$ so $F = \pi r^2(A+B) + \frac{2CV}{r}$.

Then $F' = 2\pi r(A+B) - \frac{2CV}{r^2} = 0$ when $r = \sqrt[3]{\frac{CV}{\pi(A+B)}}$. Now you can find h and F.

Section 3.6

1. (a) h has a root at $x = 1$.

(b) limits of $h(x) = f(x)/g(x)$: as $x \to 1^+$ is 0: as $x \to 1^-$ is 0: as $x \to 3^+$ is $-\infty$: as $x \to 3^-$ is $+\infty$

(c) h has a vertical asymptote at $x = 3$.

3. limits of $h(x) = f(x)/g(x)$: as $x \to 2^+$ is $+\infty$: as $x \to 2^-$ is $-\infty$: as $x \to 4^+$ is 0 : as $x \to 4^-$ is 0

5. 0 7. –3 9. 0 11. DNE

13. 2/3 15. 0 17. –7 19. 0

21. $\cos(0) = 1$ 23. $\ln(1) = 0$

25. (a) $V(t) = 50 + 4t$ gallons, and $A(t) = 0.8t$ pounds of salt

(b) $C(t) = \dfrac{\text{amount of salt}}{\text{total amount of liquid}} = \dfrac{A(t)}{V(t)} = \dfrac{0.8t}{50 + 4t}$

(c) "after a long time" (as $t \to \infty$), $C(t) \to 0.8/4 = 0.2$ pounds of salt per gallon.

(d) $V(t) = 200 + 4t$, $A(t) = 0.8t$, $C(t) = \dfrac{0.8t}{200 + 4t} \to 0.8/4 = 0.2$ pounds of salt per gallon.

27. $+\infty$ 29. $-\infty$ 31. $-\infty$ 33. $-\infty$

35. $+\infty$ 37. $-\infty$ 39. 1 41. $-\infty$

42. Horizontal: $y = 1$. Vertical: $x = 1$. 43. Horizontal: $y = 0$. Vertical: $x = 0$.

44. Horizontal: $y = 0$. Vertical: $x = 0$ (a "hole" at $x = 1$). 45. Horizontal: $y = 0$. Vertical: $x = 3$ and $x = 1$.

46. Horizontal: $y = 1/3$. Vertical: $x = 1$. 47. Horizontal: $y = 1$.

48. Horizontal: $y = 0$. Vertical: $x = 0$. 49. Horizontal: $y = 1$. Vertical: $x = 1$.

50. Horizontal: $y = 0$. Vertical: $x = 1$. (and a "hole" at $x = 0$)

51. $y = 2x + 1$. $x = 0$ 52. $y = x$ 53. $y = \sin(x)$. $x = 2$ 54. $y = x$

55. $y = x^2$ 56. $y = x^2 + 1$. $x = -1$ 57. $y = \cos(x)$. $x = 3$. 58. $y = x^2 + 2$. $x = 1$.

59. $y = \sqrt{x}$. $x = -3$.

Section 3.7

1. 3/2 3. 3/5 5. -1 7. 0

9. 0 11. 9/2

13. For $a \neq 0$: $\dfrac{f'}{g'} = \dfrac{m}{n} x^{(m-n)} \to \dfrac{m}{n} a^{(m-n)}$.

For $a = 0$, $\displaystyle\lim_{x \to a} \frac{x^m - a^m}{x^n - a^n} = \begin{cases} 0 & \text{if } m > n \\ 1 & \text{if } m = n \\ +\infty & \text{if } m < n \text{ and } (m - n) \text{ is even} \\ \text{DNE} & \text{if } m < n \text{ and } (m - n) \text{ is odd} \end{cases}$

15. 0 17. $\dfrac{f'}{g'} = \dfrac{pe^{px}}{3} \to \dfrac{p}{3}$ so $p = 3(5) = 15$.

19. (a) All three limits are $+\infty$.

(b) After applying L'Hopital's Rule d times, $\dfrac{f^{(d)}}{g^{(d)}} = \dfrac{a \cdot b^n \cdot e^{bn}}{c(d)(d-1)(d-2) \ldots (2)(1)} = \dfrac{\text{constant} \cdot e^{bn}}{\text{another constant}} \to +\infty$.

21. 0 23. 0 24. 1 25. 1 26. 1

27. 0 28. $+\infty$ 29. 1 30. 1

Chapter 4: THE INTEGRAL

Previous chapters dealt with **Differential Calculus**. They started with the "simple" geometrical idea of the **slope of a tangent line** to a curve, developed it into a combination of theory about derivatives and their properties, techniques for calculating derivatives, and applications of derivatives. This chapter begins the development of **Integral Calculus** and starts with the "simple" geometric idea of **area**. This idea will be developed into another combination of theory, techniques, and applications.

One of the most important results in mathematics, The Fundamental Theorem of Calculus, appears in this chapter. It unifies the differential and integral calculus into a single grand structure. Historically, this unification marked the beginning of modern mathematics, and it provided important tools for the growth and development of the sciences.

The chapter begins with a look at area, some geometric properties of areas, and some applications. First we will see ways of approximating the areas of regions such as tree leaves that are bounded by curved edges and the areas of regions bounded by graphs of functions. Then we will find ways to calculate the areas of some of these regions exactly. Finally, we will explore more of the rich variety of uses of "areas".

The primary purpose of this introductory section is to help develop your intuition about areas and your ability to reason using geometric arguments about area. This type of reasoning will appear often in the rest of this book and is very helpful for applying the ideas of calculus.

AREA

The basic shape we will use is the rectangle; the area of a rectangle is (base)•(height). If the units for each side of the rectangle are "meters," then the area will have the units ("meters")•("meters") = "square meters" = m^2. The only other area formulas needed for this section are for triangles, area = b•h/2, and for circles, area = $\pi \cdot r^2$. Three other familiar properties of area are assumed and will be used:

Addition Property: The total area of a region is the sum of the areas of the non–overlapping pieces which comprise the region. (Fig. 1)

Inclusion Property: If region B is on or inside region A, then the area of region B is less than or equal to the area of region A. (Fig. 2)

Location–Independence Property: The area of a region does not depend on its location. (Fig. 3)

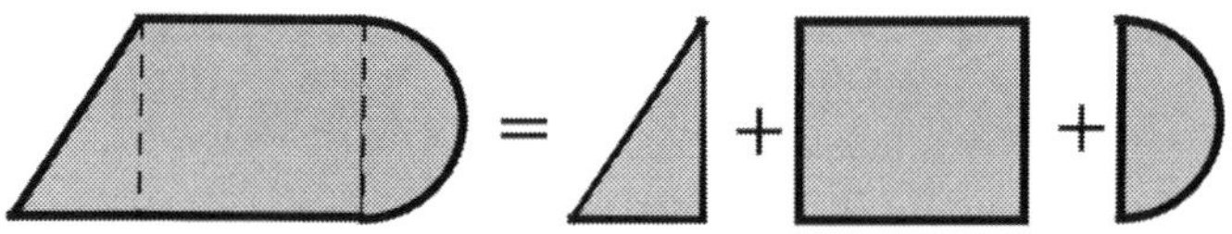

Fig. 1

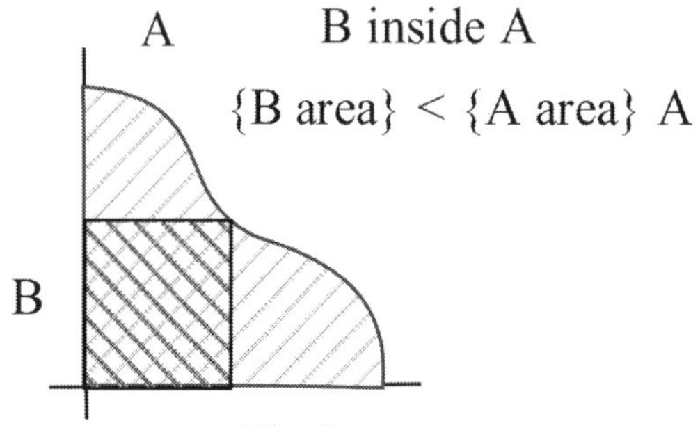

Fig. 2

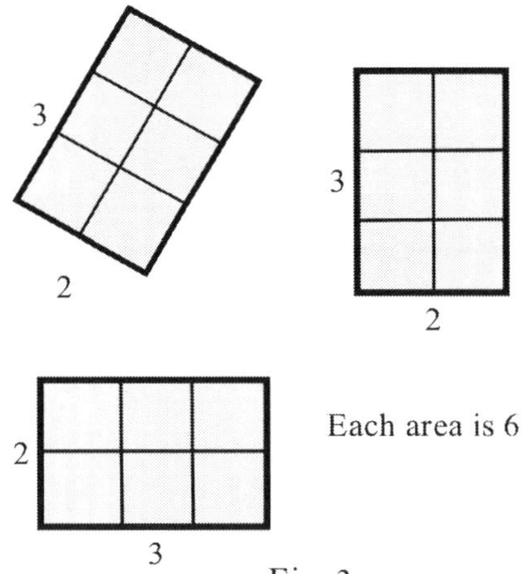

Fig. 3

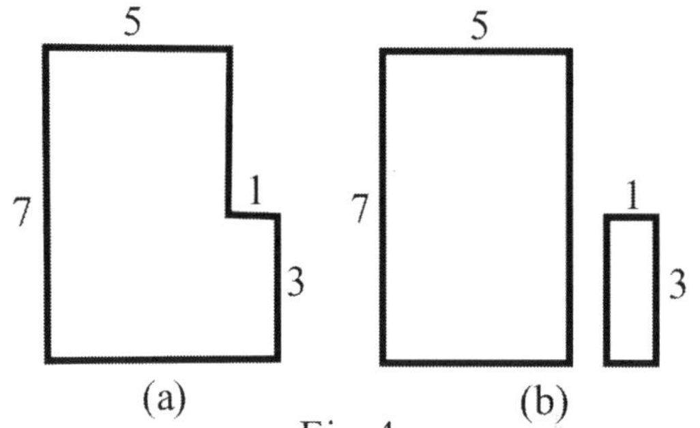

Fig. 4

Example 1: Determine the area of the region in Fig. 4a

Solution: The region can easily be broken into two rectangles, Fig. 4b, with areas 35 square inches and 3 square inches respectively, so the area of the original region is 38 square inches.

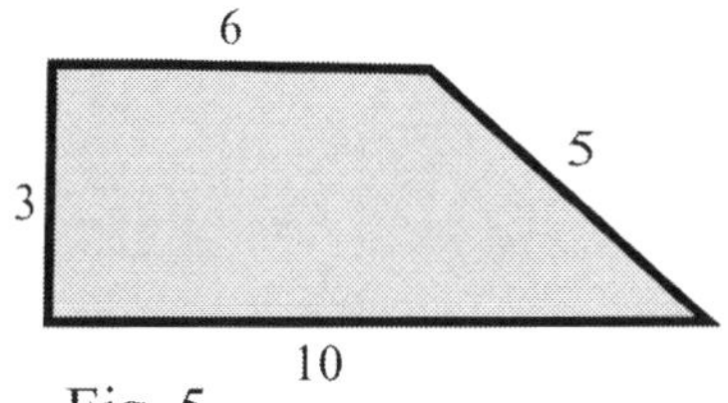

Fig. 5

Practice 1: Determine the area of the region in Fig. 5 by cutting it in two ways:
(a) into a rectangle and triangle and (b) into two triangles.

We can use the three properties of area to get information about areas that are difficult to calculate exactly. Let A be the region bounded by the graph of $f(x) = 1/x$, the x–axis, and vertical lines at $x = 1$ and $x = 3$. Since the two rectangles in Fig. 6 are inside the region A and do not overlap each, the area of the rectangles, $1/2 + 1/3 = 5/6$, is less than the area of region A.

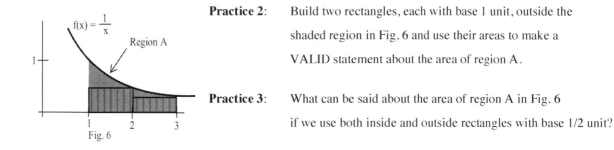

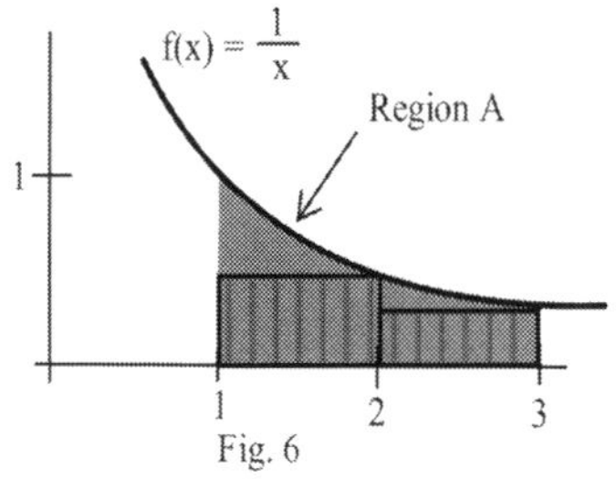

Fig. 6

Practice 2: Build two rectangles, each with base 1 unit, outside the shaded region in Fig. 6 and use their areas to make a VALID statement about the area of region A.

Practice 3: What can be said about the area of region A in Fig. 6 if we use both inside and outside rectangles with base 1/2 unit?

Example 2: In Fig. 7, there are 32 dark squares, 1 centimeter on a side, and 31 lighter squares of the same size. We can be sure that the area of the leaf is smaller than what number?

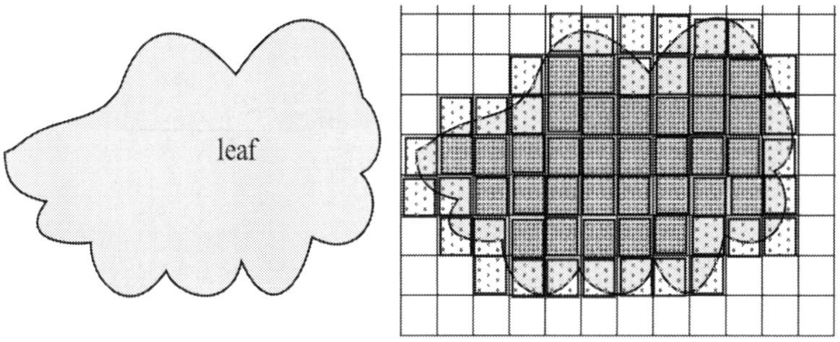

Fig. 7

Solution: The area of the leaf is smaller than $32 + 31 = 63\ \text{cm}^2$.

Practice 4: We can be sure that the area of the leaf is at least how large?

Functions can be defined in terms of areas. For the constant function $f(t) = 2$, define $A(x)$ to be the area of the rectangular region bounded by the graph of f, the t–axis, and the vertical lines at $t=1$ and $t=x$ (Fig. 8a). A(2) is the area of the shaded region in Fig. 8b, and A(2) =2. Similarly, A(3)= 4 and A(4)=6. In general, $A(x)= (\text{base})(\text{height}) = (x-1)(2) = 2x - 2$ for any $x \geq 1$. The graph of $y=A(x)$ is shown in Fig. 8c, and $A'(x) = 2$ for every value of $x > 1$.

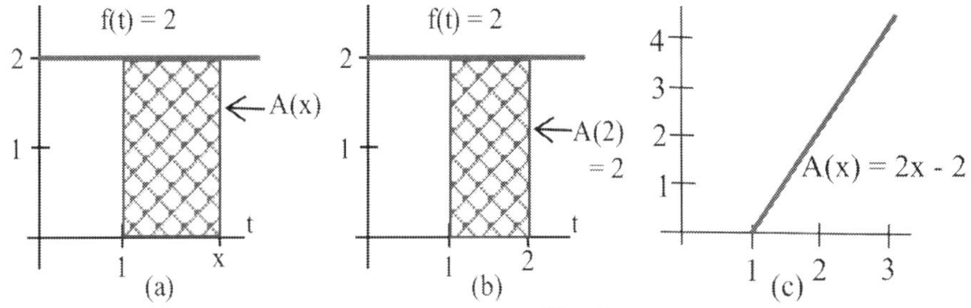

Fig. 8

Practice 5: For $f(t)=2$, define $B(x)$ to be the area of the region bounded by the graph of f, the t–axis, and vertical lines at $t = 0$ and $t = x$. Fill in the table in Fig. 9 with the values of B. How are the graphs of $y = A(x)$ and $y = B(x)$ related?

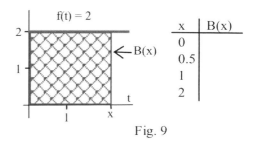

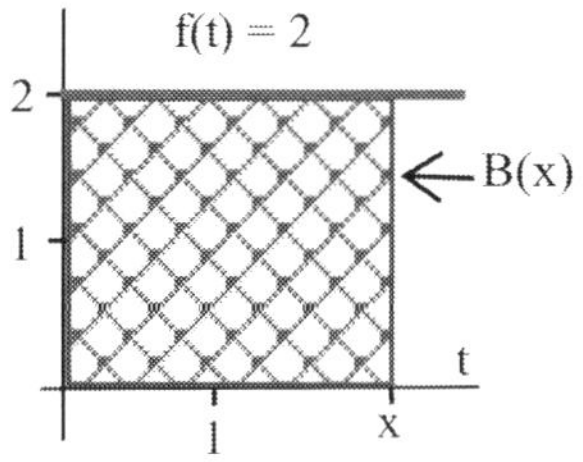

x	B(x)
0	
0.5	
1	
2	

Fig. 9

Sometimes it is useful to move regions around. The area of a parallelogram is obvious if we move the triangular region from one side of the parallelogram to fill the region on the other side and ending up with a rectangle (Fig. 10). At first glance, it is difficult to estimate the total area of the shaded regions in Fig. 11a . However, if we slide all of them into a single column (Fig. 11b), then it is easy to determine that the shaded area is less than the area of the enclosing rectangle = (base)(height) = (1)(2) = 2.

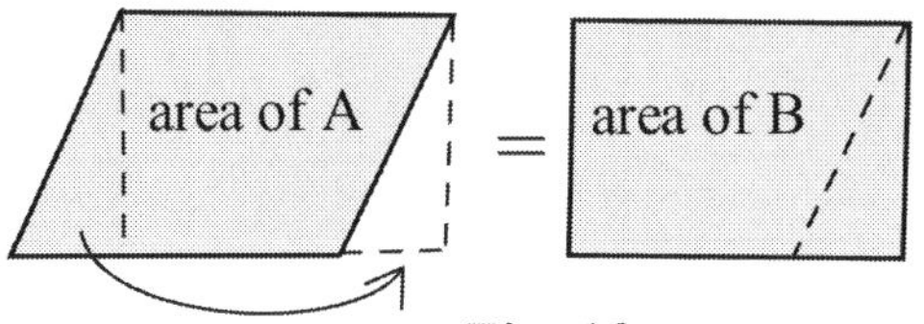

Fig. 10

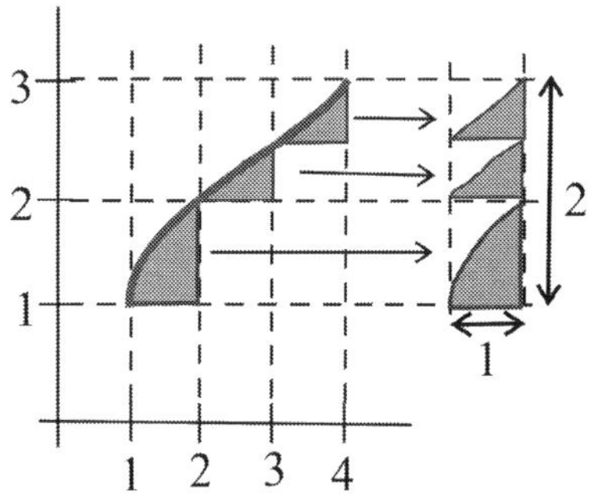

Fig. 11

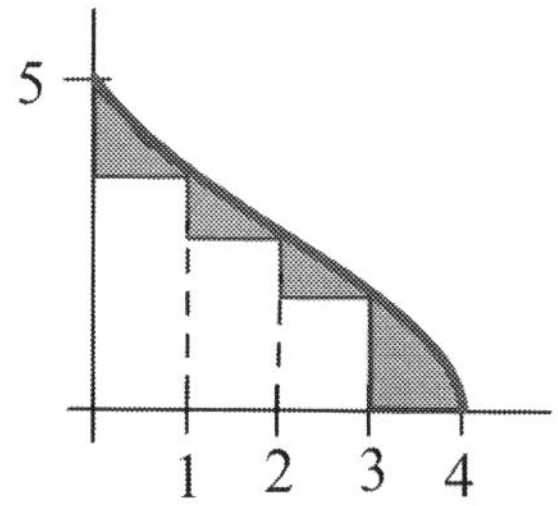

Fig. 12

Practice 6: The total area of the shaded regions in Fig. 12 is less than what number?

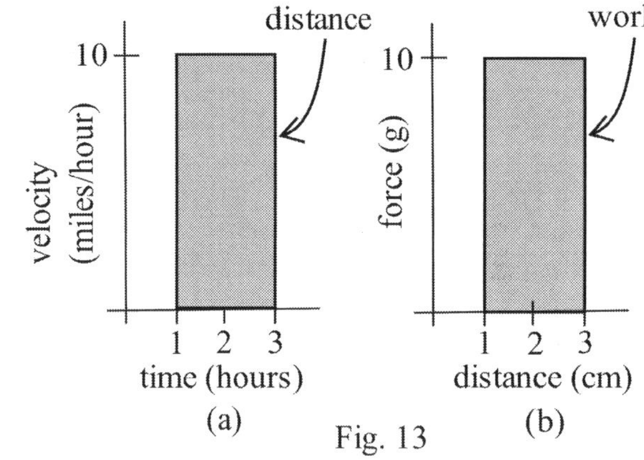

Fig. 13

SOME APPLICATIONS OF "AREA"

One reason "areas" are so useful is that they can represent quantities other than simple geometric shapes. For example, if the units of the base of a rectangle are "hours" and the units of the height are "miles/hour", then the units of the "area" of the rectangle are ("hours")•("miles/hour") = "miles," a measure of distance (Fig. 13a). Similarly, if the base units are "centimeters" and the height units are "grams" (Fig. 13b), then the "area" units are "gram•centimeters," a measure of work.

Distance as an "Area"

In Fig. 14, f(t) is the velocity of a car in "miles per hour", and t is the time in "hours." Then the shaded "area" will be (base)•(height) = (3 hours)•(20 miles/hour) = 60 miles, the distance traveled by the car in the 3 hours from 1 o'clock until 4 o'clock.

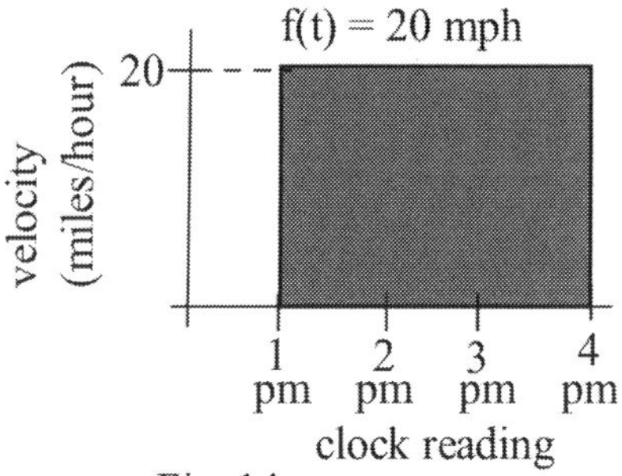

Fig. 14

Distance as an "Area"

If f(t) is the (positive) forward velocity of an object at time t, then the "area" between the graph of f and the t–axis and the vertical lines at times t = a and t = b will be the distance that the object has moved forward between times a and b.

This "area as distance" can make some difficult distance problems much easier.

Example 3: A car starts from rest (velocity = 0) and steadily speeds up so that 20 seconds later it's speed is 88 feet per second (60 miles per hour). How far did the car travel during those 20 seconds?

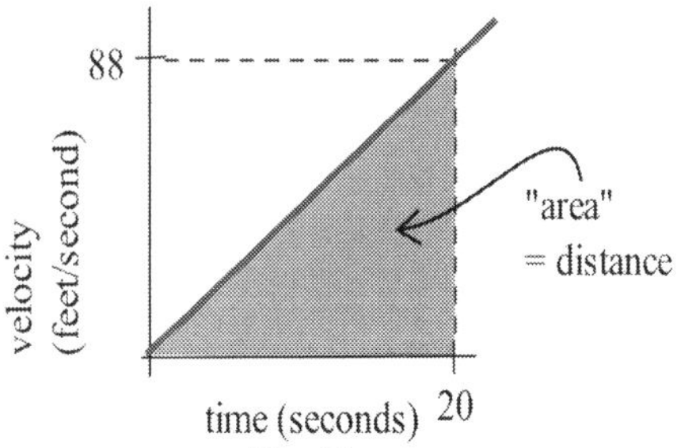

Fig. 15

Solution: We can answer the question using the techniques of chapter 3 (try it). But if "steadily" means that the velocity increases linearly, then it is easier to use Fig. 15 and the idea of "area as distance."

The "area" of the triangular region represents the distance traveled, so

$$\text{distance} = \frac{1}{2}(\text{base}) \cdot (\text{height}) = \frac{1}{2}(20 \text{ seconds}) \cdot (88 \text{ feet/second})$$

$$= 880 \text{ feet.}$$

Practice 7: A train traveling at 45 miles per hour (66 feet/second) takes 60 seconds to come to a complete stop. If the train slowed down at a steady rate (the velocity decreased linearly), how many feet did the train travel while coming to a stop?

Practice 8: You and a friend start off at noon and walk in the same direction along the same path at the rates shown in Fig. 16.

(a) Who is walking faster at 2 pm? Who is ahead at 2 pm?

(b) Who is walking faster at 3 pm? Who is ahead at 3 pm?

(c) When will you and your friend be together? (Answer in words.)

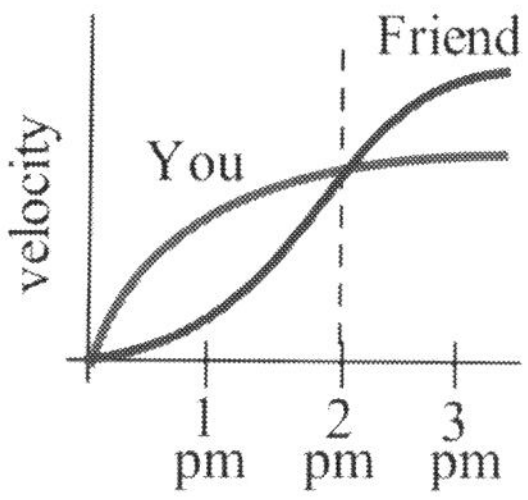

Fig. 16

Total Accumulation as "Area"

In the previous examples, the function represented a **rate** of travel (miles per hour), and the area represented the **total** distance traveled. For functions representing other **rates** such as the production of a factory (bicycles per day), or the flow of water in a river (gallons per minute) or traffic over a bridge (cars per minute), or the spread of a disease (newly sick people per week), the area will still represent the **total** amount of something.

> **"Area" as a Total Accumulation**
>
> If f(t) represents a positive **rate** (in units per time interval) at time t, then the "area" between the graph of f and the t–axis and the vertical lines at times t=a and t=b will be the total units which accumulate between times a and b. (Fig. 17)

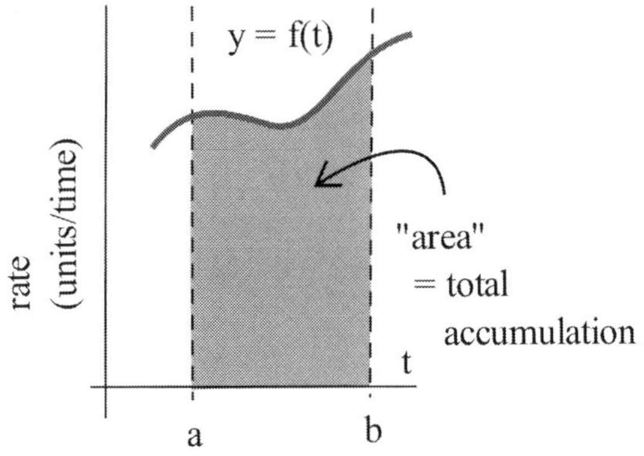

Fig. 17

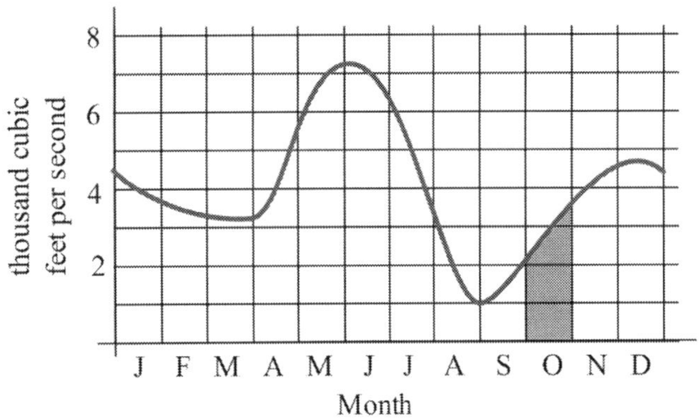

Fig. 18

For example, Fig. 18 shows the flow rate (cubic feet per second) of water in the Skykomish river at the town of Goldbar in Washington state. The area of the shaded region represents the total volume (cubic feet) of water flowing past the town during the month of October:

Total water = "area"

= area of rectangle + area of the triangle

$\approx$ (2000 cubic feet/sec)(30 days) + $\frac{1}{2}$(1500 cf/s)(30 days) = (2750 cubic feet/sec)(30 days)

= (2750 cubic feet/sec)(2,592,000 sec) $\approx 7.128 \times 10^9$ cubic feet

(For comparison, the flow over Niagara Falls is about 2.12×10^5 cf/s.)

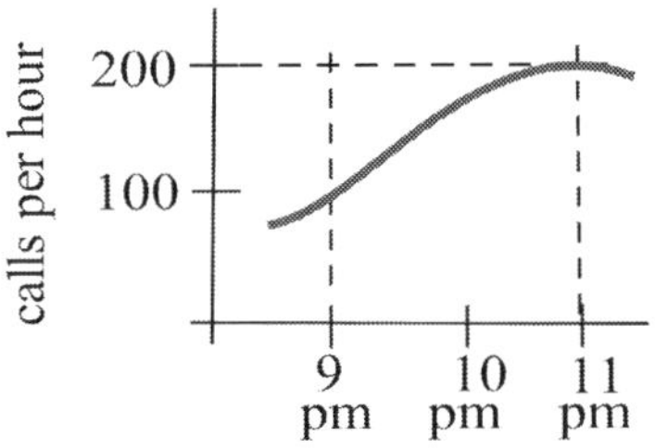

Fig. 19

Practice 9: Fig. 19 shows the number of telephone calls made per hour on a Tuesday. Approximately how many calls were made between 9 am and 11 am?

PROBLEMS

1. Calculate the areas of the shaded regions in Fig. 20.

2. Calculate the area of the trapezoidal region in Fig. 21 by breaking it into a triangle and a rectangle.

3. Break Fig. 22 into a triangle and rectangle and verify that the total area of the trapezoid is $b \cdot (\frac{h + H}{2})$.

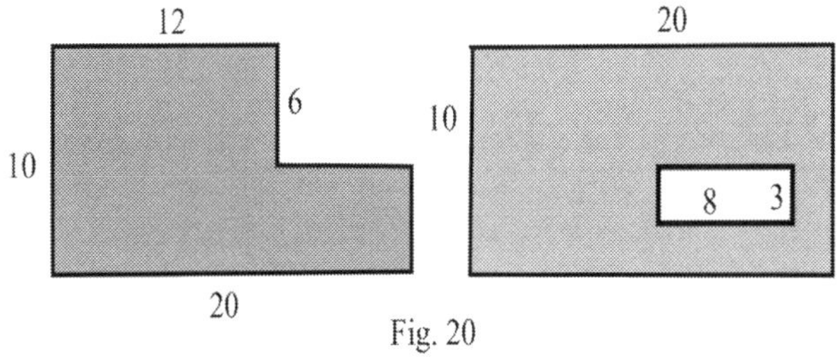

Fig. 20

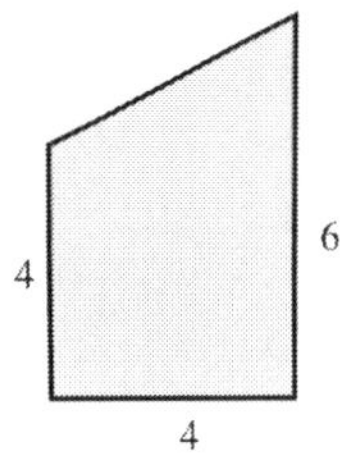

Fig. 21

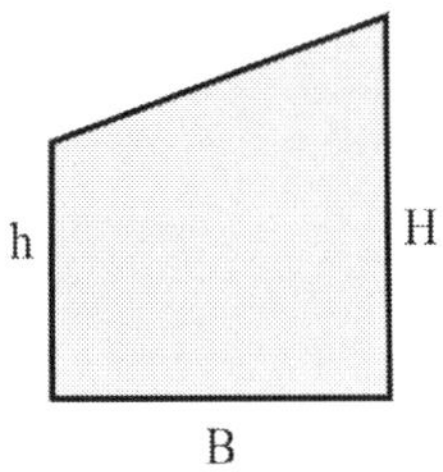

Fig. 22

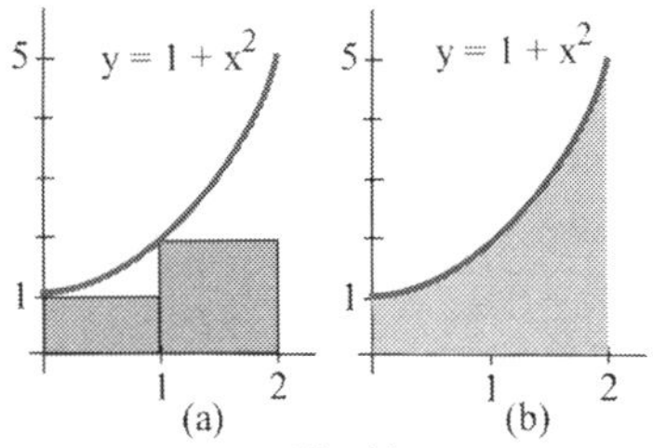

Fig. 23

4. (a) Calculate the sum of the rectangular areas in Fig. 23a.
   (b) What can we say about the area of the shaded region in Fig. 23b?

5. (a) Calculate the sum of the rectangular areas in Fig. 24a.

(b) From part (a), what can we say about the area of the shaded region in Fig. 24b?

6. (a) Calculate the sum of the areas of the shaded regions in Fig. 24c.

(b) From part (a), what can we say about the area of the shaded region in Fig. 24b?

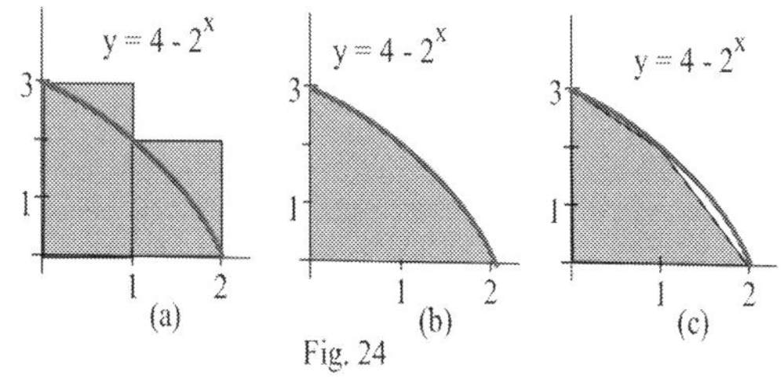

Fig. 24

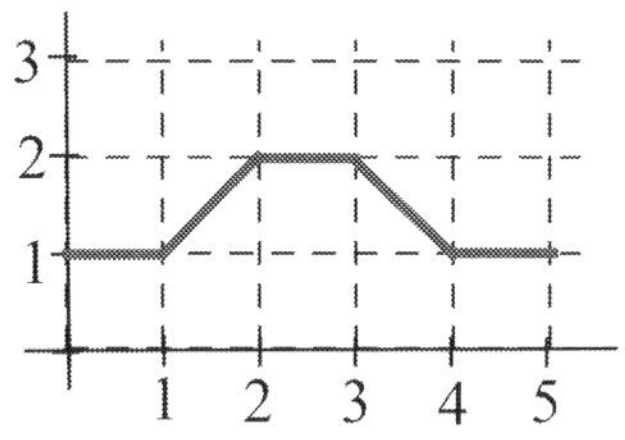

Fig. 25

7. Let A(x) represent the area bounded by the graph and the horizontal axis and vertical lines at t=0 and t=x for the graph in Fig. 25. Evaluate A(x) for x = 1, 2, 3, 4, and 5.

8. Let B(x) represent the area bounded by the graph and the horizontal axis and vertical lines at t=0 and t=x for the graph in Fig. 26. Evaluate B(x) for x = 1, 2, 3, 4, and 5.

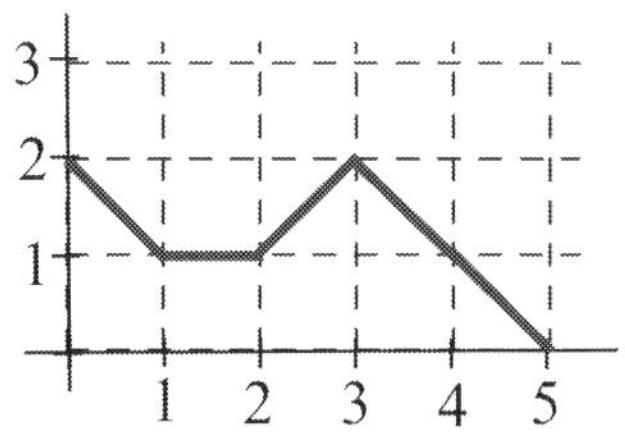

Fig. 26

9. Let C(x) represent the area bounded by the graph and the horizontal axis and vertical lines at t=0 and t=x for the graph in Fig. 27. Evaluate C(x) for x = 1, 2, and 3 and find a formula for C(x).

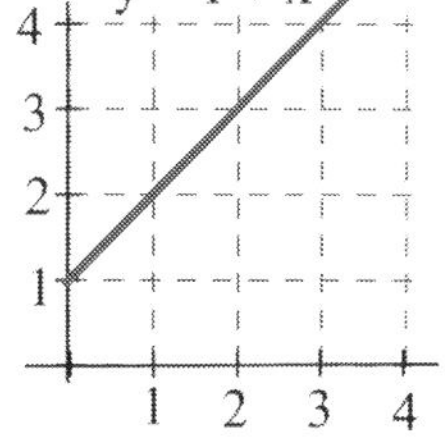

Fig. 27

10. Let A(x) represent the area bounded by the graph and the horizontal axis and vertical lines at t=0 and t=x for the graph in Fig. 28. Evaluate A(x) for x = 1, 2, and 3 and find a formula for A(x).

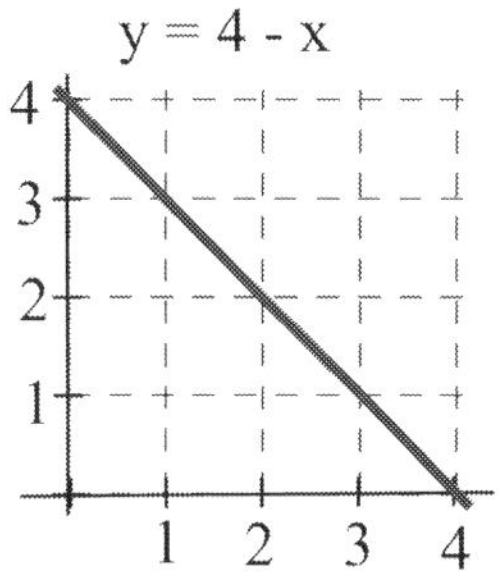

Fig. 28

11. A car had the velocity given in Fig. 29. How far did the car travel from t=0 to t=30 seconds?

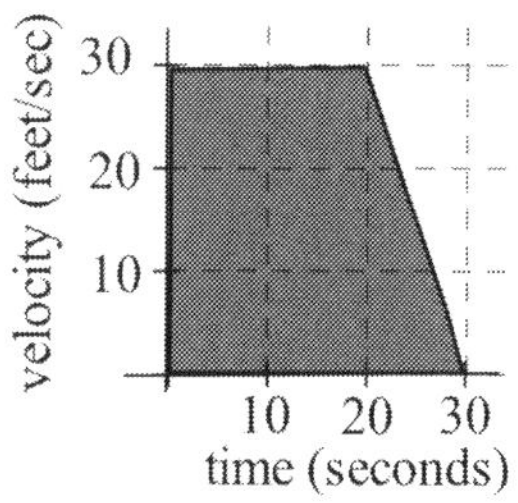

Fig. 29

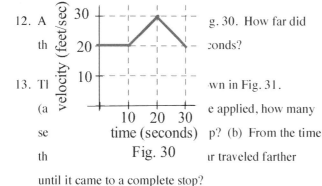

12. A g. 30. How far did
th conds?

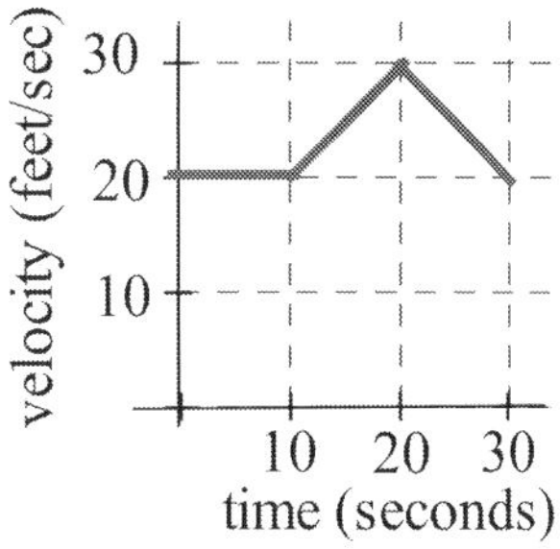

Fig. 30

13. Tl wn in Fig. 31.
(a e applied, how many
se p? (b) From the time
th ır traveled farther
until it came to a complete stop?

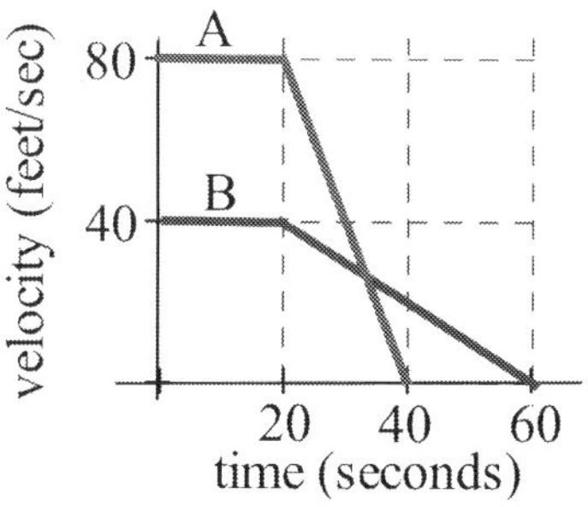

Fig. 31

14. Police chase: A speeder traveling 45 miles per hour (in a 25 mph zone) passes a stopped police car which immediately takes off after the speeder. If the police car speeds up steadily to 60 miles/hour in 20 seconds and then travels at a steady 60 miles/hour, **how long** and **how far** before the police car catches the speeder who continued traveling at 45 miles/hour? (Fig. 32)

15. What are the units for the "area" of a rectangle with the given base and height units?

Base units	Height units	"Area" units
miles per second	seconds	
hours	dollars per hour	
square feet	feet	
kilowatts	hours	
houses	people per house	
meals	meals	

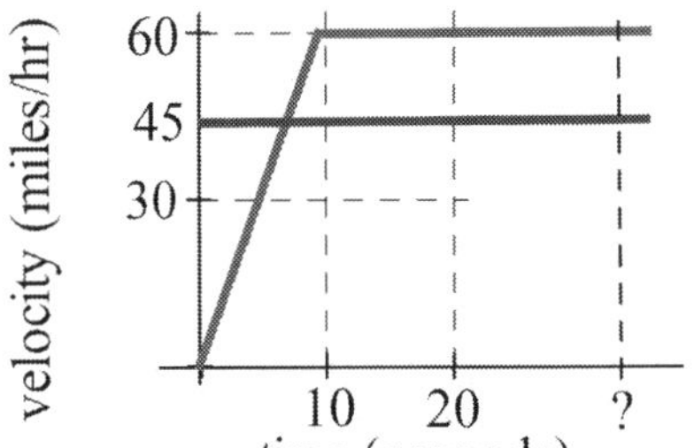

Fig. 32

Section 4.0 **PRACTICE Answers**

Practice 1: area $= 3(6) + \frac{1}{2}(4)(3) = 24$ and area $= \frac{1}{2}(3)(10) + \frac{1}{2}(6)(3) = 24$.

Practice 2: outside rectangular area $= (1)(1) + (1)(\frac{1}{2}) = 1.5$

Practice 3: Using rectangles with base = 1/2. Inside: area $= \frac{1}{2}(\frac{2}{3} + \frac{1}{2} + \frac{2}{5} + \frac{1}{3}) = \frac{57}{60} \approx 0.95$

Outside: area $= \frac{1}{2}(1 + \frac{2}{3} + \frac{1}{2} + \frac{2}{5}) = \frac{72}{60} \approx 1.2$

The area of the region is between 0.95 and 1.2 .

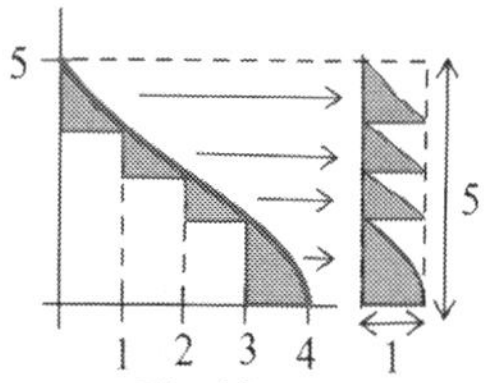

Fig. 33

Practice 4: The area of the leaf is larger than the area of the dark rectangles, 32 cm^2.

Pr 5:

x	B(x)
0	0
1/2	1
1	2
2	4

$y = B(x) = 2x$ is a line with slope 2 so it is parallel to $y = A(x) = 2x - 2$.

Practice 6: See Fig. 33. Area < area of the rectangle enclosing the shifted regions = 5

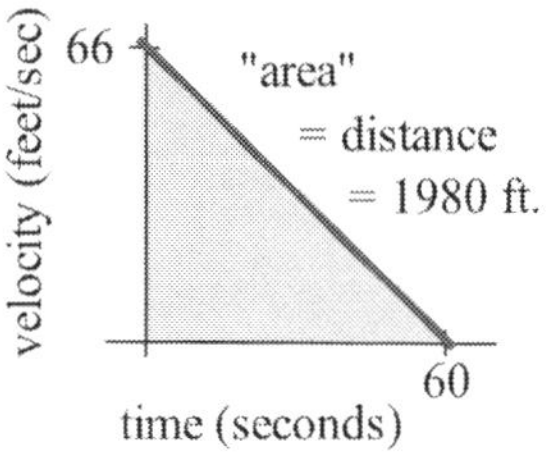

Fig. 34

Practice 7: See Fig. 34. Distance = area of shaded region $= \frac{1}{2}$(base)(height)

$= \frac{1}{2}$(60 seconds)(66 feet/second) = 1980 feet.

Practice 8: (a) At 2 pm both are walking at the same velocity. You are ahead.

(b) At 3 pm your friend is walking faster than you, but you are still ahead. (The "area" under your velocity curve is larger than the "area" under your friend's.)

(c) You and your friend will be together on the trail when the "areas" (distances) under the two velocity graphs are equal.

Practice 9: Total calls = "area" under rate curve from 9 am to 11 am ≈ 300 calls.

4.1 SIGMA NOTATION AND RIEMANN SUMS

One strategy for calculating the area of a region is to cut the region into simple shapes, calculate the area of each simple shape, and then add these smaller areas together to get the area of the whole region. We will use that approach, but it is useful to have a notation for adding a lot of values together: the sigma (Σ) notation.

Summation	Sigma notation	A way to read the sigma notation
$1^2 + 2^2 + 3^2 + 4^2 + 5^2$	$\sum_{k=1}^{5} k^2$	the sum of k squared for k equals 1 to k equals 5
$\frac{1}{3} + \frac{1}{4} + \frac{1}{5} + \frac{1}{6} + \frac{1}{7}$	$\sum_{k=3}^{7} \frac{1}{k}$	the sum of 1 divided by k for k equals 3 to k equals 7
$2^0 + 2^1 + 2^2 + 2^3 + 2^4 + 2^5$	$\sum_{j=0}^{5} 2^j$	the sum of 2 to the jth power for j equals 0 to j equals 5
$a_2 + a_3 + a_4 + a_5 + a_6 + a_7$	$\sum_{i=2}^{7} a_i$	the sum of a sub i from i equals 2 to i equals 7

The variable (typically i, j, or k) used in the summation is called the **counter** or **index variable**. The function to the right of the sigma is called the **summand**, and the numbers below and above the sigma are called the **lower and upper limits of the summation**. (Fig. 1)

Practice 1: Write the summation denoted by each of the following:

(a) $\sum_{k=1}^{5} k^3$, (b) $\sum_{j=2}^{7} (-1)^j \frac{1}{j}$, (c) $\sum_{m=0}^{4} (2m+1)$.

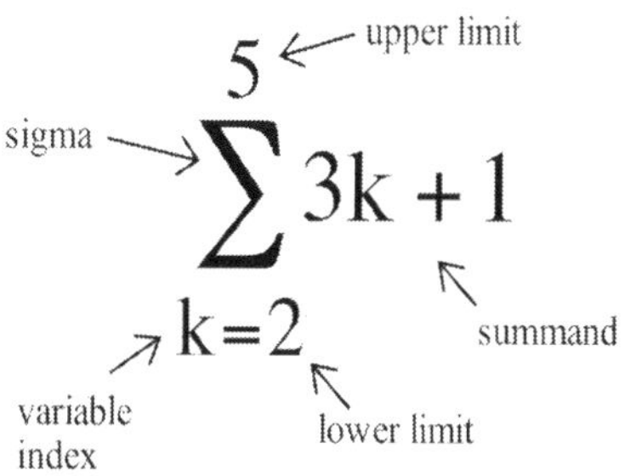

Fig. 1

In practice, the sigma notation is frequently used with the standard function notation:

$$\sum_{k=1}^{3} f(k+2) = f(1+2) + f(2+2) + f(3+2) = f(3) + f(4) + f(5) \text{ and}$$

$$\sum_{i=1}^{4} f(x_i) = f(x_1) + f(x_2) + f(x_3) + f(x_4)$$

x	f(x)	g(x)	h(x)
1	2	4	3
2	3	1	3
3	1	–2	3
4	0	3	3
5	3	5	3

Table 1

Example 1: Use the values in Table 1 to evaluate $\sum_{k=2}^{5} 2\cdot f(k)$ and $\sum_{j=3}^{5} (5 + f(j-2))$.

Solution: $\sum_{k=2}^{5} 2\cdot f(k) = 2\cdot f(2) + 2\cdot f(3) + 2\cdot f(4) + 2\cdot f(5) = 2\cdot(3)+2\cdot(1)+2\cdot(0)+2\cdot(3) = 14$

$\sum_{j=3}^{5} (5 + f(j-2)) = (5+f(3-2))+(5+f(4-2))+(5+f(5-2)) = (5+f(1))+(5+f(2))+(5+f(3))$

$= (5+2) + (5+3) + (5+1) = 21$.

Practice 2: Use the values of f, g and h in Table 1 to evaluate the following:

(a) $\sum_{k=2}^{5} g(k)$ (b) $\sum_{j=1}^{4} h(j)$ (c) $\sum_{k=3}^{5} (g(k) + f(k-1))$

Example 2: For $f(x) = x^2 + 1$, evaluate $\sum_{k=0}^{3} f(k)$.

Solution: $\sum_{k=0}^{3} f(k) = f(0) + f(1) + f(2) + f(3)$

$= (0^2 +1) + (1^2 +1) + (2^2 +1) + (3^2 +1) = 1 + 2 + 5 + 10 = 18$.

Practice 3: For $g(x) = 1/x$, evaluate $\sum_{k=2}^{4} g(k)$ and $\sum_{k=1}^{3} g(k+1)$.

The summand does not have to contain the index variable explicitly: a sum from k=2 to k=4 of the constant function f(k) = 5 can be written as

$\sum_{k=2}^{4} f(k)$ or $\sum_{k=2}^{4} 5 = 5 + 5 + 5 = 3\cdot 5 = 15$. Similarly, $\sum_{k=3}^{7} 2 = 2 + 2 + 2 + 2 + 2 = 5\cdot 2 = 10$.

Since the sigma notation is simply a notation for addition, it has all of the familiar properties of addition.

Summation Properties

Sum of Constants: $\sum_{k=1}^{n} C = C + C + C + .. + C \text{ (n terms)} = n\cdot C$

Addition: $\sum_{k=1}^{n} (a_k + b_k) = \sum_{k=1}^{n} a_k + \sum_{k=1}^{n} b_k$

Subtraction: $\sum_{k=1}^{n} (a_k - b_k) = \sum_{k=1}^{n} a_k - \sum_{k=1}^{n} b_k$

Constant Multiple: $\sum_{k=1}^{n} C\cdot a_k = C\cdot \sum_{k=1}^{n} a_k$

Problems 16 and 17 illustrate that similar patterns for sums of products and quotients are **not valid**.

Sums of Areas of Rectangles

In Section 4.2 we will approximate the areas under curves by building rectangles as high as the curve, calculating the area of each rectangle, and then adding the rectangular areas together.

Example 3: Evaluate the sum of the rectangular areas in Fig. 2, and write the sum using the sigma notation.

Solution:

{ sum of the rectangular areas } = { sum of (base)·(height) for each rectangle }

$$= (1)\cdot(1/3) + (1)\cdot(1/4) + (1)\cdot(1/5) = 47/60.$$

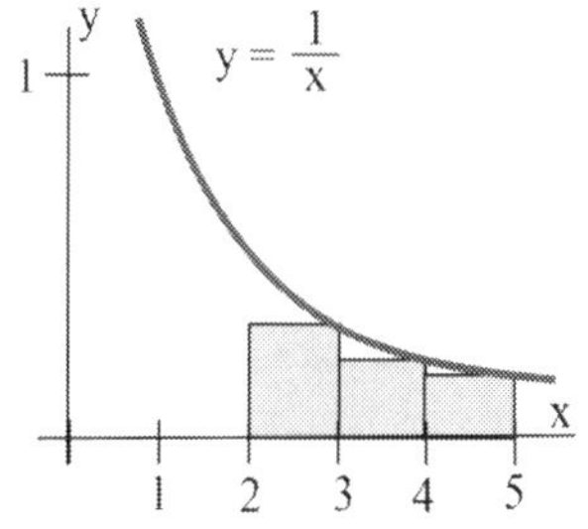

Fig. 2

Using the sigma notation,

$$(1)\cdot(1/3) + (1)\cdot(1/4) + (1)\cdot(1/5) = \sum_{k=3}^{5} \frac{1}{k} .$$

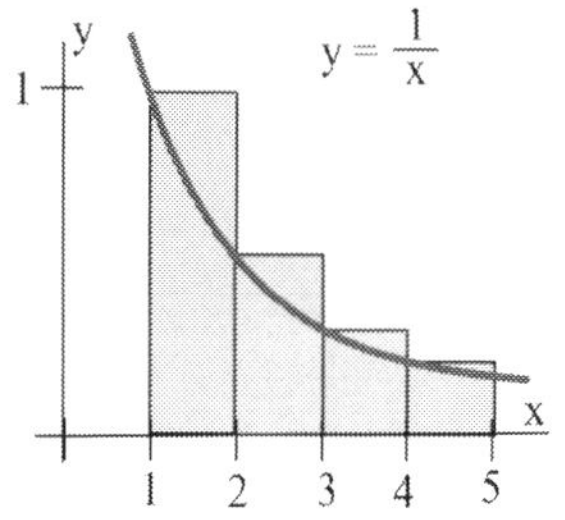

Practice 4: Evaluate the sum of the rectangular areas in Fig. 3, and write the sum using the sigma notation.

The bases of the rectangles do not have to be equal. For the rectangular areas in Fig. 4 ,

rectangle	base	height	area
1	3–1=2	f(2)=4	2·4 = 8
2	4–3=1	f(4)=16	1·16 = 16
3	6–4=2	f(5)=25	2·25 = 50

so the sum of the rectangular areas is 8 + 16 + 50 = 74.

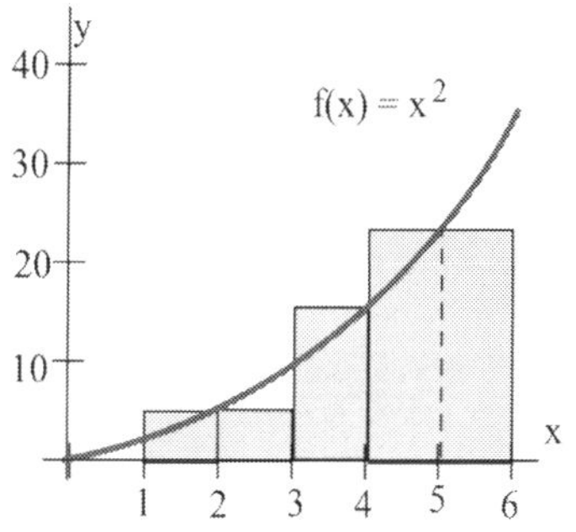

Fig. 4

Example 4: Write the sum of the areas of the rectangles in Fig. 5 using the sigma notation.

Solution: The area of each rectangle is (base)·(height).

rectangle	base	height	area
1	$x_1 - x_0$	$f(x_1)$	$(x_1 - x_0)\cdot f(x_1)$
2	$x_2 - x_1$	$f(x_2)$	$(x_2 - x_1)\cdot f(x_2)$
3	$x_3 - x_2$	$f(x_3)$	$(x_3 - x_2)\cdot f(x_3)$

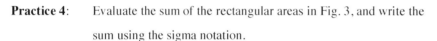

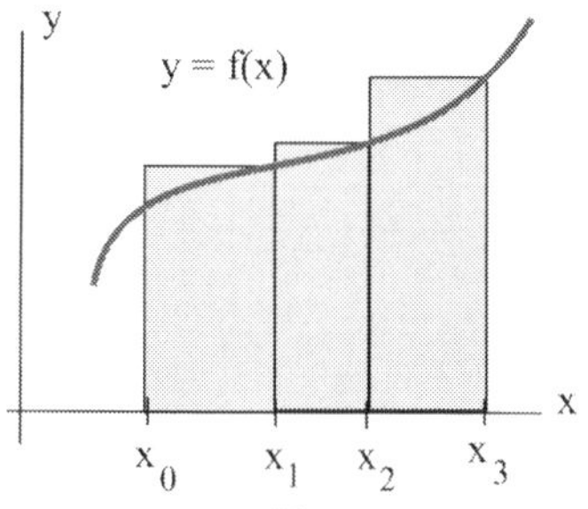

Fig. 5

The area of the kth rectangle is $(x_k - x_{k-1})\cdot f(x_k)$, and the total area of the rectangles is the sum $\sum_{k=1}^{3} (x_k - x_{k-1}) \cdot f(x_k)$.

Practice 5: Write the sum of the areas of the shaded rectangles in Fig. 6 using the sigma notation.

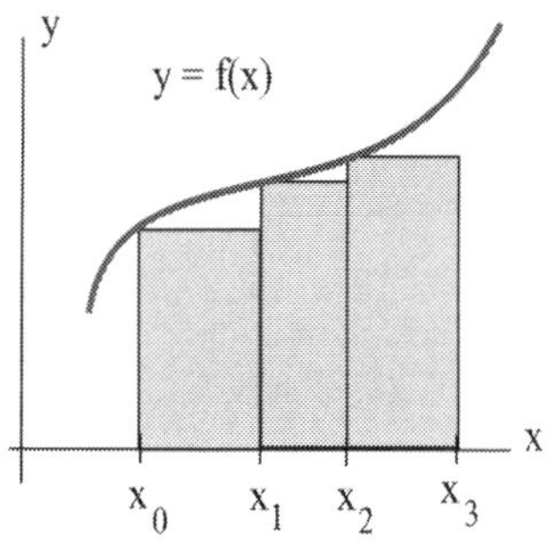

Fig. 6

Area Under A Curve — Riemann Sums

Suppose we want to calculate the area between the graph of a positive function f and the interval [a, b] on the x–axis (Fig. 7). The Riemann Sum method is to build several rectangles with bases on the interval [a, b] and sides that reach up to the graph of f (Fig. 8). Then the areas of the rectangles can be calculated and added together to get a number called a Riemann Sum of f on [a, b]. The area of the region formed by the rectangles is an **approximation** of the area we want.

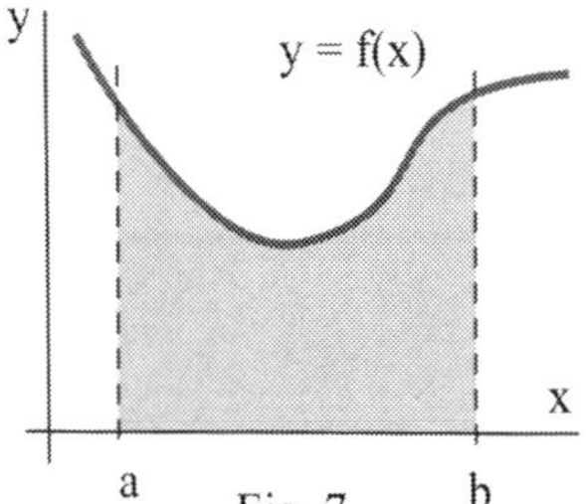

Fig. 7

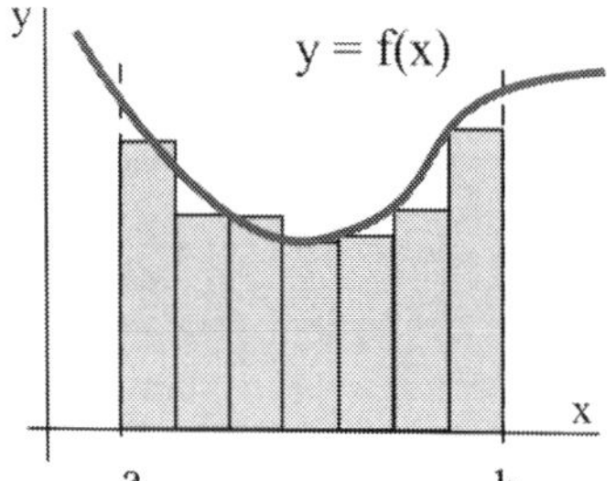

Fig. 8

Example 5: Approximate the area in Fig. 9a between the graph of f and the interval [2, 5] on the x–axis by summing the areas of the rectangles in Fig. 9b.

Solution: The total area of rectangles is (2)(3) + (1)(5) = 11 square units.

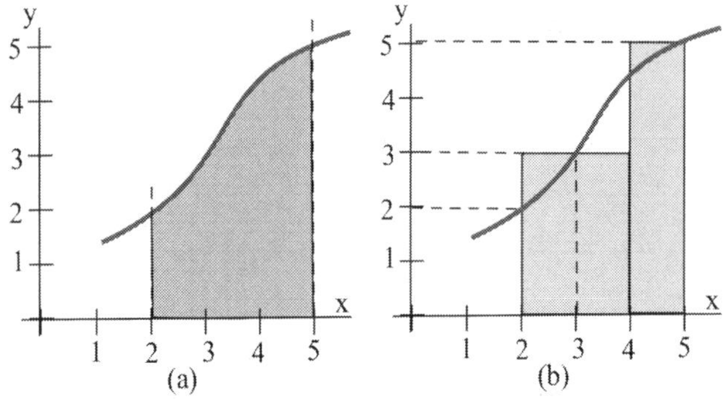

Fig. 9

In order to effectively describe this process, some new vocabulary is helpful: a **partition** of an interval and the **mesh** of the partition.

A **partition** P of a closed interval [a,b] into n subintervals is a set of n+1 points $\{ x_0 = a, x_1, x_2, x_3, \ldots, x_{n-1}, x_n = b\}$ in increasing order, $a = x_0 < x_1 < x_2 < x_3 < \ldots < x_{n-1} < x_n = b$.

(A partition is a collection of points on the axis and it does not depend on the function in any way.)

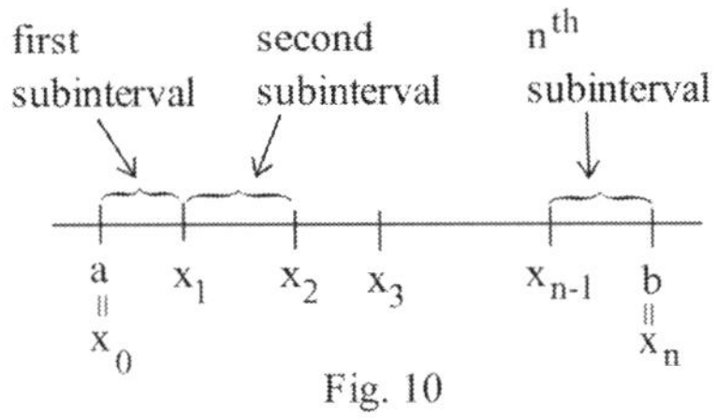

Fig. 10

The points of the partition P divide the interval into n subintervals (Fig. 10): $[x_0, x_1]$, $[x_1, x_2]$, $[x_2, x_3]$, . . . , and $[x_{n-1}, x_n]$ with lengths $\Delta x_1 = x_1 - x_0$, $\Delta x_2 = x_2 - x_1$, $\Delta x_3 = x_3 - x_2$, . . . , and $\Delta x_n = x_n - x_{n-1}$. The points x_k of the partition P are the locations of the vertical lines for the sides of the rectangles, and the bases of the rectangles have lengths Δx_k for k = 1, 2, 3, . . . , n.

The **mesh** or **norm** of partition P is the length of the longest of the subintervals $[x_{k-1}, x_k]$, or, equivalently, the maximum of Δx_k for k = 1, 2, 3, . . . , n.

For example, the set P = {2, 3, 4.6, 5.1, 6} is a partition of the interval [2,6] (Fig. 11) and divides the interval [2,6] into 4 subintervals with lengths $\Delta x_1 = 1$, $\Delta x_2 = 1.6$, $\Delta x_3 = .5$ and $\Delta x_4 = .9$. The mesh of this partition is 1.6, the maximum of the lengths of the subintervals. (If the mesh of a partition is "small," then the length of each one of the subintervals is the same or smaller.)

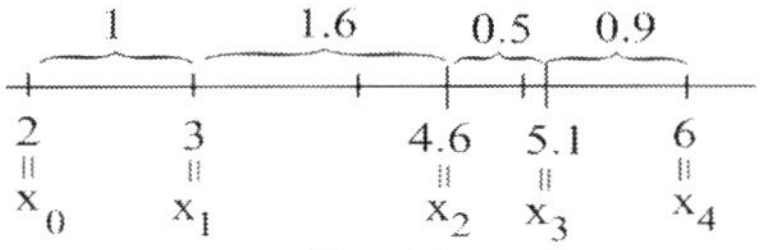

Fig. 11

Practice 6: P = {3, 3.8, 4.8, 5.3, 6.5, 7, 8} is a partition of what interval? How many subintervals does it create? What is the mesh of the partition? What are the values of x_2 and Δx_2 ?

A function, a partition, and a point in each subinterval determine a Riemann sum.

Suppose f is a positive function on the interval [a,b]

$P = \{ x_0 = a, x_1, x_2, x_3, \ldots, x_{n-1}, x_n = b\}$ is a partition of [a,b]

$\mathbf{c_k}$ is an x–value in the kth subinterval $[x_{k-1}, x_k]$: $x_{k-1} \le c_k \le x_k$.

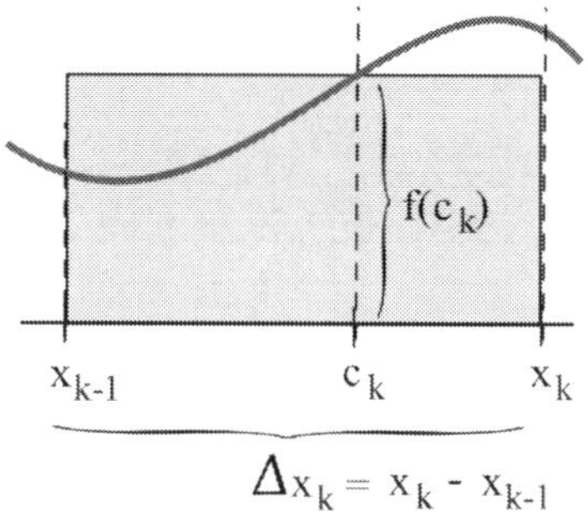

Fig. 12

Then the area of the kth rectangle is $f(c_k)\cdot(x_k - x_{k-1}) = f(c_k)\cdot\Delta x_k$. (Fig. 12)

Definition: A summation of the form $\displaystyle\sum_{k=1}^{n} \mathbf{f(c_k)\cdot\Delta x_k}$

is called a **Riemann Sum** of f for the partition P.

This Riemann sum is the total of the areas of the rectangular regions and is an approximation of the area between the graph of f and the x–axis.

Example 6: Find the Riemann sum for $f(x) = 1/x$ and the partition {1, 4, 5} using the values $c_1 = 2$ and $c_2 = 5$. (Fig. 13)

Solution: The 2 subintervals are [1,4] and [4,5] so $\Delta x_1 = 3$ and $\Delta x_2 = 1$.

Then the Riemann sum for this partition is

$$\sum_{k=1}^{n} f(c_k)\cdot\Delta x_k = \sum_{k=1}^{2} f(c_k)\cdot\Delta x_k = f(c_1)\cdot\Delta x_1 + f(c_2)\cdot\Delta x_2 = f(2)\cdot(3) + f(5)\cdot(1)$$

$$= \frac{1}{2}(3) + \frac{1}{5}(1) = 1.7 \ .$$

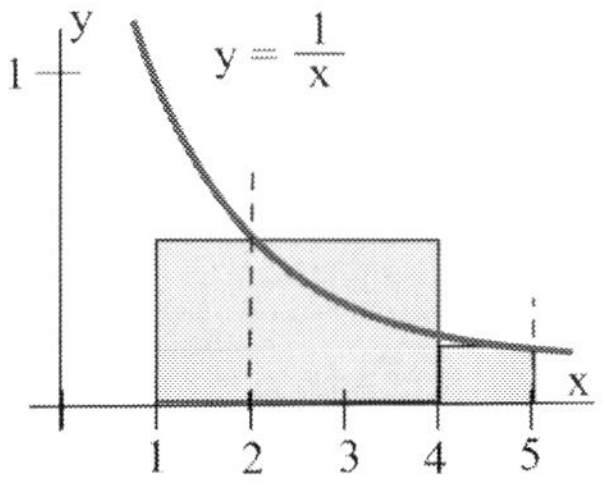

Fig. 13

Practice 7: Calculate the Riemann sum for $f(x) = 1/x$ on the partition $\{1, 4, 5\}$ using the values $c_1 = 3, c_2 = 4$.

Practice 8: What is the smallest value a Riemann sum for $f(x) = 1/x$ and the partition $\{1, 4, 5\}$ can have? (You will need to select values for c_1 and c_2.) What is the largest value a Riemann sum can have for this function and partition?

Table 2 shows the results of a computer program that calculated Riemann suns for the function $f(x) = 1/x$ with different numbers of subintervals and different ways of selecting the points c_i in each subinterval. When the mesh of the partition is small (and the number of subintervals large), all of the ways of selecting the c_i lead to approximately the same number for the Riemann sums. For this decreasing function, using the left endpoint of the subinterval always resulted in a sun that was larger than the area. Choosing the right end point gave a value smaller that the area. Why?

Table 2: Riemann sums for $f(x) = 1/x$ on the interval $[1,5]$

number of subintervals	mesh	Values of the Riemann sum for different choices of c_k: c_k = left edge $= x_{k-1}$	c_k = "random" point in $[x_{k-1}, x_k]$	c_k = right edge $= x_k$
4	1	2.083333	1.473523	1.283333
8	.5	1.828968	1.633204	1.428968
16	.25	1.714406	1.577806	1.514406
40	.1	1.650237	1.606364	1.570237
400	.01	1.613446	1.609221	1.605446
4000	.001	1.609838	1.609436	1.609038

As the mesh gets smaller, all of the Riemann Sums seem to be approaching the same value, approximately 1.609 . (ln 5 = 1.609437912)

Example 7: Find the Riemann sum for the function $f(x) = \sin(x)$ on the interval $[0, \pi]$ using the partition $\{0, \pi/4, \pi/2, \pi\}$ with $c_1 = \pi/4, c_2 = \pi/2, c_3 = 3\pi/4$.

Solution: The 3 subintervals (Fig. 14) are $[0, \pi/4]$, $[\pi/4, \pi/2]$, and $[\pi/2, \pi]$ so $\Delta x_1 = \pi/4$, $\Delta x_2 = \pi/4$ and $\Delta x_3 = \pi/2$. The Riemann sum for this partition is

$$\sum_{k=1}^{3} f(c_k)\cdot\Delta x_k = \sin(\pi/4)\cdot(\pi/4) + \sin(\pi/2)\cdot(\pi/4) + \sin(3\pi/4)\cdot(\pi/2)$$

$$= \frac{\sqrt{2}}{2}\cdot\frac{\pi}{4} + 1\cdot\frac{\pi}{4} + \frac{\sqrt{2}}{2}\cdot\frac{\pi}{2} \approx 2.45148 .$$

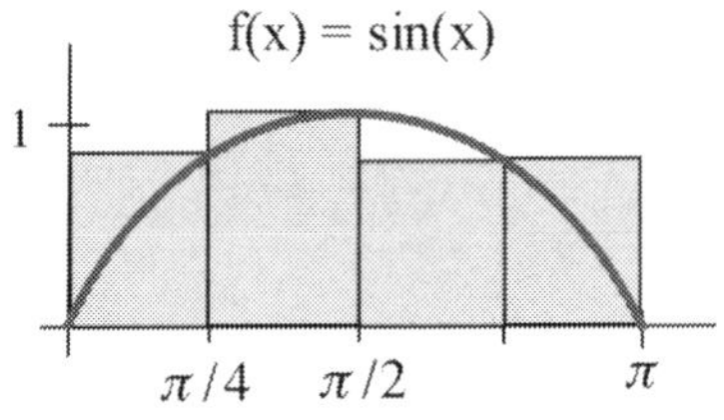

Fig. 14

Practice 9: Find the Riemann sum for the function and partition in the previous example, but use $c_1 = 0, c_2 = \pi/2, c_3 = \pi/2$.

Two Special Riemann Sums: Lower & Upper Sums

Two particular Riemann sums are of special interest because they represent the extreme possibilities for Riemann sums for a given partition.

> Definition: Suppose f is a positive function on [a,b], and P is a partition of [a,b].
> Let $\mathbf{m_k}$ be the x–value in the kth subinterval so that f($\mathbf{m_k}$) is the minimum value of f in that interval, and let $\mathbf{M_k}$ be the x–value in the kth subinterval so that f($\mathbf{M_k}$) is the maximum value of f in that interval.
>
> LS_P: $\sum_{k=1}^{n} f(\mathbf{m_k})\cdot\Delta x_k$ is the **lower sum** of f for the partition P.
>
> US_P: $\sum_{k=1}^{n} f(\mathbf{M_k})\cdot\Delta x_k$ is the **upper sum** of f for the partition P.

Geometrically, the lower sum comes from building rectangles under the graph of f (Fig. 15a), and the lower sum (every lower sum) is less than or equal to the exact area A: $LS_P \le A$ for every partition P. The upper sum comes from building rectangles over the graph of f (Fig. 15b), and the upper sum (every upper sum) is greater than or equal to the exact area A:

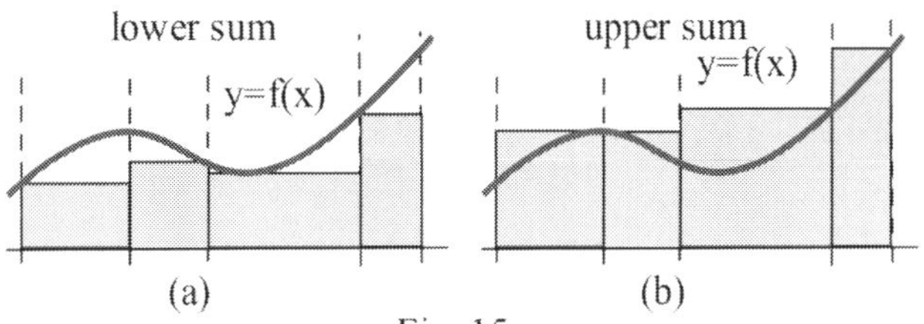

Fig. 15

$US_P \ge A$ for every partition P. The lower and upper sums provide bounds on the size of the exact area:

$$LS_P \le A \le US_P.$$

For any c_k value in the kth subinterval,

$f(\mathbf{m_k}) \le f(c_k) \le f(\mathbf{M_k})$; so, for any choice of the c_k values, the Riemann sum $RS_P = \sum_{k=1}^{n} f(c_k)\cdot\Delta x_k$ satisfies

$$\sum_{k=1}^{n} f(\mathbf{m_k})\cdot\Delta x_k \le \sum_{k=1}^{n} f(c_k)\cdot\Delta x_k \le \sum_{k=1}^{n} f(\mathbf{M_k})\cdot\Delta x_k \text{ or, equivalently, } LS_P \le RS_P \le US_P.$$

The lower and upper sums are bounds on the size of all Riemann sums.

The exact area A and every Riemann sum RS_P for partition P both lie between the lower sum and the upper sum for P (Fig. 16). Therefore, if the lower and upper sums are close together then the area and any Riemann sum for P must also be close together. If we know that the upper and lower sums for a partition P are within 0.001 units of each other, then we can be sure that every Riemann sum for partition P is within 0.001 units of the exact area.

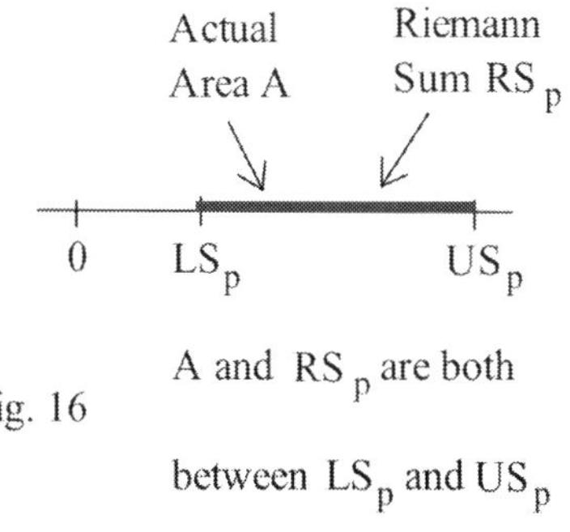

Fig. 16

Unfortunately, finding minimums and maximums can be a time–consuming business, and it is usually not practical to determine lower and upper sums for "wiggly" functions. If f is monotonic, however, then it is easy to find the values for m_k and M_k , and sometimes we can explicitly calculate the limits of the lower and upper sums.

For a monotonic bounded function we can guarantee that a Riemann sum is within a certain distance of the exact value of the area it is approximating.

Theorem: If f is a positive, **monotonically** increasing, bounded function on [a,b],
then for any partition P and any Riemann sum for P,

$$\left\{\begin{array}{l}\text{distance between the}\\ \text{Riemann sum and}\\ \text{the exact area}\end{array}\right\} \le \left\{\begin{array}{l}\text{distance between the}\\ \text{upper sum and}\\ \text{the lower sum}\end{array}\right\} \le \{f(b) - f(a)\}\cdot(\text{mesh of P}).$$

Proof: The Riemann sum and the exact area are both between the upper and lower sums so the distance between the Riemann sum and the exact area is less than or equal to the distance between the upper and lower sums. Since f is monotonically increasing, the areas representing the difference of the upper and lower sums can be slid into a rectangle (Fig. 17) whose height equals f(b)–f(a) and whose base equals the mesh of P. Then the total difference of the upper and lower sums is less than or equal to the area of the rectangle, {f(b)–f(a)}·(mesh of P) .

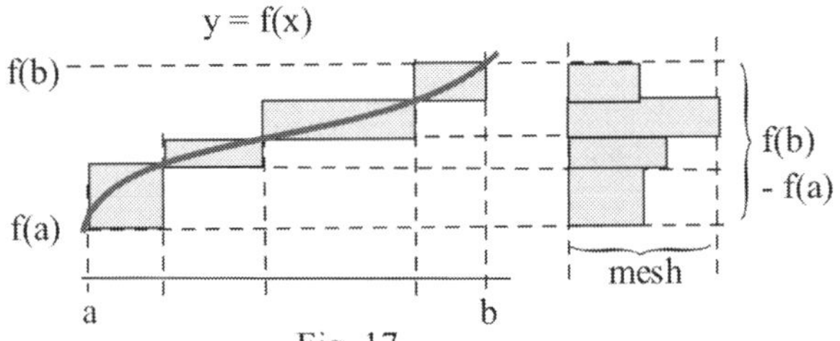

Fig. 17

PROBLEMS

In problems 1 – 6 , rewrite the sigma notation as a summation and perform the indicated addition.

1. $\sum_{k=2}^{4} k^2$ 2. $\sum_{j=1}^{5} (1+j)$ 3. $\sum_{n=1}^{3} (1+n)^2$

4. $\sum_{k=0}^{5} \sin(\pi k)$ 5. $\sum_{j=0}^{5} \cos(\pi j)$ 6. $\sum_{k=1}^{3} 1/k$

In problems 7 – 12, rewrite each summation using the sigma notation. Do not evaluate the sums.

7. $3 + 4 + 5 + \ldots + 93 + 94$ 8. $4 + 6 + 8 + \ldots + 24$ 9. $9 + 16 + 25 + 36 + \ldots + 144$

10. $\frac{3}{4} + \frac{3}{9} + \frac{3}{16} + \ldots + \frac{3}{100}$ 11. $1\cdot2^1 + 2\cdot2^2 + 3\cdot2^3 + \ldots + 7\cdot2^7$ 12. $3 + 6 + 9 + \ldots + 30$

In problems 13 – 15 , use the values of a_k and b_k in Table 3 and verify that the value in part (a) does equal the value in part (b).

Table 3

k	a_k	b_k
1	1	2
2	2	2
3	3	2

13. (a) $\sum_{k=1}^{3} (a_k + b_k)$ (b) $\sum_{k=1}^{3} a_k + \sum_{k=1}^{3} b_k$

14. (a) $\sum_{k=1}^{3} (a_k - b_k)$ (b) $\sum_{k=1}^{3} a_k - \sum_{k=1}^{3} b_k$

15. (a) $\sum_{k=1}^{3} 5a_k$ (b) $5 \cdot \sum_{k=1}^{3} a_k$

For problems 16 – 18 , use the values of a_k and b_k in Table 3 and verify that the value in part (a) does **not** equal the value in part (b).

16. (a) $\sum_{k=1}^{3} a_k \cdot b_k$ (b) $\sum_{k=1}^{3} a_k \cdot \sum_{k=1}^{3} b_k$

17. (a) $\sum_{k=1}^{3} (a_k^{\,2})$ (b) $\left(\sum_{k=1}^{3} a_k \right)^2$

18. (a) $\sum_{k=1}^{3} a_k / b_k$ (b) $\left(\sum_{k=1}^{3} a_k \right) / \left(\sum_{k=1}^{3} b_k \right)$

For problems 19 – 30 , $f(x) = x^2$, $g(x) = 3x$, and $h(x) = 2/x$. Evaluate each sum.

19. $\sum_{k=0}^{3} f(k)$

20. $\sum_{k=0}^{3} f(2k)$

21. $\sum_{j=0}^{3} 2f(j)$

22. $\sum_{i=0}^{3} f(1+i)$

23. $\sum_{m=1}^{3} g(m)$

24. $\sum_{k=1}^{3} g(\, f(k)\,)$

25. $\sum_{j=1}^{3} g^2(j)$

26. $\sum_{k=1}^{3} k \cdot g(k)$

27. $\sum_{k=2}^{4} h(k)$

28. $\sum_{i=1}^{4} h(3i)$

29. $\sum_{n=1}^{3} f(n) \cdot h(n)$

30. $\sum_{k=1}^{7} g(k) \cdot h(k)$

For problems 31 – 36 , write out each summation and simplify the result. These are examples of "telescoping sums".

31. $\sum_{k=1}^{7} (\, (k)^2 - (k-1)^2 \,)$

32. $\sum_{k=1}^{6} (\, (k)^3 - (k-1)^3 \,)$

33. $\sum_{k=1}^{5} \left(\frac{1}{k} - \frac{1}{k+1} \right)$

34. $\sum_{k=0}^{4} (\, (k+1)^3 - (k)^3 \,)$

35. $\sum_{k=0}^{8} (\, \sqrt{k+1} - \sqrt{k} \,)$

36. $\sum_{k=1}^{5} (x_k - x_{k-1})$

For problems 37– 43 , (i) list the subintervals determined by the partition P, (ii) find the values of Δx_i , (iii) find the mesh of P, and (iv) calculate $\sum_{i=1}^{n} \Delta x_i$.

37. $P = \{ 2, 3, 4.5, 6, 7 \}$

38. $P = \{ 3, 3.6, 4, 4.2, 5, 5.5, 6 \}$

39. $P = \{-3, -1, 0, 1.5, 2 \}$

40. P is given in Fig. 18.

41. P is given in Fig. 19.

42. P is given in Fig. 20.

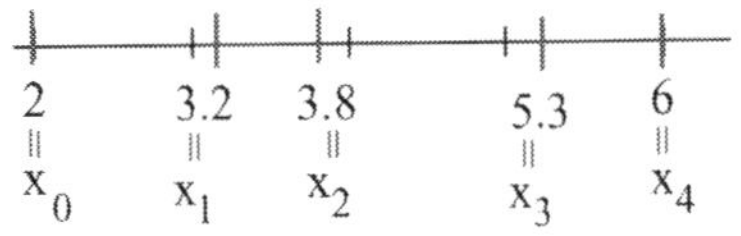

Fig. 18

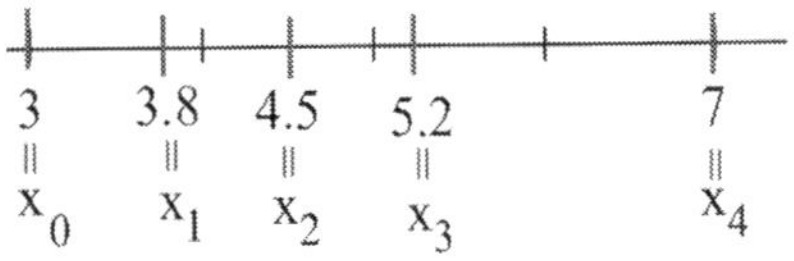

Fig. 19

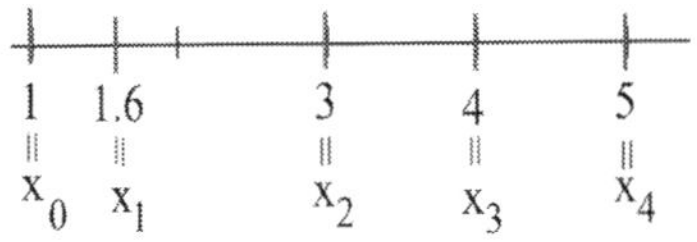

Fig. 20

43. For $\Delta x_i = x_i - x_{i-1}$, verify that $\sum_{i=1}^{n} \Delta x_i$ = length of the interval $[a, b]$.

For problems 44 – 48 , (i) sketch the graph of f on the given interval, (ii) draw vertical lines at each point of the given partition, (iii) evaluate each $f(c_i)$ and sketch the corresponding rectangle, and (iv) calculate and add together the areas of the rectangles.

44. $f(x) = x + 1$, $P = \{ 1, 2, 3, 4 \}$ (a) $c_1 = 1, c_2 = 3$, and $c_3 = 3$. (b) $c_1 = 2, c_2 = 2$, and $c_3 = 4$.

45. $f(x) = 4 - x^2$, $P = \{ 0, 1, 1.5, 2 \}$ (a) $c_1 = 0, c_2 = 1$, and $c_3 = 2$. (b) $c_1 = 1, c_2 = 1.5$, and $c_3 = 1.5$.

46. $f(x) = \sqrt{x}$, $P = \{ 0, 2, 5, 10 \}$ (a) $c_1 = 1, c_2 = 4$, and $c_3 = 9$. (b) $c_1 = 0, c_2 = 3$, and $c_3 = 7$.

47. $f(x) = \sin(x)$, $P = \{0, \pi/4, \pi/2, \pi \}$ (a) $c_1 = 0, c_2 = \pi/4$, and $c_3 = \pi/2$. (b) $c_1 = \pi/4, c_2 = \pi/2$, and $c_3 = \pi$.

48. $f(x) = 2^x$, $P = \{ 0, 1, 3 \}$ (a) $c_1 = 0, c_2 = 2$. (b) $c_1 = 1, c_2 = 3$.

For problems 49 – 52 , sketch the function and find the smallest possible value and the largest possible value for a Riemann sum of the given function and partition.

49. $f(x) = 1 + x^2$ (a) $P = \{ 1, 2, 4, 5 \}$ (b) $P = \{ 1, 2, 3, 4, 5 \}$ (c) $P = \{ 1, 1.5, 2, 3, 4, 5 \}$

50. $f(x) = 7 - 2x$ (a) $P = \{ 0, 2, 3 \}$ (b) $P = \{ 0, 1, 2, 3 \}$ (c) $P = \{ 0, .5, 1, 1.5, 2, 3 \}$

51. $f(x) = \sin(x)$ (a) $P = \{ 0, \pi/2, \pi \}$ (b) $P = \{ 0, \pi/4, \pi/2, \pi \}$ (c) $P = \{ 0, \pi/4, 3\pi/4, \pi \}$

52. $f(x) = x^2 - 2x + 3$ (a) $P = \{ 0, 2, 3\}$ (b) $P = \{ 0, 1, 2, 3 \}$ (c) $P = \{ 0, .5, 1, 2, 2.5, 3 \}$

Upper and Lower Sum Problems

53. Suppose $LS_P = 7.362$ and $US_P = 7.402$ for a positive function f and a partition P of the interval $[1, 5]$. We can be certain that every Riemann sum for the partition P is within what distance of the exact value of the area between the graph of f and the interval $[1,5]$? (b) What if $LS_P = 7.372$ and $US_P = 7.390$?

54. Suppose we divide the interval [1,4] into 100 equally wide subintervals and calculate a Riemann sum for $f(x) = 1 + x^2$ by randomly selecting a point c_i in each subinterval. We can be certain that the value of the Riemann sum is within what distance of the exact value of the area between the graph of f and the interval [1,4]? (b) What if we take 200 equally long subintervals?

55. If we divide the interval [2,4] into 50 equally wide subintervals and calculate a Riemann sum for $f(x) = 1 + x^3$ by "randomly" selecting a point c_i in each subinterval (any point c_i in the subinterval), then we can be certain that the Riemann sum is within what distance of the exact value of the area between f and the interval [2,4]?

56. If f is monotonic decreasing on [a, b] and we divide the interval [a, b] into n equally wide subintervals (Fig. 21), then we can be certain that the Riemann sum is within what distance of the exact value of the area between f and the interval [a, b].

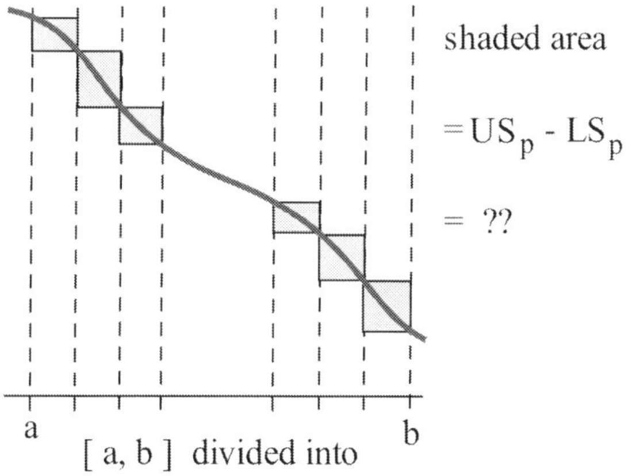

[a, b] divided into n equal subintervals
Fig. 21

Summing Powers of Consecutive Integers

Explicit formulas for some commonly encountered summations are known and are useful for explicitly evaluating some Riemann sums and their limits. The formulas below are included here for your reference. They will not be used or needed in the following sections.

The summation formula for the first n positive integers is relatively well–known, has several easy but clever proofs, and even has an interesting story.

$$1 + 2 + 3 + \ldots + (n-1) + n = \sum_{k=1}^{n} k = \frac{n(n+1)}{2}$$

Proof: Let S represent the sum $1 + 2 + 3 + \ldots + (n-2) + (n-1) + n$. Rearranging the summands, we also know $S = n + (n-1) + (n-2) + \ldots + 3 + 2 + 1$. Adding these 2 representations of S together,

$$\begin{array}{llllllll} S & = 1 & +2 & +3 & +\ldots & +(n-2) & +(n-1) & +n \\ +\ S & = n & +(n-1) & +(n-2) & +\ldots & +3 & +2 & +1 \\ \hline 2S = & (n+1) & +(n+1) & +(n+1) & +\ldots & +(n+1) & +(n+1) & +(n+1) = n\cdot(n+1) \end{array}$$

so $S = \dfrac{n(n+1)}{2}$, the desired formula.

It is said that Gauss discovered this formula for himself at the age of 5. His teacher, planning on keeping the class busy for a while, asked the students to add the integers from 1 to 100. Gauss thought a few minutes, wrote his answer on his slate, and turned it in. According to the story, he then sat smugly while his classmates struggled with the problem.

57. Find the sum of the first 100 positive integers in 2 ways: (1) using Gauss' formula, and (2) using Gauss' method.

58. Find the sum of the first 10 odd integers.
(Each odd integer can be written in the form $2k - 1$ for $k = 1, 2, 3, \ldots$)

59. Find the sum of the integers from 10 to 20.

Formulas for other integer powers of the first n integers have been discovered:

$$\sum_{k=1}^{n} k = \frac{1}{2} n^2 + \frac{1}{2} n = \frac{n(n+1)}{2} \qquad \sum_{k=1}^{n} k^2 = \frac{1}{3} n^3 + \frac{1}{2} n^2 + \frac{2}{12} n = \frac{n(n+1)(2n+1)}{6}$$

$$\sum_{k=1}^{n} k^3 = \frac{1}{4} n^4 + \frac{1}{2} n^3 + \frac{3}{12} n^2 + 0 \cdot n = \left(\frac{n(n+1)}{2} \right)^2$$

$$\sum_{k=1}^{n} k^4 = \frac{1}{5} n^5 + \frac{1}{2} n^4 + \frac{4}{12} n^3 + 0 \cdot n^2 - \frac{1}{30} n = \frac{n(n+1)(2n+1)(3n^2+3n-1)}{30}$$

In problems 60 – 62 , use the properties of summation and the formulas for powers to evaluate each sum.

60. $\sum_{k=1}^{10} (3 + 2k + k^2)$ 61. $\sum_{k=1}^{10} k \cdot (k^2 + 1)$ 62. $\sum_{k=1}^{10} k^2 \cdot (k - 3)$

Section 4.1 **PRACTICE Answers**

Practice 1: (a) $\sum_{k=1}^{5} k^3 = 1 + 8 + 27 + 64 + 125$. (b) $\sum_{j=2}^{7} (-1)^j \frac{1}{j} = \frac{1}{2} - \frac{1}{3} + \frac{1}{4} - \frac{1}{5} + \frac{1}{6} - \frac{1}{7}$

(c) $\sum_{m=0}^{4} (2m+1) = 1 + 3 + 5 + 7 + 9$.

Practice 2: (a) $\sum_{k=2}^{5} g(k) = g(2) + g(3) + g(4) + g(5) = 1 + (-2) + 3 + 5 = 7$

(b) $\sum_{j=1}^{4} h(j) = h(1) + h(2) + h(3) + h(4) = 3 + 3 + 3 + 3 = 12$

(c) $\sum_{k=3}^{5} (g(k) + f(k-1)) = (g(3) + f(2)) + (g(4) + f(3)) + (g(5) + f(4)) = (-2+3)+(3+1)+(5+0) = 10$

Practice 3: For $g(x) = 1/x$, $\sum_{k=2}^{4} g(k) = g(2) + g(3) + g(4) = \frac{1}{2} + \frac{1}{3} + \frac{1}{4} = \frac{13}{12}$ and

$\sum_{k=1}^{3} g(k+1) = g(2) + g(3) + g(4) = \frac{13}{12}$.

Practice 4: Rectangular areas $= 1 + \frac{1}{2} + \frac{1}{3} + \frac{1}{4} = \frac{25}{12} = \sum_{j=1}^{4} \frac{1}{j}$.

Practice 5: $f(x_0)(x_1 - x_0) + f(x_1)(x_2 - x_1) + f(x_2)(x_3 - x_2)$

$= \sum_{j=1}^{3} f(x_{j-1}) (x_j - x_{j-1})$ or $\sum_{k=0}^{2} f(x_k) (x_{k+1} - x_k)$.

Practice 6: Interval is [3, 8]. 6 subintervals. mesh = length of longest subinterval = 1.2 .
$x_2 = 4.8$ and $\Delta x_2 = x_2 - x_1 = 4.8 - 3.8 = 1$.

Practice 7: $RS = (3)(\frac{1}{3}) + (1)(\frac{1}{4}) = 1.25$

Practice 8: smallest $RS = (3)(\frac{1}{4}) + (1)(\frac{1}{5}) = 0.95$ largest $RS = (3)(1) + (1)(\frac{1}{4}) = 3.25$

Practice 9: $RS = (0)(\frac{\pi}{4}) + (1)(\frac{\pi}{4}) + (1)(\frac{\pi}{2}) \approx 2.356$.

4.2 THE DEFINITE INTEGRAL

Definition of The Definite Integral

Each particular Riemann sum depends on several things: the function f, the interval [a,b], the partition P of the interval, and the values chosen for c_k in each subinterval. Fortunately, for most of the functions needed for applications, as the approximating rectangles get thinner (as the mesh of P approaches 0 and the number of subintervals gets bigger) the values of the Riemann sums approach the same value independently of the particular partition P and the points c_k. For these functions, the LIMIT (as the mesh approaches 0) of the Riemann sums is the same number no matter how the c_k are chosen.

This limit of the Riemann sums is the next big topic in calculus, the definite integral. Integrals arise throughout the rest of this book and in applications in almost every field that uses mathematics.

> Definition: **The Definite Integral**
>
> If $\lim_{mesh\to 0}\left(\sum_{k=1}^{n} f(c_k)\cdot\Delta x_k\right)$ equals a finite number I
>
> then f is **integrable** on the interval [a, b].
>
> The number I is called the **Definite Integral of f on [a,b]** and is written $\int_a^b f(x)\,dx$.

The symbol $\int_a^b f(x)\,dx$ is read as "the integral from a to b of eff of x dee x" or "the integral from a to b of f(x) with respect to x." The name of each piece of the symbol is shown in Fig. 1.

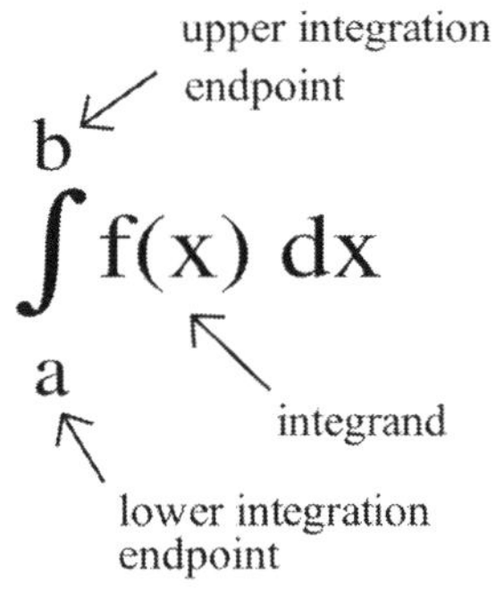

Fig. 1

Example 1: Describe the area between the graph of f(x) = $1/x$, the x–axis, and the vertical lines at $x = 1$ and $x = 5$ as a limit of Riemann sums and as a definite integral.

Solution: Area = $\lim_{mesh\to 0}\left(\sum_{k=1}^{n} \frac{1}{c_k}\Delta x_k\right) = \int_1^5 \frac{1}{x}\,dx \approx 1.609$

(from Table 2 in Section 4.1).

Practice 1: Describe the area between the graph of f(x)= sin(x), the x–axis, and the vertical lines at $x = 0$ and $x = \pi$ as a limit of Riemann sums and as a definite integral.

Example 2: Using the idea of area, determine the values of

(a) $\lim_{mesh \to 0} \left(\sum_{k=1}^{n} (1 + c_k)\Delta x_k \right)$ on the interval [1,3] (b) $\int_0^4 (5 - x)\, dx$ (c) $\int_{-1}^{1} \sqrt{1 - x^2}\, dx$

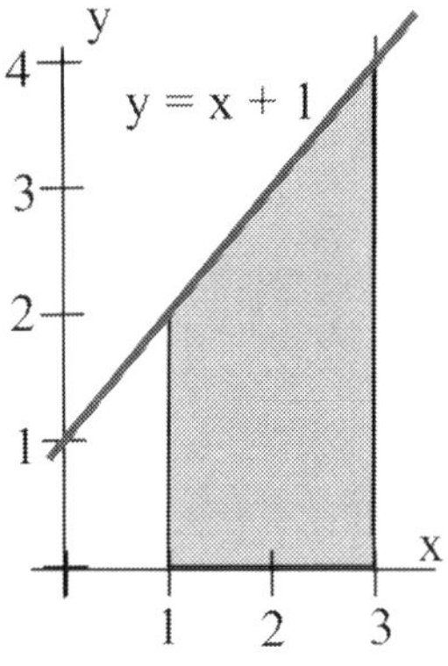

Fig. 2

Solution: (a) represents the area between the graph of f(x) = 1+x, the x–axis, and the vertical lines at 1 and 3 (Fig. 2), and this area equals 6 square units.

(b) represents the area between f(x) = 5 – x , the x–axis and the vertical lines at 0 and 4, so the integral equals 12 square units.

(c) represents the area of 1/2 of the circle $x^2 + y^2 = 1$ with radius 1 and center at (0,0), and the integral equals (circle area)/2 = $(\pi r^2)/2 = \pi/2$.

Practice 2: Using the area idea, determine the values of

(a) $\lim_{mesh \to 0} \left(\sum_{k=1}^{n} (2c_k)\Delta x_k \right)$ on the interval [1,3] and (b) $\int_3^8 4\, dx$.

Example 3: Represent the limit of each Riemann sum as a definite integral.

(a) $\lim_{mesh \to 0} \left(\sum_{k=1}^{n} (3 + c_k)\Delta x_k \right)$ on [1,4] (b) $\lim_{mesh \to 0} \left(\sum_{k=1}^{n} \sqrt{c_k} \cdot \Delta x_k \right)$ on [0,9]

Solution: (a) $\int_1^4 (3 + x)\, dx$ (b) $\int_0^9 \sqrt{x}\, dx$

Example 4: Represent each shaded area in Fig. 3 as a definite integral. (Do not evaluate the definite integral, just translate the picture into symbols.)

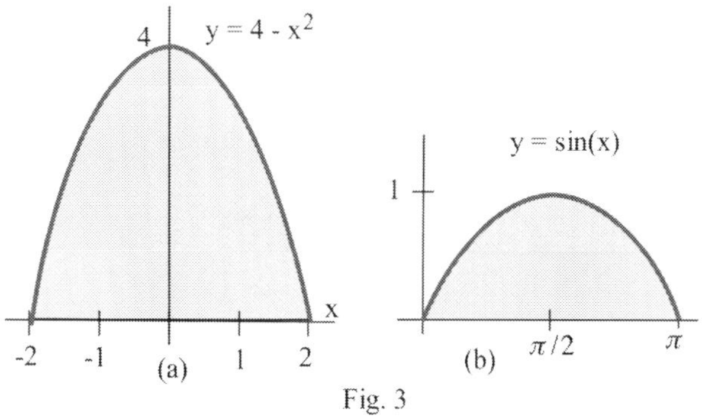

Fig. 3

Solution: (a) $\int_{-2}^{2} (4 - x^2)\, dx$ (b) $\int_{\pi/2}^{\pi} \sin(x)\, dx$

The value of a definite integral $\int_a^b f(x)\, dx$ depends only on the function f being integrated and on the interval [a, b]. The variable $\boldsymbol{x}$ in $\int_a^b f(\boldsymbol{x})\, d\boldsymbol{x}$ is a "dummy variable" and replacing it with another variable does not change the value of the integral. The following integrals each represent the integral of the function f on the interval [a,b], and they are all equal:

$$\int_a^b f(\boldsymbol{x})\, d\boldsymbol{x} = \int_a^b f(\boldsymbol{t})\, d\boldsymbol{t} = \int_a^b f(\boldsymbol{w})\, d\boldsymbol{w} = \int_a^b f(\boldsymbol{z})\, d\boldsymbol{z} .$$

Definite Integrals of Negative Functions

A definite integral is a limit of Riemann sums, and Riemann sums can be made from any integrand function f, positive or negative, continuous or discontinuous. The definite integral still has a geometric meaning even if the function is sometimes (or always) negative, and definite integrals of negative functions also have interpretations in applications.

Example 5: Find the definite integral of $f(x) = -2$ on the interval $[1,4]$.

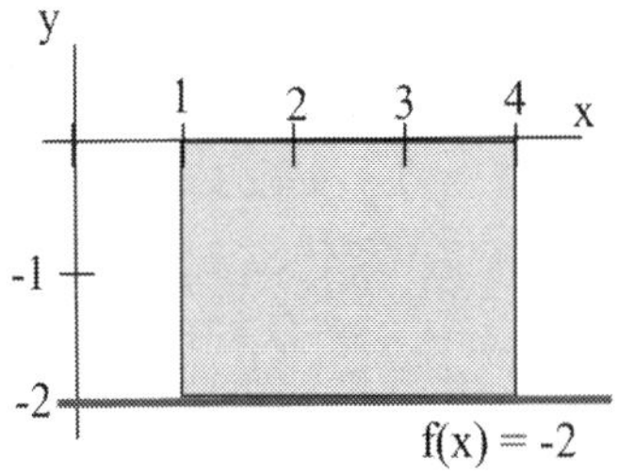

Fig. 4

Solution: $\sum_{k=1}^{n} f(c_k)\Delta x_k = \sum_{k=1}^{n} (-2)\cdot\Delta x_k = -2\cdot\sum_{k=1}^{n}\Delta x_k = -2\cdot(3) = -6$

for every partition P and every choice of values for c_k so

$$\int_1^4 -2\,dx = \lim_{mesh\to 0}\left(\sum_{k=1}^{n} f(c_k)\cdot\Delta x_k\right) = \lim_{mesh\to 0}(-6) = -6 .$$

The **area** of the region in Fig. 4 is 6 units, but because the region is **below** the x–axis, the value of the **integral** is -6 .

If	the graph of f(x) is **below** the x–axis for $a \le x \le b$ (f is **negative**),
then	$\int_a^b f(x)\,dx = -$ { area below the x–axis for $a \le x \le b$ } , a negative number.

If $f(t)$ is the rate of population change (people/year) for a town, then negative values of f would indicate that the population of the town was getting smaller, and the definite integral (now a negative number) would be the **change** in the population, a decrease, during the time interval.

Example 6: In 1980 there were 12,000 ducks nesting around a lake, and the **rate** of population change is shown in Fig. 5. Write a definite integral to represent the total change in the duck population from 1980 to 1990, and estimate the population in 1990.

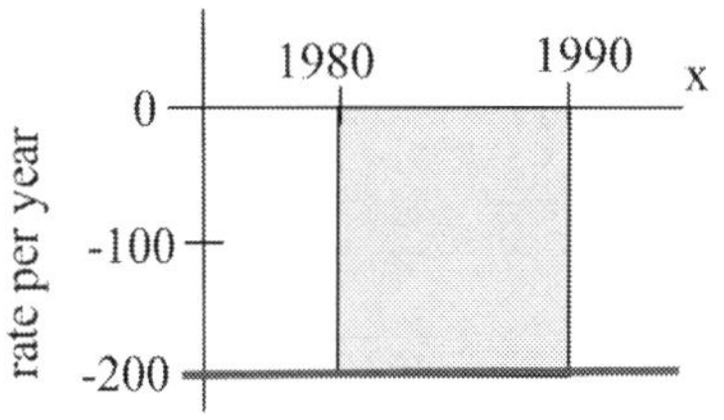

Fig. 5

Solution: Total **change** in population

$$= \int_{1980}^{1990} f(t)\,dt = -\{\text{area between f and axis}\}$$

$$\approx -\{200 \text{ ducks/year}\}\cdot\{10 \text{ years}\} = -2000 \text{ ducks.}$$

Then {1990 duck population} = {1980 population} + {change from 1980 to 1990}

= {12,000} + { –2000} = 10,000 ducks.

If $f(t)$ is the velocity of a car in the positive direction along a straight line at time t (miles/hour), then negative values of f indicate that the car is traveling in the negative direction. The definite integral of f (the integral is a negative number) is the change in position of the car during the time interval, how far the car traveled in the negative direction.

Practice 3: A bug starts at the location x = 12 on the x–axis at 1 pm walks along the axis with the velocity shown in Fig. 6. How far does the bug travel between 1 pm and 3 pm, and where is the bug at 3 pm?

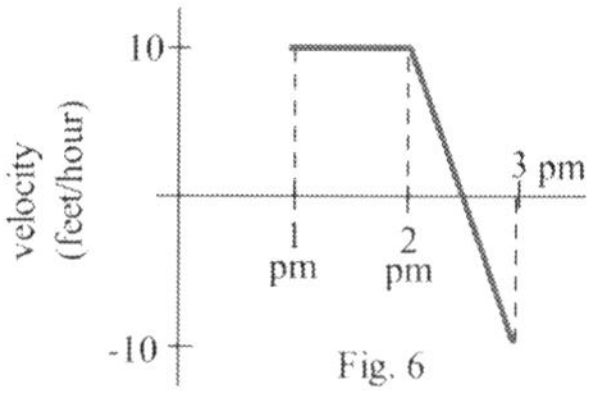

Fig. 6

Frequently our integrand functions will be positive some of the time and negative some of the time. If f represents the rate of population increase, then the integral of the positive parts of f will be the increase in population and the integral of the negative parts of f will be the decrease in population. Altogether, the integral of f over the whole time interval will be the **total (net) change** in the population.

$$\int_a^b f(x)\,dx = \{\text{ area above axis }\} - \{\text{ area below axis }\}$$

Example 7: Use Fig. 7 to calculate $\int_0^2 f(x)\,dx$, $\int_2^4 f(x)\,dx$, $\int_4^5 f(x)\,dx$, $\int_0^5 f(x)\,dx$.

Solution: $\int_0^2 f(x)\,dx = 2$, $\int_2^4 f(x)\,dx = \mathbf{-5}$, $\int_4^5 f(x)\,dx = 2$, and

$$\int_0^5 f(x)\,dx = \{\text{area above}\} - \{\text{area below}\} = \{2+2\} - \{5\} = -1.$$

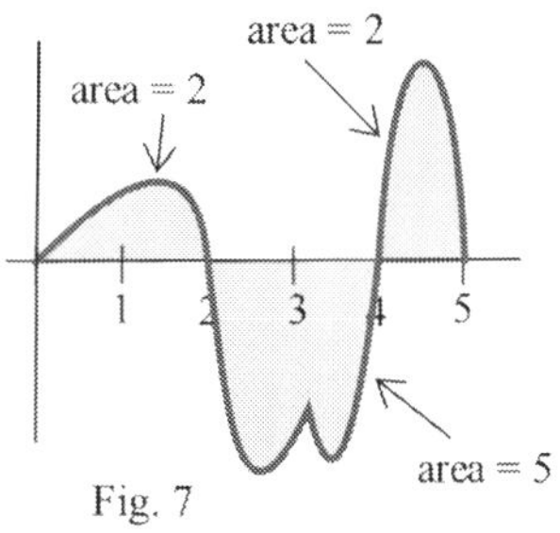

Fig. 7

Practice 4: Use geometric reasoning to evaluate $\int_0^{2\pi} \sin(x)\,dx$.

If f is a velocity, then the integrals on the intervals where f is positive measure the distances moved forward; the integrals on the intervals where f is negative measure the distances moved backward; and the integral over the whole time interval is the **total (net) change** in position, the distance moved forward minus the distance moved backward.

Practice 5: A car is driven with the velocity west shown in Fig. 8.

(a) Between noon and 6 pm how far does the car travel?

(b) At 6 pm, where is the car relative to its starting point (its position at noon)?

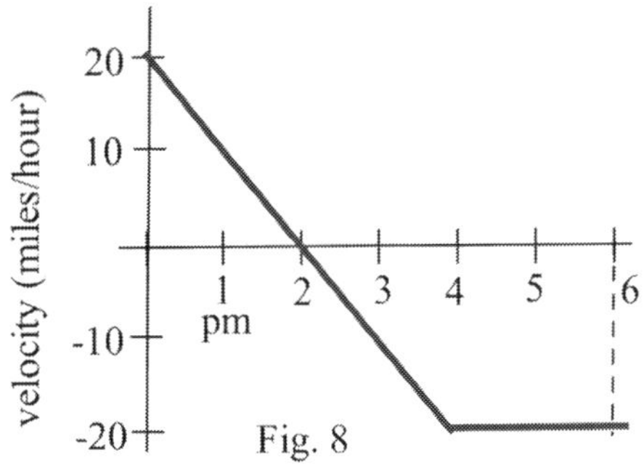

Fig. 8

Units For the Definite Integral

We have already seen that the "area" under a graph can represent quantities whose units are not the usual geometric units of square meters or square feet. For example, if x is a measure of time in "seconds" and $f(x)$ is a velocity with units "feet/second", then Δx has the units "seconds" and $f(x)\cdot\Delta x$ has the units ("feet/second")("seconds") = "feet," a measure of distance. Since each Riemann sum $\sum f(x)\cdot\Delta x$ is a sum of "feet" and the definite integral is the limit of the Riemann sums, the definite integral, has the same units, "feet".

If the units of $f(x)$ are "square feet" and the units of x are "feet", then $\int_a^b f(x)\,dx$ is a number with the units ("square feet")·("feet") = "cubic feet," a measure of volume. If $f(x)$ is a force in grams, and x is a distance in centimeters, then $\int_a^b f(x)\,dx$ is a number with the units "gram·centimeters," a measure of work.

In general, the units for the definite integral $\int_a^b f(x)\,dx$ are (units for $f(x)$)·(units for x). A quick check of the units can help avoid errors in setting up an applied problem.

PROBLEMS

In problems 1 – 4 , rewrite the limit of each Riemann sum as a definite integral.

1. $\lim_{mesh \to 0} \left(\sum_{k=1}^{n} (2 + 3c_k)\Delta x_k \right)$ on the interval [0, 4]

2. $\lim_{mesh \to 0} \left(\sum_{k=1}^{n} \cos(5c_k)\Delta x_k \right)$ on [0,11]

3. $\lim_{mesh \to 0} \left(\sum_{k=1}^{n} (c_k)^3\Delta x_k \right)$ on [2, 5]

4. $\lim_{mesh \to 0} \left(\sum_{k=1}^{n} \sqrt{c_k}\,\Delta x_k \right)$ on [1, 4]

In problems 5 – 10, represent the area of each bounded region as a definite integral. (Do **not** evaluate the integral, just translate the area into an integral.)

5. The region bounded by $y = x^3$, the x–axis, the line $x = 1$, and $x = 5$.

6. The region bounded by $y = \sqrt{x}$, the x–axis, and the line $x = 9$.

7. The region bounded by $y = x \cdot \sin(x)$, the x–axis, the line $x = 1/2$, and $x = 2$.

8. The shaded region in Fig. 9.

9. The shaded region in Fig. 10.

10. The shaded region in Fig. 10 for $2 \le x \le 3$.

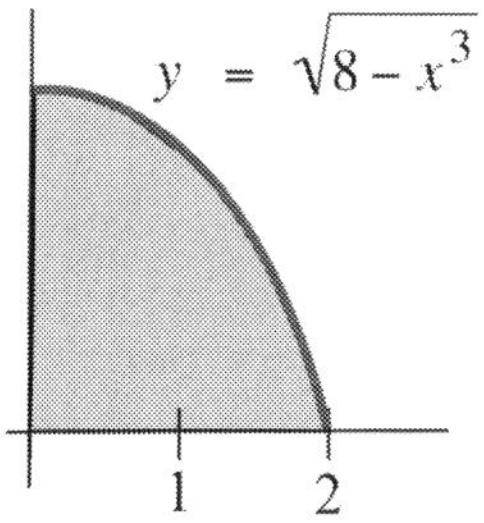

Fig. 9

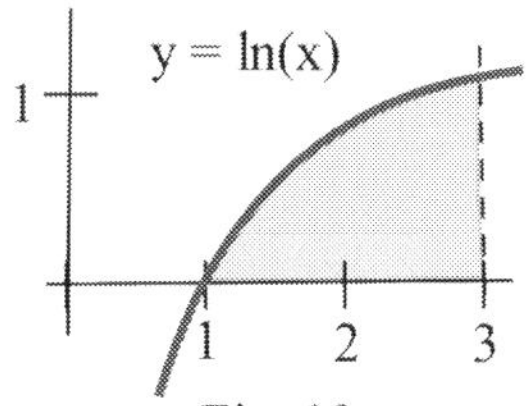

Fig. 10

In problems 11 – 15, represent the area of each bounded region as a definite integral, and use geometry to determine the value of the definite integral.

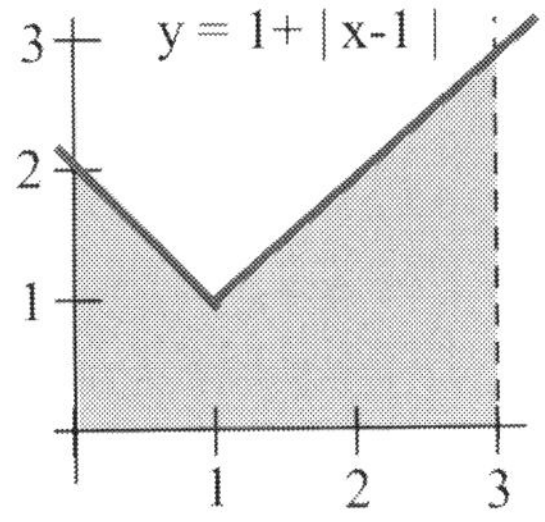

Fig. 11

11. The region bounded by $y = 2x$, the x–axis, the line $x = 1$, and $x = 3$.

12. The region bounded by $y = 4 - 2x$, the x–axis, and the y–axis.

13. The region bounded by $y = | x |$, the x–axis, and the line $x = -1$.

14. The shaded region in Fig. 11.

15. The shaded region in Fig. 12.

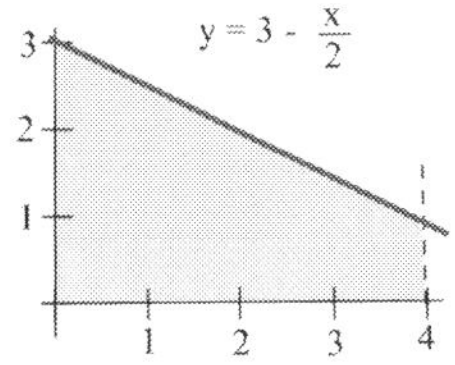

Fig. 12

16. Fig. 13 shows the graph of f and the areas of several regions. Evaluate:

(a) $\int_0^3 f(x)\,dx$ (b) $\int_3^5 f(x)\,dx$ (c) $\int_5^7 f(x)\,dx$

(d) $\int_0^5 | f(x) |\,dx$ (e) $\int_3^7 f(x)\,dx$

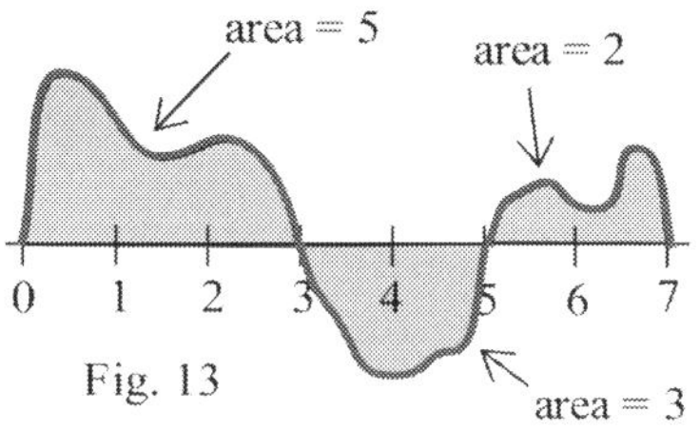

Fig. 13

17. Fig. 14 shows the graph of g and the areas of several regions.

Evaluate : (a) $\int_1^3 g(x)\,dx$ (b) $\int_3^4 g(x)\,dx$

(c) $\int_4^8 g(x)\,dx$ (d) $\int_1^8 g(x)\,dx$ (e) $\int_3^8 | g(x) |\,dx$

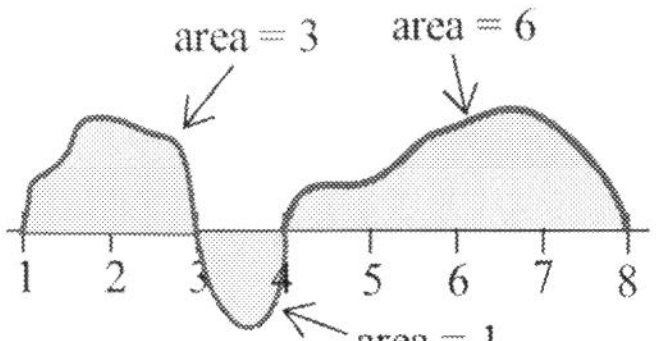

Fig. 14

18. Fig. 15 shows the graph of h and the areas of several regions. Evaluate:

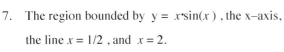

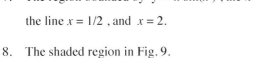

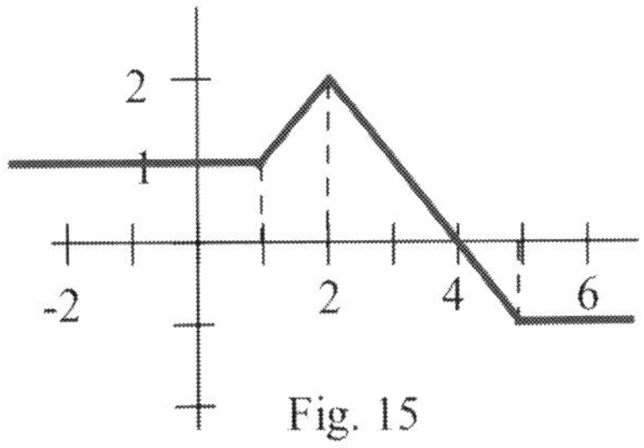

Fig. 15

(a) $\int_{-2}^1 h(x)\,dx$ (b) $\int_4^6 h(x)\,dx$

(c) $\int_{-2}^6 h(x)\,dx$ (d) $\int_{-2}^4 h(x)\,dx$ (e) $\int_{-2}^4 | h(x) |\,dx$

In problems 19 – 20, your velocity (in feet per minute) along a straight path is shown. (a) Sketch the graph of your location.

(b) How many feet did you walk in 8 minutes?

(c) Where, relative to your starting location, are you after 8 minutes?

19. Your velocity is shown in Fig. 16.

20. Your velocity is shown in Fig. 17.

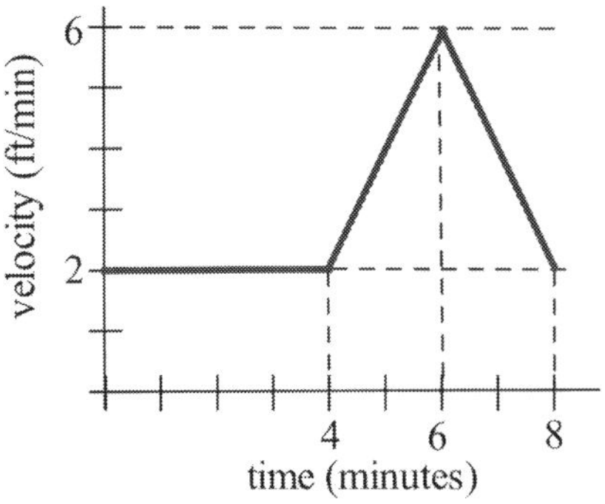

Fig. 16

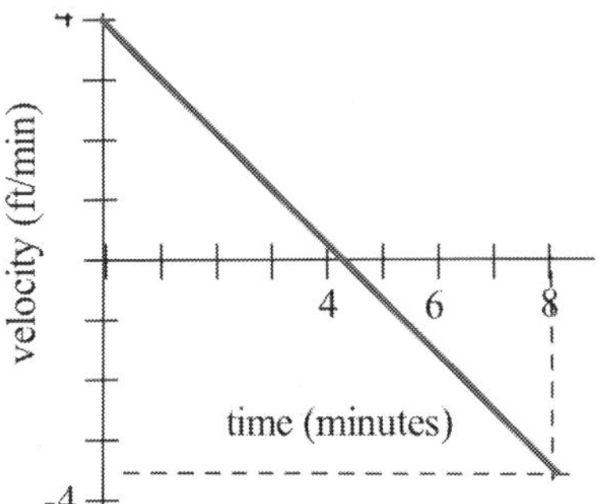

Fig. 17

In problems 21 – 27, the units are given for x and a $f(x)$. Give the units of $\int_a^b f(x)\,dx$.

21. x is time in "seconds", and $f(x)$ is velocity in "meters per second."

22. x is time in "hours", and $f(x)$ is a flow rate in "gallons per hour."

23. x is a position in "feet", and $f(x)$ is an area in "square feet."

24. x is a time in "days", and $f(x)$ is a temperature in "degrees Celsius."

25. x is a height in "meters", and $f(x)$ is a force in "grams."

26. x is a position in "inches", and $f(x)$ is a density in "pounds per inch."

27. x is a time in "seconds", and $f(x)$ is an acceleration in "feet per second per second (ft/s^2)."

Example 8: For $f(x) = x^2$, divide the interval [0,2] into n equally–wide subintervals and evaluate the lower sum and the limit of the lower sum as $n \to \infty$.

Solution: $\Delta x = \frac{b-a}{n} = \frac{2}{n}$, $x_i = \frac{2}{n}\, i$ (for i = 0, 1, 2, . . . , n) and $f(x_i) = (x_i)^2 = \frac{4}{n^2}\, i^2$. Since f is increasing on [0, 2], the minimum value of f on each subinterval occurs at the left edge of that subinterval, so $m_i = x_{i-1}$. Then

$$LS = \sum_{i=1}^{n} f(m_i)\Delta x = \sum_{i=1}^{n} (m_i)^2 \frac{2}{n} = \sum_{i=1}^{n} (x_{i-1})^2 \frac{2}{n}\,) = \sum_{i=1}^{n} (\frac{2}{n}(i-1))^2 \frac{2}{n}$$

$$= \sum_{i=1}^{n} \frac{4}{n^2}(i^2 - 2i + 1)\frac{2}{n} = \frac{8}{n^3} \sum_{i=1}^{n} (i^2 - 2i + 1)\;)$$

$$= \frac{8}{n^3}\left\{ \sum_{i=1}^{n} i^2 - 2\sum_{i=1}^{n} i + \sum_{k=1}^{n} 1 \right\} = \frac{8}{n^3}\left\{ \frac{n(n+1)(2n+1)}{6} - 2\frac{n(n+1)}{2} + n \right\}$$

$$= \frac{8}{n^3}\left\{ \frac{2n^3 - 3n^2 + n}{6} \right\} = \frac{8}{6}\left\{ \frac{2n^3 - 3n^2 + n}{n^3} \right\} = \frac{8}{6}\left\{ 2 - \frac{3}{n} + \frac{1}{n^2} \right\}$$

$$\longrightarrow \frac{8}{6}\{2\} = \frac{8}{3}. \quad \text{We can be certain that } \int_0^2 x^2\,dx \ \geq 8/3.$$

Practice 6: For $f(x) = x^2$, divide the interval $[0,2]$ into n equally–wide subintervals. Verify that $\Delta x = 2/n$, M_i = right edge $= x_i = \frac{2}{n}\, i$, $f(M_i) = \frac{4}{n^2}\, i^2$, and that the upper sum is $\sum_{i=1}^{n} \frac{4}{n^2}\, i^2\, \frac{2}{n}$. Show that the limit of the upper sum, as $n \to \infty$, is $8/3$.

From the previous Example and Practice problem, we know that

$$\frac{8}{3} \leq \int_0^2 x^2\,dx \leq \frac{8}{3} \quad \text{so we can conclude that} \quad \int_0^2 x^2\,dx = \frac{8}{3}.$$

A much easier method for evaluating this integral will be presented in Section 4.4 .

28. For $f(x) = 3 + x$, partition the interval $[0,2]$ into n equally wide subintervals of length $\Delta x = 2/n$. Write the lower sum for this function and partition, and calculate the limit of the lower sum as $n \to \infty$.

    (b) Write the upper sum for this function and partition and find the limit of the upper sum as $n \to \infty$.

29. For $f(x) = x^3$, partition the interval $[0,2]$ into n equally wide subintervals of length $\Delta x = 2/n$. Write the lower sum for this function and partition, and calculate the limit of the lower sum as $n \to \infty$.

    (b) Write the upper sum for this function and partition and find the limit of the upper sum as $n \to \infty$.

30. For $f(x) = \sqrt{x}$, partition the interval $[0, 9]$ by taking $x_i = \frac{9}{n^2}\, i^2$ for $i = 0, 1, 2, \ldots, n$. Then $\Delta x_i = x_i - x_{i-1} = \frac{9}{n^2}\, i^2 - \frac{9}{n^2}(i-1)^2 = \frac{9}{n^2}(2i + 1)$ and $f(x_i) = \sqrt{x_i} = \frac{3}{n}\, i$. Write the upper sum for this function and partition (take $c_i = x_i$ for $i = 1, 2, \ldots, n$), use the summation formulas in the previous section for sums of powers to simplify the summation, and take the limit of the result as $n \to \infty$.

    (b) Follow the same steps for the lower sum (take $c_i = x_i$ for $i = 0, 2, \ldots, n-1$).

Section 4.2 **PRACTICE Answers**

Practice 1: Area = $\lim_{mesh \to 0} \left(\sum_{k=1}^{n} \sin(c_k)\Delta x_k \right) = \int_0^{\pi} \sin(x)\,dx$.

Practice 2: $\lim_{mesh \to 0} \left(\sum_{k=1}^{n} (2c_k)\Delta x_k \right) =$ Shaded area in Fig. 18 = 8

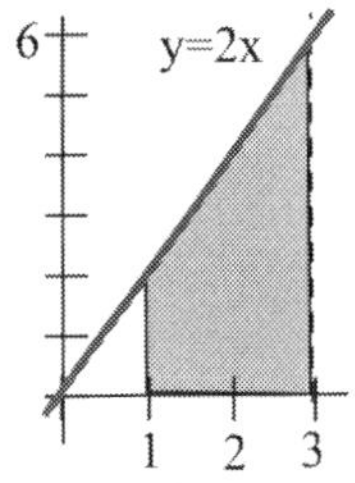

Fig. 18

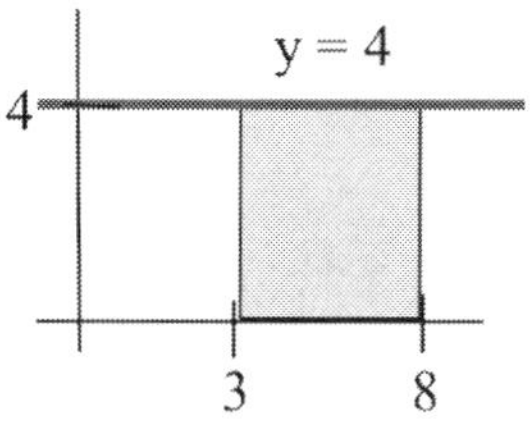

Fig. 19

$\int_3^8 4\,dx$ = shaded area in Fig. 19 = 20 .

Practice 3: (a) Total distance = 12.5 feet forward and 2.5 feet backward = 15 feet total travel.

(b) The bug ends up 10 feet forward of it's starting position at x = 12 so the bug's final location is at x = 22.

Practice 4: Between x = 0 and x = 2π, the graph of y = sin(x) (Fig. 20) has the same area above the x–axis as below the x–axis so the definite integral is 0: $\int_0^{2\pi} \sin(x)\,dx = 0$.

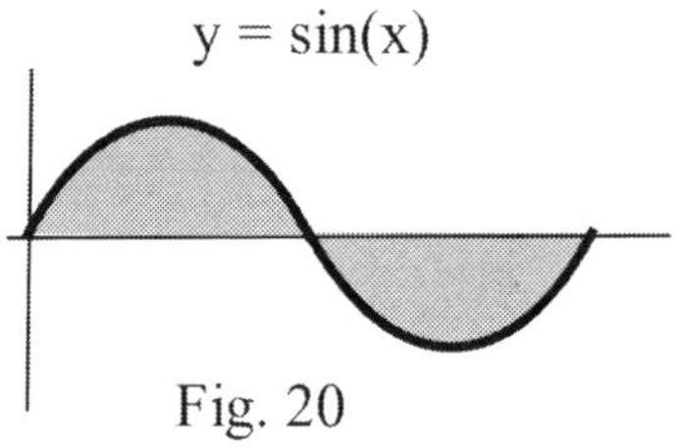

Fig. 20

Practice 5: (a) 20 miles west (from noon to 2 pm) plus 60 miles east (from 2 to 6 pm) is a total travel distance of 80 miles. (At 4 pm the driver is back at the starting position after driving 40 miles = 20 miles west and then 20 miles east.)

(b) The car is 40 miles **east** of the starting location. (East is the "negative" of west.)

Practice 6: $\Delta x = \frac{2-0}{n} = \frac{2}{n}$. $M_i = \frac{2}{n} i$ so $f(M_i) = \{ \frac{2}{n} i \}^2 = \frac{4}{n^2} i^2$. Then

$$US = \sum_{i=1}^{n} f(M_i)\Delta x = \sum_{i=1}^{n} \frac{4}{n^2} i^2 \frac{2}{n} = \frac{8}{n^3} \{ \sum_{i=1}^{n} i^2 \}$$

$$= \frac{8}{n^3} \{ \frac{1}{3} n^3 + \frac{1}{2} n^2 + \frac{2}{12} n \} = \frac{8}{3} + \frac{4}{n} + \frac{16}{12} \frac{1}{n^2} \longrightarrow \frac{8}{3} \text{ as n approaches infinity.}$$

4.3 PROPERTIES OF THE DEFINITE INTEGRAL

Definite integrals are defined as limits of Riemann sums, and they can be interpreted as "areas" of geometric regions. These two views of the definite integral can help us understand and use integrals, and together they are very powerful. This section continues to emphasize this dual view of definite integrals and presents several properties of definite integrals. These properties are justified using the properties of summations and the definition of a definite integral as a Riemann sum, but they also have natural interpretations as properties of areas of regions. These properties are used in this section to help understand functions that are defined by integrals. They will be used in future sections to help calculate the values of definite integrals.

Properties of the Definite Integral

As you read each statement about definite integrals, examine the associated Figure and interpret the property as a statement about areas.

1. $\int_a^a f(x)\,dx = 0$ (a definition)
2. $\int_b^a f(x)\,dx = -\int_a^b f(x)\,dx$ (a definition)
3. $\int_a^b k\,dx = k\cdot(b-a)$ (Fig. 1)
4. $\int_a^b k\cdot f(x)\,dx = k\cdot\int_a^b f(x)\,dx$
5. $\int_a^b f(x)\,dx + \int_b^c f(x)\,dx = \int_a^c f(x)\,dx$ (Fig. 2)

Justification of Property 3: Using area: If $k > 0$ (Fig. 1), then $\int_a^b k\,dx$ represents the area of the rectangle with base = b–a and height = k, so $\int_a^b k\,dx = (\text{height})\cdot(\text{base}) = k\cdot(b-a)$.

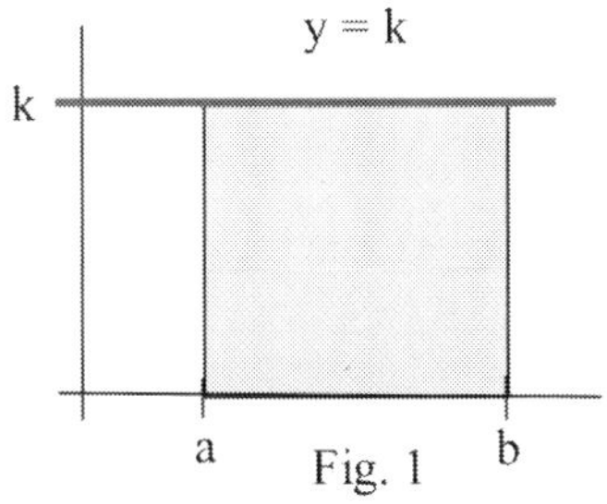

Fig. 1

Using Riemann Sums: For every partition $P = \{ x_0 = a, x_1, x_2, x_3, \ldots, x_{n-1}, x_n = b\}$ of the interval [a, b], and every choice of c_k, the Riemann sum is

$$\sum_{k=1}^{n} f(c_k)\cdot\Delta x_k = \sum_{k=1}^{n} k\cdot\Delta x_k = k \sum_{k=1}^{n}\Delta x_k$$

$$= k \sum_{k=1}^{n}(x_k - x_{k-1}) = k\{(x_1-x_0) + (x_2-x_1) + (x_3-x_2) + \ldots + (x_{n-1} - x_n)\}$$

$$= k\cdot(x_n - x_0) = k\cdot(b - a)$$

so the limit of the Riemann sums, as the mesh approaches zero, is $k\cdot(b - a)$. $\int_a^b k\,dx = k\cdot(b - a)$.

Justification of Property 4:

$$\int_a^b k\cdot f(x)\,dx = \lim_{mesh\varnothing 0}\left(\sum_{k=1}^{n} k\cdot f(c_k)\Delta x_k\right) = k\cdot\lim_{mesh\varnothing 0}\left(\sum_{k=1}^{n} f(c_k)\Delta x_k\right) = k\cdot\int_a^b f(x)\,dx .$$

Property 5 can be justified using Riemann sums, but Fig. 2 graphically illustrates why it is true.

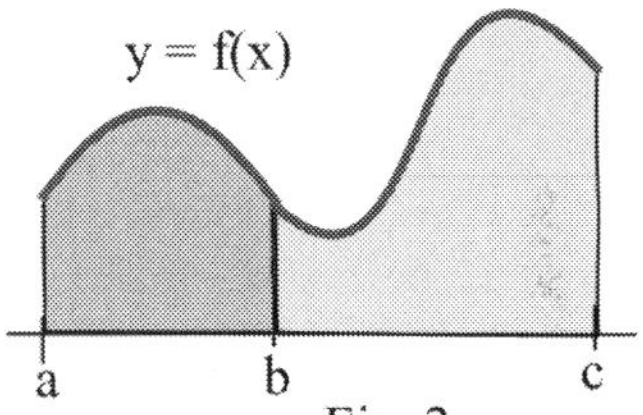

Fig. 2

Properties of Definite Integrals of Combinations of Functions

Properties 6 and 7 relate the values of integrals of sums and differences of functions to the sums and differences of integrals of the individual functions. These two properties will be very useful when we need the integral of a function which is the sum or difference of several terms: we can integrate each term and then add or subtract the individual results to get the integral we want. Both of these new properties have natural interpretations as statements about areas of regions.

6. $\int_a^b f(x) + g(x)\,dx = \int_a^b f(x)\,dx + \int_a^b g(x)\,dx$ (Fig. 3)

7. $\int_a^b f(x) - g(x)\,dx = \int_a^b f(x)\,dx - \int_a^b g(x)\,dx$

Justification of Property 6: $\int_a^b f(x) + g(x)\,dx$

$$= \lim_{mesh\to 0}\left(\sum_{k=1}^{n}(f(c_k) + g(c_k))\Delta x_k\right)$$

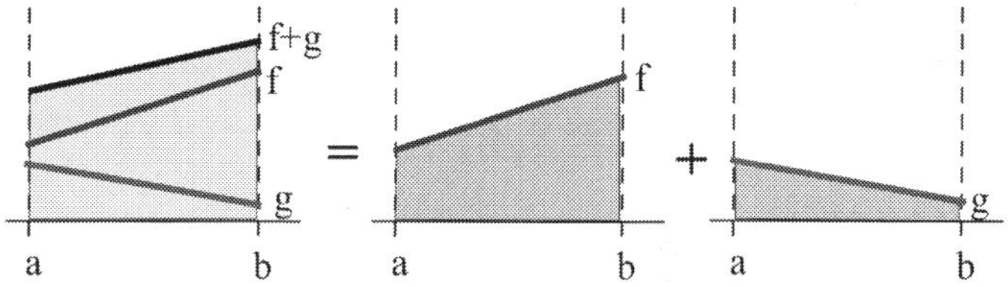

Fig. 3

=

$$\lim_{mesh \to 0} \left(\sum_{k=1}^{n} f(c_k)\Delta x_k + \sum_{k=1}^{n} g(c_k)\Delta x_k \right)$$

$$= \lim_{mesh \to 0} \left(\sum_{k=1}^{n} f(c_k) \cdot \Delta x_k \right) + \lim_{mesh \to 0} \left(\sum_{k=1}^{n} g(c_k) \cdot \Delta x_k \right) = \int_a^b f(x)\, dx + \int_a^b g(x)\, dx .$$

Practice 1: $\int_1^4 f(x)\, dx = 7$, and $\int_1^4 g(x)\, dx = 3$. Evaluate $\int_1^4 f(x)-g(x)\, dx$.

Property 8 says that if one function is larger than another function on an interval, then the definite integral of a larger function on that interval is bigger than the definite integral of the smaller function. This property then leads to Property 9 which provides a quick method for determining bounds on how large and small a particular integral can be.

8.	If $f(x) \le g(x)$ for all x in $[a,b]$, then $\int_a^b f(x)\, dx \le \int_a^b g(x)\, dx$.	(Fig. 4)
9.	$(b-a)\cdot(\text{min of } f \text{ on } [a,b]) \le \int_a^b f(x)\, dx \le (b-a)\cdot(\text{max of } f \text{ on } [a,b])$. (Fig. 5)	

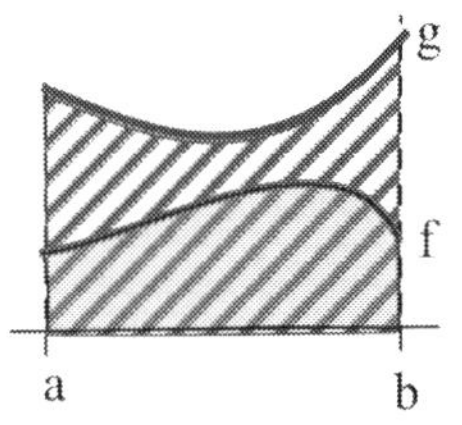

Fig. 4

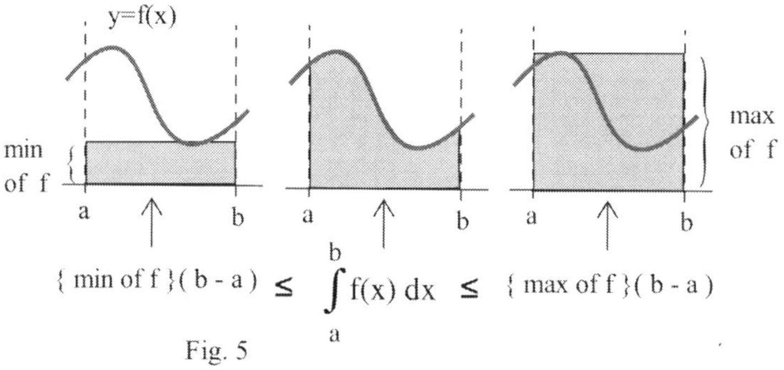

Fig. 5

Justification of Property 8:

Fig. 4 illustrates that if f and g are both positive and $f(x) \le g(x)$ for all x in $[a,b]$, then the area of region F is smaller than the area of region G and $\int_a^b f(x)\, dx \le \int_a^b g(x)\, dx$.

Similar sketches for the situations when f or g are sometimes or always negative illustrate that Property 9 is always true, but we can avoid all of the different cases by using Riemann sums.

Using Riemann Sums: If the same partition and sampling points c_k are used to get Riemann sums for f and g, then $f(c_k) \le g(c_k)$ for each k and

$$\sum_{k=1}^{n} f(c_k)\Delta x_k \le \sum_{k=1}^{n} g(c_k)\Delta x_k \text{ so } \lim_{mesh\to 0}\left(\sum_{k=1}^{n} f(c_k)\Delta x_k\right) \le \lim_{mesh\to 0}\left(\sum_{k=1}^{n} g(c_k)\Delta x_k\right).$$

Justification of Property 9: Property 9 follows easily from Property 8.

Let $g(x) = M = (\text{max of } f \text{ on } [a,b])$. Then $f(x) \le M = g(x)$ for all x in $[a,b]$ so

$$\int_a^b f(x)\,dx \le \int_a^b g(x)\,dx = \int_a^b M\,dx = (b-a)\cdot M .$$

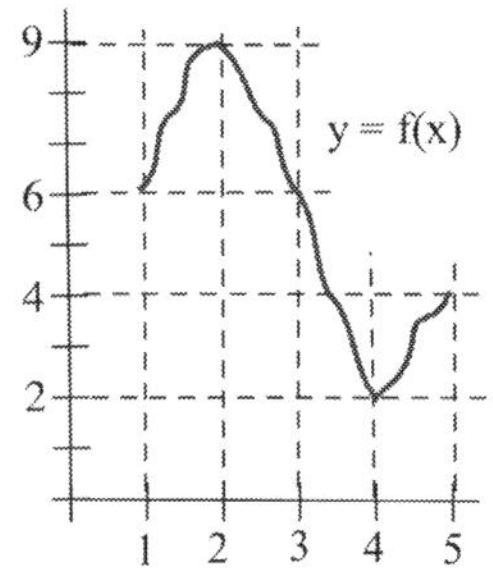

Fig. 6

Example 1: Determine lower and upper bounds for the value of $\int_1^5 f(x)\,dx$ in Fig. 6.

Solution: If $1 \le x \le 5$, then $2 \le f(x) \le 9$ so a lower bound is $(b-a)\cdot(\text{min of } f \text{ on } [a,b]) = (4)(2) = 8$.

An upper bound is $(b-a)\cdot(\text{max of } f \text{ on } [a,b]) = (4)(9) = 36$: $8 \le \int_1^5 f(x)\,dx \le 36$. This range, from 8 to 36 is rather wide. Property 9 is not useful for finding the exact value of the integral, but it is very easy to use and it can help us avoid an unreasonable value for an integral.

Practice 2: Determine a lower bound and an upper bound for the value of $\int_3^5 f(x)\,dx$ in Fig. 6.

Functions Defined by Integrals

If one of the endpoints a or b of the interval $[a, b]$ changes, then the value of the integral $\int_a^b f(t)\,dt$ typically changes. A definite integral of the form $\int_a^x f(t)\,dt$ defines a function of x, and functions defined by definite integrals in this way have interesting and useful properties. The next examples illustrate one of them: the derivative of a function defined by an integral is closely related to the integrand, the function "inside" the integral.

Example 2: For the function $f(t) = 2$, define $A(x)$ to be the area of the region bounded by f, the t–axis, and vertical lines at $t = 1$ and $t = x$ (Fig. 7).

(a) Evaluate $A(1), A(2), A(3), A(4)$.

(b) Find an algebraic formula for $A(x)$ for $x \ge 1$.

(c) Calculate $\frac{d}{dx} A(x)$.

(d) Describe $A(x)$ as a definite integral.

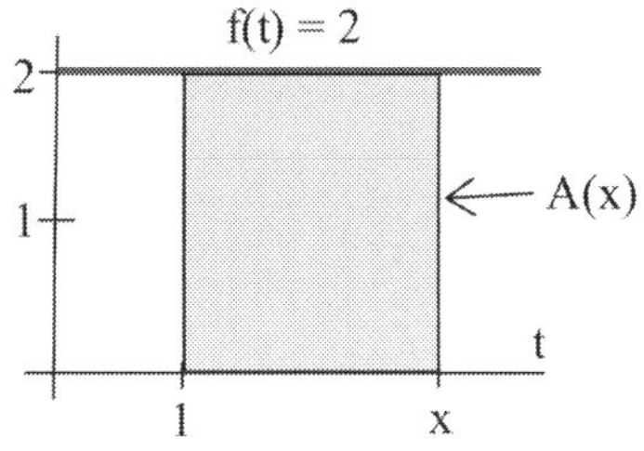

Fig. 7

Solution : (a) $A(1) = 0, A(2) = 2, A(3) = 4, A(4) = 6$.

(b) A(x) = area of a rectangle = (base)·(height) = $(x - 1)\cdot(2) = 2x - 2$.

(c) $\frac{d}{dx} A(x) = \frac{d}{dx}(2x - 2) = 2$. (d) $A(x) = \int_1^x 2\, dt$.

Practice 3: Answer the questions in the previous Example for $f(x) = 3$.

Example 3: For the function $f(t) = 1+t$, define $B(x)$ to be the area of the region bounded by the graph of f, the t–axis, and vertical lines at $t = 0$ and $t = x$ (Fig. 8).

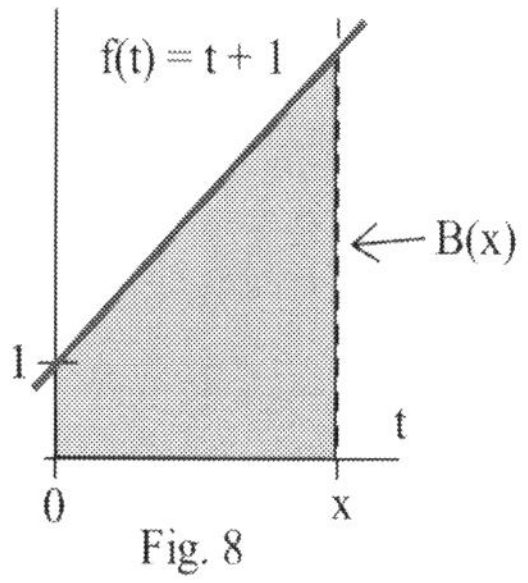

Fig. 8

(a) Evaluate B(0), B(1), B(2), B(3).

(b) Find an algebraic formula for $B(x)$ for $x \ge 0$.

(c) Calculate $\frac{d}{dx} B(x)$.

(d) Describe $B(x)$ as a definite integral.

Solution: (a) B(0) = 0, B(1) = 1.5, B(2) = 4, B(3) = 7.5

(b) B(x) = area of trapezoid = (base)·(average height) = $(x)\cdot\left(\frac{1+(1+x)}{2}\right) = x + \frac{x^2}{2}$.

(c) $\frac{d}{dx} B(x) = \frac{d}{dx}\left(x + \frac{x^2}{2}\right) = 1 + x$. (d) $B(x) = \int_0^x 1+t\, dt$

Practice 4: Answer the questions in the previous Example for $f(t) = 2t$.

A curious "coincidence" appeared in each of these Examples and Practice problems: the derivative of the function defined by the integral was the same as the integrand, the function "inside" the integral. Stated another way, the function defined by the integral was an "antiderivative" of the function "inside" the integral. In section 4.4 we will see that this "coincidence" is a property of functions defined by the integral. And it is such an important property that it is called The Fundamental Theorem of Calculus, part I. Before we go on to the Fundamental Theorem of Calculus, however, there is an "existence" question to consider: Which functions can be integrated?

Which Functions Are Integrable?

This important question was finally answered in the 1850s by Georg Riemann, a name that should be familiar by now. Riemann proved that a function must be badly discontinuous to not be integrable.

Every continuous function is integrable.

If f is **continuous** on the interval [a,b],

then $\lim_{mesh \to 0}\left(\sum_{k=1}^{n} f(c_k)\cdot\Delta x_k\right)$ is always the same finite number , $\int_a^b f(x)\, dx$,

so f is **integrable** on [a,b] .

In fact, a function can even have any finite number of breaks and still be integrable.

Every bounded, piecewise continuous function is integrable.

If f is defined and bounded ($-M \le f(x) \le M$ for some M) for all x in [a,b] and continuous except at a finite number of points in [a,b] ,

then $\lim_{mesh \to 0} \left(\sum_{k=1}^{n} f(c_k)\Delta x_k \right)$ is always the same finite number, $\int_a^b f(x)\,dx$,

so f is integrable on [a,b] .

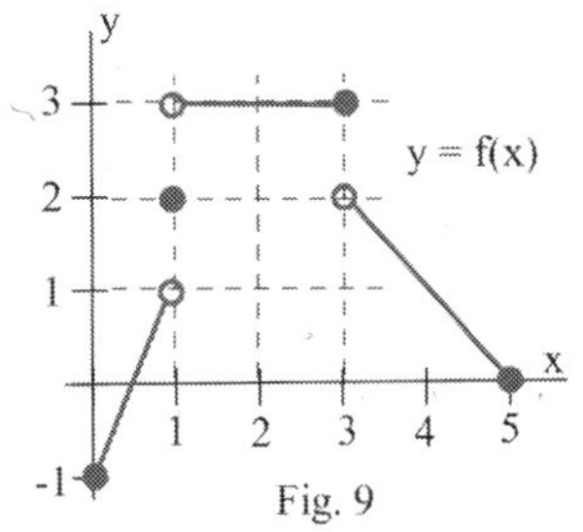

Fig. 9

The function f in Fig. 9 is always between –3 and 3 (in fact, always between –1 and 3) so it is bounded , and it is continuous except at 2 and 3. As long as the values of f(2) and f(3) are finite numbers, their actual values will not effect the value of the definite integral, and

$$\int_0^5 f(x)\,dx = 0 + 3 + 2 = 5.$$

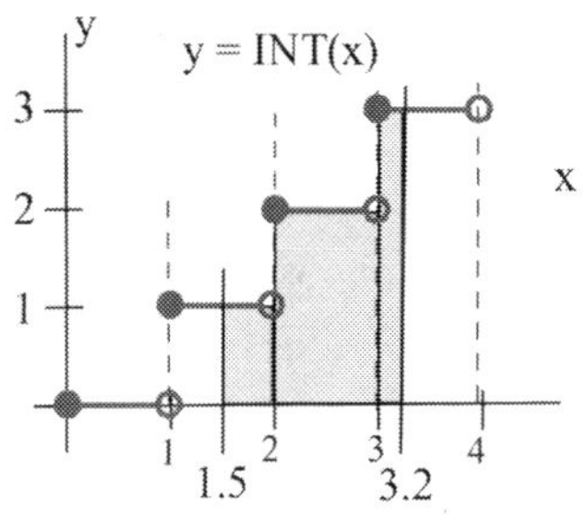

Fig. 10

Practice 5: Evaluate $\int_{1.5}^{3.2} \text{INT}(x)\,dx$. (Fig. 10)

Fig. 11 summarizes the relationships among differentiable, continuous, and integrable functions:

- Every differentiable function is continuous, but there are continuous functions which are not differentiable. (example: $| x |$ is continuous but not differentiable at $x = 0$.)
- Every continuous function is integrable, but there are integrable functions which are not continuous. (example: the function in Fig. 9 is integrable on [0, 5] but is not continuous at 2 and 3.)
- Finally, as shown in the optional part of this section, there are functions which are not integrable.

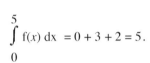

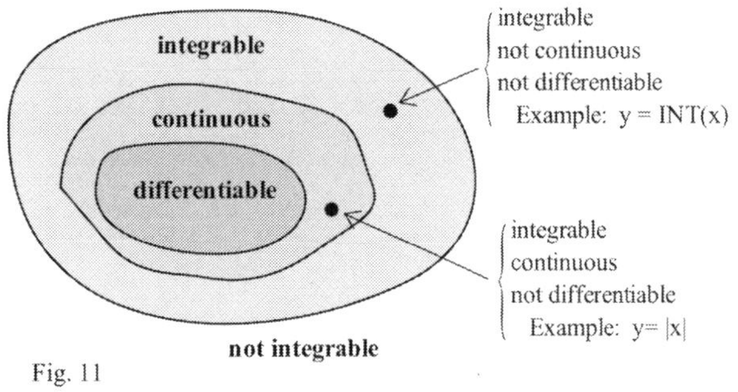

Fig. 11

A Nonintegrable Function

If f is continuous or piecewise continuous on [a,b], then f is integrable on [a,b]. Fortunately, the functions we will use in the rest of this book are all integrable as are the functions you are likely to need for applications. However, there are functions for which the limit of the Riemann sums does not exist, and those functions are not integrable.

A nonintegrable function:

The function $f(x) = \begin{cases} 1 & \text{if } x \text{ is a rational number} \\ 2 & \text{if } x \text{ is an irrational number} \end{cases}$ (Fig. 12)

is not integrable on [0,3].

Proof: For any partition P, suppose that you, a very rational person, always select values of c_k which are rational numbers. (Every subinterval contains rational numbers and irrational numbers, so you can always pick c_k to be a rational number.)

Then $f(c_k) = 1$, and your Riemann sum, YS, is always

$$YS_P = \sum_{k=1}^{n} f(c_k)\Delta x_k = \sum_{k=1}^{\mathbf{n}} \mathbf{1}\Delta x_k = 3.$$

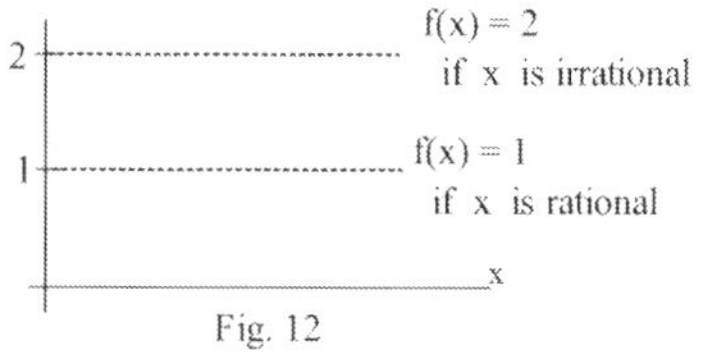

Fig. 12

Suppose your friend, however, always selects values of c_k which are irrational numbers. Then $f(c_k) = 2$, and your friend's Riemann sum, FS, is always

$$FS_P = \sum_{k=1}^{n} f(c_k)\Delta x_k = \sum_{k=1}^{\mathbf{n}} \mathbf{2}\Delta x_k = \mathbf{2}\sum_{k=1}^{n} \Delta x_k = 6.$$

Then $\lim_{mesh\to 0} YS_P = 3$ and $\lim_{mesh\to 0} FS_P = 6$ so $\lim_{mesh\to 0}\left(\sum_{k=1}^{n} f(c_k)\cdot\Delta x_k\right)$

does not exist, and this f is not integrable on [0,3] or on any other interval either.

PROBLEMS:

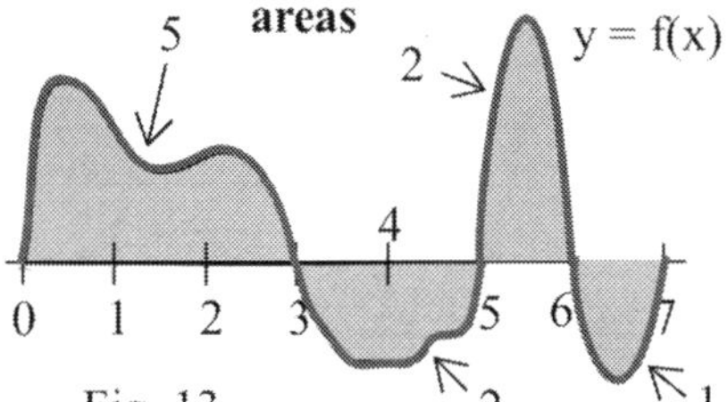

Fig. 13

Problems 1 – 20 refer to the graph of f in Fig. 13. Use the graph to determine the values of the definite integrals. (The bold numbers represent the **area** of each region.)

1. $\int_0^3 f(x)\,dx$ 2. $\int_3^5 f(x)\,dx$ 3. $\int_2^2 f(x)\,dx$ 4. $\int_6^7 f(w)\,dw$ 5. $\int_0^5 f(x)\,dx$

6. $\int_0^7 f(x)\,dx$ 7. $\int_3^6 f(t)\,dt$ 8. $\int_5^7 f(x)\,dx$ 9. $\int_3^0 f(x)\,dx$ 10. $\int_5^3 f(x)\,dx$

11. $\int_6^0 f(x)\,dx$ 12. $\int_0^3 2f(x)\,dx$ 13. $\int_4^4 f^2(s)\,ds$ 14. $\int_0^3 1+f(x)\,dx$ 15. $\int_0^3 x+f(x)\,dx$

16. $\int_3^5 3+f(x)\,dx$ 17. $\int_0^5 2+f(x)\,dx$ 18. $\int_3^5 |\,f(x)\,|\,dx$ 19. $\int_0^5 |\,f(x)\,|\,dx$ 20. $\int_7^3 1+|\,f(x)\,|\,dx$

Problems 21 – 30 refer to the graph of g in Fig. 14. Use the graph to evaluate the integrals.

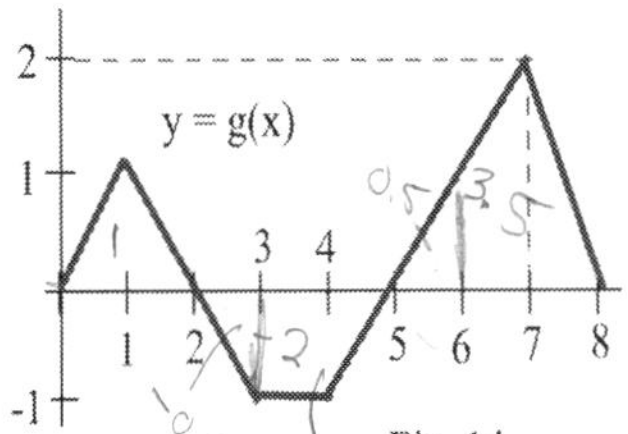

Fig. 14

21. $\int_0^2 g(x)\,dx$ 22. $\int_1^3 g(t)\,dt$ 23. $\int_0^5 g(x)\,dx$ 24. $\int_4^2 g(x)\,dx$

25. $\int_0^8 g(s)\,ds$ 26. $\int_1^4 |\,g(x)\,|\,dx$ 27. $\int_0^3 2g(t)\,dt$ 28. $\int_5^8 1+g(x)\,dx$

29. $\int_6^3 g(u)\,du$ 30. $\int_0^8 t+g(t)\,dt$

For problems 31 – 34 , use the constant functions $f(x) = 4$ and $g(x) = 3$ on the interval $[0,2]$. Calculate each integral and verify that the value obtained in part (a) is **not** equal to the value in part (b).

31.(a) $\int_0^2 f(x)dx \cdot \int_0^2 g(x)dx$ (b) $\int_0^2 f(x)\cdot g(x)\,dx$ 32.(a) $\int_0^2 f(x)dx \Big/ \int_0^2 g(x)dx$ (b) $\int_0^2 f(x)/g(x)\,dx$

33.(a) $\int_0^2 f^2(x)dx$ (b) $\left(\int_0^2 f(x)dx\right)^2$ 34.(a) $\int_0^2 \sqrt{f(x)}\,dx$ (b) $\sqrt{\int_0^2 f(x)\,dx}$

For problems 35 – 42 , sketch the graph of the integrand function and use it to help evaluate the integral.

35 $\int_0^4 |x|\,dx$ 36. $\int_0^4 1 + |t|\,dt$ 37. $\int_{-1}^2 |x|\,dx$ 38. $\int_0^2 |x| - 1\,dx$

39. $\int_1^3 \text{INT}(u)\,du$ 40. $\int_1^{3.5} \text{INT}(x)\,dx$ 41. $\int_1^3 2 + \text{INT}(t)\,dt$ 42. $\int_3^1 \text{INT}(x)\,dx$

For problems 43 – 46, (a) Sketch the graph of $y = A(x) = \int_0^x f(t)\,dt$ and (b) sketch the graph of $y = A'(x)$.

43. $f(x) = x$. 44. $f(x) = x - 2$. 45. f in Fig. 15. 46. f in Fig. 16.

For problems 47 – 50, state whether the function is
(a) continuous on [1,4], (b) differentiable on [1,4] , and (c) integrable on [1,4].

47. f in Fig. 15, 48. f in Fig. 16. 49. f in Fig. 17. 50. f in Fig. 18.

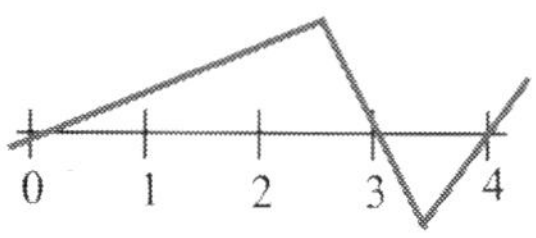

Fig. 15

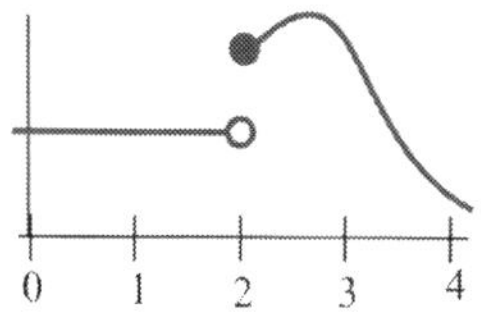

Fig. 16

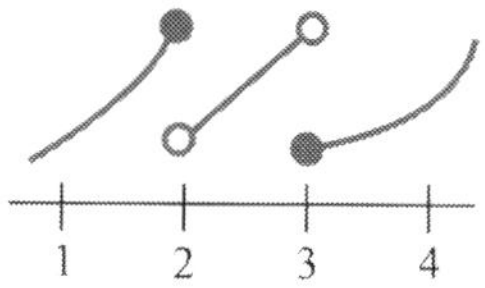

Fig. 17

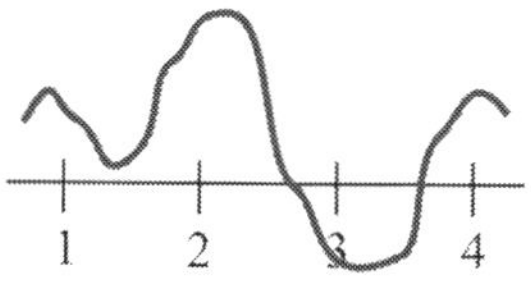

Fig. 18

51. Write the total distance traveled by the car in Fig. 19 between 1 pm and 4 pm as a definite integral and estimate the value of the integral.

52. Write the total distance traveled by the car in Fig. 19 between 3 pm and 6 pm as a definite integral and estimate the value of the integral.

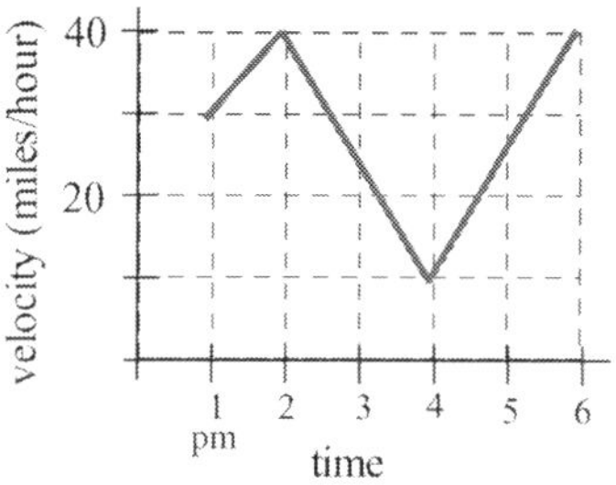

Fig. 19

Section 4.3 **PRACTICE Answers**

Practice 1: $\int_1^4 f(x) - g(x)\,dx = 7 - 3 = 4$.

Practice 2: $(2)(\text{min. of f on } [3,5]) = 4 \le \int_3^5 f(x)\,dx \le 2(\text{max. of f on } [3,5]) = 12$.

Practice 3: (a) $A(1) = 0, A(2) = 3, A(3) = 6, A(4) = 9$ (b) $A(x) = (x - 1)(3) = 3x - 3$

(c) $\frac{d}{dx} A(x) = 3$ (d) $A(x) = \int_1^x 3\,dx$

Practice 4: (a) $B(0) = 0, B(1) = 1, B(2) = 4, B(3) = 9$ (b) $B(x) = \frac{1}{2}(\text{base})(\text{height}) = \frac{1}{2}(x)(2x) = x^2$.

(c) $\frac{d}{dx} B(x) = \frac{d}{dx} x^2 = 2x$ (d) $B(x) = \int_0^x 2t\,dt$

Practice 5: The integral = the shaded area in Fig. 10 = (0.5)(1) + (1)(2) + (0.2)(3) = 3.1 .

4.4 AREAS, INTEGRALS AND ANTIDERIVATIVES

This section explores properties of functions defined as areas and examines some of the connections among areas, integrals and antiderivatives. In order to focus on the geometric meaning and connections, all of the functions in this section are nonnegative, but the results are generalized in the next section and proved true for all continuous functions. This section also introduces examples to illustrate how areas, integrals and antiderivatives can be used.

When f is a continuous, **nonnegative** function, then the "area function" $A(x) = \int_a^x f(t)\,dt$ represents the area between the graph of f, the t–axis, and between the vertical lines at t=a and t=x (Fig. 1), and the derivative of $A(x)$ represents the rate of change (growth) of $A(x)$. Examples 2 and 3 of Section 4.3 showed that for some functions f, the derivative of A(x) was equal to f so A(x) was an antiderivative of f. The next theorem says the result is true for every continuous, nonnegative function f.

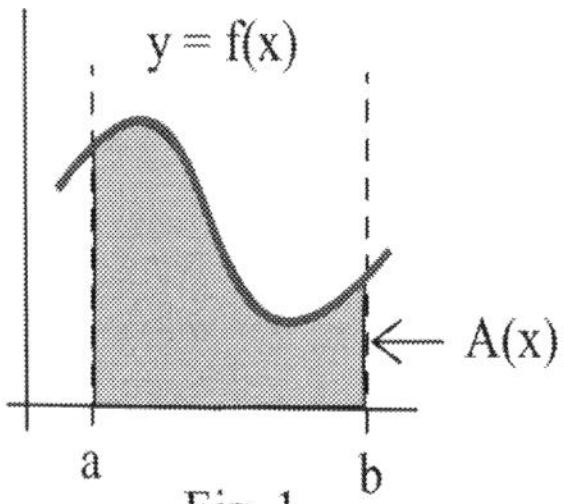

Fig. 1

The Area Function is an Antiderivative

If f is a continuous nonnegative function, $x \ge a$, and $A(x) = \int_a^x f(t)\,dt$

then $\frac{d}{dx}\left(\int_a^x f(t)\,dt \right) = \frac{d}{dx} A(x) = f(x)$, so $A(x)$ is an antiderivative of $f(x)$.

This result relating integrals and antiderivatives is a special case (for nonnegative functions f) of the Fundamental Theorem of Calculus (Part 1) which is proved in Section 4.5 . This result is important for two reasons:

(i) it says that a large collection of functions have antiderivatives, and
(ii) it leads to an **easy** way of **exactly** evaluating definite integrals.

Example 1: $A(x) = \int_1^x f(t)\,dt$ for the function $f(t)$ shown in Fig. 2.

Estimate the values of $A(x)$ and $A'(x)$ for x = 2, 3, 4 and 5 and use these values to sketch the graph of $y = A(x)$.

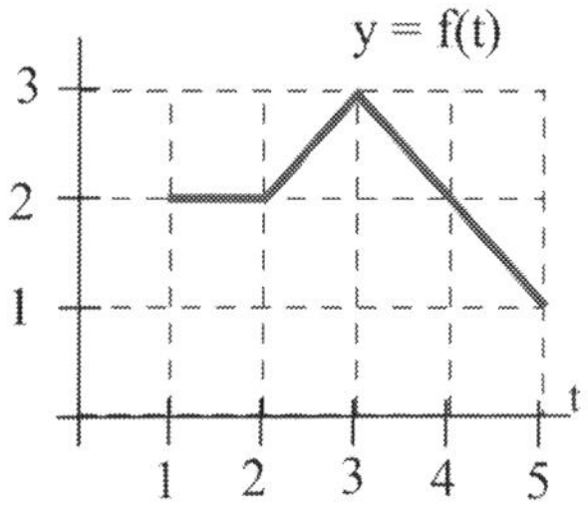

Fig. 2

Solution: Dividing the region into squares and triangles, it is easy to see that A(2) = 2, A(3) = 4.5, A(4) = 7 , and A(5) = 8.5 . Since $A'(x) = f(x)$, we know that A'(2) = f(2) = 2 , A'(3) = f(3) = 3 , A'(4) = f(4) = 2 ,

and A'(5) = f(5) = 1. The graph of $y = A(x)$ is shown in Fig. 3.

It is important to recognize that f is not differentiable at $x = 2$ and $x = 3$. However, the values of A change smoothly near 2 and 3, and the function A is differentiable at those points and at every other point from 1 to 5. Also, $f'(4) = -1$ (f is clearly decreasing near $x = 4$), but $A'(4) = f(4) = 2$ is positive (the area A is growing even though f is getting smaller).

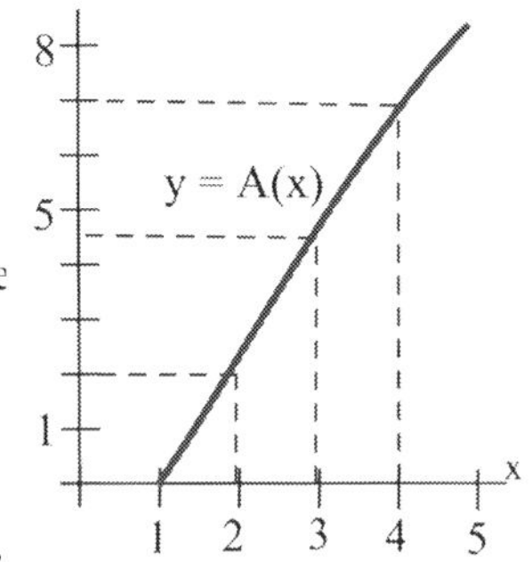

Fig. 3

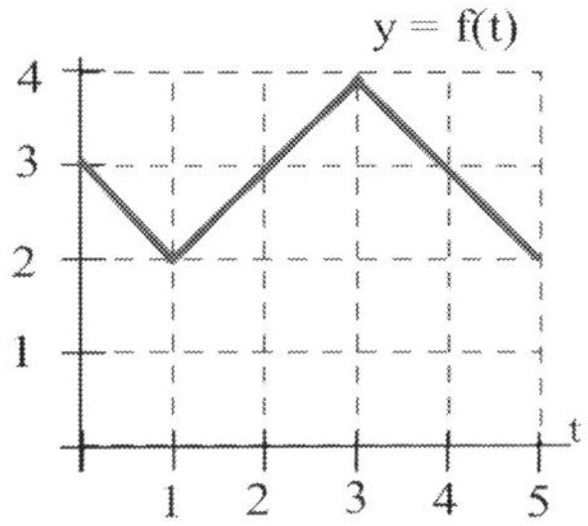

Fig. 4

Practice 1: $B(x)$ is the area bounded by the horizontal axis vertical lines at $t = 0$ and $t = x$, and the graph of $f(t)$ shown in Fig. 4. Estimate the values of $B(x)$ and $B'(x)$ for $x = 1, 2, 3, 4$ and 5.

Example 2: Let $G(x) = \frac{d}{dx}\left(\int_0^x \sin(t)\,dt\right)$. Evaluate $G(x)$ for $x = \pi/4, \pi/2$, and $3\pi/4$.

Solution: Fig. 5(a) shows the graph of $A(x) = \int_0^x \sin(t)\,dt$, and $G(x)$ is the derivative of $A(x)$. By the theorem, $A'(x) = \sin(x)$ so $A'(\pi/4) = \sin(\pi/4) \approx .707$, $A'(\pi/2) = \sin(\pi/2) = 1$, and $A'(3\pi/4) = \sin(3\pi/4) \approx .707$. Fig. 5(b) shows the graph of $y = A(x)$ and 5(c) is the graph of $y = A'(x) = G(x)$

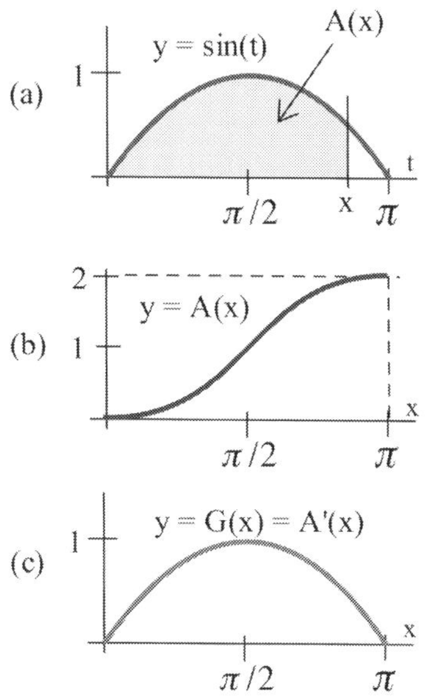

Fig. 5

Using Antiderivatives to Evaluate $\int_a^b f(x)\,dx$

Now we can put the ideas of areas and antiderivatives together to get a way of evaluating definite integrals that is exact and often easy.

If $A(x) = \int_a^x f(t)\,dt$, then $A(a) = \int_a^a f(t)\,dt = 0$, $A(b) = \int_a^b f(t)\,dt$, and $A(x)$ is an antiderivative of f, $A'(x) = f(x)$. We also know that if $F(x)$ is **any** antiderivative of f, then $F(x)$ and $A(x)$ have the same derivative so $F(x)$ and $A(x)$ are "parallel" and differ by a constant, $F(x) = A(x) + C$ for all x.

Then $F(b) - F(a) = \{ A(b) + C \} - \{ A(a) + C \} = A(b) - A(a) = \int_a^b f(t)\,dt - \int_a^a f(t)\,dt = \int_a^b f(t)\,dt$.

To evaluate a definite integral $\int_a^b f(t)\,dt$, we can find any antiderivative F of f and evaluate F(b) – F(a).

This result is a special case of Part 2 of the Fundamental Theorem of Calculus, and it will be used hundreds of times in the next several chapters. The Fundamental Theorem is stated and proved in Section 4.5.

Antiderivatives and Definite Integrals

If f is a continuous, nonnegative function
and **F is any antiderivative** of f ($F'(x) = f(x)$) on the interval [a,b] ,

then $\left\{ \begin{array}{l} \text{area bounded between the graph} \\ \text{of f and the } x\text{–axis and} \\ \text{vertical lines at } x = a \text{ and } x = b \end{array} \right\} = \int_a^b f(x)\,dx = F(b) - F(a).$

The problem of finding the exact value of a definite integral reduces to finding some (any) antiderivative F of the integrand and then evaluating F(b) – F(a). Even finding one antiderivative can be difficult, and, for now, we will stick to functions which have easy antiderivatives. Later we will explore some methods for finding antiderivatives of more difficult functions.

The evaluation F(b) – F(a) is represented by the symbol $F(x) \Big|_a^b$.

Example 3: Evaluate $\int_1^3 x\,dx$ in two ways:

(i) By sketching the graph of $y = x$ and geometrically finding the area.

(ii) By finding an antiderivative of F(x) of f and evaluating F(3)–F(1).

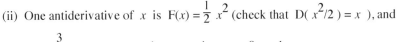

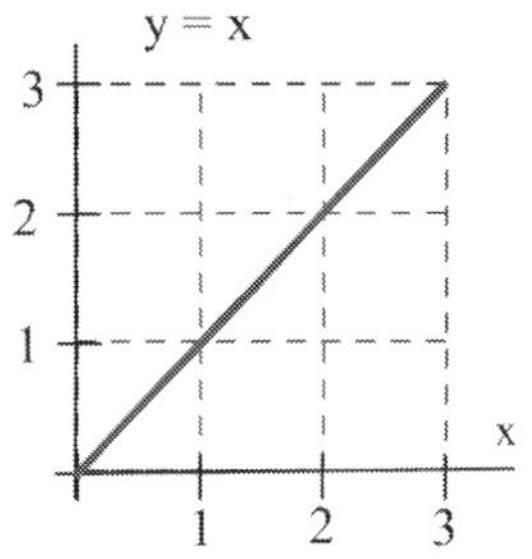

Fig. 6

Solution: (i) The graph of $y = x$ is shown in Fig. 6, and the shaded region has area 4.

(ii) One antiderivative of x is $F(x) = \frac{1}{2}x^2$ (check that $D(x^2/2) = x$), and

$F(x)\Big|_1^3 = F(3)–F(1) = \frac{1}{2}(3^2) - \frac{1}{2}(1^2) = \frac{9}{2} - \frac{1}{2} = 4$ which agrees with (i).

If someone chose another antiderivative of x, say $F(x) = \frac{1}{2}x^2 + 7$ (check that $\mathbf{D}(x^2/2 + 7) = x$),

then $F(x)\Big|_1^3 = F(3) - F(1) = \left\{ \frac{1}{2}(3^2) + 7 \right\} - \left\{ \frac{1}{2}(1^2) + 7 \right\} = \frac{23}{2} - \frac{15}{2} = 4$.

No matter which antiderivative F is chosen, F(3) – F(1) equals 4.

Practice 2: Evaluate $\int_1^3 (x - 1)\,dx$ in the two ways of the previous example.

This antiderivative method is an extremely powerful way to evaluate some definite integrals, and it is used often. However, it can only be used to evaluate a definite integral of a function defined by a formula.

Example 4: Find the area between the graph of the cosine and the horizontal axis for x between 0 and $\pi/2$.

Solution: The area we want (Fig. 7) is $\int_0^{\pi/2} \cos(x)\, dx$ so we need an antiderivative of $f(x) = \cos(x)$.

$F(x) = \sin(x)$ is one antiderivative of $\cos(x)$ (Check that $\mathbf{D}(\sin(x)) = \cos(x)$). Then

$$\text{area} = \int_0^{\pi/2} \cos(x)\, dx = \sin(x) \Big|_0^{\pi/2} = \sin(\pi/2) - \sin(0) = 1 - 0 = 1 .$$

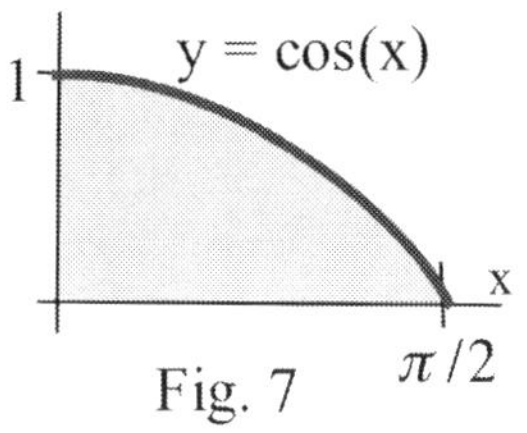

Fig. 7

Practice 3: Find the area between the graph of $y = 3x^2$ and the horizontal axis for x between 1 and 2.

Integrals, Antiderivatives, and Applications

The antiderivative method of evaluating definite integrals can also be used when we need to find an "area", and it is useful for solving applied problems.

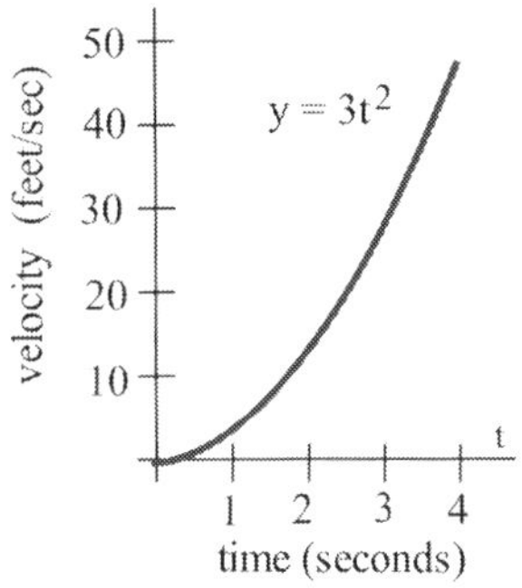

Fig. 8

Example 5: A robot has been programmed so that when it starts to move, its velocity after t seconds will be $3t^2$ feet/second.

(a) How far will the robot travel during its first 4 seconds of movement?

(b) How far will the robot travel during its next 4 seconds of movement?

(c) How many seconds before the robot is 729 feet from its starting place?

Solution: (a) The distance during the first 4 seconds will be the area under the graph (Fig. 8) of velocity , $f(t) = t$, from $t = 0$ to $t = 4$, and that area is the definite integral

$$\int_0^4 3t^2\, dt$$. An antiderivative of $3t^2$ is t^3 so $\int_0^4 3t^2\, dt = t^3 \Big|_0^4 = (4)^3 - (0)^3 = 64$ feet.

(b) $\int_4^8 3t^2\, dt = t^3 \Big|_4^8 = (8)^3 - (4)^3 = 512 - 64 = 448$ feet.

(c) This part is different from the other two parts. Here we are told the lower integration endpoint, $t = 0$, and the total distance, 729 feet, and we are asked to find the upper endpoint. Calling the upper endpoint T, we know that

$$729 = \int_0^T 3t^2\, dt = t^3 \Big|_0^T = (T)^3 - (0)^3 = T^3 , \text{ so } T = \sqrt[3]{729} = 9 \text{ seconds.}$$

Practice 4: (a) How far will the robot move between $t = 1$ second and $t = 5$ seconds?

(b) How many seconds before the robot is 343 feet from its starting place?

Example 6: Suppose that t minutes after putting 1000 bacteria on a Petri plate the rate of growth of the population is $6t$ bacteria per minute. (a) How many new bacteria are added to the population during the first 7 minutes? (b) What is the total population after 7 minutes? (c) When will the total population be 2200 bacteria?

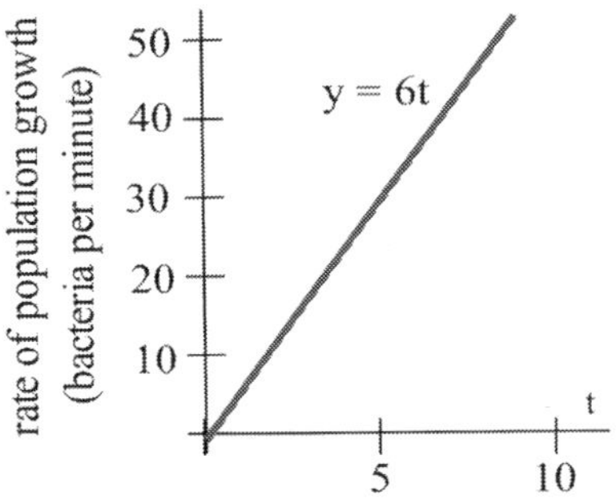

Fig. 9

Solution: (a) The number of new bacteria is the area under the rate of growth graph (Fig. 9), and one antiderivative of $6t$ is $3t^2$ (check that $D(3t^2) = 6t$) so

$$\text{new bacteria} = \int_0^7 6t\,dt = 3t^2 \Big|_0^7 = 3(7)^2 - 3(0)^2 = 147 .$$

(b) The new population = {old population} + {new bacteria} = 1000 + 147 = 1147 bacteria.

(c) If the total population is 2200 bacteria, then there are 2200 – 1000 = 1200 new bacteria, and we need to find the time T needed for that many new bacteria to occur.

$$1200 \text{ new bacteria} = \int_0^T 6t\,dt = 3t^2 \Big|_0^T = 3(T)^2 - 3(0)^2 = 3\,T^2 \quad \text{so } T^2 = 400 \text{ and}$$

T = 20 minutes. After 20 minutes, the total bacteria population will be 1000 + 1200 = 2200.

Practice 5: (a) How many new bacteria will be added to the population between $t = 4$ and $t = 8$ minutes?

(b) When will the total population be 2875 bacteria? (Hint: How many are new?)

PROBLEMS

In problems 1 – 4, the function f is given by a graph, and $A(x) = \int_1^x f(t)\,dt$.

(a) Graph $y = A(x)$ for $1 \le x \le 5$. (b) Estimate the values of $A(1), A(2), A(3)$, and $A(4)$.

(c) Estimate the values of $A'(1), A'(2), A'(3)$, and $A'(4)$.

1. f in Fig. 11.
2. f in Fig. 12.
3. f in Fig. 13
4. f in Fig. 14.

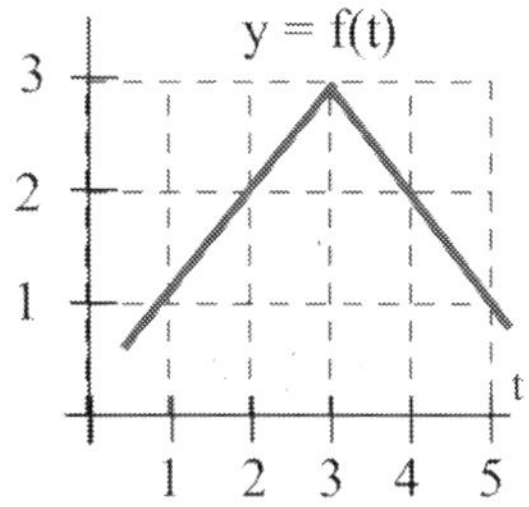

Fig. 11

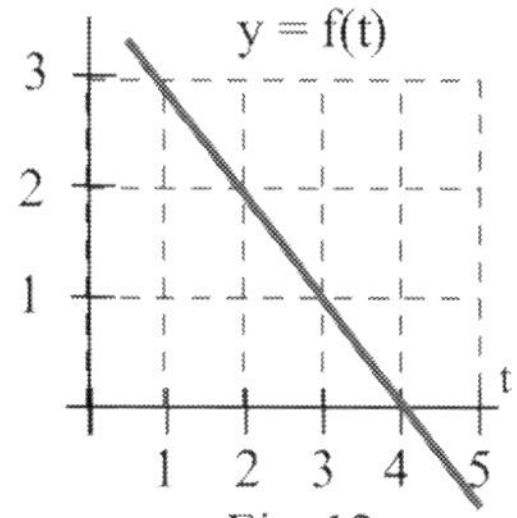

Fig. 12

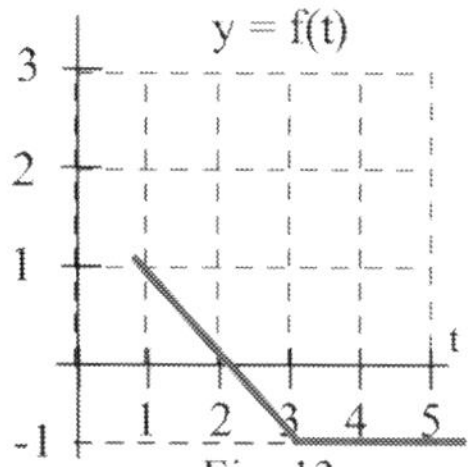

Fig. 13

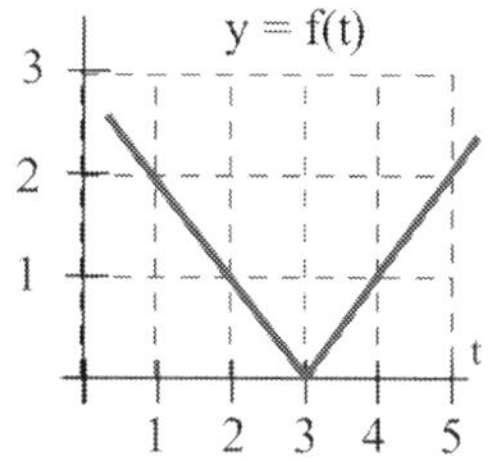

Fig. 14

In problems 5 – 8, the function f is given by a formula, and $A(x) = \int_1^x f(t)\,dt$.

(a) Graph $y = A(x)$ for $1 \le x \le 5$. (b) Calculate the values of $A(1), A(2), A(3)$, and $A(4)$.

(c) Determine the values of $A'(1), A'(2), A'(3)$, and $A'(4)$.

5. $f(t) = 2$
6. $f(t) = 1 + t$
7. $f(t) = 6 - t$
8. $f(t) = 1 + 2t$

In problems 9 – 18, use the **Antiderivatives and Definite Integrals** Theorem to evaluate the integrals.

9. $\int_0^3 2x\,dx$, $\int_1^3 2x\,dx$, $\int_0^1 2x\,dx$

10. $\int_0^2 4x^3\,dx$, $\int_0^1 4x^3\,dx$, $\int_1^2 4x^3\,dx$

11. $\int_1^3 6x^2\,dx$, $\int_1^2 6x^2\,dx$, $\int_0^3 6x^2\,dx$

12. $\int_{-2}^{2} 2x\,dx$, $\int_{-2}^{-1} 2x\,dx$, $\int_{-2}^{0} 2x\,dx$

13. $\int_0^3 4x^3\,dx$, $\int_1^3 4x^3\,dx$, $\int_0^1 4x^3\,dx$

14. $\int_0^5 4x^3\,dx$, $\int_0^2 4x^3\,dx$, $\int_2^5 4x^3\,dx$

15. $\int_{-3}^{3} 3x^2\,dx$, $\int_{-3}^{0} 3x^2\,dx$, $\int_0^3 3x^2\,dx$

16. $\int_0^3 5\,dx$, $\int_0^2 5\,dx$, $\int_2^3 5\,dx$

17. $\int_0^2 3x^2\,dx$, $\int_1^3 3x^2\,dx$

18. $\int_{-2}^{2} 12 - 3x^2 \, dx$, $\int_{0}^{2} 12 - 3x^2 \, dx$, $\int_{1}^{2} 12 - 3x^2 \, dx$

19. The velocity of a car after t seconds is 2t feet per second. (a) How far does the car travel during its first 10 seconds? (b) How many seconds does it take the car to travel half the distance in part (a)?

20. The velocity of a car after t seconds is $3t^2$ feet per second. (a) How far does the car travel during its first 10 seconds? (b) How many seconds does it take the car to travel half the distance in part (a)?

21. The velocity of a car after t seconds is $4t^3$ feet per second. (a) How far does the car travel during its first 10 seconds? (b) How many seconds does it take the car to travel half the distance in part (a)?

22. The velocity of a car after t seconds is $20 - 2t$ feet per second. (a) How many seconds does it take for the car to come to a stop (velocity = 0)? (b) How far does the car travel while coming to a stop? (c) How many seconds does it take the car to travel half the distance in part (b)?

23. The velocity of a car after t seconds is $75 - 3t^2$ feet per second. (a) How many seconds does it take for the car to come to a stop (velocity = 0)? (b) How far does the car travel while coming to a stop? (c) How many seconds does it take the car to travel half the distance in part (b)?

24. Find the exact area under half of one arch of the sine curve: $\int_{0}^{\pi/2} \sin(x) \, dx$. (Note: **D**($-\cos(x)$) = $\sin(x)$)

25. An artist you know wants to make a figure consisting of the region between the curve $y = x^2$ and the x–axis for $0 \le x \le 3$.
   (a) Where should the artist divide the region with a vertical line (Fig. 15) so each piece has the same area?
   (b) Where should the artist divide the region with vertical lines to get 3 pieces with equal areas?

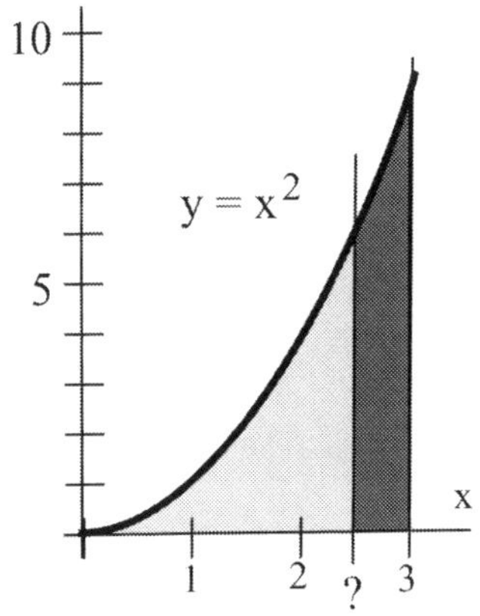

Fig. 15

Section 4.4 **PRACTICE Answers**

Practice 1: B(1) = 2.5, B(2) = 5, B(3) = 8.5, B(4) = 12, B(5) = 14.5

$B(x) = \int_0^x f(t)\,dt$ so $B'(x) = \frac{d}{dx}\left(\int_0^x f(t)\,dt \right) = f(x)$ (by The Area Function is an Antiderivative theorem): then $B'(1) = f(1) = 2$, $B'(2) = f(2) = 3$, $B'(3) = 4$, $B'(4) = 3$, and $B'(5) = 2$.

Practice 2: (a) As an area, $\int_1^3 x - 1\,dx$ is the area of the triangular region between $y = x - 1$ and the x–axis for $1 \le x \le 3$: area $= \frac{1}{2}$(base)(height) $= \frac{1}{2}(2)(2) = 2$.

(b) $F(x) = \frac{x^2}{2} - x$ is an antiderivative of $f(x) = x - 1$ so

area $= \int_1^3 x - 1\,dx = F(3) - F(1) = \left(\frac{9}{2} - 3\right) - \left(\frac{1}{2} - 1\right) = 2.$

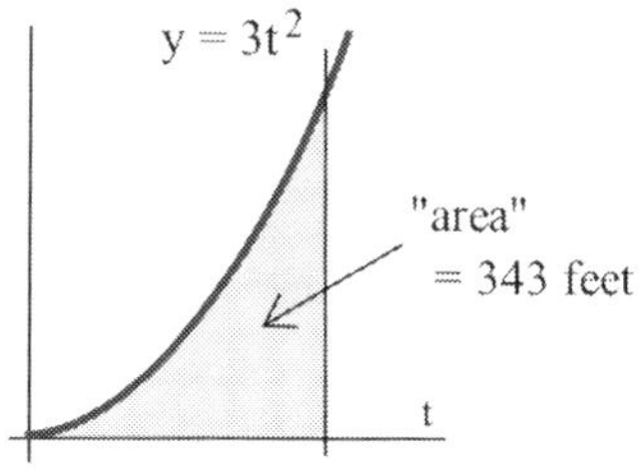

Fig. 16

Practice 3: Area $= \int_1^2 3x^2\,dx = x^3 \Big|_1^2 = 8 - 1 = \mathbf{7}$.

Practice 4: (a) distance $= \int_1^5 3t^2\,dt = t^3 \Big|_1^5 = 125 - 1 =$ **124 feet**.

(b) In this problem we know the starting point is x = 0, and the total distance ("area") is 343 feet. Our problem is to find the time T (Fig. 16) so 343 feet $= \int_0^T 3t^2\,dt$.

$343 = \int_0^T 3t^2\,dt = t^3 \Big|_0^T = T^3 - 0 = T^3$ so T = **7 seconds**.

Practice 5: (a) number of new bacteria $= \int_4^8 6t\,dt$

$= 3t^2 \Big|_4^8 = 3\cdot 64 - 3\cdot 16 = 144$ bacteria.

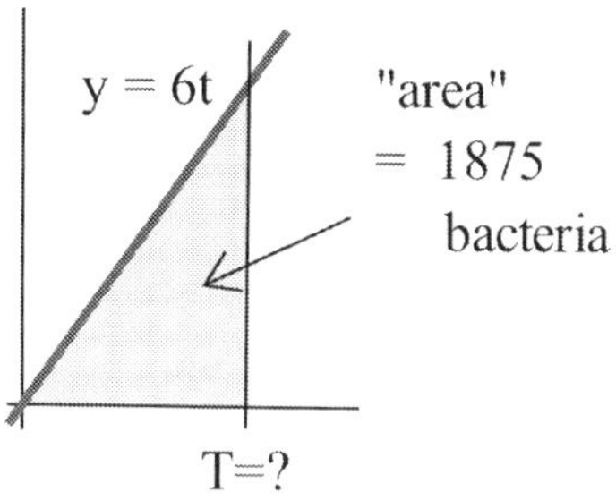

Fig. 17

(b) We know the total new population ("area" in Fig. 17) is 2875 – 1000 = 1875 so

$1875 = \int_0^T 6t\,dt = 3t^2 \Big|_0^T = 3T^2 - 0 = 3T^2$ so T = **25 minutes**.

4.5 THE FUNDAMENTAL THEOREM OF CALCULUS

This section contains the most important and most used theorem of calculus, THE Fundamental Theorem of Calculus. Discovered independently by Newton and Leibniz in the late 1600s, it establishes the connection between derivatives and integrals, provides a way of easily calculating many integrals, and was a key step in the development of modern mathematics to support the rise of science and technology. Calculus is one of the most significant intellectual structures in the history of human thought, and the Fundamental Theorem of Calculus is a most important brick in that beautiful structure.

The previous sections emphasized the meaning of the definite integral, defined it, and began to explore some of its applications and properties. In this section, the emphasis is on the Fundamental Theorem of Calculus. You will use this theorem **often** in later sections.

There are two parts of the Fundamental Theorem. They are similar to results in the last section but more general. Part 1 of the Fundamental Theorem of Calculus says that **every** continuous function has an antiderivative and shows how to differentiate a function defined as an integral. Part 2 shows how to evaluate the definite integral of any function if we know an antiderivative of that function.

Part 1: Antiderivatives

Every continuous function has an antiderivative, even those nondifferentiable functions with "corners" such as absolute value.

> **The Fundamental Theorem of Calculus** (Part 1)
>
> If f is continuous and $A(x) = \int_a^x f(t)\,dt$
>
> then $\frac{d}{dx}\left(\int_a^x f(t)\,dt \right) = \frac{d}{dx} A(x) = f(x)$. $A(x)$ is an antiderivative of $f(x)$.

Proof: Assume f is a continuous function and let $A(x) = \int_a^x f(t)\,dt$. By the definition of derivative of A,

$$\frac{d}{dx} A(x) = \lim_{h\to 0} \frac{A(x+h) - A(x)}{h} = \lim_{h\to 0} \frac{1}{h}\left\{ \int_a^{x+h} f(t)dt - \int_a^{x} f(t)dt \right\} = \lim_{h\to 0} \frac{1}{h} \int_x^{x+h} f(t)\,dt\,.$$

By Property 6 of definite integrals (Section 4.3), for $h > 0$

$$\{ \text{ min of f on } [x, x+h] \}\cdot h \le \int_{x}^{x+h} f(t)\ dt \le \{ \text{ max of f on } [x, x+h] \}\cdot h \ . \ \text{(Fig. 1)}$$

Dividing each part of the inequality by h, we have that $\dfrac{1}{h}\displaystyle\int_{x}^{x+h} f(t)\ dt$ is

between the minimum and the maximum of f on the interval [x, x+h].

The function f is continuous (by the hypothesis) and the interval

[x,x+h] is shrinking (since h approaches 0), so

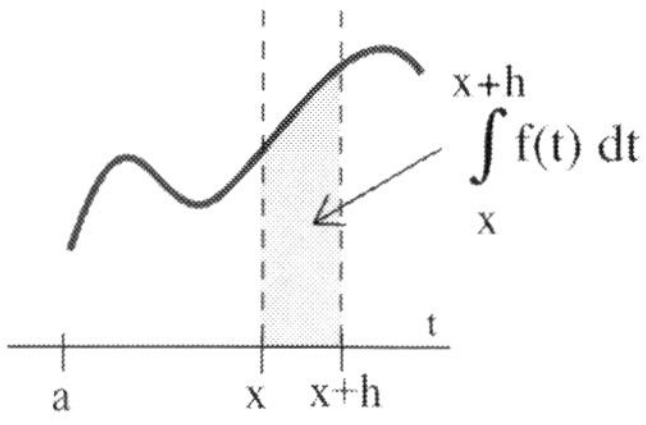

Fig. 1

$$\lim_{h\to 0} \{ \text{ min of f on } [x, x+h] \} = f(x) \text{ and}$$

$$\lim_{h\to 0} \{ \text{ max of f on } [x, x+h] \} = f(x). \text{ Therefore, } \frac{1}{h}\int_{x}^{x+h} f(t)\ dt \text{ is stuck between two}$$

$$\{ \textbf{min} \text{ of } f \text{ on } [x, x+h] \} \le \frac{1}{h}\int_{x}^{x+h} f(t)\ dt \le \{\textbf{max} \text{ of } f \text{ on } [x, x+h] \}$$

as h → 0   as h → 0   as h → 0

$$f(x) \qquad \le \qquad ? \qquad \le \qquad f(x)$$

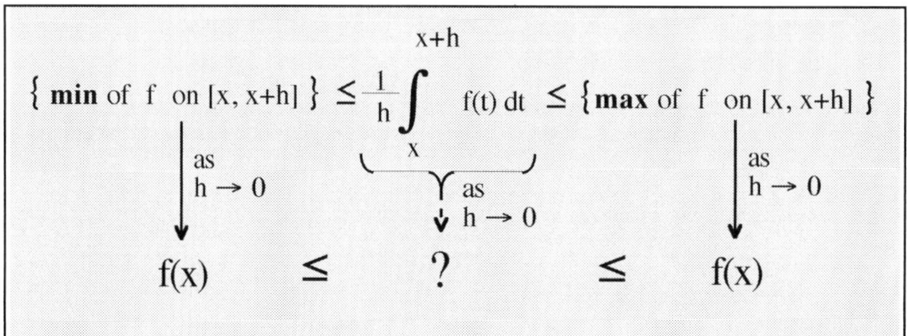

Fig. 2

quantities (Fig. 2) which both approach f(x).

Then $\dfrac{1}{h}\displaystyle\int_{x}^{x+h} f(t)\ dt$ must also approach f(x), and $\dfrac{d}{dx}\ A(x) = \lim_{h\to 0}\dfrac{1}{h}\displaystyle\int_{x}^{x+h} f(t)\ dt = f(x).$

Example 1: $A(x) = \displaystyle\int_{0}^{x} f(t)\ dt$ for f in Fig. 3. Evaluate A(x) and A'(x) for

x= 1, 2 , 3 and 4.

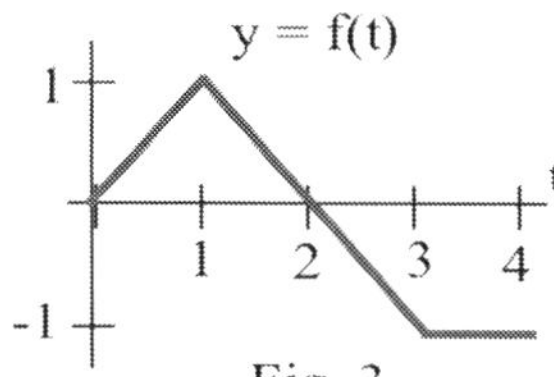

Fig. 3

Solution: $A(1) = \displaystyle\int_{0}^{1} f(t)\ dt = 1/2, \ A(2) = \int_{0}^{2} f(t)\ dt = 1, \ A(3) = \int_{0}^{3} f(t)\ dt = 1/2,$

$A(4) = \displaystyle\int_{0}^{4} f(t)\ dt = -1/2.$ Since f is continuous, A '(x) = f(x) so

$A'(1) = f(1) = 1, A'(2) = f(2) = 0, A'(3) = f(3) = -1, A'(4) = f(4) = -1.$

(1, 1)
(2, 0)
(3, -1)
(4, 1)

Fig. 4

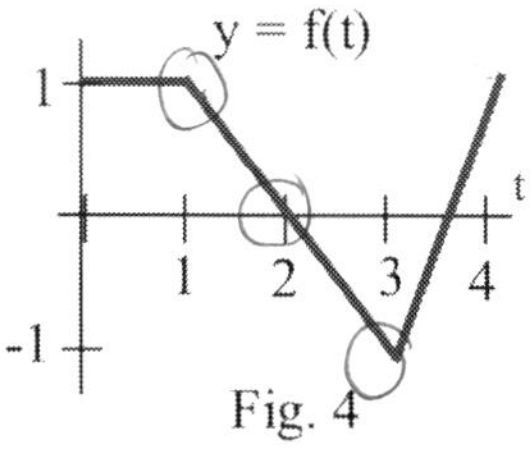

Practice 1: $A(x) = \int_0^x f(t)\,dt$ for f in Fig. 4. Evaluate $A(x)$ and $A'(x)$ for $x = 1, 2, 3$ and 4.

Example 2: $A(x) = \int_0^x f(t)dt$ for the function f shown in Fig. 5.

For which value of x is $A(x)$ maximum?

For which x is the rate of change of A maximum?

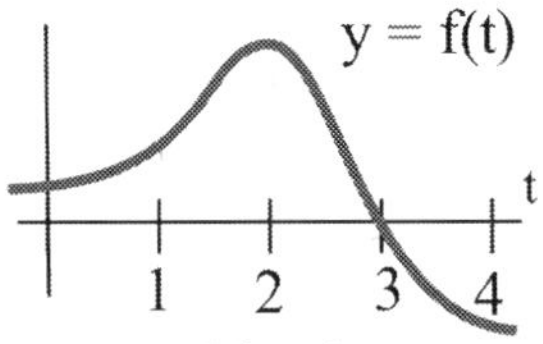

Fig. 5

Solution: Since A is differentiable, the only critical points are where $A'(x) = 0$ or at endpoints. $A'(x) = f(x) = 0$ at $x=3$, and A has a maximum at $x=3$. Notice that the values of $A(x)$ increase as x goes from 0 to 3 and then the A values decrease. The rate of change of $A(x)$ is $A'(x) = f(x)$, and $f(x)$ appears to have a maximum at $x=2$ so the rate of change of $A(x)$ is maximum when $x=2$. Near $x=2$, a slight increase in the value of x yields the maximum increase in the value of $A(x)$.

Part 2: Evaluating Definite Integrals

If we know and can evaluate some antiderivative of a function, then we can evaluate any definite integral of that function.

> **The Fundamental Theorem of Calculus** (Part 2)
>
> If $f(x)$ is continuous and $F(x)$ is **any** antiderivative of f ($F'(x) = f(x)$),
>
> then $\int_a^b f(x)\,dx = F(x)\Big|_a^b = F(b) - F(a)$.

Proof: If F is an antiderivative of f, then $F(x)$ and $A(x) = \int_a^x f(t)\,dt$ are both antiderivatives of f,

$F'(x) = f(x) = A'(x)$, so F and A differ by a constant: $A(x) - F(x) = C$ for all x.

At $x=a$, we have $C = A(a) - F(a) = 0 - F(a) = -F(a)$ so $C = -F(a)$ and the equation $A(x) - F(x) = C$ becomes $A(x) - F(x) = -F(a)$. Then $A(x) = F(x) - F(a)$ for all x so

$A(b) = F(b) - F(a)$ and $\int_a^{\mathbf{b}} f(x)\,dx = A(\mathbf{b}) = F(\mathbf{b}) - F(a)$, the formula we wanted.

The definite integral of a continuous function f can be found by finding an antiderivative of f (any antiderivative of f will work) and then doing some arithmetic with this antiderivative. The theorem does not tell us how to find an antiderivative of f, and it does not tell us how to find the definite integral of a discontinuous function. It is possible to evaluate definite integrals of some discontinuous functions (Section 4.3), but the Fundamental Theorem of Calculus can not be used to do so.

Example 3: Evaluate $\int_0^2 (x^2 - 1)\ dx$.

Solution: $F(x) = \frac{x^3}{3} - x$ is an antiderivative of $f(x) = x^2 - 1$ (check that $D(\frac{x^3}{3} - x) = x^2 - 1$), so

$$\int_0^2 (x^2 - 1)\ dx = \frac{x^3}{3} - x \Big|_0^2 = \{ \frac{2^3}{3} - 2 \} - \{ \frac{0^3}{3} - 0 \} = 2/3 - 0 = 2/3.$$

If friends had picked a different antiderivative of $x^2 - 1$, say $F(x) = \frac{x^3}{3} - x + \mathbf{4}$, then their calculations would be slightly different but the result would be the same:

$$\int_0^2 (x^2 - 1)\ dx = \frac{x^3}{3} - x + \mathbf{4} \Big|_0^2 = (\frac{2^3}{3} - 2 + \mathbf{4}) - (\frac{0^3}{3} - 0 + \mathbf{4}) = 14/3 - 4 = 2/3.$$

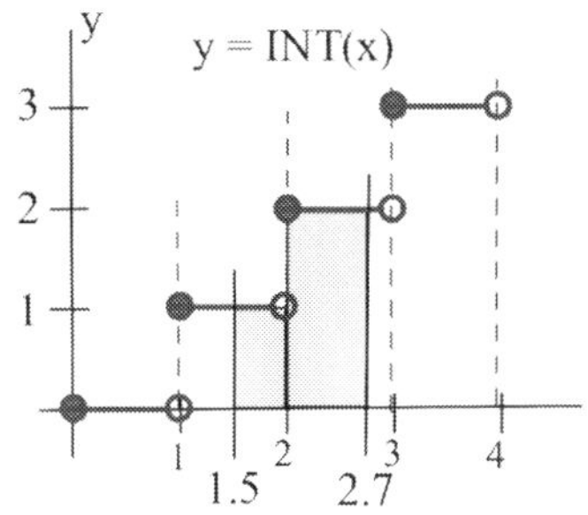

Fig. 6

Practice 2: Evaluate $\int_1^3 (3x^2 - 1)\ dx$.

Example 4: Evaluate $\int_{1.5}^{2.7} INT(x)\ dx$. (INT(x) is the largest integer less than or equal to x. Fig. 6)

Solution: $f(x) = INT(x)$ is not continuous at $x = 2$ in the interval $[1.5, 2.7]$ so the Fundamental Theorem of Calculus can not be used. We can, however, use our understanding of the meaning of an integral to get

$$\int_{1.5}^{2.7} INT(x)\ dx = (\text{area for } x \text{ between 1.5 and 2}) + (\text{area for } x \text{ between 2 and 2.7})$$

$$= (\text{base})(\text{height}) + (\text{base})(\text{height}) = (.5)(1) + (.7)(2) = 1.9\ .$$

Practice 3: Evaluate $\int_{1.3}^{3.4} INT(x)\ dx$.

Calculus is the study of derivatives and integrals, their meanings and their applications. The Fundamental Theorem of Calculus shows that differentiation and integration are closely related and that integration is really antidifferentiation, the inverse of differentiation.

Applications — The Future

Calculus is important for many reasons, but students are usually required to study calculus because it is needed for understanding concepts and doing applications in a variety of fields. The Fundamental Theorem of Calculus is very important to both pursuits.

Most applied problems in integral calculus require the following steps to get from the problem to a numerical answer:

$$\text{Applied problem} \xrightarrow{\ 1\ } \text{Riemann sum (or area)} \xrightarrow{\ 2\ } \text{definite integral} \xrightarrow{\ 3\ } \text{number}$$

In some cases, the path from the problem to the answer may be abbreviated, but the three steps are commonly used.

Step 1 is absolutely vital. If we can not translate the ideas of an applied problem into an area or a Riemann sum or a definite integral, then we can not use integral calculus to solve the problem. For a few types of applied problems, we will be able to go directly from the problem to an integral, but usually it will be easier to first break the problem into smaller pieces and to build a Riemann sum. Section 4.8 and all of chapter 5 will focus on translating different types of applied problems into Riemann sums and definite integrals. **Computers and calculators are seldom of any help with Step 1.**

Step 2 is usually easy. If we have a Riemann sum $\sum_{k=1}^{n} f(c_k)\Delta x_k$ on the interval $[a,b]$, then the limit of the sum is simply the definite integral $\int_a^b f(x)\,dx$.

Step 3 can be handled in several ways.

If the function f is relatively simple, there are several ways to find an antiderivative of f (sections 4.6, parts of chapter 6 and others), and then Part 2 of the Fundamental Theorem of Calculus can be used to get a numerical answer.

If the function f is more complicated, then integral tables (section 4.8) or computers (symbolic manipulators such as Maple or Mathematica) can be used to find an antiderivative of f . Then Part 2 of the Fundamental Theorem of Calculus can be used to get a numerical answer.

If an antiderivative of f cannot be found, approximate numerical answers for the definite integral can be found by various summation methods (section 4.9). These summation methods are typically done on computers, and program listings are included in an Appendix.

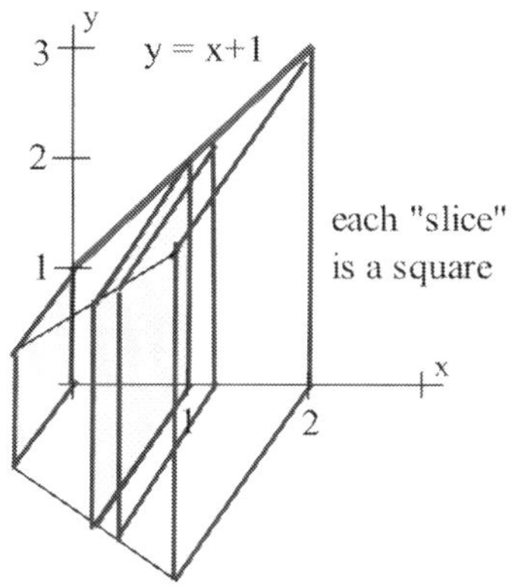

Fig. 7

Usually the difficulties in solving an applied problem come with the 1st and 3rd steps, and the most time will be spent working with them. There are techniques and details to master and understand, but it is also important to keep in mind where these techniques and details fit into the bigger picture.

The next Example illustrates these steps for the problem of finding a volume of a solid. Problems of finding volumes of solids will be examined in more detail in Section 5.1.

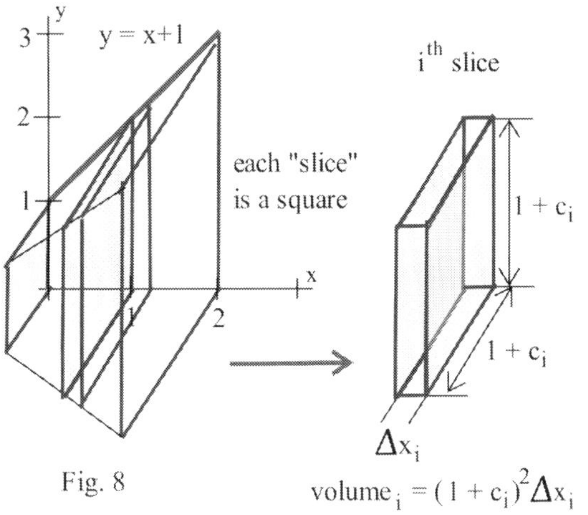

Fig. 8

Example 5: Find the volume of the solid in Fig. 7 for $0 \le x \le 2$. (Each perpendicular "slice" through the solid is a square.)

Solution:

Step 1: Going from the figure to a Riemann sum.

If we break the solid into n "slices" with cuts perpendicular to the x–axis, at $x_1, x_2, x_3, \ldots, x_{n-1}$ (like cutting a loaf of bread), then the volume of the original solid is the sum of the volumes of the "slices" (Fig. 8):

$$\text{Total Volume} = \sum_{i=1}^{n} (\text{volume of the } i^{th} \text{ slice}) .$$

The volume of the ith slice is approximately equal to the volume of a box:

(height of the slice)·(base of the slice)·(thickness)

$\approx (c_i + 1)\cdot(c_i + 1)\cdot\Delta x_i$

where c_i is any value between x_{i-1} and x_i .

Therefore,

$$\text{Total Volume} \approx \sum_{i=1}^{n} (c_i + 1)\cdot(c_i + 1)\cdot\Delta x_i$$

which is a Riemann sum.

Step 2: Going from the Riemann sum to a definite integral.

The Riemann sum approximation of the total volume in Step 1 is improved by taking thinner slices (making all of the Δx_i small), and

$$\text{Total volume} = \lim_{mesh \to 0} \left\{ \sum_{i=1}^{N} (c_i + 1) \cdot (c_i + 1) \cdot \Delta x_i \right\}$$

$$= \int_0^2 (x+1)(x+1)\,dx = \int_0^2 (x^2 + 2x + 1)\,dx \,.$$

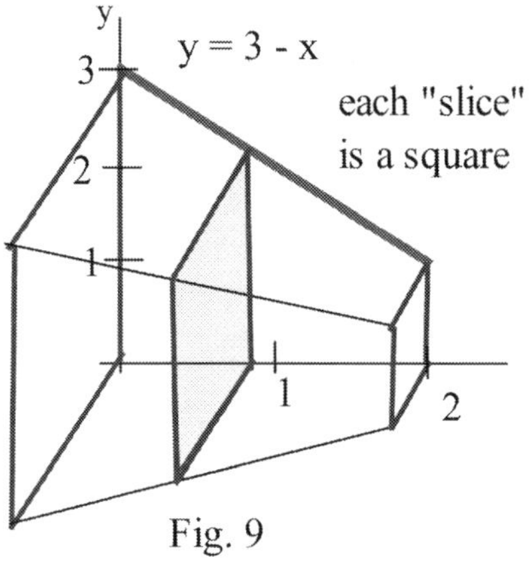

Fig. 9

Step 3: Going from the definite integral to a numerical answer.

We can use Part 2 of the Fundamental Theorem of Calculus to evaluate the integral.

$F(x) = \frac{1}{3} x^3 + x^2 + x$ is an antiderivative of $x^2 + 2x + 1$

(check by differentiating $F(x)$), so

$$\int_0^2 (x^2 + 2x + 1)\,dx = F(2) - F(0) = \left\{ \tfrac{1}{3} \mathbf{2}^3 + \mathbf{2}^2 + \mathbf{2} \right\} - \left\{ \tfrac{1}{3} \mathbf{0}^3 + \mathbf{0}^2 + \mathbf{0} \right\}$$

$$= \left\{ \tfrac{26}{3} \right\} - \left\{ 0 \right\} = \tfrac{26}{3} = 8\tfrac{2}{3} \,.$$

The volume of the solid shape in Fig. 7 is exactly $8\frac{2}{3}$ cubic inches.

Practice 4: Find the volume of the solid shape in Fig. 9 for $0 \le x \le 2$. (Each "slice" through the solid perpendicular to the x–axis is a square.)

Leibniz' Rule For Differentiating Integrals

If the endpoint of an integral is a function of x rather than simply x,
then we need to use the Chain Rule together with part 1 of the Fundamental Theorem of Calculus to calculate the derivative of the integral. According to the Chain Rule, if

$\frac{d}{dx} A(x) = f(x)$, then $\frac{d}{dx} A(x^2) = f(x^2) \cdot \mathbf{2}x$ and, applying the Chain Rule to the derivative of the integral,

$$\frac{d}{dx}\left(\int_a^{g(x)} f(t)\,dt \right) = \frac{d}{dx} A(\, g(x)\,) = f(\, g(x)\,) \cdot \mathbf{g}'(x).$$

If f is a continuous function and $A(x) = \int_a^x f(t)dt$

then $\frac{d}{dx}\left(\int_a^x f(t)dt \right) = \frac{d}{dx} A(x) = f(x)$ **(Fundamental Theorem, Part I)**

and, if g is differentiable, $\frac{d}{dx}\left(\int_a^{g(x)} f(t)dt \right) = \frac{d}{dx} A(\, g(x)\,) = f(\, g(x)\,)\cdot g'(x)$ **(Leibniz' Rule)**

Example 6: Calculate $\frac{d}{dx}\left(\int_a^{5x} t^2\, dt \right)$, $\frac{d}{dx}\left(\int_a^{x^2} \cos(u)\, du \right)$, $\frac{d}{dw}\left(\int_a^{\sin(w)} z^3\, dz \right)$.

Solution: $\frac{d}{dx}\left(\int_a^{5x} t^2\, dt \right) = (5x)^2 \cdot 5 = 125x^2$. $\frac{d}{dx}\left(\int_a^{x^2} \cos(u)\, du \right) = \cos(x^2)\cdot 2x = 2x\cdot\cos(x^2)$

$\frac{d}{dw}\left(\int_a^{\sin(w)} z^3\, dz \right) = (\sin(w))^3 \cdot\cos(w) = \sin^3(w)\cos(w)$.

Practice 5: Find $\frac{d}{dx}\left(\int_0^{x^3} \sin(t)\, dt \right)$.

PROBLEMS:

1. $A(x) = \int_0^x 3t^2\, dt$ (a) Use part 2 of the Fundamental Theorem to find a formula for $A(x)$ and then differentiate $A(x)$ to obtain a formula for $A'(x)$. Evaluate $A'(x)$ at $x = 1, 2,$ and 3.

 (b) Use part 1 of the Fundamental Theorem to evaluate $A'(x)$ at $x = 1, 2,$ and 3.

2. $A(x) = \int_1^x (\,1 + 2t\,)\, dt$ (a) Use part 2 of the Fundamental Theorem to find a formula for $A(x)$ and then differentiate $A(x)$ to obtain a formula for $A'(x)$. Evaluate $A'(x)$ at $x = 1, 2,$ and 3.

 (b) Use part 1 of the Fundamental Theorem to evaluate $A'(x)$ at $x = 1, 2,$ and 3.

In problems 3 – 8 , evaluate $A'(x)$ at $x = 1, 2,$ and 3.

3. $A(x) = \int_0^x 2t \, dt$

4. $A(x) = \int_1^x 2t \, dt$

5. $A(x) = \int_{-3}^x 2t \, dt$

6. $A(x) = \int_0^x (3 - t^2) \, dt$

7. $A(x) = \int_0^x \sin(t) \, dt$

8. $A(x) = \int_1^x | t - 2 | \, dt$

In problems 9 – 13, $A(x) = \int_0^x f(t) \, dt$ for the functions in Figures 10 – 14. Evaluate $A'(1), A'(2), A'(3)$.

9. f in Fig. 10

10. f in Fig. 11

11. f in Fig. 12

12. f in Fig. 13

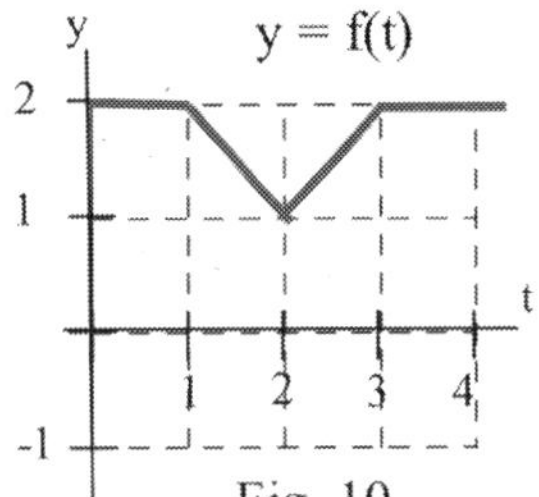

Fig. 10

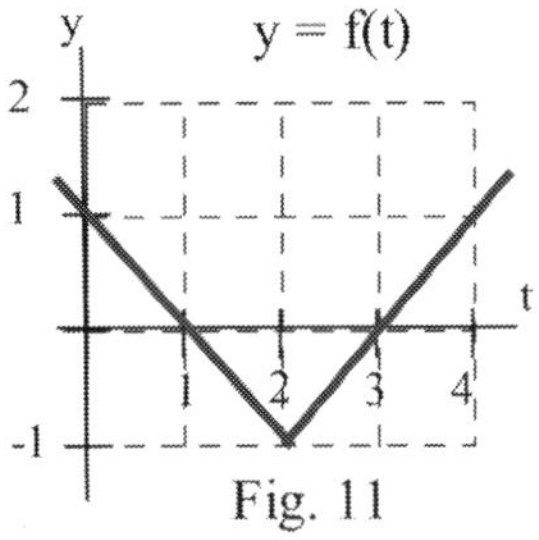

Fig. 11

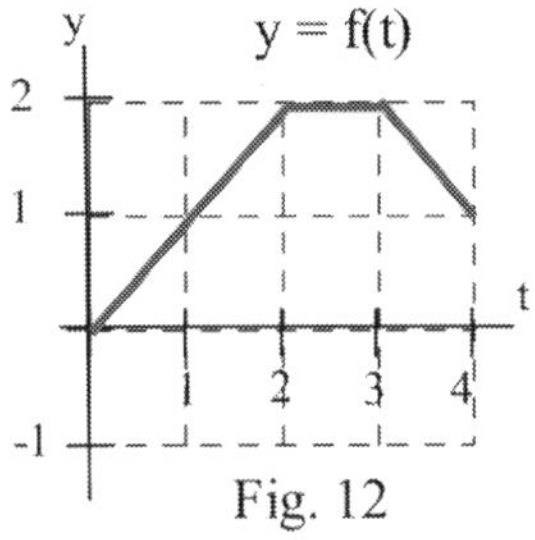

Fig. 12

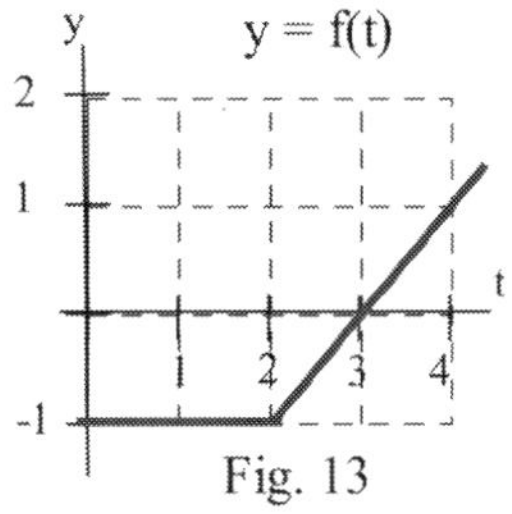

Fig. 13

In problems 13 – 33, verify that $F(x)$ is an antiderivative of the integrand $f(x)$ and use Part 2 of the Fundamental Theorem to evaluate the definite integrals.

13. $\int_0^1 2x \, dx$, $F(x) = x^2 + 5$

14. $\int_1^4 3x^2 \, dx$, $F(x) = x^3 + 2$

15. $\int_1^3 x^2 \, dx$, $F(x) = \frac{1}{3} x^3$

16. $\int_0^3 (x^2 + 4x - 3) \, dx$, $F(x) = \frac{1}{3} x^3 + 2x^2 - 3x$

17. $\int_1^5 \frac{1}{x} \, dx$, $F(x) = \ln(x)$

18. $\int_2^5 \frac{1}{x} \, dx$, $F(x) = \ln(x) + 4$

19. $\int_{1/2}^3 \frac{1}{x} \, dx$, $F(x) = \ln(x)$

20. $\int_1^3 \frac{1}{x} \, dx$, $F(x) = \ln(x) + 2$

21. $\int_0^{\pi/2} \cos(x) \, dx$, $F(x) = \sin(x)$

22. $\int_0^{\pi} \sin(x) \, dx$, $F(x) = -\cos(x)$

23. $\int_0^1 \sqrt{x} \, dx$, $F(x) = \frac{2}{3} x^{3/2}$

24. $\int_1^4 \sqrt{x} \, dx$, $F(x) = \frac{2}{3} x^{3/2}$

25. $\int_1^7 \sqrt{x} \, dx$, $F(x) = \frac{2}{3} x^{3/2}$

26. $\int_1^4 \frac{1}{2\sqrt{x}} \, dx$, $F(x) = \sqrt{x}$

27. $\int_1^9 \frac{1}{2\sqrt{x}}\,dx$, $F(x) = \sqrt{x}$

28. $\int_2^5 \frac{1}{x^2}\,dx$, $F(x) = -\frac{1}{x}$

29. $\int_{-2}^3 e^x\,dx$, $F(x) = e^x$

30. $\int_0^3 \frac{2x}{1+x^2}\,dx$, $F(x) = \ln(1+x^2)$

31. $\int_0^{\pi/4} \sec^2(x)\,dx$, $F(x) = \tan(x)$

32. $\int_1^e \ln(x)\,dx$, $F(x) = x \cdot \ln(x) - x$

33. $\int_0^3 2x\sqrt{1+x^2}\,dx$, $F(x) = \frac{2}{3}(1+x^2)^{3/2}$

For problems 34 – 48, find an antiderivative of the integrand and use Part 2 of the Fundamental Theorem to evaluate the definite integral.

34. $\int_2^5 3x^2\,dx$

35. $\int_{-1}^2 x^2\,dx$

36. $\int_1^3 (x^2 + 4x - 3)\,dx$

37. $\int_1^e \frac{1}{x}\,dx$

38. $\int_{\pi/4}^{\pi/2} \sin(x)\,dx$

39. $\int_{25}^{100} \sqrt{x}\,dx$

40. $\int_3^5 \sqrt{x}\,dx$

41. $\int_1^{10} \frac{1}{x^2}\,dx$

42. $\int_1^{1000} \frac{1}{x^2}\,dx$

43. $\int_0^1 e^x\,dx$

44. $\int_{-2}^2 \frac{2x}{1+x^2}\,dx$

45. $\int_{\pi/6}^{\pi/4} \sec^2(x)\,dx$

46. $\int_0^1 e^{2x}\,dx$

47. $\int_3^3 \sin(x) \cdot \ln(x)\,dx$

48. $\int_2^4 (x-2)^3\,dx$

In problems 49 – 54 , find the area of each shaded region.

49. Region in Fig. 14.

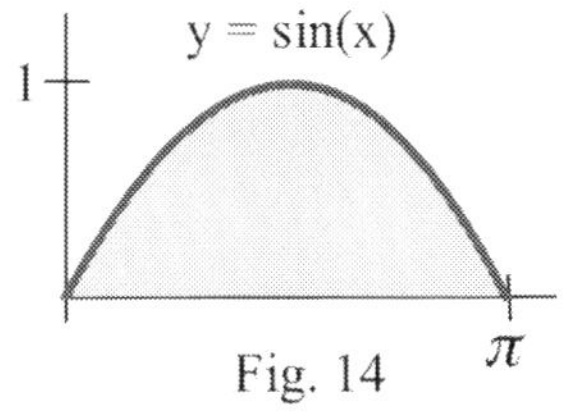

Fig. 14

50. Region in Fig. 15.

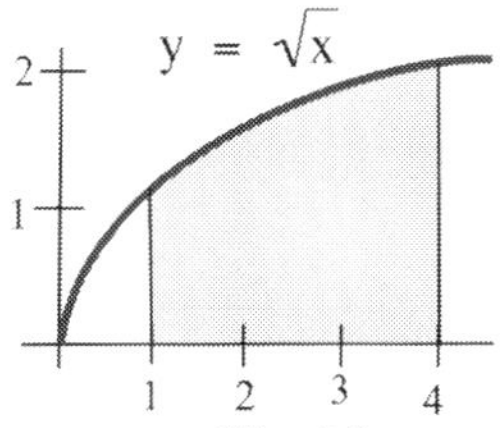

Fig. 15

51. Region in Fig. 16.

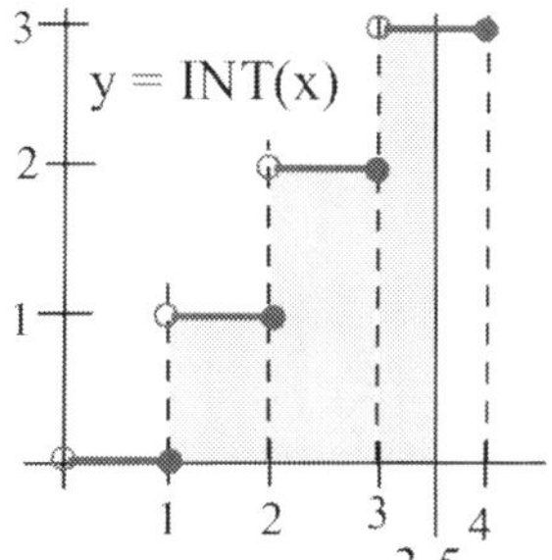

Fig. 16

52. Region in Fig. 17.

53. Region in Fig. 18.

54. Region in Fig. 19.

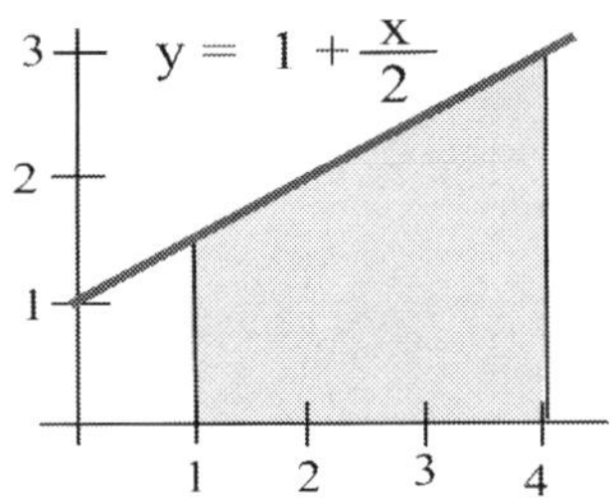

Fig. 17

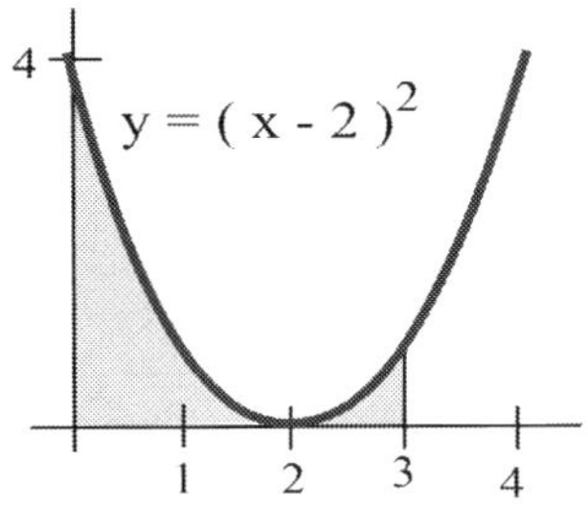

Fig. 18

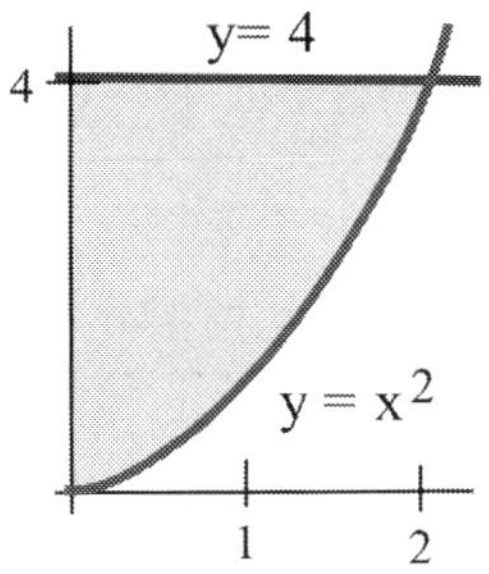

Fig. 19

Leibniz' Rule

55. If **D**(A(x)) = tan(x) , then find **D**(A(3x)) , **D**(A(x^2)) , and **D**(A(sin(x))) .

56. If **D**(B(x)) = sec(x) , then find **D**(B(3x)) , **D**(B(x^2)) , and **D**(B(sin(x))) .

57. $\frac{d}{dx}\left(\int_1^{5x} \sqrt{1+t}\, dt \right)$

58. $\frac{d}{dx}\left(\int_2^{x^2} \sqrt{1+t}\, dt \right)$

59. $\frac{d}{dx}\left(\int_0^{\sin(x)} \sqrt{1+t}\, dt \right)$

60. $\frac{d}{dx}\left(\int_1^{2+3x} t^2 + 5\, dt \right)$

61. $\frac{d}{dx}\left(\int_0^{1-2x} 3t^2 + 2\, dt \right)$

62. $\frac{d}{dx}\left(\int_x^{9} 3t^2 + 2\, dt \right)$

63. $\frac{d}{dx}\left(\int_x^{\pi} \cos(3t)\, dt \right)$

64. $\frac{d}{dx}\left(\int_{7x}^{\pi} \cos(2t)\, dt \right)$

65. $\frac{d}{dx}\left(\int_x^{x^2} \tan(t)\, dt \right)$

66. $\frac{d}{dx}\left(\int_0^{\pi} \cos(3t)\, dt \right)$

67. $\frac{d}{dx}\left(\int_2^{\ln(x)} 5t \cdot \cos(3t)\, dt) \right)$

68. $\frac{d}{dx}\left(\int_0^{\pi} \tan(7t)\, dt \right)$

Very Optional Problems

a. $\int_0^{ice} 3x^2\, dx$ What a calculus student puts in a drink.

b. $\int_1^{cabin} \frac{1}{x}\, dx$ Where Abe Lincoln was born.

c. $\int_0^{cerely} \cos(x)\, dx$ How the calculus student ended a letter.

d. $\int_1^{jam} \frac{1}{x}\, dx$ What a forester puts on toast.

Section 4.5 **PRACTICE Answers**

Practice 1: A(1) = 1, A(2) = 1.5, A(3) = 1, A(4) = 0.5

A '(x) = f(x) so A '(1) = f(1) = 1, A '(2) = f(2) = 0, A '(3) = –1, A '(4) = 0.

Practice 2: $F(x) = x^3 - x$ is one antiderivative of $f(x) = 3x^2 - 1$ (F ' = f) so

$$\int_1^3 3x^2 - 1\,dx = x^3 - x \Big|_1^3 = (3^3 - 3) - (1^3 - 1) = 24.$$

$F(x) = x^3 - x + \mathbf{7}$ is another antiderivative of $f(x) = 3x^2 - 1$ so

$$\int_1^3 3x^2 - 1\,dx = x^3 - x + \mathbf{7} \Big|_1^3 = (3^3 - 3 + \mathbf{7}) - (1^3 - 1 + \mathbf{7}) = 24.$$

No matter which antiderivative of $f(x) = 3x^2 - 1$ you use, the value of the

definite integral $\int_1^3 3x^2 - 1\,dx$ is **24.**

Practice 3: $\int_{1.3}^{3.4} \text{INT}(x)\,dx = \mathbf{3.9}$. Since f(x) = INT(x) is not continuous on the interval [1.3, 3.4] so we can not use the Fundamental Theorem of Calculus. Instead, we can think of the definite integral as an area (Fig. 20).

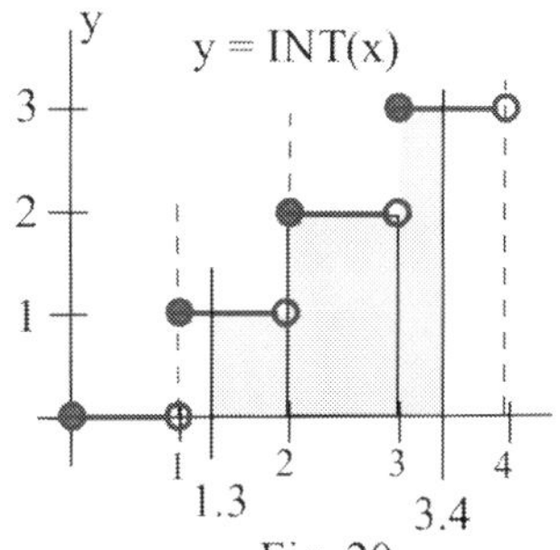

Fig. 20

Practice 4: Total volume $= \lim_{mesh \to 0} \left\{ \sum_{i=1}^{N} (3 - c_i) \cdot (3 - c_i) \cdot \Delta x_i \right\}$

$$= \int_0^2 (3 - x)(3 - x)\,dx$$

$$= \int_0^2 (9 - 6x + x^2)\,dx$$

$$= 9x - 3x^2 + \frac{x^3}{3} \Big|_0^2 = (18 - 12 + \frac{8}{3}) - (0–0+0) = \mathbf{8\frac{2}{3}} .$$

Practice 5: $\frac{d}{dx}\left(\int_0^{x^3} \sin(t)\,dt \right) = \sin(x^3)\,\frac{d\,x^3}{dx} = \mathbf{3x^2 \cdot \sin(x^3)}$.

4.6 FINDING ANTIDERIVATIVES

In order to use Part 2 of the Fundamental Theorem of Calculus, an antiderivative of the integrand is needed, but sometimes it is not easy to find one. This section collects some of the information we already have about the general properties of antiderivatives and about antiderivatives of particular functions. It shows how to use the information you already have to find antiderivatives of more complicated functions and introduces a "change of variable" technique to make this job easier.

INDEFINITE INTEGRALS AND ANTIDERIVATIVES

Antiderivatives appear so often that there is a notation to indicate the antiderivative of a function:

Definition: $\int f(x)\,dx$, read as "the **indefinite integral** of f" or as "the **antiderivatives** of f," represents the collection (or family) of functions whose derivatives are f.

If F is an antiderivative of f, then every member of the family $\int f(x)\,dx$ has the form $F(x) + C$ for some constant C. We write $\int f(x)\,dx = F(x) + C$, where C represents an arbitrary constant.

There are no small families in the world of antiderivatives: if f has one antiderivative F, then f has an infinite number of antiderivatives and every one of them has the form $F(x) + C$. This means that there are many ways to write a particular indefinite integral and some of them may look very different. You can check that $F(x) = \sin^2(x)$, $G(x) = -\cos^2(x)$, and $H(x) = 2\sin^2(x) + \cos^2(x)$ all have the same derivative $f(x) = 2\sin(x)\cos(x)$, so the indefinite integral of $2\sin(x)\cos(x)$, $\int 2\sin(x)\cos(x)\,dx$, can be written in several ways: $\sin^2(x) + C$, or $-\cos^2(x) + C$, or $2\sin^2(x) + \cos^2(x) + C$.

Practice 1: Verify that $\int 2\tan(x)\cdot\sec^2(x)\,dx = \tan^2(x) + C$ and $\int 2\tan(x)\cdot\sec^2(x)\,dx = \sec^2(x) + C$.

Properties of Antiderivatives (Indefinite Integrals)

The following properties are true of all antiderivatives.

General Properties of Antiderivatives

If f and g are integrable functions, then

1.	Constant Multiple	$\int k\cdot f(x)\,dx$	=	$k\cdot\int f(x)\,dx$
2.	Sum	$\int f(x) + g(x)\,dx$	=	$\int f(x)\,dx + \int g(x)\,dx$
3.	Difference	$\int f(x) - g(x)\,dx$	=	$\int f(x)\,dx - \int g(x)\,dx$

There are general rules for derivatives of products and quotients. Unfortunately, there are no easy general patterns for antiderivatives of products and quotients, and only one more general property will be added to the list (in Chapter 6).

We have already found antiderivatives for a number of important functions.

Particular Antiderivatives

1. Constant Function: $\int k\,dx = kx + C$

2. Powers of x: $\int x^n\,dx = \frac{x^{n+1}}{n+1} + C$ if $n \neq -1$

$$\int x^{-1}\,dx = \int \frac{1}{x}\,dx = \ln|x| + C$$

Common special cases: $\int \sqrt{x}\,dx = \frac{2}{3}x^{3/2} + C$

$$\int \frac{1}{\sqrt{x}}\,dx = 2x^{1/2} + C = 2\sqrt{x} + C$$

3. Trigonometric Functions:

$\int \cos(x)\,dx = \sin(x) + C$ $\int \sin(x)\,dx = -\cos(x) + C$

$\int \sec^2(x)\,dx = \tan(x) + C$ $\int \csc^2(x)\,dx = -\cot(x) + C$

$\int \sec(x)\cdot\tan(x)\,dx = \sec(x) + C$ $\int \csc(x)\cdot\cot(x)\,dx = -\csc(x) + C$

4. Exponential Functions: $\int e^x\,dx = e^x + C$

All of these antiderivatives can be verified by differentiating. The list of antiderivatives of particular functions will continue to grow in the coming chapters and will eventually include antiderivatives of additional trigonometric functions, the inverse trigonometric functions, logarithms, and others.

Antiderivatives of More Complicated Functions

Antiderivatives are very sensitive to small changes in the integrand, and we need to be very careful. Fortunately, an antiderivative can always be checked by differentiating, so even though we may not find the correct antiderivative, we should be able to determine if an antiderivative is incorrect.

Example 1: We know one antiderivative of $\cos(x)$ is $\sin(x)$, so $\int \cos(x)\,dx = \sin(x)+C$. Find (a) $\int \cos(2x+3)\,dx$, (b) $\int \cos(5x-7)\,dx$, and (c) $\int \cos(x^2)\,dx$.

Solution: (a) Since $\sin(x)$ is an antiderivative of $\cos(x)$, it is reasonable to hope that $\sin(2x+3)$ will be an antiderivative of $\cos(2x+3)$. When we differentiate $\sin(2x+3)$, however, we see that $\mathbf{D}(\sin(2x+3)) = \cos(2x+3)\cdot\mathbf{2}$, exactly twice the result we want. Let's try again by modifying our "guess" and differentiating. Since the result of differentiating was twice the function we wanted, lets try one half the original function: $\mathbf{D}(\frac{1}{2}\sin(2x+3)) = \frac{1}{2}\cos(2x+3)\cdot\mathbf{2} = \cos(2x+3)$, exactly what we want. Therefore, $\int \cos(2x+3)\,dx = \frac{1}{2}\sin(2x+3) + C$.

(b) $\mathbf{D}(\sin(5x-7)) = \cos(5x-7)\cdot\mathbf{5}$, but $\mathbf{D}(\frac{1}{5}\sin(5x-7)) = \frac{1}{5}\cos(5x-7)\cdot\mathbf{5} = \cos(5x-7)$ so $\int \cos(5x-7)\,dx = \frac{1}{5}\sin(5x-7) + C$.

(c) $\mathbf{D}(\sin(x^2)) = \cos(x^2)\cdot\mathbf{2x}$. It was easy enough in parts (a) and (b) to modify our "guesses" to eliminate the constants 2 and 5, but the x is much harder to eliminate.

$\mathbf{D}(\frac{1}{2x}\sin(x^2)) = \mathbf{D}(\frac{\sin(x^2)}{2x})$ requires the quotient rule (or the product rule), and

$$\mathbf{D}\left(\frac{\sin(x^2)}{2x}\right) = \frac{2x\cdot\mathbf{D}(\sin(x^2)) - \sin(x^2)\cdot\mathbf{D}(2x)}{(2x)^2} = \cos(x^2) - \frac{\sin(x^2)}{2x^2}$$

which is not the result we want. A fact which can be proved using more advanced mathematics is that $\cos(x^2)$ does not have an "elementary" antiderivative composed of polynomials, roots, trigonometric functions, or exponential functions or their inverses. The value of a definite integral of $\cos(x^2)$ could still be approximated as accurately as needed by using Riemann sums or one of the numerical techniques in Section 4.9 . The point of part (c) is that even a simple looking integrand can be very difficult. At this point, there is no easy way to tell.

Getting the Constants Right

One method for finding antiderivatives is to "guess" the form of the answer, differentiate your "guess" and then modify your original "guess" so its derivative is exactly what you want it to be. Parts (a) and (b) of the previous example illustrate that technique.

Example 2: We know $\int \sec^2(x)\,dx = \tan(x) + C$ and $\int \frac{1}{\sqrt{x}}\,dx = 2\sqrt{x} + C$. Find (a) $\int \sec^2(3x+7)\,dx$ and (b) $\int \frac{1}{\sqrt{5x+3}}\,dx$.

Solution: (a) If we "guess" an answer of tan(3x+7) and then differentiate it, we get $\mathbf{D}(\tan(3x+7)) = \sec^2(3x+7)\cdot\mathbf{D}(3x+7)) = \sec^2(3x+7)\cdot\mathbf{3}$ which is 3 times too large. If we divide our original guess by 3 and try again, we have

$$\mathbf{D}(\tfrac{1}{3}\tan(3x+7)) = \tfrac{1}{3}\mathbf{D}(\tan(3x+7)) = \tfrac{1}{3}\sec^2(3x+7)\cdot\mathbf{3} = \sec^2(3x+7) \text{ so } \int \sec^2(3x+7)\,dx = \tfrac{1}{3}\tan(3x+7) + C.$$

(b) If we "guess" an answer of $2\sqrt{5x+3}$ and then differentiate it, we get

$$\mathbf{D}(2\sqrt{5x+3}) = 2\cdot\tfrac{1}{2}(5x+3)^{-1/2}\,\mathbf{D}(5x+3) = \mathbf{5}\cdot(5x+3)^{-1/2} = \frac{\mathbf{5}}{\sqrt{5x+3}}$$ which is 5 times too large.

Dividing our guess by 5 and differentiating, we have

$$\mathbf{D}(\tfrac{2}{5}\sqrt{5x+3}) = \tfrac{1}{5}\cdot\mathbf{D}(2\sqrt{5x+3}) = \tfrac{1}{5}\cdot\mathbf{5}\cdot(5x+3)^{-1/2} = \frac{1}{\sqrt{5x+3}} \text{ so } \int \frac{1}{\sqrt{5x+3}}\,dx = \tfrac{2}{5}\sqrt{5x+3} + C.$$

Practice 2: Find $\int \sec^2(7x)\,dx$ and $\int \frac{1}{\sqrt{3x+8}}\,dx$.

The "guess and check" method is a very effective technique if you can make a good first guess, one that misses the desired result only by a constant multiple. In that situation just divide the first guess by the unwanted constant multiple. If the derivative of your guess misses by more than a constant multiple, then more drastic changes are needed. Sometimes the next technique can help.

Making Patterns More Obvious — Changing the Variable

Successful integration is mostly a matter of recognizing patterns. The "change of variable" technique can make some underlying patterns of an integral easier to recognize. Basically, the technique involves rewriting an integral which is originally in terms of one variable, say x, in terms of another variable, say u. The hope is that after doing the rewriting, it will be easier to find an antiderivative of the new integrand.

For example, $\int \cos(5x+1)\,dx$ can be rewritten by setting $u = 5x+1$. Then $\cos(5x+1)$ becomes $\cos(u)$ and $\mathbf{du} = d(5x+1) = \mathbf{5}\,dx$ so the differential dx is $\frac{1}{5}$ du. Finally, $\int \cos(5x+1)\,dx = \int \cos(u)\frac{\mathbf{1}}{\mathbf{5}}\mathbf{du}$

$= \frac{1}{5}\sin(u) + C = \frac{1}{5}\sin(5x+1) + C$.

The steps of the "change of variable" method can be summarized as

(1) set a new variable, say u , equal to some function of the original variable x (usually u is set equal to some part of the original integrand function)

(2) calculate the differential du as a function of dx

(3) rewrite the original integral in terms of u and du

(4) integrate the new integral to get an answer in terms of u

(5) replace the u in the answer to get an answer in terms of the original variable

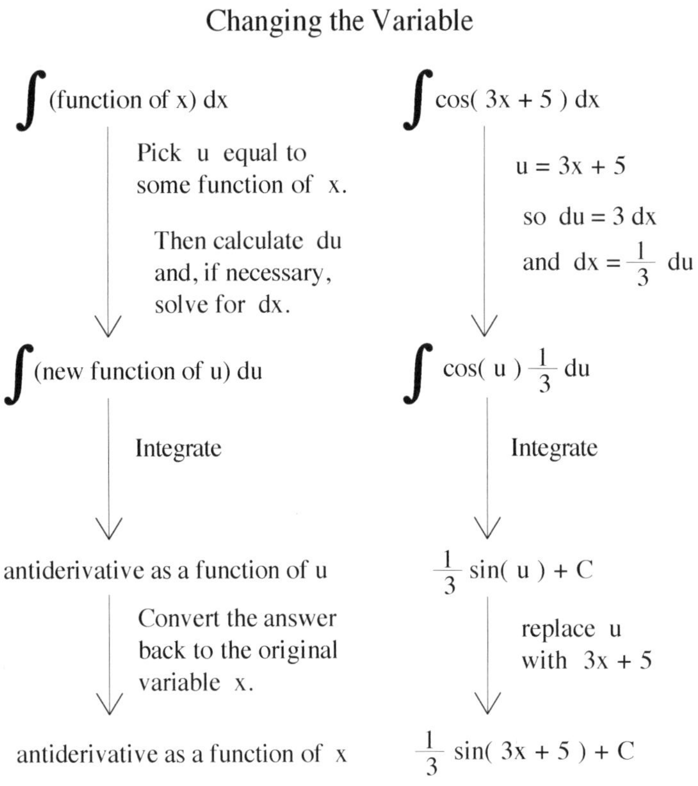

Fig. 1

Fig. 1 shows the general flow of this algorithm and another example.

Example 3: Make the suggested change of variable and rewrite each integral in terms of u and du.

(a) $\int \cos(x)\, e^{\sin(x)}\, dx$ with $u = \sin(x)$(b) $\int \frac{2x}{5+x^2}\, dx$ with $u = 5+x^2$

Solution: (a) Since $u = \sin(x)$, then $du = d(\sin(x)) = \cos(x)\, dx$ so $e^{\sin(x)} = e^u$ and $\mathbf{\cos(x)\, dx = du}$.

Then $\int \mathbf{\cos(x)}\, e^{\sin(x)}\, \mathbf{dx} = \int e^u\, \mathbf{du} = e^u + C = e^{\sin(x)} + C$.

(b) If $u = 5+x^2$, then $\mathbf{du} = d(5+x^2) = \mathbf{2x\ dx}$, so $\int \frac{\mathbf{2x}}{5+x^2}\, \mathbf{dx} = \int \frac{1}{u}\, \mathbf{du} = \ln|\, u\, | + C = \ln|\, 5+x^2\, | + C$.

In each example, the change of variable did not find the antiderivative, but it did make the pattern of the integrand more obvious so it was easier to determine an antiderivative.

Practice 3: Make the suggested change of variable and rewrite each integral in terms of u and du.

(a) $\int (7x+5)^3\, dx$ with $u = 7x+5$ (b) $\int \sin(x^3-1)\, 3x^2\, dx$ with $u = x^3-1$.

In the future, you will need to decide what u should equal, and there are no rules which guarantee that your choice will lead to an easier integral. There is, however, a "rule of thumb" which frequently results in easier integrals. Even though the following suggestion is not guaranteed, it is often worth trying.

A "Rule of thumb" for changing the variable

If part of the integrand is a composition of functions, $f(g(x))$,

then try setting $u = g(x)$, the "inner" function.

If part of the integrand is being raised to a power, try setting u equal to the part being raised to the power. For example, if the integrand includes $(3 + \sin(x))^5$, try $u = 3 + \sin(x)$. If part of the integrand involves a trigonometric (or exponential or logarithmic) function of another function, try setting u equal to the "inside" function. If the integrand includes the function $\sin(3+x^2)$, try $u = 3+x^2$.

The key to becoming skilled at selecting a good u and correctly making the substitution is **practice**.

Example 4: Select a function for u for each integral and rewrite the integral in terms of u and du.

(a) $\int \cos(3x)\sqrt{2+\sin(3x)}\, dx$ (b) $\int \frac{5e^x}{2+e^x}\, dx$ (c) $\int e^x \cdot \sin(e^x)\, dx$

Solution: (a) Put $u = 2+\sin(3x)$. Then $\mathbf{du = 3\cos(3x)\, dx}$, and the integral becomes $\int \frac{1}{3}\sqrt{u}\, du$.

(b) Put $u = 2 + e^x$. Then $\mathbf{du = e^x\, dx}$, and the integral becomes $\int \frac{5}{u}\, du$.

(c) Put $u = e^x$. Then $\mathbf{du = e^x\, dx}$, and the integral becomes $\int \sin(u)\, du$.

Changing the Variable and Definite Integrals

Once an antiderivative in terms of u is found, we have a choice of methods. We can

(a) rewrite our antiderivative in terms of the original variable x, and then evaluate the antiderivative at the integration endpoints and subtract, or

(b) change the integration endpoints to values of u, and evaluate the antiderivative in terms of u before subtracting.

If the original integral had endpoints $x = a$ and $x = b$, and we make the substitution $u = g(x)$ and $du = g'(x)dx$, then the new integral will have endpoints $u = g(a)$ and $u = g(b)$ and

$$\int_{x=a}^{x=b} (\text{original integrand})\, dx \quad \text{becomes} \quad \int_{u=g(a)}^{u=g(b)} (\text{new integrand})\, du .$$

To evaluate $\int_0^1 (3x-1)^4\,dx$, we can put $u = 3x-1$. Then **du** $= d(3x-1) =$ **3 dx** so the integral becomes

$\int u^4 \frac{1}{3}\,du = \frac{1}{15}u^5 + C.$

(a) Converting our antiderivative back to the variable x and evaluating with the original endpoints:

$$\int_0^1 (3x-1)^4\,dx = \frac{1}{15}(3x-1)^5 \Big|_0^1 = \frac{1}{15}(3\cdot1-1)^5 - \frac{1}{15}(3\cdot0-1)^5 = \frac{1}{15}(2)^5 - \frac{1}{15}(-1)^5 = \frac{32}{15} - \frac{-1}{15} = \frac{33}{15}.$$

(b) Converting the integration endpoints to values of u : when x = 0, then $u = 3x-1 = 3\cdot0 - 1 = -1$,

and when $x = 1$, then $u = 3x-1 = 3\cdot1-1 = 2$ so

$$\int_{x=0}^{x=1} (3x-1)^4\,dx = \int_{u=-1}^{u=2} u^4 \frac{1}{3}\,du = \frac{1}{15}u^5 \Big|_{-1}^{2} = \frac{1}{15}(2)^5 - \frac{1}{15}(-1)^5 = \frac{32}{15} - \frac{-1}{15} = \frac{33}{15}.$$

Both approaches typically involve about the same amount of work and calculation.

Practice 4: If the original integrals in Example 4 had endpoints (a) $x = 0$ to $x = \pi$, (b) $x = 0$ to $x = 2$, and (c) $x = 0$ to $x = \ln(3)$, then the new integrals should have what endpoints?

Special Transformations for $\int \sin^2(x)dx$ and $\int \cos^2(x)dx$

The integrals of $\sin^2(x)$ and $\cos^2(x)$ occur relatively often, and we can find their antiderivatives with the help of two trigonometric identities for $\cos(2x)$:

$$\cos(2x) \equiv 1 - 2\sin^2(x) \quad \text{and} \quad \cos(2x) \equiv 2\cos^2(x) - 1$$

Solving the identies for $\sin^2(x)$ and $\cos^2(x)$, we get $\sin^2(x) \equiv \frac{1-\cos(2x)}{2}$ and $\cos^2(x) \equiv \frac{1+\cos(2x)}{2}$.

Then $\int \sin^2(x)\,dx = \int \frac{1-\cos(2x)}{2}\,dx = \int \frac{1}{2}\,dx - \int \frac{1}{2}\cos(2x)\,dx$

$$= \frac{x}{2} - \frac{1}{2}\frac{\sin(2x)}{2} + C = \frac{x}{2} - \frac{\sin(2x)}{4} + C \quad \text{or} \quad \frac{1}{2}\{x - \sin(x)\cos(x)\} + C .$$

Similarly, $\int \cos^2(x)\,dx = \int \frac{1+\cos(2x)}{2}\,dx = \int \frac{1}{2}\,dx + \int \frac{1}{2}\cos(2x)\,dx$

$$= \frac{x}{2} + \frac{1}{2}\frac{\sin(2x)}{2} + C = \frac{x}{2} + \frac{\sin(2x)}{4} + C \quad \text{or} \quad \frac{1}{2}\{x + \sin(x)\cos(x)\} + C .$$

PROBLEMS

For problems 1 – 4, put $f(x) = x^2$, $g(x) = x$ and verify that

1. $\int_1^2 f(x)\cdot g(x)\,dx \neq \int_1^2 f(x)\,dx \cdot \int_1^2 g(x)\,dx$

2. $\int_1^2 f(x)/g(x)\,dx \neq \int_1^2 f(x)\,dx \,/ \int_1^2 g(x)\,dx$

3. $\int_0^1 f(x)\cdot g(x)\,dx \neq \int_0^1 f(x)\,dx \cdot \int_0^1 g(x)\,dx$

4. $\int_1^4 f(x)/g(x)\,dx \neq \int_1^4 f(x)\,dx \,/ \int_1^4 g(x)\,dx$

For problems 5 – 14 , use the suggested u to find du and rewrite the integral in terms of u and du. Then find an antiderivative in terms of u , and, finally, rewrite your answer in terms of x .

5. $\int \cos(3x)\,dx$ $\quad u = 3x$

6. $\int \sin(7x)\,dx$ $\quad u = 7x$

7. $\int e^x \sin(2 + e^x)\,dx$ $\quad u = 2 + e^x$

8. $\int e^{5x}\,dx$ $\quad u = 5x$

9. $\int \cos(x) \sec^2(\sin(x))\,dx$ $\quad u = \sin(x)$

10. $\int \frac{\cos(x)}{\sin(x)}\,dx$ $\quad u = \sin(x)$

11. $\int \frac{5}{3 + 2x}\,dx$ $\quad u = 3 + 2x$

12. $\int x^2 (5 + x^3)^7\,dx$ $\quad u = 5 + x^3$

13. $\int x^2 \sin(1 + x^3)\,dx$ $\quad u = 1 + x^3$

14. $\int \frac{e^x}{1 + e^x}\,dx$ $\quad u = 1 + e^x$

For problems 15 – 26 , use the change of variable technique to find an antiderivative in terms of x .

15. $\int \cos(4x)\,dx$

16. $\int e^{3x}\,dx$

17. $\int x^3 (5 + x^4)^{11}\,dx$

18. $\int x \cdot \sin(x^2)\,dx$

19. $\int \frac{3x^2}{2 + x^3}\,dx$

20. $\int \frac{\sin(x)}{\cos(x)}\,dx$

21. $\int \frac{\ln(x)}{x}\,dx$

22. $\int x\sqrt{1 - x^2}\,dx$

23. $\int (1 + 3x)^7\,dx$

24. $\int \frac{1}{x} \cdot \sin(\ln(x))\,dx$

25. $\int e^x \cdot \sec(e^x) \cdot \tan(e^x)\,dx$

26. $\int \frac{1}{\sqrt{x}} \cos(\sqrt{x})\,dx$

For problems 27 – 38 , evaluate the definite integrals.

27. $\int_0^{\pi/2} \cos(3x)\,dx$

28. $\int_0^{\pi} \cos(4x)\,dx$

29. $\int_0^1 e^x \cdot \sin(2 + e^x)\,dx$

30. $\int_0^1 e^{5x}\,dx$

31. $\int_{-1}^{1} x^2(1+x^3)^5\,dx$

32. $\int_0^1 x^4(x^5-1)^{10}\,dx$

33. $\int_0^2 \frac{5}{3+2x}\,dx$

34. $\int_0^{\ln(3)} \frac{e^x}{1+e^x}\,dx$

35. $\int_0^1 x\sqrt{1-x^2}\,dx$

36. $\int_2^5 \frac{2}{1+x}\,dx$

37. $\int_0^1 \sqrt{1+3x}\,dx$

38. $\int_0^1 \frac{1}{\sqrt{1+3x}}\,dx$

39. $\int \sin^2(5x)\,dx$

40. $\int \cos^2(3x)\,dx$

41. $\int \frac{1}{2} - \sin^2(x)\,dx$

42. $\int e^{3x} + \sin^2(x)\,dx$

43. Find the area under one arch of the $y = \sin^2(x)$ graph.

44. Evaluate $\int_0^{2\pi} \sin^2(x)\,dx$.

Problems 45 – 53 , expand the integrand and then find an antiderivative.

45. $\int (x^2+1)^3\,dx$

46. $\int (x^3+5)^2\,dx$

47. $\int (e^x+1)^2\,dx$

48. $\int (x^2+3x-2)^2\,dx$

49. $\int (x^2+1)(x^3+5)\,dx$

50. $\int (7+\sin(x))^2\,dx$

51. $\int e^x(e^x+e^{3x})\,dx$

52. $\int (2+\sin(x))\sin(x)\,dx$

53. $\int \sqrt{x}\,(x^2+3x-2)\,dx$

Problems 54 – 64 , perform the division and then find an antiderivative.

54. $\int \frac{x+1}{x}\,dx$ $\left(\frac{x+1}{x} = 1 + \frac{1}{x}\right)$

55. $\int \frac{3x}{x+1}\,dx$ $\left(\frac{3x}{x+1} = 3 - \frac{3}{x+1}\right)$

56. $\int \frac{x-1}{x+2}\,dx$

57. $\int \frac{x^2-1}{x+1}\,dx$

58. $\int \frac{2x^2-13x+15}{x-1}\,dx$

59. $\int \frac{2x^2-13x+18}{x-1}\,dx$

60. $\int \frac{2x^2-13x+11}{x-1}\,dx$

61. $\int \frac{x+2}{x-1}\,dx$

62. $\int \frac{e^x+e^{3x}}{e^x}\,dx$

63. $\int \frac{x+4}{\sqrt{x}}\,dx$

64. $\int \frac{\sqrt{x}+3}{x}\,dx$

The definite integrals in problems 65 – 70 involve areas associated with parts of circles (Fig. 2). Use your knowledge of circles and their areas to evaluate the integrals. (Suggestion: Sketch a graph of the integrand function.)

65. $\int_{-1}^{1} \sqrt{1-x^2}\, dx$

66. $\int_{0}^{1} \sqrt{1-x^2}\, dx$

67. $\int_{-3}^{3} \sqrt{9-x^2}\, dx$

68. $\int_{-4}^{0} \sqrt{16-x^2}\, dx$

69. $\int_{-1}^{1} 2+\sqrt{1-x^2}\, dx$

70. $\int_{0}^{2} 3-\sqrt{4-x^2}\, dx$

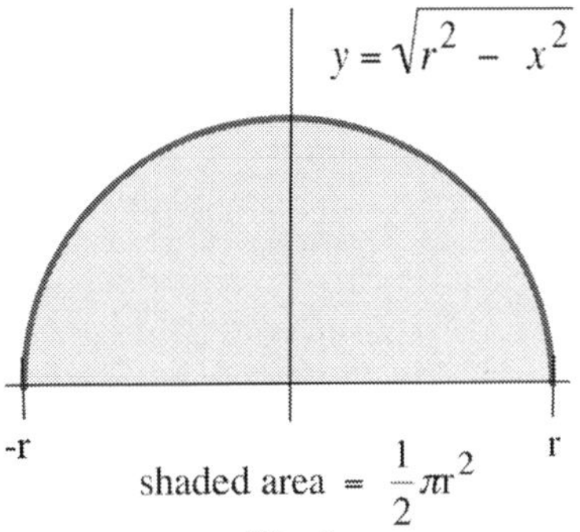

Fig. 2

Section 4.6 **PRACTICE Answers**

Practice 1: $\mathbf{D}(\tan^2(x) + C) = 2\tan^1(x)\,\mathbf{D}(\tan(x)) = 2\tan(x)\sec^2(x)$

$\mathbf{D}(\sec^2(x) + C) = 2\sec^1(x)\,\mathbf{D}(\sec(x)) = 2\sec(x)\sec(x)\tan(x) = 2\tan(x)\sec^2(x)$

Practice 2: We know $\mathbf{D}(\tan(x)) = \sec^2(x)$, so it is reasonable to try $\tan(7x)$.

$\mathbf{D}(\tan(7x)) = \sec^2(7x)\,\mathbf{D}(7x) = 7\cdot\sec^2(7x)$, a result 7 times the function we want, so divide the original "guess" by 7 and try it. $\mathbf{D}(\frac{1}{7}\tan(7x)) = \frac{1}{7}\cdot 7\cdot\sec^2(7x) = \sec^2(7x)$.

$$\int \sec^2(7x)\,dx = \frac{1}{7}\tan(7x) + C.$$

$\mathbf{D}((3x+8)^{1/2}) = \frac{1}{2}\cdot(3x+8)^{-1/2}\,\mathbf{D}(3x+8) = \frac{\mathbf{3}}{2}\cdot(3x+8)^{-1/2}$ so let's multiply our original "guess" by 2/3:

$$\mathbf{D}(\tfrac{2}{3}(3x+8)^{1/2}) = \frac{2}{3}\,\frac{1}{2}\cdot(3x+8)^{-1/2}\,\mathbf{D}(3x+8) = \frac{2}{3}\,\frac{\mathbf{3}}{2}\cdot(3x+8)^{-1/2} = (3x+8)^{-1/2}$$

$$\int \frac{1}{\sqrt{3x+8}}\,dx = \frac{2}{3}(3x+8)^{1/2} + C.$$

Practice 3: (a) $\int (7x+5)^3\,dx$ with $u = 7x+5$. Then $du = 7\,dx$ so $dx = \frac{1}{7}\,du$ and

$$\int (7x+5)^3\,dx = \int u^3\,\frac{1}{7}\,du = \frac{1}{7}\cdot\frac{1}{4}\,u^4 + C = \frac{1}{28}(7x+5)^4 + C.$$

(b) $\int \sin(x^3-1)\,3x^2\,dx$ with $u = x^3-1$. Then $du = 3x^2\,dx$ so

$$\int \sin(x^3-1)\,3x^2\,dx = \int \sin(u)\,du = -\cos(u) + C = -\cos(x^3-1) + C$$

Practice 4: (a) $u = 2+\sin(3x)$ so when $x = 0$, $u = 2+\sin(3\cdot 0) = 2$. When $x = \pi$, $u = 2+\sin(3\cdot\pi) = 2$. (This integral is now very easy to evaluate. Why?)

(b) $u = 2 + e^x$ so when $x = 0$, $u = 2 + e^0 = 3$. When $x = 2$, $u = 2 + e^2$.

The new integration endpoints are $u = 3$ to $u = 2 + e^2$.

(c) $u = e^x$ so when $x = 0$, $u = e^0 = 1$. When $x = \ln(3)$, $u = e^{\ln(3)} = 3$.

The new integration endpoints are $u = 1$ to $u = 3$.

4.7 FIRST APPLICATIONS OF DEFINITE INTEGRALS

The development of calculus by Newton and Leibniz was a vital step in the advancement of pure mathematics, but Newton also advanced the applied sciences and mathematics. Not only did he discover theoretical results, but he immediately used those results to answer important applied questions about gravity and motion. The success of these applications of mathematics to the physical sciences helped establish what we now take for granted: mathematics can and should be used to answer questions about the world.

Newton applied mathematics to the outstanding problems of his day, problems primarily in the field of physics. In the intervening 300 years, thousands of people have continued these theoretical and applied traditions and have used mathematics to help develop our understanding of all of the physical and biological sciences as well as the behavioral sciences and business. Mathematics is still used to answer new questions in physics and engineering, but it is also important for modeling ecological processes, for understanding the behavior of DNA, for determining how the brain works, and even for devising strategies for voting effectively. The mathematics you are learning now can help you become part of this tradition, and you might even use it to add to our understanding of different areas of life.

It is important to understand the successful applications of integration in case you need to use those particular applications. It is also important that you understand the **process** of building models with integrals so you can apply it to new problems. Conceptually, converting an applied problem to a Riemann sum (Fig. 1) is the most valuable step. Typically, it is also the most difficult.

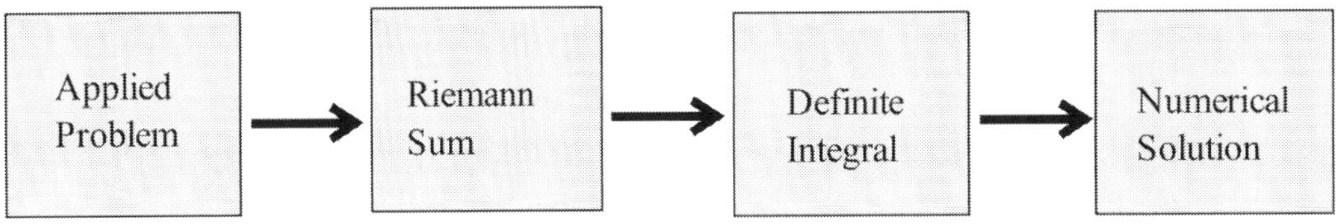

Fig. 1

Area between f and g

We have already used integrals to find the area between the graph of a function and the horizontal axis. Integrals can also be used to find the area between two graphs. If $f(x) \ge g(x)$ for all x in $[a,b]$, then we can approximate the area between f and g by partitioning the interval $[a,b]$ and forming a Riemann sum (Fig. 2). The height of each rectangle is $f(c_i) - g(c_i)$ so the area of the ith rectangle is $(\text{height})\cdot(\text{base}) = \{f(c_i) - g(c_i)\}\cdot\Delta x_i$. This approximation of the total area is

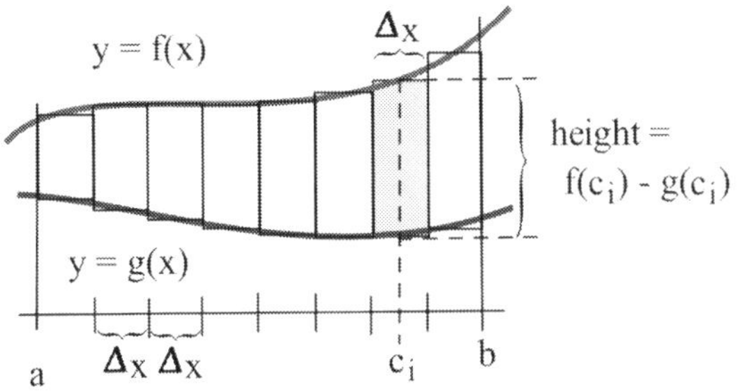

Fig. 2

$$\text{area} \approx \sum_{i=1}^{n} \{f(c_i) - g(c_i)\}\cdot\Delta x_i \text{ , a Riemann sum.}$$

The limit of this Riemann sum, as the mesh of the partitions approaches 0, is the definite integral

$$\int_a^b \{ f(x) - g(x) \} \, dx \ .$$

We will sometimes use an arrow to indicate "the limit of the Riemann sum as the mesh of the partitions approaches zero," and will write

$$\sum_{i=1}^{n} \{f(c_i) - g(c_i)\} \cdot \Delta x_i \longrightarrow \int_a^b \{ f(x) - g(x) \} \, dx \ .$$

> If $f(x) \ge g(x)$ on the interval $[a,b]$,
>
> then $\left\{ \begin{array}{l} \text{area bounded by the graphs} \\ \text{of f and g and vertical} \\ \text{lines at } x = a \text{ and } x = b \end{array} \right\} = \int_a^b \{ f(x) - g(x) \} \, dx \ .$

Example 1: Find the area bounded between the graphs of $f(x) = x$ and $g(x) = 3$ for $1 \le x \le 4$. (Fig. 3)

Solution: It is clear from the figure that the area between f and g is 2.5 square inches. Using the theorem,

area between f and g for $1 \le x \le 3$ is

$$\int_1^3 \{ 3 - x \} \, dx = 3x - \frac{x^2}{2} \Big|_1^3 = (\frac{9}{2}) - (\frac{5}{2}) = 2 \text{, and}$$

area between f and g for $3 \le x \le 4$ is

$$\int_3^4 \{ x - 3 \} \, dx = \frac{x^2}{2} - 3x \Big|_3^4 = (\frac{-8}{2}) - (\frac{-9}{2}) = \frac{1}{2} \ .$$

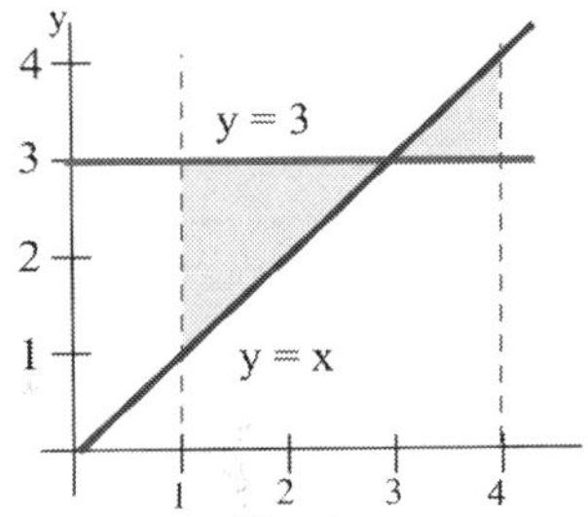

Fig. 3

The two integrals also tell us that the total area between f and g is 2.5 square inches.

The single integral $\int_1^4 \{ 3 - x \} \, dx = 1.5$ which is not the **area** we want in this problem. The value of the **integral is 1.5**, and the value of the **area is 2.5** .

Practice 1: Use integrals and the graphs of $f(x) = 1 + x$ and $g(x) = 3 - x$ to determine the area between the graphs of f and g for $0 \le x \le 3$.

Example 2: Two objects start from the same location and travel along the same path with velocities $v_A(t) = t + 3$ and $v_B(t) = t^2 - 4t + 3$ meters per second (Fig. 4).

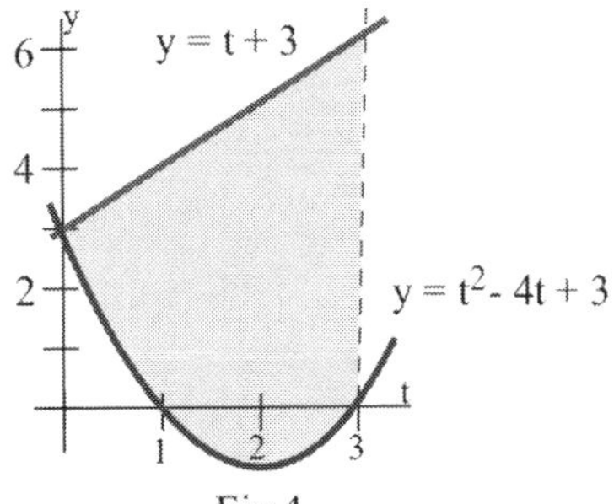

Fig.4

How far ahead is A after 3 seconds? After 5 seconds?

Solution: Since $v_A(t) \ge v_B(t)$, the "area" between the graphs of v_A and v_B represents the distance between the objects.

After 3 seconds, the distance apart = $\int_0^3 v_A(t) - v_b(t)\, dt = \int_0^3 (t+3) - (t^2-4t+3)\, dt$

$= \int_0^3 5t - t^2\, dt = \frac{5}{2}t^2 - \frac{t^3}{3} \Big|_0^3 = (\frac{5}{2}\cdot 9 - \frac{27}{3}) - (\frac{5}{2}\cdot 0 - \frac{0}{3}) = 13\frac{1}{2}$ meters.

After 5 seconds, the distance apart = $\int_0^5 v_A(t) - v_b(t)\, dt = \frac{5}{2}t^2 - \frac{t^3}{3} \Big|_0^5 = 20\frac{5}{6}$ meters.

If $f(x) \ge g(x)$, we can use the simpler argument that the area of region A is $\int_a^b f(x)\, dx$ and the area of region B is $\int_a^b g(x)\, dx$, so the area of region C, the area between f and g , is

area of C = (area of A) – (area of B) = $\int_a^b f(x)\, dx - \int_a^b g(x)\, dx = \int_a^b f(x) - g(x)\, dx$.

If the same function is not always greater, then we need to be very careful and find the intervals where $f \ge g$ and the intervals where $g \ge f$.

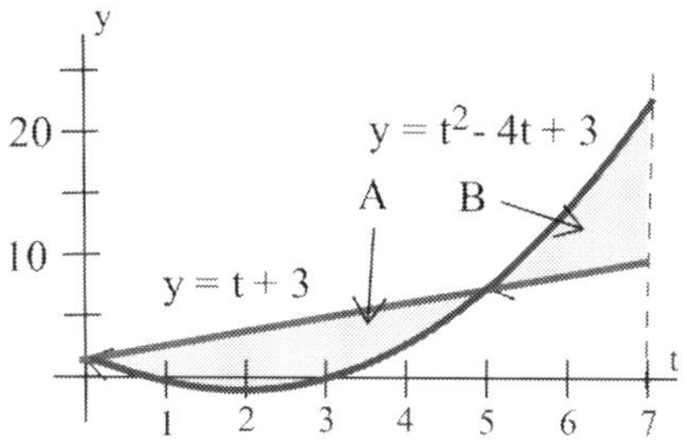

Fig. 5

Example 3: Find the **area** of the shaded region in Fig. 5.

Solution: For $0 \le x \le 5$, $f(x) \ge g(x)$ so the **area** of A is

$$\int_0^5 f(x) - g(x)\, dx = \int_0^5 (x+3) - (x^2-4x+3)\, dx$$

$$= \int_0^5 5x - x^2\, dx = \frac{5}{2}x^2 - \frac{x^3}{3} \Big|_0^5 = 20\frac{5}{6} .$$

For $5 \le x \le 7$, $g(x) \ge f(x)$ so the **area** of B is

$$\int_5^7 g(x) - f(x) dx = \int_5^7 (x^2-4x+3) - (x+3)\, dx = \int_5^7 x^2 - 5x\, dx = \frac{x^3}{3} - \frac{5}{2}x^2 \Big|_5^7 = 12\frac{4}{6} .$$

Altogether, the total **area** between f and g for $0 \le x \le 7$ is

$$\int_0^5 f(x) - g(x)\, dx + \int_5^7 g(x) - f(x)\, dx = 20\frac{5}{6} + 12\frac{4}{6} = 33\frac{1}{2} .$$

Average Value of a Function

We know the average of n numbers, $a_1, a_2, \ldots, a_n$, is their sum divided by n: average (mean) = $\frac{1}{n}\sum_{k=1}^{n} a_k$.

Finding the average of a function on an interval, an infinite number of values, requires an integral.

To find a Riemann sum approximation of the average value of f on the interval [a,b], we can partition [a,b] into n equally long subintervals of length $\Delta x = (b-a)/n$, pick a value c_i of x in each subinterval, and find the average of the numbers $f(c_i)$. Then

$$\text{average of } f \approx \frac{f(c_1)+f(c_2)+\ldots+f(c_n)}{n} = \frac{1}{n}\sum_{k=1}^{n} f(c_i) = \sum_{k=1}^{n} f(c_i)\cdot\frac{1}{n}$$

This last sum is not a Riemann sum since it does not have the form $\sum f(c_i)\cdot\mathbf{\Delta x_i}$, but it can be manipulated into one:

$$\sum_{i=1}^{n} f(c_i)\cdot\frac{1}{n} = \sum_{i=1}^{n} f(c_i)\cdot\frac{\mathbf{b-a}}{n}\cdot\frac{1}{\mathbf{b-a}} = \frac{1}{\mathbf{b-a}}\sum_{i=1}^{n} f(c_i)\cdot\frac{\mathbf{b-a}}{n} = \frac{1}{b-a}\sum_{i=1}^{n} f(c_i)\cdot\Delta x.$$

$$\text{Then } \{\text{average of } f\} \approx \frac{1}{b-a}\sum_{k=1}^{n} f(c_i)\cdot\Delta x \longrightarrow \frac{1}{b-a}\int_a^b f(x)\,dx = \{\text{average of } f\}$$

as the number of points n gets larger and the mesh, (b–a)/n, approaches 0.

> Definition: **Average (Mean) Value of a Function**
>
> For an integrable function f on the interval [a,b],
>
> **the average value of f on [a,b]** is $\frac{1}{b-a}\int_a^b f(x)\,dx$

The average value of a positive f has a nice geometric interpretation. Imagine that the area under f (Fig. 6a) is a liquid that can "leak" through the graph to form a rectangle with the same area (Fig. 6b). If the height of the rectangle is H, then the area of the rectangle is H·(b–a). We know the area of the rectangle is the same as the area under f so

$$H\cdot(b-a) = \int_a^b f(x)\,dx. \text{ Then}$$

$$H = \frac{1}{b-a}\int_a^b f(x)\,dx, \text{ the average value of f on [a,b].}$$

The average value of positive f is the height H of the rectangle whose area is the same as the area under f.

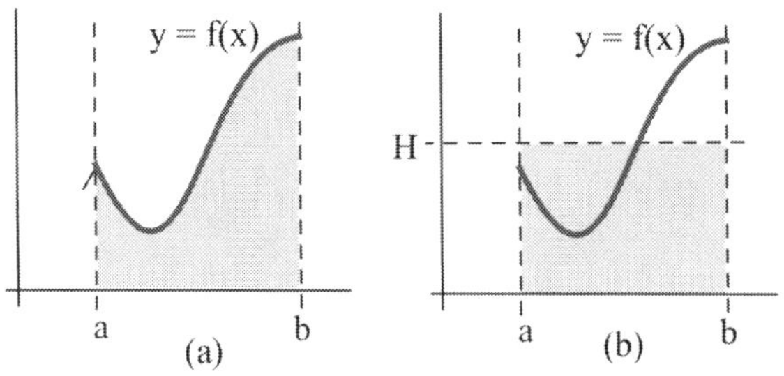

Fig. 6

Example 4: Find the average value of $f(x) = \sin(x)$ on the interval $[0,\pi]$. (Fig. 7)

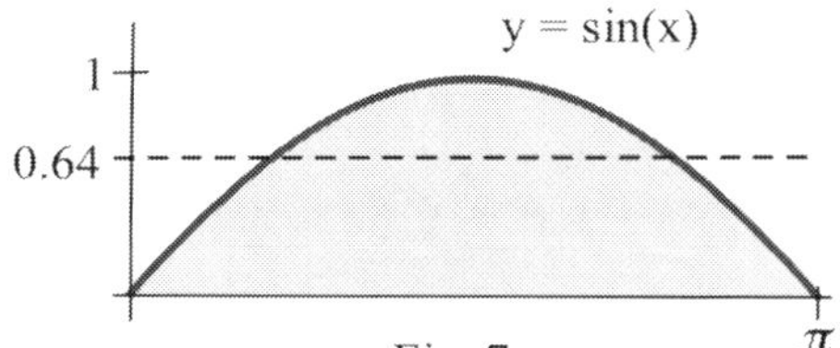

Fig. 7

Solution: Average value $= \frac{1}{\pi-0}\int_0^{\pi} \sin(x)\,dx$

$$= \frac{1}{\pi}(-\cos(x))\Big|_0^{\pi} = \frac{1}{\pi}\{-(-1)-(-1)\} = \frac{2}{\pi} \approx 0.6366 .$$

A rectangle with height $2/\pi \approx 0.64$ on the interval $[0,\pi]$ encloses the same area as one arch of the sine curve. The average value of $\sin(x)$ on the interval $[0, \mathbf{2\pi}]$ is 0 since $\frac{1}{2\pi}\int_0^{2\pi} \sin(x)dx = 0$.

Practice 2: During a 9 hour work day, the production rate at time t hours was $r(t)= 5 + \sqrt{t}$ cars per hour. Find the average hourly production rate.

Function averages, involving means and more complicated averages, are used to "smooth" data so that underlying patterns are more obvious and to remove high frequency "noise" from signals. In these situations, the original function f is replaced by some "average of f." If f is rather jagged time data, then the ten year average of f is the integral $g(x) = \frac{1}{10}\int_{x-5}^{x+5} f(t)\,dt$ .an average of f over 5 units on each side of x. For example, Fig. 9 shows the graphs of a Monthly Average (rather "noisy" data) of surface temperature data, an Annual Average (still rather "jagged), and a Five Year Average (a much smoother function). Typically the average function reveals the pattern much more clearly than the original data. This use of a "moving average" value of "noisy" data (weather information, stock prices) is a very common.

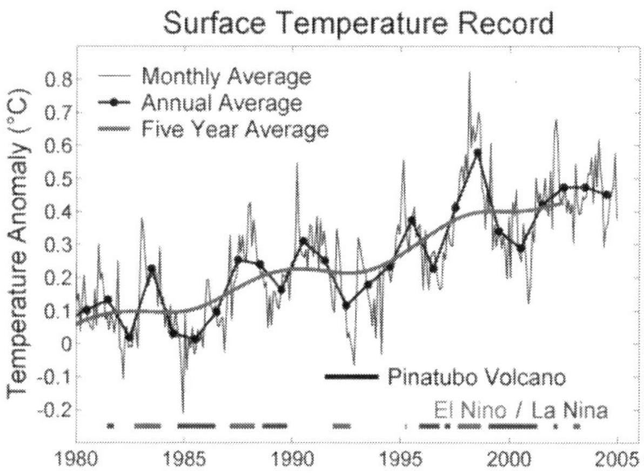

Fig. 9

Work

The amount of work done on an object is the force applied to the object times the distance the object is moved while the force is applied (Fig. 10) or, more succinctly, **work = (force)·(distance)** .

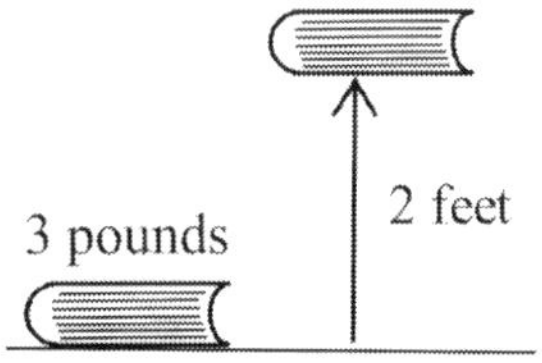

Fig. 10

If you lift a 3 pound book 2 feet, then the force is 3 pounds, the weight of the book, and the distance moved is 2 feet, so you have done (3 pounds)·(2 feet) = 6 foot–pounds of work. When the applied force and the distance are both constants, then calculating work is simply a matter of multiplying.

Practice 3: How much work is done lifting a 10 pound object from the ground to the top of a 30 foot building (assume the cable is weightless).

If either the force or the distance is variable, then integration is needed.

Example 5: How much work is done lifting a 10 pound object from the ground to the top of a 30 foot building if the cable weighs 2 pounds per foot. (Fig. 11)

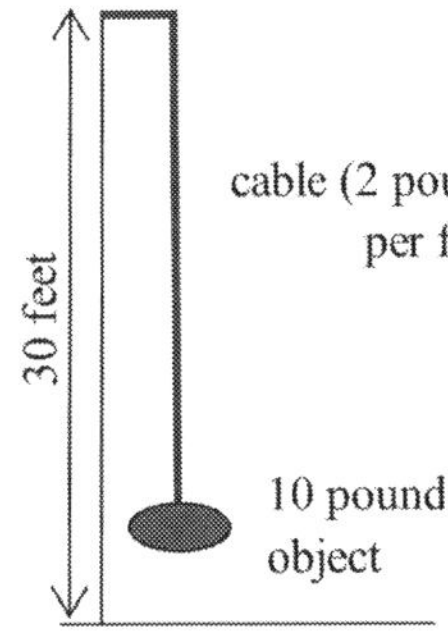

Fig. 11

Solution: This is more difficult than the Practice problem. When the object is at ground level, a force of 70 pounds (10 pounds plus the weight of 30 feet of cable) must be applied, but when the object is 29 feet above the ground, only 12 pounds of force are needed. In general, if the object is x feet above the ground (Fig. 12), then $30 - x$ feet of cable, weighing $2(30-x)$ pounds, is used so the required force is

$$f(x) = 10 + 2(30-x) = 70 - 2x \text{ pounds.}$$

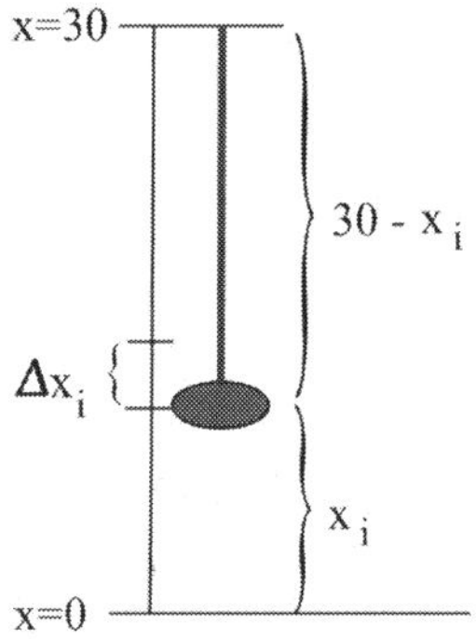

Fig. 12

Let's partition the height of the building into small increments so the force needed in each subinterval does not change much. The force in the ith subinterval will be approximately $f(c_i)$ for some c_i in the subinterval, and the distance moved will be the length of the subinterval, Δx_i. The work done to move the object through the subinterval will be $f(c_i) \cdot \Delta x_i$, and the total work will be the sum of the work on each subinterval:

$$\text{work} \approx \sum (\text{subinterval work}) = \sum f(c_i)\,\Delta x_i = \sum \{70 - 2c_i\}\,\Delta x_i \quad \text{(a Riemann sum)}$$

$$\longrightarrow \int_0^{30} \{70 - 2x\}\,dx \quad \text{as the mesh approaches 0.}$$

We have approximated the solution to an applied problem by a Riemann sum, and obtained an exact solution by taking the limit of the Riemann sum to get a definite integral. Now we just need to evaluate the definite integral:

$$\int_0^{30} (70 - 2x)\,dx = 70x - x^2 \Big|_0^{30} = \{70 \cdot 30 - (30)^2\} - \{70 \cdot 0 - (0)^2\} = (2100-900) - (0) = 1200 \text{ foot-pounds.}$$

Practice 4: Suppose the building in Example 5 is 50 feet tall and the cable weighs 3 pounds per foot.

(a) How much force is needed when the 10 pound object is x feet above the ground?

(b) Write an integral for the work done raising the object from the ground to a height of 10 feet. From a height of 10 feet to a height of 20 feet.

In the previous Example and Practice problem, the force was variable and the distance was Δx. In later sections we will examine situations where the force is constant and the distance changes.

Summary

These area, average and work applications simply introduce a few of the applications of definite integrals and illustrate the pattern of going from an applied problem to a Riemann sum, on to a definite integral and, finally, to a number. More applications will be explored in Chapter 5. The rest of this chapter will examine additional ways of finding antiderivatives and of finding the values of definite integrals when an antiderivative can not be found.

PROBLEMS

In problems 1 – 4, use the values in Table 1 to estimate the areas.

x	f(x)	g(x)	h(x)
0	5	2	5
1	6	1	6
2	6	2	8
3	4	2	6
4	3	3	5
5	2	4	4
6	2	5	2

Table 1

1. Estimate the **area** between f and g for $1 \le x \le 4$.
2. Estimate the **area** between f and g for $1 \le x \le 6$.
3. Estimate the **area** between f and h for $0 \le x \le 4$.
4. Estimate the **area** between g and h for $0 \le x \le 6$.
5. Estimate the area of the island in Fig. 13.
6. Estimate the area of the island in Fig. 13 if the distances between the lines is 50 feet instead of 40 feet,.

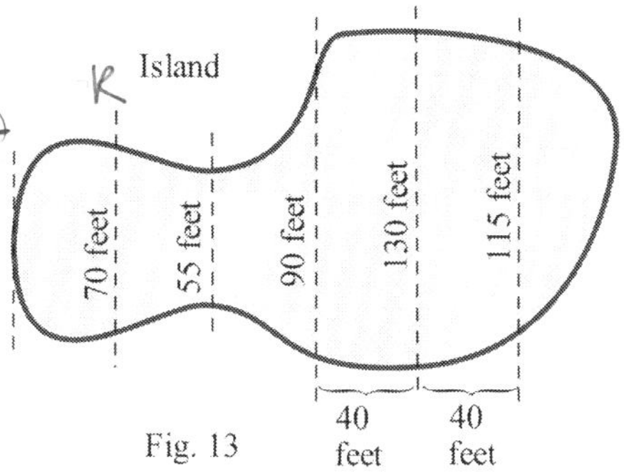

Fig. 13

In problems 7 – 18, sketch the graph of each function and find the **area** between the graphs of f and g for x in the given interval.

7. $f(x) = x^2 + 3$, $g(x) = 1$ and $-1 \le x \le 2$.
8. $f(x) = x^2 + 3$, $g(x) = 1 + x$ and $0 \le x \le 3$.
9. $f(x) = x^2$, $g(x) = x$ and $0 \le x \le 2$.
10. $f(x) = 4 - x^2$, $g(x) = x + 2$ and $0 \le x \le 2$.
11. $f(x) = \frac{1}{x}$, $g(x) = x$ and $1 \le x \le e$.
12. $f(x) = \sqrt{x}$, $g(x) = x$ and $0 \le x \le 4$.
13. $f(x) = x + 1$, $g(x) = \cos(x)$ and $0 \le x \le \pi/4$.
14. $f(x) = (x-1)^2$, $g(x) = x + 1$ and $0 \le x \le 3$.
15. $f(x) = e^x$, $g(x) = x$ and $0 \le x \le 2$.
16. $f(x) = \cos(x)$, $g(x) = \sin(x)$ and $0 \le x \le \pi/4$.
17. $f(x) = 3$, $g(x) = \sqrt{1 - x^2}$ and $0 \le x \le 1$.
18. $f(x) = 2$, $g(x) = \sqrt{4 - x^2}$ and $-2 \le x \le 2$.

In problems 19 – 22, use the values in Table 1 to estimate the average values.

19. Estimate the average value of f on the interval $[0.5, 4.5]$.

20. Estimate the average value of f on the interval $[0.5, 6.5]$.

21. Estimate the average value of f on the interval $[1.5, 3.5]$.

22. Estimate the average value of f on the interval $[3.5, 6.5]$.

In problems 23 – 32, find the **average value** of f on the given interval.

23. $f(x)$ in Fig. 14 for $0 \le x \le 2$.

24. $f(x)$ in Fig. 14 for $0 \le x \le 4$.

25. $f(x)$ in Fig. 14 for $1 \le x \le 6$.

26. $f(x)$ in Fig. 14 for $4 \le x \le 6$.

27. $f(x) = 2x + 1$ for $0 \le x \le 4$.

28. $f(x) = x^2$ for $0 \le x \le 2$.

29. $f(x) = x^2$ for $1 \le x \le 3$.

30. $f(x) = \sqrt{x}$ for $0 \le x \le 4$.

31. $f(x) = \sin(x)$ for $0 \le x \le \pi$.

32. $f(x) = \cos(x)$ for $0 \le x \le \pi$.

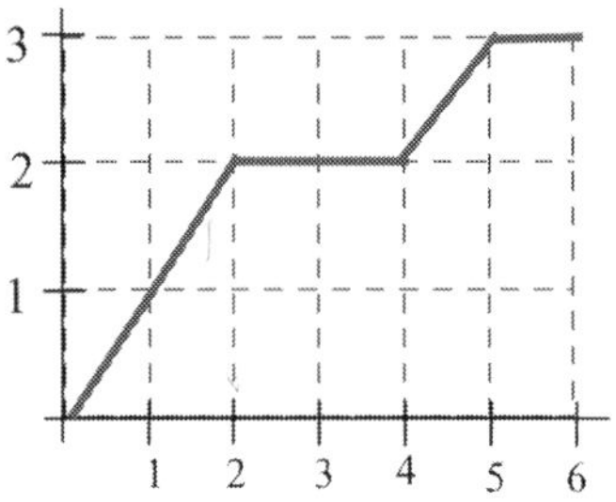

Fig. 14

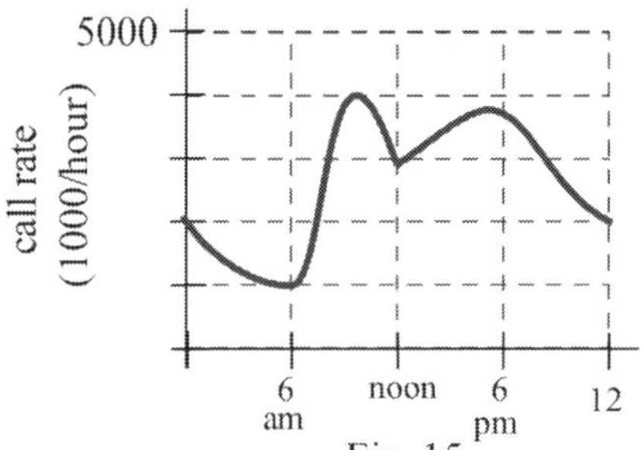

Fig. 15

33. Calculate the average value of $f(x) = \sqrt{x}$ on the interval $[0, C]$ for $C = 1, 9, 81, 100$. What is the pattern?

34. Calculate the average value of $f(x) = x$ on the interval $[0, C]$ for $C = 1, 10, 80, 100$. What is the pattern?

35. Fig. 15 shows the number of telephone calls per minute at a large company. (a) Estimate the average number of calls per minute from 8 am to 5 pm. (b) From 9 am to 1 pm.

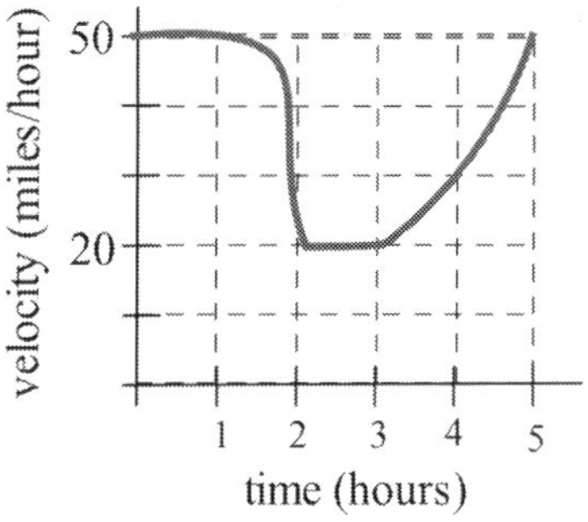

Fig. 16

36. Fig. 16 shows the velocity of a car during a 5 hour trip.

(a) Estimate how far the car traveled during the 5 hours.

(b) At what **constant** velocity should you drive in order to travel the same distance in 5 hours?

37. (a) How much work is done lifting a 20 pound bucket from the ground to the top of a 30 foot building with a cable which weighs 3 pounds per foot?

(b) How much work is done lifting the same bucket from the ground to a height of 15 feet with the same cable?

38. (a) How much work is done lifting a 60 pound chair from the ground to the top of a 20 foot building with a cable which weighs 1 pound per foot?

(b) How much work is done lifting the same chair from the ground to a height of 5 feet with the same cable?

39. (a) How much work is done lifting a 10 pound calculus book from the ground to the top of a 30 foot building with a cable which weighs 2 pounds per foot?

(b) From the ground to a height of 10 feet? (c) From a height of 10 feet to a height of 20 feet?

40. How much work is done lifting an 80 pound injured child to the top of a 20 foot hole using a stretcher weighing 14 pounds and a cable which weighs 1 pound per foot?

41. How much work is done lifting an 60 pound injured child to the top of a 15 foot hole using a stretcher weighing 10 pounds and a cable which weighs 2 pound per foot?

42. How much work is done lifting an 120 pound injured adult to the top of a 30 foot hole using a stretcher weighing 10 pounds and a cable which weighs 2 pound per foot?

Section 4.7 **PRACTICE Answers**

Practice 1: Using geometry (Fig. 17): $A = \frac{1}{2}(2)(1) = 1$ and $B = \frac{1}{2}(4)(2) = 4$ so total area = A + B = **5**.

Using integrals:

$$A = \int_0^1 (3-x) - (1+x)\,dx = \int_0^1 (2-2x)\,dx = 2x - x^2 \Big|_0^1 = (2-1) - (0) = 1.$$

$$B = \int_1^3 (1+x) - (3-x)\,dx = \int_1^3 (2x-2)\,dx = x^2 - 2x \Big|_1^3 = (9-6) - (1-2) = 4.$$

The single integral $\int_0^3 (1+x) - (3-x)\,dx$ is **not** correct: $\int_0^3 (1+x) - (3-x)\,dx = 3$.

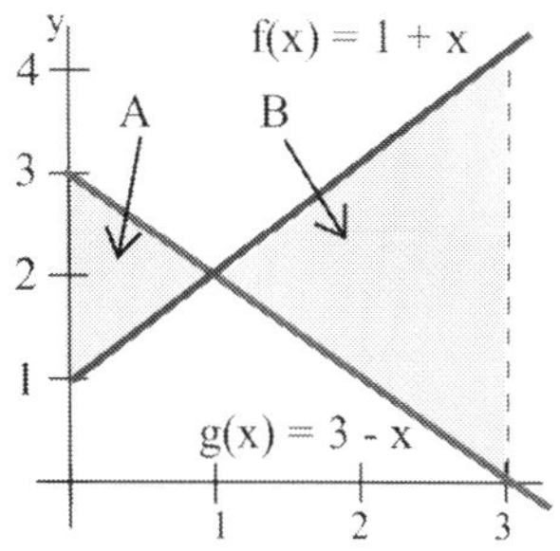

Fig.17

Practice 2: Average value = $\frac{1}{b-a}\int_a^b f(x)\,dx = \frac{1}{9-0}\int_0^9 5+\sqrt{t}\,dt$

$= \frac{1}{9}\int_0^9 5+t^{1/2}\,dt = \frac{1}{9}(\,5t + \frac{2}{3}t^{3/2}\,)\Big|_0^9$

$= \frac{1}{9}(45+\frac{2}{3}9^{3/2}\,) - \frac{1}{9}(0) = \frac{1}{9}(45+18) =$ **7 cars per hour.**

Practice 3: Work = (force)·(distance) = (10 pounds)·(30 feet) = **300 foot·pounds**.

Practice 4: (a) force = (force for cable) + (force for object)

= (length of cable)(density of cable) + 10 pounds

= (50 – x feet)(3 pounds/foot) + 10 pounds = **160 – 3x pounds**

(b) "from the ground to a height of 10 feet:"

Work $\approx \sum$ (subinterval work) $= \sum f(c_i)\,\Delta x_i = \sum \{160-3c_i\}\,\Delta x_i$ (a Riemann sum)

$\longrightarrow \int_0^{10} \{160-3x\}\,dx$ as the mesh approaches 0.

"From a height of 10 feet to a height of 20 feet:" work = $\int_{10}^{20} \{160-3x\}\,dx$.

4.8 APPROXIMATING DEFINITE INTEGRALS

The Fundamental Theorem of Calculus tells how to calculate the exact value of a definite integral IF the integrand function is continuous and IF we can find an antiderivative of the integrand. In practice, however, we may need the definite integral of a function defined by a table of measurements or a graph, or of a function which does not have an elementary antiderivative. This section includes several techniques for getting approximate numerical values for definite integrals without using antiderivatives. Mathematically, exact answers are preferable and satisfying, but for most applications, a numerical answer with several digits of accuracy is just as useful.

The ideas behind the approximation methods are geometrical and rather simple, but using the methods to get good approximations typically requires lots of arithmetic, something calculators are very good and quick at doing. All of these approximate methods can be easily programmed, and program listings for two of these methods are included after the Practice Answers.

The General Approach

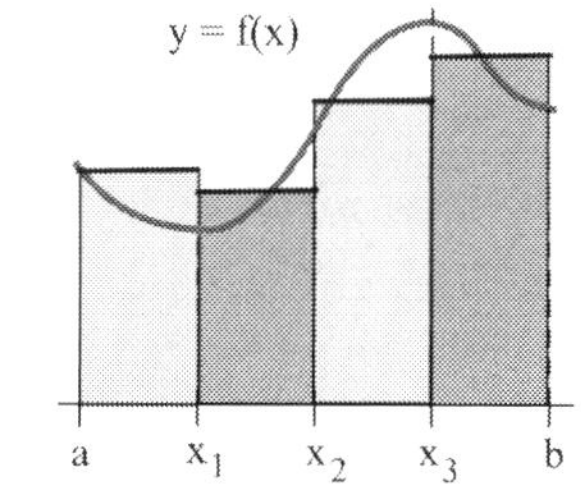

Rectangles: Fitting horizontal lines
Fig. 1

The methods in this section approximate the definite integral of a function f by building "easy" functions close to f and then exactly evaluating the definite integrals of the "easy" functions. If the "easy" functions are close enough to f, then the sum of the definite integrals of the "easy" functions will be close to the definite integral of f. The Left, Right and Midpoint approximations fit horizontal lines to f, the "easy" functions are constant functions, and the approximating regions are rectangles (Fig. 1). The Trapezoidal Rule fits slanted lines to f, the "easy" functions are linear, and the approximating regions are trapezoids (Fig. 2). Finally, Simpson's Rule fits parabolas to f, and the "easy" functions are quadratics (Fig. 3).

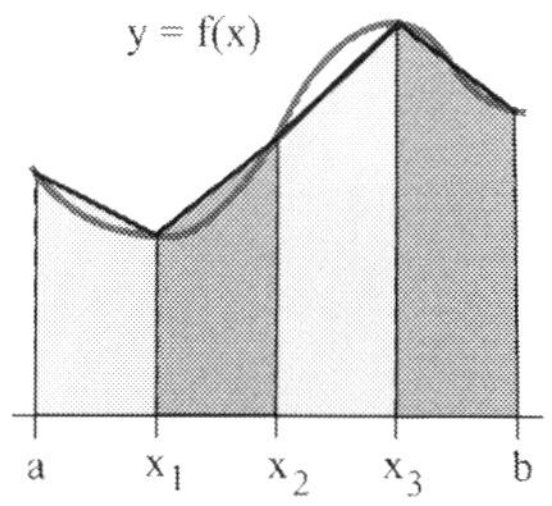

Trapezoids: Fitting slanted lines
Fig. 2

The Left and Right approximation rules are simply Riemann sums with the point c_i in each subinterval chosen to be the left or right endpoint of that subinterval. They typically require a large number of computations to get even mediocre approximations and are seldom used in practice. They and the Midpoint rule are discussed at the end of the problem set.

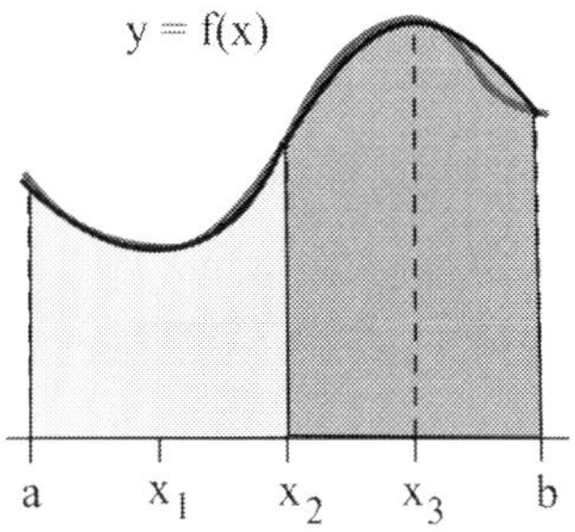

Parabolas: Fitting parabolas
Fig. 3

All of the methods divide the interval [a,b] into **n** **equally–long** subintervals. Each subinterval has length $\mathbf{h = \Delta x_i = \frac{b-a}{n}}$, and the points of the partition are

$x_0 = a$, $x_1 = a + h$, $x_2 = a + 2 \cdot h$, . . . , $\mathbf{x_i = a + i \cdot h}$, . . . , $x_n = a + n \cdot h = a + n(\frac{b-a}{n}) = b$.

Approximating A Definite Integral Using Trapezoids

If the graph of f is curved, then slanted lines typically come closer to the graph of f than horizontal ones do, and the slanted lines lead to trapezoidal regions (Fig. 2).

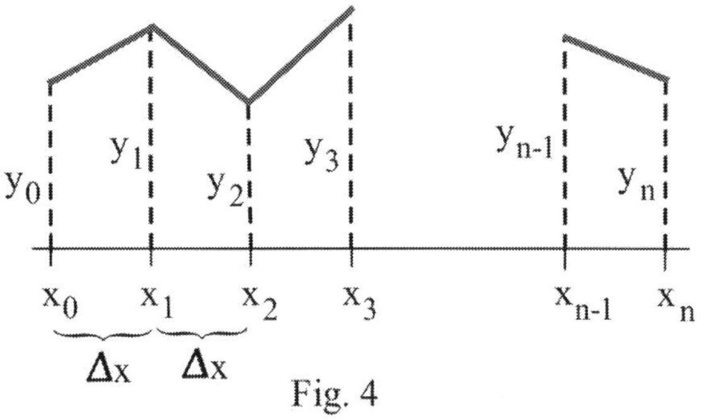

Fig. 4

The area of a trapezoid is (base)•(average height) so the area of the first trapezoid in Fig. 4 is

$$(\Delta x)\cdot\frac{y_0+y_1}{2} = \frac{\Delta x}{2}(y_0+y_1) .$$

Similarly, the areas of the other trapezoids are

$$\frac{\Delta x}{2}(y_1+y_2) , \quad \frac{\Delta x}{2}(y_2+y_3) , \quad \ldots ,$$

$$\frac{\Delta x}{2}(y_{n-1}+y_n) .$$

The sum of the trapezoidal areas is

$$T_n = \frac{\Delta x}{2}(y_0+y_1) + \frac{\Delta x}{2}(y_1+y_2) + \frac{\Delta x}{2}(y_2+y_3) + \ldots + \frac{\Delta x}{2}(y_{n-1}+y_n)$$

$$= \frac{\Delta x}{2} \{y_0 + 2y_1 + 2y_2 + \ldots + 2y_{n-1} + y_n\} \text{ or, equivalently,}$$

$$\frac{\Delta x}{2} \left\{ f(x_0) + 2f(x_1) + 2f(x_2) + \ldots + 2f(x_{n-1}) + f(x_n) \right\}.$$

Each $f(x_i)$ value, except the first (i = 0) and the last (i = n), is the right–endpoint height of one trapezoid and the left–endpoint height of the next trapezoid so it shows up in the calculation for two trapezoids and is multiplied by two in the formula for the trapezoidal approximation.

> **Trapezoidal Approximation Rule**
>
> If f is integrable on [a,b], and [a,b] is partitioned into n subintervals of length $h = \frac{b-a}{n}$,
>
> then the Trapezoidal approximation of $\int_a^b f(x)\,dx$ is
>
> $$T_n = \frac{h}{2} \left\{ f(x_0) + \mathbf{2}f(x_1) + \mathbf{2}f(x_2) + \ldots + \mathbf{2}f(x_{n-1}) + f(x_n) \right\}$$

Example 1: Calculate T_4 , the Trapezoidal approximation of $\int_1^3 f(x)\,dx$, for the function values in Table 1.

x	f(x)
1.0	4.2
1.5	3.4
2.0	2.8
2.5	3.6
3.0	3.2

Table 1

Solution: The step size is h = (b–a)/n = (3–1)/4 = 1/2 . Then

$$T_4 = \frac{h}{2} \left\{ f(x_0) + 2f(x_1) + 2f(x_2) + 2f(x_3) + f(x_4) \right\}$$

$$= \frac{.5}{2} \left\{ 4.2 + 2(3.4) + 2(2.8) + 2(3.6) + (3.2) \right\} = (.25)(27) = 6.75 .$$

Let's see how well the trapezoidal rule approximates an integral whose exact value we know, $\int_1^3 x^2\,dx = 8\frac{2}{3}$.

Example 2: Calculate T_4 , the Trapezoidal approximation of $\int_1^3 x^2\,dx$ for $n = 4$.

Solution: As in Example 1, $h = .5$ and $x_0 = 1$, $x_1 = 1.5$, $x_2 = 2$, $x_3 = 2.5$, and $x_4 = 3$. Then

$$T_4 = \frac{h}{2}\left\{f(x_0) + 2f(x_1) + 2f(x_2) + 2f(x_3) + f(x_4)\right\} = \frac{.5}{2}\left\{f(1) + 2f(1.5) + 2f(2) + 2f(2.5) + f(3)\right\}$$

$$= (.25)\left\{1 + 2(2.25) + 2(4) + 2(6.25) + 9\right\} = 8.75 .$$

Larger values for n give better approximations: $T_{20} = 8.67$ and $T_{100} = 8.6668$.

Practice 1: On a summer day, the level of the pond in Fig. 5 went down 0.1 feet because of evaporation. Use the trapezoidal rule to approximate the surface area of the pond and then calculate how much water evaporated.

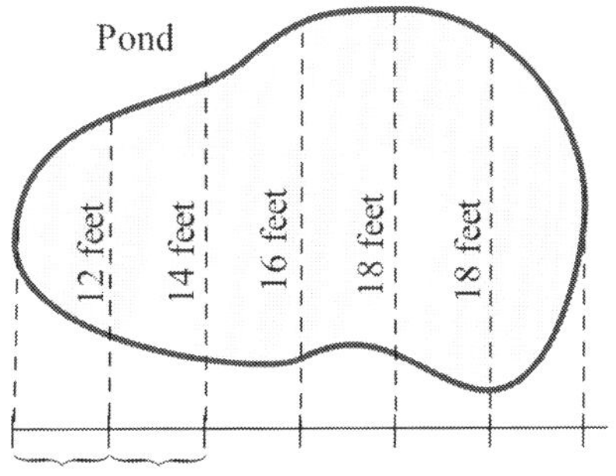

Fig. 5

Approximating A Definite Integral Using Parabolas

If the graph of f is curved, even the slanted lines may not fit the graph of f as closely as we would like, and a large number of subintervals may still be needed with the Trapezoidal rule to get a good approximation of the definite integral. Curves typically fit the graph of f better than straight lines, and the easiest nonlinear curves are parabolas.

$$\text{Area} = (2\Delta x)\left\{\frac{y_0 + 4y_1 + y_2}{6}\right\}$$

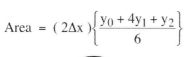

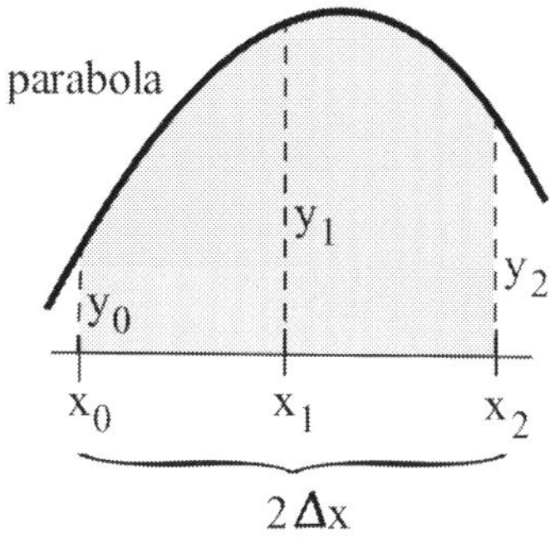

Fig/ 6

Three points $(x_0, y_0), (x_1, y_1), (x_2, y_2)$ are needed to determine the equation of a parabola, and the area under a parabolic region with evenly spaced x_i values (Fig. 6) is

$$(2\Delta x)\cdot\left\{\frac{y_0 + 4y_1 + y_2}{6}\right\} = \frac{\Delta x}{3}\cdot\{ y_0 + 4y_1 + y_2 \}.$$

(The steps to verify this formula for parabolas are outlined in problem 32.)

Taking the subintervals in pairs, the areas of the other parabolic regions are

$$\frac{\Delta x}{3}\cdot\{ y_2 + 4y_3 + y_4 \},\ \frac{\Delta x}{3}\cdot\{ y_4 + 4y_5 + y_6 \},\ldots,\ \frac{\Delta x}{3}\cdot\{ y_{n-2} + 4y_{n-1} + y_n \}$$

so the sum of the parabolic areas (Fig. 7) is

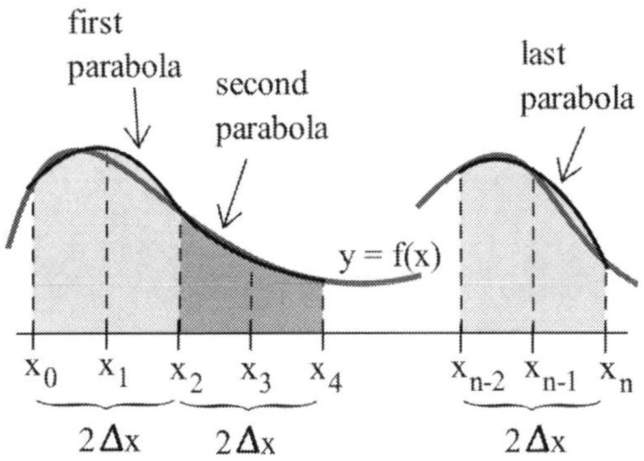

Fig. 7

$$S_n = \frac{\Delta x}{3} \cdot \{ y_0 + 4y_1 + y_2 \}$$

$$+ \frac{\Delta x}{3} \cdot \{ y_2 + 4y_3 + y_4 \}$$

$$+ \ldots + \frac{\Delta x}{3} \cdot \{ y_{n-2} + 4y_{n-1} + y_n \}$$

$$= \frac{\Delta x}{3} \cdot \{ y_0 + 4y_1 + 2y_2 + 4y_3 + 2y_4 + \ldots + 2y_{n-2} + 4y_{n-1} + y_n \}$$ or, equivalently,

$$\frac{\Delta x}{3} \left\{ f(x_0) + \mathbf{4}f(x_1) + \mathbf{2}f(x_2) + \mathbf{4}f(x_3) + \mathbf{2}f(x_4) + \ldots + \mathbf{2}f(x_{n-2}) + \mathbf{4}f(x_{n-1}) + f(x_n) \right\}.$$

In order to use **pairs** of subintervals, the number n of subintervals must be **even**. The coefficient pattern for a single parabola is 1–4–1, but when we put several parabolas next to each other, they share some edges and the pattern becomes 1–4–**2**–4–**2**– . . . –**2**–4–1 with the shared edges getting counted twice.

Parabolic Approximation Rule (Simpson's Rule)

If f is integrable on [a,b], and [a,b] is partitioned into an **even number** n of subintervals of length $h = \frac{b-a}{n}$, then the Parabolic approximation of $\int_a^b f(x)\,dx$ is

$$S_n = \frac{h}{3} \cdot \{ f(x_0) + \mathbf{4}f(x_1) + \mathbf{2}f(x_2) + \mathbf{4}f(x_3) + \mathbf{2}f(x_4) + \ldots + \mathbf{4}f(x_{n-1}) + f(x_n) \}$$

Example 3: Calculate S_4, Simpson's parabolic approximation of $\int_1^3 f(x)\,dx$, for the function in Table 1.

Solution: The step size is $h = (b-a)/n = (3-1)/4 = 1/2$. Then

$$S_4 = \frac{h}{3} \left\{ f(x_0) + 4f(x_1) + 2f(x_2) + 4f(x_3) + f(x_4) \right\}$$

$$= \frac{1/2}{3} \left\{ 4.2 + 4(3.4) + 2(2.8) + 4(3.6) + (3.2) \right\} = \frac{1}{6}(41) \approx 6.833 .$$

Example 4: Calculate S_4, Simpson's parabolic approximation of $\int_1^3 2^x\,dx$ for n = 4.

Solution: As in the previous Examples, h = (b–a)/n = .5 and $x_0 = 1$, $x_1 = 1.5$, $x_2 = 2$, $x_3 = 2.5$, and $x_4 = 3$.

$$S_4 = \frac{h}{3}\left\{f(x_0) + 4f(x_1) + 2f(x_2) + 4f(x_3) + f(x_4)\right\} = \frac{.5}{3}\left\{f(1) + 4f(1.5) + 2f(2) + 4f(2.5) + f(3)\right\}$$

$$= \left(\frac{1}{6}\right)\left\{2 + 4(2.828427) + 2(4) + 4(5.656854) + 8\right\} = \frac{1}{6}(51.941124) = 8.656854 .$$

Larger values for n give better approximations: $S_{20} = 8.656171$ and $S_{100} = 8.656170$.

Practice 2: Use Simpson's Rule to estimate the area of the pond in Fig. 5.

Which Method Is Best?

The hardest and slowest part of these approximations, whether by hand or by computer, is the evaluation of the function at the x_i values. For n subintervals, all of the methods require about the same number of function evaluations. Table 2 illustrates how closely each method approximates the definite integral of 1/x using several values of n. The values in Table 2 also show how quickly the actual error shrinks as the values of n increase: just doubling n from 4 to 8 cut the actual error of the parabolic approximation of this definite integral by a factor of 9 — a good reward for our extra work. The rest of this section discusses "error bounds" of the approximations so we can know how close our approximation is to the exact value of the integral even if we don't know the exact value.

Table 2: Approximating $\int_1^5 \frac{1}{x} dx$ = ln 5 = 1.609437912

Using n=4 (h= (5–1)/4 = 1)

method	approximation	M	error bound	actual error
T_4	1.6833333	2	.6666666	.07389542
S_4	1.6222222	24	.5333333	.01278431
L_4	2.083333333	1	2	.47389542
R_4	1.283333333	1	2	.32610458
M_4	1.574603175	2	.33333333	.03483474

Using n=8 (h= (5–1)/8 = 1/2)

T_8	1.628968254	2	.1666666	.01953034
S_8	1.610846561	24	.0333333	.00140865
L_8	1.828968254	1	1	.21953034
R_8	1.428968254	1	1	.18046966
M_8	1.599844394	2	.08333333	.00959352

Using n=20 (h= (5–1)/20 = 1/5)

T_{20}	1.612624844	2	.0266667	.00318693
S_{20}	1.609486789	24	.0008533	.00004888
L_{20}	1.692624844	1	.4	.08318693
R_{20}	1.532624844	1	.4	.07681307
M_{20}	1.607849324	2	.01333333	.00158859

How Good Are the Approximations?

The approximation rules are valuable by themselves, but they are particularly useful because there are "error bound" formulas that guarantee how close the approximations are to the exact values of the integrals. It is useful to know that an integral is "about 3.7," but we can have more confidence if we know that the integral is "within .0001 of 3.7 ." Then we can decide if our approximation is good enough for the job at hand or if we need to improve it. The formulas for the error bounds can also be solved to determine how many subintervals are needed to guarantee that our approximation is within some specified distance of the exact answer. There is no reason to use 1000 subintervals if 18 will give the needed accuracy. Unfortunately, the formulas for the error bounds require information about the derivatives of the integrands, so we can not use these formulas to determine error bounds for the approximations of integrals of functions defined by tables of values.

Error Bound for Trapezoidal Approximation

If the second derivative of f is continuous on [a,b] and $M_2 \geq \{$ maximum of $|f''(x)|$ on [a,b] $\}$,

then the "error" of the T_n approximation is $\left| \int_a^b f(x)\,dx - T_n \right| \leq \frac{(b-a)^3}{12\,n^2} \cdot M_2$ = "error bound."

The "error bound" formula $\frac{(b-a)^3}{12\,n^2} \cdot M_2$ for the Trapezoidal approximation is a "guarantee:" the actual error is guaranteed to be no larger than the error bound. In fact, the actual error is usually much smaller than the error bound. The word "error" does not indicate a mistake, it means the deviation or distance from the exact answer.

Example 5: We can be certain that the T_{10} approximation of $\int_0^2 \sin(x^2)\,dx$ is **within** what distance of the exact value of the integral?

Solution: $b - a = 2$, $n = 10$, $f(x) = \sin(x^2)$, and $f''(x) = -4x^2 \cdot \sin(x^2) + 2 \cdot \cos(x^2)$ is continuous on [0, 2]. The graph of $f''(x)$ is given in Fig. 8 . Even though we may not know the exact maximum value M_2 of $|f''(x)|$ on [0, 2], it is clear from the graph that $M_2 \leq 11$. Then

$$\text{"actual error"} \leq \text{"error bound"} = \frac{(b-a)^3}{12\,n^2} \cdot M_2$$

$$= \frac{(2)^3}{12\,(10)^2} \cdot (11) = \frac{88}{1200} < 0.074$$

so we can be certain that our T_{10} approximation of the definite integral is within .074 of the exact value:

$$T_{10} - 0.074 \leq \int_0^2 \sin(x^2)\,dx \leq T_{10} + 0.074 .$$

$T_{10} = 0.7959247$, so we can be certain that the value of the integral is between 0.722 and 0.870 .

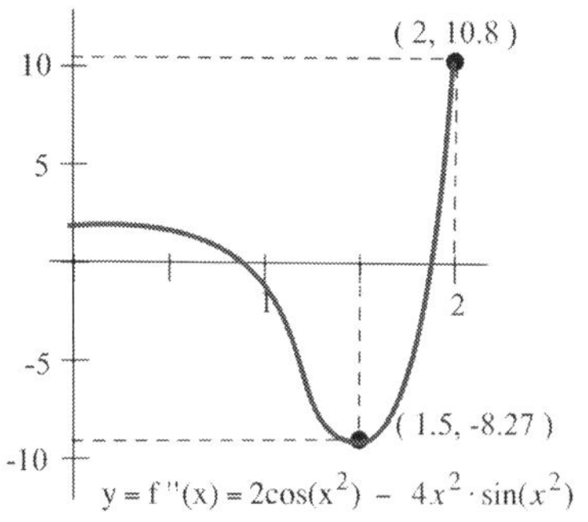

Fig. 8

Practice 3: Find an error bound for the T_{12} approximation of $\int_2^5 \frac{1}{x}\,dx$.

Example 6: How large must n be to be certain that T_n is within 0.001 of $\int_0^2 \sin(x^2)\,dx$?

Solution: The "allowable error" of 0.001 is given, and we are asked to find n. From Example 5 we know that $M_2 \le 11$, so we want the error bound to be less than the allowable error of 0.001. Then

$.001 \ge$ "error bound" $= \frac{(2)^3}{12\,n^2} \cdot (11) = \frac{88}{12}\frac{1}{n^2} = \frac{22}{3n^2}$. Solving for n, we have $n^2 \ge \frac{22}{0.003} > 7334$

so $n \ge \sqrt{7334} \approx 85.6$. Since n must be an integer, we can be certain that T_{86} is within 0.001 of $\int_0^2 \sin(x^2)\,dx$. $T_{86} \approx 0.80465$, so we can be certain that the exact value of the integral is between 0.80**365** and 0.80**565** . As is usually the case, T_{86} is even closer than 0.001 to the exact value, $|\,T_{86} - \text{exact value}\,| \approx 0.00012$.

Practice 4: How large must n be to be certain that T_n is within 0.001 of $\int_2^5 \frac{1}{x}\,dx$?

Error Bound for Simpson's Parabolic Approximation

If the fourth derivative of f is continuous on [a,b] , and $M_4 \ge$ {maximum of $|\,f^{(4)}(x)\,|$ on [a,b]},

then the "error" of the S_n approximation is $\left|\int_a^b f(x)\,dx - S_n\right| \le \frac{(b-a)^5}{180\,n^4} \cdot M_4$ = "error bound."

Example 7: Find an error bound for the S_{10} approximation of $\int_0^2 \sin(x^2)\,dx$.

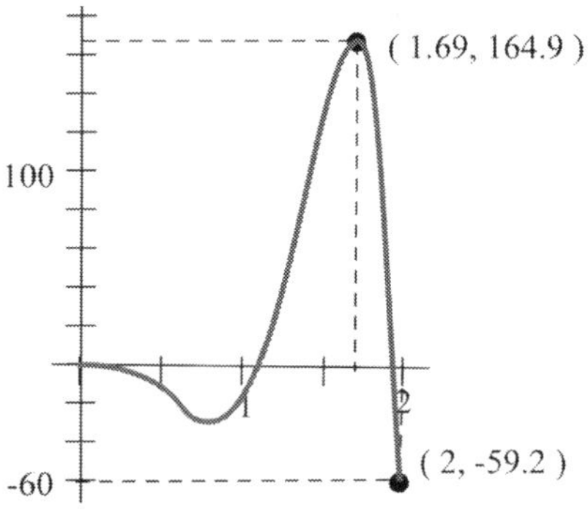

$y = f^{(4)}(x) = (16x^4 - 12)\sin(x^2) - 48x^2\cos(x^2)$

Fig. 9

Solution: $b - a = 2$, $n = 10$, $f(x) = \sin(x^2)$, and $f^{(4)}(x) = (16x^4 - 12)\sin(x^2) - 48x^2\cos(x^2)$ is continuous on [0, 2] . From Fig. 9, the graph of $f^{(4)}(x)$ on [0, 2] , we know that $M_4 \le 165$. Then

$$\text{"error"} \le \frac{(b-a)^5}{180\,n^4} \cdot M_4 = \frac{(2)^5}{180\,(10)^4}(165) = \frac{5280}{1800000} < 0.003$$

so we can be certain that our S_{10} approximation of the definite integral is within 0.003 of the exact value:

$$S_{10} - 0.003 \le \int_0^2 \sin(x^2)\,dx \le S_{10} + 0.003 .$$

$S_{10} = 0.80537615$, so we are certain that the exact value of the integral is between 0.80237615 and 0.80837615 . Notice that we got a much narrower guarantee using S_{10} compared to using T_{10} to approximate the integral .

Example 8: How large must n be to be certain that S_n is within 0.001 of $\int_0^2 \sin(x^2)\,dx$?

Solution: We are given an "allowable error" of 0.001 and are asked to find n. From Fig. 9 we know that $M_4 \le 165$, so we want the error bound to be less than the allowable error of 0.001. Then

$0.001 \ge \text{"error bound"} = \frac{(2)^5}{180\,n^4}(165) = \frac{5280}{180\,n^4}$. Solving for n, we have

$n^4 \ge \frac{5280}{(.001)180} \ge 29{,}333.34$ so $n \ge \sqrt[4]{29333.34} \approx 13.08$. Since n must be an **even** integer, we can take n = 14 and be certain that S_{14} is **within** 0.001 of $\int_0^2 \sin(x^2)\,dx$. In fact, $S_{14} = 0.8049239$ is even closer than 0.001 to the exact value, $|\,S_{14} - \text{exact value}\,| \approx 0.00015$.

A variety of other methods for approximating definite integrals can be found in most books on Numerical Analysis. Definite integrals occur often in applied problems, and these approximation methods can get us the numerical answers we need even if we can't find an antiderivative of the integrand. If you have a programmable calculator, program Simpson's rule. It will be useful in Chapter 5.

PROBLEMS

For problems 1 and 2, use the values given in Table 3 to approximate the value of $\int_2^6 f(x)\,dx$.

1. Calculate T_4 and S_4 .

2. Calculate T_8 and S_8 .

For problems 3 and 4, use the values given in Table 4 to approximate the value of $\int_{-3}^{1} g(x)\,dx$.

3. Calculate T_8 and S_8 .

4. Calculate T_4 and S_4 .

x	f(x)
2.0	2.1
2.5	2.7
3.0	3.8
3.5	2.3
4.0	0.3
4.5	−1.8
5.0	−0.9
5.5	0.5
6.0	2.2

Table 3

x	g(x)
−3.0	4.2
−2.5	1.8
−2.0	0.7
−1.5	1.5
−1.0	3.4
−0.5	4.3
0	3.5
0.5	−0.3
1.0	−4.6

Table 4

For problems 5 – 10, calculate (a) T_4, (b) S_4, and (c) the exact value of the integral.

5. $\int_1^3 x \, dx$ 6. $\int_0^2 (1-x)\, dx$ 7. $\int_{-1}^1 x^2 \, dx$

8. $\int_2^6 \frac{1}{x} \, dx$ 9. $\int_0^\pi \sin(x)\, dx$ 10. $\int_0^1 \sqrt{x} \, dx$

For problems 11 – 16, calculate (a) T_6 and (b) S_6.

11. $\int_0^2 \frac{1}{1+x^3} \, dx$ 12. $\int_1^2 2^x \, dx$ 13. $\int_{-1}^1 \sqrt{4-x^2} \, dx$

14. $\int_0^1 e^{-x^2} \, dx$ 15. $\int_1^4 \frac{\sin(x)}{x} \, dx$ 16. $\int_0^1 \sqrt{1+\sin(x)} \, dx$

For problems 17 – 23, calculate (a) the error bound for T_4, (b) the error bound for S_4, (c) the value of n so the error bound for T_n is less than 0.001, and (d) the value of n so the error bound for S_n is less than 0.001.

17. $\int_1^3 x \, dx$ 18. $\int_0^2 (1-x)\, dx$ 19. $\int_{-1}^1 x^3 \, dx$

20. $\int_2^6 \frac{1}{x} \, dx$ 21. $\int_0^\pi \sin(x)\, dx$ 22. $\int_1^4 \sqrt{x} \, dx$

23. A friend has asked you to help calculate the area of a piece of land located between a river and a road (Fig. 10). Estimate the area.

24. Estimate the area of the island in Fig. 11.

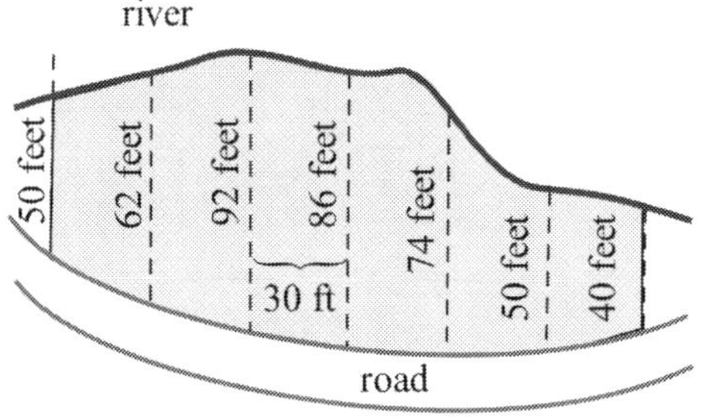

Fig. 10

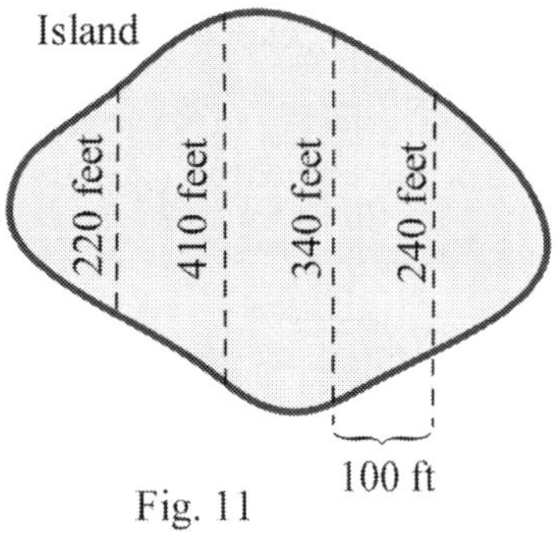

Fig. 11

25. The average depth of the reservoir in Fig. 12 is 22 feet. Estimate the amount of water in the reservoir.

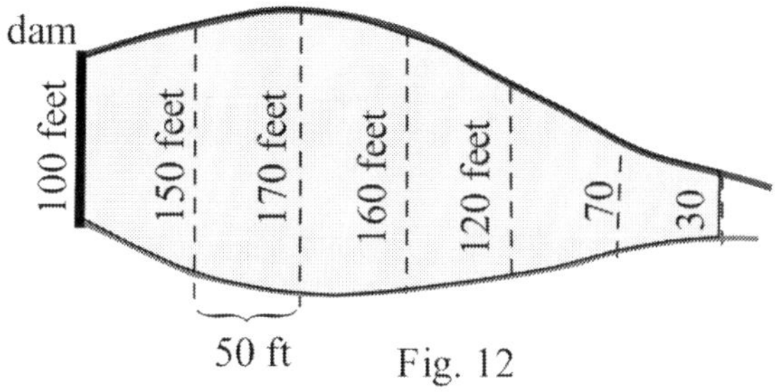

Fig. 12

26. Table 5 shows the speedometer readings for a car at one minute intervals. Estimate how far the car traveled (a) during the first 5 minutes of the trip and (b) during the first 10 minutes of the trip.

Table 5: Time (minutes) and Velocity (feet/minute) for a car.

Time	0	1	2	3	4	5	6	7	8	9	10
Velocity	0	2000	3000	5000	5000	6000	5200	4400	3000	2000	1200

27. Table 6 shows the speed of a jogger at one minute intervals. Estimate how far the jogger ran during the workout.

Table 6: Time (minutes) and Velocity (feet/minute) for a jogger.

Time	0	1	2	3	4	5	6	7	8	9	10
Velocity	0	420	540	300	500	580	520	440	360	260	180

28. Use the error formula for Simpson's rule to show that the parabolic approximation is the exact value of the integral if the integrand is a polynomial of degree 3 or less, $ax^3 + bx^2 + cx + d$.

29. A trapezoidal region (Fig. 13) with base b and heights h_1 and h_2 (assume $h_1 \le h_2$) can be cut into a rectangle with base b and height h_1 and a triangle with base b and height $h_1 - h_2$. Show that the sum of the area of the rectangle and the area of the triangle is $b \cdot \{ \frac{h_1 + h_2}{2} \}$.

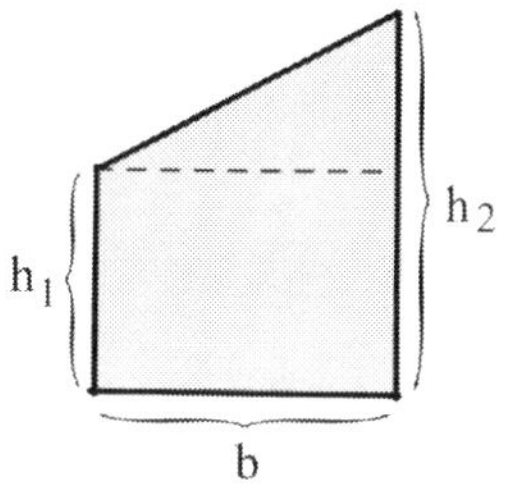

Fig. 13

30. Let f(m) be the minimum value of f on the interval $[x_0 , x_1]$. Let f(M) be the maximum value of f on $[x_0 , x_1]$. And let $h = x_1 - x_0$. Show that the trapezoidal value, $h \cdot \{ \frac{f(x_0) + f(x_1)}{2} \}$, is between $h \cdot f(m)$ and $h \cdot f(M)$. From this result, it can be shown that the trapezoidal approximation is between the lower and upper Riemann sums for f. Since the limit (as h approaches 0) of these Riemann sums is the definite integral of f , we can conclude that the limit of the trapezoidal sums is the value of the definite integral.

31. Let $f(m)$ be the minimum value of f on the interval $[x_0, x_2]$. Let $f(M)$ be the maximum of f on $[x_0, x_2]$. And let $h = x_1 - x_0 = x_2 - x_1$. Show that the parabolic value,

$2h \cdot \{ \frac{f(x_0) + 4f(x_1) + f(x_2)}{6} \}$ is between $2h \cdot f(m)$ and $2h \cdot f(M)$. From this result, it can be shown that the parabolic approximation is between the lower and upper Riemann sums for f. Since the limit (as h approaches 0) of these Riemann sums is the definite integral of f, we can conclude that the limit of the parabolic sums is the value of the definite integral.

32. This problem leads you through the steps to show that the area under a parabolic region with evenly spaced x_i values ($x_0 = m-h, x_1 = m, x_2 = m+h$) as in Fig. 14 is

$$\frac{h}{3} \cdot \{ f(x_0) + 4f(x_1) + f(x_2) \} = \frac{h}{3} \cdot \{ y_0 + 4y_1 + y_2 \} .$$

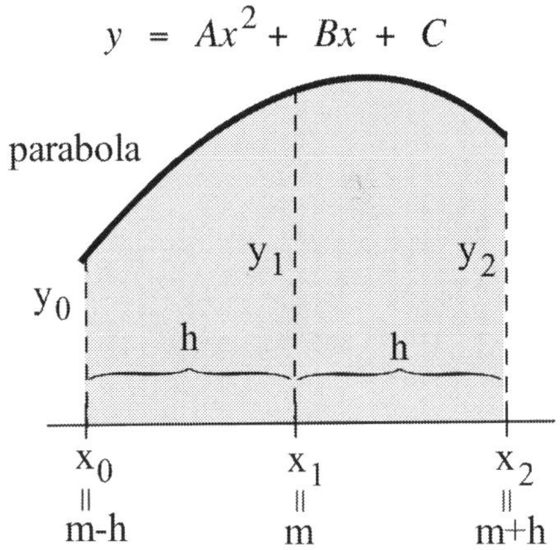

Fig. 14

(a) For $f(x) = Ax^2 + Bx + C$, a parabola, verify that

$$\int_{m-h}^{m+h} f(x)\,dx = \frac{A}{3}x^3 + \frac{B}{2}x^2 + Cx \Big|_{x=m-h}^{x=m+h}$$

$$= \mathbf{2Am^2h + \frac{2}{3}Ah^3 + 2Bmh + 2Ch}.$$

(b) Expand $y_0 = f(m-h) = A(m-h)^2 + B(m-h) + C$,

$y_1 = f(m) = Am^2 + Bm + C$, and

$y_2 = f(m+h) = A(m+h)^2 + B(m+h) + C$. Then verify that

$$\frac{h}{3} \cdot \{ y_0 + 4y_1 + y_2 \} = 2h \left\{ \frac{f(m-h) + 4f(m) + f(m+h)}{6} \right\}$$

$$= \mathbf{2Am^2h + \frac{2}{3}Ah^3 + 2Bmh + 2Ch} .$$

(c) Compare the results of parts (a) and (b) to conclude that for any parabola $f(x) = Ax^2 + Bx + C$,

$$\int_{m-h}^{m+h} f(x)\,dx = 2h \left\{ \frac{f(m-h) + 4f(m) + f(m+h)}{6} \right\} = \frac{h}{3} \cdot \{ y_0 + 4y_1 + y_2 \}.$$

Rectangular Approximations: Left Endpoint, Right Endpoint, and Midpoint Rules

The rectangular approximation methods fit horizontal lines to the integrand. The approximating regions are rectangles, and the sum of the areas of the rectangular regions is a Riemann sum. The Left and Right Endpoint Rules are easy to understand and use, but they typically require a very large number of subintervals to provide good approximations of a definite integral. The Midpoint Rule uses the value of the function at the midpoint of each subinterval. If these midpoint values of f are available, for example when f is given by a formula, then the Midpoint Rule is often more efficient than the Trapezoidal rule. The rectangular approximation rules and their error bounds are given below.

Left endpoint: $L_n = h\cdot\{ f(x_0) + f(x_1) + f(x_2) + \ldots + f(x_{n-1}) \}$

Right endpoint: $R_n = h\cdot\{ f(x_1) + f(x_2) + f(x_3) + \ldots + f(x_n) \}$

Midpoint Rule: $M_n = h\cdot\{ f(A) + f(A + h) + f(A + 2h) + \ldots + f(A + (n-1)h) \}$ where $A = x_0 + \frac{h}{2}$.

(The points A, A + h, A + 2h, . . . are the **midpoints** of the subintervals.)

The "error bound" for L_n and R_n is $\frac{(b-a)^2}{2\,n}\cdot M_1$ where $M_1 \geq \{\text{maximum of } | f'(x) | \text{ on } [a,b]\}$.

The "error bound" for M_n is $\frac{(b-a)^3}{\mathbf{24}\, n^2}\cdot M_2$ where $M_2 \geq \{\text{maximum of } | f''(x) | \text{ on } [a,b]\}$. This is half the error bound of T_n , the trapezoidal approximation.

For problems 33 – 38, calculate (a) L_4 , (b) R_4 , (c) M_4 , and (d) the exact value of the integral.

33. $\int_1^3 x \, dx$

34. $\int_0^2 (1 - x) \, dx$

35. $\int_{-1}^1 x^2 \, dx$

36. $\int_2^6 \frac{1}{x} \, dx$

37. $\int_0^\pi \sin(x) \, dx$

38. $\int_0^1 \sqrt{x} \, dx$

39. Show that the trapezoidal approximation is the average of the left and right endpoint approximations: $T_n = (L_n + R_n)/2$.

40. Which endpoint rule will give a better approximation of $\int_a^b f(x) \, dx$ if f is concave up on [a, b]?

Calculator Problems

The following definite integrals arise in applications, but they do not have easy antiderivatives. Use Simpson's Rule with n = 10 and n = 40 to approximate their values. (Is S_{40} very different from S_{10} ?)

41. $\int_{-1}^2 \sqrt{1 + 4\cdot x^2} \, dx$. This is the length of the curve $y = x^2$ from (–1, 1) to (2, 4).

42. $\int_0^\pi \sqrt{1 + \cos^2(x)} \, dx$. This is the length of one arch of the curve y = sin(x).

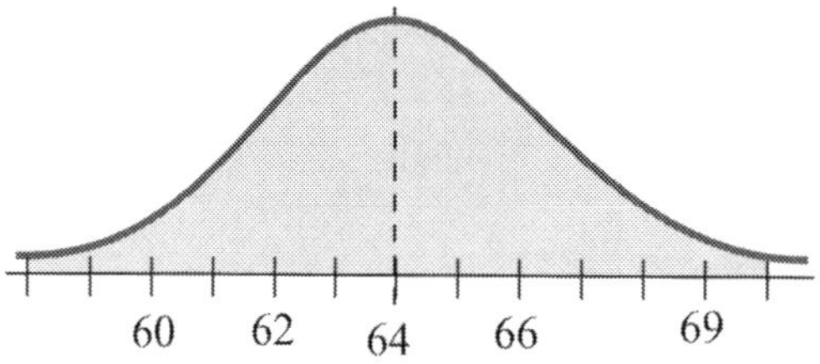

Fig. 15

43. $\int_0^{2\pi} \sqrt{16 \cdot \sin^2(x) + 9 \cdot \cos^2(x)} \; dx$.

This is the length of the ellipse $\frac{x^2}{16} + \frac{y^2}{9} = 1$.

44. $\frac{100}{\sqrt{2\pi}} \int_{60}^{69} \text{EXP}\left(-\{(x-64)/2.5\}^2/2 \right) dx$. $\text{EXP}(x) = e^x$.

This is the percentage of adult females who are between 60 and 69 inches tall (Fig. 15). Approximate the value of this integral.

45. Approximate the percentage of adult females who are between 61 and 64 inches tall.

Section 4.9 **Practice Answers**

Practice 1: Using the Trapezoidal rule to approximate the surface area of the pond in Fig. 5,

$$T \approx \frac{5 \text{ feet}}{2} \cdot \{ 0 + 2 \cdot 12 + 2 \cdot 14 + 2 \cdot 16 + 2 \cdot 18 + 2 \cdot 18 + 0 \text{ feet} \} = 390 \text{ ft}^2.$$

Then volume = (surface area)(depth) $\approx (390 \text{ ft}^2)(0.1 \text{ ft}) = 39 \text{ ft}^3$.

Practice 2: Using Simpson's Rule to estimate the area of the pond in Fig. 5,

$$S \approx \frac{5 \text{ feet}}{3} \{ 0 + 4 \cdot 12 + 2 \cdot 14 + 4 \cdot 16 + 2 \cdot 18 + 4 \cdot 18 + 0 \text{ feet}\} = 413.3 \text{ ft}^2.$$

Practice 3: $f(x) = \frac{1}{x}$, $b - a = 3$, $n = 12$, $f''(x) = \frac{2}{x^3}$. On the interval [2,5], $| f''(x) | = | \frac{2}{x^3} | \le \frac{2}{2^3} = \frac{1}{4}$

so we can take $M_2 = \frac{1}{4}$. Then $| \text{error} | \le \frac{(b-a)^3}{12n^2} \cdot M_2 = \frac{3^3}{12(12)^2} \frac{1}{4} = \frac{27}{6912} \approx \mathbf{0.004}$.

Practice 4: "error bound" $= \frac{(b-a)^3}{12n^2} \cdot M_2 = \frac{(3)^3}{12n^2} \cdot \frac{1}{4} = \frac{27}{48n^2}$. Setting this "error bound" equal to 0.001

and solving for n, we get $n = \sqrt{\frac{27}{48 \cdot (0.001)}} = \sqrt{562.5} \approx 23.7$. Put $n = 24$.

We can be certain that T_{24} is within 0.001 of the exact value of the integral. (We can not guarentee that T_{23} is within 0.001 of the exact value of the integral, but it probably is.)

PROBLEM ANSWERS Chapter Four

Section 4.0

1. (a) $(10)(12) + (8)(4) = 152$ (b) $(10)(20) - (3)(8) = 176$

3. $bh + \frac{1}{2}\, b(H - h) = bh + \frac{1}{2}\, bH - \frac{1}{2}\, bh = b\left(\frac{h + H}{2}\right)$

5. (a) $(1)(3) + (1)(2) = 5$ (b) {area of shaded region in Fig. 24b} < 5

7. $A(1) = 1$, $A(2) = 2.5$, $A(3) = 4.5$, $A(4) = 6$, $A(5) = 7$

9. $C(1) = 1.5$, $C(2) = 4$, $C(3) = 7.5$, and $C(x)$ = rect. + triangle areas $= x + \frac{1}{2}\, x \cdot x = x + \frac{1}{2}\, x^2$

11. Distance = "area" $= (20)(30) + \frac{1}{2}(10)(30) = 600 + 150 = 750$ feet.

13. (a) A: 20 seconds to stop. B: 40 seconds to stop.
(b) A: $\frac{1}{2}(20)(80) = 800$ feet to stop. B: $\frac{1}{2}(40)(40) = 800$ feet to stop.

15. miles, dollars, cubic feet, kilowatt•hours, people, square meals

Section 4.1

1. $2^2 + 3^2 + 4^2 = 29$ 3. $(1+1)^2 + (1+2)^2 + (1+3)^2 = 29$

5. $\cos(0) + \cos(\pi) + \cos(2\pi) + \cos(3\pi) + \cos(4\pi) + \cos(5\pi) = 1 + (-1) + 1 + (-1) + 1 + (-1) = 0$

7. $\sum_{k=3}^{94} k$ 9. $\sum_{k=3}^{12} k^2$ 11. $\sum_{k=1}^{7} k \cdot 2^k$

13. (a) $(1+2) + (2+2) + (3+2) = 3 + 4 + 5 = 12$ (b) $(1+2+3) + (2+2+2) = 12$

15. (a) $5 \cdot 1 + 5 \cdot 2 + 5 \cdot 3 = 5 + 10 + 15 = 30$ (b) $5 \cdot (1 + 2 + 3) = 5 \cdot 6 = 30$

17. (a) $1^2 + 2^2 + 3^2 = 1 + 4 + 9 = 14$ (b) $(1 + 2 + 3)^2 = 6^2 = 36$

19. $f(0) + f(1) + f(2) + f(3) = 0^2 + 1^2 + 2^2 + 3^2 = 14$

21. $2 \cdot f(0) + 2 \cdot f(1) + 2 \cdot f(2) + 2 \cdot f(3) = 2 \cdot 0 + 2 \cdot 1 + 2 \cdot 4 + 2 \cdot 9 = 28$

23. $g(1) + g(2) + g(3) = 3 + 6 + 9 = 18$

25. $g^2(1) + g^2(2) + g^2(3) = 3^2 + 6^2 + 9^2 = 126$

27. $h(2) + h(3) + h(4) = \frac{2}{2} + \frac{2}{3} + \frac{2}{4} = \frac{13}{6}$

29 $f(1)h(1) + f(2)h(2) + f(3)h(3) = (1)(2) + (4)(1) + (9)(2/3) = 12$

31. $(1^2 - 0^2) + (2^2 - 1^2) + (3^2 - 2^2) + (4^2 - 3^2) + \ldots + (7^2 - 6^2) = 7^2 - 0^2 = 49$

33. $(\frac{1}{1} - \frac{1}{2}) + (\frac{1}{2} - \frac{1}{3}) + (\frac{1}{3} - \frac{1}{4}) + (\frac{1}{4} - \frac{1}{5}) + (\frac{1}{5} - \frac{1}{6}) = 1 - \frac{1}{6} = \frac{5}{6}$

35. $(\sqrt{1} - \sqrt{0}) + (\sqrt{2} - \sqrt{1}) + (\sqrt{3} - \sqrt{2}) + \ldots + (\sqrt{9} - \sqrt{8}) = \sqrt{9} - \sqrt{0} = 3$

37. (i) $[2,3], [3,4.5], [4.5,6], [6,7]$ (ii) $1, 1.5, 1.5, 1$
(iii) mesh = 1.5 (iv) $1 + 1.5 + 1.5 + 1 = 5$

39. (i) $[-3,-1], [-1,0], [0,1.5], [1.5,2]$ (ii) $2, 1, 1.5, 0.5$
(iii) mesh = 2 (iv) $2 + 1 + 1.5 + 0.5 = 5$

41. (i) $[3,3.8], [3.8,4.5], [4.5,5.2], [5.2,7]$ (ii) $0.8, 0.7, 0.7, 1.8$ (iii) mesh = 1.8 (iv) $0.8+0.7+0.7+1.8=4$

43. $\Delta x_1 + \Delta x_2 + \Delta x_3 + \ldots + \Delta x_n = (x_1 - x_0) + (x_2 - x_1) + (x_3 - x_2) + \ldots + (x_n - x_{n-1}) = x_n - x_0$

Fig. for 45(a)

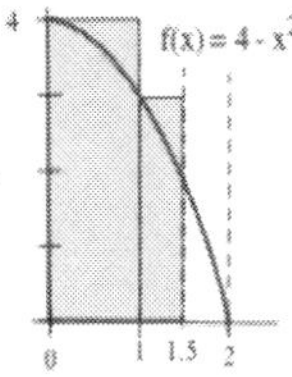

45. (a) $f(0)(1) + f(1)(0.5) + f(2)(0.5) = (4)(1) + (3)(0.5) + (0)(0.5) = 5.5$
(b) $f(1)(1) + f(1.5)(0.5) + f(1.5)(0.5) = (3)(1) + (1.75)(0.5) + (1.75)(0.5) = 4.75$

47a. (i) and (ii) See the graph.
(iii) $f(0) = 0$, $f(\pi/4) \approx 0.707$, $f(\pi/2) = 1$
(iv) $(\pi/4)(0) + (\pi/4)(0.707) + (\pi/2)(1) \approx 2.13$

Fig. for 47(a)

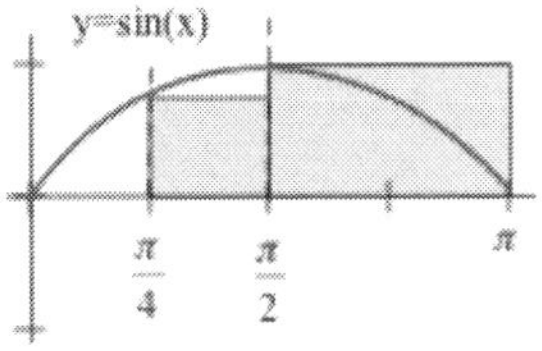

49. (a) $(2)(1) + (5)(2) + (17)(1) \le RS \le (5)(1) + (17)(2) + (26)(1)$ so $29 \le RS \le 65$.
(b) $(2)(1) + (5)(1) + (10)(1) + (17)(1) \le RS \le (5)(1) + (10)(1) + (17)(1) + (26)(1)$ so $34 \le RS \le 58$.
(c) $(2)(0.5) + (3.25)(0.5) + (5)(1) + (10)(1) + (17)(1) \le RS \le (3.25)(0.5) + (5)(0.5) + (10)(1) + (17)(1) + (26)(1)$ so $34.625 \le RS \le 57.125$.

51. (a) $(0)(\pi/2) + (0)(\pi/2) \le RS \le (1)(\pi/2) + (1)(\pi/2)$ so $0 \le RS \le \pi$
(b) $(0)(\pi/4) + (0.707)(\pi/4) + (0)(\pi/2) \le RS \le (0.707)(\pi/4) + (1)(\pi/4) + (1)(\pi/2)$ so $0.56 \le RS \le 2.91$
(c) $(0)(\pi/4) + (0.707)(\pi/\ ) + (0.707)(\pi/4) + (0)(\pi/4) \le RS \le (0.707)(\pi/4) + (1)(\pi/4) + (1)(\pi/4) + (0.707)(\pi/4)$ so $1.11 \le RS \le 2.68$

53. (a) $|\,7.402 - 7.362\,| = 0.04$ (b) $|\,7.390 - 7.372\,| = 0.018$

55. $|\text{error}| \le (\text{base})(\text{height}) = \frac{4-2}{50}(65-9) = \frac{2}{50}(56) = \frac{112}{50} = 2.24$

57. (a) $\frac{100(101)}{2} = 5050$ (b)

$$\begin{array}{r} 1 + 2 + 3 + \ldots + 100 \\ \underline{100 + 99 + 98 + \ldots + 1} \\ 101 + 101 + 101 + \ldots + 101 \end{array}$$

$= 100(101) = 10100.\ \frac{1}{2}(10100) = 5050$

59. $10 + 11 + 12 + \ldots + 20 = (1+2+3+\ldots+20) - (1+2+3+\ldots+9) = \frac{20(21)}{2} - \frac{9(10)}{2} = 210 - 45 = 165$

61. $\sum_{k=1}^{10} (k^3 + k) = \sum_{k=1}^{10} k^3 + \sum_{k=1}^{10} k = \left(\frac{10(11)}{2} \right)^2 + \frac{10(11)}{2} = (55)^2 + 55 = 3080$

Section 4.2

1. $\int_0^4 2 + 3x \, dx$
3. $\int_2^5 x^3 \, dx$
5. $\int_1^5 x^3 \, dx$

7. $\int_{0.5}^2 x \sin(x) \, dx$
9. $\int_1^3 \ln(x) \, dx$
11. $\int_1^3 2x \, dx = 8$

13. $\int_{-1}^0 |x| \, dx = 1/2$
15. $\int_0^4 3 - \frac{x}{2} \, dx = 8$

Fig. for 19(a)

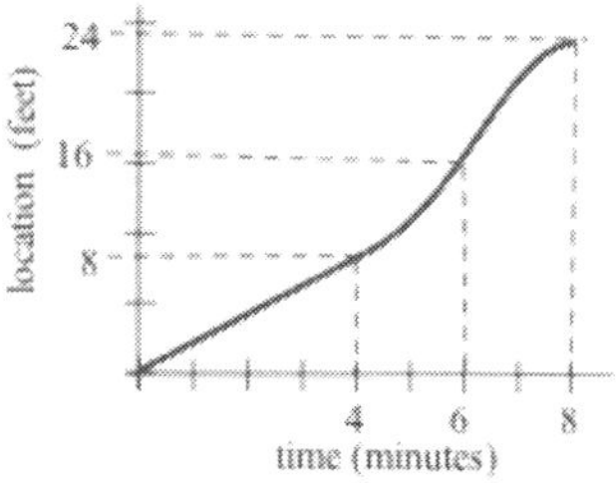

17. (a) 3 (b) –1 (c) 6 (d) 8 (e) 7

19. (a) see the graph
(b) 24 feet.
(c) 24 feet from the starting point.

21. meters
23. feet^3 = cubic feet
25. gram•meters
27. feet/second = feet per second

29. $\Delta x = \frac{2-0}{n} = \frac{2}{n}$. $m_i = \frac{2}{n}(i-1)$ and $M_i = \frac{2}{n} i$ so $f(m_i) = \{ \frac{2}{n}(i-1) \}^3$ and $f(M_i) = \{ \frac{2}{n} i \}^3$.

(a) $LS = \sum_{i=1}^{n} f(m_i)\Delta x = \sum_{i=1}^{n} \{ \frac{2}{n}(i-1) \}^3 \frac{2}{n} = \frac{2}{n} \frac{8}{n^3} \left\{ \sum_{i=1}^{n} i^3 - 3\sum_{i=1}^{n} i^2 + 3\sum_{i=1}^{n} i - \sum_{i=1}^{n} 1 \right\}$

$= \frac{16}{n^4} \{ (\frac{1}{4} n^4 + \frac{1}{2} n^3 + \frac{3}{12} n^2) - 3(\frac{1}{3} n^3 + \frac{1}{2} n^2 + \frac{2}{12} n) + 3(\frac{1}{2} n^2 + \frac{1}{2} n) - n \}$

$= \frac{16}{n^4} \{ \frac{1}{4} n^4 - \frac{1}{2} n^3 + \frac{1}{4} n^2 \} = 4 - \frac{8}{n} + \frac{4}{n^2} \longrightarrow 4$.

(b) $US = \sum_{i=1}^{n} f(M_i)\Delta x = \sum_{i=1}^{n} \{ \frac{2}{n} i \}^3 \frac{2}{n} = \frac{2}{n} \frac{8}{n^3} \left\{ \sum_{i=1}^{n} i^3 \right\} = \frac{16}{n^4} \{ \frac{1}{4} n^4 + \frac{1}{2} n^3 + \frac{3}{12} n^2 \}$

$= 4 + \frac{8}{n} + \frac{4}{n^2} \longrightarrow 4$.

Section 4.3

1. 5
3. 0
5. 3
7. 0
9. –5
11. –5
13. 0
15. 4.5+5 = 9.5
17. 10+3 = 13
19. 5+2 = 7
21. 1
23. –1

25. 2.5   27. 1   29. 1   31. (a) $8 \cdot 6 = 48$ (b) 24

33. (a) 32 (b) $8^2 = 64$ 35. 8   37. 2.5

39. 3   41. 7

43. (a) graph $y = A(x) = x^2/2$ (b) graph $y = A'(x) = x$

For 43

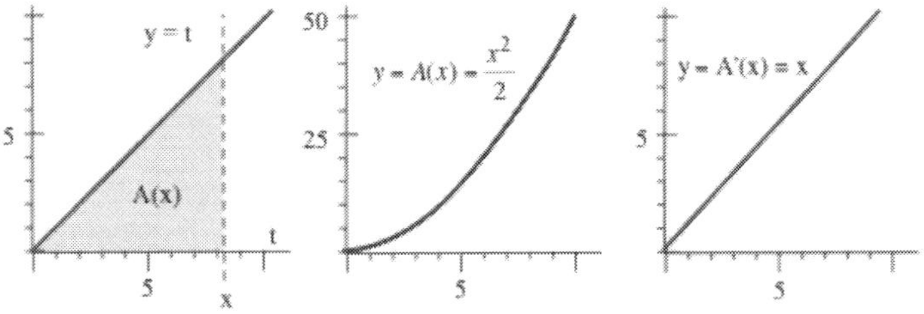

45. (a) graph of $y = A(x)$ in Figure 45a
(b) graph of $y = A'(x)$ in Figure 45b

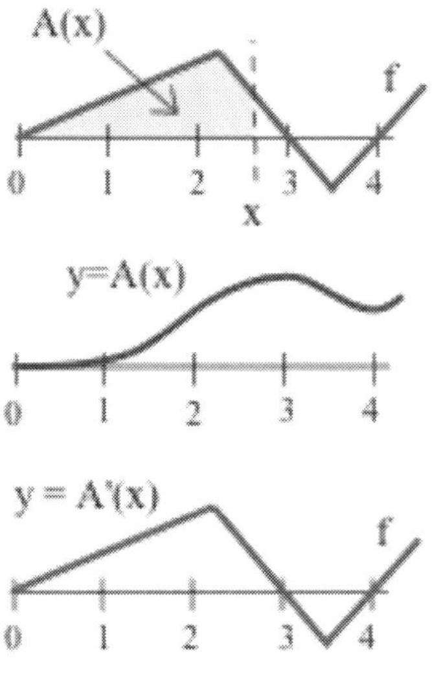

Graphs for problem 45

47. (a) f is continuous on [1,4]
(b) f is not differentiable on [1,4] (not diff. at $x \approx 2.3$ and 3.3)
(c) f is integrable on [1,4]

49. (a) f is not continuous on [1,4] (not cont. at x = 2 and 3)
(b) f is not differentiable on [1,4] (not diff. at x = 2 and 3)
(c) f is integrable on [1,4].

51. $\int_1^4 \text{velocity } dt = \int_1^2 \text{velocity } dt + \int_2^4 \text{velocity } dt = 35 + 50 = 85$ miles

Section 4.4

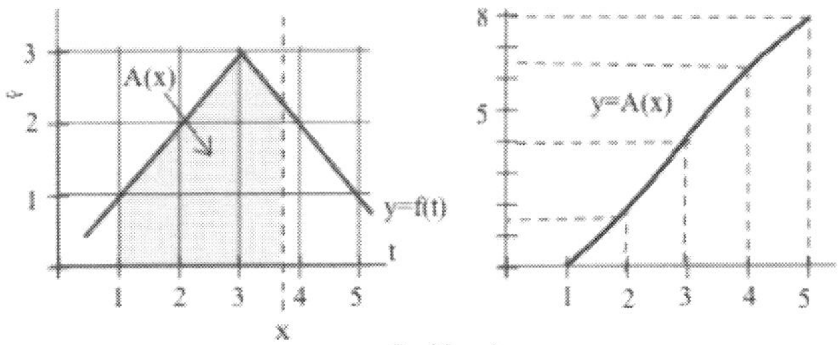

Problem 1

1. (a) see Figure
(b) A(1) = 0 , A(2) = 1.5 , A(3) = 4 , A(4) = 6.5
(c) A '(1) = 1 , A '(2) = 2 , A '(3) = 3 , A '(4) = 2

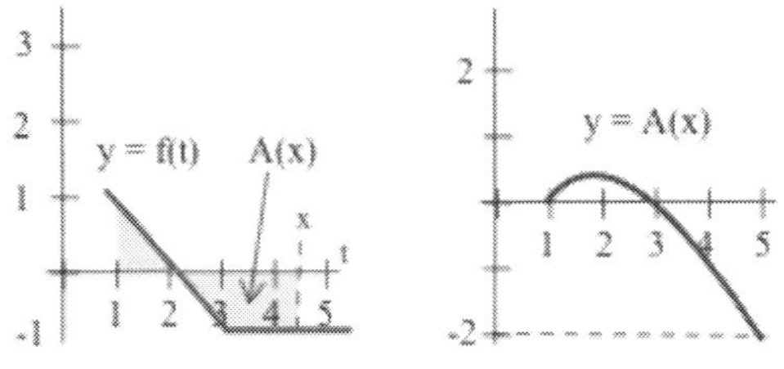

Problem 3

3. (a) see Figure
(b) A(1) = 0 , A(2) = 0.5 , A(3) = 0 , A(4) = –1
(c) A '(1) = 1 , A '(2) = 0 , A '(3) = –1 , A '(4) = –1

5. (a) see Figure
(b) $A(1) = 0$, $A(2) = 2$, $A(3) = 4$, $A(4) = 6$
(c) $A'(1) = 2$, $A'(2) = 2$, $A'(3) = 2$, $A'(4) = 2$

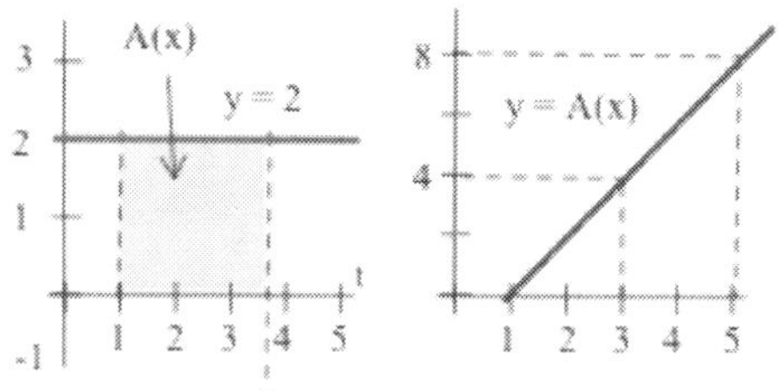

Problem 5

7. (a) see Figure
(b) $A(1) = 0$, $A(2) = 4.5$, $A(3) = 8$, $A(4) = 10.5$
(c) $A'(1) = 5$, $A'(2) = 4$, $A'(3) = 3$, $A'(4) = 2$

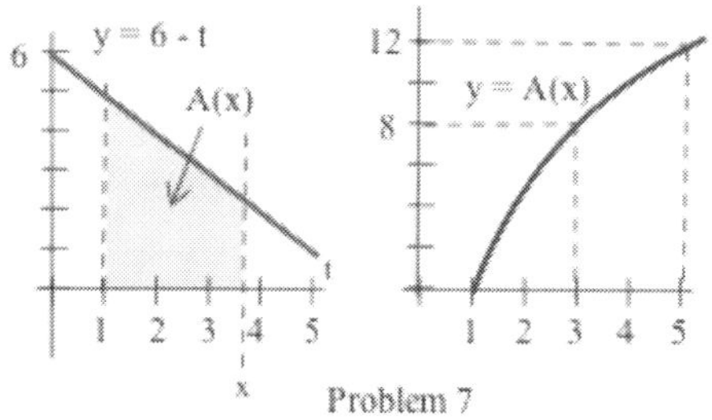

Problem 7

9. $x^2 \Big|_0^3 = 9$, $x^2 \Big|_1^3 = 8$, $x^2 \Big|_0^1 = 1$

11. $2x^3 \Big|_1^3 = 52$, $2x^3 \Big|_1^2 = 14$, $2x^3 \Big|_0^3 = 54$

13. $x^4 \Big|_0^3 = 81$, $x^4 \Big|_1^3 = 80$, $x^4 \Big|_0^1 = 1$

15. $x^3 \Big|_{-3}^3 = 54$, $x^3 \Big|_{-3}^0 = 27$, $x^3 \Big|_0^3 = 27$

17. $x^3 \Big|_0^2 = 8$, $x^3 \Big|_1^3 = 26$

19. (a) distance = $\int_0^{10} 2t \, dt = t^2 \Big|_0^{10} = 100$ feet.

(b) Find T so $50 = \int_0^T 2t \, dt = t^2 \Big|_0^T = T^2$. $T = \sqrt{50} \approx 7.07$ seconds.

21. (a) distance = $\int_0^{10} 4t^3 \, dt = t^4 \Big|_0^{10} = 10{,}000$ feet.

(b) $5000 = \int_0^T 4t^3 \, dt = T^4$. $T = \sqrt[4]{5000} \approx 8.41$ seconds.

23. (a) velocity $= 75 - 3t^2 = 0$ when $t = 5$ seconds.(b) distance = $\int_0^5 75 - 3t^2 \, dt = 75t - t^3 \Big|_0^5 = 250$ feet.

(c) $125 = \int_0^T 75 - 3t^2 \, dt = 75t - t^3 \Big|_0^T = 75T - T^3$ so $T^3 - 75T + 125 = 0$ (solve using

Newton's method or by examining the graph of $y = x^3 - 75x + 125$) and $T \approx 1.74$ seconds.

25. The total area is $\int_0^3 x^2\,dx = \frac{1}{3}x^3 \Big|_0^3 = 9$.

(a) Find T so $\frac{1}{2} \cdot 9 = \frac{9}{2} = \int_0^T x^2\,dx = \frac{1}{3}x^3 \Big|_0^T = \frac{1}{3}T^3$. T = $\sqrt[3]{27/2} \approx 2.38$.

(b) Find T so $\frac{1}{3} \cdot 9 = 3 = \int_0^T x^2\,dx = \frac{1}{3}x^3 \Big|_0^T = \frac{1}{3}T^3$. T = $\sqrt[3]{9} \approx 2.08$.

Then find T so $\frac{2}{3} \cdot 9 = 6 = \int_0^T x^2\,dx = \frac{1}{3}x^3 \Big|_0^T = \frac{1}{3}T^3$. T = $\sqrt[3]{18} \approx 2.62$.

Section 4.5

1. (a) $A(x) = x^3$. Then $A'(x) = 3x^2$, and $A'(1) = 3$, $A'(2) = 12$, and $A'(3) = 27$.

(b) $A'(x) = \mathbf{D}\left(\int_0^x 3t^2\,dt \right) = 3x^2$. $A'(1) = 3$, $A'(2) = 12$, and $A'(3) = 27$.

3. $A'(x) = 2x$ so $A'(1) = 2$, $A'(2) = 4$, $A'(3) = 6$.

4. $A'(x) = 2x$ so $A'(1) = 2$, $A'(2) = 4$, $A'(3) = 6$.

5. $A'(x) = 2x$ so $A'(1) = 2$, $A'(2) = 4$, $A'(3) = 6$.

6. $A'(x) = 3 - x^2$ so $A'(1) = 2$, $A'(2) = -1$, $A'(3) = -6$.

7. $A'(x) = \sin(x)$ so $A'(1) \approx 0.84$, $A'(2) \approx 0.91$, $A'(3) \approx 0.14$.

8. $A'(x) = | x - 2 |$ so $A'(1) = 1$, $A'(2) = 0$, $A'(3) = 1$.

9. $A'(x) = f(x)$ so $A'(1) = 2$, $A'(2) = 1$, $A'(3) = 2$.

10. $A'(x) = f(x)$ so $A'(1) = 0$, $A'(2) = -1$, $A'(3) = 0$.

11. $A'(x) = f(x)$ so $A'(1) = 1$, $A'(2) = 2$, $A'(3) = 2$.

12. $A'(x) = f(x)$ so $A'(1) = -1$, $A'(2) = -1$, $A'(3) = 0$.

13. $F(1) - F(0) = 6 - 5 = 1$

15. $F(3) - F(1) = 9 - \frac{1}{3} = \frac{26}{3}$

17. $F(5) - F(1) \approx 1.61 - 0 = 1.61$

19. $F(3) - F(1/2) \approx 1.10 - (-0.69) = 1.79$

21. $F(\pi/2) - F(0) = 1 - 0 = 1$

23. $F(1) - F(0) \approx 0.67 - 0 = 0.67$

25. $F(7) - F(1) = \frac{2}{3}(7)^{3/2} - \frac{2}{3} \approx 11.68$

27. $F(9) - F(1) = 3 - 1 = 2$

29. $F(3) - F(-2) \approx 20.09 - 0.14 = 19.95$

31. $F(\pi/4) - F(0) = 1 - 0 = 1$

33. $F(3) - F(0) = \frac{2}{3}(10)^{3/2} - \frac{2}{3} \approx 20.42$

35. $F(x) = \frac{1}{3}x^3$. $F(2) - F(-1) = \frac{8}{3} - \left(-\frac{1}{3}\right) = 3$.

37. $F(x) = \ln(x)$. $F(e) - F(1) = 1 - 0 = 1$.

39. $F(x) = \frac{2}{3}x^{3/2}$. $F(100) - F(25) = \frac{2000}{3} - \frac{250}{3} = \frac{1750}{3} \approx 583.33$

41. $F(x) = -1/x$. $F(10) - F(1) = -0.1 - (-1) = 0.9$

43. $F(x) = e^x$. $F(1) - F(0) = e - 1 \approx 1.718$

45. $F(x) = \tan(x)$. $F(\pi/4) - F(\pi/6) \approx 1 - 0.577 = 0.423$

47. The integral goes from 3 to 3 so even without knowing an antiderivative, $\int_3^3 \sin(x)\cdot\ln(x)\,dx = 0$.

49. area $= \int_0^{\pi} \sin(x)\,dx = -\cos(x)\Big|_0^{\pi} = -(-1) - (-1) = 2$

51. area$= \int_0^{3.5} \text{INT}(x)\,dx = 0 + 1 + 2 + \frac{1}{2}(3) = 4.5$.

53. area $= \int_0^3 (x-2)^2\,dx = \int_0^3 x^2 - 4x + 4\,dx = \frac{1}{3}x^3 - 2x^2 + 4x\Big|_0^3 = 3 - 0 = 3$.

55. $\mathbf{D}(\,A(3x)\,) = \mathbf{3}\cdot\tan(\,3x\,)$, $\mathbf{D}(\,A(x^2)\,) = \mathbf{2x}\cdot\tan(\,x^2\,)$, $\mathbf{D}(\,A(\,\sin(x)\,)\,) = \mathbf{cos(x)}\cdot\tan(\,\sin(x)\,)$

57. $\sqrt{1+5x}\cdot(\,5\,)$

59. $\sqrt{1+\sin(x)}\cdot\cos(x)$

61. $\{\,3(1-2x)^2 + 2\,\}\cdot(-2)$

63. $-\cos(3x)$

65. $\tan(\,x^2\,)\cdot 2x - \tan(\,x\,)$

67. $5\cdot\ln(x)\cdot\cos(\,3\cdot\ln(x)\,)\cdot\frac{1}{x}$

Section 4.6

1. Left side $= \frac{1}{4} x^4 \Big|_1^2 = \frac{15}{4}$. Right side $= \{ \frac{1}{3} x^3 \Big|_1^2 = \frac{7}{3} \}\cdot\{ \frac{1}{2} x^2 \Big|_1^2 = \frac{3}{2} \} = \frac{7}{2} \neq$ left side.

3. Left side $= \frac{1}{4}$. Right side $= (\frac{1}{3})\cdot(\frac{1}{2}) = \frac{1}{6} \neq$ left side.

5. $\frac{1}{3} \sin(3x) + C$ 7. $-\cos(2 + e^x) + C$ 9. $\tan(\sin(x)) + C$ 11. $\frac{5}{2} \ln| 3 + 2x | + C$

13. $-\frac{1}{3} \cos(1 + x^3) + C$ 15. $\frac{1}{4} \sin(4x) + C$ 17. $\frac{1}{48} (5 + x^4)^{12} + C$ 19. $\ln| 2 + x^3 | + C$

21. $\frac{1}{2} (\ln(x))^2 + C$ 23. $\frac{1}{24} (1 + 3x)^8 + C$ 25. $\sec(e^x) + C$ 27. $\frac{1}{3} \sin(3x) \Big|_0^{\pi/2} = -\frac{1}{3}$

29. $-\cos(2 + e^x) \Big|_0^1 = \cos(3) - \cos(2 + e) \approx -0.996$

31. $\frac{1}{18} (1 + x^3)^6 \Big|_{-1}^1 = \frac{32}{9}$ 33. $\frac{5}{2} \ln| 3 + 2x | \Big|_0^2 = \frac{5}{2} \ln(\frac{7}{3})$ 35. $-\frac{1}{3} (1 - x^2)^{3/2} \Big|_0^1 = \frac{1}{3}$

37. $\frac{2}{9} (1 + 3x)^{3/2} \Big|_0^1 = \frac{16}{9} - \frac{2}{9} = \frac{14}{9}$ 39. $\frac{1}{2} x - \frac{1}{20} \sin(10x) + C$

41. $\frac{1}{4} \sin(2x) + C$ 43. $\frac{1}{2} x - \frac{1}{4} \sin(2x) \Big|_0^{\pi} = \frac{\pi}{2}$ 45. $\frac{1}{7} x^7 + \frac{3}{5} x^5 + x^3 + x + C$

47. $\frac{1}{2} e^{2x} + 2e^x + x + C$ 49. $\frac{1}{6} x^6 + \frac{1}{4} x^4 + \frac{5}{3} x^3 + 5x + C$ 51. $\frac{1}{2} e^{2x} + \frac{1}{4} e^{4x} + C$

53. $\frac{2}{7} x^{7/2} + \frac{6}{5} x^{5/2} - \frac{4}{3} x^{3/2} + C$ 55. $3x - 3\cdot\ln| x + 1 | + C$ 57. $\frac{1}{2} x^2 - x + C$

59. (divide first) $x^2 - 11x + 7\cdot\ln | x - 1 | + C$ 61. (divide first) $x + 3\cdot\ln | x - 1 | + C$

63. $\frac{2}{3} x^{3/2} + 8 x^{1/2} + C$ 65. (area of semicirle with radius 1) $= \frac{1}{2} \pi(1)^2 = \frac{\pi}{2}$

67. (area of semicirle with radius 3) $= \frac{1}{2} \pi(3)^2 = \frac{9}{2} \pi$

69. (area of rectangle) + (area of semicircle of radius 1) $= (2)(2) + \frac{1}{2}(\pi(1)^2) = 4 + \frac{\pi}{2}$

Section 4.7

1. between 11 (using left endpoints of intervals) and 6 (using right endpoints)

3. between 4 (using left endpoints of intervals) and 6 (using right endpoints)

5. Using left endpoint widths: (0)(40)+(70)(40)+(55)(40)+(90)(40)+(130)(40)+(115)(40) = 18,400 ft^2.
Right endpoint widths (70, 55, ...) and average widths (70/2, 125/2, ...) give the same result, 18,400 ft^2.
All of these are reasonable methods for estimating the area of the island.

7. 9 9. 1 11. $\frac{1}{2} \cdot e^2 - \frac{3}{2}$ 13. $\frac{1}{32}\pi^2 + \frac{1}{4}\pi - \frac{\sqrt{2}}{2}$

15. $e^2 - 3$ 17. $3 - \frac{\pi}{4}$

19. Estimate using midpoints of unit intervals: $\frac{1}{4}\{f(1)(1)+f(2)(1)+f(3)(1)+f(4)(1)\} = \frac{19}{4}$. About $\frac{19}{4}$.

21. Estimate using midpoints of unit intervals: $\frac{1}{2}\{f(2)(1)+f(3)(1)\} = 5$. About 5 .

23. average ≈ 1 25. average $\approx \frac{11}{5}$ 27. average = 5 29. average $= \frac{13}{3}$ 31. average $= \frac{2}{\pi}$

33. (a) C = 1: average $= \frac{2}{3}$ (b) C = 9: average = 2 (c) C = 81: average = 6 (d) C = 100: average $= \frac{20}{3}$
In general, average $= \frac{2}{3}\sqrt{C}$.

35. (a) Graphically, average $\approx 3000 \cdot 1000 \frac{\text{calls}}{\text{hour}} = \frac{3000000}{60} \frac{\text{calls}}{\text{min}} \approx 50{,}000 \frac{\text{calls}}{\text{min}}$. (b) About $58{,}333 \frac{\text{calls}}{\text{min}}$.

37. (a) Similar to Example 5: work = 1,950 foot–pounds (b) work = 1,312.5 foot–pounds

39. (a) work = 1,200 foot–pounds (b) work = 600 foot–pounds (c) work = 400 foot–pounds

41. work = 1,275 foot–pounds

Section 4.8

1. $h = \frac{(6-2)}{4} = 1$. $T_4 = \frac{1}{2}\{ 2.1 + 2(3.8) + 2(0.3) + 2(-0.9) + 2.2 \} = 5.35$
$S_4 = \frac{1}{3}\{ 2.1 + 4(3.8) + 2(0.3) + 4(-0.9) + 2.2 \} = 5.5$

3. $h = \frac{1-(-3)}{8} = 0.5$.
$T_8 = \frac{0.5}{2}\{ 4.2 + 2(1.8) + 2(0.7) + 2(1.5) + 2(3.4) + 2(4.3) + 2(3.5) + 2(-0.3) + -4.6 \} = 7.35$
$S_8 = \frac{0.5}{3}\{ 4.2 + 4(1.8) + 2(0.7) + 4(1.5) + 2(3.4) + 4(4.3) + 2(3.5) + 4(-0.3) + -4.6 \} = 22/3 = 7.3333$

5. $T_4 = 47$. $T_4 = 0.75$
$S_4 = 4$ $S_4 = 2/3 = 0.666$
exact = 4 exact = 2/3

9. $T_4 = 1.896118898$
$S_4 = 2.004559755$
exact = 2

11. $T_6 = 1.088534906$
$S_6 = 1.090560447$

13. $T_6 = 3.815780054$
$S_6 = 3.826350295$

15. $T_6 = 0.8159928163$
$S_6 = 0.8120491229$

17. (a) $M_2 = 0$ so the error bound = 0 (the trapezoidal approximation is exact)
(b) $M_4 = 0$ so the error bound = 0 (the Simpson's rule approximation is exact)
(c) n = 1 (d) n = 2 (must be an even integer)

19. $M_2 = \{$ max. of 6x on $[-1,1]$ $\} = 6$. $f''''(x) = 0$ so $M_4 = 0$.

(a) $|\text{ error }| \le \dfrac{2^3}{(12)(4^2)}(6) = 0.25$

(b) $M_4 = 0$ so the error bound = 0 (the Simpson's rule approximation is exact)

(c) Set $0.001 = \dfrac{2^3}{(12)(n^2)}(6)$ and solve for $n^2 = \dfrac{48}{(12)(0.001)} = 4{,}000$ so $n = 63.25$.

Take n = 64 (to be certain that we have enough subintervals, always round UP).

(d) n = 2 (must be an even integer)

21. $M_2 = \{$ max. of $|-\sin(x)|$ on $[0,\pi]$ $\} = 1$. $M_4 = \{$ max. of $|\sin(x)|$ on $[0,\pi]$ $\} = 1$.

(a) $|\text{ error }| \le \dfrac{\pi^3}{(12)(4^2)}(1) = 0.1614910244$ (b) $|\text{ error }| \le \dfrac{\pi^5}{(180)(4^4)}(1) = 0.0066410522$

(c) Set $0.001 = \dfrac{\pi^3}{(12)(n^2)}(1)$ so $n^2 = \dfrac{\pi^3}{(12)(0.001)} = 2{,}583.9$ and $n = 50.83$. **n = 51**.

(d) Set $0.001 = \dfrac{\pi^5}{(180)(n^4)}(1)$ so $n^4 = \dfrac{\pi^5}{(180)(0.001)} = 1{,}700.1$ and $n = 6.42$. **n = 8**.

23. h = 30. $S_6 = \dfrac{30}{3}\{ 50 + 4(62) + 2(92) + 4(86) + 2(74) + 4(50) + 40 \} =$ 12,140 square feet.
($T_6 =$ 12,270 square feet.)

25. h = 50.

$S_6 = \dfrac{50}{3}\{ 100 + 4(150) + 2(170) + 4(160) + 2(120) + 4(70) + 30 \} = 37{,}166.7$ square feet area.

volume = (area)(depth) = (37,166.7 ft^2)(22 ft) = 817,667 ft^3 .

$T_6 = \dfrac{50}{2}\{ 100 + 2(150) + 2(170) + 2(160) + 2(120) + 2(70) + 30 \} =$ 36,750 square feet area.

volume = (area)(depth) = (36,750 ft^2)(22 ft) = 808,500 ft^3 .

27. Distance traveled by jogger is area under the graph of velocity (f(x)) vs. time (x) .

area $\approx T_{10} = \dfrac{1}{2}\{ 0+840+1080+600+1000+1160+1040+880+720+520+180 \} = 4{,}010$ ft.

29. – 32. on your own

33.	(a) $L_4 = 3.5$	(b) $R_4 = 4.5$	(c) $M_4 = 4$	(d) exact = 4
35.	(a) $L_4 = 0.75$	(b) $R_4 = 0.75$	(c) $M_4 = 0.625$	(d) exact ≈ 0.6667
37.	(a) $L_4 = 1.8961$	(b) $R_4 \doteq 1.8961$	(c) $M_4 = 2.0523$	(d) exact = 2

Calculus Reference Facts

Derivatives

Notation: $\mathbf{D}(f) = \mathbf{D}f$ represents the derivative with respect to x of the function f(x)

$\mathbf{D}(f) = \mathbf{D}f = \frac{d}{dx} f(x) = f'(x)$

$\mathbf{D}(k) = 0$ k a constant

$\mathbf{D}(f + g) = \mathbf{D}f + \mathbf{D}g$

$\mathbf{D}(f \cdot g) = f \cdot \mathbf{D}g + g \cdot \mathbf{D}f$

$\mathbf{D}(k \cdot f) = k \cdot \mathbf{D}f$ k a constant

$\mathbf{D}(f - g) = \mathbf{D}f - \mathbf{D}g$

$\mathbf{D}(f/g) = \dfrac{g \cdot \mathbf{D}f - f \cdot \mathbf{D}g}{g^2}$

$\mathbf{D}(f^n) = n f^{n-1} \cdot \mathbf{D}f$

$\mathbf{D}(f(g(x))) = f'(g(x)) \cdot \mathbf{D}g$ (Chain Rule)

$\mathbf{D}(\sin(f)) = \cos(f) \cdot \mathbf{D}f$

$\mathbf{D}(\tan(f)) = \sec^2(f) \cdot \mathbf{D}f$

$\mathbf{D}(\sec(f)) = \sec(f) \cdot \tan(f) \cdot \mathbf{D}f$

$\mathbf{D}(\cos(f)) = -\sin(f) \cdot \mathbf{D}f$

$\mathbf{D}(\cot(f)) = -\csc^2(f) \cdot \mathbf{D}f$

$\mathbf{D}(\csc(f)) = -\csc(f) \cdot \cot(f) \cdot \mathbf{D}f$

$\mathbf{D}(e^f) = e^f \cdot \mathbf{D}f$

$\mathbf{D}(a^f) = a^f \ln(a) \cdot \mathbf{D}f$

$\mathbf{D}(\ln|f|) = \frac{1}{f} \cdot \mathbf{D}f$

Logrithmic Differentiation: $\mathbf{D}f = f \cdot \mathbf{D}(\ln|f|)$

$\mathbf{D}(\log_a |f|) = \dfrac{1}{f \cdot \ln(a)} \cdot \mathbf{D}f$

$\mathbf{D}(\arcsin(f)) = \dfrac{1}{\sqrt{1 - f^2}} \cdot \mathbf{D}f$

$\mathbf{D}(\arctan(f)) = \dfrac{1}{1 + f^2} \cdot \mathbf{D}f$

$\mathbf{D}(\text{arcsec}(f)) = \dfrac{1}{|f|\sqrt{f^2 - 1}} \cdot \mathbf{D}f$

$\mathbf{D}(\arccos(f)) = \dfrac{-1}{\sqrt{1 - f^2}} \cdot \mathbf{D}f$

$\mathbf{D}(\text{arccot}(f)) = \dfrac{-1}{1 + f^2} \cdot \mathbf{D}f$

$\mathbf{D}(\text{arccsc}(f)) = \dfrac{-1}{|f|\sqrt{f^2 - 1}} \cdot \mathbf{D}f$

Integrals — General Properties

$\int k \cdot f \, dx = k \cdot \int f \, dx$

$\int (f - g) \, dx = \int f \, dx - \int g \, dx$

$\int (f + g) \, dx = \int f \, dx + \int g \, dx$

$\int f \cdot (g') \, dx = f \cdot g - \int g \cdot (f') \, dx$

Integrals — Particular Functions

1. $\int k\,dx = k \cdot x + C$

2. $\int x^n\,dx = \frac{1}{n+1} \cdot x^{n+1} + C$ if $n \neq -1$ $\int x^{-1}\,dx = \int \frac{1}{x}\,dx = \ln|x| + C$

3. $\int (ax+b)^n\,dx = \frac{1}{n+1} \cdot \frac{1}{a} \cdot (ax+b)^{n+1} + C$ $n \neq -1$

   (a) $\int (ax+b)^{-1}\,dx = \int \frac{1}{ax+b}\,dx = \frac{1}{a} \cdot \ln|ax+b| + C$

   (b) $\int \sqrt{ax+b}\,dx = \frac{2}{3a} \cdot (ax+b)^{3/2} + C$

   (c) $\int \frac{1}{\sqrt{ax+b}}\,dx = \frac{2}{a} \cdot (ax+b)^{1/2} + C$

4. $\int \frac{1}{x(ax+b)}\,dx = \frac{1}{b} \cdot \ln\left|\frac{x}{ax+b}\right| + C$

5. $\int \frac{1}{(x+a)(x+b)}\,dx = \frac{1}{b-a}\left\{\ln|x+a| - \ln|x+b|\right\} + C = \frac{1}{b-a} \ln\left|\frac{x+a}{x+b}\right| + C$ $a \neq b$

   (a) $\int \frac{1}{(x+a)(x+a)}\,dx = \int \frac{1}{(x+a)^2}\,dx = -\frac{1}{x+a} + C$

6. $\int x(ax+b)^n\,dx = \frac{(ax+b)^{n+1}}{a} \cdot \left\{ \frac{ax+b}{n+2} - \frac{b}{n+1} \right\} + C$ $n \neq -1, -2$

   (a) $\int x(ax+b)^{-1}\,dx = \int \frac{x}{ax+b}\,dx = \frac{x}{a} - \frac{b}{a^2} \cdot \ln|ax+b| + C$

   (b) $\int x(ax+b)^{-2}\,dx = \int \frac{x}{(ax+b)^2}\,dx = \frac{1}{a^2} \cdot \left\{ \ln|ax+b| + \frac{b}{ax+b} \right\} + C$

7. $\int \sin(ax)\,dx = -\frac{1}{a}\cos(ax) + C$

8. $\int \cos(ax)\,dx = \frac{1}{a}\sin(ax) + C$

9. $\int \tan(ax)\,dx = \int \frac{\sin(ax)}{\cos(ax)}\,dx = -\frac{1}{a}\ln|\cos(ax)| + C = \ln|\sec(ax)| + C$

10. $\int \cot(ax)\,dx = \int \frac{\cos(ax)}{\sin(ax)}\,dx = \frac{1}{a}\ln|\sin(x)| + C$

11. $\int \sec(ax)\,dx = \frac{1}{a}\ln|\sec(ax) + \tan(ax)| + C$

12. $\int \csc(ax)\,dx = \frac{1}{a}\ln|\csc(ax) - \cot(ax)| + C$

13. $\int \sin^2(ax)\,dx = \frac{x}{2} - \frac{\sin(2ax)}{4a} + C = \frac{x}{2} - \frac{\sin(ax)\cos(ax)}{2a} + C$

14. $\int \cos^2(ax)\,dx = \frac{x}{2} + \frac{\sin(2ax)}{4a} + C = \frac{x}{2} + \frac{\sin(ax)\cos(ax)}{2a} + C$

15. $\int \tan^2(ax)\,dx = \frac{1}{a}\tan(ax) - x + C$

16. $\int \cot^2(ax)\,dx = -\frac{1}{a}\cot(ax) - x + C$

17. $\int \sec^2(ax)\,dx = \frac{1}{a}\tan(ax) + C$

18. $\int \csc^2(ax)\,dx = -\frac{1}{a}\cot(ax) + C$

19. $\int \sin^n(ax)\,dx = \frac{-\sin^{n-1}(ax)\cos(ax)}{na} + \frac{n-1}{n}\int \sin^{n-2}(ax)\,dx$

(a) $\int \sin^3(ax)\,dx = \frac{-\sin^2(ax)\cos(ax)}{3a} - \frac{2}{3a}\cos(ax) + C$

(b) $\int \sin^4(ax)\,dx = \frac{-\sin^3(ax)\cos(ax)}{4a} + \frac{3x}{8} - \frac{3\sin(ax)\cos(ax)}{8a} + C$

20. $\int \cos^n(ax)\,dx = \frac{\cos^{n-1}(ax)\sin(ax)}{na} + \frac{n-1}{n}\int \cos^{n-2}(ax)\,dx$

21. $\int \tan^n(ax)\,dx = \frac{\tan^{n-1}(ax)}{(n-1)a} - \int \tan^{n-2}(ax)\,dx \qquad n \neq 1$

(a) $\int \tan^3(ax)\,dx = \frac{1}{2a}\tan^2(ax) + \frac{1}{a}\ln|\cos(ax)| + C$

(b) $\int \tan^4(ax)\,dx = \frac{1}{3a}\tan^3(ax) - \frac{1}{a}\tan(ax) + C$

22. $\int \cot^n(ax)\,dx = -\frac{\cot^{n-1}(ax)}{n-1} - \int \cot^{n-2}(ax)\,dx \qquad n \neq 1$

23. $\int \sec^n(ax)\,dx = \frac{\sec^{n-2}(ax)\tan(ax)}{(n-1)a} + \frac{n-2}{n-1}\int \sec^{n-2}(ax)\,dx \qquad n \neq 1$

(a) $\int \sec^3(ax)\,dx = \frac{1}{2a}\sec(ax)\tan(ax) + \frac{1}{2a}\ln|\sec(ax) + \tan(ax)| + C$

(b) $\int \sec^4(ax)\,dx = \frac{1}{3a}\sec^2(ax)\tan(ax) + \frac{2}{3a}\tan(ax) + C$

24. $\int \csc^n(ax)\,dx = -\frac{\csc^{n-2}(ax)\cot(ax)}{(n-1)a} + \frac{n-2}{n-1}\int \csc^{n-2}(ax)\,dx \qquad n \neq 1$

25. $\int \sin(ax)\sin(bx)\,dx = \frac{\sin((a-b)x)}{2(a-b)} - \frac{\sin((a+b)x)}{2(a+b)} + C \qquad a^2 \neq b^2$

26. $\int \cos(ax)\cos(bx)\,dx = \frac{\sin((a-b)x)}{2(a-b)} + \frac{\sin((a+b)x)}{2(a+b)} + C \qquad a^2 \neq b^2$

27. $\int \sin(ax)\cos(bx)\,dx = \frac{-\cos((a+b)x)}{2(a+b)} - \frac{\cos((a-b)x)}{2(a-b)} + C \qquad a^2 \neq b^2$

28. $\int x^n \sin(ax)\,dx = -\frac{1}{a} x^n \cos(ax) + \frac{n}{a^2} \int x^{n-1} \cos(ax)\,dx$

29. $\int x^n \cos(ax)\,dx = \frac{1}{a} x^n \sin(ax) - \frac{n}{a^2} \int x^{n-1} \sin(ax)\,dx$

30. $\int e^{ax}\,dx = \frac{1}{a} e^{ax} + C$

31. $\int b^{ax}\,dx = \frac{b^{ax}}{a \cdot \ln(b)} + C$

32. $\int x^n e^{ax}\,dx = \frac{1}{a} x^n e^{ax} - \frac{n}{a} \int x^{n-1} e^{ax}\,dx + C$

(a) $\int x e^{ax}\,dx = \frac{1}{a} x e^{ax} - \frac{1}{a^2} e^{ax} = \frac{e^{ax}}{a^2}(ax - 1) + C$

(b) $\int x^2 e^{ax}\,dx = \frac{1}{a} x^2 e^{ax} - \frac{2}{a^2} x e^{ax} + \frac{2}{a^3} e^{ax} = \frac{e^{ax}}{a^3}(a^2x^2 - 2ax + 2) + C$

33. $\int x^n b^{ax}\,dx = \frac{1}{\ln(b)} \left\{ \frac{1}{a} x^n b^{ax} - \frac{n}{a} \int x^{n-1} b^{ax}\,dx \right\} + C$

34. $\int \frac{1}{\sqrt{a^2 - x^2}}\,dx = \arcsin\left(\frac{x}{a}\right) + C$

35. $\int \frac{1}{a^2 + x^2}\,dx = \frac{1}{a} \cdot \arctan\left(\frac{x}{a}\right) + C$

36. $\int \frac{1}{|x|\sqrt{x^2 - a^2}}\,dx = \frac{1}{a} \cdot \text{arcsec}\left(\frac{x}{a}\right) + C$

37. $\int \frac{1}{a^2 - x^2}\,dx = \frac{1}{2a} \ln\left|\frac{x + a}{x - a}\right| + C$

38. $\int \ln(x)\,dx = x \cdot \ln(x) - x + C$

39. $\int x^n \cdot \ln(x)\,dx = \frac{x^{n+1}}{(n+1)^2} \{(n+1) \cdot \ln(x) - 1\} + C$

40. $\int \frac{1}{x \cdot \ln(x)}\,dx = \ln|\ln(x)| + C$

41. $\int e^{ax} \sin(nx)\,dx = \frac{e^{ax}}{a^2 + n^2} \{ a \cdot \sin(nx) - n \cdot \cos(nx) \} + C$

42. $\int e^{ax} \cos(nx)\,dx = \frac{e^{ax}}{a^2 + n^2} \{ a \cdot \cos(nx) + n \cdot \sin(nx) \} + C$

43. $\int \frac{1}{\sqrt{x^2 \pm a^2}}\,dx = \ln\left| x + \sqrt{x^2 \pm a^2} \right| + C$

44. $\int \sqrt{x^2 \pm a^2}\,dx = \frac{x}{2}\sqrt{x^2 \pm a^2} + \frac{1}{2} a^2 \ln\left| x + \sqrt{x^2 \pm a^2} \right| + C$

45. $\int x^2 \sqrt{x^2 \pm a^2}\,dx = \frac{1}{8} x \cdot (2x^2 \pm a^2)\sqrt{x^2 \pm a^2} - \frac{1}{8} a^4 \ln\left| x + \sqrt{x^2 \pm a^2} \right| + C$

TRIGONOMETRY FACTS

Right Angle Trigonometry

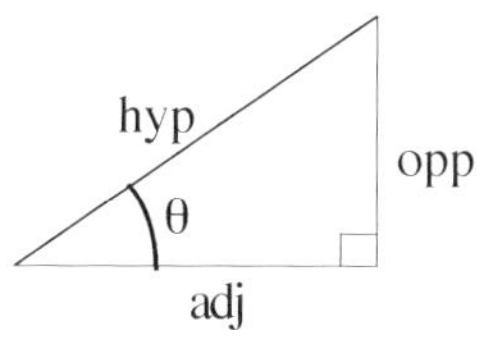

$\sin(\theta) = \frac{opp}{hyp}$ $\cos(\theta) = \frac{adj}{hyp}$

$\tan(\theta) = \frac{opp}{adj}$ $\cot(\theta) = \frac{adj}{opp}$ $\sec(\theta) = \frac{hyp}{adj}$ $\csc(\theta) = \frac{hyp}{opp}$

Trigonometric Functions

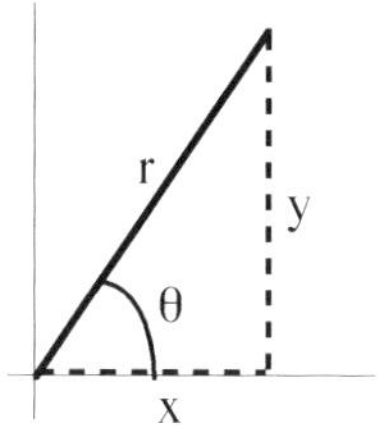

$\sin(\theta) = \frac{y}{r}$ $\cos(\theta) = \frac{x}{r}$

$\tan(\theta) = \frac{y}{x}$ $\cot(\theta) = \frac{x}{y}$ $\sec(\theta) = \frac{r}{x}$ $\csc(\theta) = \frac{x}{y}$

Fundamental Identities

$\tan(\theta) = \frac{\sin(\theta)}{\cos(\theta)}$ $\cot(\theta) = \frac{\cos(\theta)}{\sin(\theta)}$ $\sec(\theta) = \frac{1}{\cos(\theta)}$ $\csc(\theta) = \frac{1}{\sin(\theta)}$

$\mathbf{\sin^2(\theta) + \cos^2(\theta) = 1}$ $\tan^2(\theta) + 1 = \sec^2(\theta)$ $1 + \cot^2(\theta) = \csc^2(\theta)$

$\sin(-\theta) = -\sin(\theta)$ $\cos(-\theta) = \cos(\theta)$ $\tan(-\theta) = -\tan(\theta)$

$\sin(\frac{\pi}{2} - \theta) = \cos(\theta)$ $\cos(\frac{\pi}{2} - \theta) = \sin(\theta)$ $\tan(\frac{\pi}{2} - \theta) = \cot(\theta)$

Law of Sines: $\frac{\sin(A)}{a} = \frac{\sin(B)}{b} = \frac{\sin(C)}{c}$

Law of Cosines: $a^2 = b^2 + c^2 - 2bc \cdot \cos(A)$

$b^2 = a^2 + c^2 - 2ac \cdot \cos(B)$

$c^2 = a^2 + b^2 - 2ab \cdot \cos(C)$

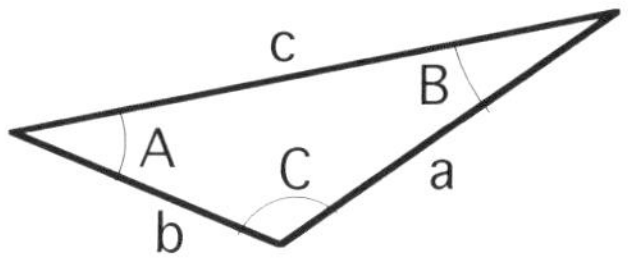

Angle Addition and Subtraction Formulas

$\sin(x + y) = \sin(x) \cdot \cos(y) + \cos(x) \cdot \sin(y)$

$\cos(x + y) = \cos(x) \cdot \cos(y) - \sin(x) \cdot \sin(y)$

$\tan(x + y) = \frac{\tan(x) + \tan(y)}{1 - \tan(x) \cdot \tan(y)}$

$\sin(x - y) = \sin(x) \cdot \cos(y) - \cos(x) \cdot \sin(y)$

$\cos(x - y) = \cos(x) \cdot \cos(y) + \sin(x) \cdot \sin(y)$

$\tan(x - y) = \frac{\tan(x) - \tan(y)}{1 + \tan(x) \cdot \tan(y)}$

Function Product Formulas

$\sin(x) \cdot \sin(y) = \frac{1}{2}\cos(x - y) - \frac{1}{2}\cos(x + y)$

$\cos(x) \cdot \cos(y) = \frac{1}{2}\cos(x - y) + \frac{1}{2}\cos(x + y)$

$\sin(x) \cdot \cos(y) = \frac{1}{2}\sin(x + y) + \frac{1}{2}\sin(x - y)$

Function Sum Formulas

$\sin(x) + \sin(y) = 2\sin(\frac{x + y}{2}) \cdot \cos(\frac{x - y}{2})$

$\cos(x) + \cos(y) = 2\cos(\frac{x + y}{2}) \cdot \cos(\frac{x - y}{2})$

$\tan(x) + \tan(y) = \frac{\sin(x + y)}{\cos(x) \cdot \cos(y)}$

Double Angle Formulas

$\sin(2x) = 2\sin(x) \cdot \cos(x)$

$\cos(2x) = \cos^2(x) - \sin^2(x) = 2\cos^2(x) - 1$

$\tan(2x) = \frac{2\tan(x)}{1 - \tan^2(x)}$

Half Angle Formulas

$\sin(\frac{x}{2}) = \pm\sqrt{\frac{1}{2}(1 - \cos(x))}$ $\pm$ depends on

$\cos(\frac{x}{2}) = \pm\sqrt{\frac{1}{2}(1 + \cos(x))}$ quadrant of x/2

$\tan(\frac{x}{2}) = \frac{1 - \cos(x)}{\sin(x)}$

21849820R00249

Made in the USA
San Bernardino, CA
09 June 2015